Multi-Sensor Information Fusion

Multi-Sensor Information Fusion

Topical Collection Editors

Xue-Bo Jin
Yuan Gao

MDPI • Basel • Beijing • Wuhan • Barcelona • Belgrade • Manchester • Tokyo • Cluj • Tianjin

Topical Collection Editors
Xue-Bo Jin
Beijing Technology and Business University
China

Yuan Gao
Heilongjiang University
China

Editorial Office
MDPI
St. Alban-Anlage 66
4052 Basel, Switzerland

This is a reprint of articles from the Topical Collection published online in the open access journal *Sensors* (ISSN 1424-8220) (available at: https://www.mdpi.com/journal/sensors/special_issues/MultiSensor).

For citation purposes, cite each article independently as indicated on the article page online and as indicated below:

LastName, A.A.; LastName, B.B.; LastName, C.C. Article Title. *Journal Name* **Year**, *Article Number*, Page Range.

ISBN 978-3-03928-302-6 (Pbk)
ISBN 978-3-03928-303-3 (PDF)

Cover image courtesy of Xue-Bo Jin and Yuan Gao

Contents

About the Topical Collection Editors

Jin Xue-bo was born in Liaoning, China, in 1972. She received a BE in industrial electrical and automation and a Master in industrial automation from Jilin University, Jilin, China, in 1994 and 1997, respectively. She received her PhD in control theory and control engineering from Zhejiang University, Zhejiang, China, in 2004. From 1997 to 2012 she was with the College of Informatics and Electronics, Zhejiang Sci-Tech University. Since 2012 she has been a professor at the College of Computer and Information Engineering of the Beijing Technology and Business University. Dr. Xue-bo is a fellow of the Information Fusion Branch of the China Society of Aeronautics and Astronautics, and a member of China Computer Society and Chinese Society of Artificial Intelligence. She is now the leader of the Intelligent Cognitive and Data Analysis Research Center of the Beijing Technology and Business University. She has published about 100 papers in various statistical, mathematics, information fusion and image processing journals. She also developed a new estimation technology known as closed-loop estimation and owns multiple patents.

Gao Yuan was born in Harbin, China, in 1978. She received a BE in Automation, a Master in Control Science and Control Engineering, and a PhD in Microelectronics and Solid State Electronics from Heilongjiang University, Harbin, China, in 2001, 2004, and 2010, respectively. Since 2004 she has been with College of Electronic Engineering of Heilongjiang University. Dr. Yuan is a fellow of the Information Fusion Branch of the China Society of Aeronautics and Astronautics, and a fellow of the Automation Society of Heilongjiang Province. She has published more than 20 papers in domestic and international academic journals.

Preface to "Multi-Sensor Information Fusion"

Around the 1970s, information fusion technology was developed for the military, purportedly because only military systems utilized such significant amounts of sensors. Half a century has now passed and sensor technology has continued to develop. With the decline in cost, it is almost impossible to find systems that do not contain multiple sensors, which has led to the development of multi-sensor technology, and more universal and diversified development.

We have found that many civil systems also have multi-sensor systems, such as unmanned vehicle technology, cyber physical system, the Internet of Things, and intelligent robot systems. In the face of these new application systems, multi-sensor information fusion technology research faces many new issues.

Our 30 paper collection includes the latest research results of current multi-sensor information fusion technology. These research papers are mainly divided into two parts, theoretical and applicational. The basis of the theoretical papers is in-depth research on methods and theories, and proposes new methods. The papers include three main aspects of theoretical and methodological research: new fusion methods based on filtering and estimation; the study of various nonlinear Klaman filters, such as CKF and UKF; and Bayesian inference—new methods for imaging, including video target follow-up and expression recognition. We were also delighted to include applicational papers for applications with extremely high reading and reference value.

From 2018 to 2019, there was an obvious new trend in the research of information fusion technology, which resulted in the emergence of more network-based fusion technologies. For example, particle swarm optimization and deep learning was used for information in this collection.

We believe that with the increasing application of sensors in the system, new technology trends will develop on the basis of classic multi-sensor fusion technology, especially that with strong processing capabilities for complex systems, such as Artificial Intelligence and optimization.

Xue-Bo Jin, Yuan Gao
Topical Collection Editors

sensors

MDPI

Article

Multi-Sensor Data Fusion for Real-Time Surface Quality Control in Automated Machining Systems

E. García Plaza [1], P. J. Núñez López [1,*] and E. M. Beamud González [2]

1 Department of Applied Mechanics & Project Engineering, Institute for Energy Research and Industrial Applications (INEI), Higher Technical School of Industrial Engineering, University of Castilla-La Mancha, Avda. Camilo José Cela, s/n, 13071 Ciudad Real, Spain; eustaquio.garcia@uclm.es

2 Department of Applied Mechanics & Project Engineering, Mining and Industrial Engineering School, University of Castilla-La Mancha, Plaza Manuel Meca 1, 13400 Almadén (Ciudad Real), Spain; elenamaria.beamud@uclm.es

* Correspondence: pedro.nunez@uclm.es; Tel.: +34-926-295-218

Received: 9 November 2018; Accepted: 7 December 2018; Published: 11 December 2018

check for updates

Abstract: Multi-sensor data fusion systems entail the optimization of a wide range of parameters related to the selection of sensors, signal feature extraction methods, and predictive modeling techniques. The monitoring of automated machining systems enables the intelligent supervision of the production process by detecting malfunctions, and providing real-time information for continuous process optimization, and production line decision-making. Monitoring technologies are essential for the reduction of production times and costs, and an improvement in product quality, discarding the need for post-process quality controls. In this paper, a multi-sensor data fusion system for the real-time surface quality control based on cutting force, vibration, and acoustic emission signals was assessed. A total of four signal processing methods were analyzed: time direct analysis (TDA), power spectral density (PSD), singular spectrum analysis (SSA), and wavelet packet transform (WPT). Owing to the nonlinear and stochastic nature of the process, two predictive modeling techniques, multiple regression and artificial neural networks, were evaluated to correlate signal parametric characterization with surface quality. The results showed a high correlation of surface finish with cutting force and vibration signals. The signal processing methods based on signal decomposition in a combined time and frequency domain (SSA and WPT) exhibited better signal feature extraction, detecting excitation frequency ranges correlated to surface finish. The artificial neural network model obtained the highest predictive power, with better behavior for the whole data range. The proposed on-line multi-sensor data fusion provided significant improvements for in-process quality control, with excellent predictive power, reliability, and response times.

Keywords: surface quality control; multi-sensor data fusion; cutting forces; vibration; acoustic emission; signal feature extraction methods; predictive modeling techniques

1. Introduction

Current quality control techniques require slow and costly measurement procedures for inspecting finished products. In industrially competitive sectors, these aspects impose serious constraints directly affecting the benefits obtained. Machining monitoring systems are an ideal tool for overcoming these deficiencies, since they permit the real-time monitoring and control of the cutting process, detect in-process malfunctions, and apply corrective measures to avoid the manufacture of defective products. Several aspects of machining processes can be supervized using monitoring techniques. In recent years, numerous publications have focused on the analysis of tool condition [1–4] and chatter [5–8], whereas other aspects such as surface finish [9–12], dimensional precision [11–14], and chip formation [15,16] have received less attention. The appropriate selection of sensors is crucial for monitoring techniques to be efficacious.

The dynamic character of machining processes, characterized by random and transitory phenomena, has prompted the need for processing on-line information captured by cutting force sensors [15,17,18], mechanical vibration [19–21], acoustic emission [22–25], sound [26–28], power consumed [29,30], among others. Frequently, the information from one single sensor has been insufficient for the accurate characterization of a process, underscoring the need for multi-sensor data fusion. The first attempts at using multi-sensor data fusion for process monitoring were developed in the 1990s to monitor tool wear in turning [31,32] and drilling operations [33,34]. Since then, numerous works with sensor fusion have been published [35–39]. The amount of information provided by a sensor depends mainly on the signal feature extraction method. Signal processing techniques in one single domain have been extensively employed, with the time direct analysis (TDA) method [40–43] being used for analysis in the time domain; and the fast Fourier transform (FFT) method [19,40,43–45], and the power spectral density (PSD) method [45–48] for frequency analysis. In many cases, these methods are not sufficiently efficacious for extracting signal information, which stresses the need for applying more complex analysis techniques that decompose a signal into an independent time series with defined frequency ranges such as: singular spectrum analysis (SSA) [19,49,50], and wavelet packet transform (WPT) [6,15,51]. Statistical parametric characterization is the most common technique used for signal information extraction [21,52]. Moreover, the selection of an efficacious predictive technique is vital for obtaining a high level of precision in predicting data from monitoring systems, with multiple regression (MR) techniques [52,53], artificial neural networks (ANN) [19,54], and support vector machines (SVM) [12,55] being the most frequent methods.

The mean deviation of the assessed profile (*Ra*) is the primary parameter used for the monitoring of surface quality in machining processes. This parameter is an indicator of the surface quality of a product and the behavior of the cutting process, since it is directly linked to machining aspects such as: cutting parameters, tool geometry, use of cutting fluids, tool wear, and chatter, among others [56,57]. In recent years, several surface finish monitoring techniques have been developed. Cutting force and vibration sensors are the most widely employed, and off-line parameters have been incorporated as an additional information source of the cutting process. When one works under ideal conditions—i.e., those recreated in the laboratory—these off-line parameters raise the performance of predictive models. However, under real working conditions characterized by the appearance of random and transitory phenomena, monitoring systems with off-line parameters are more rigid and may mask malfunctions or severe process deficiencies that go undetected by the system.

TDA is the signal analysis method most extensively used by researchers for the monitoring of surface finish. Hessainia [41] used TDA processed vibration signals and cutting conditions for the monitoring of the parameter *Ra*, using a small sample of 27 data built for regression models and validation with the same data. Kirby and Chen [58] used a single component of vibration signals and cutting conditions to monitor the parameter *Ra*. The vibration signal was processed by TDA, using 87 data for fuzzy logic predictive models and validating only with seven workpieces selected under non-random cutting conditions. Upadhay et al. [59] also monitored surface finish using TDA processed vibration signals. In this study 15 workpieces were simultaneously used for the building and validation of the predictive models. Risbood et al. [13] evaluated the parameter *Ra* using TDA processed radial vibration signals and cutting conditions on 20 workpieces to validate ANN models. Özel et al. [60] monitored surface finish using a combination of cutting forces, cutting parameters, cutting time, consumed electrical power, and specific force. The cutting force signals were processed by the TDA method, with only 18 workpieces for building ANN models and 9 for validation.

Signal analysis in the frequency domain is seldom used for the monitoring of surface finish. Abouleta and Mádl [47] calculated the *Ra* parameter with vibration signals, cutting conditions, and tool and workpiece geometry features. The signals were processed by the PSD method using a total of 480 workpieces to build MR models without validating them. Wang et al. [45] calculated the vibration modes of cutting tools by applying the PSD method to cutting forces, and found high frequency vibration (14 kHz) had a significant impact on surface roughness (*Ra*). Moreover, Krolczyk et al. [48] applied the PSD method to 3D measurement surfaces images to analyze the performance of surface morphology in turning operations. Botcha et al. [61] applied frequency analysis to determine the frequency ranges with significant information correlated to surface finish in cylindrical plunge grinding processes.

In contrast, advanced processing methods working simultaneously in a combined time and frequency domain (hereafter, time–frequency domain) are not commonly used owing to their greater complexity. Salgado et al. [49] used SSA processed vibration signals, cutting conditions, and tool angle and radius to estimate the parameter *Ra*, with 35 data for the SVM predictive models, and 20 for validation. García and Núñez [50] applied SSA to vibration signals to monitor surface finish using 270 data test to build MR predictive models, and 90 for validation. Likewise, García and Núñez [52,53] applied the WPT method to vibration and cutting force signals, respectively.

In the last three decades, acoustic emission and sound signals have been commonly applied to the monitoring of tool condition [31,32,62,63]; however, only a few studies have analyzed the monitoring of surface finish. Azouzi and Guillot [14] estimated the parameter *Ra* with the fusion of cutting force, vibration, acoustic emission signals, and cutting parameters. The signals were processed with the TDA method with only 16 workpieces for the ANN models and 5 for validation. Acoustic emission were found to have no significant impact on surface finish. Carou et al. [27] estimated surface finish using TDA processed sound signals and cutting parameters by applying regression models with 18 experimental and 18 validation replicated trials. Frigieri et al. [64] undertook a similar study, except the sound signal was processed in the time-frequency domain using the mel-frequency cepstral coeffiecients, with 15 trials replicated 10 times using 80 for Gaussian mixture models, and validation on non-independent replicated trials. Certain studies have sought to determine the behavior patterns of the acoustic emission signal in relation to surface finish without predictive models. The study of Bhuiyan et al. [42] found the *AE* signal had a moderate correlation with the *Ra* parameter, whilst Pawade and Joshi [65] observed a moderate correlation between the *AE* signal and surface finish with high quality obtained in this study. Rao et al. [66] processed vibration, cutting forces, and acoustic emission signals for the monitoring of surface roughness in ultraprecision diamond turning with a nonparametric clustering technique called, the mean-shift algorithm, and found only the force and vibration signals in the feed direction were adequate for detecting changes in process dynamics and were sensitive to surface variations.

The present study assessed a novel multi-sensor data fusion system for surface quality control in automated machining processes using cutting force, vibration, and acoustic emission sensors. An exhaustive analysis of signal feature extraction methods was performed using two signal processing methods in one single time (TDA) or frequency (PSD) domain; and two processing methods (SSA and WPT) working in the time–frequency domain. In order to correlate the signal characterization parameters with surface finish, multiple regression, and artificial neural network predictive models were analyzed. The original contribution of this study was to determine the optimum multi-sensor data system configuration, signal feature extraction method, and predictive modeling technique in terms of predictive power, reliability, and processing times for the real-time quality control of surface finish using only information from the sensors without off-line parameters.

2. Experimental Design and Methodology

The machining trials were performed on a numerical control lathe (Goratu G Crono4S). Titanium carbonitride Ti(C,N) cutting inserts for finishing operations (0.4 mm corner radius) were used. Each workpiece was machined on a new cutting edge to avoid variability in cutting edge wear. The workpiece material was AISI 1045 steel, 80 mm in diameter and 130 mm in length, with 80 mm in cantilever (Figure 1). The experimental design was based on a factorial design with three factors at different levels: feed (f) six levels (0.08, 0.11, 0.14, 0.17, 0.20, 0.23 mm/rev), cutting speed (v) five levels

(250, 275, 300, 325, 350 m/min), and cutting depth (d) four levels (0.5, 0.8, 1.1, 1.4 mm), with a total of 120 trial combinations. The machined length of each workpiece was subdivided into three 20 mm sampling areas (Figure 1), and signals were registered independently in each sampling area ($SA1$, $SA2$, $SA3$), with a total of 360 monitoring trials.

The parameter selected for the characterization of surface finish was the mean deviation of the assessed profile (Ra), as measured with a Talysurf Intra 50 profilometer. The cut-off (λ_c) was 0.8 mm with an evaluation length (l_t) of 4 mm. The arithmetic mean roughness value (Ra) was calculated for each sampling area (Ra_{SA1}, Ra_{SA2}, Ra_{SA3}) as the average of the three equidistant measures in 120° rotation (Figure 1): 0° ($Ra_{0^{\circ}}$), 120° ($Ra_{120^{\circ}}$), and 240° ($Ra_{240^{\circ}}$). The roughness measurements obtained for each sampling area were correlated to the signals registered in the machining trials.

The multi-sensor data system was designed in the LabVIEW virtual platform to process simultaneously cutting force (F_p, F_f, F_c), mechanical vibration (a_p, a_f, a_c), and acoustic emission (AE) signals with a dynamometer Kistler 9021, a triaxial accelerometer Kistler 8763B500BB, and a piezotron acoustic emission sensor 8152B111, respectively (Figure 1). In order to obtain adequate signal resolution, signals were captured with a sample frequency (f_s) 5 times higher than the maximum frequency range of each sensor. The mechanical vibration signal was sampled with a data acquisition card NI PCI 6110 using a sample frequency f_s = 50 kSamples/s, whereas cutting forces and acoustic emission (in RMS mode) were jointly sampled with a card NI PCI 6133 with a sample frequency f_s = 5 kSamples/s.

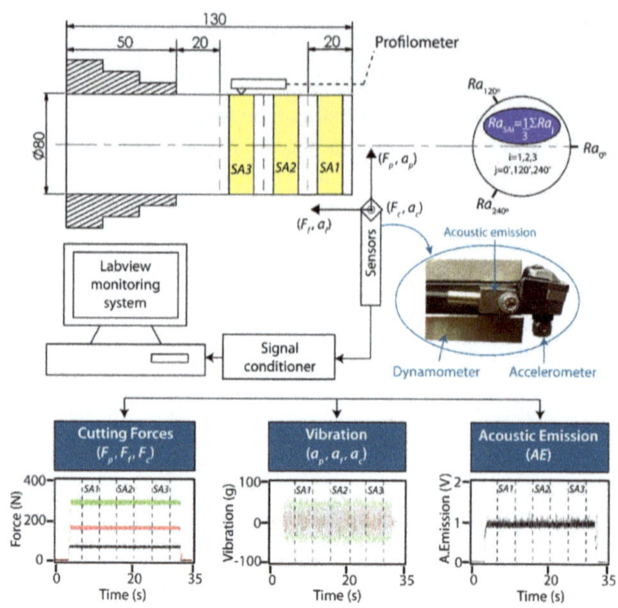

Figure 1. Experimental setup.

The methodology in this study involved the analysis of captured signals (F_p, F_f, F_c, a_p, a_f, a_c, AE) using the four signal feature extraction methods shown in Figure 2. The processed signals with TDA, PSD, SSA, and WPT methods were characterized by statistical and non-statistical parameters (Table 1). The signal feature extraction methods were evaluated in terms of each individual sensor, and multi-sensor. To correlate surface roughness with the signal characterization parameters, multiple regression and the artificial neural network were used as predictive modeling techniques. Finally, an optimum multi-sensor data fusion system was developed for real-time surface quality control.

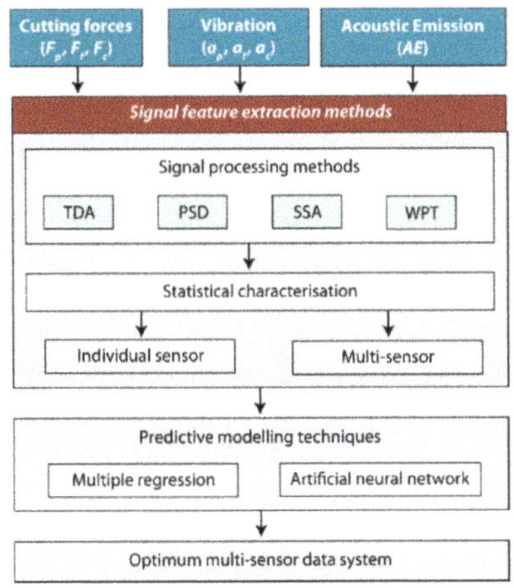

Figure 2. Methodology.

Table 1. Signal characterization parameters.

Features	Signal Components						
	F_p	F_f	F_c	a_p	a_f	a_c	AE
Mean	$X^j_{F_p}$	$X^j_{F_f}$	$X^j_{F_c}$	$X^j_{a_p}$	$X^j_{a_f}$	$X^j_{a_c}$	X^j_{AE}
Standard deviation	$\sigma^j_{F_p}$	$\sigma^j_{F_f}$	$\sigma^j_{F_c}$	$\sigma^j_{a_p}$	$\sigma^j_{a_f}$	$\sigma^j_{a_c}$	σ^j_{AE}
Peak to peak amplitude	$PP^j_{F_p}$	$PP^j_{F_f}$	$PP^j_{F_c}$	$PP^j_{a_p}$	$PP^j_{a_f}$	$PP^j_{a_c}$	PP^j_{AE}
Skewness	$S^j_{F_p}$	$S^j_{F_f}$	$S^j_{F_c}$	$S^j_{a_p}$	$S^j_{a_f}$	$S^j_{a_c}$	S^j_{AE}
Kurtosis	$K^j_{F_p}$	$K^j_{F_f}$	$K^j_{F_c}$	$K^j_{a_p}$	$K^j_{a_f}$	$K^j_{a_c}$	K^j_{AE}
Energy	$E^j_{F_p}$	$E^j_{F_f}$	$E^j_{F_c}$	$E^j_{a_p}$	$E^j_{a_f}$	$E^j_{a_c}$	E^j_{AE}
Shannon entropy	$SE^j_{F_p}$	$SE^j_{F_f}$	$SE^j_{F_c}$	$SE^j_{a_p}$	$SE^j_{a_f}$	$SE^j_{a_c}$	SE^j_{AE}

Where, $j = t$ is time for TDA method, SSA $j = \lambda_1, \ldots, \lambda_L$ are number of the eigenvalues for SSA method, PSD $j = f1, \ldots, f4$ are frequency ranges for PSD, and $j = A, \ldots, ADADA$ are frequency ranges (packets) for the WPT.

Multiple regression and artificial neural network predictive models were evaluated in four ways: (1) the goodness of fit to experimental data by the adjusted determination coefficient R^2_{adj}; (2) the predictive power by the mean relative error $\bar{e}_r$ (Equation (1)) in the prediction of the experimental validation data, and the variability of $\bar{e}_r$ by the standard deviation associated to the mean value $\sigma_{\bar{e}_r}$; (3) the reliability in the prediction by the $R_p^{0.25}$ coefficient [53], and the percentage in the distribution error; and (4) the correlation of the data estimated by the predictive models versus the experimental validation data (R). All of the models under analysis reached the minimum requirement of a mean relative prediction error of $\bar{e}_r \leq 25\%$, and reliability of $R_p^{0.25} \geq 75\%$ [53]. Of the 360 experimental data obtained, 75% were used for building the models, and the remaining 25.0% were randomly selected for model validation. The multiple regression predictive models were adjusted stepwise to include only the significant (p-value < 0.05) characterization parameters (Table 1). All regression models were diagnosed by analyzing atypical values, multicollinearity, independence, and normality of the residuals, homoscedasticity, and contrasts and hypothesis tests.

$$\bar{e}_r(\%) = \frac{1}{n}\sum_{i=1}^{n}\left|\frac{\left(Ra_i^{exp} - Ra_i^{pred}\right)}{Ra_i^{exp}}\right|100 \quad (1)$$

The design of the neural network was performed using a *feedforward* network structure with back-propagation training methodology. To determine the optimum network configuration, several training and transference functions widely used for the monitoring of machining processes were tested. As with the regression models, network validation was undertaken independently, with 25.0% of all the trials.

3. Results

3.1. Time Direct Analysis

The TDA method directly analyses the signal registered by the sensor in the time domain, with no transformation or decomposition, for fast processing at a low analytical-computational cost. This method is based on signal definition as an amplitude-time function $x(t)$ discretized by the succession $[x_i]$, with $i = 0, 1, 2, \ldots, N-1$, where N is the total number of points in the sample (Figure 3). The term N depends on sampling frequency (f_s) and sampling time (t), which has a direct impact on processing times. Thus, cutting force (F_p, F_f, F_c) and acoustic emission (AE) signals, both with $f_s = 5$ kHz, required shorter processing times than vibration signals (a_p, a_f, a_c) with $f_s = 50$ kHz. The TDA method performs signal feature extraction using parametric characterization of the original signal captured by the sensor, defined in the time domain. For parametric characterization, statistical measurements (arithmetic mean, standard deviation variance, kurtosis Shannon entropy, etc.) or non-statistical measurements (energy, maximum and minimum peak amplitude, etc.) can be used [50]. The efficiency of the TDA method depends on the type of signal to be processed, and the information to be extracted.

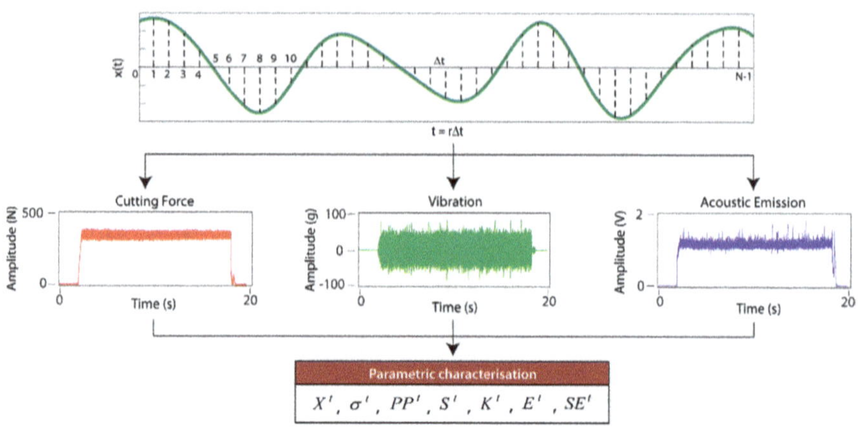

Figure 3. Time direct analysis signal processing method.

The signal feature extraction of cutting forces, vibration and acoustic emission with the TDA method, involved direct parametric signal characterization using the parameters outlined in Table 1. In order to correlate the signal information with surface finish, the variables were related using a multiple regression predictive model with higher R^2_{adj}, lower mean relative error $\overline{e}_r$, and higher reliability $R_p^{0.25}$. For the optimum characterization of the triaxial signal sensors (dynamometer and accelerometer), the signals were analyzed independently for each component, and the fusion of the three components. This methodology identified correlations between components from each sensor to avoid information overloading of the predictive model that would undermine the fit.

The results obtained for cutting force signals with the TDA method (Figure 4) showed that the back force F_p was the cutting force component most correlated to surface roughness (*Ra*), and provided the best results in all of the indices (R^2_{adj} = 79.4%, $\overline{e}_r$ = 17.8 ± 3.1%, $R_p^{0.25}$ = 71.1%). The back force F_p is responsible for tool–workpiece contact stability, and flexing of the workpiece machined in cantilever, given that this force was perpendicular to the axis of rotation of the workpiece. These results underscore the impact of this component (F_p) on surface finish, due to the influence of the tool–workpiece interaction, and the dynamic behavior of the rotated workpiece. The feed force F_f, and tangential force F_c showed weaker correlations to roughness, with an R^2_{adj} percentages of 43.4% and 34.0%, respectively. The model combining the three force components $(F_p + F_f + F_c)$, slightly improved the results of the F_p force model, increasing reliability to 76.7%, and improving the data fit to 82.5%, but with no improvement in predictive power with an $\bar{e}_r$ of 17.9 ± 3.4%. This corroborated that the back force F_p explained a larger percentage of the variability in the experimental data, indicating it was the component with the greatest impact on roughness (Ra). The feed force F_f and tangential force F_c complemented the information provided by the back force F_p, but the improvement in the fused model was relatively small.

With the TDA method, the individual analysis of vibration signal models (a_i) exhibited a moderate fit to data, with an R^2_{adj} below 63% for all of the components, and an $\bar{e}_r$ close to the critical value of 25% but never surpassing it. The three vibration components provided similar percentages of information, with the feed vibration a_f (R^2_{adj} = 62.2%) and radial vibration a_p (R^2_{adj} = 62.9%) explaining most variability of the experimental data, and to a lesser extent the a_p percentage (54.1%). The fused vibration model $(a_p + a_f + a_c)$ significantly improved model prediction, both in terms of the fit to data (81.6%), predictive power (20.5 ± 5.6%), and reliability (76.7%). This implied that the three vibration components (a_p, a_f, a_c) had a similar impact on surface finish, with a lower correlation among them than for the cutting force components.

The acoustic emission (AE) model obtained very poor results in all of the indicators analyzed, having little impact on the parameter Ra. This model explained only 17.1% of the variability of the experimental data, which indicated a very poor correlation to roughness (Ra). This implies information extracted for the AE signal and processed with the TDA method did not permit the correlation between this signal and surface roughness.

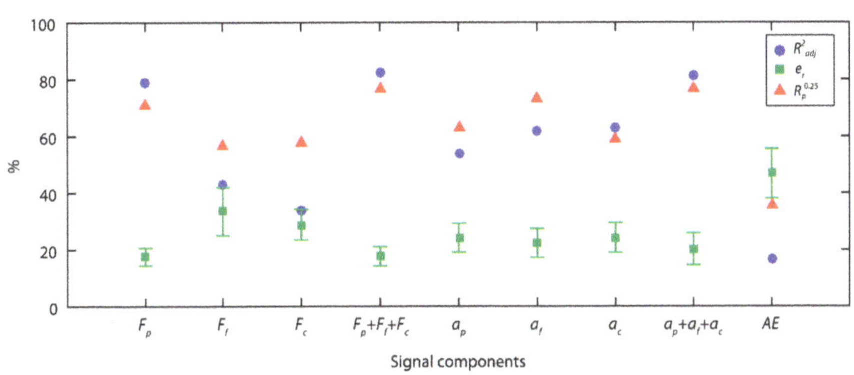

Figure 4. Signal analysis results with the TDA method.

The correlations of the estimated data versus the validation data for the best model obtained with each sensor are shown in Figure 5. Cutting force signals exhibited the best behavior (Figure 5a) with a correlation $R = 0.94$, and uniform behavior in all of the data ranges. The model tended to slightly overestimate, given that most of the estimated values were higher than the experimental validation values. As shown in Figure 5b, in the vibration model greater dispersion was observed in all of the value ranges, which weakened the correlation of $R = 0.87$ in comparison to cutting forces. As for acoustic emission (Figure 5c), a very poor correlation $R = 0.08$ was obtained, indicating a very poor correlation between this signal AE and surface finish with the TDA method. In relation to model reliability in terms of the distribution error of the validation data (Figure 5d), a similar behavior was observed between the cutting force model and the vibration model. In both models, from 56.0% to 58.0% of the validation trials were optimally estimated with an $e_r \leq 15\%$, and 76.7% of trials were satisfactorily predicted with error $e_r \leq 25\%$. The poor behavior observed with acoustic emission signal generated a model with 75% of estimated data out of the acceptable range.

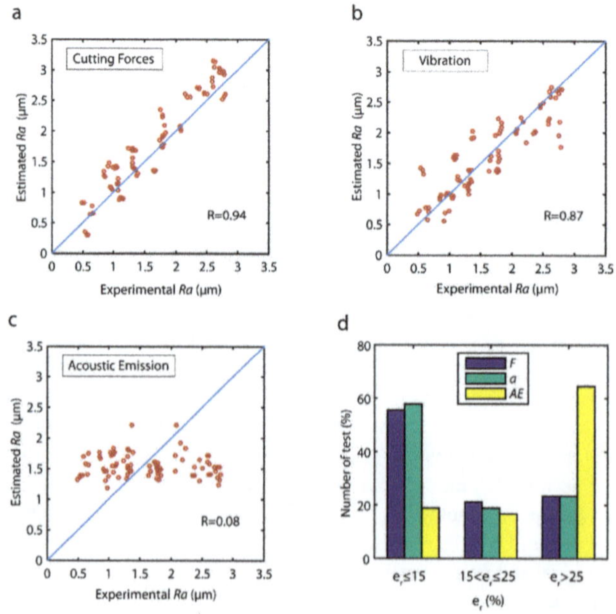

Figure 5. Estimated values versus experimental validation values for the parameter *Ra* with the TDA method: (**a**) cutting forces, (**b**) vibration, and (**c**) acoustic emission. (**d**) Prediction reliability with the TDA method.

3.2. Power Spectral Density

The PSD is a real positive function to calculate the distribution of the power of an original signal $x(t)$ along the entire frequency range registered by the signal. The PSD can be calculated according to Wiener–Khintchin's theorem as the Fourier transform of the autocorrelation function (Equation (2)).

$$S_x(f) = \int_{-T}^{T} R_x(\tau)e^{-j2\pi ft}dt \text{ where } R_x(\tau) = E\{x(t)x^*(t+\tau)\} \tag{2}$$

As mentioned in Section 2, the frequency ranges of the sensors were different: for cutting force and acoustic emission signals maximum frequency was ~1 kHz, whereas the maximum frequency for vibration was ~10 kHz. The broad bandwidth of the vibration signal entailed certain frequency ranges with significant information failed to be adequately characterized. Thus, the frequency analysis of the vibration signal was undertaken using two methods: a complete analysis of the entire bandwidth and a fractioned analysis by discretizing the bandwidth into four independent frequency ranges as shown in Figure 6.

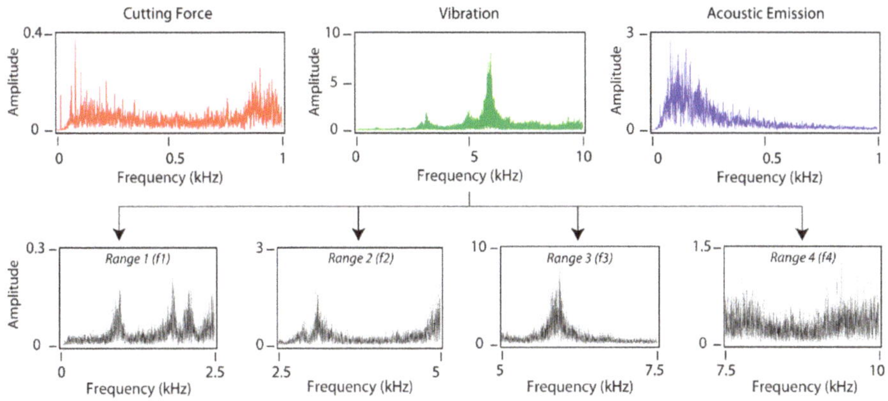

Figure 6. PSD signal processing method with four frequency ranges.

Following signal feature extraction and the building of the predictive models, Figure 7 shows the results obtained with the PSD method for each signal. For cutting force signals, neither the models built for each force component (F_p, F_f, F_c), nor the models of the fused components $(F_p + F_f + F_c)$, obtained satisfactory results, and neither reached the minimum criteria for a model to be considered acceptable. Vibration signal analysis in four frequency ranges (4R) improved the results obtained as compared to signal analysis in one single frequency range, but the mean relative errors and reliability were deficient. The feed vibration a_f was the most correlated to surface roughness with a R^2_{adj} of 70.1 %, with a lower correlation for the back vibration a_p, and tangential vibration a_c components of 46.4% and 47.5% of R^2_{adj}, respectively. The fused vibration model $(a_p + a_f + a_c)$ obtained the best results $(R^2_{adj} = 84.0\%$, $\bar{e}_r = 22.0 \pm 5.5\%$, $R_p^{0.25} = 71.1\%)$, underscoring it was the best predictive model of those analyzed. Similar to TDA method, the acoustic emission signals analyzed with the PSD method also obtained poor results, with a low correlation between this signal and surface finish.

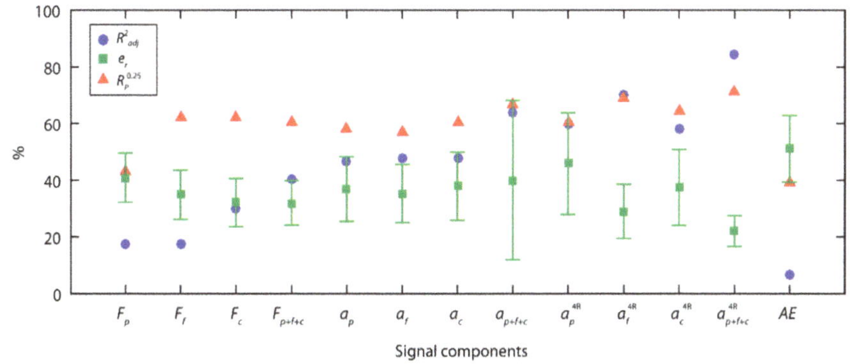

Figure 7. Signal analysis results with the PSD method.

With reference to estimated validation data (Figure 8), the best cutting force model $(F_p + F_f + F_c)$ obtained a low correlation of $R = 0.68$ (Figure 8a), with inaccurate estimates in high and low roughness values, and a slightly better behavior in mid-range values. The fused vibration model $(a_p + a_f + a_c)$ with four frequency ranges (4R) (Figure 8b) obtained the highest correlation (R = 0.76), with uniform estimated data in all of the ranges of surface finish (Figure 8b). The acoustic emission signal was not correlated in anyway $(R = 0.06)$ with very deficient results being obtained.

The analysis of model reliability in terms of the distribution error in the prediction of the validation data (Figure 8d), revealed the best model was fused vibration signals $(a_p + a_f + a_c)$ in four frequency ranges with an $R_p^{0.25}$ of 71.1%, followed by fused cutting forces model $(F_p + F_f + F_c)$ with an $R_p^{0.25}$ of 60.0%, and acoustic emission with very poor results ($R_p^{0.25}$ = 38.9%).

The analysis revealed the PSD method failed to provide adequate signal feature extraction for the prediction of surface roughness, given that none of the models analyzed reached a prediction reliability of 75%.

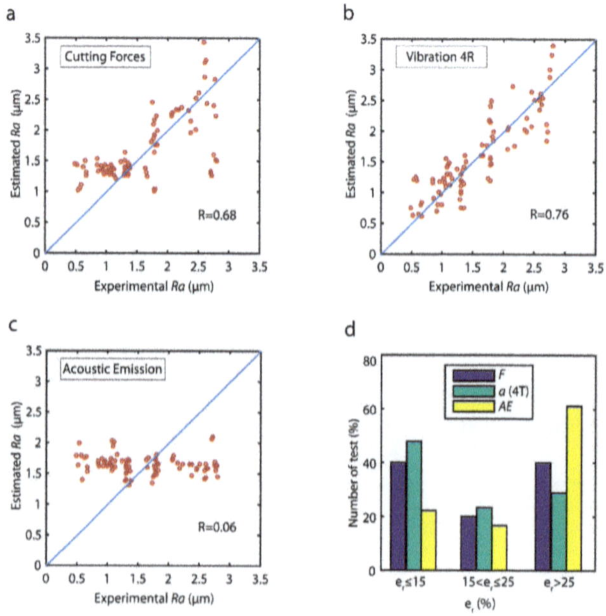

Figure 8. Estimated values versus experimental validation values of the parameter *Ra* for the PSD method: (**a**) cutting forces, (**b**) vibration (4R), and (**c**) acoustic emission. (**d**) Prediction reliability with the PSD method.

3.3. *Singular Spectrum Analysis*

The SSA method is a non-parametric time series analysis technique, based on statistical multivariability, multivariante geometry, and dynamic signal processing systems. The SSA method decomposes a signal into independent time series (with defined frequency ranges) referred to as principal components (PC_i). The SSA method builds a Hankel matrix termed the trajectory matrix X, calculated through a sliding windows (L) applied to the succession of data $[x_i]$ from the original signal. The next step is singular value decomposition (SVD) of the trajectory matrix by decomposing the X matrix into a series of elementary matrixes X_i obtained by calculating the eigenvalues and eigenvectors of the matrix $S = XX^T$. Finally, the principal components are obtained from the reconstruction of the elementary matrixes X_i [50]. The number of principal components obtained with the SSA method depends directly on the parameter L, which is a significant factor conditioning the results. To determine the principal components containing information of the original signal, it is standard practice to show in descending order the weight of each principal component (or eigenvalue) on a graph commonly referred to as Singular Spectrum (*SS*), which give its name to the SSA method itself.

The application of *SS* to cutting force, vibration, and acoustic emission signals for the window lengths L = 5 and L = 10, showed significant differences (Figure 9). A similar behavior was observed in cutting forces and acoustic emission (Figure 9a), where the eigenvalue associated to the first principal component (PC_1) contained approximately 100% of the weight of the total signal, and the remaining principal component values obtained were almost negligent. The application of the SSA method to vibration signals exhibited a different behavior (Figure 9b) with non-zero eigenvalues, resulting in principal components with different levels of information.

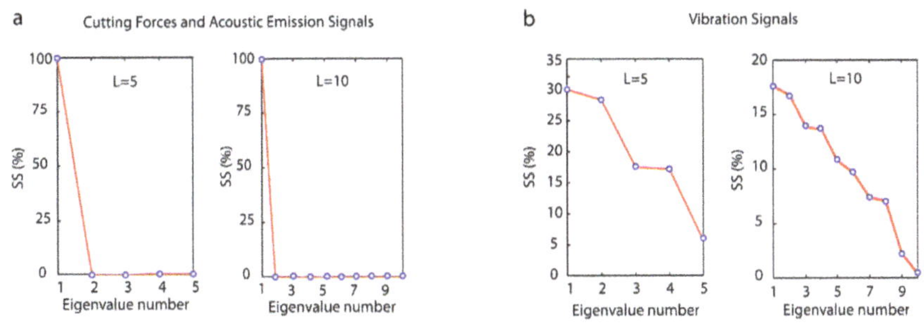

Figure 9. Signal analysis with the SSA method using two window lengths $L = 5$ and $L = 10$: (**a**) cutting forces and acoustic emission, and (**b**) vibration.

Figure 10 shows the principal components obtained with the SSA method for a window length of $L = 5$. Cutting forces and acoustic emission only contained information in the first principal component (PC_1), the remaining components being negligible, which corroborated the *SS* results (Figure 9). The first principal component (PC_1) was the original signal, thus the results obtained for the signals with the SSA method were equivalent to those obtained with the TDA method. This phenomenon occurred in signals with a characteristic function $x(t) = A\sin t + b$ where the factor b represented signal amplitude, and the term $A\sin t$ the signal oscillation. With SSA decomposition, this amplitude and the oscillation range were contained in the first principal component, and the remaining components provided little or negligible additional information. Moreover, the vibration signal obtained significant information from the original signal in the five principal components as these signals had a value of $b \approx 0$. The selection of the vibration signal configuration parameters for the SSA method was determined on the basis of the results obtained by García and Núñez [50].

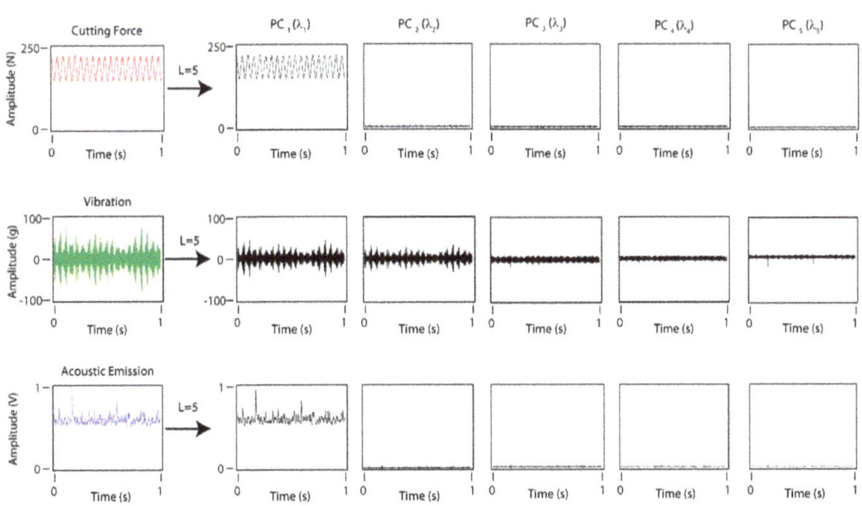

Figure 10. Principal components obtained with the WPT method for cutting forces, vibration, and acoustic emission, and a window length of $L = 5$.

The individual analysis of the vibration signals with the SSA method (Figure 11) showed the feed vibration a_f explained most of the variability in the experimental data with a fit of data of 82.8%, a predictive power of 16.7 ± 4.3%, and 76.7% of reliability. The radial vibration a_p and tangential vibration a_c components explained less variability (65.1% and 68.4%, respectively) with high relative errors (29.0% and 24.5%, respectively) and low reliabilities (61.1% and 64.4%, respectively) that failed to reach acceptable levels. In contrast, the combination of the vibration components $(a_p + a_f + a_c)$ improved the model in all of the evaluation indices of the individual analyses, with a R^2_{adj} of 87.8%, sharp fall in the relative error $\overline{e}_r$ reaching a value of 14.6%, and a significant increase in reliability to 91.1%. These results revealed that feed vibration a_f was the component with the greatest impact on surface finish (*Ra*), and that the radial a_p and tangential a_c vibration complemented the information provided by a_f.

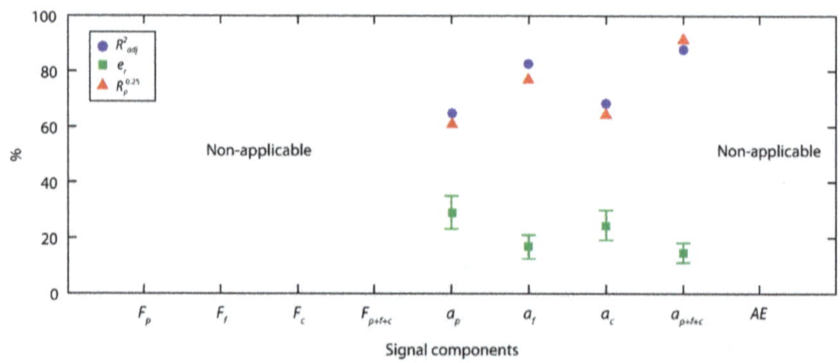

Figure 11. Signal analysis results with the SSA method.

The prediction of the validation data versus the experimental values for the best vibration model obtain with SSA method $(a_p + a_f + a_c)$ is shown in Figure 12a. The results showed the model had a good predictive power with a high correlation ($R = 0.93$), without any significant bias, except for a slight underestimation in the prediction of *Ra* values above 2.5 µm. The analysis of model reliability in terms of the distribution error in the prediction of the validation data showed the model had very good reliability (Figure 12b), with 67.8% optimum estimation of the data ($e_r \leq 15\%$), and 91.1% acceptable predictions ($e_r \leq 25\%$).

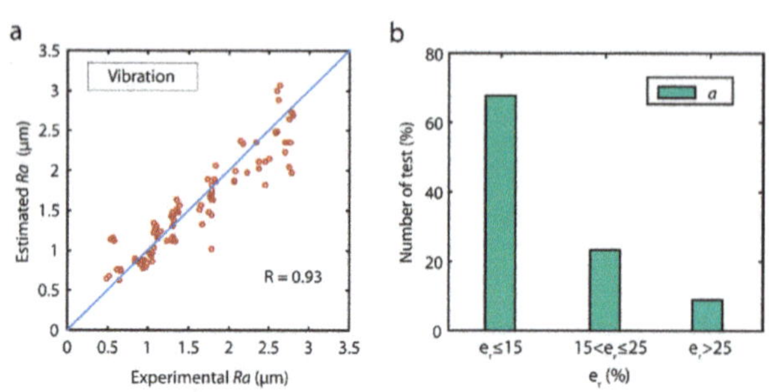

Figure 12. (**a**) Estimated data versus experimental validation data of the parameter *Ra*. (**b**) Reliability of predictive models.

3.4. Wavelet Packet Transform

The WPT method decomposes a signal into scaled and shifted series (packets) of a prototype function referred to as the mother wavelet, which is characterized on a time-frequency scale.

The method applies a pyramidal algorithm, where the original signal is successively split into approximation A_j (low frequencies) and detail D_j (high frequency) signals until the desired wavelet decomposition level is achieved. The approximation and detail signals are calculated by the coefficients described in Equations (3) and (4),

$$A_j\,(k) = \sum_n h(n-2k)c_{j-1}(n) \quad (3)$$

$$D_j\,(k) = \sum_n g(n-2k)c_{j-1}(n) \quad (4)$$

where $A_j\,(k)$ and $D_j\,(k)$ are approximation and detail coefficients, j is the number of transformation levels with $j = 1, 2, \ldots$; k is the number of scaled and wavelet coefficients with $k = 1, 2, \ldots, Nx2^{-j}$, where N is the total number of samples of the original signal; h and g are low-pass and high-pass coefficients of the scaled function and wavelet function, respectively, based on a chosen mother wavelet; and n is the filter length.

For the satisfactory application of the WPT, three fundamental factors should be borne in mind: (1) selection of the appropriate mother wavelet for each specific type of signal; (2) to determine the number of decomposition levels (L_d) needed to divide the signal into effective frequency ranges; and (3) selection of the information packets of significance to the parameter under analysis. The optimum selection of the configuration parameters applied in the WPT method to cutting force and vibration signals for the monitoring of surface finish was determined by García and Núñez in two previous studies [52,53], respectively, where the mother wavelet *bior4.4* with three decomposition levels provided the best results for vibration signals, and the mother wavelet *db06* with four decomposition levels provided the best results for cutting forces. As there are no studies published in the literature determining the optimum configuration for acoustic emission signals, this study replicated the methodology employed by García y Núñez, establishing the best configuration for mother wavelet *coifflet4.4* (*coif4.4*), and a number of decomposition levels $L_d = 5$.

The sampling frequency (f_s), and decomposition level (L_d) determined the frequency ranges of the original signal (Figure 13). For cutting force signals, seven frequency ranges of 156.25 Hz were obtained, for vibration signals four frequency ranges of 3125 Hz, and for acoustic emission 13 frequency ranges of 78.125 Hz (Figure 13).

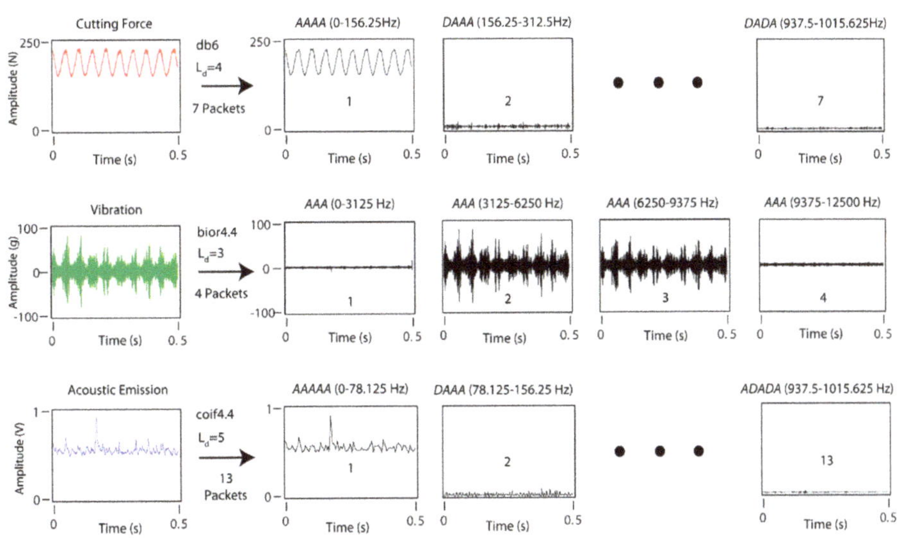

Figure 13. WPT method applied to cutting force, vibration, and acoustic emission signals.

As shown in Figure 14, in the individual analysis of cutting forces, the back force F_p was the principal information source explaining most of the variability in the experimental data, with a data fit of 86.4% and 14.1 ± 2.5% of predictive power. As with the TDA method, the back force F_p was the cutting force component having the greatest impact on surface finish, with better prediction indices than in the TDA method. Nevertheless, in this case the tangential force F_c also obtained a good fit to the experimental data with an R^2_{adj} of 77.7%, and an $\bar{e}_r$ of 20.7 ± 3.6%. As shown in the analysis, the WPT method only analyzed time series in effective frequency ranges with relevant information, excluding series that masked the original signal, and impeded adequate signal feature extraction in methods such as the TDA. This enhanced the analysis of cutting force signals, eliminated noise from the signal, with the tangential force F_c exhibiting a greater influence on roughness.

The fusion of the three orthogonal cutting force components ($F_p + F_f + F_c$) hardly improved the fit of the model with an R^2_{adj} of 88.0%, indicating a strong correlation among the cutting force components. In contrast, the fused model ($F_p + F_f + F_c$) improved the predictive power with an $\bar{e}_r$ of 11.9 ± 1.9%, and an 86.7% of reliability. These results revealed that the back force F_p, and tangential F_c were had the greatest influence on surface finish (*Ra*), and the feed force F_f complemented the information, improving the predictive power with the fusion of the cutting force components.

The results obtained for the vibration signals were similar to those obtained with the SSA method. In the individual analysis of the vibration components (a_p, a_f, a_c), once again the feed vibration a_f provided most of the information, with a data fit of 75.7%, a predictive power of 18.0 ± 4.3%, and a reliability of 75.6%. The results for the back a_p (R^2_{adj} = 62.0%) and tangential a_c (R^2_{adj} = 61.2%) components were poorer than those obtained for a_f with high relative errors and low reliabilities that failed to reach minimum acceptable levels. The combination of the three vibration components ($a_p + a_f + a_c$) improved the model in all of the evaluation indices (R^2_{adj} = 88.5%, $\bar{e}_r$ = 14.2 ± 3.8%, $R_p^{0.25}$ = 93.3%) The acoustic emission signal (*AE*) failed to obtain good results with the WPT method, with all of the indicators of predictive power being deficient.

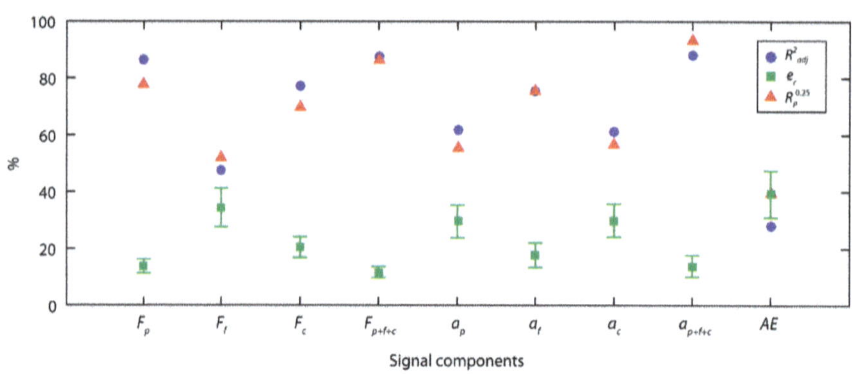

Figure 14. Signal analysis results with the WPT method.

The correlation between estimated data versus the validation data for the model obtained for each sensor is shown in Figure 15. The cutting force and vibration models obtained the best results with very high *R* correlations of 0.95 and 0.94, respectively. Though the data distribution in both models was fairly uniform in all of the ranges of surface finish (*Ra*), a slight overestimation was observed in all of the data ranges in the cutting force model (Figure 15a), which tended to increase at higher roughness values (2.6 μm ≤ *Ra* ≤ 3.0 μm). As for acoustic emission (Figure 15c), the correlation obtained was very deficient (*R* = 0.37). The analysis of model reliability (Figure 15d) revealed a similar behavior, with cutting force and vibration models obtaining the best results. Both models obtained similar results in the optimum prediction range, where 67.0–69.0% of the validation trials had an estimated error $e_r \leq 15\%$; however, in the interval of acceptable predictions, the vibration model obtained the best results, with an error $e_r \leq 25\%$ in 93.3% of the validation data, compared to 86.7% for the cutting force model. The poor behavior obtained with the acoustic emission signal produced a model where 60.0% of estimated data was outside the acceptable range.

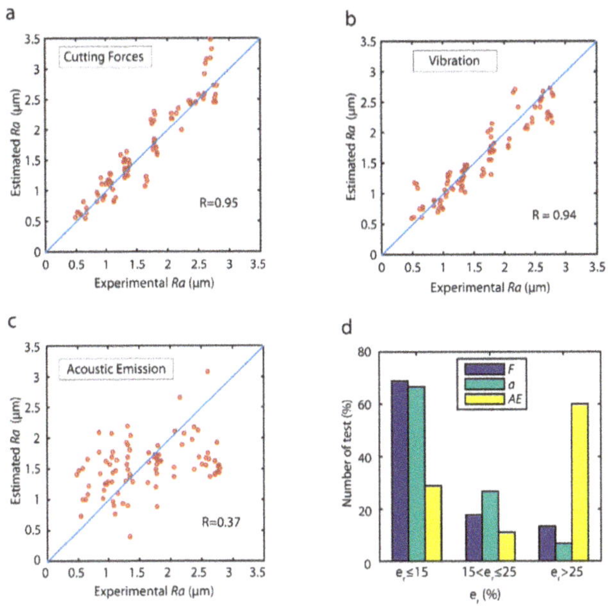

Figure 15. Estimated data versus experimental validation data for the parameter *Ra*: (**a**) cutting forces, (**b**) vibration, and (**c**) acoustic emission. (**d**) Reliability of predictive models.

3.5. Comparison of Methods

A comparison of the best predictive models obtained in the previous section, classified according to processing method and the type of signal under analysis, is shown in Figure 16. In order to draw a more accurate comparison of the processing methods in a single domain (TDA and PSD) versus methods of analysis in the time-frequency domain (SSA and WPT), the predictive models obtained with the fusion of the TDA and PSD methods were compared. To evaluate the computational cost of each method, the monitoring system response time in the processing of a second signal was analyzed (Figure 17).

In the analysis of cutting forces, the WPT method presented the best prediction results with an $\bar{e}_r$ of $12.0 \pm 2.0\%$, a reliability of 88.7%, and a 24 ms response time. As shown in Figure 16, the results obtained with the other methods fell far short of the method WPT. It should be noted that the characterization of cutting force signals in the frequency domain presented the worst results, and the combination with the TDA method worsened the predictive power $\bar{e}_r$ of the TDA method.

In the analysis of the vibration signals, the SSA and WPT methods presented the best prediction results with an $\bar{e}_r$ of $14.6 \pm 3.5\%$ and $14.2 \pm 3.8\%$, and reliability of 91.1% and 93.3%, respectively. In comparison, the prediction results obtained with the other methods fell far short of the SSA and WPT methods. In this case, PSD analysis provided complementary information to the TDA analysis, which improved the prediction and reliability results. For the vibration signals, system response times increased significantly as sampling frequency was five times greater than in cutting force signals. In spite of this increase, the processing time of the WPT method was sufficiently low for real-time monitoring (101 ms), but the SSA method presented a very long response time (10,750 ms), which discarded it as feasible for the real-time prediction of surface finish.

Acoustic emission failed to provide good results in all of the methods under analysis, with a slight improvement in the WPT method, but failing to obtain satisfactory results. The lowest computational costs were for acoustic emission in comparison to the other signals, owing to the smaller amount of information processed in one single signal.

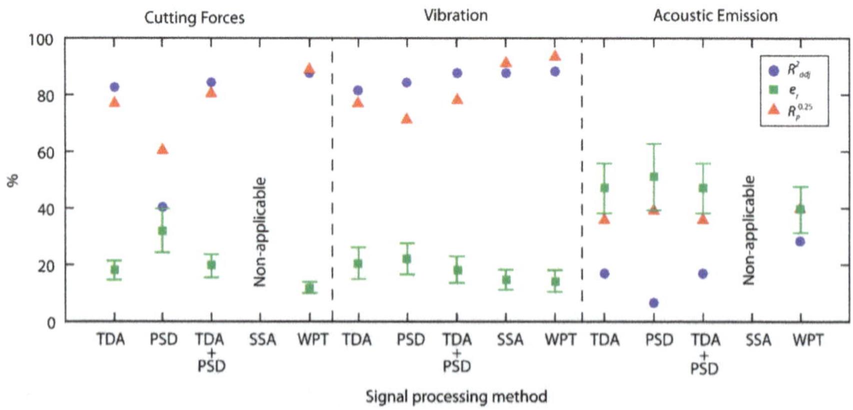

Figure 16. Comparison of the predictive results according to processing method and signal type.

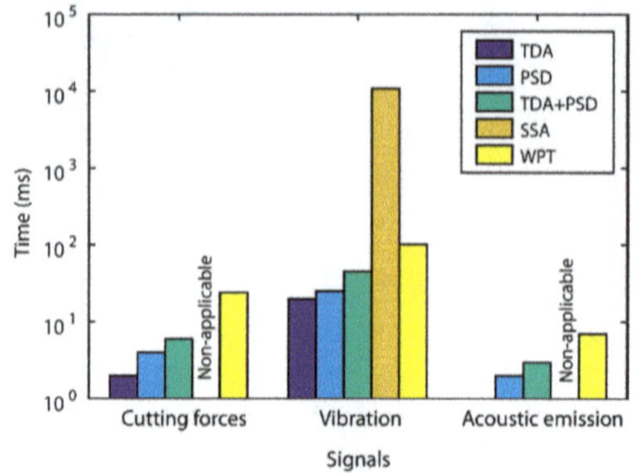

Figure 17. Response times obtained for the TDA, PSD, TDA + PSD, SSA, and WPT processing methods.

4. Multi-Sensor Data Fusion Analysis

4.1. Comparative Analysis of Sensor Fusion

Having determined the behavior of the processing methods individually for each signal, the next step was to analyze the multi-sensor fusion data by building a fused model of each processing method, using the significant characterization parameters obtained in each individual analysis. It should be borne in mind that the acoustic emission signal provided poor prediction results in each individual analysis, which indicated the *AE* signal contained no information correlated to surface finish. Thus, two types of analysis of the fusion of sensors were performed: the fusion of all the sensors, and the fusion of all the sensors except the acoustic emission signal. It should be noted that the SSA method can only be applied efficaciously to vibration signals as explained previously in Section 3.3. Nevertheless, the SSA method was applied to all of the signals in order to compare the fusion of sensors, taking into account that cutting force and acoustic emission signals were equivalent to applying the TDA method.

As shown in Figure 18, with exception of PSD analysis, the results obtained with the other methods and the fusion of three sensors provided predictive models with low relative errors $\bar{e}_r < 12\%$, and high reliability over 88.0%, which improved the results obtained with the individual analysis of each sensor. The best result was obtained for the SSA method, with excellent predictive power $\bar{e}_r = 8.2 \pm 1.6\%$, and very high reliability of 93.3%. Similar results were obtained for the WPT method with a mean relative error of 10.8 ± 2.0% and a reliability of 91.1%.

When the *AE* signal was eliminated from multi-sensor data fusion, with the exception of the model of the PSD method, the predictive power of the other models improved, indicating the *AE* signal provided no significant information for the prediction of surface finish, or even negatively affected the prediction. The results for the fusion of cutting force and vibration signals were similar with hardly any considerable differences between them, with the exception of the PSD method. All of the methods exhibited an excellent predictive power, particularly the WPT and SSA methods that obtained an excellent mean relative error of 8.5 ± 1.5% and 8.8 ± 1.8%, respectively, with an optimum reliability of 95.5% in both. Even the TDA method and TDA+PSD fusion obtained excellent results with the fusion of sensors.

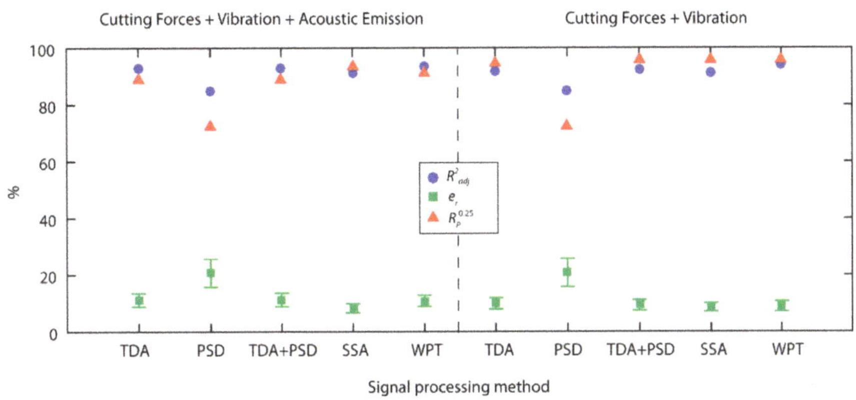

Figure 18. Comparison of the prediction results obtained with multi-sensor data fusion with and without the acoustic emission signal.

The correlations between the estimated data and the validation data of the models obtained with the fusion of sensors are shown in Figure 19. The three models had very strong correlations (Figure 19), with the SSA (Figure 19b) and WPT (Figure 19b) methods exhibiting a uniform behavior in the entire range of experimental data. However, the TDA + PSD method underperformed in data prediction at the 2.5 ÷ 3.0 µm interval. The analysis of model reliability (Figure 19d), in the range of optimum predictions ($e_r \leq 15\%$), showed the SSA model had the best performance with 85.5% of the data, followed by the WPT model with 80.0% of the data, and the TDA + PSD model with 75.5%. In the acceptable prediction range ($e_r \leq 25\%$), excellent results were obtained for the three models with a 95.5% reliability in all three models.

Bearing in mind these data and the aforementioned computational costs, the WPT model with the fusion of cutting force and vibration sensors was the best option for the time-real prediction of surface quality in CNC automated machining processes. Moreover, the WPT method was applicable to all of the signals analyzed and enabled the determination of effective frequency ranges correlated to surface finish. The significant characterization parameters obtained with the WPT model and the corresponding frequency ranges, the sum of squares (type III), and the *p*-values are shown in Table 2. The most relevant information of cutting force signals was found at very low frequency ranges *AAAA* (0–156.25 Hz), with a small contribution from low *DAAA* (156.25–312.50 Hz), and very high *DADA* frequencies (937.50–1093.75 Hz), and negligible information from the other frequency ranges analyzed. The behavior of the vibration signals was entirely different with high frequency *DDA* (6250–9375 Hz) providing most of the relevant information, followed by the very high *ADA* frequency (9375–12,500 Hz), and no information provided by the other frequency ranges.

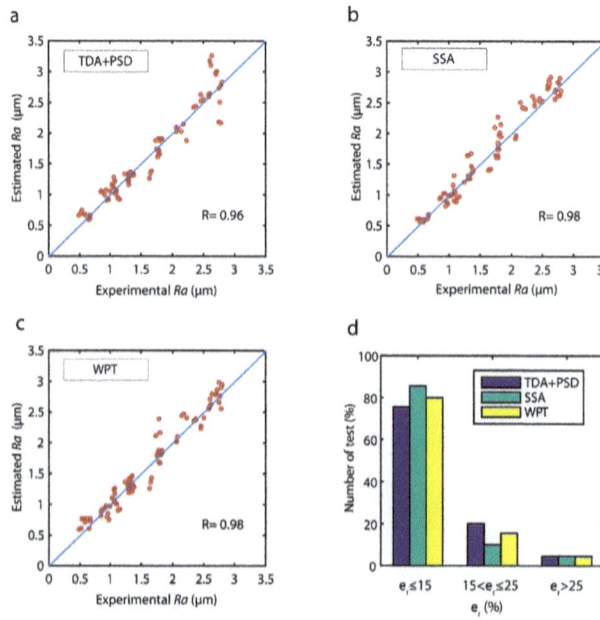

Figure 19. Estimated data versus experimental validation data for the parameter *Ra* with multi-sensor data fusion for the (**a**) TDA + PSD, (**b**) SSA, (**c**) WPT methods, and (**d**) model reliability.

Table 2. Significant signal feature extraction of the optimum model obtained with WPT.

Feature	Frequency (Hz)	Sum of Sq (Type III)	*p*-Value
$X_{F_p}^{AAAA}$	0–156.25	3.30	3.10×10^{-18}
$\sigma_{F_p}^{DAAA}$	156.25–312.50	0.22	1.39×10^{-2}
$K_{F_f}^{DADA}$	937.50–1093.75	0.19	2.17×10^{-2}
$SE_{F_c}^{AAAA}$	0–156.25	5.37	1.26×10^{-26}
$SE_{a_p}^{DDA}$	6250–9375	1.42	2.41×10^{-9}
$\sigma_{a_f}^{DDA}$	6250–9375	3.82	1.77×10^{-20}
$PP_{a_f}^{DDA}$	6250–9375	0.14	5.38×10^{-2}
$SE_{a_f}^{DDA}$	6250–9375	1.41	2.74×10^{-9}
$X_{a_f}^{ADA}$	9375–12,500	1.45	1.81×10^{-9}
$SE_{a_f}^{ADA}$	9375–12,500	0.67	3.03×10^{-5}
$\sigma_{a_c}^{DDA}$	6250–9375	3.81	1.90×10^{-20}
$SE_{a_c}^{ADA}$	9375–12,500	1.22	2.72×10^{-8}

As for the level of information provided by the cutting forces, the mean of the back force $X_{F_p}^{AAAA}$, and Shannon entropy of the tangential force $SE_{F_c}^{AAAA}$ were the primary information sources. Both components (F_p and F_c) were responsible for the load on the tool in the direction perpendicular to the axis of rotation, and flexing of the workpiece at the cantilever. This aspect led to displacement and eccentric rotation of the workpiece, which altered the dynamic behavior and causes vibrations in workpiece-tool contact areas. The parameter $X_{F_p}^{AAAA}$ measured the static component of the back force F_p, and $SE_{F_c}^{AAAA}$ measured the dynamic component of tangential force F_c.

In relation to the vibration signal, the parameters measuring the dynamic behavior of the signal were the most significant, with the feed $\sigma_{a_f}^{DDA}$ and tangential $\sigma_{a_c}^{DDA}$ standard deviation being the principal sources of information, complemented by mean feed vibration $X_{a_f}^{ADA}$ and the entropy of the three components $SE_{a_p}^{DDA}$, $SE_{a_f}^{DDA}$, $SE_{a_f}^{ADA}$, $SE_{a_c}^{ADA}$. The feed component of the vibration signal provided most of the variables with the greatest impact on the prediction of surface roughness. Feed vibration a_f was the vibration component most affecting surface roughness, and was directly correlated to the parameter *Ra*.

4.2. Comparative Analysis of Predictive Techniques

The ANN optimized for the prediction of surface finish had a feedforward structure with back-propagation training methodology. In order to obtain the optimum network configuration, the training and transference functions available in Matlab 2018 were analyzed using a pyramidal criterion [67] to determine the number of layers and neurons providing the best results. Similar to the multiple regression models, network validation was undertaken independently with 25.0% of the randomly selected experimental data. The input variables were restricted to significant signal characterization parameters obtained with the regression models to ensure the fit of the network was not weakened due to information overloading or correlations between variables. The optimum networks obtained for the TDA + PSD and SSA methods had a structure 13 × 6 × 2 × 1 and 10 × 3 × 1, respectively, with the *tansig* transference function and *trainlm* training. The network for the WPT method had a structure 12 × 3 × 1 with the *purelin* transfer function and *trainlm* training.

The results obtained for the prediction of surface finish with ANN and multiple regression are shown in Figure 20. The two prediction techniques analyzed obtained excellent prediction results with $\bar{e}_r$ below 10.0 ± 2.0%, and reliability over 95.0% for the three methods analyzed (TDA + PSD, SSA, and WPT). The ANN predictive modeling technique improved the $\bar{e}_r$ from 1.0% to 1.2 %, obtaining 8.2 ± 1.7% for TDA + PSD, 7.5 ± 1.3% for SSA, and 7.7 ± 1.6% for the WPT method. As for reliability, a 1.1% to 2.2% improvement was observed, obtaining 96.7% for the TDA + PSD, 97.8% for SSA, and 96.7% for the WPT method. Though the improvement in prediction and reliability may initially appear to be insignificant, it entailed a substantial increase in the performance of the predictive model, owing to the low prediction errors in both models, and the precision requirements of surface quality control systems.

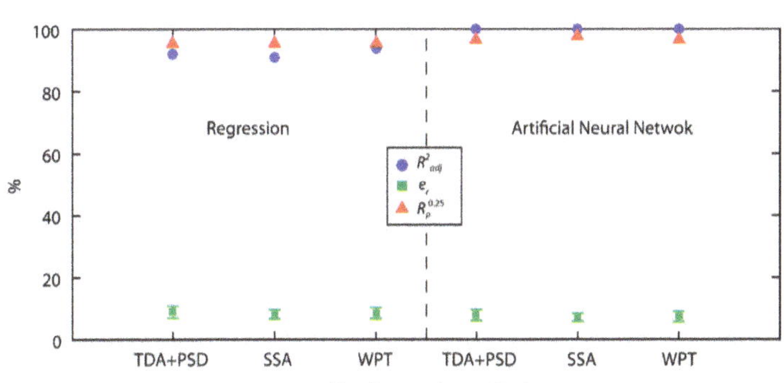

Figure 20. Prediction results for the MR and ANN models with multi-sensor data fusion of cutting force and vibration signals.

The correlation of estimated data versus experimental validation data (Figure 21) of the ANN models was very high, with excellent behavior in all of the value ranges. The reliability of the models (Figure 21d) was very similar, with the SSA method showing the best results with 86.6% of the predictive data in the optimum range ($e_r \leq 15\%$) versus the TDA + PSD and WPT methods that reached a reliability of 83.3%. As for the range of acceptable predictions ($e_r \leq 25\%$), the three models presented similar results with 97.8% for the SSA, and 96.7% for the TDA + PSD and WPT methods, indicating only from 2.2% to 3.3% of the data was predicted with an e_r greater than 25.0%.

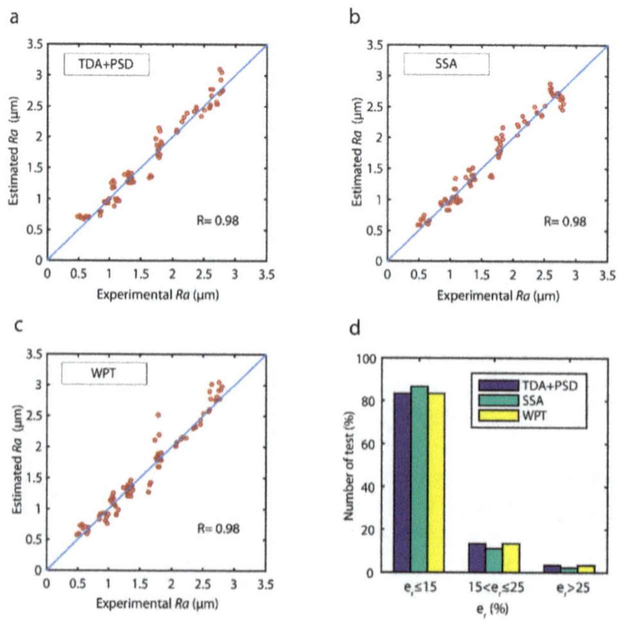

Figure 21. Correlation of estimated data versus validation experimental data obtained with ANN predictive model and multi-sensor data fusion: (**a**) TDA + PSD, (**b**) SSA, and (**c**) WPT methods. (**d**) Prediction reliability with the ANN model.

The best prediction and reliability results were obtained with the ANN model for the SSA and WPT methods, reaching excellent precision and reliability indices. The WPT method, with short responses times was the most adequate signal processing method for real-time signal feature extraction in surface quality control. It is worth noting that multiple regression enabled the determination of the significant characterization parameters of signals, eliminating factors correlated to or with little significant impact on the response variable (*Ra*), which improved the fit of ANN. In comparison to the ANN technique, multiple regression permitted greater control of the method, without random events. Thus, the use of the ANN predictive modeling technique is recommended, but with the prior multiple regression analysis of significant variables to ensure only input variables providing significant information correlated to surface quality were used in the ANN.

5. Conclusions

In this study, multi-sensor data fusion for the real-time intelligent control of surface quality in CNC machining systems was examined. The sensors most widely used for the monitoring of the machining processes based on cutting force, vibration, and acoustic emission signals were analyzed. A total of four signal processing methods were compared: two global analysis of the signal in one single domain, time (TDA) or frequency (PSD), and two signal analysis methods in the combined time-frequency domain (SSA and WPT). Owing to the nonlinear and stochastic nature of the process, two predictive modeling techniques were evaluated: multiple regression and artificial neural networks. The information provided by each individual sensor and the fusion of sensors was evaluated in terms of predictive power, reliability, and response time.

In the analysis of individual signals, both cutting forces and vibration were correlated to surface finish, allowing for the building of models with the best predictive results. In general terms, the back force F_p, responsible of the contact stability between tool and workpiece, was the cutting force component with the greatest influence on roughness. As for the vibration signals, the tool dynamic behavior in feed direction, registered by a_f, was the component with most impact on surface finish. Regarding the fusion of signals, the combination of cutting force and vibration signals improved predictive power and reliability. No correlation was observed between the *AE* signal and surface finish in either the individual or the fusion of other signals, and both even worsened the results obtained.

In the global signal analysis methods in the single domain (TDA and PSD), frequency analysis exhibited the worst behavior, with low prediction and reliability levels in all the signals analyzed, proving to be inadequate for signal feature extraction on its own. The TDA method improved the results of the PSD, which reached the minimum levels required, but with prediction and reliability below the SSA and WPT methods, particularly for individual signals. In most cases, the TDA + PSD combination slightly improved the results of the TDA method, indicating this combination could be applicable according to the precision of system requirements. The analytical-computational cost of these methods was very low, with the best response times.

The decomposition of the signals into a time series with defined frequency ranges improved signal feature extraction. The WPT method maximized information extraction in cutting force and vibration signals, whereas the SSA method was only applicable to vibration signals. The SSA method obtained long response times, rendering it inadequate for real-time monitoring systems. However, the WPT had shorter response times, making it the most adequate signal feature extraction method for use in real-time surface quality control systems.

Frequency bandwidths with information correlated to surface finish in both cutting force and vibration signals, and the sampling frequencies for each type of signal for the monitoring of surface finish were determined. The excitation ranges of cutting force signals that most influenced surface finish were found at very low frequencies (0–156.25 Hz), whilst vibration signals were at high frequency (6250–9375 Hz). Moreover, the vibration frequencies most affecting surface finish were determined for the selection of cutting speeds that did not excite the system in these frequency ranges.

The ANN models obtained the best prediction and reliability results, and was the most efficient predictive technique, but the prior elimination of non-significant characterization parameters using multiple regression is recommended.

Furthermore, the most adequate characterization parameters were determined for signal feature extraction of the three types of signals analyzed. This allowed for optimum parametric signal characterization by monitoring only parameters correlated to surface finish.

The present study proposes a multi-sensor data fusion method for on-line monitoring of surface quality in automated machining systems. Multi-sensor data fusion enabled highly accurate and reliable real-time monitoring of surface finish similar to post-process methods. This allows for in-situ decision-making based on the predictions of intelligent surface quality control systems that analyze only on-line cutting process information, which translates into greater flexibility and the real-time detection of malfunctions. The results obtained underscored that the selection of sensors, signal feature extraction method, and predictive modeling technique were crucial aspects for optimizing monitoring systems.

Author Contributions: Conceptualization, E.G.P. and P.J.N.L.; Methodology, E.G.P. and P.J.N.L.; Software, E.G.P.; Validation, E.G.P. and P.J.N.L.; Formal Analysis, E.G.P., P.J.N.L. and E.M.B.G.; Investigation, E.G.P., P.J.N.L. and E.M.B.G.; Resources, E.G.P. and P.J.N.L.; Data Curation, E.G.P.; Writing-Original Draft Preparation, E.G.P. and P.J.N.L.; Writing-Review & Editing, E.G.P., P.J.N.L. and E.M.B.G.; Visualization, E.G.P.; Supervision, P.J.N.L.; Project Administration, P.J.N.L.; Funding Acquisition, E.G.P. and P.J.N.L.

Funding: This research was funded by the Government of the Autonomous Community of Castilla-La Mancha (Spain) grant number PPII-2014-010-A, and the University of Castilla-La Mancha grant number GI20163590.

Conflicts of Interest: The authors declare no conflicts of interest.

References

1. Hocheng, H.; Tseng, H.; Hsieh, M.; Lin, Y. Tool wear monitoring in single-point diamond turning using laserscattering from machined workpiece. *J. Manuf. Process.* **2018**, *31*, 405–415. [CrossRef]
2. Barreiro, J.; Fernández-Abia, A.; González-Laguna, A.; Pereira, O. TCM system in contour milling of very thick-very large steel plates based on vibration and AE signals. *J. Mater. Process. Technol.* **2017**, *246*, 144–157. [CrossRef]
3. Nouri, M.; Fussell, B.; Ziniti, B.; Linder, E. Real-time tool wear monitoring in milling using a cutting condition. *Int. J. Mach. Tool Manuf.* **2015**, *89*, 1–13. [CrossRef]
4. Zhu, K.; Wong, Y.S.; Hong, G.S. Multi-categorymicro-milling tool wear monitoring with continuous hidden Markov models. *Mech. Syst. Signal Process.* **2009**, *23*, 547–560. [CrossRef]
5. Khasawneh, F.; Munch, E. Chatter detection in turning using persistent homology. *Mech. Syst. Signal Process.* **2016**, *70–71*, 527–541. [CrossRef]
6. Cao, H.; Lei, Y.; He, Z. Chatter identification in end milling process using wavelet packets and Hilbert-Huang transform. *Int. J. Mach. Tool Manuf.* **2013**, *69*, 11–19. [CrossRef]
7. Siddhpura, M.; Paurobally, R. A review of chatter vibration research in turning. *Int. J. Mach. Tool Manuf.* **2012**, *61*, 27–47. [CrossRef]
8. Liu, Y.; Li, T.; Liu, K.; Zhang, Y. Chatter reliability prediction of turning process system with uncertainties. *Mech. Syst. Signal Process.* **2016**, *66–67*, 232–247. [CrossRef]
9. Segreto, T.; Karam, S.; Teti, R. Signal processing and pattern recognition for surface roughness assessment in multiple sensor monitoring of robot-assisted polishing. *Int. J. Adv. Manuf. Technol.* **2017**, *9*, 1023–1033. [CrossRef]
10. Segreto, T.; Karam, S.; Teti, R.; Ramsing, J. Cognitive decision making in multiple sensor monitoring of robot assisted polishing. *Proc. CIRP* **2015**, *33*, 333–338. [CrossRef]
11. Nath, C.; Kapoor, S.; Srivastava, A. Finish turning of Ti-6Al-4V with the atomization-based cutting fluid (ACF) spray system. *J. Manuf. Process.* **2017**, *28*, 464–471. [CrossRef]
12. Niaki, F.; Mears, L.A. Comprehensive study on the effects of tool wear on surface roughness, dimensional integrity and residual stress in turning IN718 hard-to-machine alloy. *J. Manuf. Process.* **2017**, *30*, 268–280. [CrossRef]
13. Risbood, K.A.; Dixit, U.S.; Sahasrabudhe, A.D. Prediction of surface roughness and dimensional deviation by measuring cutting forces and vibration in turning process. *J. Mater. Process. Technol.* **2003**, *132*, 203–214. [CrossRef]
14. Azouzi, R.; Guillot, M. On-line prediction of surface finish and dimensional deviation in turning using neural network based sensor fusion. *Int. J. Mach. Tool Manuf.* **1997**, *37*, 1201–1217. [CrossRef]
15. Karam, S.; Teti, R. Wavelet transform feature extraction for chip form recognition during, carbon steel turning. *Proc. CIRP* **2013**, *12*, 97–102. [CrossRef]
16. Liao, Z.; Axinte, D.A. On monitoring chip formation, penetration depth and cuttingmalfunctions in bone micro-drilling via acoustic emission. *J. Mater. Process. Technol.* **2016**, *229*, 82–93. [CrossRef]
17. Wang, G.; Guo, Z.; Yang, Y. Force sensor based online tool wear monitoring using distributed Gaussian ARTMAP network. *Sens. Actuator A Phys.* **2013**, *192*, 111–118. [CrossRef]
18. Tangjitsitcharoen, S. Analysis of Chatter in Ball End Milling by Wavelet Transform. *Int. J. Ind. Manuf. Eng. (WASET)* **2012**, *6*, 2438–2444.
19. Kilundu, B.; Dehombreux, P.; Chiementin, X. Tool wear monitoring by machine learning techiques and singular spectrum analysis. *Mech. Syst. Signal Process.* **2011**, *25*, 400–415. [CrossRef]
20. Yao, Z.; Mei, D.; Chen, Z. On-line chatter detection and identification based on wavelet and support vector machine. *J. Mater. Process. Technol.* **2010**, *210*, 713–719. [CrossRef]
21. Salgado, D.R.; Alonso, F.J. Analysis of the structure of vibration signals for tool wear detection. *Mech. Syst. Signal Process.* **2008**, *22*, 735–748.
22. Diniz, A.; Liu, J.; Dornfeld, D. Correlating tool life, tool wear and surface roughness by monitoring acoustic emission in finish turning. *Wear* **1992**, *152*, 395–407. [CrossRef]
23. Andrade, L.H.; Mendes, A.; Vasconcelos, W.L.; Falco, W.; Rocha, A. A new approach for detection of wear mechanisms and determination of tool life in turning using acoustic emission. *Tribol. Int.* **2015**, *92*, 519–532.

24. Chen, X.; Li, B. AE Method for Tool Condition Monitoring Based on Wavelet Analysis. *Int. J. Adv. Manuf. Technol.* **2007**, *33*, 968–976. [CrossRef]
25. Griffin, J.M.; Chen, X. Multiple classification of the acoustic emission signals extracted during burn and chatter anomalies using genetic programming. *Int. J. Adv. Manuf. Technol.* **2009**, *45*, 1152–1168. [CrossRef]
26. Salgado, D.R.; Alonso, F.J. An approach based on current and sound signals for in-process tool wear monitoring. *Int. J. Mach. Tool Manuf.* **2007**, *47*, 2140–2452. [CrossRef]
27. Carou, D.; Rubio, E.; Lauro, C.; Cardoso, L.; Davim, P. Study based on sound monitoring as a means for superficial quality control in termittent turning of magnesium workpieces. *Proc. CIRP* **2017**, *62*, 262–268. [CrossRef]
28. Weingaertner, W.L.; Schroeter, R.B.; Polli, M.L.; Oliveira Gomes, J. Evaluation of high-speed end-milling dynamic stability through audio signal measurements. *J. Mater. Process. Technol.* **2006**, *179*, 133–138. [CrossRef]
29. Li, X.; Li, H.-X.; Guan, X.-P.; Du, R. Fuzzy Estimation of Feed-Cutting Force From Current Measurement- A Case Study on Intelligent. *IEEE Trans. Syst. Man. Cybern. C Appl. Rev.* **2004**, *34*, 506–512. [CrossRef]
30. Al-Sulaiman, F.A.; Abdul, M.; Sheikh, A.K. Use of electrical power for online monitoring of tool condition. *J. Mater. Process. Technol.* **2005**, *166*, 364–371. [CrossRef]
31. Rangwala, S.; Dornfeld, D. Sensor integration using neural networks for intelligent tool condition monitoring. *J. Eng. Ind.* **1990**, *112*, 219–228. [CrossRef]
32. Dornfeld, D.; DeVries, M. Neural networks sensor fusion for tool condition monitoring. *CIRP Ann. Manuf. Technol.* **1990**, *39*, 101–105. [CrossRef]
33. Noori-Khajavi, A.; Komaduri, R. Frequency and time domain analyses of sensor signals in drilling-I. Correlation with drill wear. *Int. J. Mach. Tool Manuf.* **1995**, *35*, 775–793. [CrossRef]
34. Noori-Khajavi, A.; Komaduri, R. Frequency and time domain analyses of sensor signals in drilling-II. Investigation on some problems associated with sensors integration. *Int. J. Mach. Tool Manuf.* **1995**, *35*, 795–815. [CrossRef]
35. Duro, J.; Padget, J.; Bowen, C.; Kim, H.; Nassehi, A. Multi-sensor data fusion framework for CNC machining monitoring. *Mech. Syst. Signal Process.* **2016**, *66–67*, 505–520. [CrossRef]
36. Liu, C.; Wang, C.F.; Li, Z.M. Incremental learning for online tool condition monitoring using Ellipsoid ARTMAP network model. *Appl. Soft Comput.* **2015**, *35*, 186–198. [CrossRef]
37. Segreto, T.; Simeone, A.; Teti, R. Sensor fusion for tool state classification in nickel superalloy high performance cutting. *Proc. CIRP* **2012**, *1*, 593–598. [CrossRef]
38. Shi, D.; Gindy, N.N. Tool Wear Predictive Model Based on Least Squares Support Vector Machines. *Mech. Syst. Signal Process.* **2007**, *21*, 1799–1814. [CrossRef]
39. Segreto, T.; Karam, S.; Simeone, A.; Teti, R. Residual stress assessment in Inconel 718 machining through wavelet sensor signal analysis and sensor fusion pattern recognition. *Proc. CIRP* **2013**, *9*, 103–108. [CrossRef]
40. Rao, P.; Bhushan, M.; Bukkapatnam, S.; Zong, Z.; Byalal, S.; Beyca, O.; Fields, A.; Komanduri, R. Process-machine interaction (PMI) modeling and monitoring of chemical mechanical planarization (CMP) process using wireless vibration sensors. *IEEE Trans. Semiconduct. Manuf.* **2014**, *27*, 1–15. [CrossRef]
41. Hessainia, Z.; Belbah, A.; Yallese, M.A.; Mabrouki, T.; Rigal, J.F. On the prediction of surface roughness in the hard turning based on cutting parameters and tool vibrations. *Measurement* **2013**, *46*, 1671–1681. [CrossRef]
42. Bhuiyan, M.H.; Choudhury, I.A.; Dahari, M. Monitoring the tool wear, surface roughness and chip formation formation occurrences using multiple sensors in turning. *J. Manuf. Syst.* **2014**, *33*, 476–487. [CrossRef]
43. Guo, Y.B.; Ammula, S.C. Real-Time acoustic emission monitoring for surface damage in hard machining. *Int. J. Mach. Tool Manuf.* **2009**, *209*, 3125–3137. [CrossRef]
44. Jeong, H.; Kim, H.; Lee, S.; Dornfeld, D. Multi-sensor monitoring system in chemical mechanical planarization (CMP) for correlations with process issues. *CIRP Ann. Manuf. Technol.* **2006**, *55*, 325–328. [CrossRef]
45. Wang, H.; To, S.; Chan, C.Y. Investigation on the influence of tool-tip vibration on surface roughness and its representative measurement in ultra-precision diamond turning. *Int. J. Mach. Tool Manuf.* **2013**, *69*, 20–29. [CrossRef]
46. Tangjitsitcharoen, S. In-process monitoring and detection of chip formation and chatter for CNC turning. *J. Mater. Process. Technol.* **2009**, *209*, 4682–4688. [CrossRef]

47. Abouelatta, O.B.; Madl, J. Surface roughness prediction based on cutting parameter and tool vibration in turning operation. *J. Mater.* **2001**, *118*, 269–277. [CrossRef]
48. Krolczyk, G.; Madura, R.; Nieslony, P.; Wieczorowski, M. Surface morphology analysis of Duplex Stainless Steel (DSS) in Clean Production using the Power Spectral Density. *Measurement* **2016**, *94*, 464–470. [CrossRef]
49. Salgado, D.R.; Alonso, F.J.; Cambero, I.; Marcelo, A. In-process surface roughness prediction system using cutting vibrations in turning. *Int. J. Adv. Manuf. Technol.* **2009**, *43*, 40–51. [CrossRef]
50. García, E.; Núñez, P.J. Surface roughness monitoring by singular spectrum analysis of vibration signals. *Mech. Syst. Signal Process.* **2017**, *84*, 516–530. [CrossRef]
51. Kunpeng, Z.; Yoke San, W.; Geok Soon, H. Wavelet analysis of sensor signal for tool condition monitoring: A review and some new results. *Int. J. Mach. Tool Manuf.* **2009**, *49*, 537–553.
52. García, E.; Núñez, P.J. Analysis of cutting force signals by wavelet packet transform for surface roughness monitoring in CNC turning. *Mech. Syst. Signal Process.* **2018**, *98*, 634–651. [CrossRef]
53. García, E.; Núñez, P.J. Application of the wavelet packet transform to vibration signals for surface roughness monitoring in CNC turning operations. *Mech. Syst. Signal Process.* **2018**, *98*, 902–919. [CrossRef]
54. Salgado, D.R.; Alonso, F.J. Tool wear detection in turning operations using singular spectrum analysis. *J. Mater. Process. Technol.* **2006**, *171*, 451–458. [CrossRef]
55. Pandiyan, V.; Caesarendra, W.; Tjahjowidodo, T.; Tan, H.H. In-process tool condition monitoring in compliant abrasive belt grinding process using support vector machine and genetic algorithm. *J. Manuf. Process.* **2018**, *31*, 199–213. [CrossRef]
56. Teti, R.; Jemielniak, K.; O'Donnel, G.; Dornfeld, D. Advanced monitoring of machining operations. *CIRP Ann. Manuf. Technol.* **2010**, *59*, 717–739. [CrossRef]
57. Lauro, C.H.; Brandão, L.C.; Baldo, D.; Reis, R.A.; Davim, J.P. Monitoring and processing signal applied in machining processes—A review. *Measurement* **2014**, *58*, 73–86. [CrossRef]
58. Kirby, E.D.; Chen, J.C. Development of a fuzzy-nets-based surface roughness prediction system in turning operations. *Comput. Ind. Eng.* **2007**, *53*, 30–42. [CrossRef]
59. Upadhyay, V.; Jain, P.K.; Mehta, N.K. In Process prediction of surface roughness in turning of Ti–6Al–4V alloy using cutting parameters and vibration signal. *Measurement* **2013**, *46*, 154–160. [CrossRef]
60. Ozel, T.; Karpat, Y.; Figueira, L.; Davim, P.J. Modelling of surface finish and tool flank wear in turning of AISI D2 steel with ceramic wiper inserts. *J. Mater. Process. Technol.* **2007**, *189*, 192–198. [CrossRef]
61. Botcha, B.; Rajagopal, V.; Bukkapatnam, S. Process-machine interactions and a multi-sensor fusion approach to predict surface roughness in cylindrical plunge grinding process. *Proc. Manuf.* **2018**, *26*, 700–711. [CrossRef]
62. Kannatey-Asibu, E., Jr.; Dornfeld, D. A study of tool wear using statistical analysis of metal-cutting acoustic emission. *Wear* **1982**, *76*, 247–261. [CrossRef]
63. Siddhpura, A.; Paurobally, R. A review of flank wear prediction methods for tool condition. *Int. J. Adv. Manuf. Technol.* **2013**, *65*, 375–393. [CrossRef]
64. Frigieri, E.; Campos, P.; Paiva, A.; Balestrassi, P.; Ferreira, J.; Ynoguti, C. A mel-frequency cepstral coefficient-based approach for surface roughness diagnosis in hard turning using acoustic signals and gaussian mixture models. *Appl. Acoust.* **2016**, *113*, 230–237. [CrossRef]
65. Pawade, R.S.; Soshi, S.S. Analysis of acoustic emission signals and surface integrity in the high speed turning of inconel 718. *J. Eng. Manuf. (Sage J.)* **2012**, *226*, 3–27. [CrossRef]
66. Rao, P.; Bukkapatnam, S.; Beyca, O.; Kong, Z.; Komanduri, R. Real-time identification of incipient surface morphology variations in ultraprecision machining process. *J. Manuf. Sci. Eng.* **2014**, *136*, 021008. [CrossRef]
67. Masters, T. *Practical Neural Networks Recipes in C++*; Academic Press, Inc.: Cambridge, MA, USA, 1993; pp. 173–180.

sensors

MDPI

Article

Observable Degree Analysis for Multi-Sensor Fusion System

Zhentao Hu [1,†], Tianxiang Chen [2], Quanbo Ge [2,*,†] and Hebin Wang [2]

1 College of Computer and Information Engineering, Henan University, Kaifeng 475004, China; hzt@henu.edu.cn

2 Institute of Systems Science and Control Engineering, School of Automation, Hangzhou Dianzi University, Hangzhou 310018, China; 18767221309@163.com (T.C.); wanghebin@hdu.edu.cn (H.W.)

* Correspondence: qbge@hdu.edu.cn

† These authors contributed equally to this work.

Received: 16 October 2018; Accepted: 14 November 2018; Published: 30 November 2018

Abstract: Multi-sensor fusion system has many advantages, such as reduce error and improve filtering accuracy. The observability of the system state is an important index to test the convergence accuracy and speed of the designed Kalman filter. In this paper, we evaluate different multi-sensor fusion systems from the perspective of observability. To adjust and optimize the filter performance before filtering, in this paper, we derive the expression form of estimation error covariance of three different fusion methods and discussed both observable degree of fusion center and local filter of fusion step. Based on the ODAEPM, we obtained their discriminant matrix of observable degree and the relationship among different fusion methods is given by mathematical proof. To confirm mathematical conclusion, the simulation analysis is done for multi-sensor CV model. The result demonstrates our theory and verifies the advantage of information fusion system.

Keywords: multi-sensor network; observable degree analysis; information fusion

1. Introduction

Multi-sensor network technology is extensively used in modern life. It has many advantages over single sensor network. However, it faces some new challenges, such as low observability and large data delay [1–3]. To some extent, observability can reflect the filtering performance of the system. The low observability caused by complex data collection and translation will deteriorate the estimator performance, and should be given more attention [4–7]. Thus, it is essential to find a way to guide the multi-sensor netting for improving the estimation performance. The most classic estimator for mobile target tracking is the Kalman filter presented by R. E. Kalman in the 1960s [8]. For the Kalman filtering theory, a basic concept is the observability of state space equation [9]. The observability is used to express the possibility of recovering the initial state by using measurement data and it is related to both state and observation models. Thus, it is important to analyze quantitatively on observability because it can guide the improvement of estimator performance.

For the modern control theory, the observability, which is a qualitative index, can generally be expressed by a variable with two values "0" and "1". Namely, the result is Boolean value. For zero value, it means that the system is unobservable, which means that the system state is not fully recovered by the measurement. As the quantitative variable, the observable degree is used to measure the observability ability [10]. In [10], an analysis on the observability and observable degree has been given, the kernel is abstracted as follows. Based on the current research work, there are some ways to evaluate the observable degree [11–14]. For this, observable degree analysis (ODA) has been presented by using estimation error covariance (EEC) of the Kalman filter in [11], which can intuitively express the observable ability of system states or linear combination of state variables. In [12], an evaluation method of observable degree with regard to singular value decomposition has also been given [10]. The observability definition has been further improved by using optimization singular value decomposition method in [13]. Another ODA method, based on pseudo-inverse with the relevant knowledge of least square, is proposed in [14].

The observable degree analysis with estimation performance measure (ODAEPM) proposed by Ma et al., [15] is a great observable degree analysis method, which considers the effect of measurement noise. It reveals the inner relation between the discriminant matrix of observability and the estimation performance of the Kalamn filtering. By synchronously defining the observable degree of state component as local observable degree (LOD) and the observable degree of system as global observable degree (GOD), it establishes the completeness of the observable degree theory.

However, the ODAEPM in [15] only discusses the observable degree problem on single sensor observation. It is necessary to establish the observable degree for multi-sensor network technology. How the fusion methods affect the estimator performance, and the advantage of information fusion should also be verified by the observable degree theory. Further research on observable degree is still needed. Motivated by these, in this paper, following the ODAEPM, we research the observable of different type fusion methods [4,7,16,17], and make the comparison among them. The main contributions of our work are: First, the observable degree discriminant matrix of three different kinds fusion methods, namely the distributed multi-sensor fusion system without feedback, the distributed multi-sensor fusion system with feedback and the centralized multi-sensor sequential fusion system, is derived, mainly based on the estimation error covariance of their fusion process. Second, the mathematical proof of the observable degree relationship between fusion center and local filter is given, along with the relation among different fusion methods, which provides evidence for the advantage of multi-sensor information fusion, and the effect of different fusion methods.

The rest of the paper is organized as follows. First, we review the ODAEPM in Section 2. Then, in Section 3, we introduce the structure of three fusion methods, and, by using ODAEPM, the observable degree analysis is performed on both fusion center and local filter. The comparison of observable degree among different fusion methods is given a mathematical proof in Section 4. In Section 5, the simulation is established to verify our research. Section 6 is the conclusion.

2. Review of ODAEPM

2.1. Problem Formulation

The associate models of linear time-varying discrete estimation system is considered as [4,18]:

$$x_{k+1} = \mathbf{\Phi}_{k+1,k} x_k + w_k \tag{1}$$

$$z_k = \mathbf{H}_k x_k + v_k \tag{2}$$

where $k(k = 1, 2, \cdots)$ is time index. $x_k \in \mathbf{R}^n$ is state variable, where n is the state dimension. $z_k \in \mathbf{R}^m$ is the observation vector, where m is the observation dimension. $\mathbf{\Phi}_{k+1,k} \in \mathbf{R}^{n\times n}$ is the linear state transition matrix and $\mathbf{H}_k \in \mathbf{R}^{m\times n}$ is the linear observation matrix. $w_k \in \mathbf{R}^n$ and $v_k \in \mathbf{R}^m$ are, respectively, n-dimensional Gaussian process noise and m-dimensional Gaussian observation noise.

2.2. The ODAEPM for Observable Degree Analysis

ODAEPM [15] is a method dealing with the observable degree analysis for linear time-varying discrete estimation system [15]. The process of ODAEPM is shown in Figure 1.

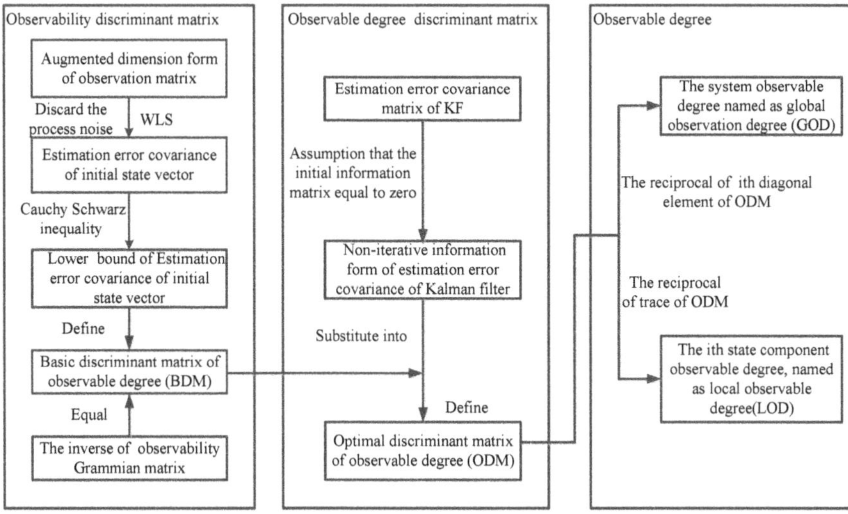

Figure 1. The process of ODAEPM.

In Figure 1, we summarize ODAPEM. Here, we need to know that the ODAEPM does not consider the process noise [15]. The ODAEPM names the observable degree of system as global observable degree (GOD):

$$\eta = \frac{1}{Trace(\mathbf{D}_{1,k}^{*})} \tag{3}$$

and names the observable degree of state component as local observable degree (LOD):

$$\Delta\eta_i = \frac{1}{(\mathbf{D}_{1,k}^{*})_i} \tag{4}$$

We define the optimal observability discriminant matrix as:

$$\mathbf{D}_{1,k}^{*} = \mathbf{\Phi}_{k,1}\mathbf{D}_{1,k}\mathbf{\Phi}_{k,1}{}^{T} \tag{5}$$

Because the non-recursive information form $Y_{k|k}$ of the EEC given the KF related knowledge is as follows:

$$Y_{k|k} = \mathbf{\Phi}_{k,0}{}^{-T}Y_{0|0}\mathbf{\Phi}_{k,0}{}^{-1} + \mathbf{\Phi}_{k,1}{}^{-T}\mathbf{D}_{1,k}{}^{-1}\mathbf{\Phi}_{k,1}{}^{-1} \tag{6}$$

Normally, if the system is stable, the system filtering accuracy is independent of the initial value. It may be assumed that $Y_{0|0} \to 0$, thus Equation (6) can be written as

$$Y_{k|k} \to \mathbf{\Phi}_{k,1}{}^{-T}\mathbf{D}_{1,k}{}^{-1}\mathbf{\Phi}_{k,1}{}^{-1} \tag{7}$$

Therefore,

$$\mathbf{P}_{k|k} \to \mathbf{\Phi}_{k,1}\mathbf{D}_{1,k}\mathbf{\Phi}_{k,1}{}^{T} \tag{8}$$

where $\mathbf{D}_{1,k}$ is

$$\mathbf{D}_{1,k} = \min\{Var[x_1 - \hat{x}_1]\} = (\boldsymbol{\vartheta}_{1,k}{}^{T}R_{1,k}^{-1}\boldsymbol{\vartheta}_{1,k})^{-1} = (\sum_{i=1}^{k}\mathbf{\Phi}_{i,1}{}^{T}H_i^{T}R_i^{-1}H_i\mathbf{\Phi}_{i,1})^{-1} \tag{9}$$

The main contributions of ODAEPM towards observable degree analysis are as follows. First, by constructing observability discriminant matrix by jointly using WLS and Cauchy Schwartz inequality, the uncertainty from the observability effects of observation noise are taken into account. Second, by substituting observability discriminant matrix into the estimation error covariance matrix of KF under the assumption that the initial value of the estimation error covariance matrix tends to be infinite, the relationship between filter estimation performance and observability analysis is clearly established. Finally, by defining the local observable degree (LOD) and global observation degree (GOD), the estimation abilities can be expressed from different scale levels.

3. Observable Degree Analysis of Multi-Sensor Observation Network System

Multi-sensor fusion system integrating signals from different sensors has the great advantage of overcoming the uncertainty and limitation under single sensor measurement condition [15,19]. To obtain performance improvements in target tracking problem, multi-sensor observation network becomes an important research field.

3.1. Problem Formulation

Distributed multi-sensor track fusion system contains L local sensors. The target moving model is

$$\boldsymbol{x}_{k+1} = \boldsymbol{\Phi}_{k+1,k}\boldsymbol{x}_k + \boldsymbol{w}_k \tag{10}$$

The ith local sensor observation model is shown as

$$\boldsymbol{z}_{i,k} = \mathbf{H}_{i,k}\boldsymbol{x}_k + \boldsymbol{v}_{i,k} \tag{11}$$

where $k(k = 1, 2, \cdot)$ is time index. The first subscript i of matrix H and v is the sensor index. $\boldsymbol{x}_k \in \mathbf{R}^n$ is state variable, where n is the state dimension. $\boldsymbol{z}_{i,k} \in \mathbf{R}^m$ is ith local sensor observation vector, where m is the observation dimension. $\boldsymbol{\Phi}_{k+1,k} \in \mathbf{R}^{n\times n}$ is the linear state transition matrix and $\mathbf{H}_{i,k} \in \mathbf{R}^{m\times n}$ is the ith local sensor linear observation matrix. $\boldsymbol{w}_k \in \mathbf{R}^n$ and $\boldsymbol{v}_{i,k} \in \mathbf{R}^m$ are, respectively, n-dimensional Gaussian process noise and m-dimensional Gaussian observation noise. The covariances of process noise $\boldsymbol{w}_k$ and observation noise $\boldsymbol{v}_{i,k}$ are $\mathbf{Q}_k$ and $\mathbf{R}_{i,k}$, respectively.

3.2. Introduction of Multi-Sensor Fusion System

There are many forms for multi-sensor fusion system; in this paper, we expound three present major fusion technologies in detail: the distributed multi-sensor fusion without feedback, the distributed multi-sensor fusion with feedback and the centralized multi-sensor sequential fusion.

3.2.1. Distributed Multi-Sensor Fusion without Feedback

The structure of distributed multi-sensor fusion without feedback is shown in Figure 2 [4,16–18,20].

The distributed multi-sensor fusion system without feedback contains several local sensors. Each sensor performs Kalman filtering on its own observation data, transmitting the state estimation value $\hat{\boldsymbol{x}}_{i,k}^{dn}$ and estimation error covariance $\mathbf{P}_{i,k}^{dn}$ to the fusion center. After the fusion center gets the filter result form all sensors, the fusion results $\hat{\boldsymbol{x}}_{f,k}^{dn}$ and $\mathbf{P}_{f,k}^{dn}$ can be calculated by fusion algorithm of distributed multi-sensor fusion system without feedback.

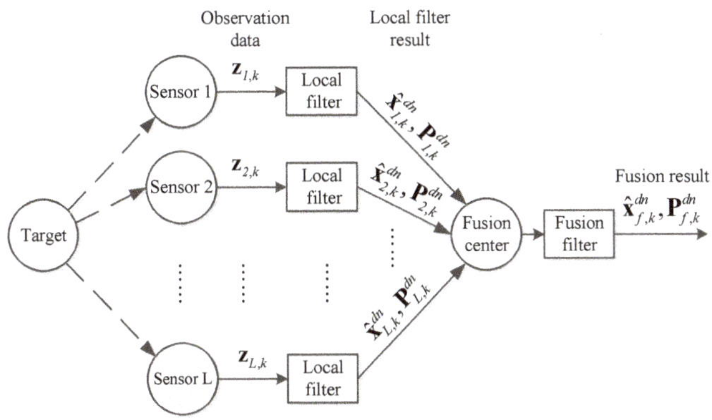

Figure 2. The process of distributed multi-sensor fusion system without feedback.

3.2.2. Distributed Multi-Sensor Fusion with Feedback

The structure of distributed multi-sensor fusion with feedback is shown in Figure 3 [4,16–18,20].

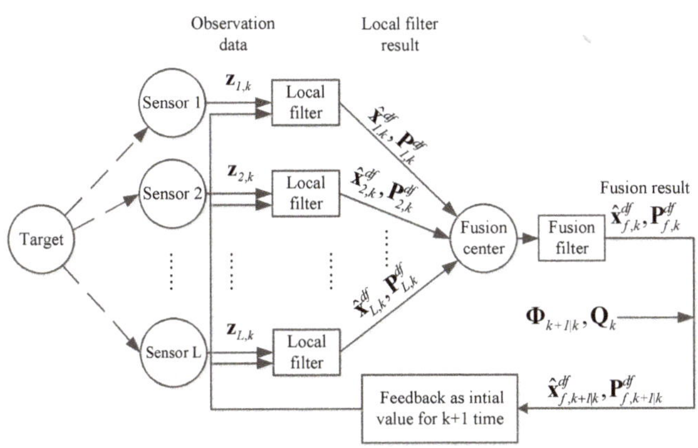

Figure 3. The process of distributed multi-sensor fusion system with feedback.

The form of distributed multi-sensor fusion with feedback is similar to the former fusion technology. The most important difference between these two methods is that the fusion with feedback needs to return the fusion result $\hat{x}^{df}_{f,k}$ and $\mathbf{P}^{df}_{f,k}$ to the local sensors as their filter initial value at time $k+1$.

3.2.3. Centralized Multi-Sensor Sequential Fusion

The structure of centralized multi-sensor sequential fusion is shown in Figure 4 [4,16–18,20].

The centralized multi-sensor sequential fusion is different from the other two fusion methods shown in this paper. This fusion technology only contains one sequential fusion filter at fusion center [21,22]. Not only the observation data $z_{i,k}$ but also the observation matrix $\mathbf{H}_{i,k}$ are necessary. The messages for same sensor are organized into same group and sequential input to the fusion filter. When each group message is received, the fusion filter will do one sequential filtering and return the result $\hat{x}^{cs}_{i,k}$, $\mathbf{P}^{cs}_{i,k}$ for next sequential step. When the step index i reaches the sensor number L, the sequential fusion is finished, and the results $\hat{x}^{cs}_{L,k}$ and $\mathbf{P}^{cs}_{L,k}$ are the final fusion result for centralized multi-sensor sequential fusion system.

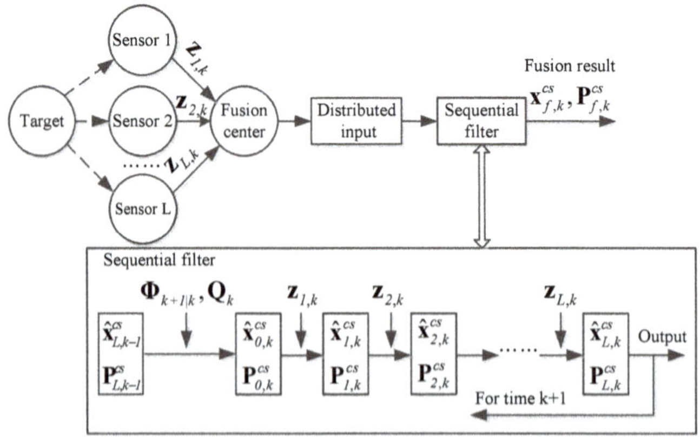

Figure 4. The structure of centralized multi-sensor sequential fusion system.

3.3. Motivation

Although the ODAEPM proposed by Ma et al., [15] provides a simple way to making observable degree analysis and presents the relationship between the observable degree and estimator accuracy, the problem in [15] is based on the single sensor observation system; the observability theory of multi sensor cooperative target tracking system and estimation of the fusion have not been solved well yet. The relationship of performance between fusion center and local sensors has only been obtained from the mutual independent simulation experiment. There is no mathematical proof yet to verify the advantage of information fusion system. Additionally, how the LOD and GOD of fusion center and local filter influenced by different fusion methods also needs a further study.

3.4. Observable Degree Analysis of Multi-Sensor Observation Network System

According to Ma et al., [15], the observable degree discriminant matrix equal to the inverse of non-iterative form of estimation error covariance under the assumption that the initial value of estimation error information matrix equals zero.

3.4.1. Observable Degree Discriminant Matrix of Distributed Multi-Sensor Fusion without Feedback

In distributed multi-sensor fusion system without feedback [23], the local sensor doing Kalman filtering is only based on individual observation data, making it the same as the single sensor observation system. Thus, its local sensors observable degree discriminant matrix at time k is the same as the ODAEPM performed on single sensor observation system, shown as

$$(\mathbf{D}_{1,k}^{*})_i^{dn} = [\mathbf{\Phi}_{k,1}^{-T}(\sum_{j=1}^{k}\mathbf{\Phi}_{j,1}^{T}\mathbf{H}_{i,j}^{T}\mathbf{R}_{i,j}^{-1}\mathbf{H}_{i,j}\mathbf{\Phi}_{j,1})\mathbf{\Phi}_{k,1}^{-1}]^{-1} \tag{12}$$

where $\Phi_{a,b}$ is the state transition matrix between time a and time b.

Its estimation error covariance matrix of fusion center is:

$$(\mathbf{P}_{f,k+1}^{dn})^{-1} = (\mathbf{P}_{f,k+1|k}^{dn})^{-1} + \sum_{i=1}^{L}[(\mathbf{P}_{i,k+1}^{dn})^{-1} - (\mathbf{P}_{i,k+1|k}^{dn})^{-1}] \tag{13}$$

where

$$(\mathbf{P}^{dn}_{i,k+1|k}) = \mathbf{\Phi}_{k+1,k}(\mathbf{P}^{dn}_{i,k})\mathbf{\Phi}^{T}_{k+1,k} + Q_k \tag{14}$$

$$(\mathbf{P}^{dn}_{i,k+1})^{-1} = (\mathbf{P}^{dn}_{i,k+1|k})^{-1} + \mathbf{H}^{T}_{i,k+1}\boldsymbol{R}^{-1}_{i,k+1}\mathbf{H}_{i,k+1} \tag{15}$$

Discarding the process noise, Equation (13) can be rewritten as

$$(\mathbf{P}^{dn}_{f,k+1})^{-1} = \mathbf{\Phi}^{-T}_{k+1,k}(\mathbf{P}^{dn}_{i,k})^{-1}\mathbf{\Phi}^{-1}_{k+1,k} + \sum_{i=1}^{L}\mathbf{H}^{T}_{i,k+1}\boldsymbol{R}^{-1}_{i,k+1}\mathbf{H}_{i,k+1} \tag{16}$$

The fusion center observable degree discriminant matrix $(\mathbf{D}^{*}_{1,k})^{dn}_{f}$ is equal to the inverse of Equation (16) at time k, thus changing Equation (16) into the non-iterative form, $(\mathbf{D}^{*}_{1,k})^{dn}_{f}$, will be

$$\begin{aligned}(\mathbf{D}^{*}_{1,k})^{dn}_{f} =& [\mathbf{\Phi}^{-T}_{k,k-1}(\mathbf{P}^{dn}_{f,k-1})^{-1}\mathbf{\Phi}^{-1}_{k,k-1} + \sum_{i=1}^{L}\mathbf{H}^{T}_{i,k}\boldsymbol{R}^{-1}_{i,k}\mathbf{H}_{i,k}]^{-1}\\ =& [\mathbf{\Phi}^{-T}_{k,k-2}(\mathbf{P}^{dn}_{f,k-2})^{-1}\mathbf{\Phi}^{-1}_{k,k-2} + \mathbf{\Phi}^{-T}_{k,k-1}\sum_{i=1}^{L}\mathbf{H}^{T}_{i,k-1}\boldsymbol{R}^{-1}_{i,k-1}\mathbf{H}_{i,k-1}\mathbf{\Phi}^{-1}_{k,k-1} + \sum_{i=1}^{L}\mathbf{H}^{T}_{i,k}\boldsymbol{R}^{-1}_{i,k}\mathbf{H}_{i,k}]^{-1}\\ &\vdots\\ =& [\sum_{j=0}^{k-1}\mathbf{\Phi}^{-T}_{k,k-j}\sum_{i=1}^{L}\mathbf{H}^{T}_{i,k-j}\boldsymbol{R}^{-1}_{i,k-j}\mathbf{H}_{i,k-j}\mathbf{\Phi}^{-1}_{k,k-j}]^{-1}\\ =& [\mathbf{\Phi}^{-T}_{k,1}\sum_{j=1}^{k}\sum_{i=1}^{L}(\mathbf{\Phi}^{T}_{j,1}\mathbf{H}^{T}_{i,j}\boldsymbol{R}^{-1}_{i,j}\mathbf{H}_{i,j}\mathbf{\Phi}_{j,1})\mathbf{\Phi}^{-1}_{k,1}]^{-1}\end{aligned} \tag{17}$$

3.4.2. Observable Degree Discriminant Matrix of Distributed Multi-Sensor Fusion with Feedback

The recursive formula of estimation error covariance of local sensor in distributed multi-sensor fusion system with feedback [23] is shown as

$$(\mathbf{P}^{df}_{i,k+1})^{-1} = (\mathbf{P}^{df}_{i,k+1|k})^{-1} + \mathbf{H}^{T}_{i,k+1}\boldsymbol{R}^{-1}_{i,k+1}\mathbf{H}_{i,k+1} \tag{18}$$

where

$$\mathbf{P}^{df}_{i,k+1|k} = \mathbf{\Phi}_{k+1,k}\mathbf{P}^{df}_{f,k}\mathbf{\Phi}^{T}_{k+1,k} + Q_k \tag{19}$$

From Equation(18), we need to get the estimation error covariance of fusion center $\mathbf{P}^{df}_{f,k}$ before calculating the observable degree of local sensor. According to [19], the result of fusion center estimation error covariance $\mathbf{P}^{df}_{f,k}$ is the same as in the fusion system without feedback. Thus, the observable degree discriminant matrix of fusion center $(\mathbf{D}^{*}_{1,k})^{df}_{f}$ is equal to $(\mathbf{D}^{*}_{1,k})^{dn}_{f}$, shown as

$$(\mathbf{D}^{*}_{1,k})^{df}_{f} = [\mathbf{\Phi}^{-T}_{k,1}\sum_{j=1}^{k}\sum_{i=1}^{L}(\mathbf{\Phi}^{T}_{j,1}\mathbf{H}^{T}_{i,j}\boldsymbol{R}^{-1}_{i,j}\mathbf{H}_{i,j}\mathbf{\Phi}_{j,1})\mathbf{\Phi}^{-1}_{k,1}]^{-1} \tag{20}$$

Then, the observable degree discriminant matrix of ith local sensor is

$$(\mathbf{D}^{*}_{1,k})^{df}_{i} = [\mathbf{\Phi}^{-T}_{k,1}\sum_{j=1}^{k-1}\sum_{i=1}^{L}(\mathbf{\Phi}^{T}_{j,1}\mathbf{H}^{T}_{i,j}\boldsymbol{R}^{-1}_{i,j}\mathbf{H}_{i,j}\mathbf{\Phi}_{j,1})\mathbf{\Phi}^{-1}_{k,1} + \mathbf{H}^{T}_{i,k}\boldsymbol{R}^{-1}_{i,k}\mathbf{H}_{i,k}]^{-1} \tag{21}$$

3.4.3. Observable Degree Discriminant Matrix of Centralized Multi-Sensor Sequential Fusion

The centralized multi-sensor sequential fusion method only operates observation data filtering at fusion center [24,25]. The process of estimation error covariance in sequential filtering is

$$\mathbf{P}^{cs}_{0,k+1} = \mathbf{\Phi}_{k+1,k}\mathbf{P}^{cs}_{L,k}\mathbf{\Phi}^{T}_{k+1,k} + Q_k \tag{22}$$

$$(\mathbf{P}^{cs}_{i,k+1})^{-1} = (\mathbf{P}^{cs}_{i-1,k+1})^{-1} + \mathbf{H}^{T}_{i,k+1}\boldsymbol{R}^{-1}_{i,k+1}\mathbf{H}_{i,k+1} \tag{23}$$

$$\mathbf{P}^{cs}_{f,k+1} = \mathbf{P}^{cs}_{L,k+1} \tag{24}$$

The observable degree discriminant matrix of fusion center is

$$\begin{aligned}
(\mathbf{D}^{*}_{1,k})^{cs}_{f} &= [(\mathbf{P}^{cs}_{L,k})^{-1}]^{-1} = [(\mathbf{P}^{cs}_{L-1,k})^{-1} + \mathbf{H}^{T}_{L,k}\boldsymbol{R}^{-1}_{L,k}\mathbf{H}_{L,k}]^{-1} \\
&= [(\mathbf{P}^{cs}_{0,k+1})^{-1} + \sum_{i=1}^{L}\mathbf{H}^{T}_{i,k}\boldsymbol{R}^{-1}_{i,k}\mathbf{H}_{i,k}]^{-1} \\
&= [\mathbf{\Phi}^{-T}_{k,k-1}(\mathbf{P}^{cs}_{0,k+1})^{-1}\mathbf{\Phi}^{-1}_{k,k-1} + \sum_{i=1}^{L}\mathbf{H}^{T}_{i,k}\boldsymbol{R}^{-1}_{i,k}\mathbf{H}_{i,k}]^{-1} \\
&\;\;\vdots \\
&= [\mathbf{\Phi}^{-T}_{k,1}\sum_{j=1}^{k}\sum_{i=1}^{L}(\mathbf{\Phi}^{T}_{j,1}\mathbf{H}^{T}_{i,j}\boldsymbol{R}^{-1}_{i,j}\mathbf{H}_{i,j}\mathbf{\Phi}_{j,1})\mathbf{\Phi}^{-1}_{k,1}]^{-1}
\end{aligned} \tag{25}$$

Although centralized multi-sensor sequential fusion does not have a local filter, the result in fusion center is computed step by step with sequential input. Define $(\mathbf{D}^{*}_{1,k})^{cs}_{i}$ as the discriminant matrix of fusion center at ith step during time k.

$$\begin{aligned}
(\mathbf{D}^{*}_{1,k})^{cs}_{i} &= [(\mathbf{P}^{cs}_{i,k})^{-1}]^{-1} \\
&= [(\mathbf{P}^{cs}_{0,k+1})^{-1} + \sum_{l=1}^{i}\mathbf{H}^{T}_{l,k}\boldsymbol{R}^{-1}_{l,k}\mathbf{H}_{l,k}]^{-1} \\
&= [\mathbf{\Phi}^{-T}_{k,1}\sum_{j=1}^{k-1}\sum_{i=1}^{L}(\mathbf{\Phi}^{T}_{j,1}\mathbf{H}^{T}_{i,j}\boldsymbol{R}^{-1}_{i,j}\mathbf{H}_{i,j}\mathbf{\Phi}_{j,1})\mathbf{\Phi}^{-1}_{k,1} + \sum_{l=1}^{i}\mathbf{H}^{T}_{l,k}\boldsymbol{R}^{-1}_{l,k}\mathbf{H}_{l,k}]^{-1}
\end{aligned} \tag{26}$$

3.5. *Brief Summary*

The observable degree analysis method for multi-sensor information fusion system is rarely mentioned because the fusion method does not have the direct observation matrix and its observability discriminant matrix for fusion center cannot be established. The traditional observable degree analysis method is only analyzed and defined on observability, which fails when studying filtering accuracy. Thus, it is hard to reveal that there is any promotion by information fusion between fusion center and local filter before operating filtering. The observable degree of ODAEPM is defined on the main body of estimation error covariance, which is used to measure the filtering accuracy, ignoring the effect of process noise, making it possible to analyze the observable degree of fusion center in multi-sensor information fusion system.

Following these considerations, in this section, we introduce the multi-sensor information fusion problem, and list the structure of three main information fusion methods. By translating the form of estimation error covariance, we are able to get the observable degree discriminant matrix for both fusion center and local filter in each fusion method.

4. The Relationship of Observable Degree among Information Fusion System

The ODAEPM makes a clear relation between filtering performance and observable degree. To study how the fusion technology affects the target tracking performance, further research on the relationship of observable degree between local filter and fusion center is needed as well as on the observable degree among different fusion methods.

4.1. The Relationship of Observable Degree between Local Filter and Fusion Center

The distributed multi-sensor fusion system contains local filter and fusion center, and the observable degree can characterize filtering performance. By studying the relationship of observable degree between local filter and fusion center, how the information fusion affects the observation system performance can be found.

For local filter and fusion center in distributed multi-sensor fusion system without feedback, the observable degree discriminant matrix of ith local sensor $(\mathbf{D}_{1,k}^{*})_{i}^{dn}$ and fusion center $(\mathbf{D}_{1,k}^{*})_{f}^{dn}$ are shown in Equations (12) and (17). To find the relationship between them, a simple way is extracting the term $(\mathbf{D}_{1,k}^{*})_{i}^{dn}$ from $(\mathbf{D}_{1,k}^{*})_{f}^{dn}$. In this thought, by employing the matrix inverse lemma, we rewrite $(\mathbf{D}_{1,k}^{*})_{f}^{dn}$ as

$$\begin{aligned}(\mathbf{D}_{1,k}^{*})_{f}^{dn} &=[\mathbf{\Phi}_{k,1}^{-T}\sum_{j=1}^{k}\sum_{l=1}^{L}(\mathbf{\Phi}_{j,1}^{T}\mathbf{H}_{l,j}^{T}R_{l,j}^{-1}\mathbf{H}_{l,j}\mathbf{\Phi}_{j,1})\mathbf{\Phi}_{k,1}^{-1}]^{-1}\\ &=[\mathbf{\Phi}_{k,1}^{-T}\sum_{j=1}^{k}(\mathbf{\Phi}_{j,1}^{T}\mathbf{H}_{i,j}^{T}R_{i,j}^{-1}\mathbf{H}_{i,j}\mathbf{\Phi}_{j,1})\mathbf{\Phi}_{k,1}^{-1}+\mathbf{\Phi}_{k,1}^{-T}\sum_{j=1}^{k}\sum_{\substack{l=1\\ l\neq i}}^{L}(\mathbf{\Phi}_{j,1}^{T}\mathbf{H}_{l,j}^{T}R_{l,j}^{-1}\mathbf{H}_{l,j}\mathbf{\Phi}_{j,1})\mathbf{\Phi}_{k,1}^{-1}]^{-1}\\ &=(\mathbf{D}_{1,k}^{*})_{i}^{dn}-(\mathbf{D}_{1,k}^{*})_{-i}^{dn}\end{aligned} \tag{27}$$

where

$$(\mathbf{D}_{1,k}^{*})_{-i}^{dn}=(\mathbf{D}_{1,k}^{*})_{i}^{dn}\mathbf{\Phi}_{k,1}^{-T}[(\sum_{j=1}^{k}\sum_{\substack{l=1\\ l\neq i}}^{L}\mathbf{\Phi}_{j,1}^{T}\mathbf{H}_{l,j}^{T}R_{l,j}^{-1}\mathbf{H}_{l,j}\mathbf{\Phi}_{j,1})^{-1}+\mathbf{\Phi}_{k,1}^{-1}(\mathbf{D}_{1,k}^{*})_{i}^{dn}\mathbf{\Phi}_{k,1}^{-T}]^{-1}\mathbf{\Phi}_{k,1}^{-1}(\mathbf{D}_{1,k}^{*})_{i}^{dn} \tag{28}$$

while the terms $(\mathbf{D}_{1,k}^{*})_{i}^{dn}$ and $\mathbf{\Phi}_{j,1}^{T}\mathbf{H}_{l,j}^{T}\mathbf{R}_{l,j}^{-1}\mathbf{H}_{l,j}\mathbf{\Phi}_{j,1}$ are positive semidefinite matrices, and, according to the discrimination method of positive semidefinite matrix [26], $(\mathbf{D}_{1,k}^{*})_{-i}^{dn}$ should also be a positive semidefinite matrix.

Following the rule of ODAEPM, the GOD of local sensor and fusion center are defined as:

$$\eta_{f}^{dn}=\frac{1}{Trace(\mathbf{D}_{1,k}^{*})_{f}^{dn}} \tag{29}$$

$$\eta_{i}^{dn}=\frac{1}{Trace(\mathbf{D}_{1,k}^{*})_{i}^{dn}} \tag{30}$$

From Equation (28) and properties of positive semi-definite matrices, we can get the following formula:

$$\begin{aligned}Trace(\mathbf{D}_{1,k}^{*})_{f}^{dn} &=Trace[(\mathbf{D}_{1,k}^{*})_{i}^{dn}-(\mathbf{D}_{1,k}^{*})_{-i}^{dn}]\\ &=Trace(\mathbf{D}_{1,k}^{*})_{i}^{dn}-Trace(\mathbf{D}_{1,k}^{*})_{-i}^{dn}\end{aligned} \tag{31}$$

$$Trace(\mathbf{D}_{1,k}^{*})_{-i}^{dn}\geq 0 \tag{32}$$

$$Trace(\mathbf{D}_{1,k}^{*})_{f}^{dn}\leq Trace(\mathbf{D}_{1,k}^{*})_{i}^{dn} \tag{33}$$

Thus, it is obvious that

$$\eta_f^{dn} \geq \eta_i^{dn} \tag{34}$$

The LOD of jth state component of local sensor and fusion center is defined as

$$\Delta(\eta_f^{dn})_j = \frac{1}{[(\mathbf{D}_{1,k}^*)_f^{dn}]_j} \tag{35}$$

$$\Delta(\eta_i^{dn})_j = \frac{1}{[(\mathbf{D}_{1,k}^*)_i^{dn}]_j} \tag{36}$$

where the subscript j of observable degree discriminant matrix means the jth diagonal element of the observable degree discriminant matrix.

To discuss the LOD, we need to extract the diagonal elements by the following equation

$$\alpha_j = [\overbrace{0 \cdots 0}^{j-1} 1 \cdots 0]^T \tag{37}$$

Then,

$$[(\mathbf{D}_{1,k}^*)_f^{dn}]_j = \alpha_j^T (\mathbf{D}_{1,k}^*)_f^{dn} \alpha_j \tag{38}$$

$$[(\mathbf{D}_{1,k}^*)_i^{dn}]_j = \alpha_j^T (\mathbf{D}_{1,k}^*)_i^{dn} \alpha_j \tag{39}$$

$$\begin{aligned} [(\mathbf{D}_{1,k}^*)_i^{dn}]_j - [(\mathbf{D}_{1,k}^*)_f^{dn}]_j &= \alpha_j^T [(\mathbf{D}_{1,k}^*)_i^{dn} - (\mathbf{D}_{1,k}^*)_f^{dn}] \alpha_j \\ &= \alpha_j^T (\mathbf{D}_{1,k}^*)_{-i}^{dn} \alpha_j \geq 0 \end{aligned} \tag{40}$$

$$[(\mathbf{D}_{1,k}^*)_f^{dn}]_j \leq [(\mathbf{D}_{1,k}^*)_i^{dn}]_j \tag{41}$$

Thus, we can get the same result of LOD between local sensor and fusion center as the GOD:

$$\Delta(\eta_f^{dn})_j \geq \Delta(\eta_i^{dn})_j \tag{42}$$

Furthermore, if the local sensor in the fusion system has the same observation matrix $\mathbf{H}_{i,j}$ for each time period, and the observation noise covariance matrix $\mathbf{R}_{i,j}$ is also the same, it means the distributed multi-sensor fusion system consists of same kind of sensors. Under this condition, Equation (17) is rewritten as

$$\begin{aligned} (\mathbf{D}_{1,k}^*)_f^{dn} &= [\mathbf{\Phi}_{k,1}^{-T} \sum_{j=1}^{k} \sum_{i=1}^{L} (\mathbf{\Phi}_{j,1}^T \mathbf{H}_{i,j}^T R_{i,j}^{-1} \mathbf{H}_{i,j} \mathbf{\Phi}_{j,1}) \mathbf{\Phi}_{k,1}^{-1}]^{-1} \\ &= L[\mathbf{\Phi}_{k,1}^{-T} \sum_{j=1}^{k} (\mathbf{\Phi}_{j,1}^T \mathbf{H}_{i,j}^T R_{i,j}^{-1} \mathbf{H}_{i,j} \mathbf{\Phi}_{j,1}) \mathbf{\Phi}_{k,1}^{-1}]^{-1} \\ &= L(\mathbf{D}_{1,k}^*)_i^{dn} \end{aligned} \tag{43}$$

which means the observation degree of fusion center in distributed multi-sensor fusion system without feedback is the sum of observation degree of its local sensors.

Next, we discuss the the relationship of observation degree between local filter and fusion center in distributed multi-sensor fusion system with feedback. According to Equations (20) and (21), we transform $(\mathbf{D}^*_{1,k})^{df}_f$ as Equation (27), making

$$\begin{aligned}(\mathbf{D}^*_{1,k})^{df}_f &= [\mathbf{\Phi}^{-T}_{k,1}\sum_{j=1}^{k}\sum_{l=1}^{L}(\mathbf{\Phi}^T_{j,1}\mathbf{H}^T_{l,j}\boldsymbol{R}^{-1}_{l,j}\mathbf{H}_{l,j}\mathbf{\Phi}_{j,1})\mathbf{\Phi}^{-1}_{k,1}]^{-1} \\ &= [\mathbf{\Phi}^{-T}_{k,1}\sum_{j=1}^{k-1}\sum_{l=1}^{L}(\mathbf{\Phi}^T_{j,1}\mathbf{H}^T_{l,j}\boldsymbol{R}^{-1}_{l,j}\mathbf{H}_{l,j}\mathbf{\Phi}_{j,1})\mathbf{\Phi}^{-1}_{k,1} + \mathbf{H}^T_{i,k}\boldsymbol{R}^{-1}_{i,k}\mathbf{H}_{i,k} + \sum_{\substack{l=1\\ l\neq i}}^{L}(\mathbf{H}^T_{l,k}\boldsymbol{R}^{-1}_{l,k}\mathbf{H}_{l,k})]^{-1} \\ &= (\mathbf{D}^*_{1,k})^{df}_i - (\mathbf{D}^*_{1,k})^{df}_{-i}\end{aligned} \tag{44}$$

where

$$(\mathbf{D}^*_{1,k})^{df}_{-i} = (\mathbf{D}^*_{1,k})^{df}_i[(\sum_{\substack{l=1\\ l\neq i}}^{L}\mathbf{H}^T_{l,k}\boldsymbol{R}^{-1}_{l,k}\mathbf{H}_{l,k})^{-1} + (\mathbf{D}^*_{1,k})^{df}_i]^{-1}(\mathbf{D}^*_{1,k})^{df}_i \tag{45}$$

Following the same operation as Equations (29)–(43), we can also get the result

$$\eta^{df}_f \geq \eta^{df}_i \tag{46}$$

$$\Delta(\eta^{df}_f)_j \geq \Delta(\eta^{df}_i)_j \tag{47}$$

From the above proof, we come to the conclusion that, in distributed multi-sensor fusion system, whether there is feedback or non-feedback in the fusion method, both LOD and GOD of fusion center are greater than them getting the local filter. That means both in state components and system level, the fusion center obtains better filter performance than local sensor. In addition, it proves that the multi-sensor information fusion can improve the observation performance of the system from the aspect of observable degree.

4.2. The Relationship of Observable Degree among Different Fusion Methods

After we get the relationship between fusion center and local filter of distributed multi-sensor fusion system, we discuss the observable degree among different fusion methods.

According to Equations (17), (20) and (25), we can find the observable discrimination matrix in different fusion methods are equal to each other. Thus, we conclude that,

$$\eta^{dn}_f = \eta^{df}_f = \eta^{cs}_f \tag{48}$$

$$\Delta(\eta^{dn}_f)_j = \Delta(\eta^{df}_f)_j = \Delta(\eta^{cs}_f)_j \tag{49}$$

the LOD and GOD of fusion center in these fusion methods are the same. Then, we discuss the observable degree of local filter in distributed multi-sensor fusion system with or without feedback and the observable degree of different steps during sequential fusion.

First, we compare the observable degree of local filter in distributed multi-sensor fusion system with or without feedback. Doing the same as Equation (27), $(\mathbf{D}_{1,k}^*)_i^{df}$ is rewritten as

$$\begin{aligned}(\mathbf{D}_{1,k}^*)_i^{df} =&[\mathbf{\Phi}_{k,1}^{-T}\sum_{j=1}^{k-1}\sum_{l=1}^{L}(\mathbf{\Phi}_{j,1}^T\mathbf{H}_{l,j}^T\boldsymbol{R}_{l,j}^{-1}\mathbf{H}_{l,j}\mathbf{\Phi}_{j,1})\mathbf{\Phi}_{k,1}^{-1}+\mathbf{H}_{i,k}^T\boldsymbol{R}_{i,k}^{-1}\mathbf{H}_{i,k}]^{-1}\\ =&[\mathbf{\Phi}_{k,1}^{-T}(\sum_{j=1}^{k}\mathbf{\Phi}_{j,1}^T\mathbf{H}_{i,j}^T\boldsymbol{R}_{i,j}^{-1}\mathbf{H}_{i,j}\mathbf{\Phi}_{j,1})\mathbf{\Phi}_{k,1}^{-1}+\mathbf{\Phi}_{k,1}^{-T}\sum_{j=1}^{k-1}\sum_{\substack{l=1\\l\neq i}}^{L}(\mathbf{\Phi}_{j,1}^T\mathbf{H}_{l,j}^T\boldsymbol{R}_{l,j}^{-1}\mathbf{H}_{l,j}\mathbf{\Phi}_{j,1})\mathbf{\Phi}_{k,1}^{-1}]^{-1}\\ =&(\mathbf{D}_{1,k}^*)_i^{dn}-(\mathbf{D}_{1,k}^*)_{-dn}^{df}\end{aligned} \tag{50}$$

where

$$(\mathbf{D}_{1,k}^*)_{-dn}^{df} =(\mathbf{D}_{1,k}^*)_i^{dn}\mathbf{\Phi}_{k,1}^{-T}[(\sum_{j=1}^{k-1}\sum_{\substack{l=1\\l\neq i}}^{L}\mathbf{\Phi}_{j,1}^T\mathbf{H}_{l,j}^T\boldsymbol{R}_{l,j}^{-1}\mathbf{H}_{l,j}\mathbf{\Phi}_{j,1})^{-1}+\mathbf{\Phi}_{k,1}^{-1}(\mathbf{D}_{1,k}^*)_i^{dn}\mathbf{\Phi}_{k,1}^{-T}]^{-1}\mathbf{\Phi}_{k,1}^{-1}(\mathbf{D}_{1,k}^*)_i^{dn} \tag{51}$$

Following the same operation as Equations (29)–(43), the conclusion is

$$\eta_i^{df} \geq \eta_i^{dn} \tag{52}$$

$$\Delta(\eta_i^{df})_j \geq \Delta(\eta_i^{dn})_j \tag{53}$$

Secondly, compare the observable degree between local filter in distributed multi-sensor fusion system with feedback and different steps during sequential fusion. Consider that the ranking of sequential input of sequential fusion system is in ascending sort order of index of the sensor number. Following the same operation as above,

$$\begin{aligned}(\mathbf{D}_{1,k}^*)_i^{cs} =&[\mathbf{\Phi}_{k,1}^{-T}\sum_{j=1}^{k-1}\sum_{l=1}^{L}(\mathbf{\Phi}_{j,1}^T\mathbf{H}_{l,j}^T\boldsymbol{R}_{l,j}^{-1}\mathbf{H}_{l,j}\mathbf{\Phi}_{j,1})\mathbf{\Phi}_{k,1}^{-1}+\sum_{l=1}^{i}\mathbf{H}_{l,k}^T\boldsymbol{R}_{l,k}^{-1}\mathbf{H}_{l,k}]^{-1}\\ =&[\mathbf{\Phi}_{k,1}^{-T}\sum_{j=1}^{k-1}\sum_{l=1}^{L}(\mathbf{\Phi}_{j,1}^T\mathbf{H}_{l,j}^T\boldsymbol{R}_{l,j}^{-1}\mathbf{H}_{l,j}\mathbf{\Phi}_{j,1})\mathbf{\Phi}_{k,1}^{-1}+\mathbf{H}_{i,k}^T\boldsymbol{R}_{i,k}^{-1}\mathbf{H}_{i,k}+\sum_{l=1}^{i-1}(\mathbf{H}_{l,j}^T\boldsymbol{R}_{l,j}^{-1}\mathbf{H}_{l,j})]^{-1}\\ =&(\mathbf{D}_{1,k}^*)_i^{df}-(\mathbf{D}_{1,k}^*)_{-df}^{cs}\end{aligned} \tag{54}$$

where

$$(\mathbf{D}_{1,k}^*)_{-df}^{cs} =(\mathbf{D}_{1,k}^*)_i^{df}[(\sum_{l=1}^{i-1}(\mathbf{H}_{l,j}^T\boldsymbol{R}_{l,j}^{-1}\mathbf{H}_{l,j})^{-1}+(\mathbf{D}_{1,k}^*)_i^{df}]^{-1}(\mathbf{D}_{1,k}^*)_i^{df} \tag{55}$$

Thus, the result is

$$\eta_i^{cs} \geq \eta_i^{df} \tag{56}$$

$$\Delta(\eta_i^{cs})_j \geq \Delta(\eta_i^{df})_j \tag{57}$$

The relationship of observable degree among the local filter in distributed multi-sensor fusion system with or without feedback and different step during sequential fusion is concluded as follows

$$\eta_i^{dn} \leq \eta_i^{df} \leq \eta_i^{cs} \tag{58}$$

$$\Delta(\eta_i^{dn})_j \leq \Delta(\eta_i^{df})_j \leq \Delta(\eta_i^{cs})_j \tag{59}$$

4.3. Brief Summary

According to [15], the observability definition of ODAEPM, the greater the observability, the better the filtering performance lead us to the the following conclusions:

In two different distributed multi-sensor fusion systems, both GOD and LOD of the fusion center are better than the local filters.That means both in state components and system level the fusion center obtains better filter performance than local sensor. In addition, it proves that the multi-sensor information fusion can improve the observation performance of the system from the aspect of observable degree.

In two different distributed multi-sensor fusion systems, the feedback information from the fusion center can improve the filtering performance of the local filter, but with or without feedback information, the fusion solution at the fusion center is equivalent. That means both in state components and system level the local filter of distributed multi-sensor fusion systems with feedback obtains better filter performance than the local filter of distributed multi-sensor fusion systems without feedback.

From the perspective of observability, in centralized multi-sensor fusion systems, the fusion solution at the fusion center is equivalent to the others. The filtering performance of the fusion center is also improved at ith step during time k.This improvement is superior to the distributed multi-sensor with feedback system for the improvement of local filter filtering performance.

5. Numerical Analysis

To validate the effectiveness of the mathematical proof of relationship of observable degree among information fusion system, computer simulation was performed to demonstrate the target tracking with three different observation sensors. The sensors used in this paper are only used to obtain the speed and displacement of CV models, so we only use the velocity sensor and the displacement sensor. The target motion is the typical two-dimensional linear discrete motion models as constant velocity (CV) model.

The state transition matrix is given as:

$$\mathbf{\Phi} = \begin{bmatrix} 1 & t & 0 & 0 \\ 0 & 1 & 0 & 0 \\ 0 & 0 & 1 & t \\ 0 & 0 & 0 & 1 \end{bmatrix}$$

The observation matrix of three different sensors are:

$$\mathbf{H}_1 = \begin{bmatrix} 1 & 0 & 0 & 0 \\ 0 & 0 & 1 & 0 \end{bmatrix}, \mathbf{H}_2 = \begin{bmatrix} 1 & 1 & 0 & 0 \\ 0 & 0 & 1 & 1 \end{bmatrix}, \mathbf{H}_3 = \begin{bmatrix} 0 & 1 & 0 & 0 \\ 0 & 0 & 0 & 1 \end{bmatrix}$$

The observation covariance matrix of them are:

$$\mathbf{R}_1 = \begin{bmatrix} 0.1 & 0 \\ 0 & 0.1 \end{bmatrix}, \mathbf{R}_2 = \begin{bmatrix} 0.2 & 0 \\ 0 & 0.2 \end{bmatrix}, \mathbf{R}_3 = \begin{bmatrix} 0.3 & 0 \\ 0 & 0.3 \end{bmatrix}$$

The subscripts of **H** and **R** correspond to same index sensor.

The simulation results are shown in Figures 5–9.

Furthermore, to make the relationship clearer, the observable degree analysis in the normalized form, as

$$\eta' = \frac{\eta}{\sum \eta} \tag{60}$$

where the numerator is the observable degree obtained by ODAEPM and the denominator is the sum of all observable degrees participating in the comparison in the same figure.

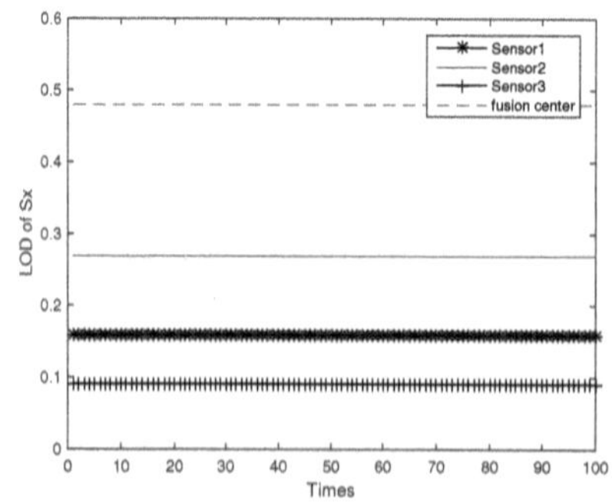

(a)

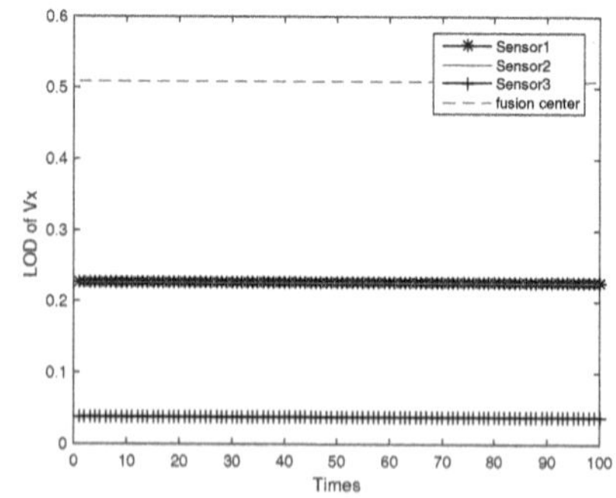

(b)

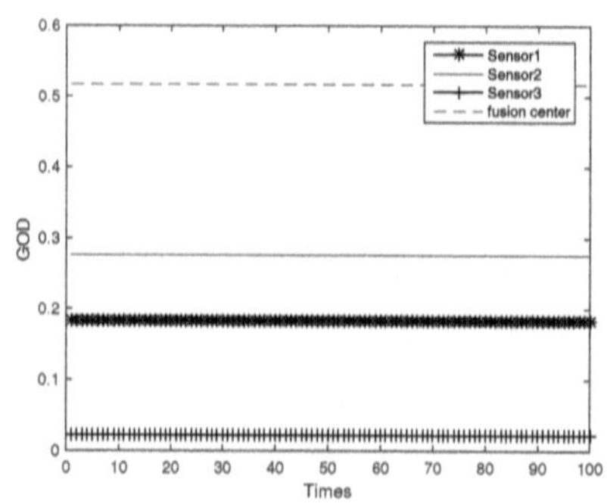

(c)

Figure 5. Observable degree comparison between local filter and fusion center in distributed multi-sensor fusion system without feedback: (**a**) LOD of Sx; (**b**) LOD of Vx; (**c**) GOD.

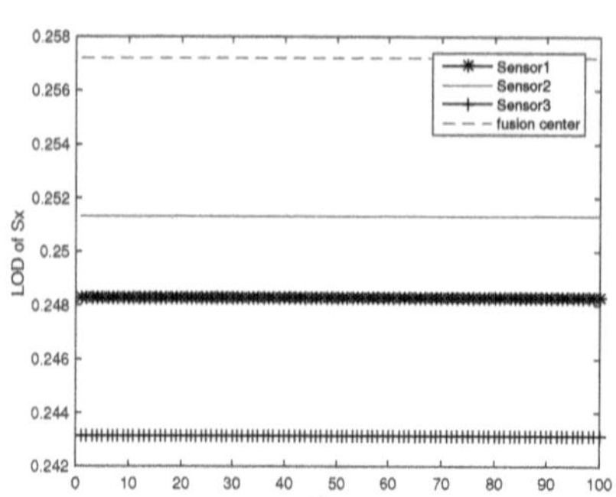

(a)

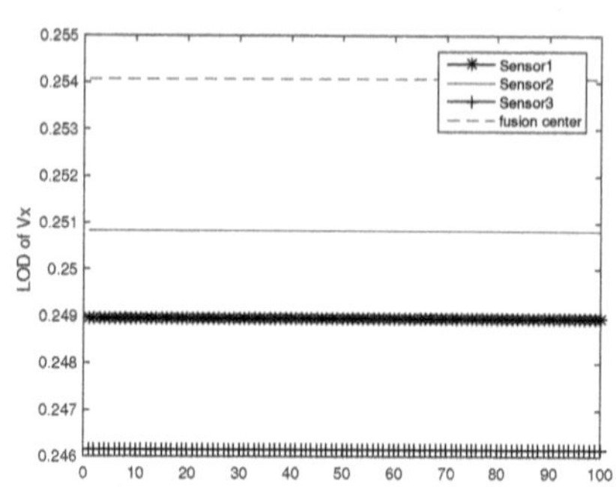

(b)

Figure 6. *Cont.*

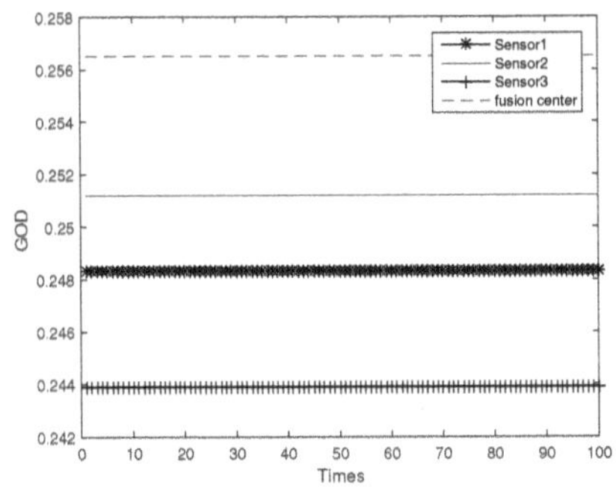

(c)

Figure 6. Observable degree comparison between local filter and fusion center in distributed multi-sensor fusion system with feedback: (**a**) LOD of Sx; (**b**) LOD of Vx; (**c**) GOD.

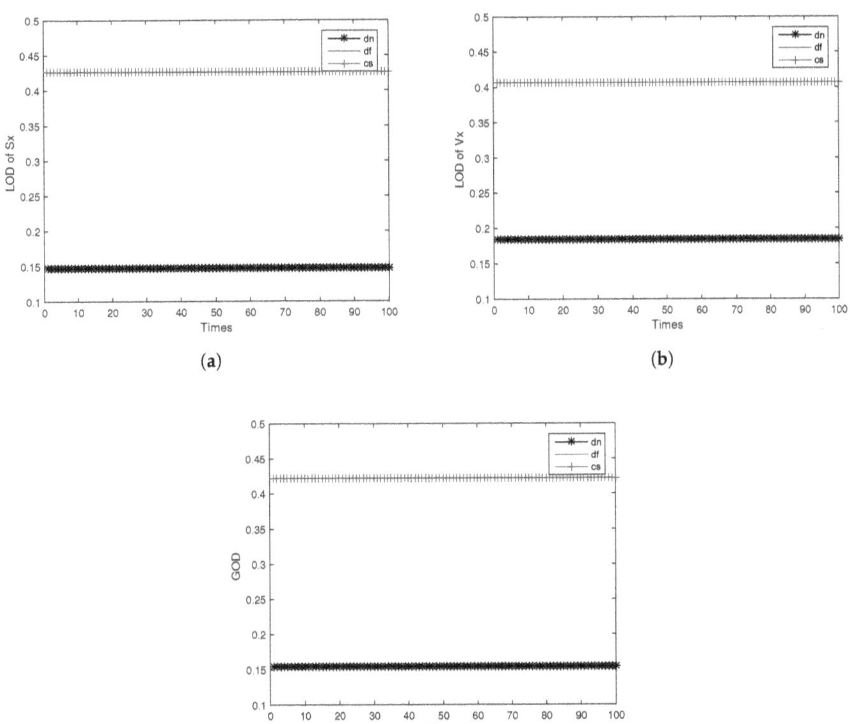

Figure 7. Observable degree comparison among different fusion methods for same index 1: (**a**) LOD of Sx; (**b**) LOD of Vx; (**c**) GOD.

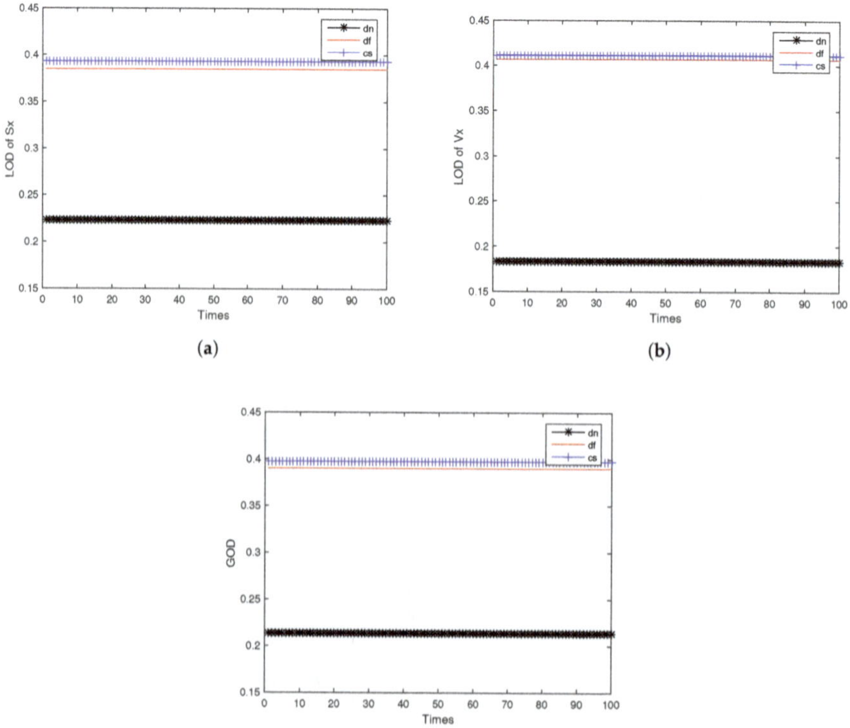

(**a**) (**b**) (**c**)

Figure 8. Observable degree comparison among different fusion methods for same index 2: (**a**) LOD of Sx; (**b**) LOD of Vx; (**c**) GOD.

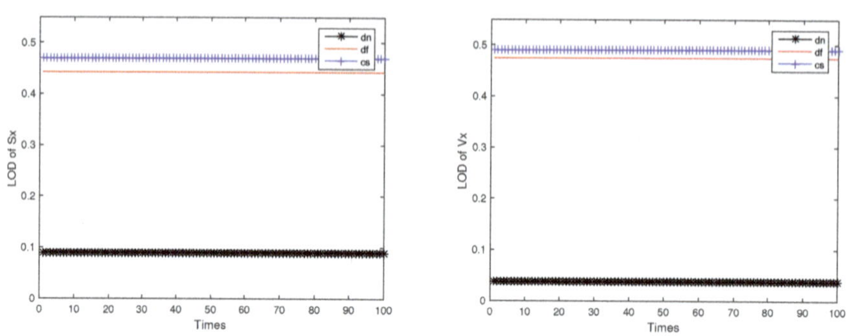

(**a**) (**b**)

Figure 9. *Cont.*

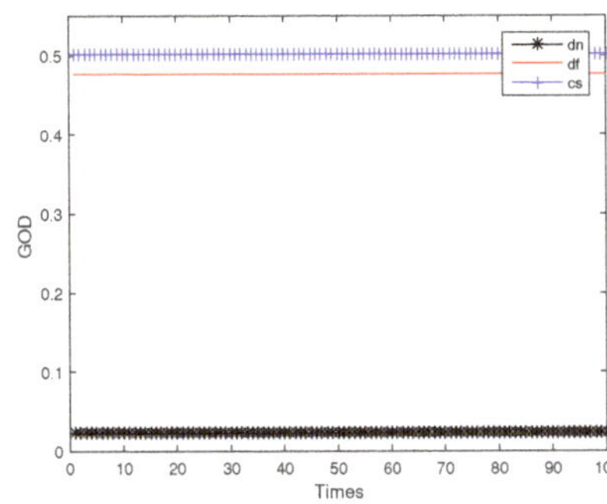

(c)

Figure 9. Observable degree comparison among different fusion methods for same index 3:(**a**) LOD of Sx; (**b**) LOD of Vx; (**c**) GOD.

Figure 5a,b shows the LOD of target position Sx and velocity Vx for ODAEPM obtained in distributed multi-sensor fusion system without feedback, both observable degree of three local filters and fusion center are contained. The GODs of local filter and fusion center using distributed multi-sensor fusion method without feedback are shown in Figure 5c. Similar to Figure 5, Figure 6 shows the condition of LOD and GOD of local filter and fusion center of distributed multi-sensor fusion system with feedback. In Figures 5 and 6, we can find that both LOD and GOD obtained from fusion center are greater than those obtained from local filter, whether the fusion method returning the fusion result to local filter as a feedback. It confirms that the information fusion can obtain better performance than single sensor observation.

Figure 7 shows the observable degree comparison among different fusion methods for same index 1, while the index 1 for two distributed multi-sensor fusion method means the observable degree obtained from the first local sensor, and for centralized multi-sensor sequential fusion system that means it is the first step of sequential fusion at this period, the fusion method only grabs the observation data from the first local sensor. Similarly, Figures 8 and 9 show the observable degree comparison among different fusion methods for same index 2 and 3, for centralized multi-sensor sequential fusion system increased acquisition the information for the two other local sensors. For each figure, both LOD and GOD are well compared. With the assumption that the sequence of sequential fusion system input is the same as the sensor index, the simulation results in Figures 7–9 show that the obtained observable degree in ascending order is the distributed multi-sensor fusion system without feedback, the distributed multi-sensor fusion system with feedback, and the centralized multi-sensor sequential fusion system. It provides the correctness of the derivation for relationship among different fusion methods in this paper. The obtained observable degree will be the same when obtained from distributed multi-sensor fusion system with feedback and centralized multi-sensor sequential fusion system only under the index equal to 1.

Then, to prove the result shown in Equation (43), if the distributed multi-sensor fusion system consists of the same kind of sensors, the observation degree of fusion center in distributed multi-sensor fusion system without feedback is the sum of observation degree of its local sensors. We established the multi-sensor fusion system without feedback with three sensors, the observation matrix and observation noise covariance matrix of which are $\mathbf{H}_1$ and $\mathbf{R}_1$. The simulation is shown in Table 1 and Figure 10.

Table 1. Observable degree of local filter and fusion center with condition that fusion system consists of the same kind of sensors.

	Sensor1	Sensor2	Sensor3	Fusion Center
Sx	253.8	253.8	253.8	761.3
Vx	8832.5	8832.5	8832.5	26,497.5
system	246.3	246.3	246.3	738.8

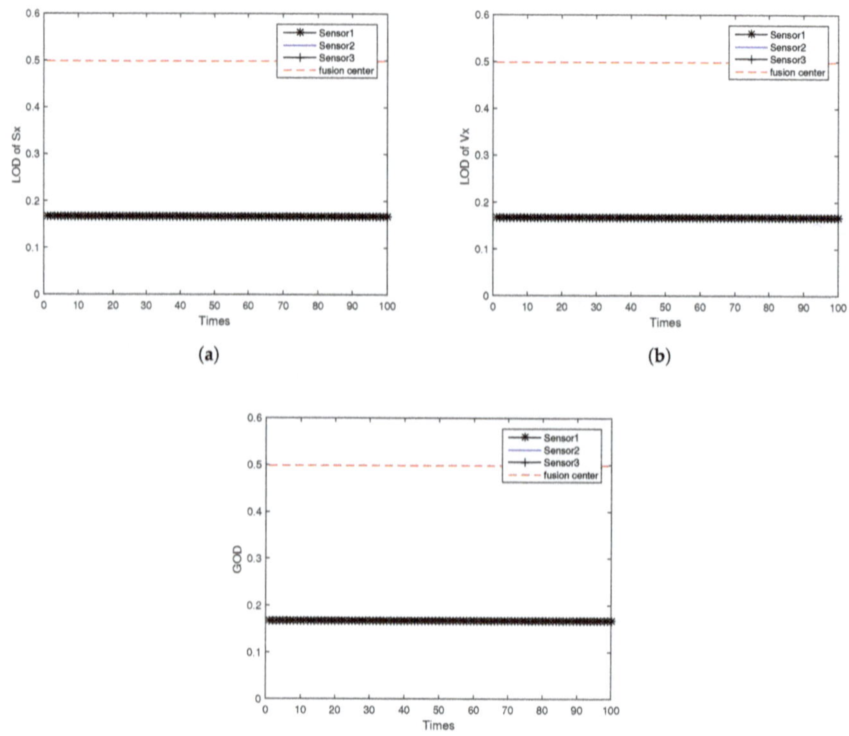

Figure 10. Observable degree of local filter and fusion center:(**a**) LOD of Sx; (**b**) LOD of Vx; (**c**) GOD.

In Table 1 and Figure 10, we can see the observable degree of fusion center is equal to the sum of its local sensors.

Based on the simulation results presented in Figures 5–10 and Table 1, our proof of observable degree relationship among multi-sensor fusion system is verified.

6. Conclusions

In this paper, observable degree analysis for multi sensor fusion system is addressed. Three different multi-sensor fusion system methods were studied, and the observability of three different fusion methods was calculated according to the definition of ODAEPM. It is also proved mathematically that, in two different distributed multi-sensor fusion systems, both GOD and LOD of their respective fusion centers are higher than their local filter, that is, the filtering performance is also better, which also illustrates the advantages of multi-sensor fusion system. The local filter performance of the distributed multi-sensor fusion system with feedback is better than without feedback. This shows that the feedback information from the fusion center can effectively improve the performance of the local filter. To confirm the mathematical proof, we simulated three different kinds of fusion methods operating on two-dimensional linear discrete motion constant velocity models and established three different observation sensors. We can verify our mathematical conclusions based on the simulation. It is clear from simulation that the feedback mechanism for distributed multi-sensor systems can significantly improve the performance of local filter. Therefore, in a multi-sensor system, the feedback mechanism should be introduced to the local nodes as much as possible, which can improve the filtering performance of the local filter.

Author Contributions: Conceptualization, Q.G., Z.H. and T.C.; Methodology, Z.H., T.C. and Q.G.; Validation, T.C. and Q.G.; Formal Analysis, T.C.; Writing, T.C.; Writing-Review-Editing, Z.H., H.W. and Q.G.

Funding: This work was supported by Zhejiang Provincial Nature Science Foundation of China (Grant no.LR17F030005), National Natural Science Foundation of China (NSFC) (Grant no.61773147, 61371064). This research received no external funding.

Conflicts of Interest: The authors declare no conflict of interest.

References

1. Raghavendra, C.S.; Sivalingam, K.M.; Znati, T. *Wireless Sensor Network*; Springer US: New York, NY, USA; 2004. [CrossRef]
2. Lersteau, C.; Rossi, A.; Sevaux, M. Robust scheduling of wireless sensor networks for target tracking under uncertainty. *Eur. J. Oper. Res.* **2016**, *10*, 407–417. [CrossRef]
3. Xu, D.; Ding, S.; Wang, Y.; Shen, B. Quadratic optimal fault tolerant control on wireless networked control systems for real-time industrial applications. *Asian J. Control* **2015**, *17*, 74–83. [CrossRef]
4. Han, C.; Zhu, H.; Duan, Z. *Multi Source Information Fusion*; Tsinghua University Press: Beijing, China, 2006; ISBN 978-7-3022-2499-0. (In Chinese)
5. Hu, Z.-T.; Hu, Y.-M.; Liu, X.X. Kalman Filter Based on Measurement Lifting Strategy. Available online: https://www.researchgate.net/publication/304942856 (accessed on 28 November 2018).
6. Hu, Z.; Liu, X.; Hu, Y. Particle filter based on the lifting scheme of observations. *IET Radar Sonar Navig.* **2015**, *9*, 48–54. [CrossRef]
7. He, Y. *Information Fusion Theory with Application*; Publishing House of Electronic Industry: Beijing, China, 2010; pp. 132–140, ISBN 978-7-1211-0323-0
8. Robert, H.; Krener, A. Nonlinear controllability and observability. *IEEE Trans. Autom. Control* **1977**, *22*, 728–740. [CrossRef]
9. Zhao, G.; Wang, J.; Jiang, W.H. *Modern Control Theory*; Zhao, G.F., Ed.; Electronic Industry Press: Beijing, China, 2010; pp. 32–58, ISBN 978-7-1112-7831-3.
10. Ma, J.; Ge, Q.; Wang, Y.; Bai, L. Comparison on system observable degree analysis methods for target tracking. In Proceedings of the IEEE International Conference on Information and Automation, Lijiang, China, 8–10 August 2015; pp. 1037–1042. [CrossRef]
11. Ham, F.M.; Brown, R.G. Observability, eigenvalues, and kalman filtering. *IEEE Trans. Aerosp. Electr. Syst.* **1983**, *19*, 269–273. [CrossRef]
12. Cheng, X.; Wan, D.; Zhong, X. Study on observability and its degree of strapdown inertial navigation system. *J. Southeast Univ.* **1997**, *27*, 6–10. [CrossRef]
13. Dai, H.D.; Chen, M.; Zhou, S.L.; Li, J. A new rapid transfer alingnment method and the analysis of observable degree. *J. Astronaut.* **2009**, *30*, 1449–1454. [CrossRef]
14. Dong, J.L.; Mo, B. The method of system observability analysis using pseudo-inverse of system observability matrix. In Proceedings of the 32nd Chinese Control Conference, Xi'an, China, 26–28 July 2013; pp. 55–59.
15. Ge, Q.; Ma, J.; Chen, S.; Wang, Y.; Bai, L. Observable Degree Analysis to Match Estimation Performance for Wireless Tracking Networks. *Asian J. Control.* **2017**, *19*, 1259–1270. [CrossRef]
16. Wen, C.; Ge, Q. A Data Fusion Algorithm of the Nonlinear System Based on Filtering Step by Step. *Int. J. Control Autom. Syst.* **2006**, *4*, 165–171.
17. Hu, Z.-T.; Zhang, J.; Hu, Y.-M.; Jin, Y. Multi-Sensor Ensemble Kalman Filtering Algorithm Based on Metropolis-Hastings Sampling. *Acta Electron. Sin.* **2017**, *45*, 868–873. [CrossRef]

18. Ge, Q.; Li, W.; Sun, R.; Xu, Z. Centralized fusion algorithms based on EKF for multisensor non-linear systems. *Acta Autom. Sin.* **2013**, *36*, 816–825. (In Chinese) [CrossRef]
19. Kong, X.; Dong, J.; Ji, Q.; Xue, J. INS observable degree analysis method based on PWCS. *J. Chin. Inert. Technol.* **2011**, *19*, 631–636.
20. Ge, Q.; Shao, T.;Yang, Q.;Shen, X.;Wen, C. Multisensor Nonlinear Fusion Methods Based on Adaptive Ensemble Fifth-degree Iterated Cubature Information Filter for Biomechatronics. *IEEE Trans. Syst. Man Cybern. Syst.* **2016**, *46*, 912–925. [CrossRef]
21. Luo, R.C.; Kay, M.G. Multisensor Integration and Fusion in Intelligent System. *IEEE Trans. Syst. Man Cyber.* **1998**, *19*, 901–931. [CrossRef]
22. Luo, R.C.; Yih, C.C.; Su, K.L. Multisensor fusion and integration: Approaches, applications, and future research directions. *IEEE Sens. J.* **2002**, *2*, 107–119. [CrossRef]
23. You, H.; Wei, X. Relationship between track fusion solutions with and without feedback information. *J. Electron. Inf. Technol.* **2003**, *6*, 47–51.
24. Zhou, G.; Xie, J.; Xu, R.; Quan, T. Sequential nonlinear tracking filter without requirement of measurement decorrelation. *J. Syst. Eng. Electron.* **2015**, *26*, 1135–1141. [CrossRef]
25. Yan, Y.H.; Li, Y.S. OCT Signal Processing Method Based on Sequential Kalman Filtering. *Adv. Mater. Res.* **2015**, *1092-1093*, 300–303. [CrossRef]
26. Greub, W.H. *Linear Algebra*; Springer-Verlag: Berlin/Heidelberg, Germany, 1981.

sensors

MDPI

Article

Multitarget Tracking Algorithm Based on Adaptive Network Graph Segmentation in the Presence of Measurement Origin Uncertainty

Tianli Ma [1,2,*], Song Gao [1,2], Chaobo Chen [1] and Xiaoru Song [1]

1 Autonomous Systems and Intelligent Control International Joint Research Center, Xi'An Technological University, Xi'an 710021, China; gaos@xatu.edu.cn (S.G.); chenchaobo@xatu.edu.cn (C.C.); songxiaoru@xatu.edu.cn (X.S.)
2 School of Mechatronic Engineering, Xi'An Technological University, Xi'an 710021, China
* Correspondence: matianli111@xatu.edu.cn; Tel.: +86-158-2908-2533

Received: 11 September 2018; Accepted: 2 November 2018; Published: 6 November 2018

check for updates

Abstract: To deal with the problem of multitarget tracking with measurement origin uncertainty, the paper presents a multitarget tracking algorithm based on Adaptive Network Graph Segmentation (ANGS). The multitarget tracking is firstly formulated as an Integer Programming problem for finding the maximum a posterior probability in a cost flow network. Then, a network structure is partitioned using an Adaptive Spectral Clustering algorithm based on the Nyström Method. In order to obtain the global optimal solution, the parallel A* search algorithm is used to process each sub-network. Moreover, the trajectory set is extracted by the Track Mosaic technique and Rauch–Tung–Striebel (RTS) smoother. Finally, the simulation results achieved for different clutter intensity indicate that the proposed algorithm has better tracking accuracy and robustness compared with the A* search algorithm, the successive shortest-path (SSP) algorithm and the shortest path faster (SPFA) algorithm.

Keywords: network flow theory; multitarget tracking; spectral clustering; A* search algorithm; RTS smoother; integer programming

1. Introduction

The purpose of multitarget tracking is to jointly estimate the number of targets and their state of motion from sensor data [1]. During the past decade, it has developed in a variety of directions, such as Air-Traffic Control [2], Marine Monitoring [3], Computer Vision [4], Autonomous Vehicle and Robot [5], etc. At present, multitarget tracking has achieved substantial advances [6]. However, the measurement origin uncertainty, for instance the unknown and time-varying number of targets, clutters, jamming signals and so forth seriously deteriorates the performance of the multitarget tracking system. Resolving the uncertainty of the measurement origin is a computationally expensive task which relied on the prior information about the target motion. To find the mapping from each measurement to its origin is often called measurement-to-track association or just data association [7,8].

Markov Chain Monte Carlo Data Association algorithm (MCMCDA) based on Bayesian Inference [9] and the Probability Hypothesis Density filter (PHD) based on finite set statistics (FISST) [10] have been proposed to cope with this problem of tracking multiple targets with measurement uncertainty. The MCMCDA algorithm can be viewed as a deferred-logic method since a track is decided based on the current and past measurements. It uses the Markov Chain Monte Carlo sampling instead of enumerating over all possible associations. In a PHD filter, it can estimate the target states by recursively computing the first-order moment of the multi-target state posterior probability distribution, without using the complex data association techniques. However, it was not designed to estimate the trajectories of targets. For this problem, Ba-Ngu Vo [11] proposed a newly labeled Random Finite Set (RFS) approach, known as the generalized labeled multi-Bernoulli (GLMB) filter, it can output trajectories and has a better performance in harsh environments.

Bar-Shalom [12] proposed an multidimensional assignment algorithm for solving the data association problem. In essence, the data association problem is converted to a combinatorial optimization problem under certain linear constraints where the total distance/benefit of assigning targets to measurements is minimized/maximized. There is a wide range of algorithms, such as Greedy algorithm [13], Genetic algorithm [14] and Lagrange relaxation theory [15], are used to find the sub-optimum solutions of the multidimensional assignment problem [16]. These approaches, while effective, need to solve the Non-deterministic Polynomial Complete (NPC) with a large amount of data. Goldberg [17] constructed an efficient min-cost flow framework. It has applied a scaling push relabel method to find the optimal solution. Under this framework, Zhang [18] formulated the multitarget data association problem as a maximum a posteriori (MAP) problem. It is mapped into the cost flow network and finds the global optimum solution by depending on the min-cost flow algorithm. An approach combining Dynamic Programming (DP) and Successive Shortest-Path algorithm (SSP) is presented by using Hidden Markov Model (HMM) in [19]. The multitarget tracking problem is formulated as an Integer Linear Programming (ILP) problem, and a greedy, successive shortest-path algorithm is used to reduce the runtime costs. For $k = 1$, this algorithm can obtain the global optimal solution. For $k > 1$, it only obtains the approximate solutions, where k is the unknown number of targets. The Shortest Path Faster algorithm (SPFA) is used to solve the Integer Programming problem of the min-cost flow network and quickly obtains the global optimal solution in [20]. The algorithm improves the robustness and tracking accuracy. In [21], an A* based tracking association algorithm is presented. The integer assumption is relaxed to the standard Linear Programming (LP) problem so that the global optimal solution can be obtained by the A* search algorithm.

The above-mentioned approaches are mainly focused on the object tracking based on video image. The available information of image targets are more than that of point targets. In addition, less clutter leads to simple network structure so that the aforementioned algorithms have a good tracking performance. For the problem of multiple point targets tracking in the presence of measurement origin uncertainty, the network may become more complicated that result in an enormous computational burden.

In this paper, a multitarget tracking algorithm based on adaptive network graph segmentation is proposed to address the problem of tracking multitarget with measurement uncertainty. Parallel search strategy is employed to solve the NP-complete problem. The network flow model of multitarget tracking is divided into different sub-graphs. The optimal trajectory is extracted by using the A* search algorithm. Our main contributions are: (1) a parallel network search framework is presented to cope with the multitarget tracking in the presence of measurement origin uncertainty; and (2) we proposed a adaptive spectral clustering algorithm based on the Nyström Method to obtain the network segmentation results for an unknown cluster number data set.

The rest of paper is organized as follows: the problem of multiple targets tracking is formulated as a cost flow network, and transform it into an Integer Programming problem in Section 2. In Section 3, the A* search algorithm is briefly reviewed. The original contribution of the paper is presented in Section 4, where we describe the adaptive spectral clustering algorithm based on the Nyström Method. Simulation results are provided in Section 5. Conclusions and possible extensions appear in Section 6.

2. Problem Formulation

The multitarget data association problem is regarded as a cost flow network. Let $\mathbf{Z} = \{\mathbf{z}_i\}$ be a set of measurements, $\mathbf{z}_i = \{\phi_i, \varphi_i, t_i\}$, where ϕ_i, φ_i is the position in x and y-axes, respectively. t_i is the time step of the measurement $\mathbf{z}_i$. $\mathbf{T}_k = \{\mathbf{z}_{k_1}, \mathbf{z}_{k_2}, \dots \mathbf{z}_{k_n}\}$ represents a trajectory. The set of trajectories is

$\mathcal{T} = \{\mathbf{T}_1, \mathbf{T}_2, \ldots, \mathbf{T}_L\}$, and the number of trajectories L is unknown. The key of the data association is to compute the maximum a posteriori (MAP) estimate of $\mathcal{T}$ given the measurement set $\mathbf{Z}$:

$$\begin{aligned}\mathcal{T}^* &= \arg\max_{\mathcal{T}} P(\mathcal{T}|\mathbf{Z}) \\ &= \arg\max_{\mathcal{T}} P(\mathbf{Z}|\mathcal{T})P(\mathcal{T}) \\ &= \arg\max_{\mathcal{T}} \prod_i P(\mathbf{z}_i|\mathcal{T})P(\mathcal{T}).\end{aligned} \tag{1}$$

Assume that tracks are independent from each other. The cost flow network framework of multitarget tracking is as follows:

$$P(\mathcal{T}) = \prod_{\mathbf{T}\in\mathcal{T}} P(\mathbf{T}), \tag{2}$$

$$P(\mathbf{T}) = P_s(\mathbf{z}_{k_0})P_l(\mathbf{z}_{k_1}|\mathbf{z}_{k_0})\ldots P_l(\mathbf{z}_{k_n}|\mathbf{z}_{k_{n-1}})P_t(\mathbf{z}_{k_n}), \tag{3}$$

where $P(\mathbf{T})$ is modeled as a Markov Chain. $P_s(\mathbf{z}_{k_0})$ is the initialization probability of a track starting at $\mathbf{z}_{k_0}$, $P_t(\mathbf{z}_{k_n})$ is the termination probability of a track ending at $\mathbf{z}_{k_n}$, and $P_l(\mathbf{z}_{k_j}|\mathbf{z}_{k_i})$ is the transition density from measurement $\mathbf{z}_{k_i}$ to measurement $\mathbf{z}_{k_j}$. $P(\mathbf{z}_i|\mathcal{T})$ denotes the likelihood function of measurement $\mathbf{z}_i$, which represents a measurement being a true target or a false alarm. In this paper, all measurements are regarded as targets; this means that $P(\mathbf{z}_i|\mathcal{T}) = 1$, and the posterior probability of trajectory set $\mathcal{T}$ is calculated as follows:

$$\mathcal{T}^* = \arg\max_{\mathcal{T}} \prod_{\mathbf{T}\in\mathcal{T}} P(\mathbf{T}). \tag{4}$$

To take advantage of the concept of network flow in Network Optimization [22], the indictor variable $f_{i,j}$ is defined as the directed flow variable that from measurement $\mathbf{z}_i$ to measurement $\mathbf{z}_j$. $f_{s,i}$ and $f_{i,t}$ represent the starting flow variable and terminated flow variable, respectively. Depending on the flow conversation method [22], for all nodes $\mathbf{z}_i$, the sum of flows arriving at node $\mathbf{z}_i$ is equal to the sum of outgoing flows from node $\mathbf{z}_i$. For any track $\mathbf{T}_k$, it is satisfies the following equation:

$$f_{s,i} + \sum_j f_{j,i} = \sum_i f_{i,j} + f_{i,t}. \tag{5}$$

Moreover, the cost flow network must guarantee that only one node is represented by a target in a moment. Let the upper bound of the sum of outgoing flows from node $\mathbf{z}_i$ is 1. For any node, the constraint is

$$\forall \mathbf{z}_i, \mathbf{z}_j \quad \sum f_{i,j} \le 1. \tag{6}$$

Taking into account that the target may appear or disappear from any location in the cost flow network, the source and sink node are introduced in [18], which is connected to each node, respectively. We transform the Network Optimization problem of Equation (4) into the IP problem, and the logarithm of the objective function can be rewritten as

$$\begin{aligned}\mathcal{T}^* &= \arg\min_{\mathcal{T}} \sum_{\mathbf{T}_k\in\mathcal{T}} -\log P(\mathbf{T}_k) \\ &= \arg\min_{\mathcal{T}} \sum_{\mathbf{T}_k\in\mathcal{T}} (c_{s,i}f_{s,i} + \sum_j c_{i,j}f_{i,j} + c_{j,t}f_{j,t}) \\ &= \arg\min_{\mathcal{T}} \sum_i c_{s,i}f_{s,i} + \sum_{i,j} c_{i,j}f_{i,j} + \sum_j c_{j,t}f_{j,t},\end{aligned} \tag{7}$$

where $c_{s,i}$ is the cost of the flow from the source node to measurement $\mathbf{z}_i$, $c_{i,j}$ is the cost of the flow from measurement $\mathbf{z}_i$ to measurement $\mathbf{z}_j$, and $c_{j,t}$ is the cost of the flow from measurement $\mathbf{z}_j$ to the sink node. Figure 1 shows an example of the cost flow network. The IP problem of multitarget tracking with measurement uncertainty can be described as

$$\min \quad \sum_i c_{s,i} f_{s,i} + \sum_{i,j} c_{i,j} f_{i,j} + \sum_j c_{j,t} f_{j,t} \tag{8}$$

$$\text{s.t.} \quad \begin{aligned} &\forall \mathbf{z}_i, \mathbf{z}_j, \quad \sum f_{i,j} \leq 1 \\ &\forall \mathbf{z}_i, \mathbf{z}_j, \quad f_{i,j} \geq 0. \end{aligned} \tag{9}$$

The cost can be defined as follows:

$$\begin{cases} c_{s,i} = -\log P_s(\mathbf{z}_i), \\ c_{i,j} = -\log P_l(\mathbf{z}_j|\mathbf{z}_i), \\ c_{j,t} = -\log P_t(\mathbf{z}_j). \end{cases} \tag{10}$$

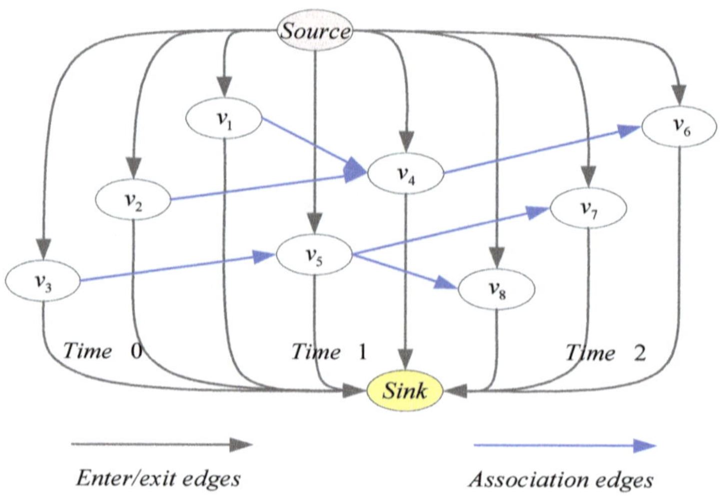

Figure 1. An example of cost flow model with three time steps.

For the IP problem of multitarget tracking, the traditional algorithms have a relatively higher computational cost due to the large number of network nodes. Hence, a parallel processing technology based on the A* search algorithm is presented to find the optimal solution.

3. Description of the A* Search Algorithm

The A* search algorithm is a heuristic graph search method which guides the search process by using the characteristic information of the problem. For the min-cost flow problem, the A* algorithm searches from the origin node, calculates and estimates the cost of each extended node, chooses the extended node that has a minimum cost and stops it when the algorithm reaches the destination node. Assuming that an evaluation function $f^*(x)$ is designed to estimate the minimum cost from origin node s_o through node x to destination node s_d. The estimated minimum cost is calculated as follows:

$$f^*(x) = g^*(x) + h^*(x), \tag{11}$$

where $g^*(x)$ is the cost from origin node s_o to the node x, $h^*(x)$ is the lower bound on the minimum cost from node x to destination node s_d, and $h(x)$ is the actual value from the node x to destination node

s_d, $h^*(x) \leq h(x)$. In order to ensure the optimality of the algorithm, Admissibility and Consistency conditions must be satisfied [23]. Admissibility condition: $f^*(x)$ never overestimates the true cost of a solution along the current path through. Consistency condition: A heuristic function $h(x)$ is consistent if, for every node x and every successor x' of x, the estimated cost of reaching the goal from x is no greater than that of the step cost of getting to x' form x plus the estimated cost of reaching the goal from x'

$$h^*(x) + e(x, x') \leq h^*(x'). \tag{12}$$

In the implementation of the A* search algorithm, two lists need to be built, named Open list and Close list. Open list is the set of nodes that have been calculated, and that are candidates for the selection of the next node. Close list is the set of nodes that have been selected, and that are not in Open list. In the initial stage, Open list contains a source node and Close list is empty. During the iterative process, the A* search algorithm calculates the evaluation function of each node in Open list chooses the node with the minimum cost and judges whether it is the termination node. If so, the algorithm is done. Otherwise, it extends all adjacent nodes and calculates the cost function of each node. If the solution exists, the A* search algorithm can guarantee obtaining the optimal solution [24].

The A* search algorithm can be expressed as follows: here, the Open list and Close list are denoted as O and C. E is the set of edges:

Step 1 Initialization:

Set $x_i = x_s, f^*(x_i) = 0; f^*(x_j) = \infty, g^*(x_j) = \infty, \forall x_j \neq x_i; O = \{x_i\}, C = O$.

Step 2 Node Selection:

Choose $x_i \in Argmin_{x_j \in O} f^*(x_j), C = C \bigcup \{x_i\}, O = O \backslash \{x_i\}$.

Step 3 Stop Rule:

If $x_i = x_d$, then stop. otherwise, continue.

Step 4 Update $f^*(x_j)$ and $g^*(x_j)$:

For each $x_j \in E(x_i)$: If $g^*(x_i) + e(x_i, x_j) + h^*(x_j) < f^*(x_j)$, then $g^*(x_j) = g^*(x_i) + e(x_i, x_j)$; $f^*(x_j) = g^*(x_i) + e(x_i, x_j) + h^*(x_j)$.

If $x_j \notin O, O = O \bigcup \{x_j\}$. Go back to Step 2.

4. Multitarget Tracking Algorithm Based on Adaptive Network Graph Segmentation

Under the multitarget tracking environment, the A* search algorithm have two obvious defects that long running time and large storage space. To solve the problem, an adaptive spectral clustering algorithm is presented to segment the cost flow network, and the A* search algorithm is used to find the optimal track of segmented sub-graph. To take advantage of track mosaic technology, the optimal track of each sub-graph is combined, and the combined track is smoothed by the Rauch–Tung–Striebel smoother. The flow chart of the proposed algorithm is depicted in Figure 2.

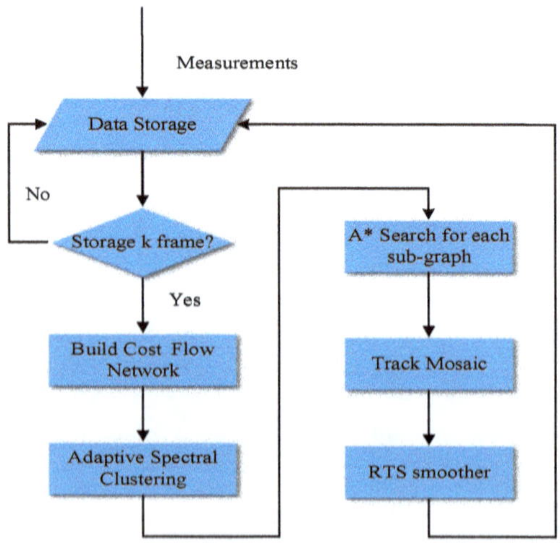

Figure 2. The flow chart of multitarget tracking algorithm based on adaptive network graph segmentation.

4.1. Adaptive Spectral Clustering

The segmentation problem of graph structure represented by the dissimilarity degree between nodes is the combinatorial optimization problem, which is NP-hard. The general solution is to consider the continuous relaxation form for this problem. Spectral clustering is an unsupervised learning method based on graph theory [25] for arbitrary image shapes. It uses the eigenvalue decomposition of graph matrix to build a spectral mapping space of the original data set, and the new space is partitioned by the K-means algorithm.

Let $G_1(\mathbf{V},\mathbf{E},\mathbf{W})$ be an undirected graph that transforms from the directed graph $G_0(\mathbf{V}_0,\mathbf{E}_0)$ with vertex set $\mathbf{V}=(v_1,v_2,\ldots,v_N)$ and edge set $\mathbf{E}=(e_{ij})_{i,j=1,2,\ldots,m}$. We assume that the edge-weighted adjacency matrix of undirected graph $\mathbf{W}=(w_{ij})_{i,j=1,2,\ldots,m}$, which is also called affinity matrix, is nonnegative, $w_{ij}=w_{ji}\geq 0$. If $w_{ij}=0$ that means the node v_i and v_j are not connected. Here, the weight of two nodes w_{ij} is the cost value c_{ij} in the directed graph model $G_0(\mathbf{V}_0,\mathbf{E}_0)$. The sets $\Psi_1 \ldots\Psi_k$ are the subsets of the graph. For sets $\Psi_1\ldots\Psi_k$, $\Psi_1\bigcup\Psi_2 .. \bigcup\Psi_k=\mathbf{V}$ and $\Psi_i\bigcap\Psi_j=\varnothing, i\neq j$. Let $cut(\Psi_1,\Psi_2,\ldots,\Psi_k)$ be the sum of the cuts between sets $\Psi_1,\Psi_2,\ldots,\Psi_k$:

$$cut(\Psi_1,\Psi_2,\ldots,\Psi_k)=\sum_{i=1}^{k} cut(\Psi_i,\bar{\Psi}_i), \tag{13}$$

where $\bar{\Psi}_i$ denotes the complement $\mathbf{V}\backslash\Psi_i$. The purpose of the spectral clustering is to find the sets $\Psi_1,\Psi_2,\ldots,\Psi_k$ such that $MNcut$ is minimized. The objective function is given as follows:

$$MNcut=\sum_{i=1}^{n}\frac{cut(\Psi_i,\bar{\Psi}_i)}{\sum\limits_{u\in\Psi_i,v\in\Psi_i} w_{u,v}}. \tag{14}$$

In [26], eigenvectors are clustered in the subspace that is generated by the first k eigenvectors of normalized Laplacian matrix. The degree matrix of the graph $\mathbf{D}$ and the normalized Laplacian matrix $\mathbf{L}_{sym}$ are defined as

$$\mathbf{D}=\begin{cases}\mathbf{D}_{ii}=\sum\limits_{j} w_{ij}, \\ \mathbf{D}_{ij}=0 \quad i\neq j,\end{cases} \tag{15}$$

$$\mathbf{L}_{sym} = \mathbf{I} - \mathbf{D}^{-1/2}\mathbf{W}\mathbf{D}^{-1/2}. \tag{16}$$

Unfortunately, $\mathbf{W}$ grows as the square of the number of elements in the grouping problem, and it quickly becomes infeasible to fit $\mathbf{W}$ in memory. Hence, an adaptive spectral clustering based on the Nyström Method is proposed to reduce the complexity of the time and space. The Nyström Method is a technique for finding numerical approximations to eigenfunction problems.

Assuming that randomly sample n points from vertex set $\mathbf{V}$ and the number of the remaining points is $N - n$. Now, partition the affinity matrix $\mathbf{W}$ as

$$\mathbf{W} = \begin{bmatrix} \mathbf{A} & \mathbf{B} \\ \mathbf{B}^T & \mathbf{C} \end{bmatrix}, \tag{17}$$

where $\mathbf{A} \in \mathbb{R}^{n \times n}$ represents the sub-block of weights among the random samples, $\mathbf{B} \in \mathbb{R}^{(N-n)\times n}$ contains the weights between the sample points and the rest of points, and $\mathbf{C} \in \mathbb{R}^{(N-n)\times(N-n)}$ denotes the weights matrix between all of the remaining points. Let $\bar{\mathbf{U}}$ represent the approximate eigenvectors of $\mathbf{W}$:

$$\bar{\mathbf{U}} = \begin{bmatrix} \mathbf{U} \\ \mathbf{B}^T\mathbf{U}\Lambda^{-1} \end{bmatrix}. \tag{18}$$

The approximation of $\mathbf{W}$, which we denote $\hat{\mathbf{W}}$, can be written as

$$\begin{aligned} \hat{\mathbf{W}} &= \bar{\mathbf{U}}\Lambda\bar{\mathbf{U}}^T \\ &= \begin{bmatrix} \mathbf{U} \\ \mathbf{B}^T\mathbf{U}\Lambda^{-1} \end{bmatrix} \Lambda \begin{bmatrix} \mathbf{U}^T & \Lambda^{-1}\mathbf{U}^T\mathbf{B} \end{bmatrix} \\ &= \begin{bmatrix} \mathbf{A} & \mathbf{B} \\ \mathbf{B}^T & \mathbf{B}^T\mathbf{A}^{-1}\mathbf{B} \end{bmatrix}. \end{aligned} \tag{19}$$

From Equations (17) and (19), $\mathbf{C}$ is approximated by $\mathbf{B}^T\mathbf{A}^{-1}\mathbf{B}$. Therefore, calculating the affinity matrix between remaining points is avoided. It is noteworthy that the columns of $\bar{\mathbf{U}}$ are not orthogonal. We need to orthogonalize $\bar{\mathbf{U}}$. If $\mathbf{A}$ is a positive definite matrix, $\mathbf{A}^{-1/2}$ represents the symmetric positive definite square root of $\mathbf{A}$. Let $\mathbf{Q} = \mathbf{A} + \mathbf{A}^{-1/2}\mathbf{B}\mathbf{B}^T\mathbf{A}^{-1/2}$, and diagonalize $\mathbf{Q}$ as $\mathbf{Q} = \mathbf{U}_s\Lambda_s\mathbf{U}_s^T$. If the matrix $\mathbf{U}_v$ is defined as

$$\mathbf{U}_v = \begin{bmatrix} \mathbf{A} \\ \mathbf{B}^T \end{bmatrix} \mathbf{A}^{-1/2}\mathbf{U}_s\Lambda_s^{-1/2}, \tag{20}$$

then the affinity matrix W is diagonalized by $\mathbf{U}_v$ and Λ_s. Without calculating $\mathbf{B}^T\mathbf{A}^{-1}\mathbf{B}$, a simple approach is proposed to calculated the row sums of $\hat{\mathbf{W}}$:

$$\hat{\mathbf{d}} = \begin{bmatrix} \mathbf{a}_r & \mathbf{b}_r \\ \mathbf{b}_c & \mathbf{B}^T\mathbf{A}^{-1}\mathbf{b}_r \end{bmatrix}, \tag{21}$$

where $\mathbf{a}_r, \mathbf{b}_r \in \mathbb{R}^{N-n}$ denote the rows sum of $\mathbf{A}$ and $\mathbf{B}$. $\mathbf{b}_c \in \mathbb{R}^n$ denotes column sum of $\mathbf{B}$. The normalized $\mathbf{A}$ and $\mathbf{B}$ can be obtained by $\hat{\mathbf{d}}$. The elements of the normalized $\mathbf{A}$ and $\mathbf{B}$ are given by

$$\mathbf{A}_{ij} = \frac{\mathbf{A}_{ij}}{\sqrt{\hat{\mathbf{d}}_i\hat{\mathbf{d}}_j}}, i, j = 1, \ldots, n, \tag{22}$$

$$\mathbf{B}_{ij} = \frac{\mathbf{B}_{ij}}{\sqrt{\hat{\mathbf{d}}_i\hat{\mathbf{d}}_{j+N}}}, i = 1, \ldots, n, j = 1, \ldots, N - n. \tag{23}$$

The number of clusters is generally determined by human experience and background knowledge. Next, the relationship between the spectrum of the weight matrix and the number of clusters can be obtained by analyzing the affinity matrix of graph.

If v_i and v_j belong to the same class, then $w_{ij} = 1$, otherwise $w_{ij} = 0$. According to perturbation analysis of spectral clustering [27], a permutation matrix $\mathbf{P}$ always exists to make elements of any node set $\mathbf{V}$ in the sequence belongs to a class after the $\mathbf{PV}$ transformation. The affinity matrix $\mathbf{W}$ is a block diagonal matrix that consists of n_k all-1 matrices. Thus, the elements of Laplacian matrix $\hat{\mathbf{L}}_{sym}$ is divided into k matrix blocks. For each matrix block $\hat{\mathbf{L}}_{n_i}$, $\lambda_k = n_i$, $k = 1, 2, \ldots n_i - 1$, where λ_k is the eigenvalues of matrix block $\hat{\mathbf{L}}_{n_i}$, and $\lambda_{n_i} = 0$. In light of Matrix Theory [28], the union of the eigenvalues of real symmetric matrix equals that of the block diagonal matrix that consists of these real symmetric matrices. Therefore, the eigenvalues of Laplacian matrix is the union of the eigenvalues of k matrices. The eigenvalues of Laplacian matrix consists of $(n-k)(n_i-1)$ nonzero eigenvalues and k zero eigenvalues. The number of clusters is the number of zero eigenvalues of Laplacian matrix $\hat{\mathbf{L}}_{sym}$.

4.2. *k-Short Paths Algorithm*

In the cost flow network of multitarget tracking, there may be multiple paths that are directed and paths are edge- and node- disjoint, which means that any two paths cannot share the same edge and node, and a path visits a node in the sub-graph at most once. Here, a path represents a possible track. We reformulated the multitarget tracking problem as an edge- and node- disjoint k-shortest paths problem on a directed acyclic graph. In order to obtain the k-shortest paths, the segmented graph is transformed into a undirected graph firstly, and then the A* search algorithm is adopted to find the single shortest path. If the maximum iteration count is not reached, remove nodes except source node and sink node on the single shortest path, and search for the next path until reaching the maximum iteration count.

4.3. *Track Mosaic*

To segment an undirected graph in the multitarget tracking environment using the adaptive spectral clustering algorithm, it may arise over segmentation. For example, a trajectory may be divided into several segments. In order to obtain the integral track, the mosaic technique is employed to deal with these segmented trajectories. Let $\mathbf{T}_i$ and $\mathbf{T}_j$ are two trajectories of different sub-graphs. x_{io} and x_{jo} are the initial position, and x_{id} and x_{jd} are the terminal position, respectively. The Euclidean distance d_D is used to decide whether to mosaic two trajectories. If $d_D < \tau$, $\mathbf{T}_i$ and $\mathbf{T}_j$ are mosaicked, where τ is the mosaic threshold.

4.4. *Rauch–Tung–Striebel Smoother*

A Rauch–Tung–Striebel (RTS) smoother [29] is used to smooth the extracted tracks. The RTS smoother consists of two parts. The first part is the Kalman filter, which calculates the state of target at each time and estimates the corresponding covariance matrix. The second part is backward recursion [30]. In this process, target state and the covariance matrix are taken as inputs to obtain the smoother output.

4.5. *Time Complexity*

The proposed algorithm is parallel processing in that each sub-graph uses the A* search algorithm to obtain the shortest path. The time complexity of the proposed algorithm is mainly composed of two parts, which are the time complexity of the adaptive spectral clustering algorithm and the worst time complexity of searching sub-graph. The time complexity of the adaptive spectral clustering algorithm is $O(n \times (N-n)) + O(n^3) + O(KNI)$. $O(n \times (N-n))$, $O(n^3)$ and $O(KNI)$ are the time complexity of calculating degree $\hat{\mathbf{d}}$, orthogonal eigenvectors $\mathbf{U}_v$ and K-means clustering algorithm, respectively. N is the number of nodes in the undirected graph, n is the sample points, K is the number of trajectories and I is the iteration number of K-means algorithm. The worst time complexity of searching sub-graph is $O(n_{\max}^2)$, and that $n_{\max}$ is less than N. The proposed algorithm is adopted to calculate the k-shortest path in $O(n \times (N-n)) + O(n^3) + O(KNI)$. The A* search algorithm, SPFA and SSP are recognized as three effective methods to solve the k-shortest path problem. The time complexity of these algorithms are $O(N^2)$, $O(NE)$ and $O(KN^2)$, respectively. E is the number of edge in the undirected graph. The complexity of these algorithms are primarily related to the value of K and N. For multitarget tracking systems, that is, N is large, and the time complexity of the proposed algorithm is far less than that of the algorithms mentioned above.

5. Experimental Results

In this section, the proposed algorithm was tested in different tracking scenarios. The optimal subpattern assignment (OSPA) [31] metric is used for performance assessment. The experiments have been performed on a computer with an Intel G840 2.8 GHz CPU and 4 GB of memory.

5.1. Clustering Quality Evaluation

To evaluate the clustering quality of the adaptive spectral clustering algorithm, the external quality measure (F-score) [32] and (MCR) [33] are used. F-score is used in spam detection for documentation as an overall assessment performance that combines the precision and recall ideas from information retrieval. F-score is defined as follows:

$$\text{F-score} = \sum_{i=1}^{K} (N_i \times \frac{2P_f(i,j)R_f(i,j)}{P_f(i,j)+R_f(i,j)})/N, \tag{24}$$

where $P_f(i,j) = N_{ij}/N_j$ represents the precision of the cluster j for the given class i. $R_f(i,j) = N_{ij}/N_i$ represents the recall of the cluster j for the given class i. N_i is the number of the members of class i. N_{ij} is the number of numbers of class i in cluster j. The MCR is given by

$$MCR = \frac{C_m}{C_t}, \tag{25}$$

where C_m is the number of misclassified targets. C_t is the total number of targets.

Figure 3 displays the comparison of classification results with different sampling rates. Since the sampling rate is 1%, clustering result is seriously affected. When the sampling rate exceeds 1%, it is clear that the classification results outperform that shown in Figure 3b. The clustering performance versus sampling rate are shown in Figure 4. It can be noted that F-score increases with increasing sample rate. Once the sample rate exceeds 30%, the data set can be accurately segmented. This is because the similarity matrix between the sampling points can be approximated by that of all points. In Figure 4b, the sampling rate is equal to 1%, MCR is 0.18. The MCR of other sampling rate is equal to 0. Based on the two evaluation criteria above, we choose the sampling rate as 10% in this paper.

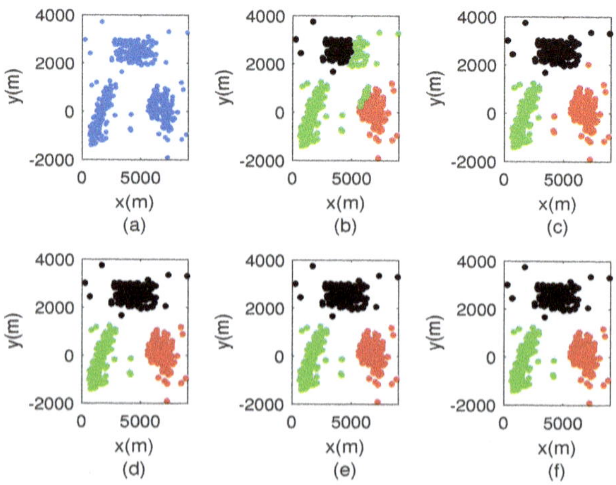

Figure 3. Comparison of classification results. (**a**) original data set; (**b**) sampling rate is 1%; (**c**) sampling rate is 10%; (**d**) sampling rate is 50%; (**e**) sampling rate is 70%; (**f**) sampling rate is 100%.

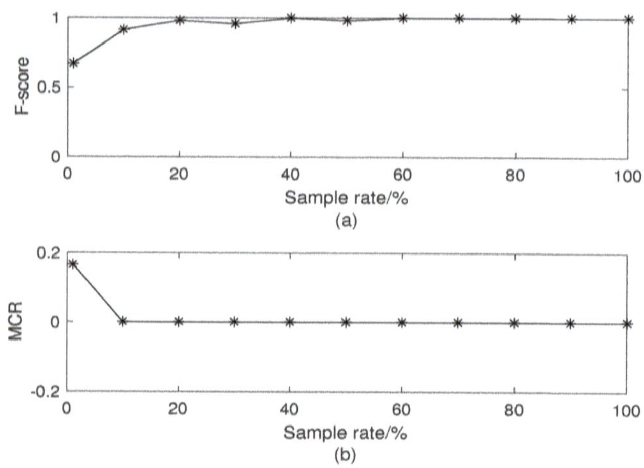

Figure 4. Clustering performance versus sampling rate. (**a**) F-score versus sampling rate; (**b**) MCR versus sampling rate.

5.2. Performance Analysis

In this subsection, we consider two scenarios for multitarget tracking. There are two types of dense clutter areas. Inside Type I dense clutter area, clutter points are uniformly distributed in the surveillance region. While Type II dense clutter area is elliptical and the position of its clutter points follows a 2D Gaussian distribution, whose mean is target position at k time and the standard deviations are determined by the major axes of the ellipse. The measurements are obtained from radar which located at [0, 0] m. The measurement model is described as

$$\mathbf{z}_t^i = \mathbf{H}\Phi_t^i + v_t^i, \tag{26}$$

where $\Phi_t^i = [\phi_t^i, \dot{\phi}_t^i, \varphi_t^i, \dot{\varphi}_t^i]^T$ is the state variable. The measurement noise v_t^i is zero-mean Gaussian random vector with covariance matrix $\mathbf{R} = diag([\delta^2, \delta^2])$, and $\delta = 10$ m in the scenarios 1 and 2.

The target motion can be modeled by combination of constant turn rate (CT) motion and constant velocity (CV) motion [34]. The dynamistic model of the target is described as follows:

$$\Phi_t^i = \mathbf{F}_t^i \Phi_{t-1}^i + w_t^i, \tag{27}$$

where $\mathbf{F}_t^i$ is the state transition matrix of the target i at the time t. Under the assumption of CV motion, it is defined as follows:

$$\mathbf{F}_t^i = \begin{bmatrix} 1 & T & 0 & 0 \\ 0 & 1 & 0 & 0 \\ 0 & 0 & 1 & T \\ 0 & 0 & 0 & 1 \end{bmatrix}, \tag{28}$$

where T denotes the sampling time. In CT motion, it is defined as

$$\mathbf{F}_t^i = \begin{bmatrix} 1 & \dfrac{\sin(\omega_t^i T)}{\omega_t^i} & 0 & \dfrac{\cos(\omega_t^i T) - 1}{\omega_t^i} \\ 0 & \cos(\omega_t^i T) & 0 & -\sin(\omega_t^i T) \\ 0 & \dfrac{1 - \cos(\omega_t^i T)}{\omega_t^i} & 1 & \dfrac{\sin(\omega_t^i T)}{\omega_t^i} \\ 0 & \sin(\omega_t^i T) & 0 & \cos(\omega_t^i T) \end{bmatrix}, \tag{29}$$

where ω_t^i denotes the turn rate. The process noise w_t^i is zero-mean Gaussian random vector with covariance matrix

$$\mathbf{Q} = \begin{bmatrix} T^3/3 & T^2/2 & 0 & 0 \\ T^2/2 & T & 0 & 0 \\ 0 & 0 & T^3/3 & T^2/2 \\ 0 & 0 & T^2/2 & T \end{bmatrix}. \tag{30}$$

Scenario 1:

A multiple non-crossing tracking scenario is considered in the surveillance region [0, 9000] m × [−2000, 5000] m. There are four manoeuvring targets whose initial position are [8000, 2500] m, [500, 3000] m, [2500, 2000] m and [2500, −1500] m, respectively. Their initial velocities are [−80, −80] m/s, [50, −130] m/s, [0, 280] m/s and [150, 0] m/s, respectively. The initial and terminal time are [1, 1, 3, 1] s and [27, 30, 30, 28] s, respectively. The mosaic threshold $\tau = 10$.

In this case, the Type I clutter intensity is $\kappa_1(z_c) = \frac{m_{Ic}}{V_I} = \frac{5}{6.3 \times 10^7} = 0.79 \times 10^{-7}$, where m_{Ic} is the expected number of clutter measurements in this Type I dense clutter area, and V_I is the surveillance region. For the Type II dense clutter, the expected number of clutter measurements in this Type II dense clutter area is 10, and $V_{II} = 4.54 \times 10^5$ m^2 is the 'volume' of the Type II dense clutter surveillance region. Therefore, the Type II clutter intensity is $\kappa_2(z_c) = \frac{m_{IIc}}{V_{II}} = \frac{10}{\pi \times 600 \times 400} = 1.33 \times 10^{-5}$. We perform a total of 100 Monte-Carlo runs to obtain the average optimal subpattern assignment (OSPA) distance [35] and average of the estimated number of targets.

Figure 5 shows the average OSPA distances of the proposed algorithm, the A* search algorithm, SSP and SPFA. It is observed that the average OSPA distance of the proposed algorithm is approximately equal to that of the A* search algorithm, which are better than that of SPFA and SSP. Figure 6 shows an average of the estimated number of targets of the proposed algorithm, the A* search algorithm, SSP and SPFA. It can be seen that the proposed algorithm presents an considerable performance in the estimation of the target numbers. To make a comparison, set the Type II clutter intensity $\kappa_2(z_c) = 1.33 \times 10^{-6}$, 6.33×10^{-6}, 19.89×10^{-6}, 26.53×10^{-6}, 39.79×10^{-6}, respectively. The average OSPA distances of the proposed algorithm, the A* search algorithm, SSP and SPFA versus the Type II clutter intensity are shown in Figure 7.

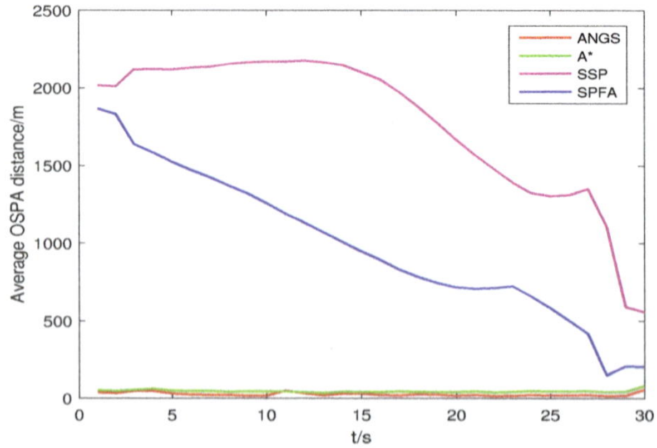

Figure 5. The average OSPA distance in scenario 1, with $c = 10$ and $p = 2$.

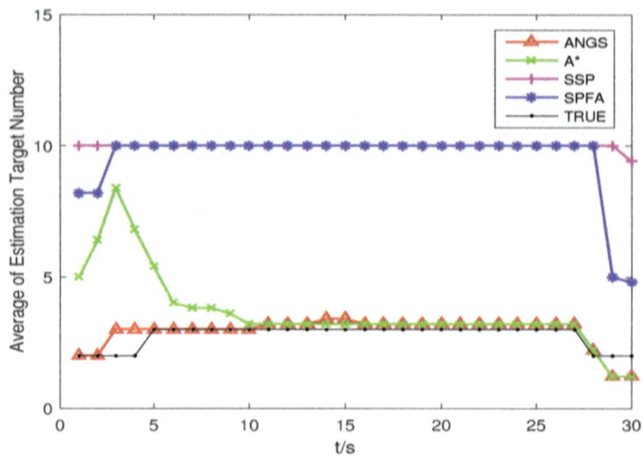

Figure 6. The average of the estimated numbers of targets in scenario 1.

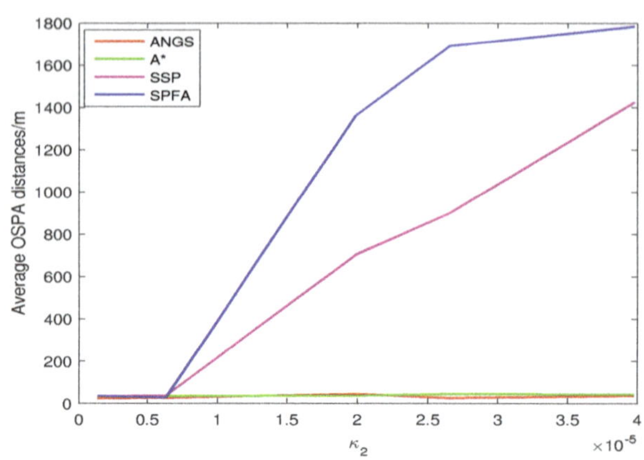

Figure 7. The average OSPA distances versus the Type II clutter intensity in scenario 1.

The average OSPA distances of the proposed algorithm have a relatively small difference than that of other algorithms at $\kappa_2(z_c) = 1.33 \times 10^{-6}$. The average OSPA distance of other algorithms increase rapidly with increasing the clutter intensity. The average OSPA distance of the proposed algorithm is basically no significant change, still at a lower value. When $\kappa_2(z_c) = 39.79 \times 10^{-6}$, it is obvious that the average OSPA distances of SSP and SPFA is about 40 and 50 times of that of the proposed algorithm, respectively.

Scenario 2:

An unknown and time-varying multiple crossing targets scenario is considered in the surveillance region [0, 10,000] m × [−4000, 5000] m. There are four manoeuvring targets whose initial positions are [7000, 4500] m, [3500, 4000] m, [1000, −500] m and [7000, −2500] m, respectively. Their initial velocities are [0, −150] m/s, [200, −20] m/s, [150, 300] m/s and [150, 300] m/s, respectively. The initial and terminal time are [1, 5, 1, 7] s and [27, 30, 30, 27] s, respectively. The mosaic threshold $\tau = 10$.

Trajectories intersect at [7478, 3618] m and [4443, −359] m. The average OSPA distances and the average of the estimated numbers of targets are shown in Figures 8 and 9. The average OSPA distances of the proposed algorithm, the A* search algorithm, SSP and SPFA are given in Figure 10. The Type I clutter intensity is $\kappa_1(z_c) = \frac{5}{9\times10^7} = 0.56 \times 10^{-7}$, and the Type II clutter intensity are $\kappa_2(z_c) = 1.33 \times 10^{-6}$, 6.33×10^{-6}, 19.89×10^{-6}, $gaihao26.53 \times 10^{-6}, 39.79 \times 10^{-6}$, respectively.

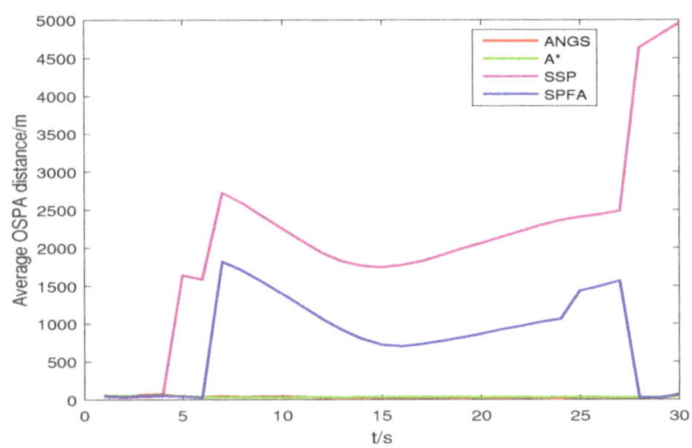

Figure 8. The average OSPA distance in scenario 2, with $c = 10$ and $p = 2$.

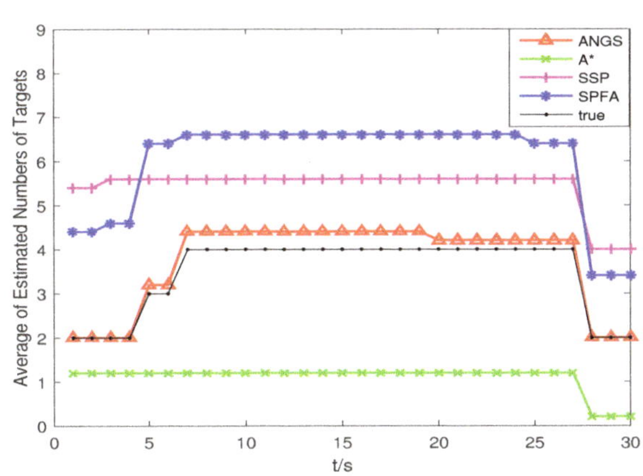

Figure 9. The average of the estimated numbers of targets in scenario 2.

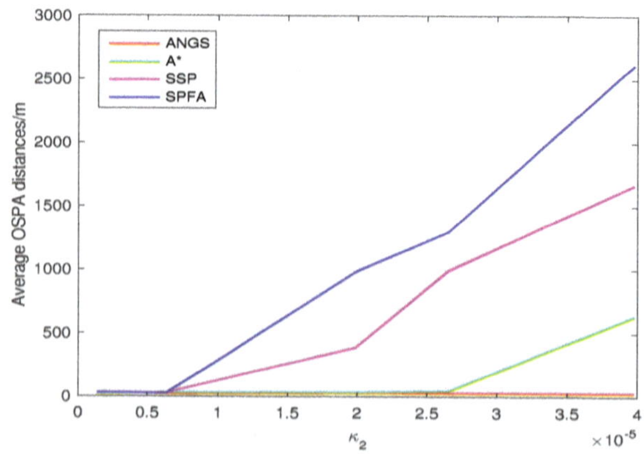

Figure 10. The average OSPA distances versus the Type II clutter intensity in scenario 2.

As seen from Figure 8, the SSP and SPFA have a larger average OSPA distances. In Figure 9, it is apparent that the average target number estimation of the proposed algorithm is exactly the same as the number of the true targets. In Figure 10, as expected, there is an overall increase of OSPA distances with with increasing the clutter intensity. The average OSPA distance of the proposed algorithm is at a lower value. When $\kappa_2(z_c) = 39.79 \times 10^{-6}$, the average OSPA distances of SSP and SPFA is about 45 and 70 times of that of the proposed algorithm, respectively.

5.3. Run Time

The average running time of different Type II clutter intensity is shown in Figure 11. It clear that the running time of the A* search algorithm, SSP, and SPFA increase exponentially. The running time of the proposed algorithm is growing slowly. When $\kappa_2(z_c) = 2.222 \times 10^{-7}$, the running time of the A* search algorithm is about 14 times that of the proposed algorithm.

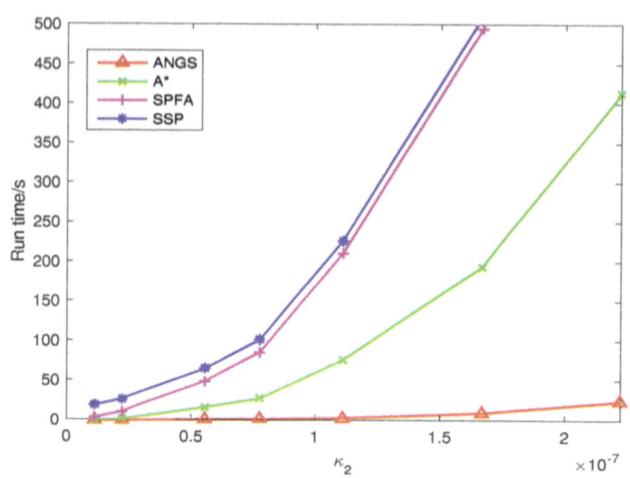

Figure 11. The comparison of average running time.

6. Conclusions

In this paper, we have presented a novel data association framework for multitarget tracking with measurement uncertainty that estimates unknown number and states of targets using the continuous multi-frame data. The multitarget tracking problem was formulated as network flow optimization problem for finding k-shortest paths, and an adaptive spectral clustering algorithm was used to segment the network structure. The optimal solution of each sub-graph can be obtained by the A* search algorithm. Experiment results indicate that the proposed algorithm is helpful in improving the accuracy of track extraction and can reduce the computational complexity. Future work will focus on tracking multitargets with low detection probability.

Author Contributions: T.M. and S.G. conceived and designed the algorithm; T.M. performed the experiments, analyzed the data, and drafted the paper; C.C. and X.S. revised the manuscript.

Funding: This research work was supported by the National Key Research and Development Program of China (Grant No. 2016YFE0111900).

Acknowledgments: We would like to thank the support received from Yuyan Cao and Xinmin Wang.

Conflicts of Interest: The authors declare no conflicts of interest.

References

1. Challa, S.; Morelande, M.R.; Mušicki, D.; Evans, R.J. *Fundamentals of Object Tracking*; Cambridge University Press: Cambridge, UK, 2011.
2. Mallick, M.; Arulampalam, S.; Yan, Y.; Ru, J. Three-Dimensional Tracking of an Aircraft Using Two-Dimensional Radars. *IEEE Trans. Aerosp. Electron. Syst.* **2018**, *54*, 585–600. [CrossRef]
3. Rodriguez, D.; Aceros, C.; Valera, J.; Anaya, E. A Framework for Multiple Object Tracking in Underwater Acoustic MIMO Communication Channels. *J. Sens. Actuator Netw.* **2017**, *6*, 2. [CrossRef]
4. Hoak, A.; Medeiros, H.; Povinelli, R.J. Image-Based Multi-Target Tracking through Multi-Bernoulli Filtering with Interactive Likelihoods. *Sensors* **2017**, *17*, 501. [CrossRef] [PubMed]
5. Banfi, J.; Guzzi, J.; Amigoni, F.; Flushing, E.F.; Giusti, A.; Gambardella, L.; Caro, G.A.D. An integer linear programming model for fair multitarget tracking in cooperative multirobot systems. *Auton. Robots* **2018**, 1–16. [CrossRef]
6. Beard, M.; Vo, B.T.; Vo, B.N. A Solution for Large-scale Multi-object Tracking. *arXiv* **2018**, arXiv:1804.06622.
7. Chen, H.; Li, X.R.; Bar-Shalom, Y. On joint track initiation and parameter estimation under measurement origin uncertainty. *IEEE Trans. Aerosp. Electron. Syst.* **2004**, *40*, 675–694. [CrossRef]
8. Huang, Y.; Zhang, Y.; Shi, P.; Wu, Z.; Qian, J.; Chambers, J.A. Robust Kalman Filters Based on Gaussian Scale Mixture Distributions with Application to Target Tracking. *IEEE Trans. Syst. Man Cybern. Syst.* **2017**, 1–15. [CrossRef]
9. Oh, S.; Russell, S.; Sastry, S. Markov Chain Monte Carlo Data Association for Multi-Target Tracking. *IEEE Trans. Autom. Control* **2009**, *54*, 481–497.
10. Vo, B.N.; Ma, W.K. The Gaussian Mixture Probability Hypothesis Density Filter. *IEEE Trans. Signal Process.* **2006**, *54*, 4091–4104. [CrossRef]
11. Vo, B.N.; Vo, B.T.; Phung, D. Labeled Random Finite Sets and the Bayes Multi-Target Tracking Filter. *IEEE Trans. Signal Process.* **2014**, *62*, 6554–6567. [CrossRef]
12. Bar-Shalom, Y.; Li, X.R. Multitarget-Multisensor Tracking: Principles and Techniques. *IEEE Aerosp. Electron. Syst. Mag.* **1995**, *11*, 41–44.
13. Jia, B.; Pham, K.D.; Blasch, E.; Chen, G. Multiple space object tracking via a space-based optical sensor. In Proceedings of the 2016 IEEE Aerospace Conference, Big Sky, MT, USA, 5–12 March 2016; pp. 1–10.
14. Wang, F.; Man, Y.; Man, L. Intelligent optimization approach for the k shortest paths problem based on genetic algorithm. In Proceedings of the International Conference on Natural Computation, Xiamen, China, 19–21 August 2014; pp. 219–224.
15. Chong, C.Y. Graph approaches for data association. In Proceedings of the International Conference on Information Fusion, Singapore, 9–12 July 2012; pp. 1578–1585.

16. Ren, X.; Huang, Z.; Sun, S.; Liu, D. An Efficient MHT Implementation Using GRASP. *IEEE Trans. Aerosp. Electron. Syst.* **2014**, *50*, 86–101. [CrossRef]
17. Goldberg, A.V. An efficient implementation of a scaling minimum-cost flow algorithm. *J. Algorithms* **1997**, *22*, 1–29. [CrossRef]
18. Zhang, L.; Li, Y.; Nevatia, R. Global data association for multi-object tracking using network flows. In Proceedings of the IEEE Conference on Computer Vision and Pattern Recognition, Anchorage, AK, USA, 23–28 June 2008; pp. 1–8.
19. Pirsiavash, H.; Ramanan, D.; Fowlkes, C.C. Globally-optimal greedy algorithms for tracking a variable number of objects. In Proceedings of the IEEE Conference on Computer Vision and Pattern Recognition, Springs, CO, USA, 20–25 June 2011; pp. 1201–1208.
20. Xi, Z.; Liu, H.; Liu, H.; Yang, B. Multiple Object Tracking Using the Shortest Path Faster Association Algorithm. *Sci. World J.* **2014**, *2014*, 481719. [CrossRef] [PubMed]
21. Zhenghao, X.I.; Hongbo, L.I.; Liu, H.; Sun, F. Multiple object tracking using A* algorithm optimization. *Qinghua Daxue Xuebao/J. Tsinghua Univ.* **2014**, *54*, 1549–1554.
22. Ahuja, R.K.; Magnanti, T.L.; Orlin, J.B. *Network Flows: Theory, Algorithms, and Applications*; Prentice Hall: Upper Saddle River, NJ, USA, 1993.
23. Russell, S.J.; Norvig, P. *Artificial Intelligence: A Modern Approach*, 3rd ed.; Pearson: London, UK, 2010.
24. Chabini, I.; Lan, S. Adaptations of the A* algorithm for the computation of fastest paths in deterministic discrete-time dynamic networks. *IEEE Trans. Intell. Transp. Syst.* **2002**, *3*, 60–74. [CrossRef]
25. Kim, J.; Min, J.; Kweon, I.S.; Lin, Z. Fusing Multiple Independent Estimates via Spectral Clustering for Robust Visual Tracking. *IEEE Signal Process. Lett.* **2012**, *19*, 527–530. [CrossRef]
26. Wang, X.; Qian, B.; Davidson, I. On constrained spectral clustering and its applications. *Data Min. Knowl. Discov.* **2014**, *28*, 1–30. [CrossRef]
27. Vautard, F.; Xu, L.; Drzal, L.T. Study on the Determination and Application of the Cluster Number of the Images Based on the Spectral Graph Theory. *Gongcheng Shuxue Xuebao/Chin. J. Eng. Math.* **2012**, *29*, 27–50.
28. Zhang, F. *Matrix Theory*; Springer: New York, NY, USA, 2011.
29. Särkkä, S. Continuous-time and continuous-discrete-time unscented Rauch–Tung–Striebel smoothers. *Signal Process.* **2010**, *90*, 225–235. [CrossRef]
30. Huang, Y.; Zhang, Y.; Zhao, Y.; Mihaylova, L.; Chambers, J. A Novel Robust Rauch–Tung–Striebel Smoother Based on Slash and Generalized Hyperbolic Skew Student's T-Distributions. In Proceedings of the 2018 21st International Conference on Information Fusion (FUSION), Cambridge, UK, 10–13 July 2018; pp. 369–376.
31. Schuhmacher, D.; Vo, B.T.; Vo, B.N. A Consistent Metric for Performance Evaluation of Multi-Object Filters. *IEEE Trans. Signal Process.* **2008**, *56*, 3447–3457. [CrossRef]
32. Kristan, M.; Matas, J. A Novel Performance Evaluation Methodology for Single-Target Trackers. *IEEE Trans. Pattern Anal. Mach. Intell.* **2016**, *38*, 2137–2155. [CrossRef] [PubMed]
33. Zhang, Y.; Brady, M.; Smith, S. Segmentation of brain MR images through a hidden Markov random field model and the expectation-maximization algorithm. *IEEE Trans. Med. Imaging* **2001**, *20*, 45–57. [CrossRef] [PubMed]
34. Li, X.R.; Jilkov, V.P. Survey of maneuvering target tracking. Part I. Dynamic models. *IEEE Trans. Aerosp. Electron. Syst.* **2003**, *39*, 1333–1364.
35. Garcia-Fernandez, A.F.; Morelande, M.R.; Grajal, J. Bayesian Sequential Track Formation. *IEEE Trans. Signal Process.* **2014**, *62*, 6366–6379. [CrossRef]

sensors

MDPI

Article

An Improved Yaw Estimation Algorithm for Land Vehicles Using MARG Sensors

Gang Shi [1,2,3], Xisheng Li [1,4,*] and Zhengfu Jiang [1]

1 School of Automation and Electrical Engineering, University of Science and Technology Beijing, Beijing 100083, China; shigang_upc@163.com (G.S.); jzf_ustb@163.com (Z.J.)
2 College of Information and Control Engineering, China University of Petroleum, Qingdao 266580, China
3 Shengli College, China University of Petroleum, Dongying 257061, China
4 Beijing Engineering Research Center of Industrial Spectrum Imaging, Beijing 10083, China
* Correspondence: lxs@ustb.edu.cn; Tel.: +86-010-6233-4885

Received: 8 August 2018; Accepted: 25 September 2018; Published: 27 September 2018

Abstract: This paper presents a linear Kalman filter for yaw estimation of land vehicles using magnetic angular rate and gravity (MARG) sensors. A gyroscope measurement update depending on the vehicle status and constraining yaw estimation is introduced. To determine the vehicle status, the correlations between outputs from different sensors are analyzed based on the vehicle kinematic model and Coriolis theorem, and a vehicle status marker is constructed. In addition, a two-step measurement update method is designed. The method treats the magnetometer measurement update separately after the other updates and eliminates its impact on attitude estimation. The performances of the proposed algorithm are tested in experiments and the results show that: the introduced measurement update is an effective supplement to the magnetometer measurement update in magnetically disturbed environments; the two-step measurement update method makes attitude estimation immune to errors induced by magnetometer measurement update, and the proposed algorithm provides more reliable yaw estimation for land vehicles than the conventional algorithm.

Keywords: attitude estimation; Kalman filter; land vehicle; magnetic angular rate and gravity (MARG) sensor; quaternion; yaw estimation

1. Introduction

As a set of Euler angles, the yaw, pitch, and roll represent the orientation of a body frame with respect to a reference frame. The pitch and roll are also referred to as attitude. Yaw and attitude estimation are widely used in vehicular technologies including driver assistance [1–3], vehicle safety [4,5], etc. In recent years, magnetic angular rate and gravity (MARG) sensors [6] are widely used in orientation estimation. A MARG sensor consists of a triaxis magnetometer, a triaxis gyroscope, and a triaxis accelerometer. Reasonable installation and calibration make it acceptable to assume that the sensor frames are aligned with the body frame. Hence, a MARG sensor can measure the geomagnetic field, angular rate of the body frame, and the gravity resolved in the body frame in undisturbed environments.

In order to obtain an orientation estimation of the body frame, we can integrate the gyroscope output based on an initial value, but the result will drift away with time because of gyroscope measurement errors [7]. Alternatively, we can solve the Wahba problem [8] using magnetometer and accelerometer outputs, but the sensor outputs are apt to be interfered by motion accelerations and magnetic disturbances [9,10]. Therefore, to make the most of the information from MARG sensors and obtain robust orientation estimation, many fusion algorithms have been studied. These algorithms can be classified into two categories: one is based on complementary filters, which realize the fusion in frequency domain [11–13], and the other is based on Kalman filters, which employ a stochastic approach [14–16]. Kalman filter based algorithms consist of two basic processing, i.e., time propagation and measurement update. In time propagation, the gyroscope output is used to predict the orientation and the result is called a priori estimation; in measurement update, the magnetometer and accelerometer outputs are used to correct the a priori estimation and the result is called a posteriori estimation.

This paper studies on the yaw estimation of land vehicles using MARG sensors and Kalman filter. Our work is based on the orientation estimation because the yaw can be extracted from orientation estimation; at the same time, we focus on special problems in yaw estimation.

Many algorithms of orientation estimation have been studied. They are different in terms of the state vector, filter structure, etc., but they are the same at one point, i.e., the magnetometer output is used to correct the yaw estimation. In fact, the correction of yaw estimation is implemented only by magnetometer measurement update because the accelerometer output only provides attitude information. Hence, one problem in yaw estimation is that it is vulnerable to magnetic disturbances. Magnetic disturbances include hard iron effects, soft iron effects, and environmental magnetic disturbances [17]. The hard and soft iron effects can be compensated through magnetometer calibration, but the environmental magnetic disturbances cannot be effectively compensated because of their nondeterminacy [18–20]. In the following, the "magnetic disturbances" refers specifically to the environmental magnetic disturbances.

Methods to handle magnetic disturbances have been proposed. A measurement vector selection scheme based on norm comparison was designed in [7]. Costanzi et al. [21] scaled down the gain associated with the magnetometer output when two particular angular constraints are violated. Wu et al. [22] did not use the magnetometer output if its norm was too big or too small. Tong et al. [23] developed a hidden Markov Model to identify the measurement disturbances and then adjust the noise covariance adaptively. Feng et al. [24] proposed a two-step correction scheme, in which the magnetometer output is used to correct the estimated direction of the magnetic field, but if the difference between the norm of the sensor output and the reference value is greater than a threshold, the correction will not be executed. In fact, the above methods apply a same strategy, i.e., detecting magnetic disturbances according to some feature, e.g., the norm, of the measured magnetic field, and reducing the measurement update weight in real time when disturbances happen. The drawback of this strategy is that the magnetometer measurement update cannot provide effective and timely correction for the yaw estimation if magnetic disturbances last for a long time. In addition, Sabatini [25] proposed an extended Kalman Filter, which compensated magnetic disturbances by including them in the state vector. The filter models magnetic disturbances by a first-order Gauss-Markov vector random process with independent components, and assume the motion acceleration is approximately zero. In fact, the actual magnetic disturbances can hardly be modelled effectively, and the assumption about the motion acceleration is not suitable for land vehicles.

Another problem in yaw estimation is its impact on attitude estimation. In some fusion algorithms, the magnetometer output is also used to correct the attitude estimation, and thus induce estimation errors caused by magnetic disturbances. To address this problem, some algorithms restrict the use of the magnetometer output to yaw estimation. In [9,10], orientation quaternion is obtained through multiplication of a series of decoupled quaternion factors, and the result can be used as the measurement for a Kalman filter with a two-layer structure [9]. Suh [26] proposed a two-step measurement update for an indirect Kalman filter, where the magnetometer measurement update only affected the yaw estimation. Afterwards, Suh et al. [27] proposed a new measurement equation for the indirect Kalman filter, which can greatly reduce the impact of the magnetometer measurement update on attitude estimation and is easier to implement than the two-step measurement update algorithm.

Obviously, both of the mentioned problems are due to magnetic disturbances. This paper aims to enhance the ability of the yaw estimation to deal with magnetic disturbances. One basic idea of this paper is that use not only the information from the sensors, but also the characters of vehicles motion to improve yaw estimation. We think motion characters can provide some independent information, which can be used as supplement to the sensors information. We note that the yaw of a vehicle running along a straight road remains basically unchanged, and this character can be utilized to improve the yaw estimation as a supplement to the magnetometer measurement update. To achieve this, it should be known whether a vehicle is running straight. A straightforward way to determine the vehicle status is comparing the gyroscope output against some preset parameters, but the gyroscope output suffers from various measurement errors and vehicle bumps. Considering the turning motion cause changes of MARG outputs, we can determine the vehicle status by analyzing the correlation between outputs from different sensors. In addition, we can reform the measurement update process of a direct Kalman filter using the condition proved in [26] and thus make the attitude estimation immune to magnetic disturbances. In fact, more accurate attitude estimation is also helpful to improve the yaw estimation because they are coupled in the time propagation. Motivated by above discussion, we propose an improved yaw estimation algorithm, and the main contributions of this paper are as follows:

(1) A new measurement update robust to magnetic disturbances is introduced, and its weight can be adjusted according to the vehicle status. The correlation coefficients between outputs from a MARG sensor are analyzed based on the vehicle kinematic model and Coriolis theorem, and a vehicle status marker is constructed.
(2) A new two-step measurement update method is designed. The method implements measurement updates in two successive steps, and make a special processing of the magnetometer measurement update to eliminate its impact on attitude estimation.

In this paper, we construct a yaw estimation algorithm using a typical linear Kalman filter to implement basic quaternion estimation and applying the conventional strategy of reducing the measurement weight to deal with magnetic disturbances. This algorithm is referred to as conventional algorithm. Then, the conventional algorithm is improved with the new measurement update and the new two-step measurement update method. Finally, the performances of the improved algorithm are evaluated through comparing its results against that of the conventional algorithm in experiments. The rest of this paper is organized as follows: Section 2 describes a conventional yaw estimation algorithm. Section 3 details the improved algorithm. Section 4 provides the experiment results and discussion. Finally, the work is concluded in Section 5.

2. Conventional Algorithm

2.1. Preliminaries

Because of not suffering from the gimbal lock, low dimension, and offering a linear formulation of the orientation dynamics [9], quaternion is widely used for orientation representation. Any orientation of a body frame with respect to a reference frame can be represented by a unit quaternion $\mathbf{q}$ defined as:

$$\mathbf{q} = \begin{bmatrix} q_0 & q_1 & q_2 & q_3 \end{bmatrix}^T = \begin{bmatrix} \cos\frac{\alpha}{2} & e_x\sin\frac{\alpha}{2} & e_y\sin\frac{\alpha}{2} & e_z\sin\frac{\alpha}{2} \end{bmatrix}^T \quad (1)$$

where q_0 is the scalar part; $\begin{bmatrix} q_1 & q_2 & q_3 \end{bmatrix}^T$ is the vector part; α is the rotation angle; $\begin{bmatrix} e_x & e_y & e_z \end{bmatrix}^T$ is the unit vector that represents the rotation axis. In quaternion estimation algorithms, the gyroscope output is used to depict the quaternion dynamics; therefore, the gyroscope bias is an important factor that affects the estimation accuracy.

In this section, we describe a Kalman filter-based yaw estimation algorithm with the unit quaternion and gyroscope bias as states. This algorithm, referred to as conventional algorithm, provides the basement and benchmark for the improved algorithm presented in the next section. The conventional algorithm is based on a typical linear Kalman filter presented in [16]. In addition, adaptive measurement weights are applied to deal with measurement disturbances, and the yaw is computed with the quaternion estimation.

The detailed derivations of the system model are presented in [16]; for conciseness, we only list the results and give necessary explanations in this section. Note that we use a different quaternion definition from that in [16] (The scalar part is the first component of a quaternion, whereas in [16], the scalar part is the last component); therefore, we rewrite the concerned equations accordingly, which will not affect the performances of the algorithm.

In this paper, the reference frame is the East, North, Up frame; the body frame is the Right, Forward, Above frame; the sensor frame is assumed aligned with the body frame; the ZXY sequence of Euler angles is chosen, and the yaw, pitch, and roll are respectively z-axis rotation angle, x-axis rotation angle, and y-axis rotation angle. For a clear writing, we define some notations that will be used throughout this paper as follows.

Vectors: $\mathbf{x}$ is the state vector; $\mathbf{b}$ is the gyroscope bias; $\mathbf{m}_r$ and $\mathbf{g}_r$ are respectively the geomagnetic filed and the gravity resolved in the reference frame; $\mathbf{m}$, $\boldsymbol{\omega}$, and $\mathbf{a}$ are respectively the output of magnetometer, gyroscope, and accelerometer.

Matrices: $\mathbf{C}_r^b$ is the rotation matrix from the reference frame to the body frame; $\mathbf{0}$ and $\mathbf{I}$ are respectively null matrices and identity matrices, and their subscripts indicate dimensions, for example, the dimensions of $\mathbf{0}_{3\times4}$, $\mathbf{0}_3$, and $\mathbf{I}_4$ are respectively 3×4, 3×3, and 4×4; $[\mathbf{v}\times]$ is the skew-symmetric matrix of vector $\mathbf{v}$.

Subscripts: x, y, and z denote respectively the x-axis, y-axis, and z-axis component of a vector; k denotes the time step.

2.2. System Model

The state vector, composed of the unit quaternion and gyroscope bias, is defined as:

$$\mathbf{x} = \begin{bmatrix} \mathbf{q}^T & \mathbf{b}^T \end{bmatrix}^T \tag{2}$$

Based on the well-known quaternion dynamics model and state augmentation, the process equation is written as:

$$\mathbf{x}_{k+1} = \boldsymbol{\Psi}_k \mathbf{x}_k + \boldsymbol{\Gamma}_k \mathbf{n}_k \tag{3a}$$

where:

$$\boldsymbol{\Psi}_k = \begin{bmatrix} \Phi(\boldsymbol{\theta}_k) & -\frac{\Delta t}{2}\boldsymbol{\Xi}_k \\ \mathbf{0}_{3\times4} & \mathbf{I}_3 \end{bmatrix} \tag{3b}$$

$$\boldsymbol{\Phi}(\boldsymbol{\theta}_k) = \exp\left(\frac{1}{2}\begin{bmatrix} 0 & -\boldsymbol{\theta}_k^T \\ \boldsymbol{\theta}_k & -[\boldsymbol{\theta}_k\times] \end{bmatrix}\right) \tag{3c}$$

$$\boldsymbol{\theta}_k = \boldsymbol{\omega}_k \Delta t \tag{3d}$$

$$\boldsymbol{\Xi}_k = \begin{bmatrix} -\mathbf{q}_{v,k}^T \\ [\mathbf{q}_{v,k}\times] + q_{0,k}\mathbf{I}_3 \end{bmatrix} \tag{3e}$$

$$\boldsymbol{\Gamma}_k = \begin{bmatrix} -\frac{1}{2}\boldsymbol{\Xi}_k & -\frac{1}{2}\boldsymbol{\Xi}_k & \mathbf{0}_{4\times3} \\ \mathbf{0}_3 & \mathbf{0}_3 & \mathbf{I}_3 \end{bmatrix} \tag{3f}$$

In (3), $\boldsymbol{\Psi}_k$ is the transition matrix; $\boldsymbol{\Gamma}_k$ is the process noise input matrix; Δt is the sample interval; $\mathbf{q}_v$ is the vector part of the quaternion state, and $\mathbf{n}_k$ is the gyroscope noise with covariance matrix $\mathrm{diag}(\sigma_\omega^2, \sigma_\omega^2, \sigma_\omega^2)$.

The linear pseudo-measurement equation of the accelerometer is:

$$\mathbf{0}_{4\times1} = \mathbf{H}_{a,k+1}\mathbf{x}_{k+1} - \frac{1}{2}\boldsymbol{\Xi}_{k+1}\delta\mathbf{a}_{k+1} \tag{4a}$$

where:

$$\mathbf{H}_{a,k+1} = \left[\begin{bmatrix} 0 & -\mathbf{d}_{a,k+1}^T \\ \mathbf{d}_{a,k+1} & -[\mathbf{s}_{a,k+1}\times] \end{bmatrix} \quad \mathbf{0}_{4\times3} \right] \tag{4b}$$

$$\mathbf{s}_{a,k+1} = \frac{1}{2}(\mathbf{a}_{k+1} + \mathbf{g}_r) \tag{4c}$$

$$\mathbf{d}_{a,k+1} = \frac{1}{2}(\mathbf{a}_{k+1} - \mathbf{g}_r) \tag{4d}$$

$$\delta\mathbf{a}_{k+1} = \mathbf{a}_{k+1} - \mathbf{C}_r^b\mathbf{g}_r \tag{4e}$$

In (4), $\mathbf{H}_{a,k+1}$ is the measurement matrix, and $-\frac{1}{2}\Xi_{k+1}\delta\mathbf{a}_{k+1}$ is the measurement noise. The covariance matrix of $\delta\mathbf{a}_{k+1}$ is $\sigma_a^2\mathbf{I}_3$.

Similarly, the linear pseudo-measurement equation of the magnetometer is:

$$\mathbf{0}_{4\times1} = \mathbf{H}_{m,k+1}\mathbf{x}_{k+1} - \frac{1}{2}\Xi_{k+1}\delta\mathbf{m}_{k+1} \tag{5a}$$

where:

$$\mathbf{H}_{m,k+1} = \left[\begin{bmatrix} 0 & -\mathbf{d}_{m,k+1}^T \\ \mathbf{d}_{m,k+1} & -[\mathbf{s}_{m,k+1}\times] \end{bmatrix} \quad \mathbf{0}_{4\times3} \right] \tag{5b}$$

$$\mathbf{s}_{m,k+1} = \frac{1}{2}(\mathbf{m}_{k+1} + \mathbf{m}_r) \tag{5c}$$

$$\mathbf{d}_{m,k+1} = \frac{1}{2}(\mathbf{m}_{k+1} - \mathbf{m}_r) \tag{5d}$$

$$\delta\mathbf{m}_{k+1} = \mathbf{m}_{k+1} - \mathbf{C}_r^b\mathbf{m}_r \tag{5e}$$

In (5), $\mathbf{H}_{m,k+1}$ is the measurement matrix, and $-\frac{1}{2}\Xi_{k+1}\delta\mathbf{m}_{k+1}$ is the measurement noise. The covariance matrix of $\delta\mathbf{m}_{k+1}$ is $\sigma_m^2\mathbf{I}_3$.

From (4) and (5), the overall measurement equation can be written as:

$$\mathbf{0}_{8\times1} = \mathbf{H}_{k+1}\mathbf{x}_{k+1} - \frac{1}{2}\mathrm{diag}(\Xi_{k+1}, \Xi_{k+1})\begin{bmatrix} \delta\mathbf{a}_{k+1}^T & \delta\mathbf{m}_{k+1}^T \end{bmatrix}^T \tag{6a}$$

where:

$$\mathbf{H}_{k+1} = \begin{bmatrix} \mathbf{H}_{a,k+1} \\ \mathbf{H}_{m,k+1} \end{bmatrix} \tag{6b}$$

The noise covariance matrices of the process and measurement equations are written as:

$$\mathrm{cov}(\mathbf{\Gamma}_k\mathbf{n}_k) = \mathbf{Q}_k \tag{7}$$

$$\mathrm{cov}\left(-\frac{1}{2}\Xi_{k+1}\delta\mathbf{a}_{k+1}\right) = \rho_{a,k+1}\mathbf{R}_{a,k+1} \tag{8}$$

$$\mathrm{cov}\left(-\frac{1}{2}\Xi_{k+1}\delta\mathbf{m}_{k+1}\right) = \rho_{m,k+1}\mathbf{R}_{m,k+1} \tag{9}$$

where ρ_a and ρ_m are adaptive weight coefficients, which can be adjusted in real time. The computation formulas for $\mathbf{Q}_k$, $\mathbf{R}_{a,k+1}$, and $\mathbf{R}_{m,k+1}$ can be found in [16]. From (8) and (9), the noise covariance matrix of (6a) can be written as:

$$\mathbf{R}_{k+1} = \mathrm{diag}\left(\rho_{a,k+1}\mathbf{R}_{a,k+1}, \rho_{m,k+1}\mathbf{R}_{m,k+1}\right) \tag{10}$$

2.3. *Adaptive Fusion Algorithm Based on Kalman Filter*

Filter Initialization: Set initial values for state vector estimation $\hat{\mathbf{x}}$, i.e., $\hat{\mathbf{q}}$ and $\hat{\mathbf{b}}$, and error covariance matrix $\mathbf{P}$.

Time Propagation: The process model, gyroscope output and posteriori estimation (or the initial estimation) at step k is used to compute the priori estimation at step $k+1$ by:

$$\hat{\mathbf{x}}_{k+1/k} = \mathbf{\Psi}_k \hat{\mathbf{x}}_k \tag{11a}$$

$$\mathbf{P}_{k+1/k} = \mathbf{\Psi}_k \mathbf{P}_k \mathbf{\Psi}_k^T + \mathbf{Q}_k \tag{11b}$$

Measurement weight adjustment: To deal with motion accelerations and magnetic disturbances, the weight coefficients ρ_a and ρ_m are adjust in real time according to the disturbance intensity, which is equivalent to adjust the measurement weight. The adjusting expressions are:

$$\rho_{a,k+1} = \exp(\lambda_a |\|\mathbf{a}_{k+1}\| - \|\mathbf{g}_r\|| / \|\mathbf{g}_r\|) \tag{12}$$

$$\rho_{m,k+1} = \exp(\lambda_m |\|\mathbf{m}_{k+1}\| - \|\mathbf{m}_r\|| / \|\mathbf{m}_r\|) \tag{13}$$

where the relative distances between the norms of measured vectors $(\mathbf{a}_{k+1}, \mathbf{m}_{k+1})$ and reference vectors $(\mathbf{g}_r, \mathbf{m}_r)$ are used to represent the disturbance intensity, and the exponential functions are used to map the disturbance intensity to weight coefficients. The function of the parameters λ_a and λ_m is to adjust the mapping relations. If the weight coefficient should be more sensitive to the disturbance intensity, the corresponding parameter should be increased, otherwise, it should be reduced. The values used for λ_a and λ_m can be determined experimentally. The exponential function instead of a linear function is applied because the former can reduce the measurement weight more quickly when the relative distance increases [28].

Measurement Update: The measurement model, magnetometer and accelerometer outputs and the priori estimation at step $k+1$ is used to compute the posteriori estimation at step $k+1$ by:

$$\mathbf{K}_{k+1} = \mathbf{P}_{k+1/k} \mathbf{H}_{k+1}^T (\mathbf{H}_{k+1} \mathbf{P}_{k+1/k} \mathbf{H}_{k+1}^T + \mathbf{R}_{k+1})^{-1} \tag{14a}$$

$$\hat{\mathbf{x}}_{k+1} = \hat{\mathbf{x}}_{k+1/k} - \mathbf{K}_{k+1} \mathbf{H}_{k+1} \hat{\mathbf{x}}_{k+1/k} \tag{14b}$$

$$\mathbf{P}_{k+1} = (\mathbf{I}_7 - \mathbf{K}_{k+1} \mathbf{H}_{k+1}) \mathbf{P}_{k+1/k} \tag{14c}$$

Unit Constraint: To preserve the unit-norm property of the quaternion estimation, the updated quaternion is normalized by:

$$\overline{\mathbf{q}}_{k+1} = \hat{\mathbf{q}}_{k+1} / \|\hat{\mathbf{q}}_{k+1}\| \tag{15}$$

2.4. *Yaw Computation*

The rotation matrix from the reference frame to the body frame can be represented as a function of either Euler angles or a unit quaternion, and the expressions are respectively:

$$\mathbf{C}_r^b(\psi, \phi, \gamma) = \begin{bmatrix} c\gamma c\psi + s\gamma s\psi s\phi & -c\gamma s\psi + s\gamma c\psi s\phi & -s\gamma c\phi \\ s\psi c\phi & c\phi c\psi & s\phi \\ s\gamma c\psi - c\gamma s\psi s\phi & -s\gamma s\psi - c\gamma c\psi s\phi & c\gamma c\phi \end{bmatrix} \tag{16}$$

$$\mathbf{C}_r^b(\mathbf{q}) = \begin{bmatrix} q_0^2 + q_1^2 - q_2^2 - q_3^2 & 2(q_1 q_3 + q_0 q_3) & 2(q_1 q_3 - q_0 q_2) \\ 2(q_1 q_2 - q_0 q_3) & q_0^2 - q_1^2 + q_2^2 - q_3^2 & 2(q_2 q_3 + q_0 q_1) \\ 2(q_1 q_3 + q_0 q_2) & 2(q_2 q_3 - q_0 q_1) & q_0^2 - q_1^2 - q_2^2 + q_3^2 \end{bmatrix} \tag{17}$$

where ψ, ϕ, and γ are respectively the yaw, pitch, and roll; s and c denote sine and cosine function, respectively.

Using the first two elements of the second row of $\mathbf{C}(\mathbf{q})$, denoted as C_{21} and C_{22} respectively, and defining the range of the yaw as $(-180°, 180°]$, we obtain the following formulas for yaw computation:

$$\hat{\psi} = \begin{cases} \arctan(C_{21}/C_{22}) & C_{22} > 0 \\ \arctan(C_{21}/C_{22}) + 180° & C_{22} < 0, C_{21} > 0 \\ \arctan(C_{21}/C_{22}) - 180° & C_{22} < 0, C_{21} < 0 \end{cases} \tag{18}$$

3. Improved Algorithm

In this section, we keep the time propagation step of the conventional algorithm unchanged, and improve its measurement update step in two ways. Firstly, to improve the accuracy of the yaw estimation in the presence of magnetic disturbances, a measurement equation of the gyroscope bias is derived and its weight can be adjusted according to the vehicle status. Secondly, we design a two-step measurement update method to eliminate the impact of the magnetometer measurement update on the attitude estimation.

3.1. New Measurement Equation and Its Adaptive Weight

The gyroscope output can be written as:

$$\boldsymbol{\omega} = \boldsymbol{\omega}_b + \mathbf{b} + \mathbf{n}_1 \tag{19a}$$

and its z-axis component is:

$$\omega_z = \omega_{bz} + b_z + n_{1z} \tag{19b}$$

where $\boldsymbol{\omega}_b$ is the angular rate resolved in the body frame, and $\mathbf{n}_1$ is a white Gaussian noise vector. When vehicles are running straight, ω_{bz} can be regarded as zero and (19b) can be rewritten as:

$$b_z = \mathbf{H}_\omega \mathbf{x} + n_{1z} \tag{20a}$$

where:

$$\mathbf{H}_\omega = \begin{bmatrix} \mathbf{0}_{1\times 6} & 1 \end{bmatrix} \tag{20b}$$

Equation (20a) is a new measurement equation named as gyroscope measurement, and its characters are as follows. Firstly, it only updates the gyroscope bias estimation in measurement update, but it will affect the quaternion estimation in time propagation through (11); specifically speaking, it will constrain the yaw estimation from changing by correcting the z-axis angular rate to zero. Secondly, it is immune to magnetic disturbances because the latter cannot impact the gyroscope output. Thirdly, the gyroscope measurement only hold true in running straight status, in other words, its validity depends on the vehicle status.

Note that, the yaw is computed using quaternion estimation, and all the gyroscope bias components can impact the quaternion propagation. But, in the fusion algorithm, the quaternion estimation is not only based on propagation, but also on measurement update. The yaw derivative [12] can be written as:

$$\dot{\psi} = \frac{\sin\gamma}{\cos\phi}\omega_{bx} - \frac{\cos\gamma}{\cos\phi}\omega_{bz} \tag{21}$$

Obviously, the yaw derivative is related to pitch, roll, ω_{bx}, and ω_{bz}. In the fusion algorithm, the estimation of pitch and roll will be corrected by the accelerometer measurement (in the form of correcting quaternion), which is immune to magnetic disturbances. The two-step measurement update method which will be introduced in Section 3.2 makes pitch and roll estimation immune to the errors induced by magnetometer measurement update. In addition, generally speaking, the pitch and roll are small for land vehicles, hence the absolute value of the coefficient of ω_{bz} is greater than that of ω_{bx}. Therefore, the z-axis bias component is of more importance, especially in magnetically disturbed environments. Equation (21) is based on Euler angles, but it should be noted that both quaternion and

Euler angles are representation of orientation, and orientation obeys a unique dynamic rule. In other words, the quaternion propagation models is different from that of Euler angles, but when they are used to represent a same orientation parameter, for example yaw, the parameter obeys a same dynamic rule no matter what representation is used.

The above characteristics suggest that the gyroscope measurement can be used to correct the yaw estimation in the presence of magnetic disturbances when vehicle is running straight. In practice, a vehicle cannot always run on a straight road, and even on a straight road the yaw may fluctuate. In order to employ this measurement correctly, we define the covariance of the measurement noise n_{1z} as $\rho_\omega \sigma_\omega^2$, where ρ_ω, similar to ρ_a and ρ_m, works as a adaptive weight coefficient that can be adjusted in real time. Obviously, the adjusting principle should be that the higher the extent of running straight is, the smaller the ρ_ω is.

Now, the problem is how to determine the extent of running straight. To address this problem, we analyze the correlations between different sensor outputs. The accelerometer output can be written as:

$$\mathbf{a} = \mathbf{a}_m - \mathbf{C}_r^b \mathbf{g}_r + \mathbf{n}_2 \tag{22}$$

where $\mathbf{a}_m$ is the motion acceleration, and $\mathbf{n}_2$ is a noise vector. According to the vehicle kinematic model [29], $\mathbf{a}_m$ is given by:

$$\mathbf{a}_m = \dot{\mathbf{v}}_b + \boldsymbol{\omega}_b \times \mathbf{v}_b \tag{23}$$

where $\mathbf{v}_b$ is the velocity of the vehicle in the body frame.

The reference fame and body frame are shown in Figure 1. For a land vehicle, the running direction can be arbitrary in reference frame, but in the body frame, which is attached to the vehicle, its running direction is always forward, although the forward direction may change in reference frame. The "forward direction" is the y direction in the body frame defined in this paper. Dissanayake et al. [30] have pointed out that, when the vehicle does not jump off the ground and does not slide on the ground, velocity of the vehicle in the plane perpendicular to the forward direction is zero, hence v_{bx} and v_{bz} can be assumed as zero. Using this assumption and substituting (16) and (23) into (22), we simplify the x-axis component of (22) as:

$$a_x = -v_{by}\omega_{bz} - g s\gamma c\phi + n_{2x} \tag{24}$$

In practical situation, the assumption is somewhat violated due to the presence of side slip during cornering and vibrations caused by the engine and suspension system [30]. When the assumption is violated, the side slips and vibrations will cause corresponding accelerations, but these accelerations can be regarded as noise and a part of n_{2x} in (24).

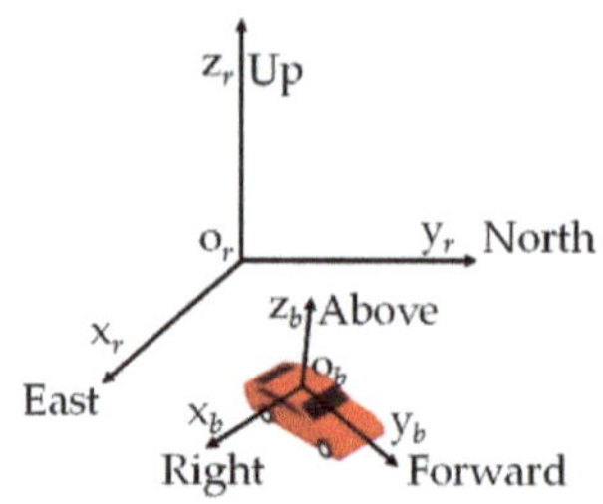

Figure 1. Reference frame $o_r x_r y_r z_r$ and body frame $o_b x_b y_b z_b$.

The magnetic field resolved in the body frame, denoted as $\mathbf{m}_b$, can be written as:

$$\mathbf{m}_b = \mathbf{C}_r^b (\mathbf{m}_r + \mathbf{d}_r) \tag{25}$$

where $\mathbf{d}_r$ is the magnetic disturbances resolved in the reference frame. According to the Coriolis theorem, the change rate of $\mathbf{m}_b$ in body frame can be written as:

$$\dot{\mathbf{m}}_b = \mathbf{C}_r^b(\dot{\mathbf{m}}_r + \dot{\mathbf{d}}_r) - \omega_b \times \mathbf{m_b} \tag{26}$$

Because $\dot{\mathbf{m}}_r = 0$, the x-axis and y-axis components of $\dot{\mathbf{m}}_b$ can be written as:

$$\dot{m}_{bx} = \mathbf{C}_{r1\cdot}^b \cdot \dot{\mathbf{d}}_r + \omega_{bz} m_{by} - \omega_{by} m_{bz} \tag{27a}$$

$$\dot{m}_{by} = \mathbf{C}_{r2\cdot}^b \cdot \dot{\mathbf{d}}_r - \omega_{bz} m_{bx} + \omega_{bx} m_{bz} \tag{27b}$$

where $\mathbf{C}_{r1\cdot}^b$ and $\mathbf{C}_{r2\cdot}^b$ denote the first row and second row of $\mathbf{C}_r^b$ respectively. We rewrite (27) as:

$$\dot{m}_{bx}/m_{by} = \omega_{bz} + \mathbf{C}_{r1\cdot}^b \cdot \dot{\mathbf{d}}_r/m_{by} - \omega_{by} m_{bz}/m_{by} \tag{28a}$$

$$-\dot{m}_{by}/m_{bx} = \omega_{bz} - \mathbf{C}_{r2\cdot}^b \cdot \dot{\mathbf{d}}_r/m_{bx} + \omega_{bx} m_{bz}/m_{bx} \tag{28b}$$

When the angle between the road plane and the horizontal plane remains constant and the vehicle do not vibrate, ω_{by} and ω_{bx} are zero. In practice, the angle between the two planes may change, however, in general, the rate of change will be small, and the duration of the vehicle attitude change is also very short. Hence, the corresponding ω_{by} and ω_{bx} will be small and close to zero for most of the time. In the case of vibration, which can be caused by bumps or ditches, ω_{by} and ω_{bx} may have large absolute value, but they will oscillate with high frequency. Hence, regarding the last item of the right side as noise and using the magnetometer output and difference to approximate the left side of (28), we obtain:

$$(m_{x,k} - m_{x,k-1})/(m_{y,k}\Delta t) = \omega_{bz} + \mathbf{C}_{r1\cdot}^b \cdot \dot{\mathbf{d}}_r/m_{by} + n_3 \tag{29a}$$

$$-(m_{y,k} - m_{y,k-1})/(m_{x,k}\Delta t) = \omega_{bz} - \mathbf{C}_{r2\cdot}^b \cdot \dot{\mathbf{d}}_r/m_{bx} + n_4 \tag{29b}$$

where n_3 and n_4 are noises. Because $m_{y,k}$ and $m_{x,k}$ may be equal or close to zero and thus cause numerical instability, we define Δm as:

$$\Delta m_k = \begin{cases} (m_{x,k} - m_{x,k-1})/(m_{y,k}\Delta t) & \left|m_{y,k}\right| > \left|m_{x,k}\right| \\ -(m_{y,k} - m_{y,k-1})/(m_{x,k}\Delta t) & \left|m_{y,k}\right| \le \left|m_{x,k}\right| \end{cases} \tag{30}$$

Equations (19b) and (24) show that the correlation between ω_z and a_x will be low if ω_{bz} is stable; otherwise, the correlation will be high. Equations (19b), (29), and (30) show that ω_z and Δm have similar relationships; more importantly, $\mathbf{d}_r$ will not affect the correlation between ω_z and Δm if it is constant and will not affect the correlation markedly except for an abrupt changing. Hence, the correlation between ω_z and Δm is robust to magnetic disturbances.

Considering ω_{bz} is stable in running straight status and unstable in turning status, we construct a vehicle status marker c, whose expression is:

$$c = \max\left(0, \frac{1}{2}\mathrm{cor}(\omega_{z,N}, \Delta m_N) - \frac{1}{2}\mathrm{cor}(\omega_{z,N}, a_{x,N})\right) \tag{31}$$

where $\mathrm{cor}(\cdot)$ estimates the correlation coefficient using the samples of the sensor outputs, and subscript N denotes the N points samples from $k - N + 1$ to k. In fact, in running straight status, MARG sensor outputs are stable except for uncorrelated noises; whereas in turning status, the outputs change, and the output from different sensor is correlative because their changes are caused by the same reason i.e., the vehicle is turning. The rationale of (31) is that use estimated correlation coefficients to distinguish status. In (31), the weighted sum combine the two correlation coefficient to make full use of the correlation between different sensors, and the second term of the sum is minus because ω_z is negatively correlated to a_x when ω_{bz} is changing. Considering the sum is theoretically nonnegative, we set c to zero when correlation coefficient estimation errors result in a negative sum. Similar to ρ_a and ρ_m, the adjusting expression for ρ_ω is:

$$\rho_{\omega,k+1} = \exp(\lambda_\omega c) \tag{32}$$

where λ_ω is a coefficient that transforms the status marker to a proper range.

3.2. Two-Step Measurement Update Method

This part presents the two-step measurement update method, and the concerned parameters of the first and second update step are respectively denoted by subscript f and s in the following.

First Step: The accelerometer and gyroscope measurement updates are implemented in this step. The measurement matrix and noise covariance matrix are respectively:

$$\mathbf{H}_{f,k+1} = \begin{bmatrix} \mathbf{H}_{a,k+1} \\ \mathbf{H}_\omega \end{bmatrix} \tag{33}$$

$$\mathbf{R}_{f,k+1} = \mathrm{diag}\left(\rho_{a,k+1}\mathbf{R}_{a,k+1}, \rho_{\omega,k+1}\sigma_\omega^2\right) \tag{34}$$

The update expressions are:

$$\mathbf{K}_{f,k+1} = \mathbf{P}_{k+1/k}\mathbf{H}_{f,k+1}^T\left(\mathbf{H}_{f,k+1}\mathbf{P}_{k+1/k}\mathbf{H}_{f,k+1}^T + \mathbf{R}_{f,k+1}\right)^{-1} \tag{35a}$$

$$\hat{\mathbf{x}}_{f,k+1} = \hat{\mathbf{x}}_{k+1/k} + \mathbf{K}_{f,k+1}(\mathbf{Z}_{f,k+1} - \mathbf{H}_{f,k+1}\hat{\mathbf{x}}_{k+1/k}) \tag{35b}$$

$$\mathbf{P}_{f,k+1} = (\mathbf{I}_7 - \mathbf{K}_{f,k+1}\mathbf{H}_{f,k+1})\mathbf{P}_{k+1/k} \tag{35c}$$

where:

$$\mathbf{Z}_{f,k+1} = \begin{bmatrix} \mathbf{0}_{4\times 1} \\ \omega_{z,k+1} \end{bmatrix} \tag{35d}$$

Second Step: In this step, the magnetometer measurement update is implemented and the ultimate quaternion estimation is computed. The measurement matrix and noise covariance matrix are respectively:

$$\mathbf{H}_{s,k+1} = \mathbf{H}_{m,k+1} \tag{36}$$

$$\mathbf{R}_{s,k+1} = \rho_{m,k+1}\mathbf{R}_{m,k+1} \tag{37}$$

The update expressions are:

$$\mathbf{K}_{s,k+1} = \mathbf{P}_{f,k+1}\mathbf{H}_{s,k+1}^T\left(\mathbf{H}_{s,k+1}\mathbf{P}_{f,k+1}\mathbf{H}_{s,k+1}^T + \mathbf{R}_{s,k+1}\right)^{-1} \tag{38a}$$

$$\hat{\mathbf{x}}_{s,k+1} = \hat{\mathbf{x}}_{f,k+1} - \mathbf{K}_{s,k+1}\mathbf{H}_{s,k+1}\hat{\mathbf{x}}_{f,k+1} \tag{38b}$$

$$\mathbf{P}_{s,k+1} = (\mathbf{I}_7 - \mathbf{K}_{s,k+1}\mathbf{H}_{s,k+1})\mathbf{P}_{f,k+1} \tag{38c}$$

Extracting quaternion part of $\hat{\mathbf{x}}_{f,k+1}$ and $\hat{\mathbf{x}}_{s,k+1}$, we obtain $\hat{\mathbf{q}}_{f,k+1}$ and $\hat{\mathbf{q}}_{s,k+1}$. Define $\mathbf{q}_\sigma$ as:

$$\mathbf{q}_\sigma = \hat{\mathbf{q}}_{f,k+1}^* \otimes \hat{\mathbf{q}}_{s,k+1} \tag{39}$$

where $\hat{\mathbf{q}}_{f,k+1}^*$ is the conjugate quaternion of $\hat{\mathbf{q}}_{f,k+1}$ and the symbol $\otimes$ is the quaternion multiplication operator. The conjugate quaternion of a unit quaternion represents the inverse rotation and a sequence of rotations can be represented by quaternion multiplication; therefore, $\mathbf{q}_\sigma$ can be viewed as the correction quaternion induced by the magnetometer measurement update. In theory, $\mathbf{q}_\sigma$

should only correct the yaw estimation; however, it also modify the attitude estimation in practice. The corresponding rotation matrix equation to Equation (39) can be written as:

$$\mathbf{C}(\hat{\mathbf{q}}_{s,k+1}) = \mathbf{C}(\mathbf{q}_\sigma)\mathbf{C}(\hat{\mathbf{q}}_{f,k+1}) \tag{40}$$

The condition for $\mathbf{C}(\hat{\mathbf{q}}_{s,k+1})$ and $\mathbf{C}(\hat{\mathbf{q}}_{f,k+1})$ to share a same attitude [26] is:

$$\mathbf{q}_{\sigma v} \times \mathbf{C}(\hat{\mathbf{q}}_{f,k+1})_{.3} = \mathbf{0}_{3\times 1} \tag{41}$$

where $\mathbf{q}_{\sigma v}$ is the vector part of $\mathbf{q}_\sigma$ and $\mathbf{C}(\hat{\mathbf{q}}_{f,k+1})_{.3}$ is the third column of $\mathbf{C}(\hat{\mathbf{q}}_{f,k+1})$. The geometric meaning of (41) is that the rotation axis of $\mathbf{q}_\sigma$ should be parallel to the z-axis of the reference frame. Based on this condition and (1), a new correction quaternion $\hat{\mathbf{q}}_\sigma$ is defined as:

$$\hat{\mathbf{q}}_\sigma = \begin{bmatrix} \cos(\alpha) & \sin(\alpha)\mathbf{C}(\hat{\mathbf{q}}_{f,k+1})_{.3}^T \end{bmatrix}^T \tag{42a}$$

where:

$$\alpha = \arccos(q_{\sigma,0})\mathbf{e} \cdot \mathbf{C}(\hat{\mathbf{q}}_{f,k+1})_{.3} \tag{42b}$$

$$\mathbf{e} = \mathbf{q}_{\sigma v} / \sin(\arccos(q_{\sigma,0})) \tag{42c}$$

Note that $\hat{\mathbf{q}}_\sigma$ is actually the projection of $\mathbf{q}_\sigma$ on the direction depicted by $\mathbf{C}(\hat{\mathbf{q}}_{f,k+1})_{.3}$, and this treatment makes $\hat{\mathbf{q}}_\sigma$ satisfy (41). Finally, the ultimate quaternion estimation is computed by:

$$\hat{\mathbf{q}}_{k+1} = \hat{\mathbf{q}}_{f,k+1} \otimes \hat{\mathbf{q}}_\sigma \tag{43}$$

3.3. Complete Improved Yaw Estimation Algorithm

Adding the adaptive gyroscope measurement update to the conventional algorithm and adopting the two-step measurement update method, we obtain the improved algorithm shown in Figure 2. Note that the improved algorithm preserves the linearity and all its measurement updates have an adaptive weight.

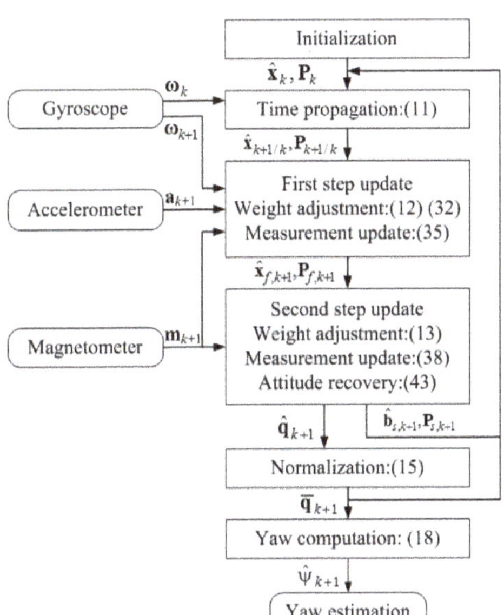

Figure 2. Improved yaw estimation algorithm.

4. Experiments

In this section, we evaluate the performances of the improved algorithm experimentally. The real data from a MARG sensor mounted on a test vehicle are processed by the conventional and improved algorithm, respectively. Then, we analyze the performances of the vehicle status marker, gyroscope measurement update, two-step measurement update method, and yaw estimation by comparing the results of the improved algorithm against that of the conventional algorithm and reference values.

4.1. Experimental Setup and Parameters

The test vehicle and experimental devices are shown in Figure 3. A Motion Tracker MTi-28A53G35 (Xsens, Enschede, The Netherlands) [31] was used as the MARG sensor. The Global Position System (GPS) unit (Unicoremm, Beijing, China) provided reference values of longitude, latitude, and forward velocity. 3-axis gyro module STIM210 (Sensonor AS, Horten, Norway) was used as the attitude and heading reference system (AHRS), which provided reference values of the yaw and attitude. The initial value of the AHRS was computed with the static accelerometer and magnetometer outputs [32]. The laptop (Lenovo, Beijing, China) supplied power for the MARG sensor, GPS unit, and the AHRS, and logged data from them at 100Hz. The GPS unit, MARG sensor, and AHRS were mounted on the test vehicle with the sensor and AHRS frames aligned with the vehicle body frame. In experiments, the vehicle was driven along test trajectories, which consisted of straight lines, corners with different angles and a circular line, involving a full range of yaw. Note that the magnetic sensor had been calibrated inline according to the sensor manual [20]; hence, we assumed that the impact of the hard and soft iron effects had been eliminated.

We analyzed the MARG sensor output in static condition, and set the noise parameters as follows: $\sigma_m = 0.0015$, $\sigma_\omega = 0.0056$ rad/s, $\sigma_a = 0.008$ m/s^2. We found the proper values for λ_m, λ_ω, and λ_a by trial and error, and set them to 20, 15, and 50 respectively. The estimation of $\mathbf{q}_0$ was computed with the initial outputs of the magnetometer and accelerometer; the estimation of $\mathbf{b}_0$ was set to the gyroscope output before the vehicle was started, and $\mathbf{P}_0 = 100\mathbf{I}_7$. The initial values of $\mathbf{Q}$, $\mathbf{R}_a$, and $\mathbf{R}_m$ were computed with $\mathbf{q}_0$ and $\mathbf{P}_0$ [16].

(a) (b)

Figure 3. Experimental setup. (**a**) Test vehicle. (**b**) Experimental devices.

4.2. Experimental Results and Discussion

4.2.1. Vehicle Status Marker

To analyze the experimental results more clearly, we intercepted a piece of the MARG sensor outputs corresponding to a segment of the test trajectories. Using these data, we computed the status marker c and the adaptive weight coefficient ρ_ω, which are shown in Figure 4. In Figure 4a, the reference yaw indicates that the vehicle undergone a turning (about 90°) between 70 s and 80 s, and mainly run straight with small fluctuation of yaw in other times. Corresponding to the reference yaw, c increases markedly between 70 s and 80 s; and oscillates with small values in other times. One key to computing c effectively is to select a proper N, because too many sample points will cause severely lagged weight adjustment, whereas too few sample points will reduce the computation accuracy. In the experiments, we set $N = 1/\Delta t$.

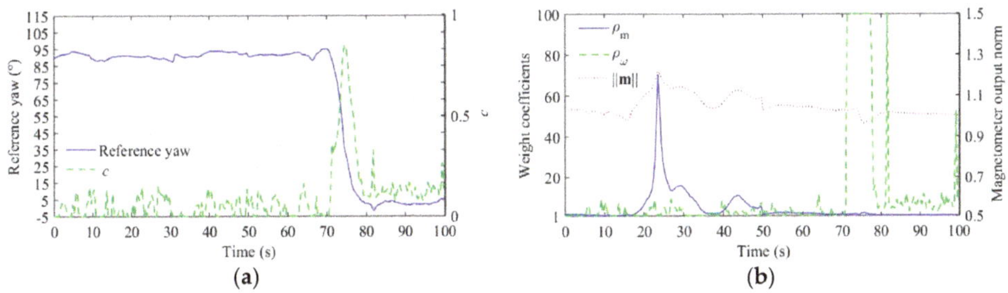

Figure 4. Adaptive weight coefficient based on the vehicle status. (**a**) Reference yaw and c. Because the reference yaw and c have different value range, two vertical axes are used for clarity with the right one corresponding to c. (**b**) ρ_m and ρ_ω. The magnetometer output norm is also plotted to show the relationship between magnetic disturbances and ρ_m, and, similar to (**a**), two vertical axes are used with the right one corresponding to $\|\mathbf{m}\|$. To obtain a proper vertical axe limit and thus the trends of curves can be showed clearly, the maximum value of the weight coefficients is set to 100.

4.2.2. Gyroscope Measurement Update

We processed the sensor outputs using the conventional and improved algorithm, respectively, and the yaw estimation results are shown in Figure 5. The results demonstrate that the improved algorithm has better estimation accuracy, and the reasons may be analyzed as follows. In the conventional algorithm, the magnetometer measurement update corrects the yaw estimation from drifting, but the measurement weight will be adjusted down in the presence of magnetic disturbances (Figure 4b shows obvious differences between the norm of magnetometer output and one from about 15 s to 50 s implying the presence of magnetic disturbances). As a result, the correction effect attenuates and thus the yaw estimation drifts. In contrast, the gyroscope measurement update provides another correction in the improved algorithm, and more importantly, the correction can hardly be attenuated by magnetic disturbances as analyzed in Section 3. The adaptive weight coefficients shown in Figure 4b verify the function of the gyroscope measurement update: ρ_ω is much lower than ρ_m in the presence of magnetic disturbances. In fact, ρ_ω represent the validity of the gyroscope measurement, which is not affected by magnetic disturbances. Hence, the gyroscope measurement can be used in magnetically disturbed environments. The real-time adjusting ρ_ω is critical for the gyroscope measurement to work effectively. Figure 5 also shows the yaw estimation result when ρ_ω is set to constant 1. The result demonstrates that the constant ρ_ω causes erroneous correction from the gyroscope measurement when the vehicle is turning, and hence results in significant estimation errors.

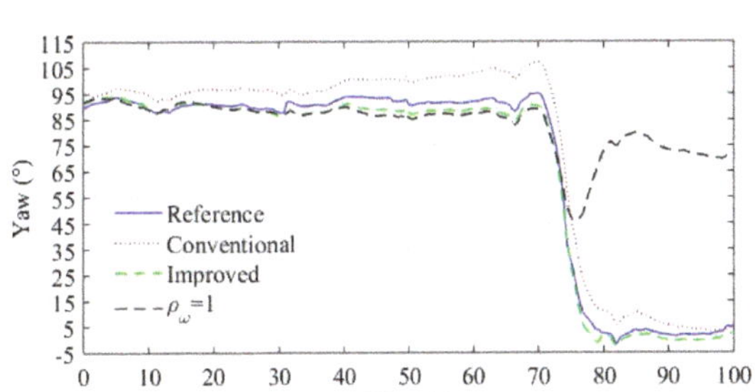

Figure 5. Yaw estimation.

The gyroscope bias estimation results are shown in Figure 6. We estimated the reference value of b_z by calculating the mean value of the gyroscope output in a not-moving interval, and the result was 0.00058 rad/s. Obviously, the b_z estimation is updated more effectively in the improved algorithm, which is verified by its better yaw estimation accuracy.

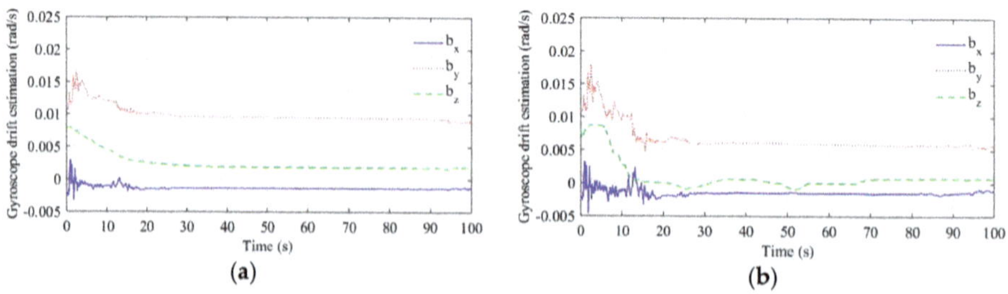

Figure 6. Gyroscope bias estimation. (**a**) Conventional algorithm. (**b**) Improved algorithm.

In Section 3.1, we draw a conclusion based on Equation (21) that b_z estimation is more important than b_x and b_y estimation in the fusion algorithm. To verify this conclusion based on experimental data, we constructed a conventional adaptive fusion algorithm using quaternion as state variables, and using Equation (6) as measurement equation. Note that, in this algorithm, the gyroscope bias is not estimated and compensated, and hence it will always impact the quaternion propagation. To examine the impact of the gyroscope bias, we added a constant bias on the reference angular rate provided by the AHRS and used the biased value to realize the quaternion propagation in the fusion algorithm.

We used this algorithm to estimate yaw for three times, and the constant bias were respectively set to [b, 0, 0], [0, b, 0], and [0, 0, b]. The results were denoted as **bx**, **by**, and **bz** respectively. Note that, in this process, all the parameters of the algorithm are same except for the constant bias. The root mean square (RMS) errors of the yaw estimation corresponding to different b are listed in Table 1. Obviously, b_z has a more significant impact on yaw estimation.

Table 1. RMS errors of the yaw estimation corresponding to different b.

Result	b = 0.005 rad/s	b = 0.01 rad/s	b = 0.015 rad/s
bx	6.8°	7.6°	7.9°
by	7.1°	8.3°	9.1°
bz	11.2°	19.4°	29.0°

It should be noted that the gyroscope measurement updates only provide a "partial" correction, in other words, they cannot provide absolute yaw information but that the yaw is unchanged to some extent, and the "extent" is indicated by the adaptive weight. Therefore, it is reasonable to employ the gyroscope measurement updates in combination with the magnetometer measurement updates. In addition, the gyroscope measurement updates only work effectively in running straight status. In practice, turning status is inevitable, but the duration of turning is relatively short, and running along a straight road is a more usual status for most land vehicles.

In practice, a running vehicle cannot avoid bumps and ditches, which will cause oscillations of a_x, ω_x, and ω_y. We assume these oscillations as part of the noise items in Equations (24) and (29) respectively, and construct the vehicle status marker c based on the correlations between different sensor outputs. To evaluate these noise assumptions, we analyzed a piece of the sensor outputs corresponding to a straight road with some bumps and ditches. The raw signals of a_x, ω_x, and ω_y are shown in Figures 7 and 8, respectively. The term c computed by Equation (31) is shown in Figure 9.

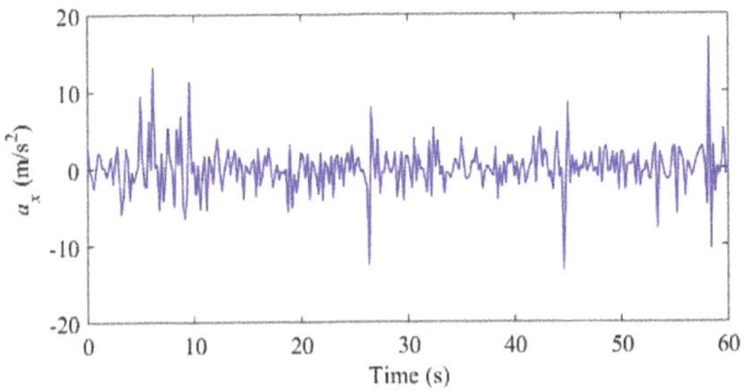

Figure 7. a_x in bumps and ditches experiment.

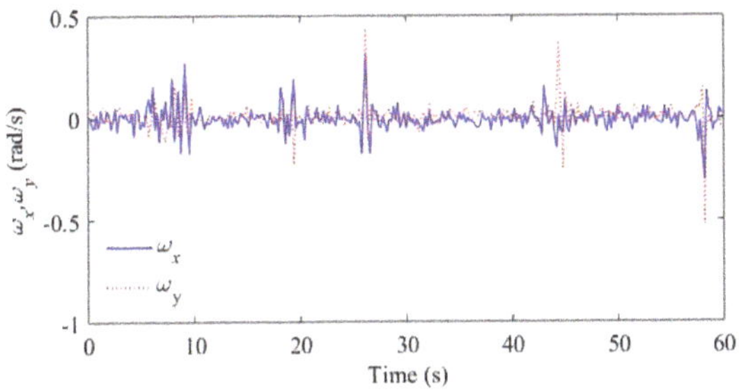

Figure 8. ω_x and ω_y in bumps and ditches experiment.

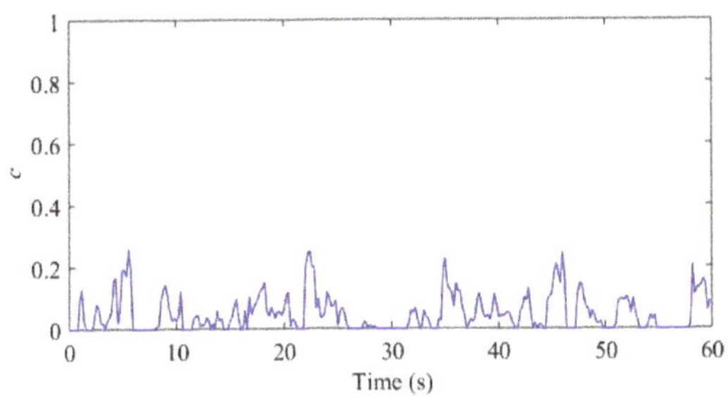

Figure 9. c in bumps and ditches experiment.

The vehicle met a bump or ditch at about 7 s, 19 s, 27 s, 45 s, and 58 s. Obviously, the bumps and ditches cause oscillations of the sensor outputs. Figure 9 demonstrate that the bumps and ditches do not affect c significantly, which displays small values and is consistent with the running straight status.

4.2.3. Two-Step Measurement Update Method

In the experiments, the test road is basically level except for some speed breaks. To evaluate the performances of the two-step measurement update method, we computed attitudes using the quaternion [33] from the first step update, the conventional algorithm and the improved algorithm respectively, and the results are shown in Figure 10. Obviously, the improved algorithm and the first step update have the same attitude estimation, which verify that the second step update do not modify the attitude estimation. In addition, the attitude estimation of the improved algorithm is more consistent with the reference values than the conventional algorithm. The comparison demonstrates that the two-step measurement update method eliminates the attitude estimation errors induced by the magnetometer measurement update. Besides magnetic disturbances, the magnetometer measurement error can also cause attitude estimation errors. For example, the attitude estimation of the conventional algorithm change incorrectly during the turning of the vehicle (70–80 s), and similar phenomena can be found in other turning processes, which, we think, is due to the dynamic errors of the magnetometer.

It should be noted that modifications of the attitude estimation made by magnetometer measurement update essentially arise from the disagreements between the accelerometer and magnetometer outputs, which can also be caused by non-gravitational acceleration and accelerometer measurement errors. The improved algorithm prevents the magnetometer output from modifying the attitude estimation because firstly the gravity resolved in the body frame can provide sufficient attitude information, and secondly the accelerometer output is assumed more reliable than the magnetometer output in magnetically disturbed environments as has been accepted and verified in many applications.

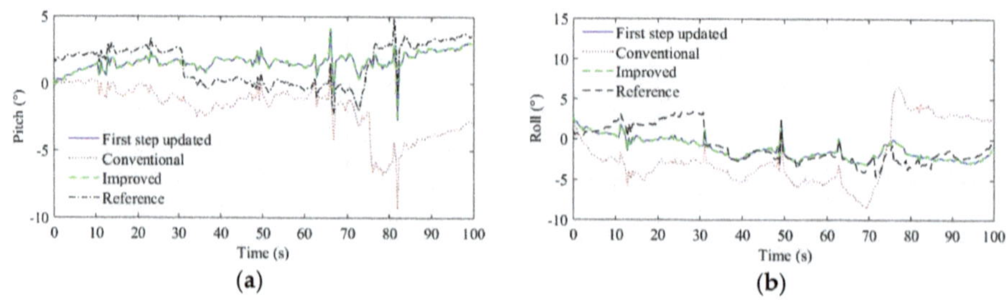

Figure 10. Attitude estimation. (**a**) Pitch estimation. (**b**) Roll estimation.

4.2.4. Yaw Estimation

We tested two trajectories, which are referred to as A and B. MARG sensor outputs corresponding to the trajectories were processed. To show the performances of the yaw estimation intuitively, we used dead reckoning [29] to reproduce the test trajectories based on the yaw estimations and the vehicle velocity data provided by the GPS unit. The results are shown in Figure 11, where the reference trajectories recorded by the GPS unit is also plotted. Clearly, the reproduced trajectories from the improved algorithm are closer to the reference than that from the conventional algorithm.

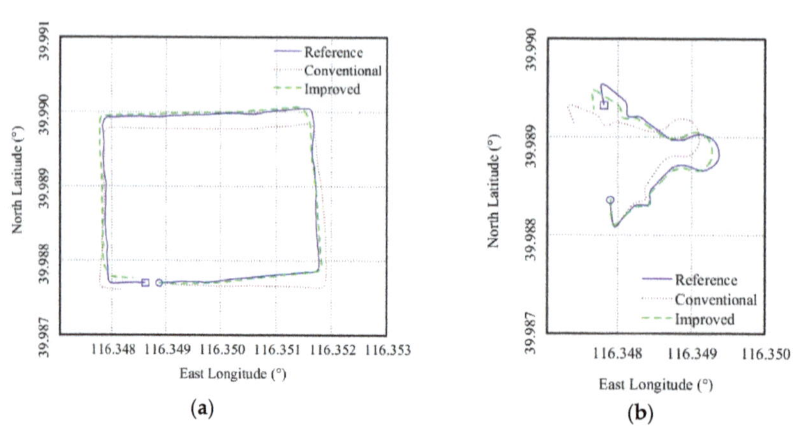

Figure 11. Dead reckoning results. The circle represents the start point, and the square represents the end point. (**a**) Trajectory A. (**b**) Trajectory B.

The root mean square (RMS) errors of the improved algorithm in trajectory A and B are respectively 1.8° and 2.9°. The reason the improved algorithm performed better in A is, as pointed out earlier, the gyroscope measurement only work effectively in running straight status. A was approximately rectangular and has longer straight roads, whereas B consisted of more turning road with different angles and curvature. Obviously, A was more suitable for the gyroscope measurement to take effect. It should be noted that vehicle status cannot affect the work of the two-step measurement update method, which is also helpful for the yaw estimation.

The RMS errors of the yaw and attitude estimation are listed in Table 2. In addition, we used the Kalman filter with vector selection [7] and the complementary filter with varying gains [21], referred to as VSKF and VGCF respectively, to process the experiment data, and the RMS errors are also listed in Table 2. Clearly, the improved algorithm performed best in the yaw and attitude estimation. In fact, as pointed out in the introduction, the VSKF and VGCF essentially use the same strategy as the conventional algorithm to handle magnetic disturbances: reducing the confidence in the magnetometer measurements, which will result in poor correction for the yaw estimation. Whereas the improved algorithm not only reduce the confidence in the disturbed measurements but also introduce new measurements based on vehicle status, which provide supplemental information for the yaw estimation. In addition, the two-step measurement update method eliminates the impacts of magnetometer errors and magnetic disturbances on attitude estimation, which is also helpful for the yaw estimation.

Table 2. RMS errors of the yaw and attitude estimation.

Algorithm	Yaw (°)	Pitch (°)	Roll (°)
VSKF	4.9	4.0	3.1
VGCF	5.6	3.5	3.3
Conventional	5.2	3.7	2.9
Improved	2.1	1.8	1.6

Note that the better yaw estimation accuracy of the improved algorithm is due to its enhanced ability to deal with magnetic disturbances; and it will have the same level of yaw estimation accuracy as the conventional algorithm in magnetically homogeneous environments.

5. Conclusions

An improved yaw estimation algorithm for land vehicle using a MARG sensor was proposed in this paper. Under running straight assumption, we derived the gyroscope measurement equation, which update the gyroscope bias and thus constrain the yaw estimation from changing. The validity of the gyroscope measurement depends on the vehicle status; to determine the vehicle status, we analyzed the correlations between different sensors based on the vehicle kinematic model and Coriolis theorem, and constructed a vehicle status marker used to adjust the weight of the gyroscope measurement. In addition, we designed a two-step measurement update method, which implements the magnetometer measurement update separately and eliminates its impact on attitude estimation. Adopting the gyroscope measurement update and the two-step measurement update method, we improved the conventional yaw estimation algorithm. The improved algorithm was tested in experiments and compared against the conventional algorithm. Based on the experiment results, the performances and characters of the improved algorithm were discussed and the conclusion is as follows. The gyroscope measurement update is an effective supplement to the magnetometer measurement update in magnetically disturbed environments; the two-step measurement update methods make attitude estimation immune to the errors induced by magnetometer measurement update; and the improved algorithm provides more reliable yaw estimation for land vehicles than the conventional algorithm. Finally, it should be noted that the vehicle status marker is based on statistic characteristics between different sensors, which make it robust to disturbances, but on the other hand, insensitive to the varying status. The improvements of its real-time performance and ability to detect the turning status with small angular rates will be the topics of further work.

Author Contributions: Conceptualization, G.S. and X.L.; Methodology, G.S.; Software, G.S.; Validation, X.L. and G.S.; Formal Analysis, G.S.; Investigation, G.S. and Z.J.; Resources, X.L.; Data Curation, G.S. and Z.J.; Writing—Original Draft Preparation, G.S.; Writing—Review & Editing, G.S.; Visualization, G.S.; Supervision, X.L.; Project Administration, X.L.; Funding Acquisition, X.L.

Funding: This research was funded by National Natural Science Foundation of China, grant number [61273082].

Conflicts of Interest: The authors declare no conflict of interest.

References

1. Vahidi, A.; Eskandarian, A. Research advances in intelligent collision avoidance and adaptive cruise control. *IEEE Trans. Intell. Transp. Syst.* **2003**, *4*, 143–153. [CrossRef]
2. McCall, J.; Trivedi, M.M. Video-based lane estimation and tracking for driver assistance: Survey, system, and evaluation. *IEEE Trans. Intell. Transp. Syst.* **2006**, *7*, 20–37. [CrossRef]
3. Wada, M.; Yoon, K.S.; Hashimoto, H. Development of advanced parking assistance system. *IEEE Trans. Ind. Electron.* **2003**, *50*, 4–17. [CrossRef]
4. Fernandez Llorca, D.; Milanes, V.; Parra Alonso, I.; Gavilan, M.; Garcia Daza, I.; Perez, J.; Sotelo, M.Á. Autonomous pedestrian collision avoidance using a fuzzy steering controller. *IEEE Trans. Intell. Transp. Syst.* **2011**, *12*, 390–401. [CrossRef]
5. Yim, S.; Jeon, K.; Yi, K. An investigation into vehicle rollover prevention by coordinated control of active anti-roll bar and electronic stability program. *Int. J. Control Autom. Syst.* **2012**, *10*, 275–287. [CrossRef]
6. Masuya, K.; Sugihara, T. A nonlinear complementary filter for attitude estimation with dynamics compensation of MARG sensor. In Proceedings of the IEEE International Conference on Advanced Intelligent Mechatronics (AIM), Banff, AB, Canada, 12–15 July 2016; pp. 976–981.
7. Zhang, Z.Q.; Meng, X.L.; Wu, J.K. Quaternion-Based Kalman filter with vector selection for accurate orientation tracking. *IEEE Trans. Instrum. Meas.* **2012**, *61*, 2817–2824. [CrossRef]
8. Wahba, G. A least squares estimate of satellite attitude. *SIAM Rev.* **1965**, *7*, 409. [CrossRef]
9. Valenti, R.G.; Dryanovski, I.; Xiao, J. A linear Kalman filter for MARG orientation estimation using the algebraic quaternion algorithm. *IEEE Trans. Instrum. Meas.* **2016**, *65*, 467–481. [CrossRef]
10. Yun, X.; Bachmann, E.R.; McGhee, R.B. A simplified quaternion-based algorithm for orientation estimation from earth gravity and magnetic field measurements. *IEEE Trans. Instrum. Meas.* **2008**, *57*, 638–650. [CrossRef]
11. Mahony, R.; Hamel, T.; Pflimlin, J.M. Nonlinear complementary filters on the special orthogonal group. *IEEE Trans. Autom. Control* **2008**, *53*, 1203–1218. [CrossRef]
12. Yoo, T.S.; Hong, S.K.; Yoon, H.M.; Park, S. Gain-scheduled complementary filter design for a MEMS based attitude and heading reference system. *Sensors* **2011**, *11*, 3816–3830. [CrossRef] [PubMed]
13. Madgwick, S.O.H.; Harrison, A.J.L.; Vaidyanathan, R. Estimation of IMU and MARG orientation using a gradient descent algorithm. In Proceedings of the IEEE International Conference on Rehabilitation Robotics, Zurich, Switzerland, 27 June–1 July 2011; pp. 1–7.
14. Markley, F.L. Attitude error representations for Kalman filtering. *J. Guid. Control Dynam.* **2003**, *26*, 311–317. [CrossRef]
15. Crassidis, J.; Markley, F.L. Unscented filtering for spacecraft attitude estimation. *J. Guid. Control Dyn.* **2003**, *26*, 536–542. [CrossRef]
16. Choukroun, D.; Bar-Itzhack, I.Y.; Oshman, Y. Novel quaternion Kalman filter. *IEEE Trans. Aerosp. Electron. Syst.* **2006**, *42*, 174–190. [CrossRef]
17. Kok, M.; Schön, T.B. Magnetometer calibration using inertial sensors. *IEEE Sens. J.* **2016**, *14*, 5679–5689. [CrossRef]
18. Shi, G.; Li, X.S.; Li, X.F.; Liu, Y.X.; Kang, R.Q.; Shu, X.Y. Equivalent two-step algorithm for the calibration of three-axis magnetic sensor in heading measurement system. *Chin. J. Sci. Instrum.* **2017**, *38*, 402–407.
19. Vasconcelos, J.F.; Elkaim, G.; Silvestre, C.; Oliveira, P.; Cardeira, B. Geometric approach to strapdown magnetometer calibration in sensor frame. *IEEE Trans. Aerosp. Electron. Syst.* **2011**, *47*, 1293–1306. [CrossRef]
20. *Magnetic Field Mapper Documentation*; Xsens Technologies B.V.: Enschede, The Netherlands, 2013.
21. Costanzi, R.; Fanelli, F.; Monni, N.; Ridolfi, A.; Allotta, B. An attitude estimation algorithm for mobile robots under unknown magnetic disturbances. *IEEE/ASME Trans. Mechatron.* **2016**, *21*, 1900–1911. [CrossRef]

22. Wu, J.; Zhou, Z.B.; Chen, J.J. Fast complementary filter for attitude estimation using low-cost MARG sensors. *IEEE Sens. J.* **2016**, *16*, 6997–7007. [CrossRef]
23. Tong, X.; Li, Z.F.; Han, G.W. Adaptive EKF based on HMM recognizer for attitude estimation using MEMS MARG sensors. *IEEE Sens. J.* **2018**, *18*, 3299–3310. [CrossRef]
24. Feng, K.; Li, J.; Zhang, X.; Shen, C.; Bi, Y.; Zheng, T.; Liu, J. A new quaternion-based Kalman Filter for real-time attitude estimation using the two-step geometrically-intuitive correction algorithm. *Sensors* **2017**, *17*, 2146. [CrossRef] [PubMed]
25. Sabatini, A.M. Kalman-Filter-based orientation determination using inertial/magnetic sensors: Observability analysis and performance evaluation. *Sensors* **2011**, *11*, 9182–9206. [CrossRef] [PubMed]
26. Suh, Y.S. Orientation estimation using a quaternion-based indirect Kalman filter with adaptive estimation of external acceleration. *IEEE Trans. Instrum. Meas.* **2010**, *59*, 3296–3305. [CrossRef]
27. Suh, Y.S.; Ro, Y.S.; Kang, H.J. Quaternion-based indirect Kalman filter discarding pitch and roll information contained in magnetic sensors. *IEEE Trans. Instrum. Meas.* **2012**, *61*, 1786–1792. [CrossRef]
28. Sun, S.; Meng, X.; Ji, L.; Wu, J.; Wong, W.C. Adaptive sensor data fusion in motion capture. In Proceedings of the 13th International Conference on Information Fusion, Edinburgh, UK, 26–29 July 2010.
29. Ahmed, H.; Tahir, M. Accurate attitude estimation of a moving land vehicle using low-cost MEMS IMU sensors. *IEEE Trans. Intell. Transp. Syst.* **2017**, *18*, 1723–1739. [CrossRef]
30. Dissanayake, G.; Sukkarieh, S.; Nebot, E.; Durrant-Whyte, H. The aiding of a low-cost strapdown inertial measurement unit using vehicle model constraints for land vehicle applications. *IEEE Trans. Robot. Autom.* **2001**, *17*, 731–747. [CrossRef]
31. *MTi and MTx User Manual and Technical Documentation*; Xsens Technologies B.V.: Enschede, The Netherlands, 2013.
32. Jurman, D.; Jankovec, M.; Kamnik, R.; Topič, M. Calibration and data fusion solution for the miniature attitude and heading reference system. *Sens. Actuator A-Phys.* **2007**, *138*, 411–420. [CrossRef]
33. Qin, Y.Y. *Inertial Navigation*, 2nd ed.; Science Press: Beijing, China, 2014; pp. 251–252, ISBN 978-7-03-039465-1.

sensors

Article

Semi-Supervised Segmentation Framework Based on Spot-Divergence Supervoxelization of Multi-Sensor Fusion Data for Autonomous Forest Machine Applications

Jian-lei Kong [1,2], Zhen-ni Wang [1], Xue-bo Jin [1,2], Xiao-yi Wang [1,2,*], Ting-li Su [1,2] and Jian-li Wang [1]

1 School of Computer and Information Engineering, Beijing Technology and Business University, Beijing 100048, China; kongjianlei@btbu.edu.cn (J.-l.K.); wzhenni@163.com (Z.-n.W.); jinxuebo@btbu.edu.cn (X.-b.J.); sutingli@btbu.edu.cn (T.-l.S.); wangjianli@btbu.edu.cn (J.-l.W.)

2 Beijing Key Laboratory of Big Data Technology for Food Safety, Beijing Technology and Business University, Beijing 100048, China

* Correspondence: sdwangxy@163.com; Tel./Fax: +86-138-1057-7889

Received: 18 July 2018; Accepted: 10 September 2018; Published: 12 September 2018

Abstract: In this paper, a novel semi-supervised segmentation framework based on a spot-divergence supervoxelization of multi-sensor fusion data is proposed for autonomous forest machine (AFMs) applications in complex environments. Given the multi-sensor measuring system, our framework addresses three successive steps: firstly, the relationship of multi-sensor coordinates is jointly calibrated to form higher-dimensional fusion data. Then, spot-divergence supervoxels representing the size-change property are given to produce feature vectors covering comprehensive information of multi-sensors at a time. Finally, the Gaussian density peak clustering is proposed to segment supervoxels into sematic objects in the semi-supervised way, which non-requires parameters preset in manual. It is demonstrated that the proposed framework achieves a balancing act both for supervoxel generation and sematic segmentation. Comparative experiments show that the well performance of segmenting various objects in terms of segmentation accuracy (F-score up to 95.6%) and operation time, which would improve intelligent capability of AFMs.

Keywords: multi-sensor joint calibration; high-dimensional fusion data (HFD); supervoxel; Gaussian density peak clustering; sematic segmentation

1. Introduction

For the foreseeable future, autonomous forest machines (AFMs) will play a central role in harvesting, tending and forest management because lots of artificial and natural forests are facing with poor forest quality on a large-scale [1]. Currently, the proportion of good quality forest resources is very small, which directly affects the economic potential and ecological value of the whole forestry industry. Scientific tending and harvesting operations, which rationally adjust the mixed structure and wood quality of forest, can improve the level of forestry production in a relatively short time [2]. However, traditional artificial technology relying on the high-cost and low-efficiency labor could not meet the urgent quality improvement of modern forestry. Therefore, replacing human beings with AFMs rapidly to engage in dangerous and heavy works including tending and harvesting have been a core focus, which will increase the efficiency and value of unit labor in handing with quality promotion of large-scale forest lands. Nevertheless, the automatic or semiautonomous ways presented that can work without operators cannot perceive the complex forest environments accurately as drivers, because so many complex factors need to be considered in the maneuvering of forest machines in harvesting operations that full automation would be extremely difficult. Many problems still remain to be addressed before the field of AFMs can be widely expanded including: which trees should be harvested or tended, where the harvester should be positioned and what suitable driving routes should be taken? To this end, an efficient environment-aware system that could facilitate decision-making in those complex work described above should be constructed before a forest machine can fully and authentically automate the harvesting and tending process, which will relieve fatigue and stress of drivers and improve overall productivity and efficiency of forestry workload [3].

In comparison to the structured environment in agriculture or the semi-structured outdoor environment in urban traffic, the forest environment is much more challenging for operation and perception of AFMs since the various objects of interest are surrounded by massive areas of dense shrubs, dead trees and fallen objects. In addition, forest trails are rarely straight or flat, and obstacles are common, the AFM itself has more problems in the process of moving due to high amount of logging residues on the ground surface and sometimes high variability of the soil-bearing capacity [4]. In order to improve the capacity of environmental perception and complex decisions in forest operations, most of the current AFMs combine several sensors to compensate for the drawbacks of each sensor and to merge various information into a single percept about the nearby environment [5]. Although multi-sensor fusion technology has become the standard techniques for AFMs to identify objects, select roads and decide to execute the best operation, separating individual objects and backgrounds from forest environments is extremely demanding and raises problems that have not yet been satisfactorily resolved, which becomes the main challenge in forestry autonomous awareness and navigation systems [6].

Many previous works dealing with this perception issue did so by solving a semantic segmentation problem, aimed at determining which objects of the input data correspond to harvesting and tending operation, and detecting which areas and trails are suitable for driving. Segmentation of images and point clouds is an important capability for AFMs in unstructured forestry scene, which is a prerequisite for solving subsequent tasks such as navigation or reconstruction. The basic process of segmentation is labeling each measurement unit, so that the pixels or points belonging to the same surface or region are given the same label [7]. However, processing unstructured and massive data (including laser point clouds, visible images, thermal infrared photos etc.) obtained by different sensors is a much harder problem.

On one hand, most of segmentation methods are proposed to handle with a single type of sensor data (images or point clouds), which does not take full advantage of higher-dimensional fusion data (HFD) captured by the multi-sensor measuring system of AFMs. Additionally, those algorithms are focusing on point-level data or patch-level supervoxel with a fixed size, which is not suitable for the size change of HFD caused by the occurrence of spot-divergence in complex forestry scenes. Thus, finding approaches that can directly operate on size-changed HFD in an effective and affordable way is still largely open in application requirements of AFMs. On the other hand, the existing segmentation works falling into the cluster category in Euclidean space are supervised-based methods, which encourages users to try many different input thresholds, therefore, increases the chance of selecting good input values for better results. Consequently, such a process usually requires numerous parameters of human intervention and can be quite time consuming. This strategy may perform well in simple and sparse datum obtained from the city vehicles or indoor robots in structured environment. However, it is difficult to deal with the noise-filled HFD of forest environment in real-time, which even lead to inaccurate results of segmentation.

Aiming to improve the segmentation performance of the multi-sensor measuring system for AFMs, we propose a semi-supervised segmentation framework based on a size-changed supervoxel, which takes the spot divergence of each HFD into account and produces the valid feature vector covering the spatial, visual and thermal information at a time. On the basis of precomputed supervoxelization, we further extend the traditional density peak clustering method in Gaussian constraint to solve the sematic segment problem of different objects in complex scenes. This framework has two major benefits: (a) it provides a patch-level process that every supervoxel can describe the variational size of HFD; (b) sematic objects can be segmented without the artificially preset of clustering central number or convergence thresholds, which gives an opportunity to promote the segment performance in the term of accuracy and operation time.

The rest of this paper is organized as follows: some related works are introduced in Section 2. Section 3 briefly presents our multi-sensor measuring system of AFMs. Section 4 presents the principle and notations of our segmentation framework. Experiments are conducted in Section 5. Finally, we conclude our work in Section 6.

2. Related Works

Focusing on HFD, there is a growing tendency of innovative methods for the treatment and analysis of these data, aimed ultimately to exploit in-depth the informative value of semantic segmentation. The early attempt to group segmentation methods followed the works of spatial transformation by converting 3D/2.5D point clouds into 2D depth images, which could be deposed with proven image segmentation techniques [8]. However, those methods lacked the geospatial information of 3D point clouds. Consequently, many further research paid attention on 3D-based segmentation methods able to understand a complex scene directly [9]. Those algorithms fell into the basic combination of the original point-level data and model-fitting method, which did take visual information and reflectivity intensity of HFD in account. In order to promote the segment performance processed in a point-wise manner, the following works of extracting supervoxels for 3D point clouds began to take multi-sensor information fusion and machine learning methods into account [10,11]. In this section, we will review some representative algorithms that are related to sematic segmentation of HFD.

2.1. Point Clouds-Based Segmentation Method

The traditional segmentation methods are dividing large amount of unstructured 3D point clouds into a certain number of independent objects with special semantics according to spatial distribution characteristics. Over the past decade, several algorithms for object extraction from 3D point clouds have been reported by researchers. Euclidean clustering segmentation was based on defining a neighbourhood of radius and all the points within the sphere of radius are belong to one cluster [12]. Although such methods allow a fast segmentation, they may produce inaccurate results in case of noise and uneven density of point clouds, which commonly occur in point clouds.

For higher accuracy, model-fitting methods were proposed with the observation that many objects could be decomposed into geometric primitives like planes, cylinders and spheres [13]. For example, the cylinder was usually fitted onto point clouds of forest scenes to distinguish the trunks which were conform to the mathematical representation of the primitive shape. As part of the model fitting-based category, two widely employed algorithms were the Hough Transform (HT) and the Random Sample Consensus (RANSAC) approach. Compared to the HT only detects fixed shapes, the RANSAC method was used to extract shapes by randomly drawing minimal data points to construct candidate shape primitives, which were checked against all points of dataset to determine the appropriate value. The model-fitting method has been adapted to segment tree stems in forestry scene. Ref. [14] proposed hierarchical minimum cut method based on that the detected trunk points are recognized according to pole-like shape. By detecting the repetitive appearance of cylindrical segment units, this method isolated individual trees from point clouds of forest scene and achieved good balance in terms of accuracy and correctness. However, the segmentation quality of the model fitting-based algorithms is sensitive to the point clouds characteristics (density, positional accuracy, and noise) and is over-reliance on predictive shapes and parameters, which lack of adaptability to segment for various objects in forest scenes.

An alternative was the region growing approach involving two stages: identification of the seed points based on the curvature of each point, and growing them based on predefined criteria such as proximity of points and planarity of surfaces. This method and several variations were presented for 3D point clouds segmentation. For example, [15] performed a marker-controlled region growing segmentation using a treetop as the seed surface and the angle and height between the neighboring triangles for the growing. However, the region growing methods strongly depends on multiple criteria, such as the location of initial seed regions and curvatures of points near region boundaries. Moreover, the high point density requires a large amount of computer resources for spatial searching if the original LiDAR points are processed directly in those methods. Thus, octree construction providing an efficient spatial index with high position accuracy was combined with the region growing methods to detect planar segments, which realized better point cloud management and provided faster refinement process [16]. Similarly, [17] proposed an initial-to-fine algorithm performed on an octree-based representation of the input point cloud to extract stem-based initial segments. Then the output was then passed through a refinement segmentation of overlapped canopy, which can reduce technical difficulties and effectively separate neighboring trees even if their canopies are overlapped. In these works, the partition of a point cloud was achieved by an octree structure. Local patches were then extracted according to the leaves of the octree. The number of local patches is related to the number of whole points and the size of octree leaves. A major limitation of this method is that the interior shape structure is discarded. Further, those methods are not particularly robust as has been shown experimentally in part because the segmentation quality strongly depends both on multiple criteria and the selection of seed points/regions.

As demonstrated above, point clouds-based algorithms are well established as robust methods for segmenting dense 3D point clouds in acquired in urban areas. However, these works have some disadvantages in dealing with large 3D data sets or scenes with complex geometries. As these algorithms only use all individual points, the computational cost and significant processing time are very high, making it impractical for real time applications. Besides, the raw point clouds from terrestrial laser scanner (TLS) or mobile laser scanner (MLS) often exhibit unorganized stripe structures due to the rotary scanning mechanism. These structures make the point clouds difficult in providing any information on local surfaces, which help extracting the inside/outside of the underlying feature for efficient segmentation. Since point clouds are unstructured and often massive, it is sought to reduce these points by grouping together or removing redundant or un-useful points for improving the segmentation quality.

2.2. Supervoxel Process

In order to accelerate the existing segmentation with 3D point clouds processed directly in a point-wise manner, the patch-level methods have been proposed by clustering the individual 3D points together to form over-segmented voxels. In order to create the voxels, a 3D point is selected as center and all 3D points in the vicinity are selected with a fixed diameter (equal to maximum voxel size) to determine an actual voxel. With the voxel representation, the point clouds can first be divided into a number of patches and the processing can then be operated in a patch-wise manner. After the voxelized process, the 3D model can not only maintain the surface shape of the object, but also effectively describe the internal distribution. Since the number of patches is much smaller than the number of points in a point cloud, the efficiency of point clouds processing can be significantly improved. In the work of [18], volumetric 3D model was proposed to explicitly representing the forestry scenes, with the details of the trees and the surrounding unknown areas represented accurately. However, the voxelization of point clouds lacks of the fusion information including color, texture, thermal and reflectivity obtained by other sensors. This make the local description capabilities of each voxel degraded and limit the segmentation performance in complex applications of forest environment.

To gaining the higher representation than the voxel, some patch-wise segmentation applied supervoxels as basic elements to cope with HFD of complex tasks. These methods are inspired by the superpixel approaches that have been widely used in image analysis and processing. A typically superpixel method was the SLIC (Simple Linear Iterative Clustering) algorithm based on gradient-ascent theory in which the relationship between the color similarity and the spatial distance was used to form the cluster centers of superpixels [19]. Another representative method was the Superpixels Extracted via Energy-Driven Sampling (SEEDS) algorithm based on graph-based theory, which started with the color distribution and discriminates edge shape of each superpixel iteratively to achieve superpixel partitioning [20].

Currently, many supervoxel-based segmentation methods are the simply extension of 2D superpixel segmentation to the domain of 3D volumes. Reference [21] proposed a Voxel Cloud Connectivity Segmentation (VCCS) method to take full advantage of 3D geometry information. In this work, the points with similar normals, colors, and Fast Point Feature Histograms (FPFHs) were clustered into a supervoxel. Similarly, the proposed SEED-3D algorithm was designed to minimise the cost of the shortest path in the weighted graph with consideration of characteristics of the sensor for complex urban environment. The performance showed the better boundary recall and under-segmentation result [22]. These kinds of supervoxel methods seem to not only be suitable for real 3D volumetric data, but also be appropriate for video with object occlusion and moving objects. Similarly, [23] proposed a novel voxel-related Gaussian mixture model for supervoxel segmentation to address the problem of new and moving objects in continuous frames. According to the experiments, the proposed method performed well in terms of segmentation accuracy while possessing a competitive computing. In particular, the supervoxels have showed as the best processing unit for the individual tree segmentation from LiDAR point clouds in urban environments. Reference [24] proposed an automatic method for the individual tree segmentation (ATS) based supervoxel generation. With the preprocessing of extracting tree points, the supervoxel was defined as a polyhedral region consisting of homogeneous points. Then assigning other points to optimize centers obtained complete supervoxels and delineate trees from complex scenes. This method overcame two main drawbacks in the commonly used tree point assignment strategy, including the low efficiency caused by assigning the index to each point and the assignment of different tree indexes for homogeneous points.

2.3. Supervoxel-Based Segmentation Method

As describing the local characteristics of point clouds effectively and reducing the processing time of segmentation, the supervoxels are selected as the basic processing unit for patch extraction of HFD in this paper. When the 3D point clouds or HFD are converted into some supervoxels, the next issue is to group these patches to segment into distinct objects. Usually for such task, [25] proposed a link-chain method instead to group these s-voxels together into segmented objects. However, this method has many features and parameters which need to be adjusted manually in order to obtain better results with very long computational time. Therefore, segmentation algorithms based on K-means clustering were applied to group set of supervoxels into different objects using few attributes/features. In [26], the feature distances between cluster centers and the neighborhood supervoxels are minimized to segment street trees from 3D point clouds. Since the choice of neighborhood strongly influences segmentation results of the K-means clustering methods, it is difficult to segment the boundary supervoxels with abundant features. Thus, a refinement phase was necessary to test whether the supervoxel was within the same cluster. The extracted segmentation based on hierarchical clustering was proposed to compute geometrical and radiometric characteristics (position, surface normals, reflectance etc.) of each supervoxel for forestry scene segmentation [27]. Similarly, a novel Density Based Spatial Clustering of Applications with Noise (DBSCAN) clustering algorithm was presented to cluster any dimensional data including terrestrial point clouds and HFD [28].

As described previously, those clustering segmentation methods require artificially determining the number of cluster centers or selecting convergence thresholds, which lacks automatic adaptability. Furthermore, these methods are all supervised model as they rely on a set of provided training examples (features) to learn how to correctly perform a task. While high-quality features can enhance algorithm performance, and can also causing computationally expensive in large datasets [29]. Hence, the partially un-supervised extraction of scene structures from 3D point clouds or HFD has been found to be an attractive approach to urban and forestry scene analysis, because it can tremendously reduce the resources and time consumption of data analyzing for subsequent tasks and other applications of AFM in forest environments. Consequently, the density peak clustering (DPC) algorithm is adopted to construct our proposed segmentation method on the basis of the idea that cluster centers are characterized by a higher density than their neighbors and by a relatively large distance from data with higher densities. The DPC has been widely applied to the problem of classification as the clusters are recognized automatically regardless of their shape and of the dimensionality of the space [30]. However, it cannot be directly applied to supervoxel segmentation because its accuracy excessively depended on the suitable threshold estimated on the basis of empirical experience [31]. What's worse, it does not encode the constraint on dimensional consistency of feature vector for each supervoxel. Thus, a new Gaussian way is proposed to automatically extract the optimal value of threshold by using the normalized feature distance. For any multi-sensor fusion data of forestry scenes to be clustered, our proposed method can extract sematic objects with semi-supervised way from the supervoxels dataset objectively instead of empirical estimation. The details of the algorithm process are explained in the following sections.

3. Multi-Sensor Measuring System

According to the environmental characteristics of AFMs' operation, the vehicle-mounted holder is designed to carry a moving 2D laser scanner, thermal infrared camera and visual camera to build the real-time measuring and perception systems as shown in Figure 1. The camera could obtain real-time visible light information in the forest environment. This sensor has a wide-angle view field with $75° \times 75°$ and a focal length with 0.1 m to 10 m, which has produced images with the resolution of 1920×1080 pixels at 20 frames/second speed. An ARTCAM-320-THERMO (ARTRAY CO., LTD, Tokyo, Japan) is selected as the thermal infrared device. Its measurement temperature range is from −40 to 150 °C. We set image resolution as 480×640 and the speed rate as 20 frames/second to detect the forestry objects. Finally, the LMS511-20100 PRO type laser scanner produced by SICK Corporation (Waldkirch, Germany) is used as a non-contact scanner. Its wavelength is 905 nm, which is safe and reliable for the human eye. To acquire abundant tree features with adequate resolution from the laser scanning measurements taken in the forest, the scanning angular resolution is set to its minimum value 0.1667°. Then the scanning angle is set to −5° to 185° and maximum scanning distance is 50 m. The measurement points corresponding to the surrounding contour is output in hexadecimal format to form the raw point clouds via the Ethernet interface at the frequency of 100 Hz. In this study, we extended the 2D scanning model to 3D scanning model combining the pan/tilt motion of the vehicle-mounted holder with the internal motor motion of 2D laser device. Therefore, the horizontal direction parameter of 3D points depended on the setting result of the laser scanner above, and the vertical direction parameter was determined by the vehicle-mounted holder. Here, we set the scanning angular resolution as 0.1°. Similarly, the scanning angle is set to −70° to 70° (0° is parallel to the ground) and the scanning frequency is set to 10 Hz.

Through multi-sensor cooperation, information such as distance, position distribution, color and surface temperature of objects in the forest area can be directly obtained and stored through the host control software in the data acquisition and processing module. This software also was used to control the working states of all sensors, vehicle-mounted holder and the system display module. The whole system was equipped on different forest machines (including forestry firefighting vehicles, forestry cutting and harvesting equipment, forestry tending and breeding equipment as shown in Figure 2) with proof level of IP67, which can effectively prevent the entry of rain or dust and be adapted to the damage conditions in the actual forestry areas.

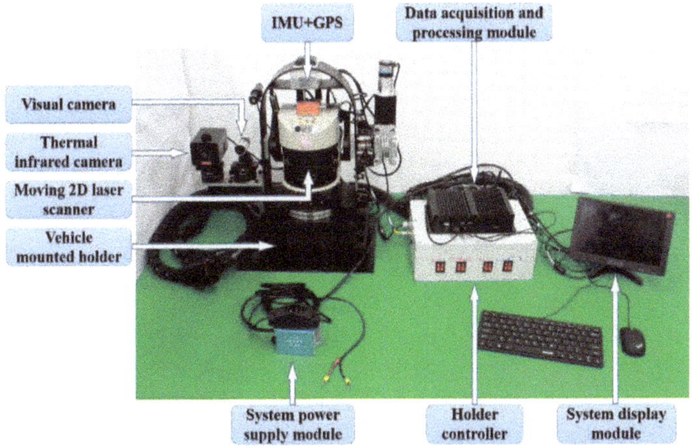

Figure 1. The AFM multi-sensor measuring system.

Figure 2. Measuring system installed at different forestry machines for operations experiment.

The multi-sensor data acquisition experiment of forest scenes was conducted in southern and northern forest farms in different seasons. The southern experiment was conducted in the artificial eucalyptus area of Qinzhou (Guangxi Province, China) where the diameter at breast height (DBH) was more than 16 cm and the height was more than 15 m. The experiment was carried out from July to November with the high temperature above 35–42 °C. In order to further enrich the measurement objects and scenes, we also selected the Jiufeng forest farm in Beijing for experimentation with various tree species including larch, fir, birch, etc. The experiments were carried in typical cold northern weather conditions with temperatures from −21 °C and 9 °C. 3D point clouds, visible images, and thermal infrared images of various objects and obstacles under different scenarios were acquired to form the multi-modal database of the forest environment. Based on the multi-sensor technology foundation, we focused on practical issues for AFM application in complex environments and carried out the sematic segmentation framework including subsequent four steps: multi-sensor joint calibration, spot-divergence supervoxelization, feature vector extraction and Gaussian density peak clustering as described in the following sections.

4. Methodology

4.1. Multi-Sensor Joint Calibration

Each sensors' data in the multi-model database has its own characteristics. Compared with the visible and thermal infrared images, the 3D point clouds have a larger field of view, but the density of point clouds is relatively sparse and noisy compared to images. To integrate the advantages of different sensors for segmentation, it is necessary to perform multi-sensor data preprocessing and fusion on the basis of analyzing the various sensors' conditions. According to the working principle of laser scanners, the measuring value of a laser beam is influenced by the reflectance of objects and the returned energy of the laser beams, which makes laser point clouds filled with discrete and systematic noise. Therefore, this paper uses a Gaussian weighted filter for removing discrete points in the laser scanner data which have the farther distance from all nearby point clouds. Then the average smoothing filter is following to suppress the systematic noise in the point clouds, making the surface of the object much smooth. Based on the noise processing, this paper conducts a joint calibration study of multiple sensors. Its purpose is to realize the original data conversion relationship between various sensor coordinate systems and achieve pixel-level data fusion. The joint calibration schematic is shown in Figure 3.

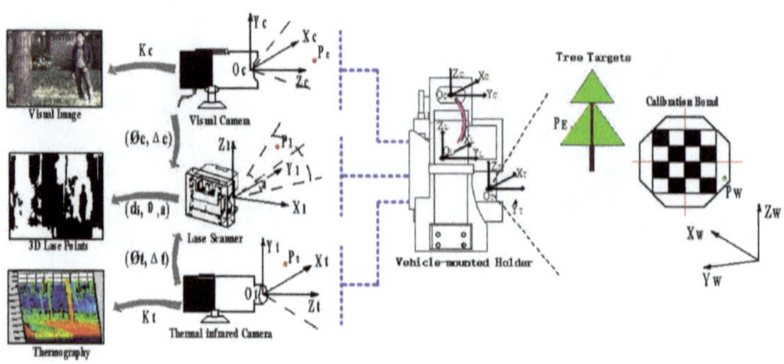

Figure 3. The joint calibration schematic of the thermal infrared camera, visible camera and 3D laser scanner for HFD.

The entire multi-sensor joint calibration process consisted of two stages: internal calibration of each sensor and external calibration between them. In the internal calibration stage, we applied the linear pinhole imaging model on the basis of the nonlinear distortion to descript the internal geometric projection process in the visible and thermal infrared cameras. We took an octagonal calibration plate with checkerboard to calculate the internal parameters including focal length, distortion etc. By using the Zhang calibration method of [33], the visible camera's coordinate $X^{cam} = [X_c, Y_c, Z_c]^T$ and the thermal infrared camera's coordinate $Y^{temp} = [X_t, Y_t, Z_t]^T$ projected by the corresponding point in the world coordinate were confirmed to construct the joint calibration model. Meanwhile, the internal coordinate relationship of 3D point clouds by combining the pan/tilt motion of the vehicle-mounted holder with the 2D laser device needs to be analyzed. Since the center of the 2D laser scanner O_l and the pan/tilt center O_L did not coincide, which leaded the laser scanning plane rotated around O_L with a certain distance d and angle δ. We defined the initial position of the 3D laser device when the plane of the O_l center is parallel to the ground. Then the internal calibration expression of 3D laser device is modeled as:

$$J^{lase} = \begin{bmatrix} X_L \\ Y_L \\ Z_L \end{bmatrix} = \begin{bmatrix} D\cos(n_1\eta + \theta_0) \bullet \sin(n_2\varepsilon + \delta)] \\ [D\sin(n_1\eta + \theta_0) - d_x] \bullet \sin(n_2\varepsilon + \delta) \\ (H + d_y) \bullet \cos(n_2\varepsilon + \delta) \end{bmatrix} \tag{1}$$

where d_x and d_y are the horizontal and vertical offset of d, D is the distance value measured by the point cloud, and $\theta = n_1\eta + \theta_0$ is the horizontal distribution angle of the laser beam formed by the horizontal angular resolution η and the horizontal laser beam sequence n_1, θ_0 represents the initial angle of the working range of the 2D laser scanner. $n_2\varepsilon$ is formed by the tilt angle resolution ε and the sequence of scan plane layers n_2. According to the parameters of the laser scanner and the vehicle-mounted holder preset in Section 3, we acquired the quantitates $\eta = 0.1667$, $\varepsilon = 0.1$, $\theta_0 = -5$, $H = 1.6$. Applied the method in [33], we used the outer edge characteristics of the calibration plate consisted of the points (D, n_1, n_2) to confirm the internal parameters d_x, d_y, δ in 3D laser scanner coordinate. With the internal parameters of three sensors determined, we established fusion coordinate system based on the 3D laser scanner, and built a joint calibration parameter solving model between multiple sensors:

$$\begin{bmatrix} L_1 \\ L_2 \\ L_3 \end{bmatrix} = \begin{bmatrix} \Phi_{lase-fuse} & 0 & 0 \\ 0 & \Phi_{cam-fuse} & 0 \\ 0 & 0 & \Phi_{temp-fuse} \end{bmatrix} \bullet \begin{bmatrix} w_1 \bullet J^{lase} \\ w_2 \bullet X^{cam} \\ w_3 \bullet Y^{temp} \end{bmatrix} + \begin{bmatrix} \Delta_{lase-fuse} \\ \Delta_{cam-fuse} \\ \Delta_{temp-fuse} \end{bmatrix} \quad (2)$$

where w_1, w_2 and w_3 respectively represent the weights of 3D point clouds, visible images and thermal infrared images in the joint calibration process under the same scene. $\Phi_{lase-fuse}$ is an 3×3 orthogonal matrix representing the rotation relationship between the laser scanning coordinate and the fusion coordinate, and $\Delta_{lase-fuse}$ is a corresponding 3×1 translation matrix. Similarly, $\Phi_{cam-fuse}$, $\Delta_{cam-fuse}$, $\Phi_{temp-fuse}$, $\Delta_{temp-fuse}$ represent the spatial rotation and translation matrix from the visible and thermal infrared camera coordinate to the fusion coordinate. After the correction of the weights of each sensor, different kinds of information in forest environment can be merged in the fusion coordinate through rotation and translation parameters of each sensor are calibrated.

In this paper, the internal calibration of each sensor and the joint external calibration process are accomplished simultaneously with calibration plate placed at different distances and locations. Then, the corresponding linear equations of multiple edges of the calibration plate are selected to model a PNP problem. In the conditional constraint of $n \geq 36$, there is a least-squares solution as the initial result for parameters of the multi-sensor joint calibration. In order to eliminate the nonlinear errors caused by human intervention, the Levenberg-Marquardt (LM) method [34] was applied to optimize the joint calibration parameters with the following expression:

$$E = \text{argmin} \sum_{i \geq 36} a_i \left\| \begin{bmatrix} L_1^i \\ L_2^i \\ L_3^i \end{bmatrix} - \begin{bmatrix} w_1 \Phi_{lase-fuse} \bullet J^{lase} + \Delta_{lase-fuse} \\ w_2 \Phi_{cam-fuse} \bullet X^{cam} + \Delta_{cam-fuse} \\ w_3 \Phi_{temp-fuse} \bullet Y^{temp} + \Delta_{temp-fuse} \end{bmatrix}^i \right\|^2 \quad (3)$$

where a_i represents the weight effect of each edge obtained by the three sensors on the of the objective function E. After the transformation relationship between different sensors are determined, the HFD dominated by 3D point clouds is obtained, in which each point has properties including distance, angle, reflectivity, as well as color and temperature.

4.2. Spot-Divergence Supervoxelization

As traditional segmentation method based on supervoxels are unsuitable for the complex application of AFMs in face with HFD, of which size is changed due to the occurrence of spot-divergence in complex forestry scenes. Therefore, this paper started with the inherent working principle of laser scanner and proposed the spot-divergence supervoxel representing size-changed character of HFD. In this work, we used 26-neighborhood region constraint to construct mutual topological relationship from the high-dimensional fusion data to supervoxels. Assume that there are n HFD in a forestry scene, which form the origin dataset $S : \{P_1, P_2, \cdots, P_n\}$. Then the dataset has been divided into K supervoxels, which constitutes the sets $V : \{V_1, \ldots, V_K\}$. The detail of supervoxelization process is divided into the four steps: supervoxel space division, spot-divergence process of HFD, center selection and adjacent partition, and extracting feature vector of supervoxel as following.

4.2.1. Supervoxel Space Division

In the fusion coordinate system, the fusion dataset $S : \{P_1, P_2, \cdots, P_n\}$ with the largest spatial coordinate value in $P_{\max}(x_{\max}, y_{\max}, z_{\max})$ and the smallest $P_{\min}(x_{\min}, y_{\min}, z_{\min})$ value are selected as the two vertices of the entire cuboid space. The length, height and height of the cuboid space are $L_x = |x_{\max} - x_{\min}|$, $L_y = |y_{\max} - y_{\min}|$, and $L_z = |z_{\max} - z_{\min}|$, respectively. According to actual requirements, we divide all HFD into presupposition spaces of supervoxel with the even length R_{super}. Set the supervoxel spaces along the x, y, z direction divided by the number of n_x, n_y, n_z, we can initially determine the number of preset supervoxels:

$$K = n_x \times n_y \times n_z = \frac{Lx}{R_{super}} \times \frac{Ly}{R_{super}} \times \frac{Lz}{R_{super}} \tag{4}$$

The minimum size of supervoxel space must satisfy the constraint conditions:

$$R_{super} > N \bullet R_{voxel}(\max) \tag{5}$$

where N is the constant coefficient, which is set to 4 in this paper. $R_{voxel}(\max)$ denotes the largest edge length of P_n in the fusion dataset, which is proposed on account of the laser beam divergence principle.

4.2.2. Spot-Divergence Process of HFD

The basic starting point is: each laser beam has a divergence angle. As surface reflectivity, texture, roughness, etc. of the object change, the spot-divergence phenomenon of multi-sensor fusion data occurs when the laser beam is reflected back over a long distance. As a result, the spot area of the measured point on the surface of the object is much larger than ever and continuously changes with increasing distance from the laser scanner as shown in Figure 4a. Therefore, describing each HFD with a fixed area size does not meet the actual situation of supervoxels, which leads to inaccurate results of segmentation. Moreover, with increasing distance from the object, the distance between the individual measured points also increases. The distance between the measured points is also dependent on the angular resolution selected. With a coarser resolution (e.g., 0.1667°), the distance is larger, with a finer resolution (e.g., 0.1°) the distance is smaller. To reliably detect an object, the valid area of laser beam with concentrated energy must be fully incident on it once. If the measured laser beam is only partially incident, less energy could be reflected by the object and be disturbed by adjacent beams as shown in see Figure 4b. The size of valid area is proportional to the degree of spot divergence, which represents a lower energy remission than the measured laser beam actually [35]. Therefore, the valid area is applied to describe the size change of each HFD due to spot divergence of 3D point clouds. Based on this idea, a novel supervoxel process was proposed to determine the supervoxel center and adjacent areas.

The distance-dependent spacing between the measured points is the tangent of the angular resolution × distance. The initial size D_1 of the laser beam launched from the emitter with the inherent divergence angle θ_1 to the surface of the object. After the transmission distance L, the diameter of spot area representing the actual size of each HFD is obtained by the principle of trigonometry as:

$$d = D_1 + \cot(\frac{\theta_1}{2})L \tag{6}$$

As shown in Figure 4b, the diameter of each spot area will increase with the distance increases, which leads to overlap of adjacent spot areas. Assume that the center distance of adjacent spot areas is H, which is calculated as:

$$H = 2 \times L \times \tan(\frac{\varepsilon}{2}) \tag{7}$$

where ε represents the smallest angle of the adjacent spot areas, which is equal to the pan/tilt angular resolution captured by 3D laser scanner. According to the laser energy distribution, the overlap causes the measurement interference of adjacent laser beam. Therefore, we choose the center area (blue area) as the valid area where there is not mutual overlap and interference of the adjacent beam. The diameter of valid area is defined as:

$$B = H - d/2 = 2 \times L \times \tan(\frac{\varepsilon}{2}) - (D_1 + \cot(\frac{\theta_1}{2})L)/2 \quad (8)$$

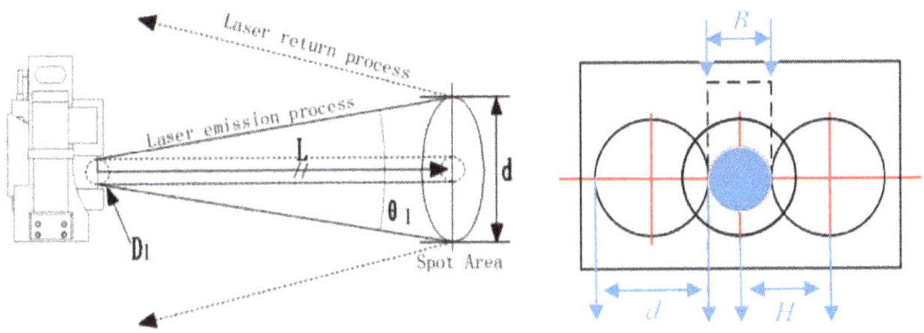

(**a**) Spot divergence principle (**b**) Valid area of laser beam

Figure 4. (**a**) Schematic of spot divergence principle with increasing distance. (**b**) The valid area of laser beam is layout of the distance between measured points at different angular resolutions.

According to space division in Section 4.2.1, the cube is the basic computation unit of supervoxel. Therefore, extend the valid area of the laser beam to the 3D space expression, which conform to the realistic geometric distribution of HFD. Then the 2D valid area of each laser beam becomes a 3D sphere with a radius $R_E = B/2$. Then, we selected the inscribed cube inside the sphere as the basic element to construct the supervoxel (as shown in Figure 5).

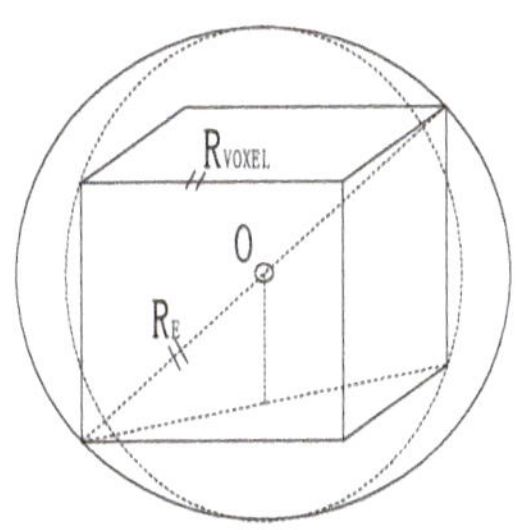

Figure 5. Inscribed cube inside the spheres represents the 3D extension of the valid area of HFD.

The edge length R_{voxel} of every cube obtained is:

$$R_{voxel} = \frac{2}{\sqrt{3}} R_E = \frac{2 \times L \times \tan(\frac{\varepsilon}{2}) - (D_1 + \cot(\frac{\theta_1}{2})L)/2}{\sqrt{3}} \quad (9)$$

Calculate the lengths of all HFD in $S : \{P_1, P_2, \cdots, P_n\}$ and select the maximum value as $R_{voxel}(\max)$. Taking Equation (9) into Equation (5), the size of each HFD and the total number of supervoxels were preset with the practical physical meaning with the spot divergence constraint, which improve the effect of supervoxelization.

4.2.3. Center Selection and Adjacent Partition

Based on spot-divergence process, the HFD near the regional center of the supervoxel space is generally selected as the initial seeds. However, in order to avoid the unreasonable situation that the selected data is a noise point or an outlier on the edge position of objects, it is necessary to calculate the size gradient function between the initial seeds and the neighborhood HFD within the search radius $R_{search} = \frac{R_{super}}{2}$ as follows:

$$G(i) = \sum_{k=1}^{N_{seed}} \sum_{j \in N_{adj}} \left(\frac{\|R^i_{seed} - R^j_{voxel}\|}{N_{adj}} + \|R^i_{seed} - R^k_{seed}\| \right) \tag{10}$$

where R^i_{seed} represents the size of the i-th initial seed, R^j_{voxel} represents the size value of j-th neighboring HFD around this seed. N_{adj} is the number of HFD available in 26-field. N_{seed} represents the number of initial seeds in this supervoxel space, and R^k_{seed} represents the k-th initial seed within the search range. When $G(i)$ is less than the preset threshold, it indicates that the i-th seed meet the constraint requirements and is selected as the central seed of this supervoxel. If the result does not satisfy the constraint, it means that the i-th initial seed is invalid. Then the gradient values of different size seeds need to be calculated sequentially until the smallest gradient is selected as the supervoxel center. Subsequently, calculate the spatial distances d^{ij}_{adj} between other HFD and different supervoxel centers for adjacent partition:

$$d^{ij}_{adj} = \sqrt{(x_i - x_j)^2 + (y_i - y_j)^2 + (z_i - z_j)^2} + \|G(i) - G(j)\| \tag{11}$$

By comparing the distance thresholds ε_1, all HFD are allocated to the nearest supervoxel. In order to facilitate the display, this paper uses a schematic diagram to show the partitioning process of two adjacent supervoxels as shown in Figure 6:

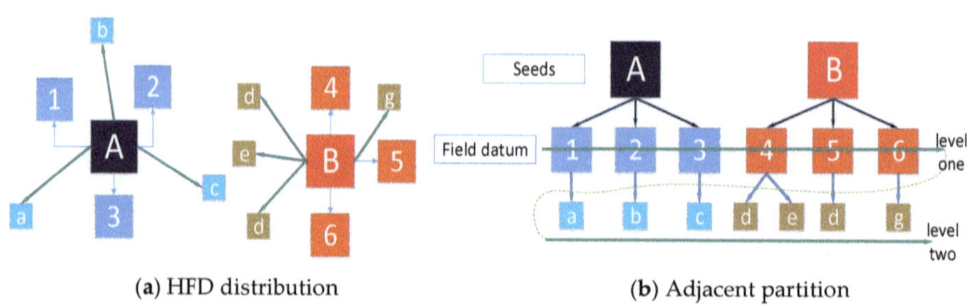

Figure 6. (**a**) Schematic of HFD distribution in two neighbouring supervoxels; (**b**) Schematic of adjacent partition and search process for supervoxels (**b**).

4.2.4. Supervoxel Feature Vector

With the seed of each supervoxel and neighbouring HFD divided, the large-volume and dense HFD can be divided into small-volume supervoxels distributed sparsely. Each supervoxel can be regarded as a cluster collection of similarity HFD with local characterization, including spatial relations, color, temperature, reflectivity, normal vector and size similarity. Those features are extracted to construct the feature vectors of K supervoxels in high dimensional space as follows:

$$F = \left[\begin{pmatrix} x^{1\sim K} \\ y^{1\sim K} \\ z^{1\sim K} \end{pmatrix}^T, \begin{pmatrix} L^{1\sim K} \\ a^{1\sim K} \\ b^{1\sim K} \end{pmatrix}^T, V_T^{1\sim K}, V_{ref}^{1\sim K}, V_S^{1\sim K}, Var_S^{1\sim K}, \begin{pmatrix} N_x^{1\sim K} \\ N_y^{1\sim K} \\ N_z^{1\sim K} \end{pmatrix}^T, T_h^{1\sim K} \right] \tag{12}$$

The detail properties of each feature mainly include:

1) Spatial coordinates of supervoxel center: $V^k_{xyz} = [x^k, y^k, z^k]$
2) CIELAB color average of n HFD in the supervoxel: $V_{Lab} = [\frac{\sum_{i=1}^{n} L_i}{n}, \frac{\sum_{i=1}^{n} a_i}{n}, \frac{\sum_{i=1}^{n} b_i}{n}]$
3) Temperature average of the supervoxel: $V_T = \sum_{i=1}^{n} T_i/n$
4) Reflectance average of the supervoxel: $V_{ref} = \sum_{i=1}^{n} r_i/n$
5) Edge length R_{voxel} mean of n HFD in the supervoxel: $V_S = \sum_{i=1}^{n} R^i_{voxel}/n$
6) Absolute range between maximum and minimum of R_{voxel}: $Var_S = |R_{voxel}(\max) - R_{voxel}(\min)|$
7) Surface normal vector of supervoxel: $\mathbf{N} = N_x v_0 + N_y v_1 + N_z v_2$ with $N_x^2 + N_y^2 + N_z^2 = 1$

In this paper, principal component analysis (PCA) is used to calculate the surface normal vector of each supervoxel. The basic principle is calculating the surface normal vector of the approximate plane by minimizing the distance from the surrounding data to the center of supervoxel:

$$d^2 = \frac{\sum_{i=1}^{n} \left\| (p_n - \overline{p})^T \mathbf{N} \right\|^2}{\|\mathbf{N}\|^2} = \frac{\mathbf{N}^T \bullet \sum_{i=1}^{n} \left| (p_i - \overline{p})^T \cdot (p_i - \overline{p}) \right| \bullet \mathbf{N}}{n \bullet \|\mathbf{N}\|^2} \tag{13}$$

where $\overline{p}$ is the local center of supervoxel, and the approximate normal is associated with the smallest eigenvalue (v_0, v_1, v_2) of the symmetric positive semi-definite matrix. Searching n HFD to determine a local surface normal vector of each supervoxel.

8) Comprehensive dissimilarity of vectors: $T_h = \eta_1 \cdot \arccos\frac{|n_0 \bullet n_k|}{|n_0||n_k|} + \eta_2 \|n_0 - n_k\|$

where $\theta_k = \arccos\frac{|n_0 \bullet n_k|}{|n_0||n_k|}$ indicates the angle between the normal vector of k-th supervoxel and the Z-axis of the fusion coordinate system. $\|n_0 - n_k\|$ indicates the numerical vector deviation of k-th supervoxel. η_1 and η_2 are weights applied to balance the relationship between angle and deviation of normal vector.

4.3. Gaussian Density Peak Clustering

Compared to traditional clustering methods requiring the artificially preset of clustering central number or convergence thresholds, the density peak clustering (DPC) accomplishes semantic object segmentation adapting to arbitrary shapes and feature types. However, the segment result of DPC excessively depended on the suitable threshold including truncation distance, local density and the minimum higher-density distance, which were all estimated on the basis of empirical experience. This was difficult to segment objects from supervoxels automatically in forestry scenes. Thus, this paper used the normalized feature to construct Gaussian density peak clustering model. With semi-supervised way for extracting threshold, the proposed method can cluster different objects in the forestry environment, which improves accuracy and timeliness of segmentation.

4.3.1. Feature Normalization

As the feature units and quantity levels of supervoxels are very different, each feature channel needs to be normalized by the central regularization process. The Euclidean distance of the supervoxels F_k and F_q in each feature space is calculated as:

$$
\begin{aligned}
d_1 &= \sqrt{(x^k - x^q)^2 + (y^k - y^q)^2 + (z^k - z^q)^2} \\
d_2 &= \sqrt{(L^k - L^q)^2 + (a^k - a^q)^2 + (b^k - b^q)^2} \\
d_3 &= \sqrt{(V_{ref}^k - V_{ref}^q)^2 + (V_T^k - V_T^q)^2} \\
d_4 &= \sqrt{(V_S^k - V_S^q)^2 + (Var_S^k - Var_S^q)^2} \\
d_5 &= \sqrt{(N_x^k - N_x^q)^2 + (N_y^k - N_y^q)^2 + (N_z^k - N_z^q)^2} \\
d_6 &= \|T_h^k - T_h^q\|_2
\end{aligned}
\tag{14}
$$

After the weights of the influence of spatial distribution, color difference, temperature and reflection difference, edge length difference, normal vector difference, and synthetic similarity are assigned, all different features are set in the range 0–1, then the high-dimensional distance D_{kq} is obtained:

$$
D_{kq} = \left\| \sum_{j=1}^{6} \tau_j {d_j}^2 \right\| \tag{15}
$$

4.3.2. Gaussian Local Density Distribution

According to the density peak clustering, we define the local density of the k-th supervoxel as ρ_k, which is obtained by the interaction between the high-dimensional distance space D_{kq} and the truncation distance D_c. The following relationship exists:

$$
\rho_k = H(D_{kq}, D_c) \tag{16}
$$

This paper assumes that the local density of all supervoxels conforms to a specific Gaussian distribution:

$$
\rho_k = \frac{1}{(K-1)D_c} \sum_{q=1}^{K-1} \frac{1}{\sqrt{2\pi}} \exp\left\{ -\frac{{D_{kq}}^T \bullet D_{kq}}{2{D_c}^2} \right\} \tag{17}
$$

When the D_c is small, the local density distribution of F_k shows the prominent form of the middle peak. Only supervoxels that are especially close to F_k can play a role, which limits the local density function performance range to a small area. With the increase of D_c, the distribution of local density function also tends to be flattened, making it possible to influence the local densities of different supervoxels. However, the smooth also inhibits the fact that the contribution degree of ρ_k on divergence of supervoxels with different feature association. Therefore, the selection of D_c affect the segmentation results and needed to be preset in fixed value [30]. In this paper, a proportional coefficient $t = D_{kq}/D_c$ is chosen to select the D_c value, which represents the proportion of the neighbors number of each supervoxel in the entire HFD dataset. While taking into account the dimension coefficient $w = 1$, the non-parametric the rule of thumb method was used to determine the D_c of Gaussian local density function. In order to meet 98% confidence, the ratio 2.58% is selected as the optimal choice according to the actual requirement of AFMs in forestry environmental. The probability strategy of determining the truncation distance through the proportional coefficient reduces the dependence of the parameter on the specific problem to some extent, and the choice of this ratio is simple and applicable to other problems.

4.3.3. Clustering Supervoxels as Objects

Assume that there are $K - k$ supervoxels with higher local density than the k-th supervoxel. Apply the expression (15) to calculate the distance between these supervoxels and F_k, and form the distance vector:

$$
W = [D_{a-k}, \cdots, D_{K-k}] \tag{18}
$$

Subsequently, the minimum value $W_{\min}$ is selected to calculate the minimum higher-density distance of F_k:

$$\delta_k = \begin{cases} W_{\min}, k < K \\ \max_{\rho_k}(D_{kq}), k = K \end{cases} \tag{19}$$

If $k = K$, the super voxel F_k is the maximum in the local density ranking, its minimum higher-density distance needs to be redefined. Calculate the higher-density distance from this supervoxel to other supervoxels and select the maximum value as δ_k. Then each supervoxel can be expressed as $F_k(\rho_k, \delta_k)$ with two novel parameters. Draw the distribution schematic of different supervoxels with δ_k as horizontal axis and ρ_k as vertical axis:

As shown in Figure 7 above, there are 22 supervoxels with two actual classifications A and B projection to 2D feature space. In the distribution schematic, the partitioning coefficient $\rho_\Delta, \delta_\Delta$ can be set according to the actual situation to determine the corresponding clustering center. When the $\rho_k > \rho_\Delta, \delta_k > \delta_\Delta$ constraints are satisfied, the supervoxel can be considered as a clustering center. If there is a case where the minimum distance is large but its density value is less than the threshold, it is defined as an outlier noise point and it needs to be eliminated. By selecting the density threshold and the higher-density distance threshold dynamically, the cluster centers are determined without the number preset of clusters in advance. However, the preferable thresholds $\rho_\Delta, \delta_\Delta$ need to be selected by human observation and intervention. Thus, a novel comprehensive evaluation expression $\gamma_k = \delta_k \bullet \rho_k$ is proposed to select the cluster centers in semi-supervised way as follows:

$$F_c = F_k(\gamma_k > \gamma_\Delta) \tag{20}$$

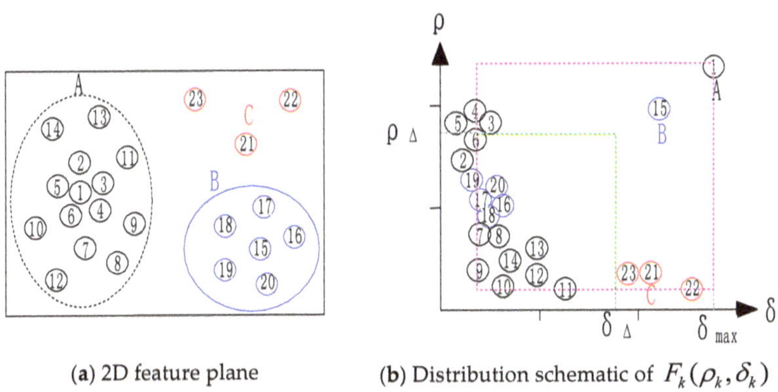

Figure 7. (**a**) Supervoxels projection to 2D feature space data with two clusters; (**b**) The corresponding distribution schematic with $F_k(\rho_k, \delta_k)$.

Through this optimization process, the γ_k of all supervoxels are calculated and ranked in descending order of density values:

$$\gamma_K > \cdots > \gamma_{K-m} > \gamma_\Delta > \cdots > \gamma_1 \tag{21}$$

where γ_Δ is judgment threshold, which equivalent to finding the number of supervoxels significant improved than other supervoxels. Through this semi-supervised method without manually observe, the number of clusters can be achieved automatically. When m supervoxels are identified as cluster centers, supervoxels close to each center are selected in density ordering and divided into different areas of several objects $C : \{C_1, \ldots, C_m\}$. For any supervoxels F_w of a non-clustered center, a cluster center with a larger density is sought in the local density arrangement. The higher-dimensional distance between F_w and these center $\{peak_1, \ldots, peak_m\}$ is calculated, and the cluster center $peak_w$

with the smallest distance is selected as its cluster center, which defines F_w as the corresponding neighborhood. In order to determine the classification of different super voxels:

$$\{peak_1, \ldots, peak_m | \rho_{center_m} > \rho_{F_w}\} \rightarrow \mathrm{D}(F_w, \{peak_1, \ldots, peak_m\}) = \min \tag{22}$$

$$C_w : near(F_1, \ldots, F_w) \rightarrow peak_w \tag{23}$$

Compared with the way that all supervoxels need to traverse the calculation relationship with all center, this method only calculates the relationship between centers and adjacent higher-density super voxels, which can reduce calculated quantity effectively and improve the speed of neighborhood division. Finally, the overall analysis flow of our segmentation framework is shown in Figure 8, and is mainly divided into three consecutive phases as following:

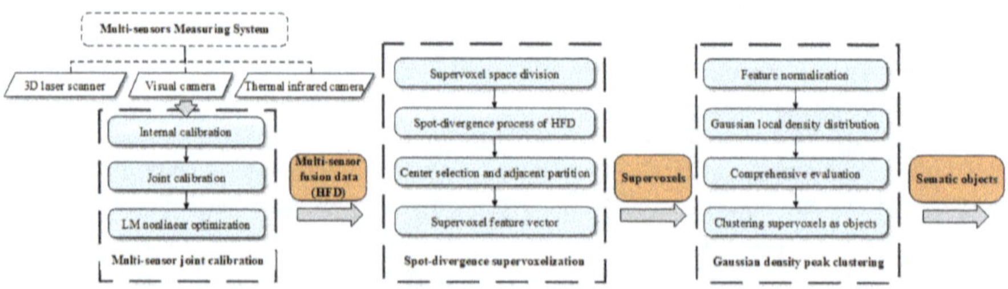

Figure 8. The analysis flow of semi-supervised segmentation framework for AFMs.

5. Results and Analysis

5.1. Multi-Sensor Fusion Evaluation

The multi-sensor joint calibration process presented in the previous section was programmed with the octagonal calibration plate. The calibration experiment was accomplished in indoor scene as shown in Figure 9a. By posing the calibration plate in different positions and distances, the internal and mutual relationship of multi-sensor coordinates were confirmed to fuse visible images and thermal infrared images with 3D laser point clouds. On this basis of joint calibration process, the coordinate relationship of the three measuring devices was kept constant, then the multi-sensor measuring system was directly mounted in various AFMs to capture high-quality HFD in urban and forest scenes without repeating calibration. As the result, objects with fused information including 3D space, color, temperature etc. could be displayed on a human-computer interface of the measuring system for AFMs operation.

In order to illustrate the performance of HFD, this study selected partial data of the urban environments in Figure 9b to define Scene A, which was captured in the Jiufeng forest farm during the cold winter season. As a comparison, Scene B was extracted from the forest environments in the artificial eucalyptus farm of Qinzhou during the hot summer and autumn. Both Scene A and Scene B contain six objects such as tree, shrub, pedestrian, stone, building and ground, which were more complicated. In general, Scene A was large-scale displayed in the range of 0.7 m to 40 m with horizontal angle ranging from -5° to 185° and vertical angle ranging from -70° to 70°. Since the measured object was relatively obvious mutual occlusion and measuring temperature is relatively low, the multi-sensor data were well fused in relatively tight form. It's proved that the measuring system constructed in this paper could cope well with the perception task of urban environment. In order to show the fusing performance of the proposed calibration work in more complex environments, fractional HDF of Scene B were selected to display in Figure 9c at extreme distances ranging 45 m to 50 m (the preset maximum distance of laser scanner). As a result, the HFD were relatively sparsely arranged subject to limitations of laser scanner with spot-divergence. And the visual information was ambiguous during to collective effects of high temperature, low reflectivity of objects and background interference.

However, such fusion effect has met the application requirements of AFMs in the forest scenes, and it also proved the necessity of subsequent supervoxels based on spot divergence in this paper.

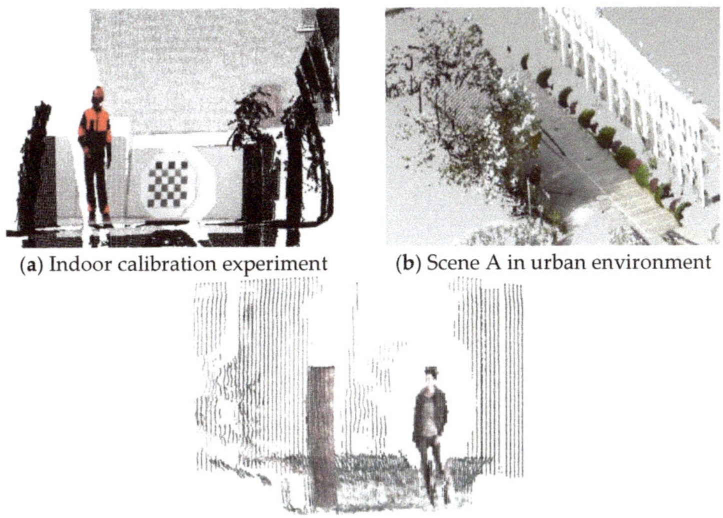

(**a**) Indoor calibration experiment (**b**) Scene A in urban environment

(**c**) Scene B in forest environment

Figure 9. (**a**) Joint calibration experiment was accomplished in indoor scene with well fusion performance; (**b**) HFD of Scene A were obtained in urban environment without extra calibration process; (**c**) HFD were obtained in Scene B of forest environment at extreme distances.

To objectively evaluate the fusion performance of calibration process optimized by the LM nonlinear method, some analysis results are presented in Table 1. Compared with the calibration method in our previous work [33], the edges relationship of the octagonal plate based on the LM method reduced the calculated value of the average calibration offset error to 2.764 cm and repressed the average angle error to 0.553° effectively with better calibration accuracy, which made HFD suitable for segment application in forest environment. Moreover, the root mean square error (RMSE) in this paper was closed to 5.126, which showed the obvious improvement of nonlinear optimization. Then the corresponding Standard Deviation (STD) was 13.032, which mean that the error distribution was not very discrete and the calibration process was much stable and robust for the following supervoxelization.

Table 1. Analysis result of multi-sensor joint calibration.

Parameters	Calibration Work in [33]	Proposed Calibration Work
Average calibration offset error	5.819 (cm)	2.764 (cm)
Average angular error	1.164°	0.553°
RMSE	8.232	5.126
STD	19.823	13.032

5.2. *Supervoxelization Evaluation*

We conducted experiments on Scene A and Scene B to evaluate the quality of the supervoxels generated by the proposed spot-divergence algorithms. There were 318,331 original HFD in Scene A and 339,547 original HFD in Scene B, which is a relatively large computation for the workload of point-level segmentation. Based on the obtained HFD, the proposed supervoxelization method based on the laser divergence scale change were applied to determine the supervoxel center and search relationship of adjacent areas. With the large-volume and dense HFD divided into small-volume supervoxels sparsely, each supervoxel contains HFD with similar properties of local features. In order to execute and run the supervoxelization method of this article, a workstation machine with 32 GB memory, 500 G SSD and the Intel Core 7 core processor was chosen for model calculation. Then the software platform Point Cloud Library (PCL) for supervoxelization [36], which was an open source programming library run on Ubuntu system.

After the changes from HFD to supervoxels, non-ground 135,369 HFD were converted to 14,240 supervoxels, and other 155,040 HFD on the ground were converted to 7602 supervoxels. The remaining 27,922 HFD were discriminated as noise and deleted in the Scene A. Similarly, the round of Scene B originally had 200,985 HFD, and after the change, 8357 supervoxels were obtained, while the non-ground was converted from 127,546 HFD to 10,714 supervoxels, and the remaining 11,016 HFD were identified as noise. Obviously, supervoxelization could reduce the amount of computation and improve the efficiency of segmentation.

To evaluate the performance of our algorithm, it is reasonable to compare the proposed method with algorithms that were also designed to generate supervoxels. We compared our method with three of these kinds of algorithms, including VCCS [21], (vSLIC), SEED-3D [22], and ATS [24], whose source codes were publicly available at their respective research websites. We used the default parameters provided by their authors for all the compared methods. Comparisons of some early methods that segment fusion data without considering the property of spot-divergence could be found in Figure 10. As shown, a further analysis on the ability adhere to object boundaries was developed. Under-segmentation error was chosen as the standard measure for boundary adherence (namely, the error between the given region from the ground truth segmentation and the set of supervoxels required to cover it in minimum number). Then the relationship between the under-segmentation error and the number of supervoxels was shown as following:

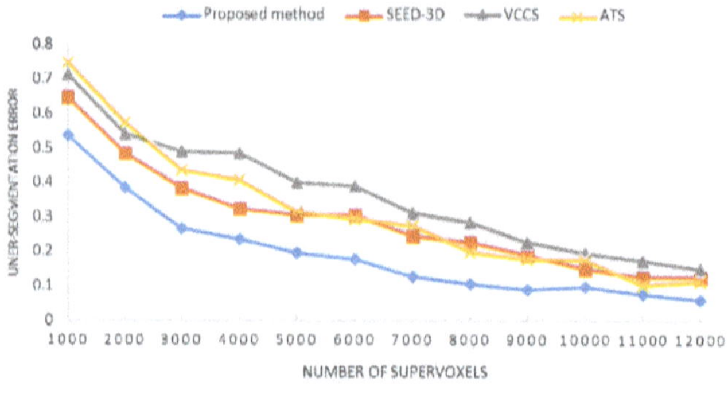

Figure 10. Relationship between under-segmentation error and the number of supervoxels for VCCS (gray), SEED-3D (orange), ATS (yellow) and the proposed method (blue).

Since the number of super voxels increased, the over-segmentation errors of the four methods show a decreasing trend. As plotted above, the blue curve repressing the proposed supervoxelization outperforms the other methods in under-segmentation error, showing the lowest undersegmentation error for most of the useful operating regime. It also means the supervoxel partitioning based on the spot divergence constraints is a better approach, which tightly fitted the ground truth result of object edge in complex scenes.

Further, supervoxels were often proposed to replace the point-wise operation to help speed up segmentation algorithms, which mean that it is important to generate lots of supervoxels efficiently in the first place. Thus, we compared the operational time required for the various methods to segment HFD with the same hardware platform in Figure 11.

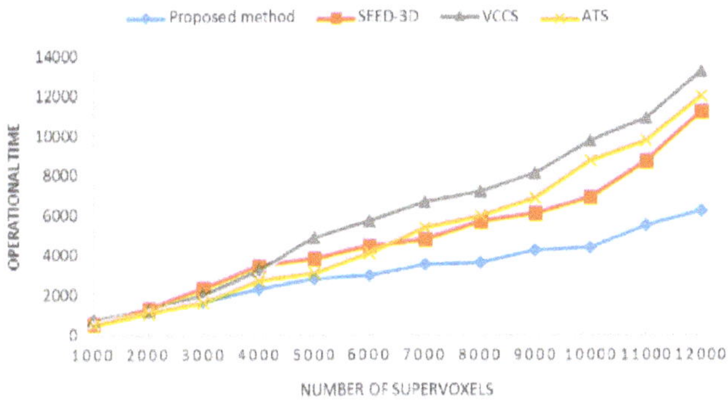

Figure 11. Relationship between the operation time and the supervoxel number.

With increasing size on number of supervoxels, the operation time curves of all methods increased with the corresponding O(N) complexity. Comparing with curves of VCCS (gray), SEED-3D (orange) and ATS (yellow), the spot-divergence-based algorithm was the fastest supervoxel method, and its advantage increased with the size of supervoxel magnitude. While the operation time of other methods were greatly affected by the increasing trend of supervoxels' number, especially in the range of 9000 to 12,000. It showed a significant gap in processing speed and memory efficient in order to handle large multi-sensor fusion data, which can not only reduce the redundancy in subsequent data processing, but also facilitate the feature extraction of complex environment.

5.3. Semantic Segmentation Evaluation

This section tested the semantic segmentation based on density peaks clustering for Scene A and Scene B. as shown in Figure 12. With the supervoxel features. Scene A was segmented semantically as 13 categories, including four trees, three shrubs, one building, two pedestrians, one stone, and two grounds respectively. Scene A was divided into 11 objects, including seven trees, one shrub, one building, one pedestrian, and one ground. Each target is randomly assigned a color to distinguish. Obviously, this algorithm can effectively segment supervoxels as different types of independent objects with small error, as shown in Figure 13.

(**a**) Scene A (**b**) Scene B

Figure 12. Semantic segmentation of independent objects in Scene A (**a**) and Scene B (**b**).

(a) Pedestrian (b) Stone (c) Tree (d) Shrub (e) Building

Figure 13. Five types of objects segmented from Scene A and B.

In order to further evaluate the segmentation performance, the artificial manual segmentation was used as the standard segmentation results, and compared with the segmentation results obtained by the proposed segmentation method. The association matrix of segmentation results for Scene A and Scene B were shown in Tables 2 and 3. Each element in the table represents the corresponding supervoxels of the actual output label. If a supervoxel is segmented to the correct target, it is called true positive *TP*; if a supervoxel is not segmented but assigned to a nearby target, it is called false negative *FN*; if a target does not exist but a supervoxel is wrongly segmented to it, it is called false positive *FP*. Calculate the precision rate *Per*, recall rate *Rec*, and F1-score value F of each scene separately to achieve the evaluation of the segmentation effect:

$$Per = \frac{TP}{TP + FN}, \ Rec = \frac{TP}{TP + FP}, \ F = \frac{2 \times Per \times Rec}{Per + Rec} \tag{24}$$

where *Per* measures the probability between the number of supervoxels correctly segmented for a certain class and the true total number belonging to that class in the artificial standard results. And *Rec* is the ratio between the number of supervoxels correctly segmented and the total number of supervoxels assigned to the class in this segmentation methods, which describes the probability of objects that can be extracted from supervoxel feature by our method. The *F* value indicates the harmonic mean evaluation of precision and recall.

Table 2. Evaluation results of six objects segmentation in Scene A.

	Ground	Pedestrian	Tree	Shrub	Building	Stone	Average
Ground	14048	12	17	63	76	24	
Pedestrian	6	410	5	1	0	11	
Tree	13	2	2208	65	3	5	
Shrub	19	4	35	1526	13	6	
Building	26	0	9	12	2672	15	
Stone	15	16	0	4	9	492	
Precision	0.987	0.947	0.962	0.952	0.977	0.918	0.957
Recall	0.994	0.923	0.971	0.913	0.964	0.890	0.943
F value	0.990	0.935	0.966	0.932	0.970	0.904	0.950

Experiments showed that the proposed algorithm achieved very competitive results in the individual objects segmentation in complicated scenes. From the result in Table 2, it was concluded that the accuracy and recall of stones were slightly poorer because supervoxels of stones were partly divided into ground and pedestrian in many cases. However, the comprehensive segmentation of all objects maintained a high value, which validated the performance and stability of the proposed segmentation method. Table 3 showed that Scene B had one less category of stone than Scene A, which improved the evaluation results of segmentation with small amplitude. Moreover, the excellent performance proved that the algorithm retained the characteristics of the original HFD, which were adaptable to both urban and forest environments. Nevertheless, the comparison result also indicated that the semantic categories and individual number of objects affected directly segmentation capability of this proposed method, which were associated with environmental distribution and various attributes of raw datum obtained by the multi-sensor measuring system. The detail results of the proposed algorithm were shown as following.

Table 3. Evaluation results of five objects segmentation in Scene B.

	Ground	Pedestrian	Tree	Shrub	Building	Average
Ground	8145	24	26	149	13	
Pedestrian	32	710	4	19	1	
Tree	66	7	5341	65	6	
Shrub	19	4	35	3476	1	
Building	44	11	29	12	832	
Precision	0.975	0.927	0.974	0.983	0.897	0.951
Recall	0.981	0.939	0.983	0.934	0.975	0.962
F value	0.978	0.933	0.978	0.958	0.934	0.956

In order to further verify the applicability and robustness of the proposed pipeline, we conducted a comparative trial on the Scene C extracted from forestry environment. The scene was relatively complex manually judged as 43 sematic objects. The proposed method automatically divided Scene C into 44 objects including four big stones, six pedestrians, one ground, 27 trees, two buildings and four shrubs, which was close to the result of manual segmentation with one more tree. Looking in the scene, we found that too many objects were obtained at once, resulting in mutual occlusion and data interference, which was the source of this problem. Wherever, this method still maintained better segmentation performance than comparison methods as shown in Figure 14:

(**a**) VCCS

(**b**) ATS

(**c**) SEED-3D

(**d**) Proposed method

Figure 14. Comparison of segmentation results with the method of VCCS, ATS, SEED-3D in K-means clustering, and the proposed method in Gaussian density peak clustering.

The K-means clustering was combined with supervoxel for sematic segmentation in VCCS, SEED-and ATS to construct the comparison methods respectively. With the preset parameters of 43 clusting centers and 2 m search radius, we could see the segmentation performance among different approaches displayed intuitively. As can be seen, the results of VCCS was presented as 39 final spectral classes different objects, resulting in severe over-segmentation and poor performance even with supervoxels. In Figure 14b, using supervoxel as the neighborhood allowed one to better discern differences inside and between tree areas. Results showed that, despite the ATS method taking only 3D point clouds for supervoxel generation, the subtle differences associated with the main trees and the other objects were properly represented. However, due to the lack of other information in HFD, the method easily divided the discrete data into more object classes or noise clusters in large-scope. With fusing various information of HDF, SEED-3D achieved a better performance of sematic segmentation presented in the complex environment. Although SEED-3D also correctly detected the various objects closing to manual results, this algorithm was easy to assign the same object with different labels in Figure 14c. The main reason was that the features of SEED-3D cannot represent the size-changed character of HFD due to the spot divergence and this process required the presetting parameters of K-mean under human intervention consequently, which needed to be tuned by experiments for achieving the optimal results in different scenes. As a consequence, for both supervoxelization and sematic segmentation, the proposed method better reflected the distribution and features of objects in the HFD, showing a notable variety in a semi-supervised way. In order to compare the overall performance among different approaches in a statistically-rigorous fashion, the statistical significance of differences in terms of accuracy and operation time were evaluated in Table 4 as following:

Table 4. The segmentation evaluation of four segmentation algorithms.

Segmentation Algorithm	Integrated Clusters	Discrete Clusters	F Value	Time (Approximate)	Effective HDF
VCCS [21] + K-mean	39	317	0.893	92 min	803,252
SEED-3D [22] + K-mean	45	382	0.938	51 min	756,328
ATS [24] + K-mean	48	426	0.920	65 min	983,174
Proposed method	44	125	0.942	34 min	1,139,829

Note: Effective HDF means Number of data for all integrated clusters (larger the value, less data is lost).

As the results summarized in Table 4 show, the proposed segmentation framework accounted for an accuracy improvement of the overall sematic segmentation performance in many forest stands. Obviously, the integrated clusters and discrete clusters showed the segmentation work can be performed reasonably according to the environmental characteristic in the semi-supervised case. Another strength of the proposed techniques was that the accuracy and quantity utilization of HFD was significantly improved (see *F* value). This is mainly due to the fact that the spot-divergence supervoxel must be more precise than the stationary supervoxel or point clouds. The performance of supervoxel-based sematic segment depends on the multi-sensor data density and the forest type. Thought the proposed approach achieve an improvement in operation time, the major limitation of our work is that the whole time cannot meet real-time applications of AFMs. Thus, however, the achieved improvement of the overall time would be the focus of subsequent work, which is necessary to the real-time perception of AFMs in forest environment.

6. Conclusions

In this paper, we have focused on a semi-supervised segmentation framework based on a spot-divergence supervoxelization of multi-sensor fusion data acquired by AFMs in complex environments. On the basis of multi-sensor measuring system, we have presented a novel three-step segmentation framework representing a semi-supervised processing workflow: Firstly, the relationship of multi-sensor coordinates was joint calibrated to form higher-dimensional fusion data. Secord is given by a spot-divergence supervoxelization instead of producing immutable supervoxels. The novel supervoxel took the size change of each HFD into account to produce feature vectors covering the valid information at a time. Finally, the Gaussian density peak clustering was proposed to segment supervoxels into sematic objects in the semi-supervised way, which non-required the artificially preset of clustering central number or convergence thresholds. Experiments demonstrated that the proposed framework performed well in terms of segmentation accuracy and operation time, which was much appropriate to applications of AFMs. For future research, we would focus on real-time improvement in sematic segmentation of objects. We would also like to extend the method to a more complex scene such as the food security of the grain & oil supply chain.

Author Contributions: J.-l.K. and X.-b.J. conceived and designed the whole structure of the paper under the supervision of X.-y.W., Z.-n.W. and T.-l.S. accomplished experimental work and wrote the paper. J.-l.W. contributed materials, especially the application part.

Funding: This work is partially supported by National Key R&D Program of China No. 2017YFC1600605, Beijing Municipal Education Commission No. KM201810011013 and KM201810011005, Research Foundation for Youth Scholars of BTBU No. QNJJ2017-15 and QNJJ2016-13, NSFC under Grant No. 61273002, Construction of Scientific and Technological Innovation and Service Capability—Basic Scientific Research Funding Project (Grant No. PXM2018_014213_000033), Construction of technological innovation and service capability—Basic scientific research service fee-innovation platform No. PXM2018_014213_000033.

Conflicts of Interest: The authors declare no conflict of interest.

References

1. Waser, L.T.; Boesch, R.; Wang, Z.; Ginzler, C. Towards Automated Forest Mapping. In *Mapping Forest Landscape Patterns*; Springer: New York, NY, USA, 2017.
2. Qian, C.; Liu, H.; Tang, J.; Chen, Y.; Kaartinen, H.; Kukko, A.; Zhu, L.; Liang, X.; Chen, L.; Hyyppä, J. An Integrated GNSS/INS/LiDAR-SLAM Positioning Method for Highly Accurate Forest Stem Mapping. *Remote. Sens.* **2016**, *9*, 3. [CrossRef]
3. Becker, R.; Keefe, R.; Anderson, N. Use of Real-Time GNSS-RF Data to Characterize the Swing Movements of Forestry Equipment. *Forests* **2017**, *8*, 44. [CrossRef]
4. Heinzel, J.; Huber, M.O. Detecting Tree Stems from Volumetric TLS Data in Forest Environments with Rich Understory. *Remote. Sens.* **2016**, *9*, 9. [CrossRef]
5. Kong, J.L.; Ding, X.K.; Liu, J.; Yan, L.; Wang, J. New Hybrid Algorithms for Estimating Tree Stem Diameters at Breast Height Using a Two Dimensional Terrestrial Laser Scanner. *Sensors* **2015**, *15*, 15661–15683. [CrossRef] [PubMed]
6. Thomas, H.; Pär, L.; Tomas, N.; Ola, R. Autonomous Forest Vehicles: Historic, envisioned, and state-of-the-art. *J. For. Eng.* **2009**, *20*, 31–38.
7. Miettinen, M.; Ohman, M.; Visala, A.; Forsman, P. Simultaneous Localization and Mapping for Forest Harvesters. In Proceedings of the IEEE International Conference on Robotics and Automation, Roma, Italy, 10–14 April 2007; pp. 517–522.
8. Engelmann, F.; Kontogianni, T.; Hermans, A.; Leibe, B. Exploring Spatial Context for 3D Semantic Segmentation of Point Clouds. In Proceedings of the IEEE International Conference on Computer Vision Workshop, Venice, Italy, 22–29 October 2017; pp. 716–724.
9. Marinello, F.; Proto, A.R.; Zimbalatti, G.; Pezzuolo, A.; Cavalli, R.; Grigolato, S. Determination of forest road surface roughness by Kinect depth imaging. *Ann. For. Res.* **2017**, *60*. [CrossRef]
10. Giusti, A.; Guzzi, J.; Dan, C.C.; He, F.-L.; Rodriguez, J.P.; Fontana, F.; Faessler, M.; Forster, C.; Schmidhuber, J.; Caro, G.D.; et al. A Machine Learning Approach to Visual Perception of Forest Trails for Mobile Robots. *IEEE Robot. Autom. Lett.* **2017**, *1*, 661–667. [CrossRef]
11. Xu, Y.; Tuttas, S.; Hoegner, L.; Stilla, U. Voxel-based segmentation of 3D point clouds from construction sites using a probabilistic connectivity model. *Pattern Recognit. Lett.* **2018**, *102*, 67–74. [CrossRef]
12. Trochta, J.; Krůček, M.; Vrška, T.; Král, K. 3D Forest: An application for descriptions of three-dimensional forest structures using terrestrial LiDAR. *PLoS ONE* **2017**, *12*, e0176871. [CrossRef] [PubMed]

13. Ramiya, A.M.; Nidamanuri, R.R.; Krishnan, R. Segmentation based building detection approach from LiDAR point cloud. *Egypt. J. Remote. Sens. Space Sci.* **2016**, *20*, 71–77. [CrossRef]
14. Yang, B.; Dai, W.; Dong, Z.; Liu, Y. Automatic Forest Mapping at Individual Tree Levels from Terrestrial Laser Scanning Point Clouds with a Hierarchical Minimum Cut Method. *Remote. Sens.* **2016**, *8*, 372. [CrossRef]
15. Hamraz, H.; Contreras, M.A.; Zhang, J. Forest understory trees can be segmented accurately within sufficiently dense airborne laser scanning point clouds. *Sci. Rep.* **2017**, *7*, 6770. [CrossRef] [PubMed]
16. Vo, A.V.; Truong-Hong, L.; Laefer, D.F.; Bertolotto, M. Octree-based region growing for point cloud segmentation. *ISPRS J. Photogramm. Remote. Sens.* **2015**, *104*, 88–100. [CrossRef]
17. Zhong, L.; Cheng, L.; Xu, H.; Wu, Y.; Chen, Y.; Li, M. Segmentation of Individual Trees from TLS and MLS Data. *IEEE J. Sel. Top. Appl. Earth Obs. Remote Sens.* **2017**, *10*, 774–787. [CrossRef]
18. Weinmann, M.; Weinmann, M.; Mallet, C.; Brédif, M. A Classification-Segmentation Framework for the Detection of Individual Trees in Dense MMS Point Cloud Data Acquired in Urban Areas. *Remote Sens.* **2017**, *9*, 277. [CrossRef]
19. Achanta, R.; Shaji, A.; Lucchi, A.; Lucchi, A.; Fua, P.; Süsstrunk, S. SLIC superpixels compared to state-of-the-art superpixel methods. *IEEE Trans. Pattern Anal. Mach. Intell.* **2012**, *34*, 2274–2282. [CrossRef] [PubMed]
20. Van, D.; Bergh, M.; Boix, X.; Roig, G.; Van Gool, L. SEEDS: Superpixels extracted via energy-driven sampling. *Int. J. Comput. Vis.* **2015**, *111*, 298–314.
21. Papon, J.; Abramov, A.; Schoeler, M.; Worgotter, F. Voxel Cloud Connectivity Segmentation—Supervoxels for Point Clouds. In Proceedings of the IEEE Conference on Computer Vision Pattern Recognition, Portland, OR, USA, 23–28 June 2013; pp. 2027–2034.
22. Kim, J.S.; Park, J.H. Weighted-graph-based supervoxel segmentation of 3D point clouds in complex urban environment. *Electron. Lett.* **2015**, *51*, 1789–1791. [CrossRef]
23. Ban, Z.; Chen, Z.; Liu, J. Supervoxel Segmentation with Voxel-Related Gaussian Mixture Model. *Sensors* **2018**, *18*, 128.
24. Xu, S.; Ye, N.; Xu, S.; Zhu, F. A supervoxel approach to the segmentation of individual trees from LiDAR point clouds. *Remote Sens. Lett.* **2018**, *9*, 515–523. [CrossRef]
25. Aijazi, A.K.; Checchin, P.; Trassoudaine, L. Segmentation Based Classification of 3D Urban Point Clouds: A Super-Voxel Based Approach with Evaluation. *Remote Sens.* **2013**, *5*, 1624–1650. [CrossRef]
26. Li, M.; Sun, C. Refinement of LiDAR point clouds using a super voxel based approach. *J. Photogramm. Remote. Sens.* **2018**. [CrossRef]
27. Yun, J.S.; Sim, J.Y. Supervoxel-based saliency detection for large-scale colored 3D point clouds. In Proceedings of the IEEE International Conference on Image Processing, Phoenix, AZ, USA, 25–28 September 2016; pp. 4062–4066.
28. Verdoja, F.; Thomas, D.; Sugimoto, A. Fast 3D point cloud segmentation using supervoxels with geometry and color for 3D scene understanding. In Proceedings of the 2017 IEEE International Conference on Multimedia and Expo (ICME), Hong Kong, China, 10–14 July 2017; pp. 1285–1290.
29. Wu, F.; Wen, C.; Guo, Y.; Wang, J.; Yu, Y.; Wang, C.; Li, J. Rapid Localization and Extraction of Street Light Poles in Mobile LiDAR Point Clouds: A Supervoxel-Based Approach. *IEEE Trans. Intell. Transp. Syst.* **2017**, *18*, 292–305. [CrossRef]
30. Alex, R.; Alessandro, L. Machine learning. Clustering by fast search and find of density peaks. *Science* **2014**, *344*, 1492–1496.
31. Wang, S.; Wang, D.; Caoyuan, L.I.; Li, Y. Clustering by Fast Search and Find of Density Peaks with Data Field. *Chin. J. Electron.* **2017**, *25*, 397–402. [CrossRef]
32. Zhang, Z. A Flexible New Technique for Camera Calibration. *IEEE Trans. Pattern Anal. Mach. Intell.* **2000**, *22*, 1330–1334. [CrossRef]
33. Kong, J.; Li, F.; Liu, J.; Yan, L.; Ding, X. New Calibration Method of Two-Dimensional Laser Scanner and Camera Based on LM-BP Neural Network. *Int. J. Signal Process. Image Process. Pattern Recognit.* **2016**, *9*, 231–244. [CrossRef]

34. Lourakis, M.I.A. *A Brief Description of the Levenberg-Marquardt Algorithm Implemened by Levmar*; Foundation of Research & Technology: Heraklion, Greece, 2005.
35. SICK Sensor Intelligence, Operating Instructions: Laser Measurement Sensors of the LMS5xx Product Family. SICK AG Waldkirch, 2012. Available online: https://www.sick.com/cn/zh/ (accessed on 11 September 2018).
36. Rusu, R.B.; Cousins, S. 3D is here: Point Cloud Library (PCL). In Proceedings of the IEEE International Conference on Robotics and Automation, Shanghai, China, 9–13 May 2011.

sensors

MDPI

Article

A KPI-Based Probabilistic Soft Sensor Development Approach that Maximizes the Coefficient of Determination

Yue Zhang [1], Xu Yang [1,*], Yuri A. W. Shardt [2], Jiarui Cui [1] and Chaonan Tong [1]

1 Key Laboratory of Knowledge Automation for Industrial Processes of Ministry of Education, School of Automation and Electrical Engineering, University of Science and Technology Beijing, Beijing 100083, China; s20160638@xs.ustb.edu.cn (Y.Z.); cuijiarui@ustb.edu.cn (J.C.); tcn@ies.ustb.edu.cn (C.T.)

2 Department of Automation Engineering, Technical University of Ilmenau, 98684 Ilmenau, Thuringia, Germany; yuri.shardt@tu-ilmenau.de

* Correspondence: yangxu@ustb.edu.cn

Received: 28 July 2018; Accepted: 7 September 2018; Published: 12 September 2018

check for updates

Abstract: Advanced technology for process monitoring and fault diagnosis is widely used in complex industrial processes. An important issue that needs to be considered is the ability to monitor key performance indicators (KPIs), which often cannot be measured sufficiently quickly or accurately. This paper proposes a data-driven approach based on maximizing the coefficient of determination for probabilistic soft sensor development when data are missing. Firstly, the problem of missing data in the training sample set is solved using the expectation maximization (EM) algorithm. Then, by maximizing the coefficient of determination, a probability model between secondary variables and the KPIs is developed. Finally, a Gaussian mixture model (GMM) is used to estimate the joint probability distribution in the probabilistic soft sensor model, whose parameters are estimated using the EM algorithm. An experimental case study on the alumina concentration in the aluminum electrolysis industry is investigated to demonstrate the advantages and the performance of the proposed approach.

Keywords: soft sensor; coefficient of determination maximization strategy; expectation maximization (EM) algorithm; Gaussian mixture model (GMM); alumina concentration

1. Introduction

With the increasing demands placed on industry, requiring a decrease in the defective rate of products, better economic efficiency, and improved safety, there has been a growing demand to develop and implement approaches that can improve the overall control strategy [1]. The first issue that needs to be solved is achieving accurate and real-time estimation of key performance indicators (KPIs) [2]. The difficulty is that these KPIs are usually not easy to measure, or the measurement has significant time delay. Even if some KPIs are measurable, due to the complexity and nonlinearity of modern industrial systems and their complex working conditions, the KPIs may be extremely unreliable [3]. One way to solve the above problems is to develop a soft sensor, which seeks to select a group of easier-to-measure secondary variables that are correlated with the required primary variables (i.e., KPIs in this paper), so that the system is capable of providing process information as often as necessary for control [4,5]. In the development of a successful soft sensor, a good process model is required. The process models can be divided into two major categories: first principles models and data-driven models [6,7]. Although it is desirable to apply mass and energy balances to build a complete first principles model, lack of process knowledge, plant–model mismatch, and nonlinear characteristics limit the applicability of such an approach to the simplest processes. As an alternative, data-driven soft sensors are developed from historical data without necessarily considering any outside process knowledge. Data-driven soft sensors, which solely use available process data to develop a model of the process, have recently attracted considerable attention and have been successfully applied in many fields [8], such as fault detection (FD) and process monitoring, that are important for many industrial processes. Serdio [9] introduced an improved fault detection approach based on residual signals extracted online from system models identified by high-dimensional measurements provided by the multisensor network. The data-driven system identification model can also be combined using multivariate orthogonal space transformations and vectorized time-series models to achieve enhanced residual-based fault detection in condition monitoring systems equipped with a multisensor network [10]. Shardt [11] proposed a data-driven design of a diagnostic-observer-based process monitoring method, which was extended to include the ability to detect changes given infrequent KPI measurements. Yan [12] and Gabrys [13] introduced the most popular data-driven soft sensor modelling techniques, as well as discussing some issues in soft sensor development and maintenance and their possible solutions. Data-driven methods can be divided into three categories: models based on statistical analysis, models based on statistical learning theory [14], and models based on artificial intelligence [15].

Of interest for this paper are models developed using statistical methods to extract the relevant information from the large amounts of industrial data that are produced by the complex processes. Statistical methods have been developed that can handle such large datasets and develop useful models. Common methods include principal component analysis (PCA) [16] and partial least squares (PLS) [17]. PCA is a powerful tool for data compression and information extraction that can simplify the model structure and improve the speed of operations. However, PCA can only deal with the correlations between vectors in the same matrix. To overcome this limitation, PLS was developed as an approach that models the correlation between independent variables and dependent variables. Since PLS only applies to linear systems or weakly nonlinear systems, many nonlinear PLS algorithms have been developed to handle nonlinear systems. The neural-network-based PLS algorithm [18] uses the nonlinear processing capability of a neural network to describe the relationship between variables. However, the determination of the network structure and the selection of network training algorithms are difficult problems. In addition, if there are too many datapoints, the model structure will be very complex and the accuracy will be difficult to guarantee.

On the other hand, considering that data-driven modeling methods use historical data for training, this raises the question of how to handle missing data. Along with issues such as the reliability of sensors and multirate sampling, missing data is common in practical industry process [19,20]. For example, in the aluminum electrolysis process, the alumina concentration is usually obtained manually by laboratory staff. Considering human factors and chemical examination equipment reliability, data loss occurs from time to time. In this case, this type of measurement has different effects on the soft sensor modeling process and state estimation performance. Therefore, in order to make the soft sensor more suitable for practical, complex industrial processes, the missing data problem needs to be taken seriously. Compared with the direct deletion of missing data, the data interpolation method [21] is better able to restore the real situation. Currently, data interpolation methods include the mean substitution method, the regression interpolation method, and the expectation maximization (EM) algorithm. Of these, the mean substitution method can cause biased estimates, and the regression interpolation method is built based on a complete data set, where the linear relationship between the variables with missing values and other variables is necessary, which, in many cases, cannot be satisfied. In fact, the EM algorithm has good practical value as an iterative algorithm for simplifying the maximum likelihood estimation when dealing with missing data in sample sets [22].

Recently, in order to evaluate the accuracy of the model output, the coefficient of determination approach has been considered. The coefficient of determination is the measurement of how well the regression model fits the data [23]. Feng [24] introduced the coefficient of determination as a criterion for comparing the best-wavelength partial least squares regression (PLSR) model with the full-wavelength model. Boyaci [25] used the coefficient of determination to evaluate the adulteration rate of coffee beans, thus ensuring coffee quality. However, these applications only consider the coefficient of determination as an evaluation index without applying it for the modeling process. In general, the coefficient of determination is a criterion that can evaluate the quality of a model and has a concise structure, so it is appropriate to apply it to the soft sensor development process to establish a simpler and more accurate model for complex industrial process.

Therefore, this paper develops a KPI-based soft sensor model with simple structure and high accuracy, using the coefficient of determination method, which also solves the missing data issue using the EM algorithm.

2. Background

2.1. The Gaussian Mixture Model

As a flexible and efficient tool for probabilistic data models, a Gaussian mixture model (GMM) can be used to define any complex probability distribution function and is, therefore widely used in many statistical data modelling applications. In this paper, GMM is used to approximate the joint probability distribution in the soft sensor probability model. The reason for introducing GMM is that, theoretically, any probability distribution can be approximated using the joint weighted Gaussian distribution [26].

If x represents a multidimensional random variable, then the joint probability distribution of the GMM is expressed as

$$p(x|\Theta) = \sum_{l=1}^{M} \alpha_l p_l(x|\theta_l) \tag{1}$$

where α_l is the mixing coefficient, which represents the prior probability of each mixed component; M is the number of mixed components; and $\sum_{l=1}^{M} \alpha_l = 1$. $\Theta = (\theta_1, \theta_2, \cdots, \theta_M)$ is the parameter vector of each mixed component, and each Gaussian probability density function $p_l(x)$ is determined by the parameter $\theta_l = (\mu_l, \Sigma_l)$, where μ_l is the mean and Σ_l is the covariance matrix. The GMM parameters α_l, μ_l, and Σ_l ($l = 1, 2, \ldots, M$) are estimated using the EM algorithm.

2.2. The Expectation Maximization Algorithm

The EM algorithm is a maximum likelihood estimation method for solving model distribution parameters from "incomplete data" and was first introduced in [27]. Each iteration of the algorithm involves two steps, called the expectation step (E-step) and the maximization step (M-step).

2.2.1. E-Step

Given the observation data set X and the current parameters $\Gamma^{(i)}$, the expectation of the log-likelihood function is called the Q-function which can be written as

$$Q(\Gamma, \Gamma^{(i)}) = E\left[\log p(X, |\Gamma) \middle| X, \Gamma^{(i)}\right] \tag{2}$$

where γ can represent missing data due to observational conditions and other reasons, and can also refer to hidden variables. Since the direct optimization of the likelihood function is usually very difficult, the relationship between X, Γ, and γ can be established by introducing an additional variable γ to achieve the purpose of simplifying the likelihood function.

2.2.2. M-Step

A new parameter $\Gamma^{(i+1)}$ is calculated by maximizing $Q(\Gamma, \Gamma^{(i)})$ which was obtained from the E-step; that is,

$$\Gamma^{(i+1)} = \underset{\Gamma}{\operatorname{argmax}} Q(\Gamma, \Gamma^{(i)}) \,. \tag{3}$$

The iteration between the E- and M-steps continues until the elements of Γ are less than a given value.

2.3. The Coefficient of Determination

Analysis of variance is an approach for determining the significance and validity of a regression model using variances obtained from the data and model. The coefficient of determination is an analysis of variance approach that seeks to decompose the total variability in the data into various orthogonal components that can then be independently analyzed [23]. For the purposes of analyzing the regression, let the total sum of squares, denoted by TSS, be defined as

$$TSS = \sum_{i=1}^{n} (y_i - \overline{y})^2 \tag{4}$$

where the real data set is represented as $y = <y_1, y_2, \ldots, y_n>$ and $\overline{y}$ refers to the average of y_i. Let the sum of squares due to regression, SSR, be defined as

$$SSR = \sum_{i=1}^{n} (\hat{y}_i - \overline{y})^2 \tag{5}$$

where $\hat{y}_i$ denotes the predicted value of the regression model for y_i. The coefficient of determination R^2 represents the ratio of SSR to TSS, that is,

$$R^2 = \frac{SSR}{TSS} \,. \tag{6}$$

Let the sum of squares due to the error, SSE, be defined as

$$SSE = \sum_{i=1}^{n} (y_i - \hat{y}_i)^2. \tag{7}$$

It can be proved that TSS = SSR + SSE [23,28], so R^2 can also be expressed as

$$R^2 = 1 - \frac{SSE}{TSS} = 1 - \frac{\sum_{i=1}^{n} (y_i - \hat{y}_i)^2}{\sum_{i=1}^{n} (y_i - \overline{y})^2} \,. \tag{8}$$

3. Development of the Probabilistic Soft Sensor Model

In this section, in order to obtain more accurate KPI estimates, a soft sensor development approach based on maximizing the coefficient of determination is proposed. In addition, the problem of missing data in the training sample set is also considered. In order to more clearly describe the soft sensor development process, Figure 1 shows the modeling flow chart.

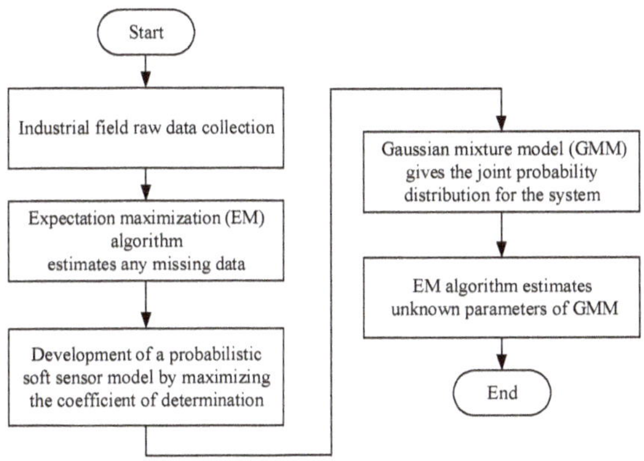

Figure 1. The flow chart of soft sensor development process.

3.1. EM Algorithm Handing Missing Data

Let $X_1, X_2, \ldots, X_n$ be a random sample from a p-variate normal population, where $X_j = (x_{j1}, x_{j2}, \ldots, x_{jp})$, $1 \leq j \leq n$, so the training sample set X can be written as

$$X = \begin{pmatrix} X_1 \\ X_2 \\ \vdots \\ X_n \end{pmatrix} = \begin{bmatrix} x_{11}, & x_{12}, & \cdots, & x_{1p} \\ x_{21}, & x_{22}, & \cdots, & x_{2p} \\ \vdots & & \ddots & \vdots \\ x_{n1}, & x_{n2}, & \cdots, & x_{np} \end{bmatrix}. \tag{9}$$

The basic steps for processing missing data using the EM algorithm are given in [29].

3.1.1. E-Step: Prediction

For each sample X_j containing missing values, $X_j = (m_j, a_j)$, where m_j is the missing value and a_j is the available values. Given the population mean and variance, $\overset{\sim i}{\mu}$ and $\overset{\sim i}{\Sigma}$, from the ith iteration and a_j, we use the expectation of the conditional normal distribution of m_j as the estimate of the missing value. The $(i+1)$th iteration is

$$\tilde{m}_j^{i+1} = E(m_j | a_j, \tilde{\mu}^i, \tilde{\Sigma}^i) = \tilde{\mu}_m^i + \tilde{\Sigma}_{ma}^i \left(\tilde{\Sigma}_{aa}^i\right)^{-1} (a_j - \tilde{\mu}_a^i) \tag{10}$$

where $\overset{\sim i}{\mu}$ is a $p \times 1$ matrix defined as $\tilde{\mu}^i = \left[\tilde{\mu}_m^i, \tilde{\mu}_a^i\right]'$, $\tilde{\mu}_m^i$ is the mean of the missing part, and $\tilde{\mu}_a^i$ is the mean of the available part. In addition, $\overset{\sim i}{\Sigma}$ can be written as

$$\tilde{\Sigma}^i = \begin{bmatrix} \tilde{\Sigma}_{mm}^i \tilde{\Sigma}_{ma}^i \\ \tilde{\Sigma}_{am}^i \tilde{\Sigma}_{aa}^i \end{bmatrix}. \tag{11}$$

3.1.2. M-Step: Estimation

We compute the maximum likelihood estimates as follows:

$$\tilde{\mu}^{i+1} = \overline{X}^{i+1} \tag{12}$$

$$\tilde{\Sigma}^{i+1} = \frac{(n-1)S^{i+1}}{n} \tag{13}$$

where $\overline{X}^{i+1}$ is the mean of the samples and S^{i+1} is the sample standard deviation, and they are all sufficient statistics. For a normal population, the importance of sufficient statistics is that the total information about μ and Σ in the data matrix X is contained in $\overline{X}$ and S, regardless of the sample size n. By transforming $\overline{X}$ and S, two new sufficient statistics T_1 and T_2 [29], given by

$$T_1 = n\overline{X} \tag{14}$$

$$T_2 = (n-1)S + n\overline{X}\overline{X}' \tag{15}$$

are obtained. Combining Equations (14) and (15) with Equations (12) and (13) gives

$$\widetilde{\mu}^{i+1} = \frac{T_1^{i+1}}{n} \tag{16}$$

$$\widetilde{\Sigma}^{i+1} = \frac{1}{n}T_2^{i+1} - \widetilde{\mu}^{i+1}(\widetilde{\mu}^{i+1})' \tag{17}$$

where

$$\widetilde{m_j m_j'}^{i+1} = E(m_j m_j' | a_j, \widetilde{\mu}^i, \widetilde{\Sigma}^i) = \widetilde{\Sigma}^i_{mm} - \widetilde{\Sigma}^i_{ma}(\widetilde{\Sigma}^i_{aa})^{-1}\widetilde{\Sigma}^i_{am} + \widetilde{m}_j^{i+1}\left(\widetilde{m}_j^{i+1}\right)' \tag{18}$$

$$\widetilde{m_j a_j'}^{i+1} = E(m_j a_j' | a_j, \widetilde{\mu}^i, \widetilde{\Sigma}^i) = \widetilde{m}_j^{i+1}\left(a_j\right)'. \tag{19}$$

The iteration between the E- and M-steps continues until the elements of $\widetilde{\mu}$ and $\widetilde{\Sigma}$ are less than a given value. Therefore, the iteration result $\widetilde{m}$ is the optimal substitution for the missing values, resulting in a complete training sample set X.

3.2. Soft Sensor Development Approach Based on the Coefficient of Determination Maximization Strategy

For the complete training sample set X obtained from Section 3.1, which can be written as

$$X = \begin{bmatrix} x_{11}, & x_{12}, & \cdots, & x_{1p} \\ x_{21}, & x_{22}, & \cdots, & x_{2p} \\ \vdots & & \ddots & \vdots \\ x_{n1}, & x_{n2}, & \cdots, & x_{np} \end{bmatrix} \tag{20}$$

let $(x_1, x_2, \cdots x_{p-1})$ denote the secondary variables, and x_p denote the KPI. Our objective is to estimate x_p from $(x_1, x_2, \cdots x_{p-1})$.

R^2 measures the fraction of the total variance in the model explained by the regression with the given variables [23]. The range of R^2 is [0,1]. Let x_p be the y mentioned in Section 2.3. Then, the coefficient of determination is

$$R^2 = 1 - \frac{\sum_{i=1}^{n}(x_{ip} - \hat{x}_{ip})^2}{\sum_{i=1}^{n}(x_{ip} - \overline{x}_p)^2}. \tag{21}$$

If the secondary variables in the soft sensor model do not account for the variance of x_p, the estimate of x_{ip}, denoted $\hat{x}_{ip}$, is exactly equal to the sample mean of x_{ip}, denoted $\overline{x}_{ip}$. In this case, SSR is 0 and SSE equal to TSS, so $R^2 = 0$. On the other hand, if $\left(x_{i1}, x_{i2}, \cdots x_{i(p-1)}\right)$ fully explains the variance of x_{ip}, for $i = 1, 2, \ldots, n$, it follows that $x_{ip} = \overline{x}_{ip}$, i.e., each error is zero and SSR = TSS, so $R^2 = 1$. In general, R^2 does not take the extreme values 0 or 1, but instead takes a certain value between the two [28]. For the case where the number of variables, p, is much smaller than the sample number n, the closer R^2 is to 1, the better the model. Therefore, when the model for the KPI maximizes R^2, it becomes the best estimate of the KPI, that is,

$$1 - \frac{\sum_{i=1}^{n} (x_{ip} - \tilde{x}_{ip})^2}{\sum_{i=1}^{n} (x_{ip} - \overline{x}_p)^2} = \max \left[1 - \frac{\sum_{i=1}^{n} (x_{ip} - K_i)^2}{\sum_{i=1}^{n} (x_{ip} - \overline{x}_p)^2} \right] \tag{22}$$

where $\tilde{x}_{ip}$ is the best estimate of x_{ip}, and K_i represents all possible estimates of x_{ip}. Simplifying the above equation gives

$$\frac{\sum_{i=1}^{n} (x_{ip} - \tilde{x}_{ip})^2}{\sum_{i=1}^{n} (x_{ip} - \overline{x}_p)^2} = \min \left[\frac{\sum_{i=1}^{n} (x_{ip} - K_i)^2}{\sum_{i=1}^{n} (x_{ip} - \overline{x}_p)^2} \right] \tag{23}$$

where x_{ip} and $\overline{x}_p$ are both computed values. Equation (23) can then be written as

$$\sum_{i=1}^{n} (x_{ip} - \tilde{x}_{ip})^2 = \min \left[\sum_{i=1}^{n} (x_{ip} - K_i)^2 \right]. \tag{24}$$

Multiplying Equation (24) on both sides by n^{-1} gives

$$\frac{1}{n}\sum_{i=1}^{n} (x_{ip} - \tilde{x}_{ip})^2 = \min \left[\frac{1}{n}\sum_{i=1}^{n} (x_{ip} - K_i)^2 \right]. \tag{25}$$

Considering that the mathematical expectation of a discrete random variable is

$$E(x) = \sum_{i} x_i p_i \tag{26}$$

where x_i represents the ith value of the random variable x and p_i represents its probability, Equation (26) can be expressed as

$$E\left\{ \|x_p - \tilde{x}_p\|^2 \right\} = \min E\left\{ \|x_p - K\|^2 \right\} \tag{27}$$

where K denotes all possible estimates of the KPI x_p, and $\tilde{x}_p$ represents the best estimate of the KPI when the coefficient of determination R^2 is maximized. Since x_p is derived from the soft sensor models and secondary variables, the above equation can be written as

$$\tilde{x}_p = \underset{K}{\operatorname{argmin}} E\left[\|x_p - K\|^2 | (x_1, x_2, \cdots x_{p-1}) \right]. \tag{28}$$

In order to establish a more direct connection between $\tilde{x}_p$ and $(x_{i1}, x_{i2}, \ldots, x_{i(p-1)})$, the left-hand side of Equation (28) will be simplified further. Firstly, it can be noted that K does not have an impact on the simplification, that is,

$$\begin{aligned}
& E\left[\|x_p - K\|^2 | (x_1, x_2, \cdots x_{p-1}) \right] \\
= \; & E\left[\|x_p - E(x_p|(x_1, x_2, \cdots x_{p-1})) + E(x_p|(x_1, x_2, \cdots x_{p-1})) - K\|^2 | (x_1, x_2, \cdots x_{p-1}) \right] \\
= \; & E\left[\|x_p - E(x_p|(x_1, x_2, \cdots x_{p-1}))\|^2 | (x_1, x_2, \cdots x_{p-1}) \right] + E\left[\|E(x_p|(x_1, x_2, \cdots x_{p-1})) - K\|^2 | (x_1, x_2, \cdots x_{p-1}) \right] \\
& + E\left[[x_p - E(x_p|(x_1, x_2, \cdots x_{p-1}))]^T [E(x_p|(x_1, x_2, \cdots x_{p-1})) - K] | (x_1, x_2, \cdots x_{p-1}) \right] \\
& + E\left[[E(x_p|(x_1, x_2, \cdots x_{p-1})) - K]^T [x_p - E(x_p|(x_1, x_2, \cdots x_{p-1}))] | (x_1, x_2, \cdots x_{p-1}) \right]
\end{aligned} \tag{29}$$

In order to minimize the above equation, the following should hold:

$$K = E[x_p|(x_1, x_2, \cdots x_{p-1})] \tag{30}$$

which can be rewritten as

$$\widetilde{x}_p = E\left[x_p\middle|(x_1, x_2, \cdots x_{p-1})\right]. \tag{31}$$

Furthermore, $E\left[x_p\middle|(x_1, x_2, \cdots x_{p-1})\right]$ can be expanded according to the definition of expectation, giving

$$\begin{aligned}\widetilde{x}_p &= E\left[x_p\middle|(x_1, x_2, \cdots x_{p-1})\right] \\ &= \int x_p p\left[x_p\middle|(x_1, x_2, \cdots x_{p-1})\right]dx_p . \\ &= \int x_p \frac{p(x_1, x_2, \cdots x_{p-1}, x_p)}{p(x_1, x_2, \cdots x_{p-1})}dx_p\end{aligned} \tag{32}$$

Thus, this establishes the basic framework of the probabilistic soft sensor model with KPI optimal estimation.

The next part is to solve the joint probability distribution in the model.

In this paper, GMM is used to approximate the joint probability distribution. Let $p(x_e) = p(x_1, x_2, \cdots x_{p-1})$; that is,

$$p(x_e) = \sum_{j=1}^{M} \alpha_j p(x_{je}|\theta_j) \tag{33}$$

$$p(x_e, x_p) = \sum_{l=1}^{M} \alpha_l p\left(x_{le}, x_{lp}\middle|\theta_l\right). \tag{34}$$

In order to deduce the specific representation of the KPI optimal estimation $\widetilde{x}_p$ under the proposed probabilistic soft sensor model, we first introduce Lemma 1.

Lemma 1. [30] *Let $G(x; \mu, \Sigma)$ be a multidimensional normal density function with mean μ and covariance matrix Σ. Let $x^T = (x_1^T, x_2^T)$, $\mu^T = (\mu_1^T, \mu_2^T)$, and $\Sigma = \begin{bmatrix} \Sigma_{11}\Sigma_{12} \\ \Sigma_{21}\Sigma_{22} \end{bmatrix}$; then, the joint probability density is*

$$p(x) = G(x_1; \mu_1, \Sigma_{11})G\left(x_2; \mu_{x_2|x_1}, \Sigma_{x_2|x_1}\right) \tag{35}$$

where

$$\mu_{x_2|x_1} = \mu_2 - \Sigma_{21}\Sigma_{11}^{-1}(\mu_1 - x_1) \tag{36}$$

$$\Sigma_{x_2|x_1} = \Sigma_{22} - \Sigma_{21}\Sigma_{11}^{-1}\Sigma_{12}. \tag{37}$$

Proof. The details of the proof can be found in [30].

Using Lemma 1, it follows that

$$\begin{aligned}&p\left(x_{le}, x_{lp}\right) \\ =\ &G(x_l; \mu_l, \Sigma_l) \\ =\ &G(x_{le}; \mu_{le}, \Sigma_{lee})G\left(x_{lp}; \mu_{lp|e}, \Sigma_{lp|e}\right)\end{aligned} \tag{38}$$

where $\mu_l = \left(\mu_{le}^T, \mu_{lp}^T\right)$ and $\Sigma_l = \begin{bmatrix} \Sigma_{lee}\Sigma_{lep} \\ \Sigma_{lpe}\Sigma_{lpp} \end{bmatrix}$. Therefore, Equations (33) and (34) can be written as

$$p(x_e) = \sum_{j=1}^{M} \alpha_j G\left(x_{je}; \mu_{je}, \Sigma_{jee}\right) \tag{39}$$

$$p(x_e, x_p) = \sum_{l=1}^{M} \alpha_l G(x_{le}; \mu_{le}, \Sigma_{lee})G\left(x_{lp}; \mu_{lp|e}, \Sigma_{lp|e}\right). \tag{40}$$

Substituting Equations (39) and (40) into Equation (32) gives

$$\begin{aligned}\tilde{x}_p &= \int x_p \frac{p(x_e,x_p)}{p(x_e)} dx_p \\ &= \int x_p \frac{\sum\limits_{l=1}^{M} \alpha_l G(x_{le};\mu_{le},\Sigma_{lee}) G(x_{lp};\mu_{lp|e},\Sigma_{lp|e})}{\sum\limits_{j=1}^{M} \alpha_j G(x_{je};\mu_{je},\Sigma_{jee})} dx_p\,.\end{aligned} \tag{41}$$

Extracting the sum in the numerator to outside the integral gives

$$\tilde{x}_p = \sum_{l=1}^{M} \int x_p \frac{\alpha_l G(x_{le};\mu_{le},\Sigma_{lee}) G\left(x_{lp};\mu_{lp|e},\Sigma_{lp|e}\right)}{\sum\limits_{j=1}^{M} \alpha_j G\left(x_{je};\mu_{je},\Sigma_{jee}\right)} dx_p\,. \tag{42}$$

In order to make the derivation more concise, the positions of some factors in the integral are changed as follows:

$$\begin{aligned}\tilde{x}_p &= \sum_{l=1}^{M} \int \frac{\alpha_l G(x_{le};\mu_{le},\Sigma_{lee})}{\sum\limits_{j=1}^{M} \alpha_j G(x_{je};\mu_{je},\Sigma_{jee})} x_p G\left(x_{lp};\mu_{lp|e},\Sigma_{lp|e}\right) dx_p \\ &= \sum_{l=1}^{M} \frac{\alpha_l G(x_{le};\mu_{le},\Sigma_{lee})}{\sum\limits_{j=1}^{M} \alpha_j G(x_{je};\mu_{je},\Sigma_{jee})} \int x_p G\left(x_{lp};\mu_{lp|e},\Sigma_{lp|e}\right) dx_p\,.\end{aligned} \tag{43}$$

When the integral part is the conditional expectation, the above equation can be simplified to

$$\tilde{x}_p = \sum_{l=1}^{M} \frac{\alpha_l G(x_{le};\mu_{le},\Sigma_{lee})}{\sum\limits_{j=1}^{M} \alpha_j G\left(x_{je};\mu_{je},\Sigma_{jee}\right)} \mu_{lp|e}\,. \tag{44}$$

Therefore, the detailed soft sensor model expression of the KPI optimal estimation is obtained.

In this paper, unknown parameters in the model are estimated using the EM algorithm. The iterative equations of the EM algorithm for estimating the GMM parameters are [31]

$$\mu_l^{(i+1)} = \frac{\sum\limits_{j=1}^{n} \gamma_{jl}^{(i+1)} X_j}{\sum\limits_{j=1}^{n} \gamma_{jl}^{(i+1)}},\ \Sigma_l^{(i+1)} = \frac{\sum\limits_{j=1}^{n} \gamma_{jl}^{(i+1)} \left(X_j - \mu^{(i)}\right)^2}{\sum\limits_{j=1}^{n} \gamma_{jl}^{(i+1)}},\ \alpha_l^{(i+1)} = \frac{\sum\limits_{j=1}^{n} \gamma_{jl}^{(i+1)}}{n} \tag{45}$$

where γ_{jl} represents the responsivity of the mixed component l on the training sample data X_j. It can be written as

$$\gamma_{jl}^{(i+1)} = \frac{\alpha_l p(X_j|\theta_l)}{\sum\limits_{l=1}^{M} \alpha_l p(X_j|\theta_l)}\,. \tag{46}$$

Consequently, the above steps give the GMM parameters, and the KPI optimal estimate $\tilde{x}_p$ follows.

4. Case Study

In this section, the effectiveness and feasibility of the proposed soft sensor model approach based on maximizing the coefficient of determination are evaluated through an industrial aluminum electrolytic production process. To show the advantages of the probabilistic soft sensor framework, the estimations are compared with the real values. For performance evaluation, the root-mean-squared error (RMSE) index is used.

4.1. Soft Sensor Development for Industrial Aluminum Electrolytic Process

Aluminum is widely used in construction and electrical industries [32]. The main method currently chosen for smelting aluminum plants is the cryolite–alumina molten salt electrolysis process, in which the electrochemical reaction process takes place in an electrolytic cell. Figure 2 shows the internal structure of the electrolytic cell.

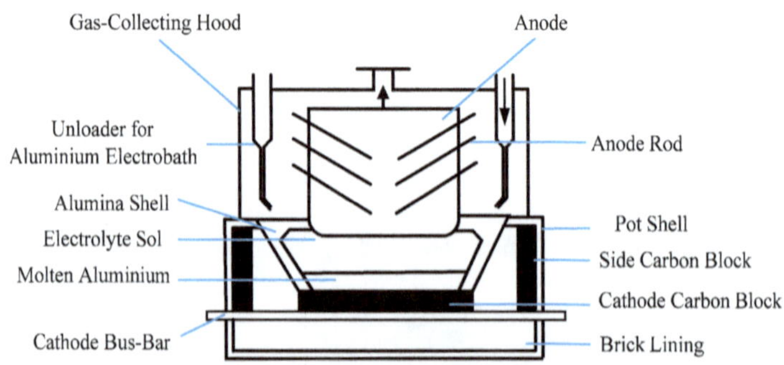

Figure 2. The internal structure of the aluminum electrolytic cell.

Molten cryolite is a solvent in which aluminum oxide is dissolved as a solute, forming a melt with good electrical conductivity. Carbon materials are used as cathodes and anodes, and a direct current is passed through them. The thermal energy of the direct current is used to melt the cryolite and maintain a constant electrolysis temperature. Furthermore, the electrochemical reaction occurs between the two electrodes, where the product at the cathode is aluminum liquid, and carbon dioxide and other gases are generated at the anode. The chemical reaction of the electrolytic process is

$$2Al_2O_3 + 3C \rightarrow 4Al + 3CO_{2\uparrow}.$$

The chemical reaction can produce gases other than carbon dioxide and carbon monoxide, as well as fluorocarbon gases. The gas purifying device uses alumina and fluorine generated in the mixed gas to produce fluorinated alumina, and the fluorinated alumina is then recycled to the electrolytic cell for chemical reaction. Figure 3 shows the process flow diagram of the aluminum electrolysis process.

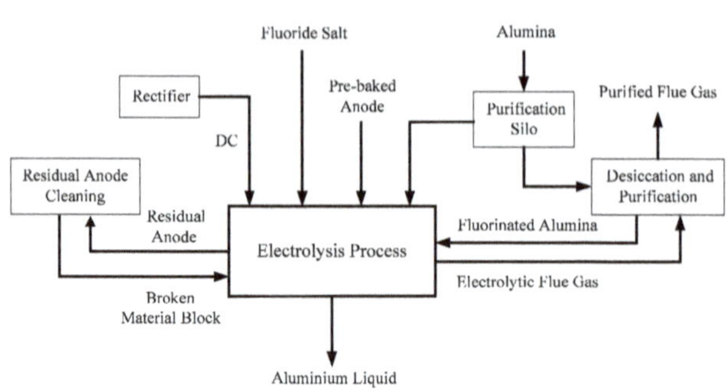

Figure 3. The process flow diagram of the aluminum electrolysis process.

The main control goal of the aluminum electrolysis process is to keep the alumina concentration in the electrolysis cell stable within a certain range, preferably between 1.5% and 3.5% [33]. The control of alumina concentration relates to energy consumption and economic benefits of the aluminum electrolytic production process. On one hand, when the alumina concentration is too low, an additional chemical reaction occurs at the anode, which can easily cause a sudden rise in the cell voltage and the energy balance of the cell is destroyed. On the other hand, when the concentration reaches saturation, if the feeder continues to add alumina at the time, the raw material will be deposited at the bottom of the cell, so that the resistance increases and the current efficiency becomes low. Therefore, it is necessary to keep the alumina concentration in the proper range.

In soft sensor development for the aluminum electrolytic process, the measurable variables, the voltage x_1 between the two electrodes obtained by the first voltage measuring instrument; the anode conductor current x_2; the voltage x_3 between the two electrodes obtained by the second voltage measuring instrument; and the alumina concentration x_4 provided by an electrochemical analyzer, were selected as the secondary variables. The interelectrode voltage refers to the voltage between the anode guide and the corresponding cathode steel bar. The alumina concentration y provided by the laboratory is the primary variable for the model. Figure 4 shows a diagram of the process measurement system.

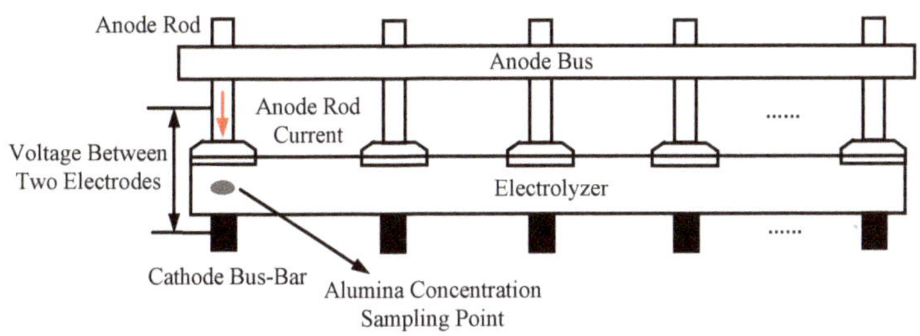

Figure 4. Schematic diagram of the variable collection system.

The variables $x_1(k)$, $x_2(k)$, $x_3(k)$, $x_4(k)$, and $y(k)$ form the joint probability distribution

$$p(x(k)) = p(x_1(k), x_2(k), x_3(k), x_4(k), y(k)) . \tag{47}$$

The soft sensor was then developed according to the process described in Section 3 of this paper. It is assumed that $M = 2$.

4.2. Experimental Results

4.2.1. EM Algorithm and Missing Values

We took 600 complete data groups from the training sample set, and deleted 10%, 20%, or 30% of the alumina concentration variable data. Then, the mean substitution method, the regression interpolation method, and the EM algorithm were used to process the sample set with missing values. Tables 1–3 show the mean and RMSE of the alumina concentration sample set for the three method simulations for missing ratios of 10%, 20%, and 30%.

Table 1. Comparison of three data interpolation methods for a 10% missing rate.

	Mean Substitution Method	Regression Interpolation Method	EM Algorithm	Real Value
Mean	2.4133	2.4225	2.4225	2.4259
RMSE	0.0867	0.4209	0.0698	0

Table 2. Comparison of three data interpolation methods for a 20% missing rate.

	Mean Substitution Method	Regression Interpolation Method	EM Algorithm	Real Value
Mean	2.4139	2.4217	2.4215	2.4259
RMSE	0.1451	0.4075	0.1361	0

Table 3. Comparison of three data interpolation methods for a 30% missing rate.

	Mean Substitution Method	Regression Interpolation Method	EM Algorithm	Real Value
Mean	2.4140	2.4204	2.41198	2.4259
RMSE	0.1700	0.4068		0

First, comparing the mean value, we can see from the above tables that the means of the regression interpolation method and the EM data interpolation method are closer to the mean of the real value set, and the mean substitution method is less effective. Obviously, the RMSE of the EM data interpolation method is much smaller than that of the regression interpolation method. Therefore, the accuracy and effectiveness of the EM data interpolation method in processing missing values is verified. Further, if there is a problem with missing values in the practical industrial process, the EM algorithm can be selected for data interpolation.

4.2.2. Experimental Results of the Soft Sensor Model Based on Maximizing the Coefficient of Determination

In order to verify the feasibility of the proposed approach, a test sample set was used to validate the designed soft sensor model. The test sample set was divided into four subsets of 100 samples. The actual alumina concentration measurement obtained from the laboratory was compared with the output of the soft sensor model to acquire an estimated performance evaluation of the model. The results are shown in Figure 5. Figure 5a–d show the estimated alumina concentrations based on the first, second, third, and fourth test subsets, respectively. Table 4 shows the root-mean-square errors (RMSE) of the four test subsets. It can be seen that, overall, the soft sensor model based on maximizing the coefficient of determination can accurately track the overall trends in the process. The alumina concentration output by the model is approximately the same as the actual laboratory measurement.

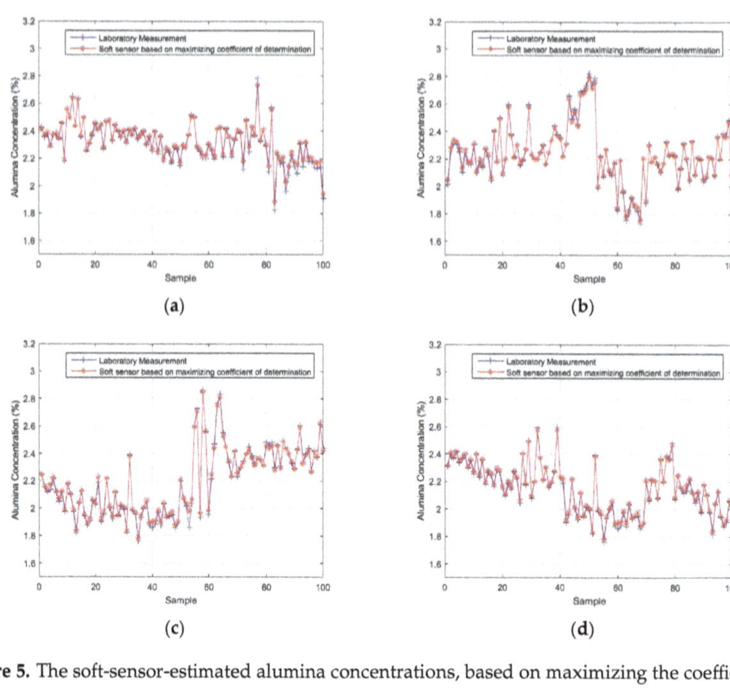

Figure 5. The soft-sensor-estimated alumina concentrations, based on maximizing the coefficient of determination, compared with the actual laboratory measurement using (**a**) the first test subset, (**b**) the second test subset, (**c**) the third test subset, and (**d**) the fourth test subset.

Table 4. The RMSE values of the four test subsets.

Test Subset	RMSE
First	0.0231
Second	0.0145
Third	0.0209
Fourth	0.0155

4.2.3. Comparison with BP and LSSVM

The backpropagation (BP) neural network and the least-squares, support vector machine (LSSVM) model were applied to the test sample set, and the first test subset was used for performance comparison. The parameters of the comparison algorithms were determined as follows: The number of hidden layer nodes in the BP neural network model was 100 and the activation function of the hidden layer was a sigmoid [34]. The kernel function of the LSSVM model was the radial basis function (RBF), and the kernel parameter and regular parameter were 1 and 20, respectively [34]. For each model, the number of secondary variables was 4, and the number of primary variables was 1. It could be seen that the two comparison models need different parameters in order to achieve an accurate estimation performance, while this is not necessary for the soft sensor model based on maximizing the coefficient of determination. The estimated results are shown in Figures 6 and 7. Figure 6 shows the estimated values of the soft sensor based on the BP neural network for the first test subset, and Figure 7 shows the estimated values of the soft sensor based on the LSSVM for the first test subset. It can be seen from Figure 6 that the soft sensor based on a BP neural network can roughly follow the trend of the laboratory measurements, but the error is still large at many points. It can be seen from Figure 7 that the overall performance of the soft sensor based on LSSVM is better than that based on a BP neural network, but compared with Figure 5a, it is obvious that the estimation of some extreme points is not as accurate as that given by the soft sensor based on maximizing the coefficient of determination.

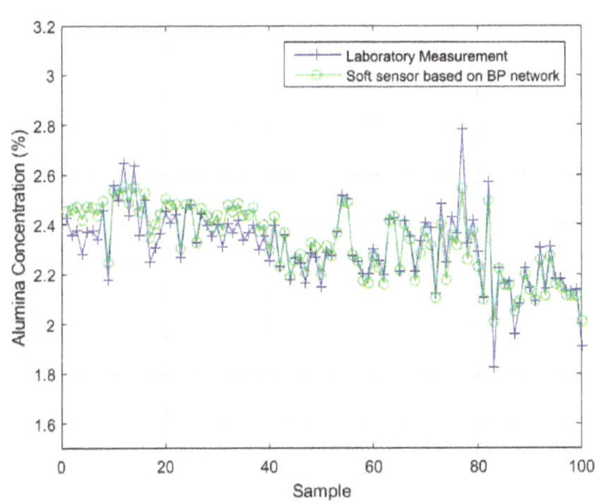

Figure 6. The estimated values of the soft sensor based on a backpropagation (BP) network compared with actual laboratory measurements.

Figures 8–10 show the soft sensor estimates based on different modelling methods as a function of the laboratory measurements. The green circles show the BP neural network model; the purple circles the LSSVM model; and the red circles the proposed coefficient of determination maximization model. In the ideal case, the circles should lie on the blue $y = x$ line. In practice, deviations from this behavior can provide information about the accuracy of the models. The BP neural network soft sensor produces a soft sensor system that has a consistent bias, since the values are consistently located above the $y = x$ line. Furthermore, the bias in the LSSVM soft sensor model is smaller, but there also seems to be a calibration issue, since the data does not lie parallel to the $y = x$ line. Finally, the proposed model has the smallest deviations and the most ideal performance.

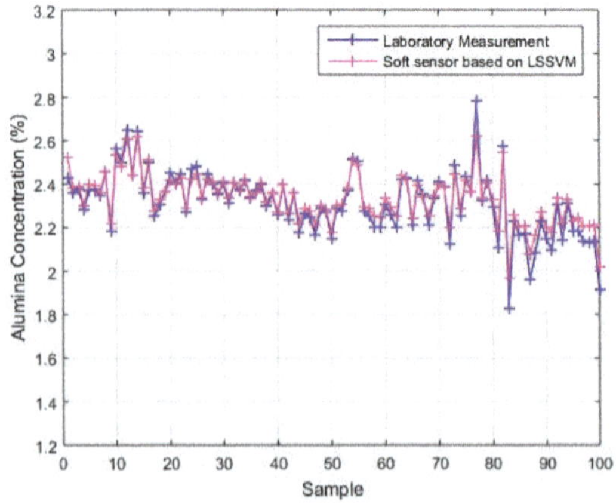

Figure 7. The estimated values of the soft sensor based on LSSVM compared with actual laboratory measurements.

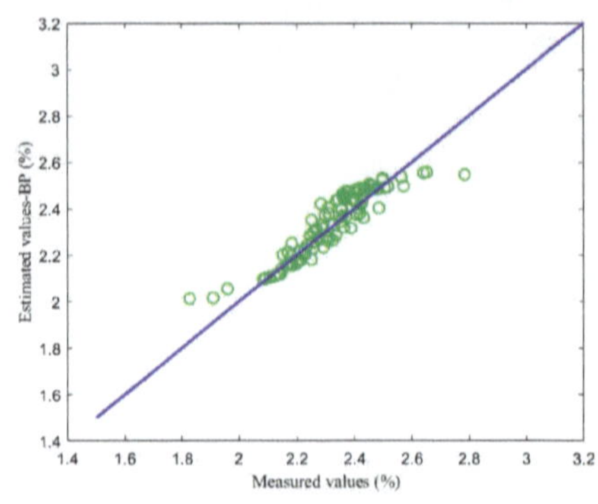

Figure 8. Comparison between the soft sensor based on a BP neural network and laboratory measurements.

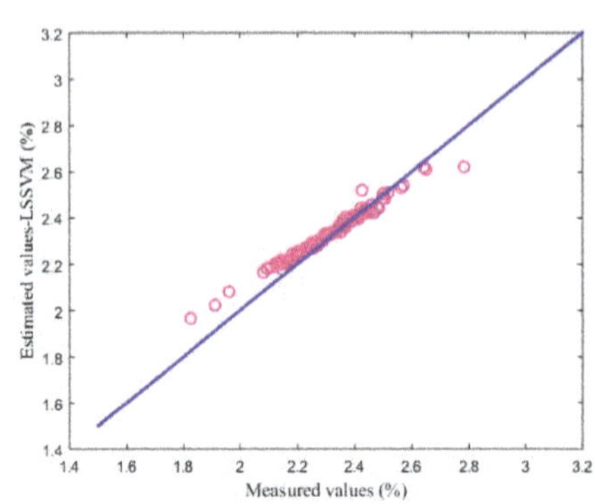

Figure 9. Comparison between the soft sensor based on LSSVM and laboratory measurements.

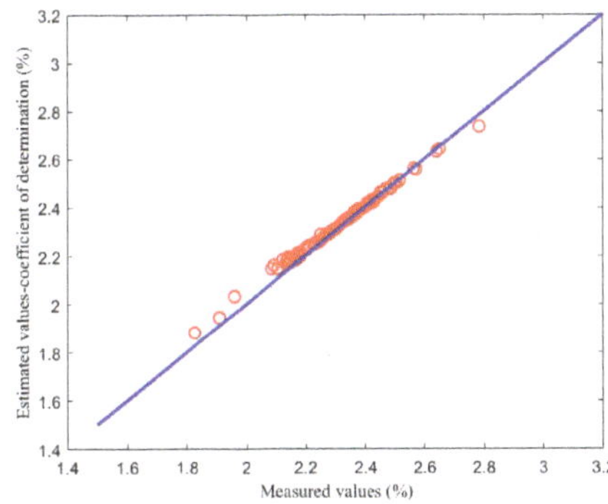

Figure 10. Comparison between the soft sensor based on maximizing the coefficient of determination and laboratory measurements.

To better illustrate the performance of the proposed soft sensor model, Table 5 shows the RMSE values for the different methods. As can be seen from Table 5, the RMSE of the proposed method is smallest, which means that the estimation effect of the proposed model is better than those of the BP neural network model and the LSSVM model.

Table 5. The comparison of the RMSE between the three modelling methods.

Method	RMSE
BP neural network	0.0616
LSSVM	0.0431
Maximizing the Coefficient of Determination	0.0231

5. Conclusions

In this paper, a new KPI estimation method for probabilistic soft sensor development is proposed based on maximizing the coefficient of determination. The joint probability distribution in the probability model is approximated using GMM, while the EM algorithm is used to estimate the GMM parameters. In addition to providing accurate, real-time estimates of the KPIs, this paper also considers the missing values that training sample sets often face and uses the EM algorithm for processing. The resulting soft sensor design method was tested on a case study of the alumina extraction process, which shows that the proposed method can provide alumina concentration estimations that are consistent with the actual measurements obtained from laboratory tests. Future work will focus on applying the proposed soft sensor development approach to solving various problems such as dealing with dynamic, non-Gaussian, or batch processes.

Author Contributions: Y.Z. and X.Y. conceived the idea, while Y.A.W.S. and X.Y. provided assistance with the development and implementation of the methods. J.C. and C.T. provided the industrial data and the experimental set-up for case study, respectively. Y.Z. performed the simulations and analysed the data, with assistance from X.Y. and Y.A.W.S. Y.Z. wrote the paper with editorial assistance from Y.A.W.S. and X.Y.

Funding: This research was funded by the National Natural Science Foundation of China, grant number [#61673053 and #61603034]; the Beijing Natural Science Foundation, grant number [#4162041 and #3182027]; and the National Key R&D Program of China, grant number [2017YFB0306403].

Conflicts of Interest: The authors declare no conflict of interest.

References

1. Shardt, Y.A.W.; Mehrkanoon, S.; Zhang, K.; Yang, X.; Suykens, J.; Ding, X.S.; Peng, K.X. Modelling the Strip Thickness in Hot Steel Rolling Mills Using Least-squares Support Vector Machines. *Can. J. Chem. Eng.* **2018**, *96*, 171–178. [CrossRef]

2. Zhang, K.; Shardt, Y.A.W.; Chen, Z.W.; Yang, X.; Ding, S.X.; Peng, K.X. A KPI-Based Process Monitoring and Fault Detection Framework for Large-Scale Processes. *ISA Trans.* **2017**, *68*, 276–286. [CrossRef] [PubMed]
3. Stanojevic, P.; Orlic, B.; Misita, M.; Tatalovic, N.; Lenkey, G.B. Online Monitoring and Assessment of Emerging Risk in Conventional Industrial Plants: Possible Way to Implement Integrated Risk Management Approach and KPI's. *J. Risk Res.* **2013**, *16*, 501–512. [CrossRef]
4. Paulsson, D.; Gustavsson, R.; Mandenius, C.F. A Soft Sensor for Bioprocess Control Based on Sequential Filtering of Metabolic Heat Signals. *Sensors* **2014**, *14*, 17864–17882. [CrossRef] [PubMed]
5. Abeykoon, C. A novel soft sensor for real-time monitoring of the die melt temperature profile in polymer extrusion. *IEEE Trans. Ind. Electron.* **2014**, *61*, 7113–7123. [CrossRef]
6. Yuan, X.F.; Ge, Z.Q.; Huang, B.; Song, Z.H. A Probabilistic Just-in-Time Learning Framework for Soft Sensor Development with Missing Data. *IEEE Trans. Control Syst. Technol.* **2017**, *25*, 1124–1132. [CrossRef]
7. Chen, K.; Liang, Y.; Gao, Z.L. Just-in-Time Correntropy Soft Sensor with Noisy Data for Industrial Silicon Content Prediction. *Sensors* **2017**, *17*, 1830. [CrossRef] [PubMed]
8. Khatibisepehr, S.; Huang, B.; Khare, S. Design of inferential sensors in the process industry: A review of Bayesian methods. *J. Process Control* **2013**, *23*, 1575–1596. [CrossRef]
9. Serdio, F.; Lughofer, E.; Zavoianu, A.C.; Pichler, K.; Pichler, M.; Buchegger, T.; Efendic, H. Improved fault detection employing hybrid memetic fuzzy modeling and adaptive filters. *Appl. Soft. Comput.* **2017**, *51*, 60–82. [CrossRef]
10. Serdio, F.; Lughofer, E.; Pichler, K.; Buchegger, T.; Pichler, M.; Efendic, H. Fault detection in multi-sensor networks based on multivariate time-series models and orthogonal transformations. *Inf. Fusion* **2014**, *20*, 272–291. [CrossRef]
11. Shardt, Y.A.W.; Hao, H.Y.; Ding, S.X. A New Soft-Sensor-Based Process Monitoring Scheme Incorporating Infrequent KPI Measurements. *IEEE Trans. Ind. Electron.* **2015**, *62*, 3843–3851. [CrossRef]
12. Yan, W.W.; Shao, H.H.; Wang, X.F. Soft sensing modeling based on support vector machine and Bayesian model selection. *Comput. Chem. Eng.* **2004**, *28*, 1489–1498. [CrossRef]
13. Kadlec, P.; Gabrys, B.; Strandt, S. Data-driven Soft Sensors in the process industry. *Comput. Chem. Eng.* **2009**, *33*, 795–814. [CrossRef]
14. Shang, C.; Gao, X.Q.; Yang, F. Novel Bayesian Framework for Dynamic Soft Sensor Based on Support Vector Machine with Finite Impulse Response. *IEEE Trans. Control Syst. Technol.* **2014**, *22*, 1550–1557.
15. Fujiwara, K.; Kano, M.; Hasebe, S. Development of correlation-based pattern recognition algorithm and adaptive soft-sensor design. In Proceedings of the IFAC Symposium on Advanced Control of Chemical Processes (ADCHEM), Istanbul, Turkey, 12–15 July 2009.
16. Yuan, X.; Ye, L.; Bao, L.; Ge, Z.; Song, Z. Nonlinear feature extraction for soft sensor modeling based on weighted probabilistic PCA. *Chemom. Intell. Lab. Syst.* **2015**, *147*, 167–175. [CrossRef]
17. Geladi, P. Notes on the history and nature of partial least squares (PLS) modelling. *J. Chemom.* **1988**, *2*, 231–246. [CrossRef]
18. Qin, S.J.; McAvoy, T.J. Nonlinear PLS modeling using neural networks. *Comput. Chem. Eng.* **1992**, *16*, 379–391. [CrossRef]
19. Khatisbisepehr, S.; Huang, B. Dealing with Irregular Data in Soft Sensors: Bayesian Method and Comparative Study. *Ind. Eng. Chem. Res.* **2008**, *47*, 8713–8723. [CrossRef]
20. Qi, F.; Huang, B.; Tamayo, E.C. A Bayesian Approach for Control Loop Diagnosis with Missing Data. *AICHE J.* **2010**, *56*, 179–195. [CrossRef]
21. Newman, D.A. Missing Data: Five Practical Guidelines. *Organ. Res. Methods* **2014**, *17*, 372–411. [CrossRef]
22. Zhang, K.K.; Gonzalez, R.; Huang, B.; Ji, G.L. Expectation-Maximization Approach to Fault Diagnosis with Missing Data. *IEEE Trans. Ind. Electron.* **2015**, *62*, 1231–1240. [CrossRef]
23. Shardt, Y.A.W. *Statistics for Chemical and Process Engineers: A Modern Approach*; Springer International Publishing: Cham, Switzerland, 2015; ISBN 978-3-319-21508-2.
24. Feng, C.H.; Makino, Y.; Yoshimura, M.; Rodriguez, F.J. Estimation of adenosine triphosphate content in ready-to-eat sausages with different storage days, using hyperspectral imaging coupled with R statistics. *Food Chem.* **2018**, *264*, 419–426. [CrossRef] [PubMed]
25. Sezer, B.; Apaydin, H.; Bilge, G.; Boyaci, I.H. Coffee arabica adulteration: Detection of wheat, corn and chickpea. *Food Chem.* **2018**, *264*, 142–148. [CrossRef] [PubMed]

26. Sun, S.L.; Zhang, C.S.; Yu, G.Q. A Bayesian network approach to traffic flow forecasting. *IEEE Trans. Intell. Transp. Syst.* **2006**, *7*, 124–132. [CrossRef]
27. Dempster, A.P.; Laird, N.M.; Rubin, D.B. Maximum likelihood from incomplete data via the EM algorithm. *J. R. Stat. Soc. B* **1977**, *39*, 1–38.
28. Stock, J.H.; Watson, M.W. *Introduction to Econometrics*, 3rd ed.; Addison-Wesley: Bosten, MA, USA, 2010; ISBN 978-0-13-800900-7.
29. Johnson, R.A.; Wichern, D.W. *Applied Multivariate Statistical Analysis*, 6th ed.; Pearson: London, UK, 2007; ISBN 978-0-13-187715-3.
30. Rao, C.R. *Linear Statistical Inference and Its Applications*; Wiley: New York, NY, USA, 1973; ISBN 978-0-47-031643-6.
31. Bilmes, J.A. A gentle tutorial of the EM algorithm and its application to parameter estimation for Gaussian mixture and hidden Markov model. *Int. Comput. Sci. Inst.* **1998**, *4*, 126.
32. Mouedhen, G.; Feki, M.; Wery, M.D.P.; Ayedi, H.F. Behavior of aluminum electrodes in electrocoagulation process. *J. Hazard. Mater.* **2008**, *150*, 124–135. [CrossRef] [PubMed]
33. Yao, Y.C.; Cheung, C.Y.; Bao, J.; Skyllas-Kazacos, M.; Welch, B.; Akhmetov, S. Estimation of spatial alumina concentration in an aluminium reduction cell using a multilevel state observer. *AICHE J.* **2017**, *63*, 2806–2818. [CrossRef]
34. Zhang, S.; Zhang, T.; Yin, Y.X.; Xiao, W.D. Alumina concentration detection based of the kernel extreme learning machine. *Sensors* **2017**, *17*, 2002. [CrossRef] [PubMed]

sensors

MDPI

Article

A Time-Distributed Spatiotemporal Feature Learning Method for Machine Health Monitoring with Multi-Sensor Time Series

Huihui Qiao [1,2], Taiyong Wang [1,2,*], Peng Wang [1,2], Shibin Qiao [3] and Lan Zhang [1,2]

1 Key Laboratory of Mechanism Theory and Equipment Design of Ministry of Education, Tianjin University, Tianjin 300350, China; huihuiqiao@tju.edu.cn (H.Q.); pengwang@tju.edu.cn (P.W.); tju_zhanglan@tju.edu.cn (L.Z.)
2 School of Mechanical Engineering, Tianjin University, Tianjin 300354, China
3 Institute for Special Steels, Central Iron and Steel Research Institute, Beijing 100081, China; zjjhxmd29@tju.edu.cn
* Correspondence: tywang@tju.edu.cn; Tel.: +86-181-0203-2867

Received: 25 July 2018; Accepted: 28 August 2018; Published: 3 September 2018

check for updates

Abstract: Data-driven methods with multi-sensor time series data are the most promising approaches for monitoring machine health. Extracting fault-sensitive features from multi-sensor time series is a daunting task for both traditional data-driven methods and current deep learning models. A novel hybrid end-to-end deep learning framework named Time-distributed ConvLSTM model (TDConvLSTM) is proposed in the paper for machine health monitoring, which works directly on raw multi-sensor time series. In TDConvLSTM, the normalized multi-sensor data is first segmented into a collection of subsequences by a sliding window along the temporal dimension. Time-distributed local feature extractors are simultaneously applied to each subsequence to extract local spatiotemporal features. Then a holistic ConvLSTM layer is designed to extract holistic spatiotemporal features between subsequences. At last, a fully-connected layer and a supervised learning layer are stacked on the top of the model to obtain the target. TDConvLSTM can extract spatiotemporal features on different time scales without any handcrafted feature engineering. The proposed model can achieve better performance in both time series classification tasks and regression prediction tasks than some state-of-the-art models, which has been verified in the gearbox fault diagnosis experiment and the tool wear prediction experiment.

Keywords: multi-sensor time series; deep learning; machine health monitoring; time-distributed ConvLSTM model; spatiotemporal feature learning

1. Introduction

Accurate and real-time monitoring of machine health status has great significance. Appropriate maintenance strategies can be adopted depending on the real-time health status of the machine to avoid catastrophic failures, shorten downtime and reduce economic losses. Machine health monitoring (MHM) is of great significance to ensure the safety and reliability of equipment operation. Modern complex machinery systems, such as CNC machining equipment and trains, are moving in the direction of large-scale, complex, high-precision, reliable and intelligent. Moreover, the features of the signal to be processed vary with different devices, different operating conditions and different fault conditions [1]. Therefore, it puts forward higher requirements for the accuracy, efficiency and versatility of condition monitoring and fault diagnosis methods.

With the rapid development of advanced sensing technology and affordable storage, it is much easier to acquire mechanical condition data, enabling large scale collection of time series data. With the massive monitoring data, data-driven methods have been greatly developed and applied in the field of MHM. The framework of the traditional data-driven MHM system includes four steps: signal acquisition, feature extraction, feature reduction and condition classification or prediction [2,3].

The types of sensor signal commonly used in MHM systems include: vibration signal, acoustic emission signal, force, rotational speed, current signal, power signal, temperature and so forth. Different types of sensors and measuring positions are sensitive to different types of damage and have different advantages and limitations. Multi-sensor fusion data that contains redundant and complementary mechanical health information can help the MHM system to achieve higher diagnostic accuracy and detect more failures than a single type of sensor data [4]. The monitoring values of each sensor channel constitute a 1D temporal sequence in chronological order. If the outputs of a plurality of sensor channels are arranged in parallel and all channels have the same sampling rate, a 2D spatiotemporal sequence data is formed. Since multiple measuring channels with high sampling frequency and long data collection period are required for each mechanical functional component, the 2D spatiotemporal sequence not only includes the local spatial-domain dependency between different channels but also includes the time-domain dependency of each channel data, which presents complex temporal correlation and spatial correlation. In addition, the 2D spatiotemporal sequence is dynamic, non-linear, multivariable, high redundant and strong noisy, which poses a huge challenge to feature extraction.

Many studies have made great efforts in handcrafted feature extraction methods and feature reduction methods. However, conventional handcrafted methods still suffer from weaknesses in the following areas: (1) The handcrafted feature extraction and feature reduction methods need to be designed according to different kinds of monitored objects and signal sources, which depend on prior domain knowledge and expert experience [2]. As a result, these methods present low efficiency and poor generalization performance [5]. Especially when facing multi-sensor based condition monitoring tasks, due to the influence of noise and a large amount of redundant information, the feature engineering is more difficult and labor-intensive. It is difficult to select the suitable data fusion method and extract sparse features, which directly affects the performance MHM models [6]. (2) Considering that feature extraction and machine learning models that work in a cascaded way are independent of each other, without considering the relationship between them, so it is impossible to jointly optimize them [7]. The extracted features of input data determine the performance of the subsequent classification or prediction models [8], therefore, it is necessary to explore an effective method for multi-sensor time series feature extraction.

Deep learning (DL) [9] provides a powerful solution to above weaknesses. Unlike traditional models that are mostly based on handcrafted features, Deep neural network (DNN) can operate directly on raw data and learn features from a low level to a higher level to represent the distributed characteristics of data [10], which doesn't require additional domain knowledge. After the layer-wise feature learning, DNN can adaptively extract deep and essential features according to the internal structure of massive data without domain knowledge. DNN has been successfully applied in speech recognition [11], image classification [12], motion recognition [13], text processing [14] and many other domains. In the past few years, the typical deep learning frameworks including deep autoencoder (DAE), deep belief network (DBN), deep convolutional neural network (DCNN), deep recurrent neural network (DRNN) and their variants have been developed in the field of machine health monitoring [15,16].

The unsupervised layer-by-layer pre-training process of DAE and DBN can reduce the need for labeled training samples and facilitate the training of DNN [15]. However, these fully-connected structures of DAE and DBN may lead to heavy computation cost and overfitting problems caused by huge model parameters, so DAE and DBN are not suitable for processing raw data, especially multi-sensor raw data.

The Convolutional neural network (CNN) is now the most prominent framework, which is usually used for learning spatial distribution of data. In the CNN model, the local connection mechanism between layers allows CNN to learn the local features of the data and the weight sharing mechanism can reduce model parameters. As a method to prevent overfitting, the spatial pooling layer of CNN can help the model learn more significant and robust features. The special structure of CNN can reduce the complication as well as the training time of the model. CNN has also been introduced to address time series data for mechanical fault diagnosis or remaining useful life estimation [17–19]. However, since the time series data is treated as static spatial data in CNN, where the sequential and temporal dependency are not taken into account, it may lead to the loss of most information between time steps [20].

The sensor data in the MHM system is usually a natural time series. Opposed to CNN, Long Short-Term Memory (LSTM) working in temporal domain is capable of sequence processing. As an advanced RNN variant, LSTM can adaptively capture long-term dependencies and nonlinear dynamics of time series data [21]. Although LSTM can directly receive raw data as input [20,22] and has been proven to be powerful for modeling time series data in MHM tasks, it does not take spatial correlation into consideration and easily leads to overfitting for multi-channel time series data containing crucial temporal and spatial dependencies.

Given the complementary strengths of CNN and LSTM, ConvLSTM is proposed for spatiotemporal sequence forecasting in [23]. Compared with LSTM, ConvLSTM preserves the spatial information [24], therefore it facilitates the spatiotemporal feature learning. Multi-sensor time series data in MHM tasks usually have high sampling rate (such as vibration signals and acoustic emission signals), so the 2D time series are sequential with long-term temporal dependency. The input sample always contains thousands of timestamps. The information of single timestamp may not be discriminative enough. Therefore, extracting local features in a short period of time can make it easier to learn long temporal dependencies between successive timestamps and often produce a better performance [7,25].

In this paper, a novel framework named Time-distributed ConvLSTM (TDConvLSTM) is proposed for intelligent MHM, which is powerful for learning spatiotemporal features of multi-sensor time series data on different time scales. TDConvLSTM is a hybrid end-to-end deep learning model, which has 5 main components: a data segmentation layer, time-distributed local spatiotemporal feature extractors, a holistic ConvLSTM layer, a fully-connected (FC) layer and a supervised learning layer. Firstly, the data segmentation layer utilizes a sliding window strategy along the temporal dimension to segment the normalized multi-sensor time series data into a collection of subsequences. Each subsequence is a 2D tensor and is taken as one time step in the holistic ConvLSTM layer. Then, all the subsequences are arranged in sequence and transformed into a 3D tensor. The local spatiotemporal feature extractor is applied to each time step to extract local spatiotemporal features inside a subsequence. The Holistic ConvLSTM layer can extract holistic spatiotemporal features between subsequences based on the local spatiotemporal features. Then a FC layer and a softmax or regression layer are stacked on the top of the model for classification or regression prediction. The main contributions of this paper are summarized as follows:

1. The ConvLSTM is first applied to extract spatiotemporal features of multi-sensor time series for real-time machine health monitoring tasks. It can learn both the complex temporal dependency and spatial dependency of multi-sensor time series, enabling the ConvLSTM to discover more hidden information than CNN and LSTM.
2. The time-distributed structure is proposed to learn both short-term and long-term features of time series. Therefore, it can make full use of information on different time scales.
3. The proposed end-to-end TDConvLSTM model directly works on raw time series data of multi-sensor and can automatically extract optimal discriminative features without any handcrafted features or expert experience. The time-distributed spatiotemporal feature learning method is not limited to a specific machine type or a fault type. Therefore, TDConvLSTM has wide applicability in MHM systems.

4. The proposed model is suitable for multisensory scenario and achieves better performance in both time series classification tasks and regression prediction tasks than some state-of-the-art models, which has been verified in the gearbox fault diagnosis experiment and the tool wear prediction experiment.

The remainder of the paper is organized as follows: In Section 2, machine health monitoring method based on CNN and LSTM are reviewed. In Section 3, the typical architecture of LSTM and ConvLSTM are briefly described. Section 4 illustrates the procedures of the proposed method. In Section 5, a gearbox fault diagnosis experiment and a tool wear prediction experiment are used to validate the effectiveness of the proposed method. Finally, conclusions are drawn in Section 6.

2. Related Work

2.1. Machine Health Monitoring Based on CNN

In some works, the raw sensor data in time domain has been transformed to frequency spectrum or time-frequency spectrum before being input to CNN models. The spectral energy maps of the acoustic emission signals are utilized as the input of CNN to automatically learn the optimal features for bearing fault diagnosis in [26]. Ding et al. proposed a deep CNN where wavelet packet energy images were used as input for spindle bearing fault diagnosis [27]. The methods presented above that indirectly processing time series data using CNN are time-consuming and limited by frequency domain and time-frequency domain transformation methods.

CNN can also directly address raw temporal signals in MHM tasks without any time-consuming preliminary frequency or time-frequency transformation. Zhang et al. presented a novel rolling element bearings fault diagnosis algorithm based on CNN, which performs all the operation on the raw temporal vibration signals without any other transformation [28]. Lee et al. addressed a CNN model for fault classification and diagnosis in semiconductor manufacturing processes with multivariate time-series data as the input [29]. In [19], CNN was first adopted as a regression approach for remaining useful life (RUL) estimation with multi-sensor raw data as model input. The raw time series data is treated as static spatial distribution data in CNN and its long temporal dependency information is lost, which makes CNN models perform poorly and error-prone.

2.2. Machine Health Monitoring Based on LSTM

A LSTM based encoder-decoder scheme was proposed in [30] for anomaly detection, which can learn to reconstruct the "normal" time-series and thereafter the reconstruction error was used to detect anomalies. Based on the work in [30], an advanced LSTM encoder-decoder was proposed to obtain a health index in an unsupervised manner using multi-sensor time series data as input and thereafter the health index was used to learn a model for estimation of remaining useful life [31]. Bruin et al. utilized a LSTM network to timely detect faults in railway track circuits [32]. They compared the LSTM network with a convolutional network on the same task. It was concluded that the LSTM network outperforms the convolutional network for the track circuit case, while the convolutional networks are easier to train. Zhao et al. applied LSTM model encoded the raw sensory data into embedding and predicted the corresponding tool wear [22].

Due to the fact that multi-sensor time series data of mechanical equipment usually have high sampling rate, the input sequence may contain thousands of timestamps. Although the LSTM can directly work on raw time series data, the high dimensionality of input data will increase model size and make the model hard to train.

2.3. Hybrid Models Based on CNN and LSTM for Machine Health Monitoring

The hybrid models connecting CNN layers and LSTM layers in order, which expressed as CNN-LSTM in this paper, have been designed to extract both spatial and temporal features for speech recognition [33], emotion recognition in video [34] and gesture recognition [35] and so forth.

A deep architecture was proposed for automatic stereotypical motor movements (SMM) detection by stacking an LSTM layer on top of the CNN architecture in [36]. Based on the work in [36], a further research that enhancing the performance of SMM detectors was presented in [37]. In the research, CNN was used for parameter transfer learning to enhance the detection rate on longitudinal data and ensemble learning was employed to combine multiple LSTM learners into a more robust SMM detector. In MHM tasks, the sensor data is often a multi-channel time series, which contains both temporal and spatial dependencies. The combination of CNN and LSTM has achieved higher performance on MHM tasks than single CNN and single LSTM [32]. Zhao et al. [38] designed a deep neural network structure named Convolutional Bi-directional Long Short-Term Memory networks (CBLSTM). One-layer CNN was applied in the model to extract local and discriminative features from raw input sequence, after which, two-layer bi-directional LSTMs were built on top of the previous CNN to encode the temporal information. The CBLSTM was able to outperform several state-of-the-art baseline methods in the tool wear estimation task.

CNN-LSTM models usually learn spatial features first and thereafter learn temporal features. However, one layer in ConvLSTM can learn the temporal features and spatial features simultaneously by using convolutions operation to replace the matrix multiplication within the LSTM unit and pay more attention to how data changes between time steps. ConvLSTM has been used to extract spatiotemporal features of weather radar maps [23] and videos [39,40] but no application of ConvLSTM in MHM tasks has been found so far.

3. Introduction of ConvLSTM

3.1. Convolutional Operation

The convolutional layer and the activation layer are the most central parts of the CNN. Input data is first convoluted with the convolution kernel and the convolutional output is added with an offset. Then the following activation unit is used to generate the output features. The convolutional operation uses a local connected and weight shared method. Compared with traditional fully-connected layers, the convolutional layer can reduce model parameters and improve model calculation speed, which is more suitable for directly processing complex input data and extracting local features.

A convolutional layer usually contains multiple convolution kernels, that is, multiple filters. Assuming that the number of convolution kernels is k, each convolution kernel is used to extract one type of feature, corresponding to one feature matrix and k convolution kernels can output a total of k feature matrices. The convolutional operation can be expressed by:

$$Z_k = f(W_K * X + b) \tag{1}$$

where X is the input data with size of $m \times n$. W_K is the Kth convolution kernel with size of $k_1 \times k_2$. b denotes the offset. '$*$' denotes the convolution operator. The stride and the padding method in convolutional operation together determine the size of the Kth feature matrix Z_k. For example, when stride is (1,1) and using no padding during convolution, the size of Z_k is $(m - k_1 + 1) \times (n - k_2 + 1)$. f is the nonlinear activation function which performs nonlinear transformation on the output of the convolutional layer. The commonly used activation functions are sigmoid, tanh and ReLu.

3.2. From LSTM to ConvLSTM

LSTM has been proven to be the most stable and powerful model to learn long-range temporal dependences in practical applications as compared to standard RNNs or other variants. The structure of the repeating module in the LSTM is shown in Figure 1. The LSTM uses three 'gate' structures to control the status of the memory cell c_t. The three gates have the ability to remove or add information to the cell state. The three gates are input gate i_t, forget gate f_t and output gate o_t, which can be understood as a way to optionally allow information to pass through [41]. The process of information passing and updating in LSTM can be described by the equations shown in (2)–(7), where $''$ denotes the Hadamard product. At each time step t, the memory cell state c_t and the hidden state h_t can be update by the current input x_t, the hidden state at previous time step h_{t-1} and the memory cell state at previous time step c_{t-1}. When a new input comes, f_t can decide how many information in c_{t-1} should be forgotten. Then, i_t and $\widetilde{c_t}$ will decide what new information can be store in the cell state. The next step is to update the old cell state c_{t-1} into the new cell state c_t. Finally, x_t, h_{t-1} and c_t determine the output h_t. The input, cell state and output of the LSTM are all 1D vectors. The LSTM uses full connections in input-to-state and state-to-state transitions.

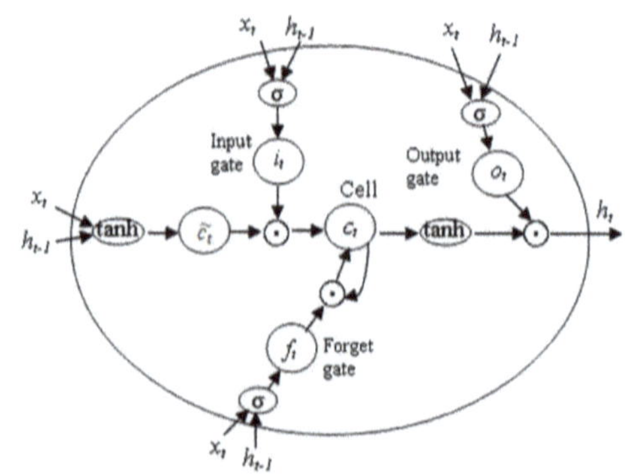

Figure 1. The structure of three gates in the Long Short-Term Memory (LSTM) cell.

$$f_t = \sigma\left(W_{xf}\, x_t + W_{hf}h_{t-1} + b_f\right) \tag{2}$$

$$i_t = \sigma(W_{xi}x_t + W_{hi}h_{t-1} + bi) \tag{3}$$

$$\widetilde{c_t} = tanh(W_{xc}x_t + W_{hc}h_{t-1} + b_c) \tag{4}$$

$$c_t = f_t c_{t-1} + i_t\widetilde{c_t} \tag{5}$$

$$o_t = \sigma(W_{xo}\, x_t + W_{ho}h_{t-1} + bo) \tag{6}$$

$$h_t = o_t tanh(c_t) \tag{7}$$

LSTM is capable of modeling time series data with long-term dependency in MHM tasks. Although LSTM can also be applied on multi-dimensional sequence by reshaping the multi-dimensional input to a 1D vector but it fails to maintain structural locality [39] and contains too much redundancy [23].

To exploit both spatial and temporal information in multi-sensor time series data, we proposed a model based on ConvLSTM. ConvLSTM is an extension of LSTM, which replaces the matrix multiplication in LSTM with convolutional operation [23]. The equations of ConvLSTM are shown in (8)–(13), where '$*$' denotes the convolution operator. The input x_t, cell state c_t and hidden output h_t are all 3D tensors, where the first two dimensions are spatiotemporal information and the last dimension is the number of convolutional filters. The convolutional operation of ConvLSTM can reduce the number of model parameters and prevent overfitting [42]. ConvLSTM retains the advantages of learning temporal dependency between different time steps, in addition to this, it can capture the local spatial information. Therefore, ConvLSTM can learn more discriminative features from multi-sensor time series data.

$$f_t = \sigma\left(W_{xf} * x_t + W_{hf} * h_{t-1} + b_f\right) \tag{8}$$

$$i_t = \sigma(W_{xi} * x_t + W_{hi} * h_{t-1} + bi) \tag{9}$$

$$\widetilde{c_t} = tanh(W_{xc} * x_t + W_{hc} * h_{t-1} + b_c) \tag{10}$$

$$c_t = f_t c_{t-1} + i_t \widetilde{c_t} \tag{11}$$

$$o_t = \sigma(W_{xo} * x_t + W_{ho} * h_{t-1} + bo) \tag{12}$$

$$h_t = o_t tanh(c_t) \tag{13}$$

4. Methods

4.1. Notation

In the multisensory MHM scenario, the time series collected from monitored machine is a sequence of real-valued data points generated by M different sensor channels. The input sample of the model can be represented as a 2D matrix, which is denoted as $X = \{x_1, x_2, \cdots, x_L\}$, where L is the length of the sample and the input data x_i at the ith timestamp is a vector with M elements. Each training sample has a corresponding target value Y. Y is a categorical value that has been encoded to a one-hot vector in the fault classification task or a real-valued data in the regression prediction task. The machine health monitoring task is defined to obtain the target value Y based on multi-sensor time series data X. In the following text, we divide X into N local subsequences, then, the input can be denoted as $X = \{P_{T1}, P_{T2}, \cdots, P_{TN}\}$, each subsequence $P_{Ti} \in R^{M \times l}$ is denoted as $P_{Ti} = \left\{x_{Ti}^1, x_{Ti}^2, \cdots, x_{Ti}^l\right\}$, where $x_{Ti}^k \in R^M$ is the kth timestamp in the ith subsequence. l is the length of each subsequence. Further, (A,B) represents the shape of a tensor with A rows and B columns.

4.2. The Proposed TDConvLSTM Model

In this section, a time-distributed ConvLSTM model (TDConvLSTM) is presented for multi-sensor time series based machine health monitoring. TDConvLSTM is a hybrid end-to-end framework that focuses on time-distributed spatiotemporal feature learning which is an extension method of basic ConvLSTM. The basic ConvLSTM model are consists of only a few ConvLSTM layers, a FC layer and a supervised learning layer, which is shown in Figure 2. ConvLSTM directly extract spatiotemporal features in the whole range of the multi-sensory input data. Although ConvLSTM can directly work on multi-sensor time series data to simultaneously capture the temporal dependencies and spatial dependencies, the input time series in the MHM task always contains thousands of timestamps, which will make the model size too large and make it difficult to train the model. The basic ConvLSTM model cannot learn long temporal dependencies well. Therefore, extracting local features in a local range of the input data before extracting features in the whole range can make it easier to learn long temporal dependencies between successive timestamps and promote the model for better performance.

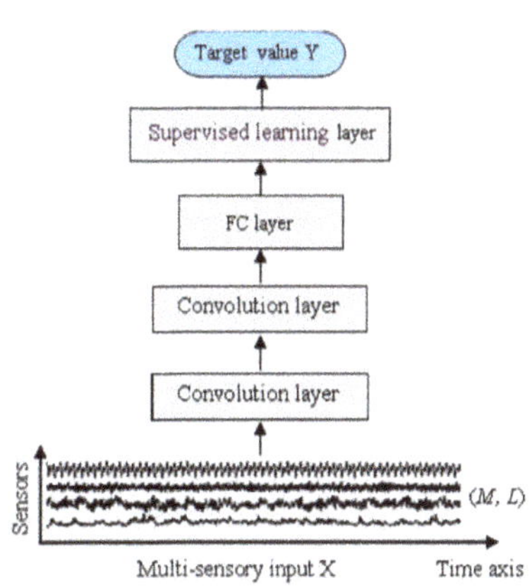

Figure 2. The framework of the basic ConvLSTM model.

Considering the above shortcomings of basic ConvLSTM, a time-distributed ConvLSTM model (TDConvLSTM) is proposed. The proposed TDConvLSTM has three major procedures: data segmentation, time-distributed local spatiotemporal features extraction and holistic spatiotemporal features extraction. The framework of the TDConvLSTM model is shown in Figure 3. Firstly, the normalized multi-sensor time series is segmented into a collection of subsequences using a sliding window along the time dimension. Then all the subsequences are reorganized into the shape that fit into the subsequent time-distributed local spatiotemporal feature extraction layers. Holistic ConvLSTM layers stacked on the top of time-distributed local spatiotemporal feature extraction layers are used to extract holistic spatiotemporal features between subsequences based on the time-distributed local spatiotemporal features. At last, a FC layer and a supervised learning layer are stacked on the top of the model to obtain the target value Y. The local spatiotemporal features extracted in each subsequence only contain the features of a part of the input data. The holistic features are extracted from local features of all subsequences to learn the long temporal spatiotemporal dependencies between subsequences. So, the holistic features contain the spatiotemporal features of the whole input data. Local spatiotemporal features are extracted before extracting holistic spatiotemporal features, which can make it easier to learn long temporal dependencies between successive timestamps and enable the TDConvLSTM to get better performance.

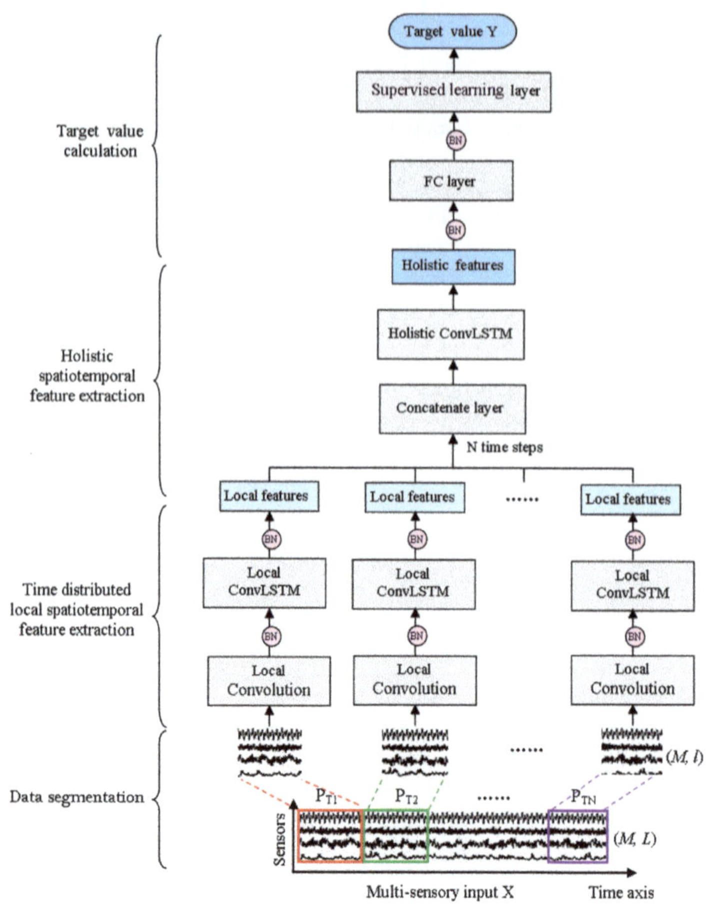

Figure 3. The framework of the proposed TDConvLSTM model (TDConvLSTM).

4.2.1. Data Normalization and Segmentation

Each channel of the multi-sensor time series data may come from different kinds of sensors and the order of magnitude of each channel may be different. If the raw multi-sensor time series data is used directly to train the model, the model will be difficult to converge. Therefore, the raw multi-sensor time series data is normalized by the z-score method before being input to the model. The main purpose of z-score is to convert data of different magnitudes into the same order of magnitude to ensure the comparability between the data. The conversion function can be expressed as:

$$x_j^z = \frac{x_j - \mu_j}{\sigma_j} \tag{14}$$

where, x_j is the time series of the jth sensor channel. μ_j and σ_j are the mean and standard deviation of x_j. x_j^z is the time series data after z-score normalization.

The input of the model $X \in R^{M \times L}$ is segmented into N local subsequences by a sliding window along the time dimension. All subsequences have the same length l *and* $N = \frac{L}{l}$. Setting l to a value that makes L divisible by l is more appropriate. If L cannot be divisible by l, the remainder part will be discarded. In other words, just the integer part of N will be retained. The demonstration of data segmentation is shown in Figure 3. Each subsequence $P_{Ti} \in R^{M \times l}$ is a window of the multi-sensor input signal and is regarded as one-time step in the holistic ConvLSTM layer.

The length l of each subsequence is a hyperparameter of the TDConvLSTM model, which controls the number of subsequences, that is, the number of time steps in the holistic ConvLSTM layer. It is obvious that a small l may not be able to obtain discriminative local features. Oppositely, if l is large, the number of time steps of the holistic ConvLSTM layer will be decreased, so that much holistic spatiotemporal information will be lost. A suitable l can be selected by comparing experiments.

4.2.2. Time-Distributed Local Spatiotemporal Feature Extraction

After data segmentation, a local feature extractor is used to extract spatiotemporal features of each subsequence. The local feature extractor is applied to each subsequence P_{Ti} simultaneously using a "TimeDistributed wrapper," which is shown in Figure 3. N local feature extraction processes are performed simultaneously and independently of each other.

Since local feature extractors of different time steps have the same structure, we focus on one local feature extractor with the input subsequence $P_{Ti} = \left\{x_{Ti}^1, x_{Ti}^2, \cdots, x_{Ti}^l\right\}$, which is a $(l \times M)$ tensor. P_{Ti} is divided into n slices, each slice $f_q, (q = 1, 2, \cdots n)$ is a $(l_0 \times M)$ tensor, where $l_0 = \frac{l}{n}$. As a result, the input P_{Ti} is transformed into a 3D tensor with shape of (n, M, l_0). We can think of the 3D tensor as a movie and f_q is a frame in the movie. Then the 3D tensor will be input to the local feature extractor.

The local feature extractor consists of a time-distributed local convolutional layer and a local ConvLSTM layer, which is shown in Figure 4. There are two types of features embedded in P_{Ti}, that is, temporal features inside a sensor channel and spatial features between different sensor channels. We applied the convolutional operation to each frame f_q simultaneously using a local "TimeDistributed wrapper." 2D kernels with shape $(k_1, 1)$ are applied in the first local convolutional layer to extract features inside a sensor channel and preserve the independence of each channel. In addition, we choose a larger convolution stride $(k_1, 1)$ than conventional $(1, 1)$ used in image recognition. The large convolutional stride can reduce the dimensionality of the input data and keep the timing unchanged. We apply c filter channels to each of the first three layers, which enable the model to get more non-linear functions and learn more information of the current sequence. The local convolutional layer returns a feature with shape of (n, l_1, M, c), which is thereafter served as the input of the local ConvLSTM layer. In the local ConvLSTM layer, 2D kernels with shape (k_2, M) and convolutional stride $(1, 1)$ are adopted to learn deeper temporal features and the dependencies between different sensor channels. The local ConvLSTM layer returns a feature Lf_{Ti} with shape of (n, l_2, c).

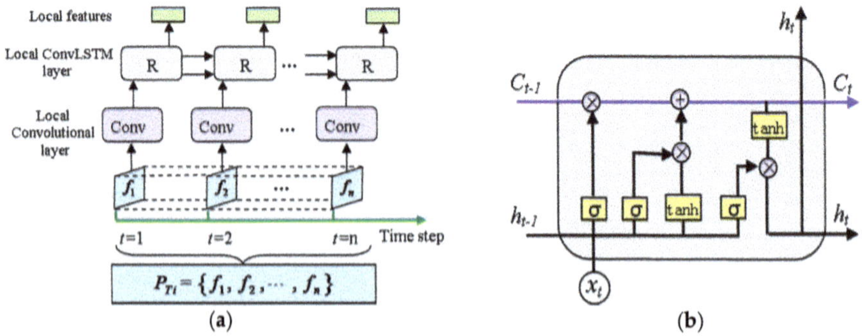

Figure 4. Local spatiotemporal feature extractor: (**a**) Structure of the local spatiotemporal feature extractor; (**b**) Diagram of the recurrent cell in ConvLSTM.

After two local feature extraction layers, local spatiotemporal features inside each subsequence P_{Ti} are extracted and the noise of the raw input data is eliminated. The time-distributed local feature extractors turn the raw input time series into a shorter sequence, which make it easier to learn long temporal dependencies.

4.2.3. Holistic Spatiotemporal Feature Extraction

After N time-distributed local spatiotemporal feature extractors, a local feature sequence with N time steps is returned with shape of (N, n, l_2, c). A holistic ConvLSTM layer is applied on the local feature sequence to extract holistic spatiotemporal features. There are N time steps in the holistic ConvLSTM layer and the local feature $Lf_{Ti} = (n, l_2, c)$ is the input at time step Ti. Small 2D kernels (k_3, k_3) and convolution stride $(1, 1)$ are adopted to further learn deeper and sparser spatiotemporal features. After holistic spatiotemporal feature extraction, holistic spatiotemporal feature at each time step is flattened into a 1D tensor, whose length is l_3. Then spatiotemporal features of N time steps are concatenated into a 1D feature vector v with length of $N \times l_3$.

4.2.4. Supervised Learning Layer

At last, the feature vector v is passed into another FC layer and a supervised learning layer. If the targets are discrete labels such as fault types, the supervised learning layer is a softmax layer, which is defined as:

$$P(y = j) = \frac{e^{\theta_j^T v}}{\sum_{k=1}^{K} e^{\theta_j^T v}} \tag{15}$$

where K is the number of labels and θ denotes parameters of softmax layer.

If the targets are continuous values such as remaining useful life (RUL) and tool wear, the supervised learning layer can be a linear-regression layer given by:

$$y = Wv + b \tag{16}$$

where W and b denote the transformation matrix and the offset in the linear regression layer.

The error between predicted values and truth values in training data can be calculated and back propagated to train the parameters of the whole model. Then, the trained model can be applied to monitor machine health condition.

4.2.5. Batch Normalization

The proposed model has a multi-layer structure. The model parameters change continuously in the training process, resulting in continuous changes in the input distribution of each subsequent layer. The learning process has to adapt each layer to the new input distribution, so the learning rate has to be reduced, resulting in a slow model converge rate. The batch normalization (BN) layer [43] is designed to reduce the shift of internal covariance and accelerate the training process of deep model by normalizing the output of each layer to obey the normal distribution. In our model, BN layers are added right after the local convolutional layer, the local ConvLSTM layer and the FC layer and before the activation unit. Assume that the input vector of the BN layer is x, $x \in R^m$, then the output of the BN layer can be calculated by:

$$y_i = \gamma x_i' + \beta \tag{17}$$

$$x_i' = \frac{x_i - \mu_B}{\sqrt{\sigma_B^2 + \epsilon}} \tag{18}$$

$$u_B = \frac{1}{m}\sum_{i=1}^{m} x_i \tag{19}$$

$$\sigma_B^2 = \frac{1}{m}\sum_{i=1}^{m} (x_i - \mu_B)^2 \tag{20}$$

where μ_B is the mean of x_i, σ_B^2 is the variance of x_i, ϵ is a small constant, γ and β are parameters that need to be learned in the model. BN can accelerate the convergence of the model and prevent overfitting. With BN layers, we can reduce the use of Dropout and adopt a large learning rate.

5. Experiments and Discussion

To verify the effectiveness of our proposed TDConvLSTM model, two experiments about gearbox fault diagnosis and real industrial milling tool wear monitoring were conducted.

5.1. Case Study 1: Gearbox Fault Diagnosis

5.1.1. Data Collection

To verify the effectiveness of the proposed TDConvLSTM model for gearbox fault diagnosis, an experiment was conducted on a gearbox test rig as shown in Figure 5a. The gearbox test rig is composed of three main units including the motor, the parallel gearbox and the magnetic powder brake. Four single-axis accelerometers were mounted vertically on the upper surface of the gearbox. Figure 5b shows the locations of sensors. The gearbox test rig has been operated under 4 different health conditions. The descriptions of different health conditions are listed in Table 1. Gearbox in each health condition has been operated at three speeds (280 rpm, 860 rpm and 1450 rpm) of the pinion. Vibration signals of 4 channels under each speed were acquired synchronously through a data acquisition box. The sampling frequency was 10.24 kHz and the sample time was 5s.

The acquired signals are first normalized as described in Section 4.2.1. Then, a sliding window with length of 5120 is used to slice the signals with overlap. There are 450 samples for each health condition under each identical operating speed. According to the operating speeds, all samples are grouped into four datasets (D1–D4) to test the performance of the proposed model respectively. D1, D2 and D3 contain samples at speeds of 280 rpm, 860 rpm and 1450 rpm, respectively. D1, D2 and D3 are brought together to form dataset D4. The samples at the three rotational speeds for each health condition are taken together as the same class in D4. There are 1800 samples in D1, D2 and D3, respectively. We first shuffle the sample order of D1, then, two-thirds of the 1800 samples are selected as the training dataset and the remaining one-third samples are selected as the testing dataset. The same processing procedure is used for D2 and D3. Finally, there are 1200 training samples and

600 testing samples in dataset D1, D2 and D3, respectively. There are 3600 training samples and 1800 testing samples in dataset D4.

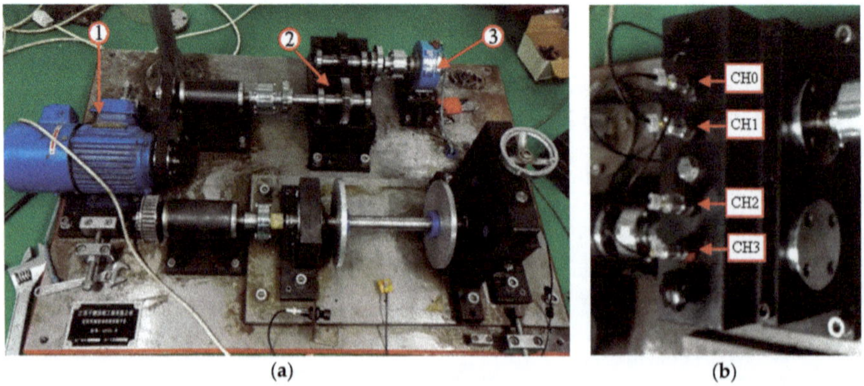

Figure 5. Gearbox test rig: (**a**) Main units: (1) Motor (2) Parallel gearbox (3) Magnetic powder brake (**b**) Locations of sensors.

Table 1. Description of four gearbox health conditions.

Label	Condition	Description	Speed (rpm)
0	FT	A root fracture tooth in the big gear	280, 860 and 1450
1	CT	A root crack tooth in the big gear	280, 860 and 1450
2	CTFT	A root crack tooth in the big gear and a half fracture tooth in the small gear	280, 860 and 1450
3	CIB	A crack on the inner race of the bearing	280, 860 and 1450

5.1.2. Parameters of the Proposed TDConvLSTM

The architecture of the proposed TDConvLSTM model used in experiments is built according the procedures described in Section 4. It should be noted that the hyperparameters of the model are selected through cross-validated experiments. The hyperparameters such as the kernels, strides and channels in main layers with the best performance are displayed in Table 2. Batch normalization is used right after each main layer to improve the performance of the model. The batch normalization axis is set to the channel axis. The activation function of the last layer is softmax and activation functions of other layers are all set to sigmoid. The categorical cross-entropy is adopted as the loss function and Adam is employed for model training. The dropout rate is set to 0.2.

As stated in Section 4.2.1, the length of subsequence l is a crucial hyperparameter of the TDConvLSTM model, so it is meaningful to research the influence of different l on the performance of the proposed model. In this research, we set l to 64, 128, 160, 256, 320, 512, 640 and 1024 respectively to test the performance of the model. Each subsequence P_{Ti} in the test model is divided into 8 slices. The other parameters of the model are same as shown before. An appropriate l is needed to fit the signals under different operating conditions, so the dataset acquired under nonstationary condition is suitable to test the performance of the model with different l. Dataset D4 is used to train the model for 15 epochs with the batch size of 20. The fault classification accuracy and the model training time are used to evaluate the model performance. The performances of the model with different l are compared and shown in Figure 6. It can be seen that when the length of the subsequence l is set to 256, the proposed model has the best performance in both fault classification accuracy and model training speed. Smaller and larger l would decline the performance of the proposed model. The model with a small l cannot learn discriminative local features. A large l would decrease the time steps of the holistic ConvLSTM layer, as a result, the model cannot obtain effective holistic spatiotemporal information. Therefore, l is set to 256 in the following experiments.

Table 2. Parameters of the proposed model used in gearbox fault diagnosis experiments.

No.	Layer Type	Kernel	Stride	Channel	BN Axis	Activation
1	Local Convolution	(4,1)	(4,1)	4	4	sigmoid
2	Local ConvLSTM	(1,4)	(1,1)	4	5	tanh
3	Holistic ConvLSTM	(2,2)	(1,1)	4	4	tanh
4	FC layer	100	-	1	−1	sigmoid
5	Supervised learning layer	4	-	1	-	softmax

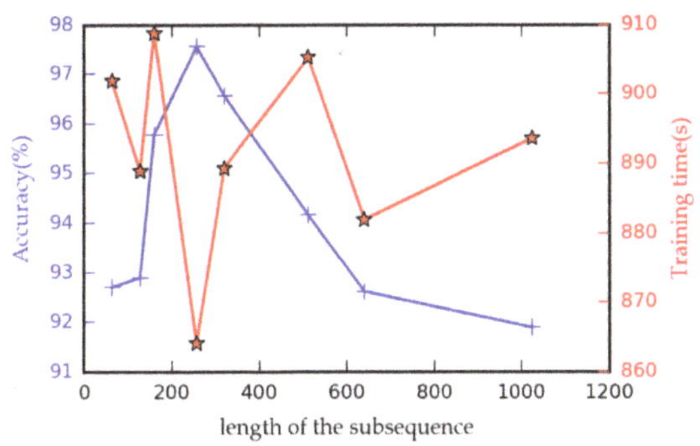

Figure 6. Accuracy and training time under different length of the subsequence.

5.1.3. Results and Discussion

To prove the advantage of the proposed TDConvLSTM, the same multi-sensor data is processed by some comparative models: empirical mode decomposition and support vector machine method (EMD-SVM), convolutional neural network (CNN), long short-term memory neural network (LSTM), a hybrid model that series connects CNN and LSTM (CNN-LSTM) and the proposed TDConvLSTM model without batch normalization (TDConvLSTM without BN).

To compare the performance of traditional machine learning models based on handcrafted features with the deep learning models based on raw sensor data, EMD-SVM is adopted as a comparative model. In EMD-SVM, the data of each sensor channel is decomposed by EMD firstly and the normalized energy, kurtosis, kurtosis and variance of the top five intrinsic mode functions are extracted as handcrafted features. A total of 80 features are obtained from four sensor channels to constitute a feature vector, which is used as the input of the SVM.

It should be noted that, all the deep learning models in this experiment are consist of five main layers. The last two layers in each model are a FC layer with size of [100] with dropout and a softmax layer with size of [4]. In CNN, three pairs of convolutional layers and pooling layers are stacked. The filter size, stride, channel and pooling size of three pairs of layers are set to [(4, 1), (4, 1), 10, (2, 1)], [(1, 4), (1, 1), 10, (2, 1)] and [(2, 1), (1, 1), 10, (2, 1)] respectively. In LSTM, the raw data with size of (5120, 4) is divided into 20 time steps firstly. Each time step is a 2D tensor with size of (256, 4). Then we flatten the 2D tensor into a 1D tensor (1024). As a result, the raw data is reshaped to (20, 1024). For the LSTM model, three LSTM layers with sizes of [500], [100] and [10] are stacked. In the CNN-LSTM model, two CNN layers with size of [(4, 1), (4, 1), 10] and [(1, 4), (1, 1), 10] are firstly designed, which is followed by a LSTM layer with size of [50]. Between the two CNN layers, a pooling layer with size of (2, 1) is adopted. Except the softmax layer, the activation function of each layer in CNN model and CNN-LSTM model are set to ReLu and activation functions in LSTM model are set to sigmoid. Parameters of the proposed TDConvLSTM are shown in Section 5.1.2. The Parameter settings of the TDConvLSTM without BN keep the same as the proposed TDConvLSTM model, except that all the batch normalization layers are removed. Dataset D1, D2, D3 and D4 are used to test all the comparative models respectively. The testing results are listed in Table 3. It is shown that our proposed TDConvLSTM model can diagnose the faults of the gearbox effectively with the highest test accuracy both under constant rotation speed and nonstationary rotation speed.

As shown in Table 3, all the deep learning models based on raw sensor data can achieve better performance than the handcrafted features based model EMD-SVM. Under constant rotation speed, the proposed TDConvLSTM model can achieve higher test accuracy than CNN and LSTM, which can be explained that the proposed TDConvLSTM model can both extract temporal features and spatial features of multi-sensor time series, which enables the TDConvLSTM layer to discover more hidden information than CNN and LSTM. The ConvLSTM structure can simultaneously learn the temporal features and spatial features and pay more attention to how data changes between time steps, so it can obtain better performance that CNN-LSTM structure. Under nonstationary rotation speed, the features that related to faults are hidden on different time scales. The proposed time-distributed structure can learn both short-term and long-term spatiotemporal features of time series, that is, it can make full use of the information on different time scales in the signal. Therefore, the proposed model can deliver better performance under nonstationary rotation speed. The comparison of TDConvLSTM and TDConvLSTM without BN proves that BN can improve the fault diagnose accuracy of the TDConvLSTM model. In addition to this, BN can improve the speed of model convergence, which is corroborated in Figure 7. The time required to calculate each sample is just 0.006s with i5-4570 CPU, which proves that the proposed TDConvLSTM model can be used for real-time mechanical fault diagnosis.

Table 3. Testing accuracy of comparative methods.

Model	Constant Rotation Speed			Nonstationary Rotation Speed
	D1	D2	D3	D4
TDConvLSTM	**100%**	**100%**	**100%**	**97.56%**
TDConvLSTM Without BN	100%	99.5%	99.78%	93.11%
CNN-LSTM	98.67%	97%	98.33%	91.89%
CNN	96.83%	99.5%	98.17%	86.78%
LSTM	96.67%	99.83%	100%	80.94%
EMD-SVM	90.67%	89.67%	91.33%	75.67%

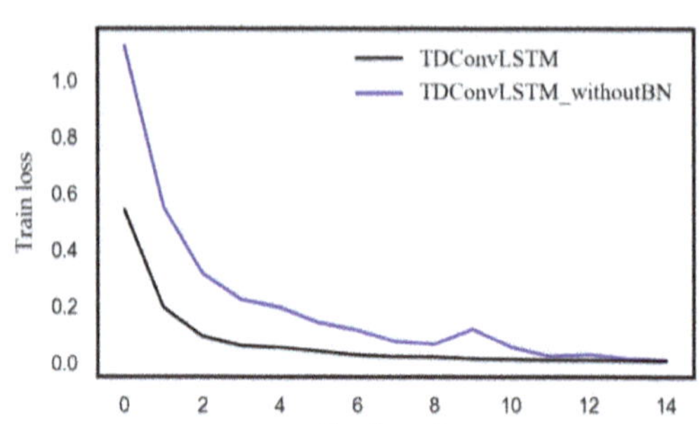

Figure 7. The effect of batch normalization (BN) on model training.

5.1.4. Feature Visualization

As we know, deep learning models work like a black box, so it is hard to understand its process of extracting features. In this section, the t-SNE method [44] is used to show the features extracted by each layer in our proposed TDConvLSTM model. T-SNE is an effective dimensionality reduction method, which can help us to visualize high-dimensional data by mapping the data from high-dimensional space to a two-dimensional space. Features extracted by each layer are respectively converted to a two-dimensional feature map. The feature maps of raw data, the local convolutional layer, the local ConvLSTM layer, the holistic ConvLSTM layer and the FC layer are shown in Figure 8, in which features of different fault types are distinguished by different colors. It can be seen that as the layers get deeper and deeper, the features of different fault types become more and more separate. As shown in the Figure 8a, raw data of four fault types are all mix together. Then the local convolutional layer disperses all features, which is shown in Figure 8b. Starting from the local ConvLSTM layer, the features of the same fault type begin to cluster. In the Figure 8c, we can see that the FT type and the CIB type start clustering first, while the CT type and FTCT type are still mix together. It is because that both the CT type and FTCT type have a root crack tooth in the big gear, so they have some same features. As we can see in Figure 8d, after the local ConvLSTM layer, the features of four fault types are almost separated. At last, the FC layer further separates the features of the four fault types and further clusters the features of the same fault type.

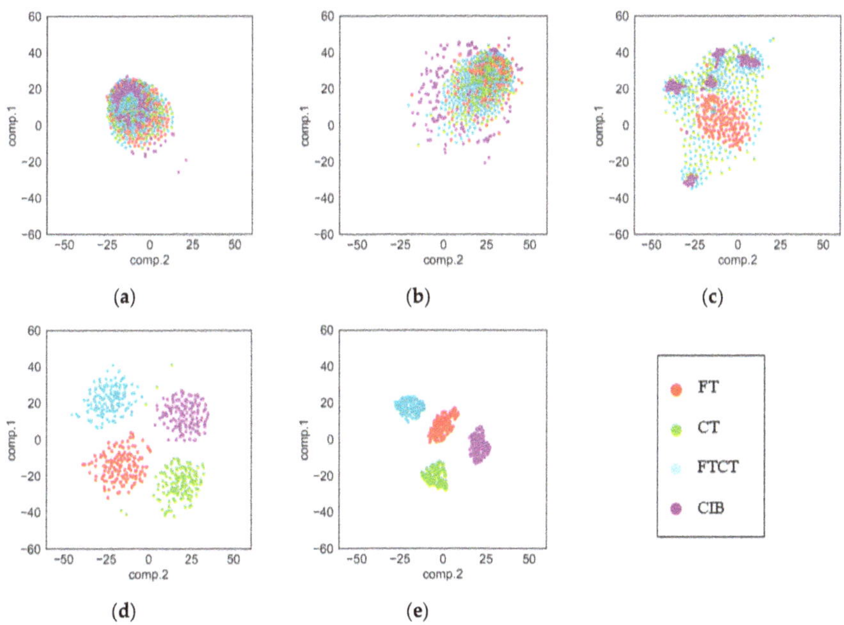

Figure 8. Feature visualization: (**a**) Raw data; (**b**) Local Convolutional layer; (**c**) Local ConvLSTM layer; (**d**) Holistic ConvLSTM layer; (**e**) Fully-connected (FC) layer.

5.2. *Case Study 2: Tool Wear Monitoring*

5.2.1. Experiment Setup and Data Description

The experiment was implemented on a high-speed CNC machine with a spindle speed of 10,400 rpm [45]. The experiment setup is illustrated in Figure 9. The material of the work-piece is Inconel 718. Ball-nose cutting tools of tungsten carbide with 3 flutes were used to mill the work-piece. The operation parameters are as follows: the feed rate in the x direction was 1555 mm/min; the depth of cut in the y direction (radial) was 0.125 mm; the depth of cut in the z direction (axial) was 0.2 mm. During the milling process, a Kistler quartz 3-component platform dynamometer, three Kistler Piezo accelerometers and a Kistler acoustic emission (AE) sensor were used to measure the cutting force, machine tool vibration and the high frequency stress wave generated by the cutting process, respectively. Seven channels of signals (force_x, force_y, force_z, vibration_x, vibration_y, vibration_z, AE_RMS) were acquired by DAQ NI PCI1200 with a sampling frequency of 50 kHz. After completing one surface milling, which was regarded as one cut, the corresponding flank wear of the three flutes were measured offline using a LEICA MZ12 microscope. The tool wear is the average wear of the three flutes. There are 300 cuts in each tool life and the multi-sensor data of one cut is regarded as a sample. The target value of a sample is the corresponding tool wear. Finally, three tool life dataset C1, C4 and C6 are selected to test our model. Due to the high sampling rate of raw data, the length of each sample is up to 200 thousand. The raw data is first downsampled; as a result, the length of each sample is reduced to 20,000. The acquired signals are first normalized as described in Section 4.2.1. In each testing case, two of the datasets are selected as the training dataset and the other one is used as the testing dataset. There are three model testing cases and according to the testing dataset, they are marked as C1, C4 and C6, respectively.

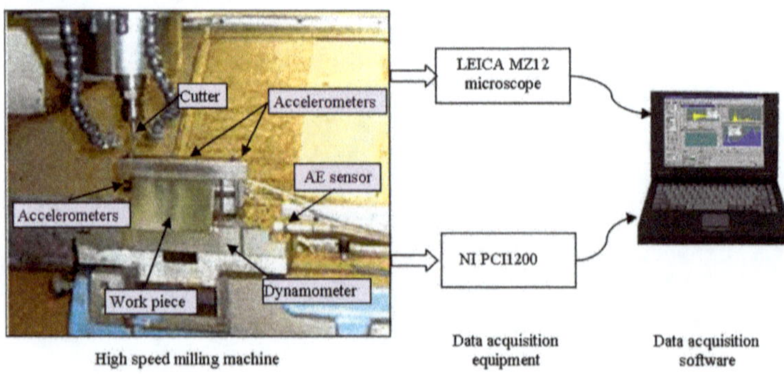

Figure 9. The experiment setup for tool wear monitoring

5.2.2. Model Settings

For our proposed TDConvLSTM model tested in tool wear monitoring experiments, the main parameters in main layers are listed in Table 4. The subsequence length l of the proposed model in this experiment is set to 500. The mean squared error is adopted as the loss function and the Nesterov-accelerated adaptive moment estimation (Nadam) algorithm [46] is employed as the optimizer for model training. It should be noted that, we formulate the tool wear prediction task as a regression prediction problem, so the supervised learning layer is a linear-regression layer and the activation in this layer is set to linear. The dropout rate of the FC layer is set to 0.5. Other parameters keep the same as stated in Section 5.1.2.

5.2.3. Results and Discussion

Three models including CNN, LSTM and CNN-LSTM are compared with the proposed model. Their structures are same as that described in Section 5.1.3, except that some parameter settings are changed. In CNN, the filter size, stride, filter number and pooling size of three pairs of convolutional and pooling layers are set to [(500, 3), (250, 3), 10, (2, 1)], [(4, 2), (1, 1), 10, (2, 1)] and [(2, 1), (1, 1), 10, (2, 1)], respectively. The activation functions of CNN models are all set to ReLu. In LSTM, the raw data with size of (20,000, 7) is divided into 40 time steps along the temporal dimension firstly. The data of each time step is reshaped to a 1D tensor and the raw data finally reshaped to (40, 3500). The output size of the three LSTM layers are [1000], [100] and [10] respectively. All the activation functions in LSTM model are set to tanh. In the CNN-LSTM model, the filter size, stride, filter number of two CNN layers are set to [(500, 3), (250, 3), 10] and [(500, 3), (1, 1), 10], respectively. The pooling size in the pooling layer is (2, 1). The size of the LSTM layer is set to [10]. The settings of the last two layers of the three models are same as the TDConvLSTM model described in Section 5.2.2.

Table 4. Parameters of the proposed model used in tool wear monitoring experiments

No.	LAYER TYPE	Kernel	Stride	Channel	BN Axis	Activation
1	Local Convolution	(10,3)	(5,3)	4	4	ReLu
2	Local ConvLSTM	(2,2)	(1,1)	4	5	tanh
3	Holistic ConvLSTM	(4,4)	(1,1)	1	4	tanh
4	FC layer	10	-	1	−1	ReLu
5	Supervised learning layer	1	-	1	-	linear

Mean absolute error (MAE) and root mean squared error (RMSE) of the true targets and the predicted targets are adopted as the indicators of model performance. The corresponding equations for the calculations of MAE and RMSE are given as follows:

$$MAE = \frac{1}{n}\sum_{i=1}^{n} |y\_test - y\_pre| \tag{21}$$

$$RMSE = \sqrt{\frac{1}{n}\sum_{i=1}^{n} (y\_test - y\_pre)^2} \tag{22}$$

where y_test is the true tool wear value in the test dataset and y_pre is the predicted tool wear value, n is the number of testing samples.

MAE and RMSE of all models in three different model testing cases are shown in Table 5. As we can see in the table, the TDConvLSTM model and the CNN-LSTM model both perform better than CNN and LSTM. The result can be explained that the TDConvLSTM model and the CNN-LSTM model can extract spatiotemporal features, while, the CNN model discards the long-term temporal correlation information in each channel data and the LSTM model discards the spatial correlation information between different channels. The hybrid models can discover more hidden information than CNN and LSTM.

It is shown that our proposed TDConvLSTM model achieves the best performance among all compared models. The most competitive CNN-LSTM model independently extracts the spatial features and the temporal features in succession, while, the TDConvLSTM model can simultaneously learn the temporal features and spatial features and pay more attention to capture the data changing features between time steps. The time-distributed structure can prompt the TDConvLSTM model make full use of information on different time scales. The above two advantages make the proposed model get better performance. The regression performances of TDConvLSTM in three different testing cases are illustrated in Figure 10. It is found that the predicted tool wear values are able to follow the trend of true tool wear values well with very small error. The testing time for each sample is 0.013s with i5-4570 CPU, which proves that the proposed TDConvLSTM model can be used for real-time tool wear monitoring.

Table 5. Mean absolute error (MAE) and root mean square error (RMSE) of models.

Model	MAE			RMSE		
	C4,C6/C1 [1]	C1,C6/C4 [2]	C1,C4/C6 [3]	C4,C6/C1	C1,C6/C4	C1,C4/C6
TDConvLSTM	**6.99**	**6.96**	**7.50**	**8.33**	**8.39**	**10.22**
CNN-LSTM	11.18	9.39	11.34	13.77	11.85	14.33
CNN	15.32	14.34	17.36	18.50	18.80	21.85
LSTM	19.09	16.00	22.61	21.42	17.78	25.81

[1] "C4,C6/C1" denote that C4 and C6 are the training datasets, C1 is the testing dataset. [2] "C1,C6/C4" denote that C1 and C6 are the training datasets, C4 is the testing dataset. [3] "C1,C4/C6" denote that C1 and C4 are the training datasets, C6 is the testing dataset.

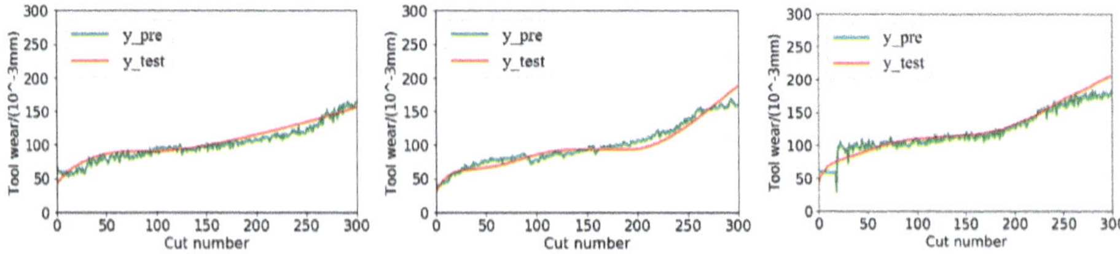

Figure 10. Regression performances of TDConvLSTM for three different testing cases: (**a**) C1; (**b**) C4; (**c**) C6.

6. Conclusions

The TDConvLSTM model has been proposed in this paper to extract spatiotemporal features of multi-sensor time series for machine health monitoring. The TDConvLSTM model is suitable for raw multi-sensor data and does not require any expert knowledge and feature engineering. In TDConvLSTM, the normalized multi-sensor time series is first segmented into a collection of subsequences using a sliding window. Then a time-distributed local feature extractor is designed with a time-distributed convolution layer and a ConvLSTM layer, which is employed in each subsequence to extract local spatiotemporal features inside a subsequence. The holistic ConvLSTM layer that stacked on the top of time-distributed local feature extractors can extract holistic spatiotemporal features between subsequences. At last, the fully-connected layer and the supervised learning layer can further reduce the feature dimension and obtain the machine health condition. The time-distributed structure can learn both short-term and long-term spatiotemporal features of multi-sensor time series. Therefore, it can make full use of information on different time scales. In the gearbox fault diagnosis experiment and the tool wear monitoring experiment, the results have confirmed the superior performance of the proposed TDConvLSTM model.

In future work, we plan to apply the proposed time-distributed spatiotemporal feature learning method in machine remaining useful life prediction tasks and continue to optimize our model to get better performance.

Author Contributions: Conceptualization, H.Q.; Data curation, H.Q. and P.W.; Formal analysis, H.Q.; Funding acquisition, T.W.; Investigation, H.Q.; Methodology, H.Q.; Project administration, T.W.; Resources, P.W.; Software, H.Q. and S.Q.; Supervision, T.W.; Validation, H.Q., S.Q. and L.Z.; Visualization, H.Q.; Writing—original draft, H.Q.; Writing—review & editing, P.W. and L.Z.

Funding: This research was funded by [National Natural Science Foundation of China] grant number [51475324], the [National Natural Science Foundation of China and Civil Aviation Administration of China jointly funded project] grant number [U1533103] and the [The Basic Innovation Project of China North Industries Group Corporation] grant number [2017CX031].

Conflicts of Interest: The authors declare no conflict of interest.

References

1. Xu, X.; Hua, Q. Industrial Big Data Analysis in Smart Factory: Current Status and Research Strategies. *IEEE Access* **2017**, *5*, 17543–17551. [CrossRef]
2. Liu, R.; Meng, G.; Yang, B. Dislocated Time Series Convolutional Neural Architecture: An Intelligent Fault Diagnosis Approach for Electric Machine. *IEEE Trans. Ind. Inform.* **2017**, *13*, 1310–1320. [CrossRef]
3. Shatnawi, Y.; Al-Khassaweneh, M. Fault diagnosis in internal combustion engines using extension neural network. *IEEE Trans. Ind. Electron.* **2014**, *61*, 1434–1443. [CrossRef]
4. Chen, Z.; Li, W. Multisensor Feature Fusion for Bearing Fault Diagnosis Using Sparse Autoencoder and Deep Belief Network. *IEEE Trans. Instrum. Meas.* **2017**, *66*, 1693–1702. [CrossRef]

5. Lei, Y.; Han, D.; Lin, J.; He, Z. Planetary gearbox fault diagnosis using an adaptive stochastic resonance method. *Mech. Syst. Signal Process.* **2013**, *38*, 113–124. [CrossRef]
6. Jing, L.; Wang, T.; Zhao, M. An Adaptive Multi-Sensor Data Fusion Method Based on Deep Convolutional Neural Networks for Fault Diagnosis of Planetary Gearbox. *Sensors* **2017**, *17*, 414. [CrossRef] [PubMed]
7. Zhao, R.; Wang, D.; Yan, R. Machine Health Monitoring Using Local Feature-based Gated Recurrent Unit Networks. *IEEE Trans. Ind. Electron.* **2018**, *65*, 1539–1548. [CrossRef]
8. Uekita, M.; Takaya, Y. Tool condition monitoring for form milling of large parts by combining spindle motor current and acoustic emission signals. *Int. J. Adv. Manuf. Technol.* **2017**, *89*, 65–75. [CrossRef]
9. Hinton, G.; Salakhutdinov, R.R. Reducing the dimensionality of data with neural networks. *Science* **2006**, *313*, 504–507. [CrossRef] [PubMed]
10. Bengio, Y.; Delalleau, O. On the expressive power of deep architectures. In Proceedings of the 14th International Conference on Discovery Science, Espoo, Finland, 5–7 October 2011; Springer: Berlin, Germany, 2011.
11. Lee, H.; Pham, P.; Largman, Y. Unsupervised feature learning for audio classification using convolutional deep belief networks. In Proceedings of the Advances in Neural Information Processing Systems, Vancouver, BC, Canada, 7–10 December 2009.
12. Krizhevsky, A.; Sutskever, I.; Hinton, G.E. Imagenet classification with deep convolutional neural networks. In Proceedings of the Advances in Neural Information Processing Systems, Carson, NV, USA, 3–8 December 2012; pp. 1097–1105.
13. Le, Q.V.; Zou, W.Y.; Yeung, S.Y. Learning hierarchical invariant spatio-temporal features for action recognition with independent subspace analysis. In Proceedings of the Computer Vision and Pattern Recognition (CVPR), Colorado Springs, CO, USA, 20–25 June 2011.
14. Singh, S.P.; Kumar, A.; Darbari, H. Building Machine Learning System with Deep Neural Network for Text Processing. In Proceedings of the International Conference on Information and Communication Technology for Intelligent Systems, Ahmedabad, India, 25–26 March 2017; Springer: Cham, Switzerland, 2017.
15. Zhao, R.; Yan, R.; Chen, Z. Deep Learning and Its Applications to Machine Health Monitoring: A Survey. *arXiv* **2016**, arXiv:1612.07460.
16. Zhao, G.; Zhang, G.; Ge, Q. Research advances in fault diagnosis and prognostic based on deep learning. In Proceedings of the Prognostics and System Health Management Conference, Harbin, China, 9–12 July 2017.
17. Ince, T.; Kiranyaz, S.; Eren, L. Real-time motor fault detection by 1-d convolutional neural networks. *IEEE Trans. Ind. Electron.* **2016**, *63*, 7067–7075. [CrossRef]
18. Abdeljaber, O.; Avci, O.; Kiranyaz, S. Real-time vibration-based structural damage detection using one dimensional convolutional neural networks. *J. Sound Vib.* **2017**, *388*, 154–170. [CrossRef]
19. Babu, G.S.; Zhao, P.; Li, X.L. Deep convolutional neural network based regression approach for estimation of remaining useful life. In Proceedings of the International Conference on Database Systems for Advanced Applications, Dallas, TX, USA, 16–19 April 2016.
20. Wielgosz, M.; Skoczeń, A.; Mertik, M. Using LSTM recurrent neural networks for monitoring the LHC superconducting magnets. *Nuclear Instrum. Methods Phys. Res. Sect. A* **2017**, *867*, 40–50. [CrossRef]
21. Gers, F.A.; Schmidhuber, J.; Cummins, F. Learning to forget: Continual prediction with lstm. *Neural Comput.* **2000**, *12*, 2451–2471. [CrossRef] [PubMed]
22. Zhao, R.; Wang, J.; Yan, R.; Mao, K. Machine health monitoring with LSTM networks. In Proceedings of the 10th International Conference on Sensing Technology (ICST), Nanjing, China, 11–13 November 2016.
23. Shi, X.; Chen, Z.; Wang, H. Convolutional LSTM Network: A Machine Learning Approach for Precipitation Nowcasting. In Proceedings of the 29th Annual Conference on Neural Information Processing Systems, Montreal, QC, Canada, 7–12 December 2015.
24. Luo, W.; Liu, W.; Gao, S. Remembering history with convolutional LSTM for anomaly detection. In Proceedings of the IEEE International Conference on Multimedia and Expo, Hong Kong, China, 10–14 July 2017.
25. Palaz, D.; Collobert, R. Analysis of cnn-based speech recognition system using raw speech as input. In Proceedings of the 16th Annual Conference of the International-Speech-Communication-Association, Dresden, Germany, 6–10 September 2015.

26. Tra, V.; Kim, J.; Khan, S.A. Bearing Fault Diagnosis under Variable Speed Using Convolutional Neural Networks and the Stochastic Diagonal Levenberg-Marquardt Algorithm. *Sensors* **2017**, *17*, 2834. [CrossRef] [PubMed]
27. Ding, X.; He, Q. Energy-Fluctuated Multiscale Feature Learning with Deep ConvNet for Intelligent Spindle Bearing Fault Diagnosis. *IEEE Trans. Instrum. Meas.* **2017**, *66*, 1926–1935. [CrossRef]
28. Zhang, W.; Peng, G.; Li, C. Rolling Element Bearings Fault Intelligent Diagnosis Based on Convolutional Neural Networks Using Raw Sensing Signal. Smart Innovation Systems and Technologies. In Proceedings of the 12th International Conference on Intelligent Information Hiding and Multimedia Signal Processing (IIH-MSP), Kaohsiung, Taiwan, 21–23 November 2016.
29. Lee, K.B.; Cheon, S.; Chang, O.K. A Convolutional Neural Network for Fault Classification and Diagnosis in Semiconductor Manufacturing Processes. *IEEE Trans. Semicond. Manuf.* **2017**, *30*, 135–142. [CrossRef]
30. Malhotra, P.; Ramakrishnan, A.; Anand, G. LSTM-based encoder-decoder for multi-sensor anomaly detection. *arXiv* **2016**, arXiv:1607.00148.
31. Malhotra, P.; Vishnu, T.V.; Ramakrishnan, A. Multi-Sensor Prognostics using an Unsupervised Health Index based on LSTM Encoder-Decoder. *arXiv* **2016**.
32. Bruin, T.D.; Verbert, K.; Babuška, R. Railway Track Circuit Fault Diagnosis Using Recurrent Neural Networks. *IEEE Trans. Neural Netw. Learn. Syst.* **2017**, *28*, 523–533. [CrossRef] [PubMed]
33. Cai, M.; Liu, J. Maxout neurons for deep convolutional and LSTM neural networks in speech recognition. *Speech Commun.* **2016**, *77*, 53–64. [CrossRef]
34. Kahou, S.E.; Michalski, V.; Konda, K. Recurrent Neural Networks for Emotion Recognition in Video. In Proceedings of the 2015 ACM on International Conference on Multimodal Interaction, New York, NY, USA, 9–13 November 2015.
35. Tsironi, E.; Barros, P.; Weber, C. An Analysis of Convolutional Long-Short Term Memory Recurrent Neural Networks for Gesture Recognition. *Neurocomputing* **2017**, *268*, 76–86. [CrossRef]
36. Rad, N.M.; Kia, S.M.; Zarbo, C.; Jurman, G.; Venuti, P.; Furlanello, C. Stereotypical Motor Movement Detection in Dynamic Feature Space. In Proceedings of the IEEE International Conference on Data Mining Workshops, New Orleans, LA, USA, 18–21 November 2017.
37. Rad, N.M.; Kia, S.M.; Zarbo, C.; Laarhoven, T.V.; Jurman, G.; Venuti, P.; Marchiori, E.; Furlanello, C. Deep learning for automatic stereotypical motor movement detection using wearable sensors in autism spectrum disorders. *Signal Process.* **2018**, *144*, 180–191. [CrossRef]
38. Zhao, R.; Yan, R.; Wang, J. Learning to Monitor Machine Health with Convolutional Bi-Directional LSTM Networks. *Sensors* **2017**, *17*, 273. [CrossRef] [PubMed]
39. Xiong, F.; Shi, X.; Yeung, D.Y. Spatiotemporal Modeling for Crowd Counting in Videos. In Proceedings of the 16th IEEE International Conference on Computer Vision (ICCV), Venice, Italy, 22–29 October 2017.
40. Qiu, Z.; Yao, T.; Mei, T. Learning Deep Spatio-Temporal Dependence for Semantic Video Segmentation. *IEEE Trans. Multimedia* **2017**, *20*, 939–949. [CrossRef]
41. Understanding LSTM Networks. Available online: http://colah.github.io/posts/2015-08-Understanding-LSTMs/ (accessed on 26 May 2018).
42. Zhang, Y.; Chan, W.; Jaitly, N. Very Deep Convolutional Networks for End-to-End Speech Recognition. In Proceedings of the IEEE International Conference on Acoustics, Speech, and Signal Processing (ICASSP), New Orleans, LA, USA, 5–9 March 2017.
43. Ioffe, S.; Szegedy, C. Batch normalization: Accelerating deep network training by reducing internal covariate shift. *arXiv* **2015**, arXiv:1502.03167.
44. Maaten, L.; Hinton, G. Visualizing data using t-SNE. *J. Mach. Learn. Res.* **2008**, *9*, 2579–2605.
45. Li, X.; Lim, B.; Zhou, J.; Huang, S.; Phua, S.; Shaw, K.; Er, M. Fuzzy neural network modelling for tool wear estimation in dry milling operation. In Proceedings of the Annual Conference of the Prognostics and Health Management Society, San Diego, CA, USA, 27–30 September 2009.
46. Dozat, T. Incorporating nesterov momentum into adam. In Proceedings of the International Conference on Learning Representations, Caribe Hilton, San Juan, Puerto Rico, 2–4 May 2016.

sensors

MDPI

Article

Hybrid Particle Swarm Optimization for Multi-Sensor Data Fusion

Hyunseok Kim [1] and Dongjun Suh [2,*]

[1] IoT Research Division, Electronics and Telecommunications Research Institute (ETRI), Daejeon 34129, Korea; hertzkim@etri.re.kr

[2] School of Convergence & Fusion System Engineering, Kyungpook National University, Sangju 37224, Korea

* Correspondence: dongjunsuh@knu.ac.kr; Tel.: +82-54-530-1482

Received: 11 July 2018; Accepted: 22 August 2018; Published: 24 August 2018

Abstract: A hybrid particle swarm optimization (PSO), able to overcome the large-scale nonlinearity or heavily correlation in the data fusion model of multiple sensing information, is proposed in this paper. In recent smart convergence technology, multiple similar and/or dissimilar sensors are widely used to support precisely sensing information from different perspectives, and these are integrated with data fusion algorithms to get synergistic effects. However, the construction of the data fusion model is not trivial because of difficulties to meet under the restricted conditions of a multi-sensor system such as its limited options for deploying sensors and nonlinear characteristics, or correlation errors of multiple sensors. This paper presents a hybrid PSO to facilitate the construction of robust data fusion model based on neural network while ensuring the balance between exploration and exploitation. The performance of the proposed model was evaluated by benchmarks composed of representative datasets. The well-optimized data fusion model is expected to provide an enhancement in the synergistic accuracy.

Keywords: multi-sensor system; multi-sensor information fusion; particle swarm optimization; sensor data fusion algorithm; distributed intelligence system

1. Introduction

A moment of evolution is now emerging toward a new paradigm known as smart convergence, which is bringing together both heterogeneous and information communication technologies. These emerging phenomena have prompted researchers to explore new possibilities for sophisticated smart devices [1] to be embedded in various real objects and to cope with various environmental changes. Recently, multiple similar and/or dissimilar sensors, as shown in Figure 1, have been widely used to provide precise sensing information from different viewpoints [2,3] and to realize the Internet of Things in a cyber-physical system [4]. Given that the accuracy of a sensor system is dictated by the degree to which repeated measurements under unchanging conditions are able to produce the same results, a multi-sensor system has typically been thought of as a way to guarantee the accuracy of a measurement system [5]. Hence, multi-sensor systems are an emerging research topic that is becoming increasingly important in various environmental perception activities. Nevertheless, challenging problems of multisensory data fusion algorithms are still far from accomplished [6]. Evolutionary computation methods [7] recently seem to be making a comeback in order to solve real-world problems concerning typically not iid (independent and identically distributed) data or sparse labeled data, and these methods are expected to help such a fusion model enhanced.

Because a single sensor usually only recognizes a limited set of partial information about the environment, multiple similar and/or dissimilar sensors are needed to provide accurate sensing information from a variety of perspective in an integrated manner [8]. Such multiple-sensing information is combined depending on data fusion algorithms to achieve synergistic effects. However, the construction of a data fusion model is not trivial because of difficulties in meeting the restricted conditions of a multi-sensor system such as its limited options for deploying sensors and nonlinear characteristics or correlation errors of multiple sensors. Such a nonlinearity optimization problem in a data fusion model can be solved by a neural network algorithm with an effective back propagation method ensuring the best performance of the network [9]. In recent multi-sensor fusion research [10,11], neural networks (including deep neural networks) play a major role in feature classification and decision making. However, difficulties for the efficient and high accurate multisensory fusion model remain.

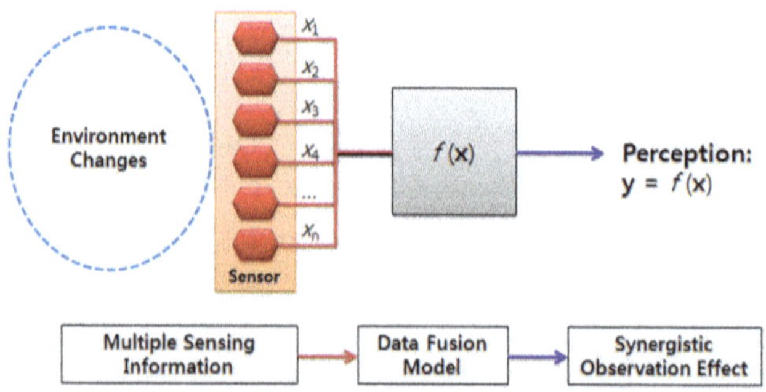

Figure 1. A multi-sensor system.

As is well known, maintaining the balance between the exploration of new possibilities and the exploitation of old certainties [12,13] has been considered a priority in designing an optimization scheme. As one of latest evolutionary approaches, particle swarm optimization (PSO) uses randomly placed particles on the search-space to explore new possibilities toward a global best position as a new solution to solve real-world problems [14–16]. PSO can be utilized in training a neural network by a population based stochastic back-propagation technique [17]. The randomly placed particles are more likely to find the global minimum than neural networks using a single particle. Venu et al. [18] showed that the parameters of feed-forward neural network converge faster using PSO than any other algorithm based on back-propagation methods, e.g., stochastic gradient descent, scaled gradient descent, or Levenberg-Marquardt (LM) [19]. They adopted PSO as a training algorithm involving adjusting the parameters (i.e., weights and biases) to optimize performance of the neural network. Kim et al. [20] suggested the PSO proportional-integral-derivative (PSOpid) which is one of the enhanced PSO algorithms through the stabilization of particle movement. Although a particle can be used as one of the solutions to regulate the parameters of neural network, it is necessary to constrain the range of search space for quicker convergence and higher fitness in PSO.

This paper proposes a hybrid PSO capable of overcoming large-scale nonlinearity or heavy correlation in the sensing data fusion and facilitates the construction of a robust data fusion model while maintaining the balance between the exploration of new possibilities and the exploitation of old certainties. The proposed algorithm was evaluated by benchmarks composed of representative datasets and the well-optimized data fusion model is expected to provide an enhancement in synergistic accuracy. In this paper, the neural network is used as a basic model and different backpropagation methods such as PSO, LM, and PSOpid are considered, thus the expression of neural network is simply omitted.

Section 2 discusses the problem of data fusion model and previous methods such as ordinary LM and PSO. We then propose a hybrid PSO, LM, and PSOpid in Section 3, and Section 4 describes how evaluation is performed for the proposed approach with the different weighing participants in diverse user-scenarios and with the different exemplary datasets from MATLAB. Section 5 concludes the research and describe possible future works.

2. Data Fusion Model Using Neural Network and PSO

Compared to a single sensor system, multiple sensor systems have the advantage of broadening the sensing range and improving the perception of environmental conditions [8]. In addition, a multi-sensor system allows information from a set of homogeneous sensors to be combined using a data fusion model. Thus, a multi-sensor system represents a proven method for enhancing the accuracy of a measurement system.

Figure 2 shows the case of using multiple Force Sensitive Resistor (FSR) sensors on a smart floor block [16]. FSRs is one of best solutions to meet multi-dimensional requirements such as maintaining visibility; holding weights of standing, moving, or jumping user; avoiding occlusion in sensing area; and reducing production cost. In this paper, an FSR is composed of two substrate layers with a conductive core, and, when pressure is applied to the FSR, the substrate moves, compressing the conductive core to detect the weight on the smart floor. The pressure changes are measured by the analog output from the FSR resistive divider. The weight is dispersed to the corner based on the approximate inverse power law which is the relation between distance and force of the FSR [21].

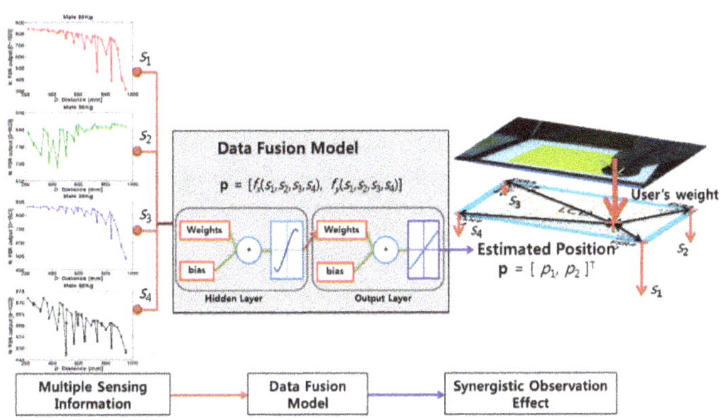

Figure 2. Example of user localization using a multi-sensor system.

In [16], a neural network adopting three layers is used, and we can clarify our notation and describe it as follows. Each four-dimensional vector is applied to the input layer as $\mathbf{s} = [\ s_1 \ s_2 \ s_3 \ s_4\ ]^T$, where $\{s \in \mathbb{R} : 0 \le s \le 2^{10}\}$. An output produced by a non-linear activation function at each hidden unit, $g(net)$, that is, a hyperbolic tangent sigmoid as,

$$g(net) = \frac{e^{net} - e^{-net}}{e^{net} + e^{-net}} \tag{1}$$

where the net is the inner product of inputs with weights at each neural network unit.

Each output of the neural network as $\mathbf{p} = [p_1\ p_2]^T$ calculates the activation function based on the hidden units described in Equation (2):

$$p_k(\mathbf{s}) \equiv \sum_{j=1}^{n_H} \theta_{kj} g\left(\sum_{i=1}^{d} \theta_{ji} s_i + \theta_{j0}\right) + \theta_{k0} \tag{2}$$

where the subscript i indexes units in the input layer and j indexes units in the hidden layer; θ_{ji} denotes the input-to-hidden layer parameters at the hidden unit j; and the dimension of input vector is d.

In addition, the subscript k indexes units in the output layer, n_H indicates the number of hidden units and θ_{kj} denotes the hidden-to-output layer weights. θ_{j0} and θ_{k0} are mathematically treated as the bias of the layer. We refer to its output, $\mathbf{p} = [p_1\ p_2]^T$, as the user's estimated 2D position (x, y) on the screen, respectively. The authors in [16] regarded the cost function as the sum of the output units of the squared difference between the target t_k and the actual output a_k.

$$J(\boldsymbol{\theta}) \equiv \frac{1}{2m}\sum_{i=1}^{m}\sum_{k=1}^{c}(t_k - a_k)^2 = \frac{1}{2m}\sum_{i=1}^{m}\| \mathbf{t} - \mathbf{a} \|^2 \tag{3}$$

where $\mathbf{t}$ and $\mathbf{a}$ are the target and actual output vectors of length c (e.g., c is 2 in this experiment), the batch size is m and $\boldsymbol{\theta}$ represents all weights and biases.

FSRs show nonlinear issues or problems of correlation errors. It can be caused by instability of the glass or metal frame, response time of the sensor, errors in manufacturing, assembly processes, the sensor placement, and human errors. Therefore, Kim et al. [16] suggested a PSO-based neural network model, namely PSO, which can be used as alternatives method to the gradient-descent algorithm by randomly spreading multiple particles capable of finding each optimum and being converged toward the global optima. However, if a system needs many parameters that should be adjusted for many epochs and other potential parameters, then the system needs to be improved.

3. A Hybrid PSO Model for Multi-Sensor Data Fusion

Since March [12] proposed the balancing between exploration and exploitation in learning, it has been extensively researched and widely applied to various domains [13]. In this paper, the PSO first explores possibilities as a global search, the LM then exploits certainties as a local search, and the PSOpid suggested in [20] lastly explores a new possibility within the range-optimized search space. The PSOpid is one of the enhanced PSO algorithms and each particle finds the global optimum securely while preventing a particle from becoming unstable or exploding. In addition, the algorithm can converge more quickly and get high fitness values compare to other algorithms in the range-optimized search space.

3.1. Improved Exploitation of Neural Network Using Ordinary PSO

The idea of PSO allows particles randomly placed on the search-space to explore new possibilities to the best global position. PSO is used to train a neural network with back propagation method considering a population based stochastic optimization technique [17]. On the other hand, the randomly placed particles will have a high probability of discovering the global minimum in comparison with the ordinary LM using a single particle. In this method, PSO can be alternatively used as the training algorithm at each iteration n, including parameter tuning to optimize the performance of the neural network. In addition, a particle is used to determine the value of parameters in the neural network. The magnitude of vectors $\boldsymbol{\theta}_i$ and $\mathbf{v}_i$ are equivalent to the dimension of the weights and biases. The velocity and position of the i-th particle after the n-th iteration are shown in Equations (4) and (5).

$$\mathbf{v}_i[n] = K\left\{\begin{array}{c} \omega\cdot\mathbf{v}_i[n-1] \\ +c_1\cdot rand_1\cdot(\mathbf{p}_{i_{best}}[n-1] - \boldsymbol{\theta}_i[n-1] \\ +c_1\cdot rand_2\cdot(\mathbf{g}_{best}[n-1] - \boldsymbol{\theta}_i[n-1] \end{array}\right\} \tag{4}$$

$$\boldsymbol{\theta}_i[n] = \boldsymbol{\theta}_i[n-1] + \mathbf{v}_i[n] \tag{5}$$

The previous best position is selected using Equation (6), and the global best position is decided by Equation (7).

$$\mathbf{p}_{i_{best}}[n] := \left\{\begin{array}{c} \boldsymbol{\theta}_i \text{ if } \mathrm{J}(\boldsymbol{\theta}_i[n]) \leq \mathrm{J}(\boldsymbol{\theta}_i[n-1]) \\ \mathbf{p}_{i_{best}}[n-1] \text{ otherwise} \end{array}\right. \tag{6}$$

$$g_{\text{best}}[n] = \{ p_{j_{\text{best}}}[n] \,|\, \forall i : J(p_{j_{\text{best}}}[n]) \leq J(p_{i_{\text{best}}}[n]) \} \quad (7)$$

where the number of particles is Z_k ($\approx 10 + 2 \times$ (the dimension of the weights and biases)$^{0.5}$) and the inertia weight is ω, which dynamically adjusts the velocity. Moreover, the cognitive component c_1 and the social component c_2 change the particles' velocity toward the previous best and global best position, respectively. The PSO uses random numbers determined from a uniform distribution $rand_1$ and $rand_2$ to avoid unfortunate states in which all particles quickly settle into an unchanging direction. Consequently, the parameters are updated in accordance with the found global best position.

This paper proposes a hybrid LM and PSO scheme, namely LM-PSO, based on the concept of two-phase evolutionary programming [22]: First, PSO is used to explore a new possibility for overcoming multiple minima, and second, the initial value of the weights and biases of LM-based neural network, namely LM, is set—in other words, the neural network is trained—to the optimal configuration derived in the PSO phase. Consequently, the proposed LM-PSO can derive more accurately a globally optimal solution compared to LM.

In the first stage, normally a PSO has the initial parameters shown in Table 1 for a multiple FSR system. As illustrated in Figure 3, PSO finds suboptimal range of parameters which can reduce the configuration space for LM's parameters.

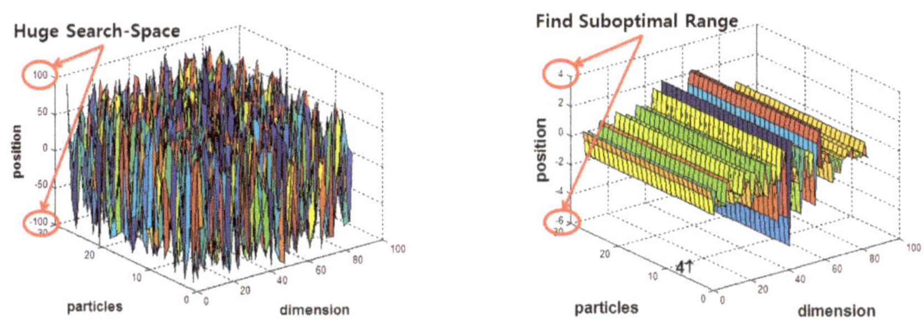

(a) (b)

Figure 3. PSO alone: (**a**) Initial position of the particles within a hypercube using a uniform random distribution; and (**b**) converged position of the particles.

Table 1. Parameter information of the PSO method alone.

Parameters	Value
Swarm size (Z_k)	28
Initial Position of Particles	Spread within a hypercube using a uniform random distribution
Minimum velocity norm	0.05
Inertial weight (ω)	1
Minimum position (min_pos)	−100
Maximum position (max_pos)	100

In the second stage, the proposed LM-PSO is trained with the optimal initial weight and bias values derived during the PSO stage. Although PSO is implausible on its own as a solution for the convergence rate and solution accuracy as illustrated in Figure 3, the proposed hybrid LM-PSO shows an enhanced performance of both accuracy and convergence rate as compared to conventional LM. However, hope remains for new possibilities for deriving an ultimate global optimum.

3.2. Exploration Toward Ultimate Goals for the Use of Enhanced PSO

Ordinary PSO alone has difficulty negotiating the tradeoff between global and local search because particles are initially deployed following a uniform random distribution in a hypercube big enough to contain a prospective global optimum. When the particle position range is not limited within the minimum and maximum positions or is too broad, the swarm can become unstable or explode, resulting in a slow convergence. Therefore, a novel approach is necessary for each particle to discover the global optimal solution securely, preventing the particles from becoming unstable or exploding.

The proportional-integral-derivative (PID) is widely applied in a feedback control loop technology [23]. The major advantages of the PID are easy to be implemented as well as only three parameters, i.e., proportional, integral and derivative terms, are required to be adjusted. The proportional gain is subject to the current error, the integral gain varies in proportion to both the magnitude and the duration error, and the derivative gain represents a prediction of future errors [24]. In [20], Kim et al. proposed a new approach using PSOpid as described in Equation (8) to change the particle's position of PSO with less oscillation.

$$\boldsymbol{\theta}_i[\mathrm{n}] = \boldsymbol{\theta}_i[\mathrm{n}-1] + \left\{ \begin{array}{c} K_P \cdot \mathbf{v}_i[\mathrm{n}] \\ +K_I \cdot \sum\limits_{\mathrm{k}=0}^{\mathrm{n}} \mathbf{v}_i[\mathrm{k}] \\ +K_D \cdot (\mathbf{v}_i[\mathrm{n}] - \mathbf{v}_i[\mathrm{n}-1]) \end{array} \right\} \quad (8)$$

where K_P is the proportional term, K_I is integral term, and K_D is the derivative term, and they are selected through trial and error operations, as shown in Table 2.

Table 2. Parameter information of the PSOpid-based method alone.

Parameters	Value
Minimum position (min_pos)	1.2 × min (LM-PSO)
Maximum position (max_pos)	1.2 × max (LM-PSO)
Proportional term (K_P)	0.5 (fixed)
Integral term (K_I)	0.5 (fixed)
Derivative term (K_D)	0.6 (fixed)

Adjustment of the velocity is calculated from the acceleration with respect to the distance error produced by both present and its previous best position compared to the global best solution. Furthermore, two random, uniformly distributed variables are utilized in preserving the diversity. The position of each particle is updated according to the multiplying the velocity with the three PID terms, which are described in Table 2. In addition, it is necessary to restrain the dynamic range of each particle position to prevent a swarm from exploding or unstable conditions.

The position of each particle is initialized to the range between maximum and minimum derived from the output of the precedent LM-PSO technique. The initial global best position is optimally configured in accordance with the LM-PSO method. Consequently, particles randomly distributed can better guarantee an accurate global optimal solution than they can through either method alone, as shown in Figure 4.

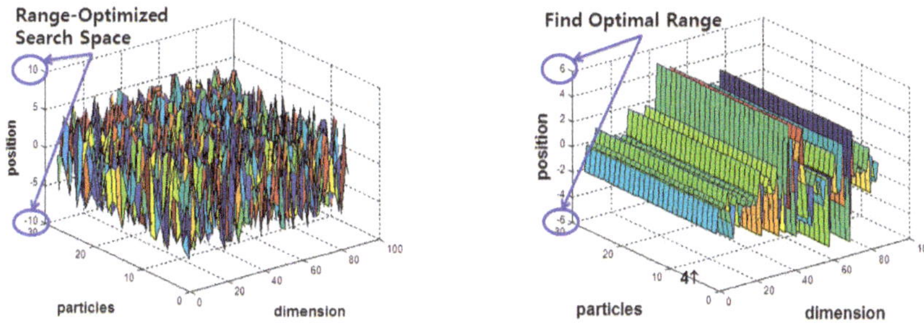

(a) (b)

Figure 4. A hybridization of PSOpid, LM and PSO, namely PSOpid-LM-PSO: (**a**) initial position of particles within a shrunk hypercube; and (**b**) convergence position of the particles.

The stabilized PSOpid, in which each particle is evaluated by using a fitness function and update, is located in an enhanced location to discover the best optimal solution without local minimum. The convergence rate of the proposed PSOpid is faster and it has much higher performance than previous methods. This technique enables efficient implementation because of its small number of parameters, which both shortens the training time and reduces overfitting compared to ordinary PSO.

3.3. Three-Phase Hybrid PSO Method Balancing between Exploration and Exploitation

The proposed PSOpid-LM-PSO, as shown in Figure 5, assigns each particle a position calculated based on the output of the LM-PSO phase to initialize an optimal configuration of a global best position. The swarm intelligence technology, in which each individual evaluates, compares, and imitates one another, is a better way to find the best optimum with no local minima. Importantly, the dynamic range of each particle position is limited to prevent a swarm from becoming unstable or exploding. In the three-phase hybrid optimization method, PSO-based exploration of new possibilities is firstly executed for the configuration space of parameters in the multi-sensor data fusion model. Secondly, ordinary LM configured with the sub-optimized range can focus on the exploitation of old certainties. Finally, PSOpid can explore the ultimate goal with more accuracy and faster convergence rate. As a result of the hybrid method, the time needed for convergence is also shorter than that of an ordinary method. Thus, the number of iterations required for convergence counts in comparing speed of convergence.

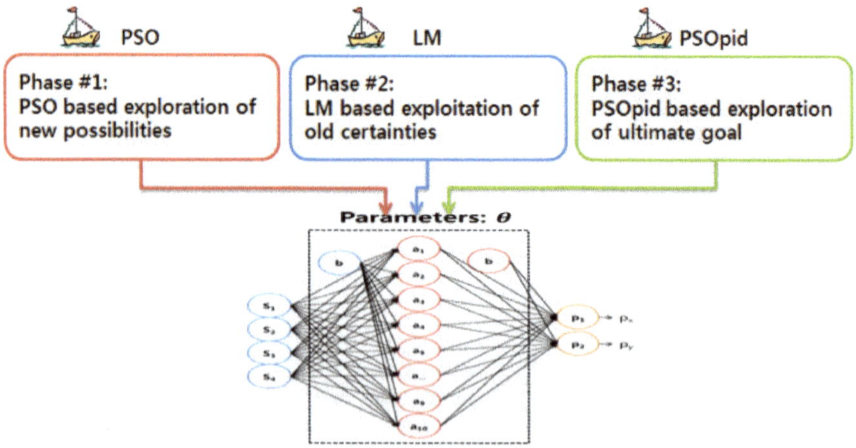

Figure 5. Three-phase hybrid optimization method.

In the case of PSO and PSOpid alone, the range of each particle position become unstable or explodes; therefore, the hybrid of three algorithms can be used to not only exploit old certainty optimum but also to explore new possibilities. Through all evaluative processes, this paper shows the enhanced performance by a hybrid scheme of ordinary PSO, LM, and PSOpid, namely PSOpid-LM-PSO. The results showed that the proposed PSOpid-LM-PSO provides faster convergence and better fitness than other algorithms within the range-optimized search space, as shown in Figure 6. Each coordinate in Figure 6 indicates the minimum and maximum of the range for parameters in multi-sensor data fusion model. This well-optimized data fusion model for the sensing information of a set of homogeneous FSRs is expected to provide an enhancement in synergistic accuracy.

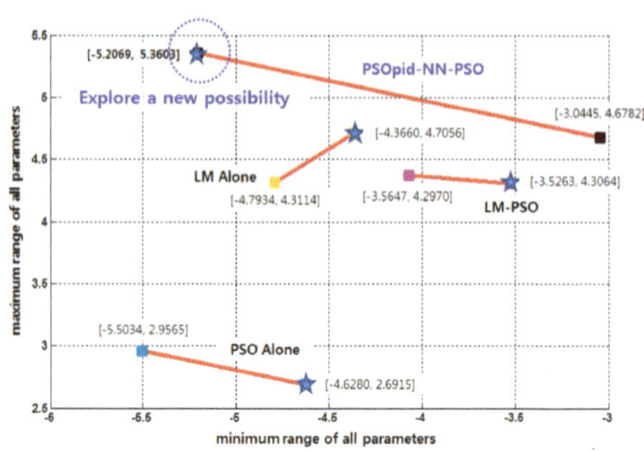

Figure 6. Exploration of a new possibility. Each (x,y) coordinate indicates the minimum and maximum of all parameters such as weights and biases, and the outcomes are from each independent trial.

4. Performance Analysis

We first evaluated the experimental dataset of the touch floor system in [16] and selected men and women from 58 kg to 90 kg as participants. While a user is standing on the floor, the user weight is distributed to four corners according to the inverse power-law in the distance vs. force relation, and the force on each FSRs is sampled through a resistive divider with noise filters. Therefore, it is reasonable to perform experiments with various weighing participants. In this experiment, we evaluated the mean distance error between prediction user location and test set location. As shown in Figure 7, five different algorithms, geometrical trilateration [25], ordinary LM-based neural network (LM), PSO-based neural network (PSO), two phases method (LM-PSO), and the proposed three-phase hybrid method (PSOpid-LM-PSO), are evaluated to investigate whether the suggested method is sufficiently robust to identify the position of differently weighing participants in realistic indoor conditions. In the evaluation, each algorithm was trained with same sized learning data consisting of 100 steps per each participant with 4 × 185 matrix. In addition, these algorithms used the same normalized input data via preprocessing with mean cancellation, principle component analysis and covariance equalization.

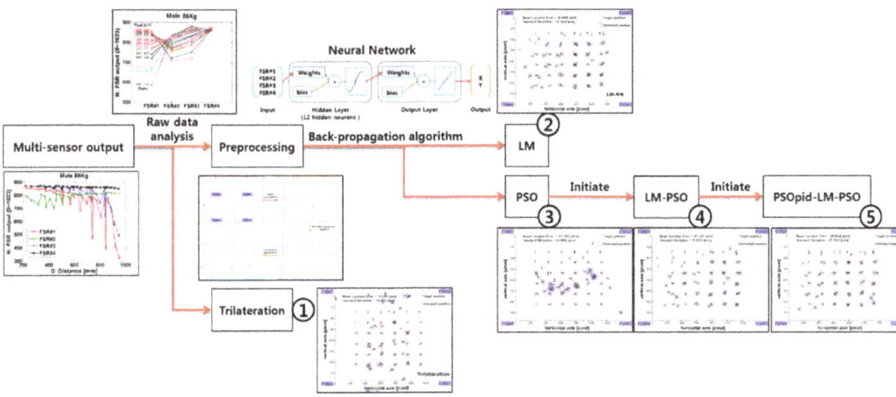

Figure 7. Performance analysis.

Table 3 presents the results from a comparison test consisting of 100 independent trials. In each trial, an inner random number generator was initialized with a nonnegative integer based on the current clock time, and, thus, the randomized functions can produce a predictable sequence of numbers. As described in Table 3, the estimation error of the proposed PSOpid-LM-PSO is reduced by approximately 88.77% in comparison with the trilateration method. (e.g., in 58 kg case, error reduction is calculated as follows: (1.0 − 23.18 mm/546.07 mm) × 100 = 95.76%).

Table 3. Comparison results of the algorithm enhancement rates (out of 100 independent runs).

Single-Subject Evaluation: Mean Distance Error [mm]					
Weight	**Trilateration**	**LM alone**	**LM-PSO**	**PSOpid-LM-PSO**	**PSOpid-LM-PSO Error Reduction**
58 kg	546.07	26.28	24.85	23.18	95.76%
64 kg	213.54	30.39	27.01	25.14	88.23%
72 kg	637.15	31.34	30.07	24.76	96.11%
85 kg	160.16	50.14	44.02	41.48	74.10%
90 kg	196.56	27.32	24.96	20.35	89.65%

Figure 8 shows the rate of enhancement under the PSOpid-LM-PSO in comparison with the classic LM. The PSOpid-LM-PSO method decreased the mean location error by about 18.57% compared to the LM alone case and the overall performance was enhanced [16].

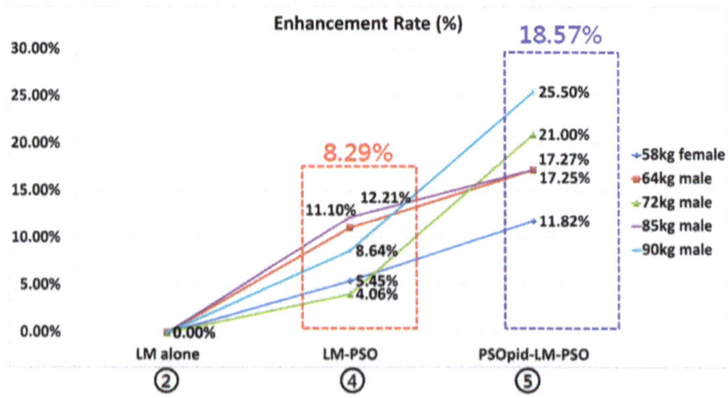

Figure 8. Enhancement graph of each algorithm as compared to the classic LM-based backpropagation algorithm.

In contrast, a performance analysis was performed using different exemplary datasets from MATLAB [26], which are widely used to evaluate the performance of machine learning, to confirm that the proposed method is robust enough to improve stability, robustness and convergence speed in terms of neural network training. In this study, we used the three values of the PID (K_P, K_I, and K_D) which are fixed to 0.5, 0.4, and 0.3, respectively to verify the reliability of performance evaluation of the proposed methodology in different exemplar datasets. Other parameters were used in the same manner with the generic PSO and the PSOpid approach. Figure 9 shows the enhancement of hybridizations among PSOpid-LM-PSO and the other algorithms. The results are summarized with the best, median, and worst results (out of 100 independent runs) reported. In the case of PSO and PSOpid, the range of each particle position becomes unstable or explodes; therefore, neither algorithm alone can be used to exploit the old certainty optimum as well as explore new possibilities. Through all evaluative processes, the proposed PSOpid-LM-PSO showed enhanced performance by using a hybrid scheme of ordinary PSO, LM, and PSOpid.

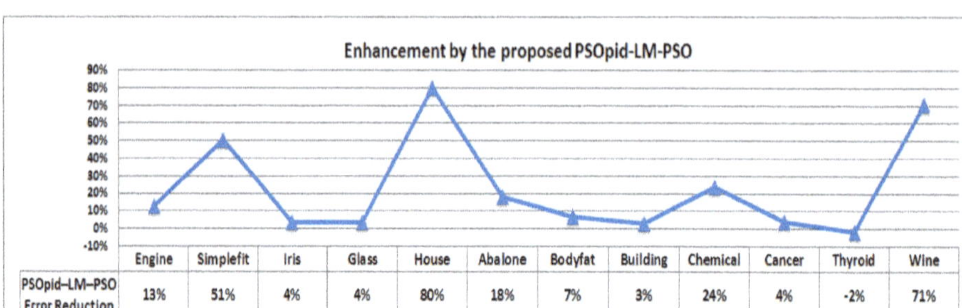

Figure 9. Performance test using a sample dataset.

5. Conclusions

This study proposed a hybridization of enhanced particle swarm optimization (PSO) and a classic neural network to build a multi-sensor data fusion model. The results show that the proposed PSOpid-LM-PSO provides a faster convergence and better fitness than other algorithms within the range-optimized search space with the system accuracy improved by approximately 18.6% compared to the classic LM algorithm. The contributions of this paper will reduce human effort in training the data fusion model using an on-line adaptation approach based on small changes from a prior trained model, or by using support vector machines for the one-to-many mapping approximation [27,28]. Another important area for further investigation is to explore the universal approximation capabilities of a standard multi-layer feed-forward neural network in most applications where numerous input samples need to be processed. Specifically, a two-hidden-layer feed-forward network using Kolmogorov's theorem can be considered for approximating the high-dimensional data fusion model [29,30]. This well-optimized model for data fusing from the sensing information of homogeneous sensors is expected to support an enhancement in the synergistic accuracy.

Author Contributions: Conceptualization, H.K.; Methodology, H.K.; Software, H.K. and D.S.; Validation, H.K. and D.S.; Formal Analysis, H.K.; Investigation, H.K.; Resources, H.K.; Data Curation, H.K.; Writing-Original Draft Preparation, H.K. and D.S.; Writing-Review & Editing, H.K. and D.S.; Visualization, H.K.; Supervision, D.S.; Project Administration, H.K.; Funding Acquisition, H.K.

Acknowledgments: This work was supported by Electronics and Telecommunications Research Institute (ETRI) grant funded by the Korean government [18ZH1100, Distributed Intelligence Core Technology of Hyper-Connected Space].

Conflicts of Interest: The authors declare no conflict of interest.

References

1. Lippi, M.; Mamei, M.; Mariani, S.; Zambonelli, F. Distributed Speaking Objects: A Case for Massive Multiagent Systems. In Proceedings of the International Workshop on Massively Multi-Agent Systems (MMAS2018), Stockholmsmässan, Stockholm, Sweden, 14 July 2018; pp. 1–18.
2. Smaiah, S.; Sadoun, R.; Elouardi, A.; Larnaudie, B.; Bouaziz, S.; Boubezoul, A.; Vincke, B.; Espié, S. A Practical Approach for High Precision Reconstruction of a Motorcycle Trajectory Using a Low-Cost Multi-Sensor System. *Sensors* **2018**, *18*, 2282. [CrossRef] [PubMed]
3. Jagannathan, S.; Mody, M.; Jones, J.; Swami, P.; Poddar, D. Multi-sensor fusion for Automated Driving: Selecting model and optimizing on Embedded platform. *Auton. Veh. Mach.* **2018**, 256-1–256-5. [CrossRef]
4. Kim, D.; Rodriguez, S.; Matson, E.T.; Kim, G.J. Special issue on smart interactions in cyber-physical systems: Humans, agents, robots, machines, and sensors. *ETRI J.* **2018**, *40*, 417–420. [CrossRef]
5. Dong, J.; Zhuang, D.; Huang, Y.; Fu, J. Advances in multi-sensor data fusion: Algorithms and applications. *Sensors* **2009**, *9*, 7771–7784. [CrossRef] [PubMed]
6. Khaleghi, B.; Khamis, A.; Karray, F.O.; Razavi, S.N. Multisensor data fusion: A review of the state-of-the-art. *Inf. Fusion* **2013**, *14*, 28–44. [CrossRef]
7. Kendall, G. Is Evolutionary Computation Evolving Fast Enough? *IEEE Comput. Intell. Mag.* **2018**, *13*, 42–51. [CrossRef]
8. Xiong, N.; Svensson, P. Multi-sensor management for information fusion: Issues and approaches. *Inf. Fusion* **2002**, *3*, 163–186. [CrossRef]
9. LeCun, Y.; Bottou, L.; Orr, G.B.; Müller, K.-R. Efficient backprop. In *Lecture Notes in Computer Science*; Springer: Berlin/Heidelberg, Germany, 1998; pp. 9–50.
10. Gravina, R.; Alinia, P.; Ghasemzadeh, H.; Fortino, G. Multi-sensor fusion in body sensor networks: State-of-the-art and research challenges. *Inf. Fusion* **2017**, *35*, 1339–1351. [CrossRef]
11. Novak, D.; Riener, R. A survey of sensor fusion methods in wearable robotics. *Rob. Auton. Syst.* **2015**, *73*, 155–170. [CrossRef]
12. March, J.G. Exploration and exploitation in organizational learning. *Organiz. Sci.* **1991**, *2*, 71–87. [CrossRef]
13. Li, Y.; Tian, X.; Liu, T.; Tao, D. On better exploring and exploiting task relationships in multitask learning: Joint model and feature learning. *IEEE Trans. Neural Netw. Learn. Syst.* **2018**, *29*, 1975–1985. [CrossRef] [PubMed]
14. Yang, Q.; Chen, W.N.; Deng, J.D.; Li, Y.; Gu, T.; Zhang, J. A Level-based Learning Swarm Optimizer for Large Scale Optimization. *IEEE Trans. Evol. Comput.* **2017**, *22*, 578–594. [CrossRef]
15. Xu, J.; Yang, W.; Zhang, L.; Han, R.; Shao, X. Multi-sensor detection with particle swarm optimization for time-frequency coded cooperative WSNs based on MC-CDMA for underground coal mines. *Sensors* **2015**, *15*, 21134–21152. [CrossRef] [PubMed]
16. Kim, H.; Chang, S. High-resolution touch floor system using particle swarm optimization neural network. *IEEE Sens. J.* **2013**, *13*, 2084–2093. [CrossRef]

17. Chen, G.; Yu, J. In Particle swarm optimization neural network and its application in soft-sensing modeling. In *International Conference on Natural Computation*; Springer: Berlin/Heidelberg, Germany, 2005; pp. 610–617.
18. Gudise, V.G.; Venayagamoorthy, G.K. Comparison of particle swarm optimization and backpropagation as training algorithms for neural networks. In Proceedings of the 2003 IEEE Swarm Intelligence Symposium (SIS'03), Indianapolis, IN, USA, 24–26 April 2003; pp. 110–117.
19. Lv, C.; Xing, Y.; Zhang, J.; Na, X.; Li, Y.; Liu, T.; Cao, D.; Wang, F.Y. Levenberg-Marquardt Backpropagation Training of Multilayer Neural Networks for State Estimation of a Safety Critical Cyber-Physical System. *IEEE Trans. Ind. Inf.* **2017**, *14*, 3436–3446. [CrossRef]
20. Kim, H.; Chang, S.; Kang, T.G. Enhancement of particle swarm optimization by stabilizing particle movement. *ETRI J.* **2013**, *35*, 1168–1171. [CrossRef]
21. Interlink Electronics, Inc. Force Sensing Resistor (FSR) Guide. Available online: http://www.interlinkelectronics.com/Force-Sensing-Resistor (accessed on 24 August 2018).
22. Kim, J.-H.; Myung, H. Evolutionary programming techniques for constrained optimization problems. *IEEE Trans. Evol. Comput.* **1997**, *1*, 129–140.
23. Willis, M. *Proportional-Integral-Derivative Control*; University of Newcastle: Newcastle, UK, 1999.
24. Araki, M. PID control. In *Control Systems, Robotics and Automation: System Analysis and Control: Classical Approaches II*; Unbehauen, H., Ed.; EOLSS Publishers Co., Ltd.: Oxford, UK, 2009; pp. 58–79. ISBN 9781848265912.
25. Cotera, P.; Velazquez, M.; Cruz, D.; Medina, L.; Bandala, M. Indoor Robot Positioning Using an Enhanced Trilateration Algorithm. *Int. J. Adv. Rob. Syst.* **2016**, *13*, 110. [CrossRef]
26. Matlab. Sample Data Sets. Available online: https://kr.mathworks.com/help/nnet/gs/neural-network-toolbox-sample-data-sets.html (accessed on 24 August 2018).
27. Smola, A.J.; Schölkopf, B. A tutorial on support vector regression. *Stat. Comput.* **2004**, *14*, 199–222. [CrossRef]
28. Gunn, S.R. Support vector machines for classification and regression. *ISIS Tech. Rep.* **1998**, *14*, 5–16.
29. Kůrková, V. Kolmogorov's theorem and multilayer neural networks. *Neural Netw.* **1992**, *5*, 501–506. [CrossRef]
30. Huang, G.-B.; Chen, L.; Siew, C.K. Universal approximation using incremental constructive feedforward networks with random hidden nodes. *IEEE Trans. Neural Netw.* **2006**, *17*, 879–892. [CrossRef] [PubMed]

sensors

MDPI

Article

Health Management Decision of Sensor System Based on Health Reliability Degree and Grey Group Decision-Making

Kai Song [1], Peng Xu [1,*], Guo Wei [1], Yinsheng Chen [2] and Qi Wang [1]

[1] School of Electrical Engineering and Automation, Harbin Institute of Technology, Harbin 150001, China; kaisong@hit.edu.cn (K.S.); weiguo@sina.com (G.W.); wangqi@hit.edu.cn (Q.W.)

[2] School of Measurement and Communication Engineering, Harbin University of Science and Technology, Harbin 150001, China; chen_yinsheng@126.com

* Correspondence: zeroloe@163.com; Tel.: +86-156-6369-2270

Received: 10 June 2018; Accepted: 10 July 2018; Published: 17 July 2018

Abstract: Metal Oxide Semiconductor (MOS) gas sensor has been widely used in sensor systems for the advantages of fast response, high sensitivity, low cost, and so on. But, limited to the properties of materials, the phenomenon, such as aging, poisoning, and damage of the gas sensitive material will affect the measurement quality of MOS gas sensor array. To ensure the stability of the system, a health management decision strategy for the prognostics and health management (PHM) of a sensor system that is based on health reliability degree (*HRD*) and grey group decision-making (*GGD*) is proposed in this paper. The health management decision-making model is presented to choose the best health management strategy. Specially, *GGD* is utilized to provide health management suggestions for the sensor system. To evaluate the status of the sensor system, a joint *HRD-GGD* framework is declared as the health management decision-making. In this method, *HRD* of sensor system is obtained by fusing the output data of each sensor. The optimal decision-making recommendations for health management of the system is proposed by combining historical health reliability degree, maintenance probability, and overhaul rate. Experimental results on four different kinds of health levels demonstrate that the *HRD-GGD* method outperforms other methods in decision-making accuracy of sensor system. Particularly, the proposed *HRD-GGD* decision-making method achieves the best decision accuracy of 98.25%.

Keywords: health management decision; grey group decision-making; health reliability degree; maintenance decision; sensor system

1. Introduction

Sensor systems are extensively used in many fields, such as industry, manufacturing, aviation, and aerospace. Metal Oxide Semiconductor (MOS) gas sensor has become the most common gas sensor in sensor system at present because it has the advantages of fast response to target gas, high sensitivity, simple structure, easy to operation, low cost, and so on. Limited to the properties of metal oxide gas sensitive materials, the phenomenon such as aging, poisoning and damage of the gas sensitive material will affect the measurement quality of MOS gas sensor array. As a result, the trained pattern recognition method greatly degrades the performance of odor detection and analysis to target gases [1]. The influence of the work state and measurement quality of MOS gas sensor array to the performance of sensor system cannot be ignored.

At present, the following three ways are used to improve the fallen performance of the odor detection and analysis for sensor system that is caused by the decrease of reliability of the measurement value of the MOS gas sensor array.

(1) Improve the material, structure and technology of gas sensor to optimize the stability of gas sensor [2,3].
(2) Take high redundancy gas sensor array for data acquisition to minimize the impact of fault sensor on the detection and analysis effect of pattern recognition methods subsequently [4].
(3) Adopt periodic calibration and maintenance for sensor to replace the gas sensor whose performance decrease obviously [5,6].

Although the above methods can improve the reliability of sensor system to some extent, there are their own application limitations still existing. Due to the inherent characteristics of MOS gas-sensing materials, the current technology cannot completely solve the problem of stability of gas-sensing materials. The high redundancy gas sensor array can only reduce the influence of the fault sensor, but it cannot completely eliminate the effect of fault gas sensor on the detection and analysis results of the sensor system. Regular calibration and maintenance not only consume a lot of manpower and material resources, but they also cannot determine the working state and measurement quality of the sensor system during the period between twice calibration and maintenance consequently [7–10]. There are numerous sensitive elements and components in a sensor system and relationships among certain components that influence each other. Sensitive elements and components often exposed to harsh environments (high temperature, high pressure, and strong corrosion), which cause the system to fail. In the past, when one or several sensors faulted in sensor arrays, changing the failure sensors is often applied. However, it is difficult to guarantee that the consistency of sensors is exactly all the same in replacement. It is necessary to rectify the parameters of the concentration output model. Sometimes, there are no standby sensors when the sensor is failure. Therefore, it is necessary to apply health management decision-making to the sensor system. To ensure the stability of the system, a suitable solution must be determined to make the health status of all the sensors in the system more clear.

Prognostics and health management (PHM) is used widely in a great number of fields [11–13]. Prognostics and health management decision is a synthesis technique that includes data acquisition, failure detection, failure diagnosis, failure recovery, health evaluation, failure prediction, maintenance decision-making, and any other aspect [14–16]. The purpose of health management decision is to improve the safety and reliability of systems. Health management decision can achieve an evaluation and prediction of system health status according to the collected data [17]. According to the health management decision-making method, maintenance recommendations are provided. In other words, choosing the corresponding measures to reduce the failure level or to prevent the occurrence of a fault. In this way, system state is clearer and the maintenance times are reduced accordingly [18–20]. The PHM structure of sensor system is shown in Figure 1. Based on some previous work [21–27], the research of condition monitoring and health evaluation has been completed. This paper will focus on health management decision-making on the basis of condition monitoring and health evaluation methods. The purpose of this paper is not to separate the other parts of PHM from health management decision-making, but to increase the reliability of the system combined with the other parts.

The current health management decision method, according to the theory and technical application in research, can be divided into three categories: model-based maintenance decision-making [28], data-based maintenance decision-making [29,30] and reliability-based decision-making [31,32].

The sensor system has a complex structure and changeable working condition and is easily affected by the environment. The outputs of sensors are greatly influenced by the environment. The baselines of the same concentrations are different at different times. It is difficult to define the failure range. As a result, it is hard to build an appropriate model for maintenance decision-making of sensor system. Data-based maintenance decision-making method is difficult to build for the same reason. The traditional reliability-based method, such as D-S evidence theory [33–36], Bayes theory [37,38], and fuzzy set theory [39–43], will face severe challenges with the uncertainty of information and variety of data types. When provided with conflicting evidence, the D-S evidence theory results tend to deviate from the understanding of the user. Under system failure, using D-S evidence theory to meet the conflict changing from health status to failure status is difficult. Priori probability is essential for the Bayes decision method. Accuracy results are easily obtained when priori probability is known. However, obtaining priori probability is difficult. The application will be limited to some extent. Fuzzy set theory is a great data fusion method, but when handling maintenance decision-making, there are many subjective factors in the description of information because of its logical reasoning. Therefore, the representation and processing of information lacks objectivity. The three types of maintenance decision-making methods are difficult to apply for such systems. In order to evaluate the work status of the sensor, Shen et al. proposed the concept of health reliability degree of multi-functional sensors [25]. Health Reliability Degree (*HRD*) is a quantitative description of heath information. However, when there are too many sensors, a single sensor failure cannot be effectively reflected [26].

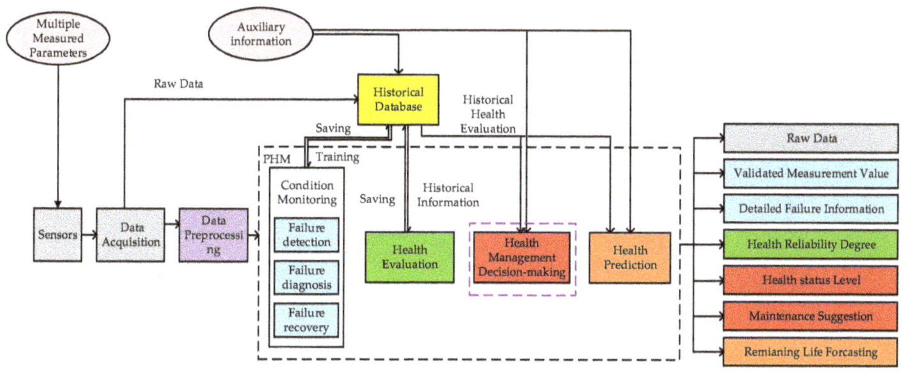

Figure 1. Prognostics and health management (PHM) structure of sensor system.

In recent years, group decision-making technology has rapidly developed [44–47]. The main research content of group decision-making is making effective decisions when multiple decision makers make decisions simultaneously. The main problem that must be solved is how to aggregate the decision information of different experts with different preferences to obtain consistent decision results. By fusing the decision objective of each expert, the accuracy of the system can be improved. However, group decision-making technology research is undeveloped. The methods of correctly obtaining decision information, including property value, property weight, and decision maker weight information have not been established.

In this paper, a method for health management and maintenance decision based on health reliability degree and grey group decision-making (*HRD-GGD*) is proposed. In this method, *HRD* of sensor system is obtained by fusing the output data of each sensor. The optimal decision-making recommendations for health management of the system is proposed by combining historical health reliability degree, maintenance probability, and overhaul rate. The *HRD-GGD* is proposed to realize the maintenance of the sensor system by comprehensively considering the decision results of multiple expert sets. Not only can the system give out the system state, but also provide the maintenance suggestion for each failure mode after the system working and give the confidence degree of each maintenance proposal.

The rest of this paper is organized as follows. In Section 2, the framework of health management and maintenance decision and the corresponding methods are presented. In Section 3, the experimental setup and analytical discussion are introduced. In Section 4, two situations are presented and 400 different health status level samples are analyzed to give the results of health management decision. Finally, the conclusion is accounted in Section 5.

2. Health Management Decision

2.1. Implementation Framework of Health Management and Maintenance Decision

The main purpose of the health management decision is to obtain the working state of sensor system quantitatively and to provide the maintenance decision for the system at this status. In order to evaluate the state of the sensor system, the historical failure information, historical maintenance records, and trends of historical health status for the system is used in order to model the health management mode. As the state of the system is clearer, it is easy to reduce the proportion of unscheduled maintenance in the maintenance plan and change the unscheduled maintenance to predictive maintenance (scheduled maintenance).

The establishment of framework is the core of health management decision theory. The health management suggestion is dynamically obtained by collecting fault information, health status, failure prediction conclusion, and historical maintenance situation. The framework is shown in Figure 2. The system input vector is composed of three parameters: historical health trend, maintenance probability, and overhaul rate. The historical health trend indicates the working state of the system during the last period of time. The parameter is acquired by fusing the historical *HRD* during this period. Maintenance probability is obtained from historical maintenance records. The value is equal to history maintenance times/total test times. The more frequent maintenance of the system, the greater the probability of failure. The value of maintenance probability is larger at this condition. Overhaul rate is the parameter of unpredictable maintenance task. The value is equal to the next inspection time/overhaul cycle. The longer the overhaul time, the greater the uncertainty of the system. The system is inclined to failure in this way.

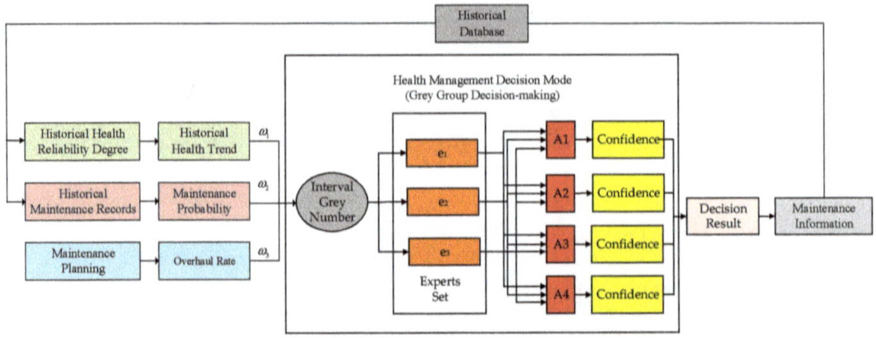

Figure 2. Framework of the health management decision.

There are three experts in the expert set. The experts can be changed when facing different problems. In the system, three algorithms are used as the experts, namely D-S evidence theory, Bayes theory, and fuzzy set theory. Three experts give their suggestion, respectively, according to the above three parameters. It can be found from the experiment that these three methods have their limitations, respectively, which will be discussed in Section 4.

The second part is the group decision-making. This part is responsible for data fusion of the decisions of the expert set to obtain the final decision result. The decision information is recorded as part of the next decision. The decision information is recorded and used as the basis for the next decision. The solution set of grey group decision-making is $\{A_1, A_2, A_3, A_4\}$, the corresponding decision frameworks are A_1 {no maintenance}, A_2 {preventive maintenance}, A_3 {corrective maintenance}, and A_4 {immediate maintenance}. The decision expert set is $\{e_1, e_2, e_3\}$, which represent three experts, respectively. The decision index set is $\{u_1, u_2, u_3\}$. The corresponding evidences are u_1 {historical health reliability degree}, u_2 {unpredictable maintenance task}, and u_3 {historical maintenance record}.

The evidence property weight vector for three evidences is $\omega = \{\omega_1, \omega_2, \omega_3\}$. The size of decision framework is four. The decision framework and its corresponding health status levels and maintenance levels are shown in Table 1.

The state description and corresponding maintenance suggestion are shown, as follows:

1. Health: The whole system is very healthy. All of the sensors are also healthy. Their measurements are close to the expected value. There is no need to repair the system.
2. Subhealth: The system is working at subhealthy status. The output of the system is within a normal range. All of the parameters may fluctuate near their expected value. It is essential to execute preventive maintenance regularly. Failure detection and failure isolation methods should be used in this situation.
3. Failure Edge: The system is nearly failure. Their actual measurements have deviated from the expected value, but they have not deviated completely. In this status some sensors may be faulty, but the system can work effectively when fault recovery is performed. Corrective maintenance is needed after experiment [25]. Failure recovery method will be applied in this status to improve the work status sometimes.
4. Failure: The system is failure. Most of sensors are failure. The actual output has completely deviated from its expected results. Immediate repaired the failure components or replacement failure components immediately may be the best choice.

Maintenance decision method can provide the maintenance suggestion for each failure mode and give the confidence degree of each maintenance proposal.

Table 1. Maintenance level of health and maintenance decision fault preventive measures.

Solution Set	Health Status Level	Health Description	Maintenance Level
A_1	Health (*HS*)	healthy condition	No maintenance
A_2	Subhealth (*SHS*)	normal range	Preventive maintenance
A_3	failure edge (*FES*)	fault edge	Corrective maintenance
A_4	Failure (*FS*)	fault condition	Immediate maintenance

2.2. Health Reliability Degree (HRD)

HRD is a novel conception to define a quantitative health level. *HRD* represents the health level of whole system. The *HRD* of the system is fused by the health level of all the sensitive elements in the system. The value ranges from 0 to 1. When the value is 0, the system works at a severe failure state. When the value is 1, the system works at 100% healthy state. The larger the *HRD*, the higher the health level. The relationship between *HRD* and health level is defined as Table 2. When to evaluate the health status of the system, the four health status levels, healthy status (*HS*), subhealthy status (*SHS*), failure edge status (*FES*), and failure status (*FS*). The specific values vary according to different application objects [26].

Table 2. Health status level.

Solution Set	Range of Health Reliability Degree	Health Status Level
A_1	$0.9 \leq HRD \leq 1$	Healthy
A_2	$0.6 \leq HRD < 0.9$	Subhealthy
A_3	$0.2 \leq HRD < 0.6$	Failure edge
A_4	$0 \leq HRD < 0.2$	Failure

HRD is fused of four belonging relationship degree (*brd*) of sensor system by applying grey theory. The four parameters *brd* are the keys to computation *HRD*. The values can be expressed in a simplified way, as shown in Figure 3. If the *brd* is equal to 1, then the current working status is completely belonged to its corresponding status completely. If the *brd* is equal to 0, then the current status is completely not belonged to its corresponding status. The *brd* is changed with the fluctuation of *HRD*. In order to map the relationship between *HRD* and *brd*, Relevance Vector Machine (RVM) is used to fuse four *brd* to *HRD*.

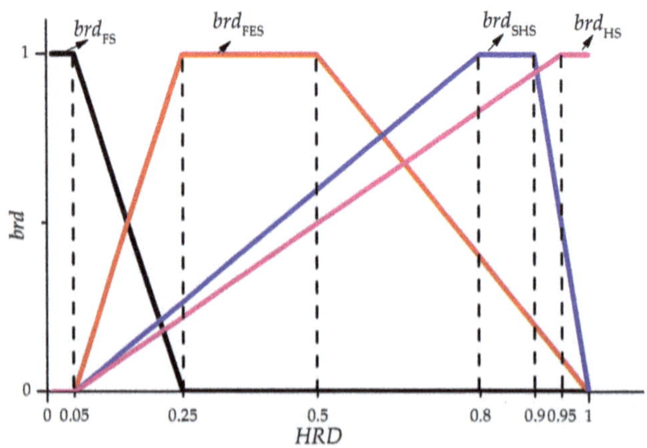

Figure 3. The relationship between Health Reliability Degree (*HRD*) and Belonging Relationship Degree (*brd*).

The relationship of *brds* and output parameters are shown in Figure 4. In summary, the whitening function of four grey sets are obtained by Equations (1)–(4).

$$f_{HS}(x) = \exp\left[-\|x-\mu\|^2/2\delta^2\right] \tag{1}$$

$$f_{SHS}(x) = \begin{cases} \exp\left[-\|x-\mu+\delta\|^2/2\delta^2\right] & x<\mu \\ \exp\left[-\|x-\mu-\delta\|^2/2\delta^2\right] & x>\mu \end{cases} \tag{2}$$

$$f_{FES}(x) = \begin{cases} \exp\left[-\|x-\mu+3\delta\|^2/2\delta^2\right] & x<\mu \\ \exp\left[-\|x-\mu-3\delta\|^2/2\delta^2\right] & x>\mu \end{cases} \tag{3}$$

$$f_{FS}(x) = \begin{cases} 1 & x<\mu-5\delta \quad or \quad x>\mu-5\delta \\ -1/\delta\cdot[x-\mu+4\delta] & \mu-5\delta<x<\mu-4\delta \\ 1/\delta\cdot[x-\mu-4\delta] & \mu+4\delta<x<\mu+5\delta \\ 0 & \mu-4\delta<x<\mu+4\delta \end{cases} \tag{4}$$

For the all of the components in sensor system, the grey sample evaluation (*GSE*) matrix at the single time point can be expressed as $GSE_j = (gse_{ijk})_{m\times n}(i = 1,2,\cdots,m;\ k = 1,2,\cdots,n)$, which is shown in (5).

$$GSE_j = \begin{matrix} \\ S_1 \\ S_2 \\ \vdots \\ S_m \end{matrix} \begin{matrix} \begin{matrix} I_1 & I_2 & I_3 & I_4 \end{matrix} \\ \begin{bmatrix} a_{1j1} & a_{1j2} & a_{1j3} & a_{1j4} \\ a_{2j1} & a_{2j2} & a_{2j3} & a_{2j4} \\ \vdots & \vdots & \vdots & \vdots \\ a_{mj1} & a_{mj2} & a_{mj3} & a_{mj4} \end{bmatrix} \end{matrix} \tag{5}$$

where j represents the time point, S_1 indicates all the elements in sensor system. I_k is the evaluating criterion.

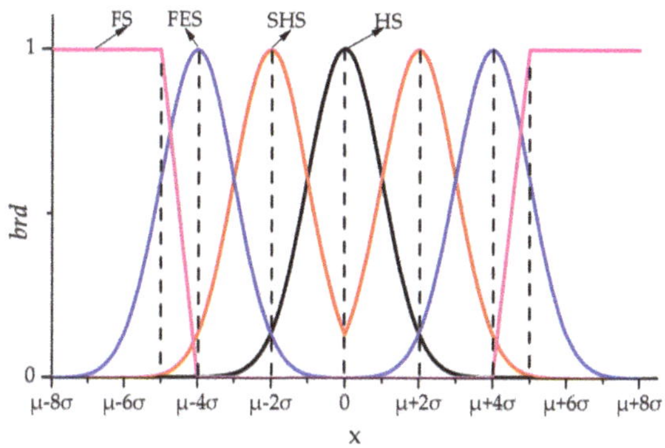

Figure 4. The relationship between the output and *brd*.

The decision weight vector of different elements in sensor system is obtained by using information entropy method. The probabilistic proportion of *k*th assessment criterion of *i*th elements is shown in (6).

$$P_{ik} = \frac{gse_{ik}}{\sum\limits_{k=1}^{n} gse_{ik}} \quad (i = 1, 2, \cdots, m) \tag{6}$$

Then compute the information the information entropy of the *i*th elements by (7).

$$E_i = -\frac{1}{\ln n}\sum_{k=1}^{n} P_{ik} \ln P_{ik} \quad (i = 1, 2, \cdots, m) \tag{7}$$

The weight vector $W_j = \{w_{1j}, w_{2j}, \cdots, u_{mj}\}$ is determined by

$$w_{ij} = \frac{1-E_i}{\sum\limits_{i=1}^{m} (1-E_i)} \quad (i = 1, 2, \cdots, m) \tag{8}$$

After obtaining the decision weight, the comprehensive grey assessment values (*CGAV*) are calculated by (9).

$$CGAV = W \times GSE \tag{9}$$

where $CGAV = [\ brd_{SH} \quad brd_{SHS} \quad brd_{FES} \quad brd_{FS}\ ]$.

The flowchart of the *HRD* methodology is shown in Figure 5 and the detail steps are shown in Table 3. The correlation among multiple parameters has been fully considered for the weight distribution of different sensors.

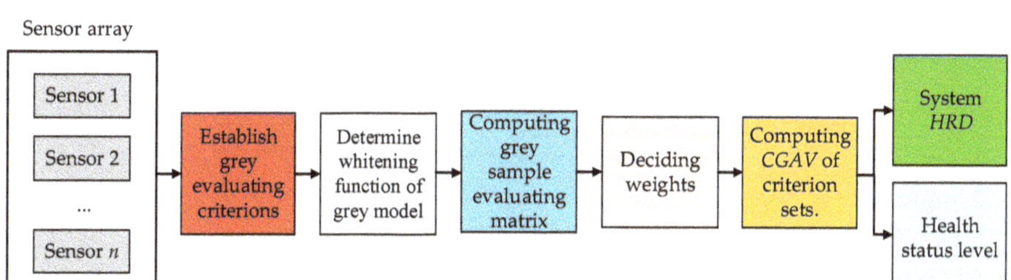

Figure 5. The flowchart of *HRD* methodology.

Table 3. *HRD* computing procedure.

***HRD* Based on Grey Theory**
Input:
Output of the sensor system
Output:
Health Reliability Degree
Procedure:
Step 1: Establish the grey evaluating criterions, which is shown as (HS, SHS, FES, FS).
Step 2: Determine the whitening function of the grey model according to Equations (1)–(4).
Step 3: Compute decision weights by using information entropy method.
Step 4: Compute Grey Sample Evaluating (*GSE*) Matrix by (5).
Step 5: Calculate the *CGAV* under evaluating criterion sets.
Step 6: Calculate *HRD* by RVM.

2.3. *Grey Group Decision-Making*

2.3.1. Grey Risk Decision-Making

The scheme-set of grey risk group decision-making problem is $(A_1, A_2, \ldots, A_n)$, the decision indicator set is $(u_1, u_2, \ldots, u_m)$, and the decision group set is $(e_1, e_2, \ldots, e_q)$ $(q \geq 2)$, where e_s represents the s-th decision maker. For each decision indicator u_j, there are l possible states $\theta = \{\theta_1, \theta_2, \ldots, \theta_l\}$. The probability of the state θ_t occurring is $p^s_{tj}(1 \leq t \leq l)$ for the decision maker e_s under the decision indicator u_j, which fits $(0 \leq p^s_{tj} \leq l)$, $\sum_{t=1}^{l} p^s_{tj} = l$. The attribute value of plan A_i is $a^s_{ijt}(\otimes) \in [\underline{a}^s_{ijt}, \overline{a}^s_{ijt}]$ [48]. The expert decision-making is divided into f-levels. The grey expert attribute set is denoted as $M = \{A_1, A_2, \ldots, A_f\}$.

For the decision expert attribute set M of level f, the decision expert fuzzy attribute value is $m_s = k(1 \leq k \leq f)$. Define the deviation coefficient a_s, which indicates the difference between m_s and the s-th expert actual importance of the expert and $a_s \in [-0.5, 0.5]$.

The index $\omega = (\omega_1, \omega_2, \ldots, \omega_m)$ weight is obtained while using the reciprocal of entropy weight method to increase the experts weight with higher accuracy. If the information entropy index is smaller, the more information it provides. The greater the role that it plays in the comprehensive evaluation, the greater the weight of the index. In order to reduce the effluence, the reciprocal was used in this way. The weight of the decision-maker s is

$$H_s = \sum_{j=1}^{m} \omega_j \sum_{t=1}^{l} (-p^s_{tj} \ln p^s_{tj})^{-1} \tag{10}$$

where p^s_{tj} is the input data. Deviation factor is

$$a_s = \frac{H_s - 0.5(\min H_s + \max H_s)}{\min H_s - \max H_s} \tag{11}$$

where $a_s \in [-0.5, 0.5]$. Let $H_s = \min H_s$, $a_s = 0.5$. If $H_s = \max H_s$, $a_s = -0.5$. Therefore, the decision-making expert attributes after recuperation are $m'_s = m_s + a_s$.

The expert weight is obtained, as follows:

$$\lambda^s = \frac{m'_s}{\sum_{s=1}^{q} m'_s} \tag{12}$$

As H_s increases, more right information included in the decision maker s increases the effectiveness and increases the weight. Thus, it is more realistic to correct the importance of the decision maker using entropy to obtain the decision maker weights.

Let $z_{ij}(\otimes) = \sum_{s=1}^{q} r_{ij}^{s}(\otimes)\lambda^{s}$, the standardized synthesis decision matrix can be obtained [49].

$$z(\otimes) = \left(z_{ij}(\otimes)\right)_{n\times m} = \left(\left[\underline{z}_{ij}, \overline{z}_{ij}\right]\right)_{n\times m} \tag{13}$$

2.3.2. Grey Group Decision Model for Decision-Making

The grey group decision-making theory is used to make the effective choice of the three decision algorithms and the final comprehensive decision is made by fusing the three types of algorithms. The specific process is shown in Figure 6. The comprehensive step of the grey group decision-making theory is shown in Table 4. In summary, the maintenance decision results and the comprehensive confidence of the system failure modes can be obtained while using the grey group decision-making method, and the dynamic maintenance plan can be formulated on this basis.

Table 4. The comprehensive decision-making step of grey group decision-making theory.

Grey Group Decision-Making Algorithm
Input: Historical Health Reliability Degree (*HHRD*): The parameter is composed of the last n *HRD*s. Maintenance Probability (*MP*): Maintenance probability is equal to history maintenance times/total test times. Overhaul Rate (*OR*): Overhaul rate is equal to the next inspection time/overhaul cycle. **Output:** Decision Result: The parameter is the level of the maintenance decision-making. The size of the framework is four: {no maintenance, preventive maintenance, corrective maintenance, immediate maintenance} Confidence: The parameter is output vector of the maintenance decision-making confidence. **Procedure:** Step 1: Calculate the interval grey numbers of each decision result for each evidence in the decision framework under each decision method. The interval grey number is expressed as $a_{ij}^{s}(\otimes)$, $a_{ij}^{s}(\otimes) \in [\underline{a}_{ij}^{s}, \overline{a}_{ij}^{s}]$. $\underline{a}_{ij}^{s}$ represents the grey number lower limit and $\overline{a}_{ij}^{s}$ is the grey number upper limit (i = 1, 2, 3, 4, j = 1, 2, 3, s = 1, 2, 3). Step 2: According to the upper-lower limit $[\underline{a}_{ij}^{s}, \overline{a}_{ij}^{s}]$ in the interval grey number $a_{ij}^{s}(\otimes)$ in Step 1, establish the comprehensive decision matrix (CDM) of each decision method as shown in Table 5. Step 3: By utilizing the interval grey number weakening transformation, the decision matrix of three decision methods is initialized and transformed to obtain the standardized decision matrix. Step 4: Calculate the weight of each decision method. First, determine the effect vector of the three decision methods for all decision frameworks according to the standardized decision matrix in Step 3. The matrix elements are the effect vectors of the three decision methods for each decision result in the decision framework. Then, according to the interval grey number vector distance formula and the weight of each evidence attribute ω_j(j = 1, 2, 3), the space projector distance of each effect vector is calculated. Finally, the ratio between the vector distance of a decision method and the sum of effect vector distance for the other decision methods is the weight coefficient λ_s(s = 1, 2, 3, 4) of the decision method. Step 5: Calculate the comprehensive decision results. According to the maintenance decision result of the experts, the confidence of each decision result corresponding to the four decision frames can be obtained. Finally, according to the weight coefficients of each decision method that were obtained in Step 4, the final decision result is obtained by applying weighted averaging to the confidence.

Table 5. Comprehensive decision matrix of decision method e_s(s = 1, 2, 3).

	u_1		u_2		u_3	
A_1	$\underline{a}_{11}^{s}$	$\overline{a}_{11}^{s}$	$\underline{a}_{12}^{s}$	$\overline{a}_{12}^{s}$	$\underline{a}_{13}^{s}$	$\overline{a}_{13}^{s}$
A_2	$\underline{a}_{21}^{s}$	$\overline{a}_{21}^{s}$	$\underline{a}_{22}^{s}$	$\overline{a}_{22}^{s}$	$\underline{a}_{23}^{s}$	$\overline{a}_{23}^{s}$
A_3	$\underline{a}_{31}^{s}$	$\overline{a}_{31}^{s}$	$\underline{a}_{32}^{s}$	$\overline{a}_{32}^{s}$	$\underline{a}_{33}^{s}$	$\overline{a}_{33}^{s}$
A_4	$\underline{a}_{41}^{s}$	$\overline{a}_{41}^{s}$	$\underline{a}_{42}^{s}$	$\overline{a}_{42}^{s}$	$\underline{a}_{43}^{s}$	$\overline{a}_{43}^{s}$

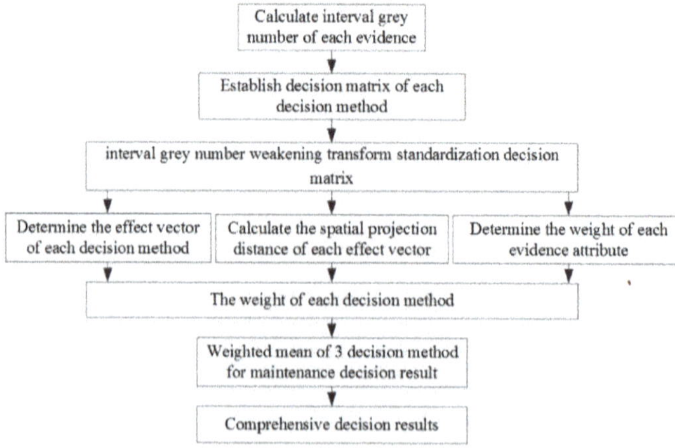

Figure 6. Decision diagram of grey group decision-making theory.

2.4. The Process of Health Management Decision Method Based on Grey Group Decision

In order to improve the accuracy of the evaluation. The health management decision method combines grey group decision making and *HRD* theory to implement health management decision to sensor system. The detail steps of *HRD-GGD* are shown, as follows.

Step 1: To obtain the system measurement point parameters of the sensor or network interface, the data acquisition device is used to collect data.

Step 2: Pre-process the parameters collected by each measuring point in the system. Pre-processing includes abnormal value elimination, filtering noise reduction, calculating average value, and 3σ standard deviation.

Step 3: Failure detection, isolation, and diagnostic algorithms should be applied to determine the location of and the type of failure [21,22].

Step 4: If there is no failure in the array, give out the best estimation value of each sensor, according to the correlation of sensor array. If failure occurs in the sensor array, the best estimation value of failure sensor can be obtained, according to the normal data and the value of normal sensor can be got from the best estimation value of failure sensor and the other normal sensors.

Step 5: Establish the health reliability degree evaluation mode. The results of *HRD* are used as the inputs of the historical health status trend. The calculation steps of *HRD* are thoroughly introduced in Section 2.2.

Step 6: Establish a health management decision model based on grey group decision-making theory. Make maintenance decision on various failure modes and give out the corresponding maintenance suggestion.

Firstly, obtain the evidences for each expert.

The number of evidence is three, expressed as u_1, u_2, u_3, known, as follows:

u_1: Analysis of health reliability degree and historical health reliability degree;

u_2: Historical failure information and corresponding maintenance records; and,

u_3: Prepared system maintenance program.

Evidence u_1 is fused according to three decision methods. Evidence u_2 and u_3 is obtained by using the whitening function as shown in Figure 7. *brd* is grey parameters. *HS* represents Health Status. *SHS* represents Subhealth Status. *FES* represents Failure Edge Status and *FS* represents Failure Status Level.

The size of the decision framework is 4. The framework is expressed with a 4-bit binary number:

A_1: no maintenance, 0001,

A_2: preventive maintenance, 0010,
A_3: corrective maintenance, 0100, and
A_4: immediate maintenance, 1000.

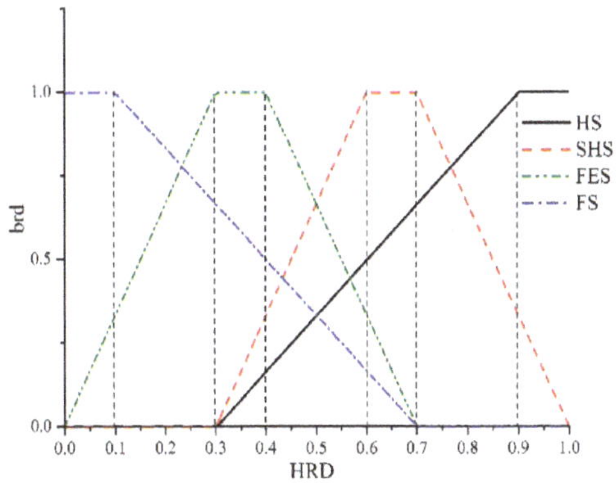

Figure 7. Whitening function of evidence u_2 and u_3.

Utilizing the historical experiment data in Step 4. Analysis and prediction of the trend of the historical health parameters is based on the set of data. Determine the weight of each evidence attribute $\omega = (\omega_1, \omega_2, \omega_3)$ based on historical failure information and maintenance records combined with the prepared maintenance program. ω_2 indicates that maintenance probability is equal to history maintenance times/total test times. ω_3 is the overhaul rate, which is equal to the next inspection time/overhaul cycle. $\omega_1 = 1 - \omega_2 - \omega_3$.

Secondly, to process the input data and turn the data to grey numbers. Calculate the confidence intervals of each evidence with a significance level of 0.05.

Thirdly, establish the comprehensive decision matrix (CDM) of each decision method. The CDM of the three experts is calculated for each failure model. Then, the decision matrix is standardized to calculate the standardization comprehensive decision matrix. Finally, the final comprehensive decision results are obtained. The matrix includes upper bound evaluation matrices, lower bound evaluation matrices, and whitening evaluation matrices. The upper bound evaluation matrices are $\mathrm{CDM}^{(1+)}$, $\mathrm{CDM}^{(2+)}$, and $\mathrm{CDM}^{(3+)}$. The lower bound evaluation matrices are $\mathrm{CDM}^{(1-)}$, $\mathrm{CDM}^{(2-)}$, and $\mathrm{CDM}^{(3-)}$. The whitening evaluation matrices are $\mathrm{CDM}^{(1)}$, $\mathrm{CDM}^{(2)}$, and $\mathrm{CDM}^{(3)}$. The upper bound evaluation matrices represent the maximum confidence of three evidences in the system. The lower bound evaluation matrices represent the minimum confidence of three evidences in the system. Whitening evaluation matrixes represent the confidence of three evidences for whitening degree in the system. The matrices are shown as (14):

$$\mathrm{CDM}^{(k)} = \begin{bmatrix} \sigma_{11} & \sigma_{12} & \sigma_{13} & \sigma_{14} \\ \sigma_{21} & \sigma_{22} & \sigma_{23} & \sigma_{24} \\ \sigma_{31} & \sigma_{32} & \sigma_{33} & \sigma_{34} \end{bmatrix} \tag{14}$$

where k represents the expert ($k = 1, e_1; k = 2, e_2; k = 3, e_3$). σ_{ij} represents the confidence of each evidence. i is the health level of the decision framework ($i = 1, A_1; i = 2, A_2 ; i = 3, A_3; i = 4, A_4$). j is the evidence ($j = 1$, historical health trends; $j = 2$, maintenance probability; $j = 3$, overhaul rate).

The steps to calculate the health status level using CDM are shown in Table 6.

Table 6. Procedure of calculating health status level.

Health Status Level
Step 1: to calculate the health level of the decision framework and the confidence of every evidence.
Step 2: In grey group decision, the weight vector of each evidence attribute is $\omega = (\omega_1, \omega_2, \omega_3)$
Step 3: calculate the decision confidence for the expert set by $\text{confidence}_i = \omega \cdot \text{CDM}$.
Step 4: the weight vector for each experts $\omega_e = (\omega_{e1}, \omega_{e2}, \omega_{e3})$ is obtained by entropy weight method.
Step 5: calculate the confidence of the final health status level by $\text{confidence} = \sum_{i=1}^{3} \omega_e \cdot \text{CDM}^{(i)}$.

Finally, ranking all of the alternative health status levels in accordance with the confidence and choosing the optimal health management suggestion with largest confidence.

3. Experimental Setup and Analytical Discussion

The detailed *HRD-GGD* process is given in Section 2.4. In this part, the problem of health management decision for an atmosphere pollution gas sensor system is taken as an example to verify the *HRD-GGD method.*

3.1. Sensor System Experimental System

The sensor system used for testing atmosphere pollution gas was mainly composed of gas source, MFCs, gas chamber, sensor array, heater driver circuit, signal conditional circuit, data acquisition circuit, power supply, and laptop PC. The sensor array that consisted of four different types of gas sensors (CO, NO_2, O_3, SO_2) and temperature, humidity, and pressure sensors, was fixed in the gas chamber. The number of each type gas sensor is three. The gas chamber temperature was maintained at 30 °C by constant temperature control. The structure of sensor system is shown in Figure 8 and the physical picture of sensor system is shown in Figure 9. The normal working ranges and units for the 15 sensors (three CO sensors, three NO_2 sensors, three O_3 sensors, three SO_2 sensors, one temperature, one humidity, and one pressure sensor) are given in the Table 7. The health management decision platform is applied to the system. In the experiment, QT is used as the experimental software platform combined with SQL server software to realize database storage function. The *HRD-GGD* algorithm is implemented using Visual Studio.

Table 7. Failure type of senor system and its form.

Failure	Name	Failure Feature and Form	Failure Place	Failure Prevention and Control Measures
F1	Sensor disconnect	Step Response. Lower than lower threshold	Target gas sensor	Check the sensor pin, change the target sensor
F2	Sensor overload	Step Response. Above upper threshold	Target gas sensor	Check the sensor pin, change the target sensor
F3	Sensor poisoned	No response or irregular fluctuation	Target gas sensor	Change the target sensor
F4	Sensor drift	Slowly varying. Baseline offset	Target gas sensor	Increase the preheating time, change the target sensor
F5	Abnormal changed	Output fluctuation	Target gas sensor	Check and replace the filter capacitor, check and replace the power supply module, change the target sensor
F6	Heater circuit failure	Sensor has no response. Heater has no input.	Target gas sensor circuit	Circuit connection check, change the chips

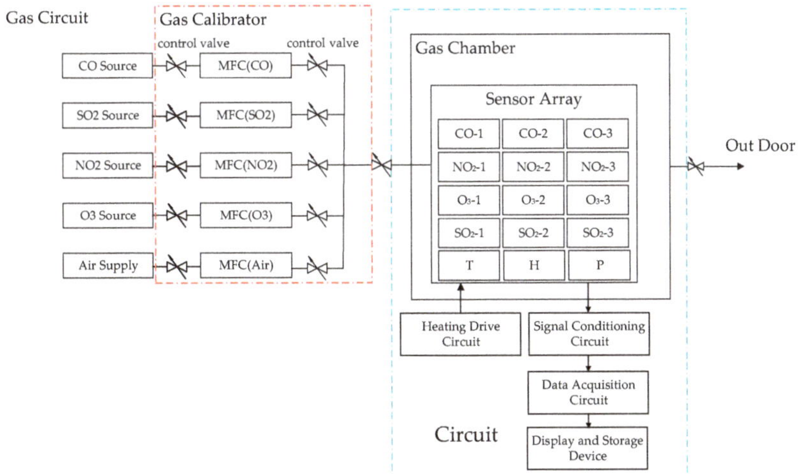

Figure 8. The model of the sensor system.

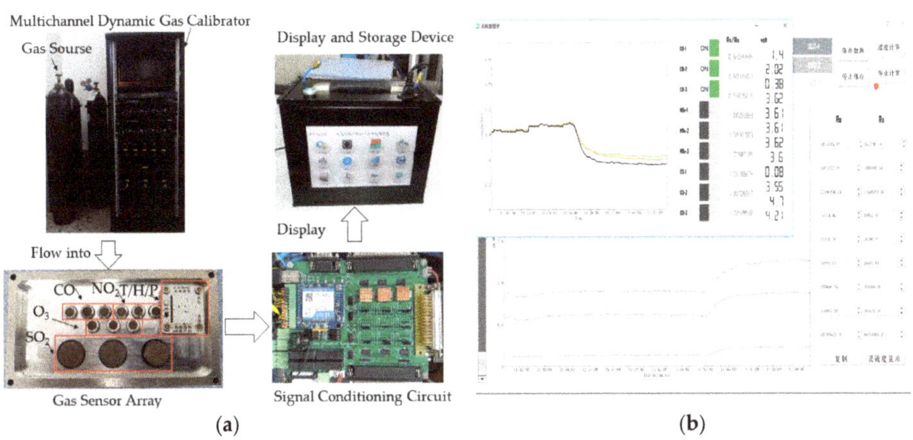

Figure 9. (**a**) Physical picture of sensor system; and, (**b**) display interface.

The system is used in the laboratory. The acquisition device utilized USB-bus cards (USB-2089, Art Technology Development Co. Ltd., Beijing, China) with 16 analog inputs at up to 400 KHz and a 14-bit A/D conversion accuracy. The sampling period is once per second. The failure data is generated by the failure simulating software. By analysis of the feature of historical failure data, the failure forms of different failures are summarized in the software. The maintenance suggestions are given according to practical experience. The failure information and maintenance suggestions are given in Table 7.

3.2. Experiment Data

The data of each sensor are the voltage values collected by the 15 sensors in Table 8. Both the test and training samples include normal samples and fault samples. The normal samples are historical experimental data, i.e., system history experimental samples. The major frequent failure part in the sensor system is the sensor array. So, the failure type is mainly aimed at sensor array and heater drive circuit. The system *HRD* is the *HRD* of sensor array. The fault samples are analysed using fault simulation software, according to the fault modes. Figure 10 illustrates the sensor response process when exposed to 50 ppm CO in experiment. Each set of data includes 15 measurement points, with a sampling time of 1 s. The experiment involved 2 min for sensor to response completely and 2 min for the sensor to recover. When the system works for a long time, the performance of the sensor system will decrease with the increasing of running time. The historical *HRD* of the sensor system in 200 h is shown in Figure 11.

F1–F5 are the failure of sensors, F6 is the failure of heater circuit. Because of the correlation among components, the output tends to be abnormal when the system fails. F1 to F6 are both sensors or circuit faults for single sensor. Every sensor fault can be diagnosed as a kind of failure. Due to the same form of expression and different location, it can be classified as a kind of failure. When multiple failures occur, which is to say that different sensors have different failures at the same time. This situation can be understood as multiple fault superposition, not as new failure. 200 sets of historical data are used for obtaining the best work state. 400 groups data of different health status level (each type of health status level contains 100 groups data) are used for testing.

Table 8. The scope of all the sensors.

Sensor	Range	Unit	Sensor	Range	Unit
CO-1	1–4	V	O_3-1	0.15–1.8	V
CO-2	1–4	V	O_3-2	0.15–0.9	V
CO-3	0.3–1.8	V	O_3-3	0.15–0.4	V
NO_2-1	0.3–5	V	SO_2-1	2–5	V
NO_2-2	0.3–5	V	SO_2-2	1.3–5	V
NO_2-3	0.1–1.5	V	SO_2-3	1.3–3.7	V
T	15–50	°C	H	20–65	%RH
P	0.09–0.12	Kpa			

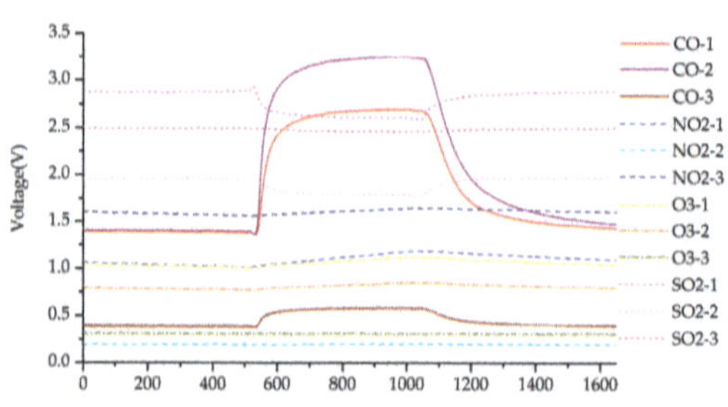

Figure 10. Response of the sensor system for 50 ppm CO.

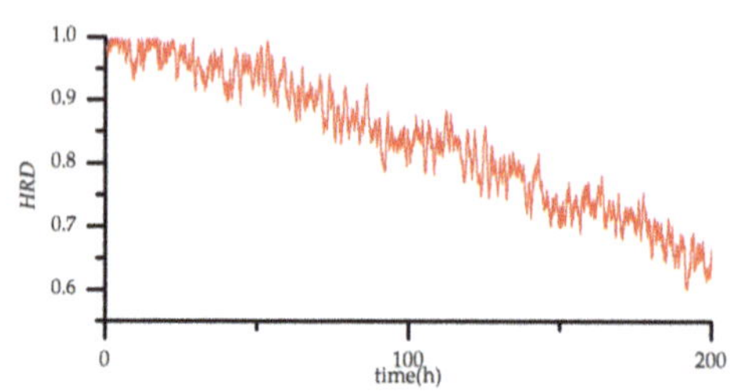

Figure 11. The historical *HRD* of the sensor system.

4. Results and Analysis

The health management decision of sensor system means that the maintenance suggestion is implemented in a global way and it refers to all of the sensors and components. In this section, two situations that represent different health levels are introduced to interpret the proposed strategy.

Situation 1: all the sensor and components are fault free

The collected data is processed using the grey data fusion method to obtain the historical health reliability degree, the details are shown in Section 2.2. The last seven health reliability degrees are used as the reference historical health parameter. The analysis and prediction of the historical health parameters trend are performed based on a data set. Historical failure information, maintenance records, and the established maintenance program are also considered.

Consider a random experiment as an example. The size of decision framework is 4. The framework is expressed as: {no maintenance}: A_1, {preventive maintenance}: A_2, {corrective maintenance}: A_3, {immediate maintenance}: A_4. Regard a series of historical health reliability degree as evidence. First, obtain the series of experimental data and the previous seven series. The historical health parameters are shown in Table 9 and are fused by three experts' decision methods. Deal with data by grey processing to obtain the grey interval and calculate separately. Deal with data by grey processing to obtain the grey interval and calculate the confidence of each level separately. The decision result of evidence 1: historical health trends are shown as (15).

Table 9. Historical health parameters for situation 1.

Evidence	A_1	A_2	A_3	A_4
1	0.8150	0.3930	0.0039	0
2	0.8648	0.3357	0.0029	0
3	0.8568	0.3447	0.0029	0
4	0.8556	0.3457	0.0031	0
5	0.7337	0.4792	0.0057	0
6	0.9388	0.2373	0.0014	0
7	0.9393	0.2449	0.0014	0

According to detection conditions, the historical maintenance number is selected as five and the total experiment time is 100 times. The next inspection time is set as 300 days and the overhaul cycle is 365 days. Attribute weights are determined as $\omega = (\omega_1, \omega_2, \omega_3) = (0.7719, 0.05, 0.1781)$. $\omega_2 = 0.05$ is the maintenance probability, which is equal to historical maintenance times/total test times. $\omega_3 = 0.1781$ is the overhaul rate, which is equal to the next inspection time/overhaul cycle. In the example, the historical maintenance times are five and total test times are 100. The next inspection time is 300 days and the overhaul cycle are 365 days. The decision matrix that was established by the fusion of each decision method is shown from Tables 10–12. The attribute weights of these three decision methods are (0.4158, 0.3467, 0.2375), which is attained by (10)~(12). The whitening degree is the mathematical expression of whitening rules in the grey set under the existing information. Whitening degree is obtained by whitening function, which is shown in Figure 7. The comprehensive decision matrixes that were obtained using the three decision methods by using (14).

Table 10. Comprehensive decision matrix of decision method e_1.

	u_1		u_2		u_3	
	Grey Interval	Whitening Degree	Grey Interval	Whitening Degree	Grey Interval	Whitening Degree
A_1	[0.8562, 0.8617]	0.8589	[0.8690, 0.8707]	0.8698	[1, 1]	1
A_2	[0.1383, 0.1438]	0.1411	[0.5853, 0.5920]	0.5937	[0.1650, 0.1683]	0.1667
A_3	[0, 0]	0	[0, 0]	0	[0, 0]	0
A_4	[0, 0]	0	[0, 0]	0	[0, 0]	0
weight	0.7719		0.05		0.1781	

Table 11. Comprehensive decision matrix of decision method e_2.

	u_1		u_2		u_3	
	Grey Interval	**Whitening Degree**	**Grey Interval**	**Whitening Degree**	**Grey Interval**	**Whitening Degree**
A_1	[0.8860, 0.9026]	0.8943	[0.8690, 0.8707]	0.8698	[1, 1]	1
A_2	[0.0974, 0.1140]	0.1057	[0.5853, 0.5920]	0.5937	[0.1650, 0.1683]	0.1667
A_3	[0, 0]	0	[0, 0]	0	[0, 0]	0
A_4	[0, 0]	0	[0, 0]	0	[0, 0]	0
weight	0.7719		0.05		0.1781	

Table 12. Comprehensive decision matrix of decision method e_3.

	u_1		u_2		u_3	
	Grey Interval	**Whitening Degree**	**Grey Interval**	**Whitening Degree**	**Grey Interval**	**Whitening Degree**
A_1	[0.6579, 0.6611]	0.6595	[0.8690, 0.8707]	0.8698	[1, 1]	1
A_2	[0.3348, 0.3381]	0.3364	[0.5853, 0.5920]	0.5937	[0.1650, 0.1683]	0.1667
A_3	[0.0040, 0.0041]	0.0040	[0, 0]	0	[0, 0]	0
A_4	[0, 0]	0	[0, 0]	0	[0, 0]	0
weight	0.7719		0.05		0.1781	

Since the results of the matrix are floating from 0 to 1, normalization is not required. The final grey group decision-making result can be obtained by directly fusing with the weight. For another decision-making method, such as D-S evidence theory, Bayes theory, and fuzzy set theory are used as the off-the-shelf maintenance decision-making method. The decision results are shown in Figure 12. According to descending order of the maintenance confidence, the ranks and maintenance suggestions that are based on grey group decision and another three comparative decision methods are shown in Table 13, the rank of four methods are shown as $A_1 > A_2 > A_3 > A_4$, the final maintenance suggestion of four methods are all A_1: no maintenance, which is suitable to the status description.

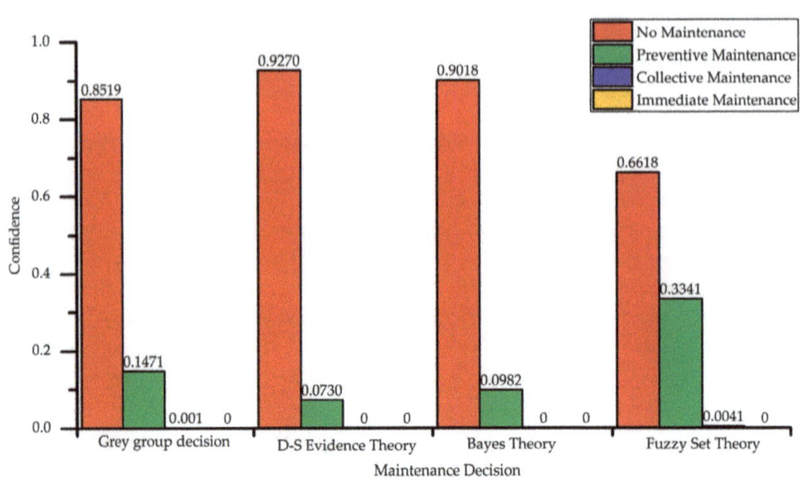

Figure 12. The confidence to 4 health status level of Situation 1 based on 4 decision methods.

In order to verify the evaluation ability of maintenance decision method for health status, 100 groups of health status data at different times were used to test the accuracy of evaluation. The confidence of the 100 groups health state data to four health status level are shown in Figure 13. False alarm occurred when using Bayes theory. The accuracy for the four methods is shown in

Table 14. Grey group decision ignores the disadvantage of Bayes theory for this situation to improve the decision accuracy.

Table 13. The rank and maintenance suggestion of Situation 1.

Method	Rank	Maintenance Suggestion
Grey Group Decision	$A_1 > A_2 > A_3 > A_4$	A_1: No Maintenance
D-S evidence Theory	$A_1 > A_2 > A_3 > A_4$	A_1: No Maintenance
Bayes Theory	$A_1 > A_2 > A_3 > A_4$	A_1: No Maintenance
Fuzzy Set Theory	$A_1 > A_2 > A_3 > A_4$	A_1: No Maintenance

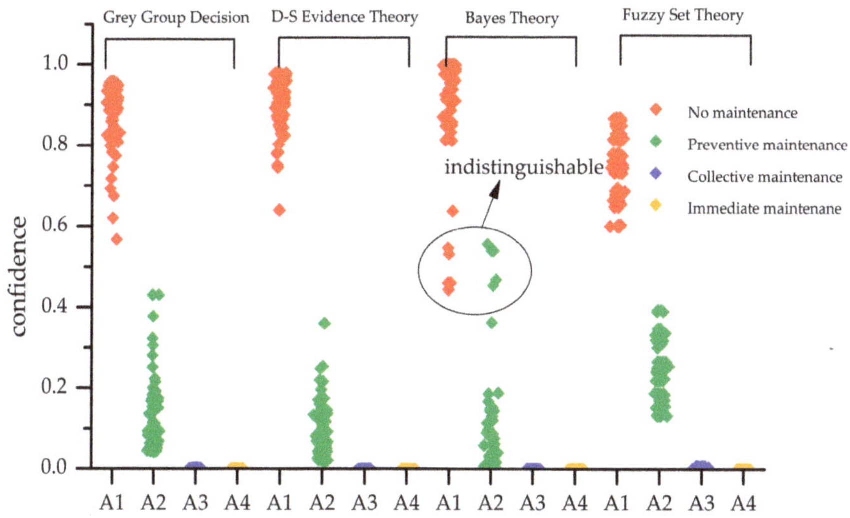

Figure 13. The confidence of 100 groups health state data to different health status level.

Situation 2: single sensor is failure

To verify the effectiveness of the grey group maintenance decision method, situation 2 is used to verify the versatility and correctness of the method. In situation 2, the data of each group are all kinds of failures simulated by fault simulation software. The attribute weights are $\omega = (\omega_1, \omega_2, \omega_3) = (0.0932, 0.4, 0.5068)$. In this example, the historical maintenance times are 40 and the total test times are 100. The next inspection time is 250 days and the overhaul cycle is 365 days.

The decision results of four methods are shown in Figure 14. All of the decision results are immediate maintenance except for the fuzzy set. The ranks and maintenance suggestions based on grey group decision and other three comparative decision methods are shown in Table 15.

Table 14. The accuracy of maintenance decision-making.

Health Status Level	Grey Group Decision	D-S Evidence Theory	Bayes Theory	Fuzzy Set Theory
A_1	100%	100%	94%	100%
A_2	100%	93%	100%	94%
A_3	95%	40%	95%	60%
A_4	98%	49%	98%	93%
average	98.25%	65.5%	96.75%	85.75%

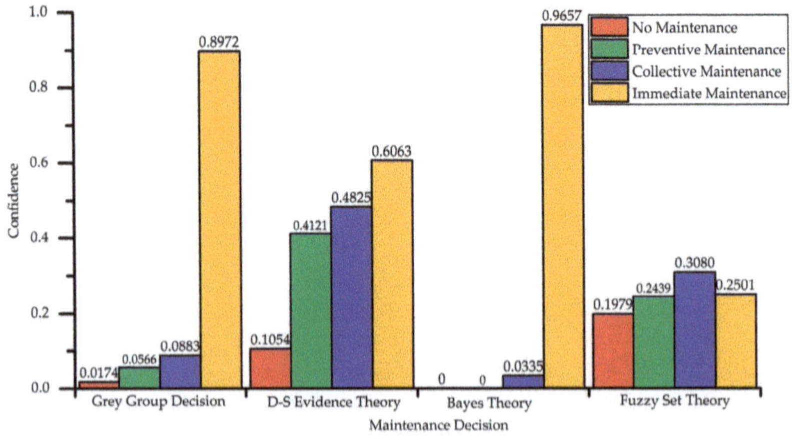

Figure 14. The confidence to 4 health status level of Situation 2 based on four decision methods.

To verify the effectiveness of the grey group maintenance decision method for single failure. 100 groups of failure status data are used for testing. The decision results are shown in Figure 15. The decision accuracies are shown in Table 14. The result of grey group decision is the same as Bayes theory and it is superior to the other algorithm. The result of D-S evidence theory is fluctuated and it is hard to distinguish the optimal maintenance result. The confidences of four health status level of fuzzy set theory are almost at the same level and it is difficult to realize the optimal maintenance result Grey group decision utilizes the advantage of Bayes theory for this situation to ignore the inaccuracy result of the other experts.

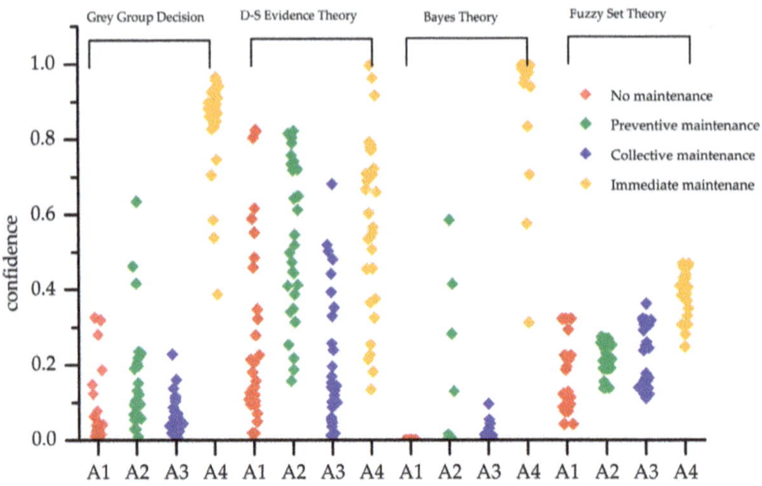

Figure 15. The decision result of 100 group failure state data.

The accuracy of maintenance decision-making which is verified by 400 groups different health status level samples (100 groups for each health status level) are shown in Table 14. It is not difficult to find out from the above test that D-S evidence theory has good detection results at health state. However, it cannot be detected effectively at the failure state because of the large range data fluctuation.

Bayes theory has a good decision-making result for the failure status, but false alarm will appear at health state. The fuzzy set theory is very good in the health state. But, in the case of failure, the health statuses often appear with a similar confidence, and cannot be effectively evaluated. According to the group decision method, the grey group decision exerts the advantage of the D-S evidence theory and fuzzy set theory in the health state and it reduces the missing alarm rate. In the case of failure, it plays the advantage of the Bayes theory and it reduces the probability of false alarm.

Table 15. The rank and maintenance suggestion of Situation 2.

Method	Rank	Maintenance Suggestion
Grey Group Decision	$A_4 > A_3 > A_2 > A_1$	A_4: Immediate Maintenance
D-S evidence Theory	$A_4 > A_3 > A_2 > A_1$	A_4: Immediate Maintenance
Bayes Theory	$A_4 > A_3 > A_2 = A_1$	A_4: Immediate Maintenance
Fuzzy Set Theory	$A_3 > A_4 > A_2 > A_1$	A_3: Collective Maintenance

5. Conclusions

In this paper, a method of health management decision strategy of a sensor system is proposed by utilizing *HRD-GGD* theory. Health reliability degree strategy is utilized to quantify system state and to provide support for decision making. The system can provide the maintenance suggestion after the system runs and give the confidence degree of each maintenance proposal. The experimental results show that this method can evaluate the system state effectively, and the accuracy rate of maintenance recommendation is 98.25%. The result proves that the accuracy is improved over 2% when compared with the other methods and the decision results are optimal under all health status levels.

In the future, we will investigate the remain life of sensor system and failure prediction by analysis of the historical trend of MOS sensor degradation in the sensor system.

Author Contributions: The manuscript was written through contributions of all authors. All authors have given approval to the final version of the manuscript.

Funding: This research was funded by the National Natural Science Foundation of China (No. 61201306 and 61327804) and National High-Tech R&D Program of China (No. 2014AA06A505).

Acknowledgments: The authors will thank Jinlei Sun, Shiming Xu and Zecheng Yu for their great help.

Conflicts of Interest: The authors declare no conflict of interest.

References

1. Fonollosa, J.; Vergara, A.; Huerta, R. Algorithmic mitigation of sensor failure: Is sensor replacement really necessary? *Sens. Actuators B Chem.* **2013**, *183*, 211–221. [CrossRef]
2. Korotcenkov, G.; Cho, B. Engineering approaches for the improvement of conductometric gas sensor parameters: Part 1. Improvement of sensor sensitivity and selectivity (short survey). *Sens. Actuators B Chem.* **2013**, *188*, 709–728. [CrossRef]
3. Korotcenkov, G.; Cho, B. Engineering approaches to improvement of conductometric gas sensor parameters. Part 2: Decrease of dissipated. (consumable) power and improvement stability and reliability. *Sens. Actuators B Chem.* **2014**, *198*, 316–341. [CrossRef]
4. Fernandez, L.; Gutierrez-Galvez, A.; Marco, S. Robustness to Sensor Damage of a Highly Redundant Gas Sensor Array. *Procedia Eng.* **2014**, *87*, 851–854.
5. Romain, A.C.; André, P.; Nicolas, J. Three years experiment with the same tin oxide sensor arrays for the identification of malodorous sources in the environment. *Sens. Actuators B Chem.* **2002**, *84*, 271–277. [CrossRef]
6. Romain, A.C.; Nicolas, J. Long term stability of metal oxide-based gas sensors for e-nose environmental applications: An overview. *Sens. Actuators B Chem.* **2009**, *146*, 502–506. [CrossRef]
7. Lazarescu, M.T. Design and Field Test of a WSN Platform Prototype for Long-Term Environmental Monitoring. *Sensors* **2015**, *15*, 9481–9518. [CrossRef] [PubMed]

8. Chen, Y.; Bouferguene, A.; Al-Hussein, M. Analytic Hierarchy Process-Simulation Framework for Lighting Maintenance Decision-Making Based on the Clustered Network. *J. Perform. Constr. Facil.* **2018**, *32*, 04017114. [CrossRef]
9. Dababneh, F.; Li, L.; Shah, R.; Haefke, C. Demand Response Driven Production and Maintenance Decision Making for Cost Effective Manufacturing. *J. Manuf. Sci. Eng.* **2018**, *140*, 061008. [CrossRef]
10. Chen, D.; Trivedi, K.S. Optimization for condition-based maintenance with semi-Markov decision process. *Reliab. Eng. Syst. Saf.* **2017**, *90*, 25–29. [CrossRef]
11. Tsui, K.L.; Chen, N.; Zhou, Q.; Hai, Y.; Wang, W. Prognostics and health management: A review on data driven approaches. *Math. Probl. Eng.* **2015**, *6*, 1–17. [CrossRef]
12. Bai, G.; Wang, P.; Hu, C. A self-cognizant dynamic system approach for prognostics and health management. *J. Power Sources* **2015**, *278*, 163–174. [CrossRef]
13. Coble, J.; Ramuhalli, P.; Bond, L.; Hines, J.W.; Upadhyaya, B. A review of prognostics and health management applications in nuclear power plants. *Int. J. Progn. Health Manag.* **2015**, *6*, 1–22.
14. Shen, Z.; He, Z.; Chen, X.; Sun, C.; Liu, Z. A Monotonic Degradation Assessment Index of Rolling Bearings Using Fuzzy Support Vector Data Description and Running Time. *Sensors* **2012**, *12*, 10109–10135. [CrossRef] [PubMed]
15. Sohaib, M.; Kim, C.H.; Kim, J.M. A Hybrid Feature Model and Deep-Learning-Based Bearing Fault Diagnosis. *Sensors* **2017**, *17*, 2876. [CrossRef] [PubMed]
16. Wang, J.; Wang, X.; Wang, L. Modeling of BN Lifetime Prediction of a System Based on Integrated Multi-Level Information. *Sensors* **2017**, *17*, 2123. [CrossRef] [PubMed]
17. Agarwal, V.; Lybeck, N.J.; Bickford, R.; Rusaw, R. *Development Of Asset Fault Signatures For Prognostic And Health Management in the Nuclear Industry*; Prognostics and Health Management: Austin, TX, USA, 2015; pp. 1–7.
18. Kumar, S.; Pecht, M. Modeling approaches for prognostics and health management of electronics. *Int. J. Perform. Eng.* **2010**, *6*, 222–229.
19. Carnero, M.C.; Gómez, A. A multicriteria decision making approach applied to improving maintenance policies in healthcare organizations. *Bmc Me. Inf. Decis. Mak.* **2016**, *16*, 47. [CrossRef] [PubMed]
20. Alaswad, S.; Xiang, Y. A review on condition-based maintenance optimization models for stochastically deteriorating system. *Reliab. Eng. Syst. Saf.* **2017**, *157*, 54–63. [CrossRef]
21. Shen, Z.; Wang, Q. Failure detection, isolation, and recovery of multifunctional self-validating sensor. *IEEE Trans. Instrum. Meas.* **2012**, *61*, 3351–3362. [CrossRef]
22. Chen, Y.; Xu, Y.; Yang, J. Fault detection, isolation, and diagnosis of status self-validating gas sensor arrays. *Rev. Sci. Instrum.* **2016**, *87*, 045001. [CrossRef] [PubMed]
23. Chen, Y.; Jiang, S.; Yang, J.; Song, K.; Wang, Q. Grey bootstrap method for data validation and dynamic uncertainty estimation of self-validating multifunctional sensors. *Chemometr. Intell. Lab. Syst.* **2015**, *146*, 63–76. [CrossRef]
24. Song, K.; Wang, Q.; Li, J.; Zhang, H. In quantitative measurement of gas component using multisensor array and NPSO-based LS-SVR. *Instrum. Meas. Technol. Conf.* **2013**, *80*, 1740–1743.
25. Shen, Z.; Wang, Q.; Zhu, F. Data-driven health evaluation of multifunctional self-validating sensor using health reliability degree. *Inf. Technol. J.* **2012**, *11*, 1597–1604. [CrossRef]
26. Shen, Z.; Wang, Q. A Novel Health Evaluation Strategy for Multifunctional Self-Validating Sensors. *Sensors* **2013**, *13*, 587–610. [CrossRef] [PubMed]
27. Feng, Z.G.; Wang, Q. Research on health evaluation system of liquid-propellant rocket engine ground-testing bed based on fuzzy theory. *Acta Astronaut.* **2007**, *61*, 840–853. [CrossRef]
28. Xia, T.; Xi, L.; Zhou, X.; Lee, J. Dynamic maintenance decision-making for series–parallel manufacturing system based on MAM–MTW methodology. *Eur. J. Oper. Res.* **2012**, *221*, 231–240. [CrossRef]
29. Berges, L.; Galar; Gustafson, A.; Tormos, B. Maintenance decision making based on different types of data fusion. *Eksploat. i Niezawodn.-Maint. Reliab.* **2012**, *14*, 135–144.
30. Cates, G.L.; Skinner, C.H.; Watson, T.S.; Meadows, T.J.; Weaver, A.; Jackson, B. Instructional effectiveness and instructional efficiency as considerations for data-based decision making: An evaluation of interspersing procedures. *Sch. Psychol. Rev.* **2003**, *32*, 601–616.
31. Aughenbaugh, J.M.; Herrmann, J.W. Reliability-based decision making: A comparison of statistical approaches. *J. Stat. Theory Pract.* **2009**, *3*, 289–303. [CrossRef]

32. Liu, W.B.; Wang, Q.F.; Gao, J.J.; Zhong, X.; Liang, G.H. Reliability-centered intelligent maintenance decision-making model. *J. Beijing Univ. Technol.* **2012**, *38*, 672–677.
33. Wang, A.; Jiang, J.; Zhang, H. Multi-sensor Image Decision Level Fusion Detection Algorithm Based on D-S Evidence Theory. In Proceedings of the Fourth International Conference on Instrumentation and Measurement, Computer, Communication and Control, Harbin, China, 18–20 September 2014; pp. 620–623.
34. He, Z.; Zhang, H.; Zhao, J.; Qian, Q. Classification of power quality disturbances using quantum neural network and DS evidence fusion. *Eur. Trans. Electr. Power* **2013**, *22*, 533–547. [CrossRef]
35. Wang, H.; Lin, D.; Qiu, J.; Du, Z.; He, B. Research on Multiobjective Group Decision-Making in Condition-Based Maintenance for Transmission and Transformation Equipment Based on D-S Evidence Theory. *IEEE Trans. Smart Grid* **2015**, *6*, 1035–1045. [CrossRef]
36. Lin, S.; Li, C.; Xu, F.; Li, W. The strategy research on electrical equipment condition-based maintenance based on cloud model and grey D-S evidence theory. *Intell. Decis. Technol.* **2018**, *3*, 1–10. [CrossRef]
37. Mehta, P.; Werner, A.; Mears, L. Condition based maintenance-systems integration and intelligence using Bayesian classification and sensor fusion. *J. Intell. Manuf.* **2015**, *26*, 331–346. [CrossRef]
38. Herrle, S.R.; Corbett Jr, E.C.; Fagan, M.J.; Moore, C.G.; Elnicki, D.M. Bayes' theorem and the physical examination: Probability assessment and diagnostic decision making. *Acad. Med. J. Assoc. Am. Med. Coll.* **2011**, *86*, 618. [CrossRef] [PubMed]
39. Lin, P.C.; Gu, J.C.; Yang, M.T. Intelligent maintenance model for condition assessment of circuit breakers using fuzzy set theory and evidential reasoning. *IET Gener. Transm. Distrib.* **2014**, *8*, 1244–1253. [CrossRef]
40. Yin, K.; Yang, B.; Li, X. Multiple Attribute Group Decision-Making Methods Based on Trapezoidal Fuzzy Two-Dimensional Linguistic Partitioned Bonferroni Mean Aggregation Operators. *Int. J. Environ. Res. Public Health* **2018**, *15*, 194. [CrossRef] [PubMed]
41. Chen, S.; Cheng, S.; Chiou, C. Fuzzy multiattribute group decision making based on intuitionistic fuzzy sets and evidential reasoning methodology. *Inf. Fusion* **2016**, *27*, 215–227. [CrossRef]
42. Efe, B. An integrated fuzzy multi criteria group decision making approach for ERP system selection. *Appl. Soft Comput.* **2016**, *38*, 106–117. [CrossRef]
43. Joshi, D.; Kumar, S. Interval-valued intuitionistic hesitant fuzzy Choquet integral based TOPSIS method for multi-criteria group decision making. *Eur. J. Oper. Res.* **2016**, *248*, 183–191. [CrossRef]
44. Pramanik, S.; Pramanik, S.; Giri, B.C. TOPSIS method for multi-attribute group decision-making under single-valued neutrosophic environment. *Neural Comput. Appl.* **2016**, *27*, 727–737.
45. Wang, H.; Lin, D.; Qiu, J.; Ao, L.; Du, Z.; He, B. Research on multiobjective group decision-making in condition-based maintenance for transmission and transformation equipment based on D-S evidence theory. *IEEE Trans. Smart Grid* **2015**, *6*, 1035–1045. [CrossRef]
46. Peng, J.; Wang, J.; Wang, J.; Zhang, H.; Chen, X. Simplified neutrosophic sets and their applications in multi-criteria group decision-making problems. *Int. J. Comput. Intell. Syst.* **2016**, *47*, 2342–2358. [CrossRef]
47. Liu, P.; Chen, S.M. Multiattribute group decision making based on intuitionistic 2-tuple linguistic information. *Inf. Sci.* **2018**, *430*, 599–619. [CrossRef]
48. Manzardo, A.; Ren, J.; Mazzi, A.; Scipioni, A. A grey-based group decision-making methodology for the selection of hydrogen technologies in life cycle sustainability perspective. *Int. J. Hydrog. Energy* **2012**, *37*, 17663–17670. [CrossRef]
49. Liu, P.; Chu, Y.; Li, Y. The multi-attribute group decision-making method based on the interval grey uncertain linguistic generalized hybrid averaging operator. *Neural Comput. Appl.* **2015**, *26*, 1395–1405. [CrossRef]

sensors

MDPI

Article

A Practical Approach for High Precision Reconstruction of a Motorcycle Trajectory Using a Low-Cost Multi-Sensor System

Sarra Smaiah [1,2,*], Rabah Sadoun [1,2], Abdelhafid Elouardi [1], Bruno Larnaudie [1], Samir Bouaziz [1], Abderahmane Boubezoul [1,3], Bastien Vincke [1] and Stéphane Espié [1,3]

1 SATIE Laboratory, University Paris Sud, 91405 Orsay, France; rabah.sadoun@g.enp.edu.dz (R.S.); abdelhafid.elouardi@u-psud.fr (A.E.); Bruno.larnaudie@u-psud.fr (B.L.); samir.bouaziz@u-psud.fr (S.B.); abderrahmane.boubezoul@ifsttar.fr (A.B.); bastien.vincke@u-psud.fr (B.V.); stephane.espie@ifsttar.fr (S.E.)

2 Signal and Communication Laboratory, National Polytechnic School, 16200 El-Harrach, Algiers, Algeria

3 IFSTTAR, Champs-sur-Marne, F-77447 Marne la Vallée, France

* Correspondence: sarra.smaiah@u-psud.fr; Tel.: +33-78-388-3572

Received: 4 June 2018; Accepted: 9 July 2018; Published: 14 July 2018

Abstract: Motorcycle drivers are considered among the most vulnerable road users, as attested by the number of crashes increasing every year. The significant part of the fatalities relates to "single vehicle" loss of control in bends. During this investigation, a system based on an instrumented multi-sensor platform and an algorithmic study was developed to accurately reconstruct motorcycle trajectories achieved when negotiating bends. This system is used by the French Gendarmerie in order to objectively evaluate and to examine the way riders take their bends in order to better train riders to adopt a safe trajectory and to improve road safety. Data required for the reconstruction are acquired using a motorcycle that has been fully instrumented (in VIROLO++ Project) with several redundant sensors (reference sensors and low-cost sensors) which measure the rider actions (roll, steering) and the motorcycle behavior (position, velocity, acceleration, odometry, heading, and attitude). The proposed solution allowed the reconstruction of motorcycle trajectories in bends with a high accuracy (equal to that of fixed point positioning). The developed algorithm will be used by the French Gendarmerie in order to objectively evaluate and examine the way riders negotiate bends. It will also be used for initial training and retraining in order to better train riders to learn and estimate a safe trajectory and to increase the safety, efficiency and comfort of motorcycle riders.

Keywords: trajectory reconstruction; low-cost sensors; embedded systems; powered two wheels (PTW); safe trajectory; data fusion

1. Introduction

Motorcycle drivers are considered the most vulnerable road users. In France, such riders account for more than 20% of all road fatalities (compared with 15% in Europe as a whole) and 43% of road injuries (ONISR, 2014). For the same distance traveled, the risk to be killed in a fatal crash is 35 times higher for a motorcyclist than for an automobilist [1]. More than 50% of individual motorcycle crashes are due to loss of control in bends (including crossroads and intersections) because of the complexity of motorcycle dynamics and the intrinsic instability of such vehicles. In 2012, in France, more than a third of all the powered two wheels (PTW) fatalities occurred in bends (248 killed).

Counter-measures are needed to improve road safety and decrease the risk of crashes in bends. In order to do so, it is necessary to better train riders to learn, estimate, and adopt a safe trajectory [2,3]. Motorcycle trajectory reconstruction represents a fairly important tool for an objective evaluation of bend taking practices and the characterization of the achieved trajectories compared to a safe trajectory.

The main idea of the safe trajectory is to "dive" towards the bend tangent point only when the driver sees the whole of the bend, up to its exit. This technique, if applied correctly, allows the rider to drive fast and safe because it allows him to anticipate dangers and potential hazards. Until today, no system exists to objectively evaluate bend-taking maneuvers. This evaluation is subject to an instructor's "imprecise" appreciation of the French Gendarmerie.

In this context, the purpose of this work is to develop an algorithm using low-cost sensors to accurately reconstruct motorcycle trajectories in order to develop a tool for the objective evaluation of the achieved trajectory and that allow PTW riders to receive better training or retraining that focuses on "safe" cornering. The challenge is to achieve a high accuracy trajectory reconstruction by using a system that can be embedded implementing low-cost sensors and proposing an alternative solution to the traditional INS/GPS systems. The powered two wheels (PTW) used for this study is a motorcycle instrumented by the VIROLO++ Project team. Thanks to experienced and inexperienced (trainee) French national gendarmerie drivers, a very large dataset was acquired at the "La Ferté Gaucher" circuit by the VIROLO++ team [3,4].

The main contributions of our work are summarized below:

- A trajectory reconstruction study involving algorithms and ad hoc sensors. Bend trajectory reconstruction is highlighted namely for driver safety.
- Identification of an optimal low-cost system further to a multi-sensor instrumentation and different approaches for trajectory reconstruction, beyond simulation.
- An alternative solution to the INS/GPS systems, that allows a high-precision trajectory reconstruction. Results are given using a real dataset provided by various drivers.

This paper is organized in two major sections. The first one is dedicated to the literature survey, the description of the multi-sensor system used, the circuit-based experiments, and the major data adjustment. In the second section, we firstly explore and evaluate the usual methods of trajectory reconstruction. Then, we demonstrate the gain obtained with our proposed fusion method comparatively to the best of the usual method of reconstruction.

2. Literature Survey

Among the existing literature about motorcycle trajectory reconstruction, there is no work that addresses the problem of "safe trajectory". In addition, motorcycle trajectory is usually reconstructed using traditional GPS/INS systems that can achieve good accuracy only if the GPS signal is available. However, during a GPS outage, the accuracy is degraded. Some works proposed an alternative solution based on vision systems, but the accuracy obtained is far from the expected objectives.

It is also worth noting that works about motorcycle trajectory reconstruction are limited compared to cars because the dynamics of a two-wheeled vehicle represents a higher level of complexity. In this section, we present research works that deal with motorcycle trajectory reconstruction.

In [5], Yuichiro Koyama presents a new algorithm (as he cited) for motorcycle trajectory reconstruction using GPS only. This algorithm is based on the interpolation of satellites positions that cannot provide adequate data within a few seconds (missing observation data) using polynomial curves calculated by the least-squares method and the calculation of the motorcycle coordinates based on interpolated pseudoranges. The obtained trajectory is smoothed using an extended Kalman filter. This method allowed obtaining a trajectory with an accuracy of less than one meter. This precision is obtained only if the observed data are absent for less than 7 s. No hardware realization was described.

Luca Gasbbaro et al. [6] presented an algorithm for the precise (as mentioned) reconstruction of the motorcycle trajectory based on vision integration and miniaturized MEMS accelerometers, using an extended Kalman filter and a dynamic model of the bike. The constraints of the model act as virtual measurements and make it possible to estimate the biases and drifts of the accelerometers. The visual reference marks were used to estimate the biases and drifts of the navigation sensors. The idea is to distribute a number, more than six, of accelerometers in specific locations on the chassis of the motorcycle and then use the shape invariance of the rigid body motion to estimate the trajectory. The data acquisition system uses a basic processing unit; a PC-104 industrial computer (266 MHz Geode, 256 MB RAM). A Compact Flash card was used to store the recorded data, as well as a Linux operating system and user's programs.

The same idea involved in [6] was applied in [7], where the authors proposed a low-cost system based on micro electro-mechanical systems (MEMS) technology coupled with images through the Whipple model [8] and a cascade of a Kalman filter and a Bayesian particle filter to reconstruct the "Vespa" scooter trajectory. The authors used an Xsens MTi-G IMU, a 1.3 megapixel progressive scan color SONY CCD camera and a notebook computer for data acquisition. The reference trajectory was determined by a Novatel DL-4 double frequency GPS receiver. Their method provides relatively acceptable accuracy (mean error: 1.033 m, max error: 10.12 m, absolute mean: 3.2 m, and standard deviation: 2.53 m). However, the application of this method depends on the environmental conditions. The roll angle estimation in this work is based on the Hough transform, which necessitates a minimal amount of linear elements in the scene, and their absence can degrade the achievable results. For instance, a complex skyline and low contrast between the road segment and neighboring object can be problematic, even if not common.

In [9], the authors proposed an experimental low-cost differential GPS/MEMS-IMU system, using an extended Kalman filter approach in a loosely coupled mode to accurately (as cited) reconstruct the trajectory and the orientation of a motorcycle. The system performance was evaluated through a set of experiments using a motorbike-embedded MEMS-IMU (Xsens MTi), rigidly fixed to the GPS antenna. A low-cost mono-frequency GPS receiver (u-blox AEK4) was used with a dual-frequency GPS receiver from Javard as a reference. The proposed system provided an absolute position accuracy of 0.5 m and an orientation accuracy of 1°–2°.

The advantage of [5] is the accuracy of the trajectory reconstruction obtained using only a GPS. However, the developed algorithm is applied only to data lost within a few seconds (7 s according to authors). Beyond the given loss time, the efficiency of the interpolation is reduced and the accuracy is degraded.

In [6,7], authors succeeded in reconstructing the motorcycle trajectory using a system based on vision (camera), MEMS technology and a computer (industrial computer in [6] and a notebook PC in [7]) for data processing. This system is complex, not embeddable, and depends on the environmental conditions. Additionally, the obtained accuracy was not mentioned in [6], the maximum error achieved in [7] is 10.11 m.

In [9], a high accuracy was achieved (50 cm) using a GPS/INS integration algorithm. The traditional GPS/INS systems can achieve great accuracies when the GPS signal is available. However, during GPS outage the accuracy is degraded.

From an application point of view, the main objective of [5] is the simulation of the motorcycle dynamics to analyze the behavior of both the motorcycle and the rider in a virtual three-dimensional space that correlates strongly with real driving tests. In [6], the authors consider the trajectory reconstruction as estimation techniques for the full reconstruction of the dynamical vehicle state. Their algorithm is used in racing applications. In [7], the authors aim to identify the vehicle position in a mapping reference frame for driving directions and best-route analysis with significant accuracy. In addition, in [9] the designed system is used in sport applications. The precise trajectory reconstruction is used to determine tire slips of a motorcycle.

The challenge in our work is to study several motorcycle trajectory reconstruction algorithms in order to design an embedded system, with an optimal set of low-cost sensors that allows the reconstruction of the motorcycle trajectory with high precision without using GPS (due to drawbacks of signal loss). This system can then be used to better train riders to adopt a safe trajectory in order to reduce the risk of crashes in bends and to improve road safety.

3. Multi-Sensor System and Circuit-Based Experiments Description

A multi-sensor architecture and an embedded data-logger have been designed by the VIROLO++ Project team [4]. The system allowed acquiring and recording data related to the rider behavior, actions, and to the motorcycle dynamics [10]. The system architecture is based on a CAN (controller area network) bus that interfaces multiple sensors and a data logger. This architecture makes it possible to add or remove one or more devices (sensors or recorders), without interrupting tasks of the others nodes. This offers an essential flexibility in the development phase of the multi-sensor system.

3.1. Embedded Sensors

The instrumented PTW (Figure 1) integrates redundant sensors, "low-cost" sensors and "reference" sensors, in order to compare data measurements and to identify the subset of sensors necessary for each reconstruction method.

- Standard GPS receiver (7): a GPS module was designed, using an A2200-A circuit [11] of Maestro Wireless Solutions, to reduce the costs and to have a completely mastered GPS (sampling time: 200 ms);
- (10Dof) Inertial navigation system (5): The low-cost INS MPU9250 [12] of InvenSense was implemented in the motorcycle in order to acquire inertial movements and data required for the trajectory reconstruction (sampling time: 10 ms);
- Handlebar sensor (3): The magnetic sensor AS5047P of AMS is used to recover the absolute angle of the handlebar. The angle is coded on 14 bits which allows a maximum resolution of 2048 steps per complete rotation (0.176°/step) [13] (sampling time: 1 ms);
- Wheel tire sensors (1): Two Hall effect sensors in quadrature are attached to the front and rear wheels to measure the distance travelled by the motorcycle, used in trajectory reconstruction (sampling time: 1 ms);
- GPS RTK (6): The position delivered by the GPS RTK is used only as a reference to evaluate the accuracy of the reconstructed trajectory. The two (rover and base) GPS "Altus APS-3" of Septentrio are used to obtain real-time kinematic positioning [14] (sampling time: 40 ms);
- Inertial navigation system "Xsens" (4): The MTi Xsens [15] is used as a (redundant) reference system in order to check the degradation induced by others sensors. (sampling time: 10 ms); and
- Tilt sensors (2): To measure the roll angle, two identical (laser) optical distance sensors are placed on both sides (right and left) of the motorbike (sampling time: 10 ms).

Figure 1. The ANR team instrumented motorcycle.

3.2. Data Logger

The data logger (recording node 8 in Figure 1) is based on a BECK programmable microcontroller and a CAN interface. The main function of the recorder is to collect messages sent on the CAN bus, put them in a specific format according to the identifier of the message, and place them in a specific file on the embedded memory storage. When receiving a CAN frame, the data logger writes a new block of data to a specific file containing the CAN frames and the reception times (time stamping task). The purpose of this instrumented architecture (Figure 2) is to have a variety of sensors in order to choose the optimal match algorithm sensors (allowing the best accuracy with low-cost sensors).

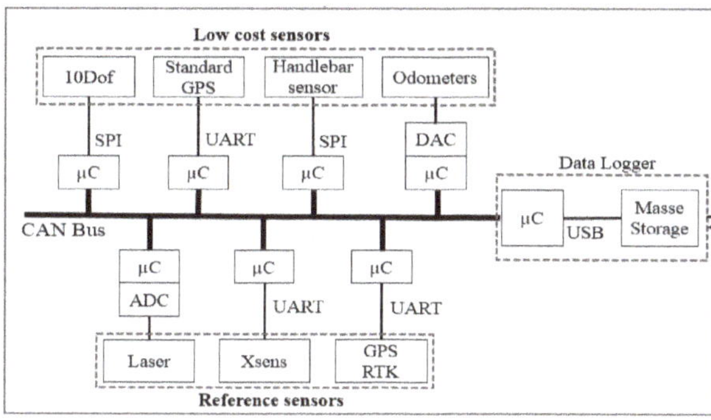

Figure 2. System architecture.

3.3. Circuit-Based Experiments

In order to evaluate the motorcycle trajectory reconstruction methods, several experiments were conducted. In this work, we have used the experiences carried out at the "La Ferté Gaucher" circuit (Figure 3) which is composed of different types of bends (right, left, 90°, and 180° bends with large or small curvature). The experimental data have been collected with a Honda CBF 1000 (Figure 1) which was driven on a 1.9 km loop. The departure point and the arrival point are identified. The motorcycle speed varies up to 110 km/h.

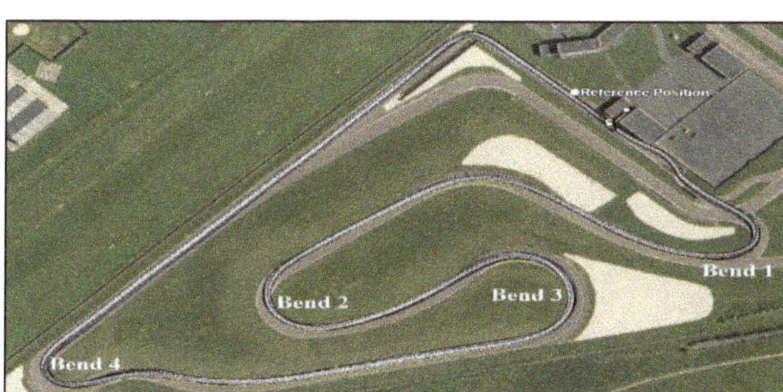

Figure 3. The "Ferté Gaucher" circuit mapped on IGN.

4. Sensor Data Correction

4.1. Reference Trajectory Correction

Sensors were embedded in different positions on the motorcycle. As a consequence, the sensors' measurements are in different references, especially in turning, as shown in Figure 4. Therefore, before evaluating the trajectory reconstruction methods, and in order to improve the accuracy, we considered the rear wheel contact point "P_r" as a reference point and we brought all sensors' data back to this reference point. Our reference frame, in this case, is the mobile triad (P_r, x, y, z), specified by the Society of Automotive Engineers (SAE) [16]. The origin is established at the rear wheel contact point "P_r" with the road plane. The "x" axis is horizontal and parallel to the rear wheel plane. The "z" axis is vertical and directed downward while the "y" axis lies on the road plane. The road surface is, therefore, represented by the plane $z = 0$.

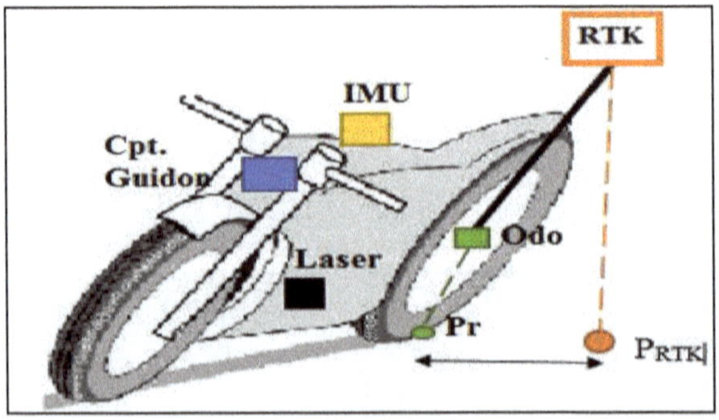

Figure 4. Sensor positioning.

Positions of the sensors installed on the motorcycle are known. The GPS *RTK* position with respect to the reference point "P_r" is given by the coordinates $(dx,\ dy,\ dz)$. Thus, the reference trajectory of the rear contact point "P_r" is calculated from the "*RTK*" trajectory using the following equation:

$$\begin{bmatrix} X_{Pr} \\ Y_{Pr} \\ Z_{Pr} \end{bmatrix} = \begin{bmatrix} X_{RTK} \\ Y_{RTK} \\ Z_{RTK} \end{bmatrix} - R_b^n. \begin{bmatrix} dx \\ dy \\ dz \end{bmatrix} \tag{1}$$

Figure 5 shows the RTK reference trajectory before and after bringing it to the reference point.

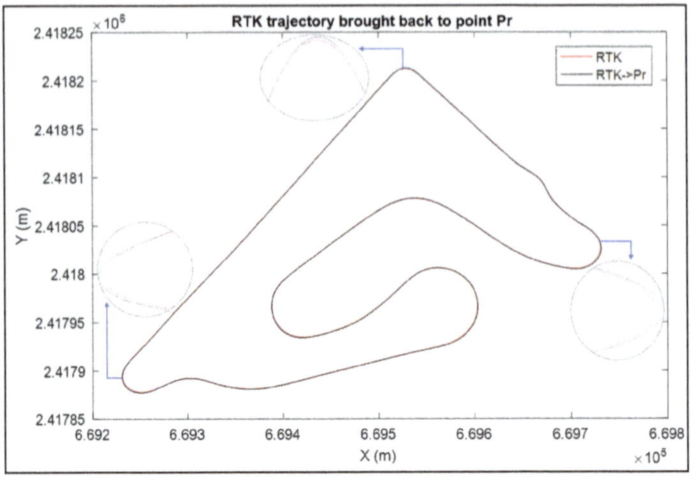

Figure 5. The reference trajectory brought back to the rear contact point.

It is clear that the main difference between the two trajectories exists in turns. When negotiating a bend, the motorcycle tilts to the inside bend direction. Thus, since the GPS RTK is installed on the bike trunk, its trajectory is always in the inside bend direction with respect to the rear contact point where the trajectory is in the outside bend direction.

4.2. Odometry Correction

Odometers are one of the sensors used to measure the motorcycle's traveled distance independently from GPS. They work by counting wheel rotations and assume that the distance traveled is the number of wheel rotations times the tire circumference (tire diameter times pi).

According to our expertise, the wheel radius is not constant; it varies according to the roll angle. Therefore, a wheel radius model is required to correct the traveled distance estimated by odometers.

4.2.1. Proposed Wheel Radius Model

When negotiating a bend, the motorcycle passes from a vertical position to a tilted position with a roll angle "φ" in order to stay balanced. Following the roll motion, the contact point of the wheel with the road plane is displaced and the wheel radius is changed. Thus, in order to accurately estimate the motorcycle traveled distance, we propose the rear wheel radius model illustrated in Figure 6.

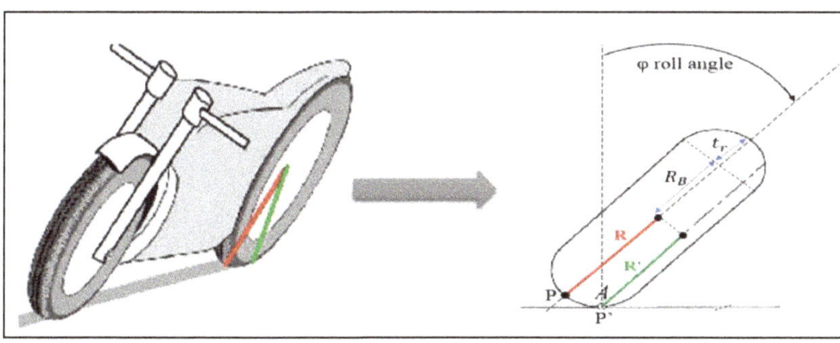

Figure 6. Lateral displacement of the rear contact point in a curve.

In a vertical position, the wheel radius of the motorcycle is R. However, when the motorcycle tilts, assuming a lateral roll without slippage on the road plane, the contact point of the rear tire "P" moves laterally, as illustrated in Figure 6, in the "Y" direction over a distance "$t_r\varphi$" which is proportional to the radius of the tire cross section "t_r" and the roll angle "φ" of the rear frame. The wheel radius in this case is R':

$$R' = R_B - t_r \cos(\varphi) \quad (2)$$

where $R_B = (R - t_r)$: is the radius of the torus center circle.

4.2.2. Validation of the Correction Model

In order to validate our correction model, we measured the difference between the real traveled distance and the one given by the odometers using a constant wheel radius, Figure 7a, and using our proposed wheel radius model, Figure 7b. From Figure 7, it is clear that our proposed model greatly ameliorates the estimation of the distance and minimizes the error from (10 m) to (1 m with zero mean error).

From Figure 8, we can clearly see that, in straight line trajectories, there is no difference between the real distance and the one estimated by the odometers using a fixed wheel radius, while in turns, where the roll angle is important, this difference increases (the colored zones, Figure 8) which confirms that the wheel radius changes according to the roll angle.

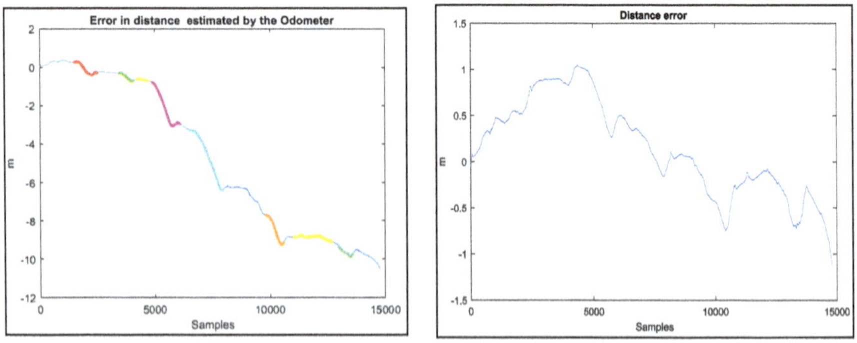

(a) (b)

Figure 7. Difference between the real traveled distance and the one given by the odometers using a fixed wheel radius (**a**) and our proposed wheel radius model (**b**).

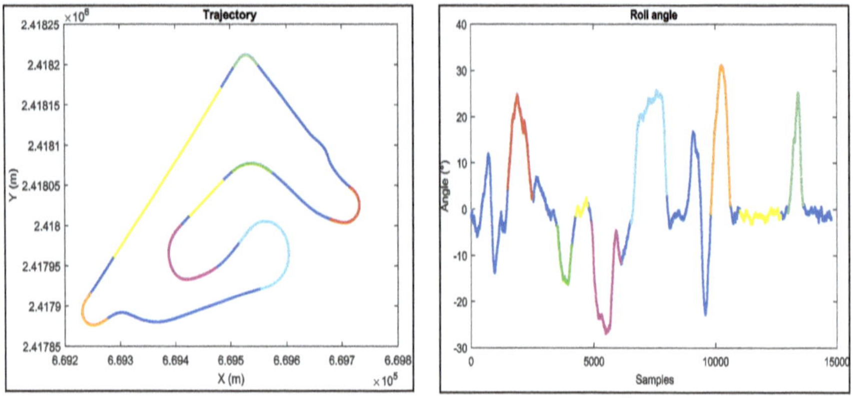

(a) (b)

Figure 8. Location of the distance error (fixed radius) of each zone of Figure 7a on the trajectory (**a**) and the corresponding roll angle (**b**).

4.3. Data Preprocessing

During the data acquisition phase, no filter is applied [17] because the aim is to achieve a real dataset that can be used for other studies. Therefore, the sensors' measurements are affected by noise (due mainly to engine vibrations) and a filtering step appeared necessary. In the first step, we proceeded to spectrally analyze the all IMU signals in static conditions (motorbike immobile and motor on). All the analyses exhibit two singular frequencies; less than 2 Hz and around 40 Hz. The two figures bellow (Figure 9a,b) illustrate our assertions.

Then, "wavelet" filter was chosen to denoise the data because of its advantage compared to conventional filters listed below.

- Little to no signal leakage or phase shifting of the original signal.
- The ability to denoise complex signals far better than conventional filters that are based on the Fourier transform.
- Wavelets are efficient for removing noise where the noise and signal spectra overlap. Conventional filters are efficient in removing out-of-band signals. However, if applied to in-band signals, wavelets will also remove the signal of interest.

To validate and confirm our choice, three filtering techniques were synthetized:

- Low-pass Butterworth filter with a 10 Hz cut-off (sixth order)
- Median filter (with a window of 20 points)
- Wavelet filter with a Daubechies mother wavelet of Db20. Six levels of decomposition have been considered as enough to provide a significant reduction of the high-frequency noise components.

Comparing the signal-to-noise ratio (SNR) of the three filtering techniques in Table 1, the "wavelet" has given better performances.

A static test was carried out in order to estimate and eliminate the static bias of sensors from the data before using them in the trajectory reconstruction algorithms.

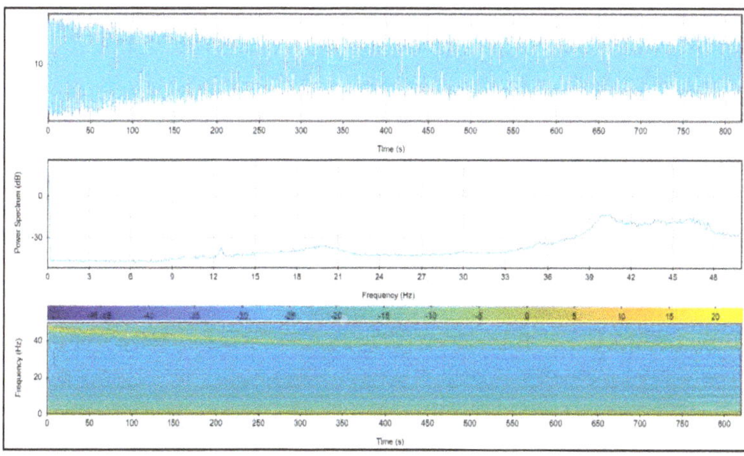

(a)

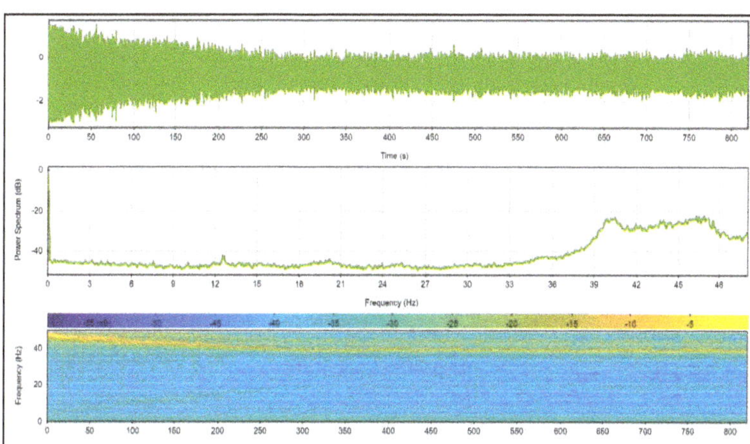

(b)

Figure 9. Illustration of spectrograms of "Ax" (**a**) and "Az" (**b**) IMU signals (signal on the top, power spectral density in the middle, and the signal spectrogram at the bottom).

Table 1. Comparison of the signal-to-noise ratio (SNR) of the three filters.

SNR (dB)	Wavelet Filter	Median Filter	Butterworth Filter
Ax	4.77	3.83	2.26
Ay	8.65	7.13	3.91
Az	27.41	19.51	11.29
Rx	1.45	0.42	0.13
Ry	3.80	2.90	1.74
Rz	0.24	0.05	0.01
Mx	1.18	1.20	1.02
My	0.27	0.18	0.19
Mz	2.19	2.04	1.83
Handelbar	0.38	0.04	−0.17
Laser	0.38	0.35	0.17

5. Evaluation of the Usual Methods of PTW Trajectory Reconstruction: A Comparative Study

The developed multi-sensor architecture allows for evaluating several models of trajectory reconstruction. Depending on the nature of sensor information and the model used to reconstruct the motorcycle trajectories, we can distinguish four methods: kinematic models, absolute localization, relative localization, and data fusion-based localization.

5.1. Kinematic Model

Several motorcycle kinematic models exist in the literature and give the position of the motorcycle according to a certain number of input data (sensor data). Among these models, we note the Cossalter model [16] and the bicycle model [17] (Figure 10).

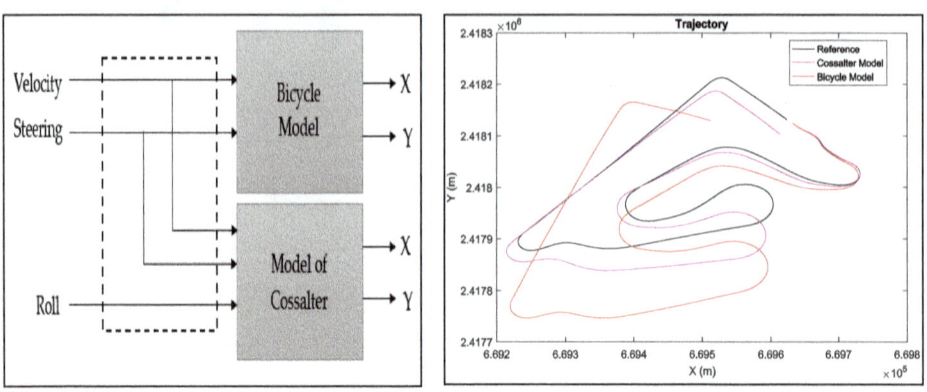

Figure 10. Trajectory reconstruction from kinematic models.

In the literature, no dynamic model allowing the passage of the motorcycle position or the trajectory reconstruction exist, a reason why we used only the kinematic models.

5.2. Data Fusion Methods

This approach consists of a fusion of sensor data that presents measurement uncertainties for a "sufficiently accurate" positioning. Different approaches exist in the literature to fuse data delivered by sensors. A very good bibliographical study is presented in [18].

This solution is based on the idea of jointly using both localization methods: relative and absolute poses in order to take advantage of the complementarity of proprioceptive and exteroceptive sensors. Indeed, the absolute localization system is generally dedicated to regularly correct the estimate of the relative localization system in order to remedy the drift problems encountered with it. Dead reckoning provides measurements at a very high frequency, but requires initialization. Absolute localization provides long-term precision, but it suffers from problems of availability, latency and, often, insufficient frequency for some applications. Therefore, to reap the benefits and complementarities of the two localization systems, researchers paired these two families, giving birth to the data fusion approach like GPS/INS systems.

5.2.1. GPS/INS Data Fusion

In the literature, several techniques are suggested to fuse GPS and INS data. The Kalman filter is the most used algorithm with, typically, three main strategies: namely loose integration, tight integration, and deep (or ultra-tight) integration. In this work, we have chosen to use a loosely-coupled implementation mode in a closed loop (Figure 11). This mode allows control of the navigation accuracy and reduces the cost of the design [19].

Due to the non-linearity of the process model, an extended Kalman filter of 15 states was built in this work using the dynamic equations of the error (the filter update is based on an error state vector which includes error vectors for position, velocity, attitude, accelerometer bias, and gyrometer bias, as explained in [20]).

In this method, the GPS measurements are used to correct data of the INS and to eliminate bias and drifts. However, the GPS accuracy is degraded because of multiple routes and for the small number of visible satellites (low availability). Sometimes GPS data are absent for a long time (if the satellites visibility conditions are degraded, the reception of the signals is blocked), which affects the accuracy of this method.

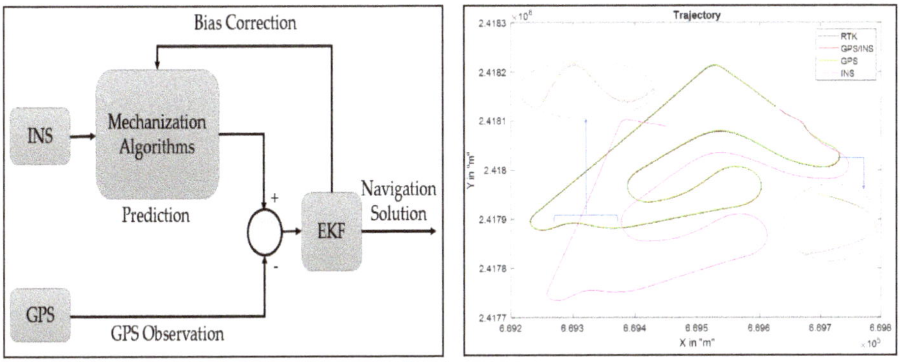

Figure 11. Loose coupling integration scheme (INS/GPS).

5.2.2. INS/Odometer Data Fusion

In the literature [21,22], different configurations are proposed to integrate odometer and INS data. In our work, we have chosen the RISS configuration (reduced inertial sensor system) (Figure 12).

The discrete form of the mechanization algorithm of this system is:

$$\begin{aligned} x(k+1) &= x(k) + T_e\, V(k)\, \cos(\theta(k)) \\ y(k+1) &= y(k) + T_e\, V(k)\, \sin(\theta(k)) \\ \theta(k+1) &= \theta(k) + T_e\, W_z \end{aligned} \tag{3}$$

where W_z is the gyroscope measurement (rad/s).

Two main errors influence the trajectory reconstruction:

- odometer errors, which come from the inaccuracy of the vehicle's geometrical parameters; and
- heading errors, which come from the drifts in the gyroscopes data (accumulation of errors during the integration of the gyrometer's data).

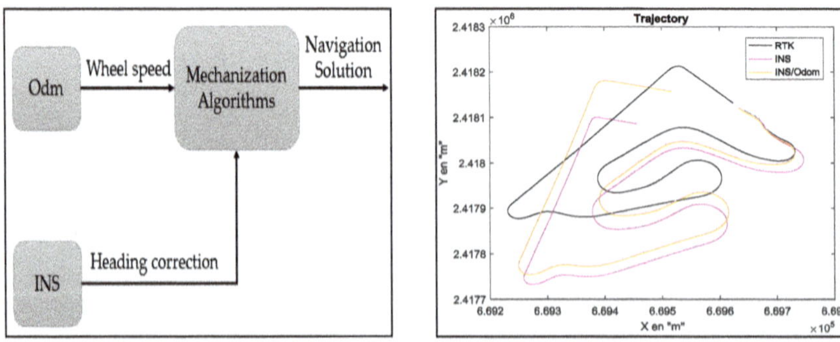

Figure 12. Schematic diagram of the INS/Odo integration.

From Table 2, it is clear that the GPS/INS data fusion solution gives a precise reconstruction and good accuracy. However, this accuracy is obtained in good conditions (no GPS outage and the number of visible satellites is more than four). Contrariwise, during GPS outage, the accuracy of the reconstructed trajectory will strongly decrease. Therefore, in order to propose an alternative to the GPS/INS solution, we propose in this work to improve the INS/odometer data fusion method rather than kinematic models for two reasons:

- The first objective is the design of a motorcycle trajectory reconstruction system with an optimal set of low-cost sensors. In other words, we want to develop an algorithm that uses as few sensors as possible: in kinematic models three sensors (steering, roll, and odometer) are used, while in the INS/odometer method, only two sensors are involved (INS and odometer).
- The second objective is the use of a non-invasive approach. In our work, we aim to propose an algorithm that can be used and implemented without changing the basic design of the motorcycle. Thus, the INS/odometer algorithm can be used either directly if the commercial PTW contains an M-ABS or MTC, or by adding low-cost sensors. While in the case of kinematic models, adding a steering sensor may require some changes to the handlebar.

Table 2. Resulting precision of trajectory reconstruction methods.

Method		Error RMS
Kinematic Models	Bicycle Model	90.86 m
	Cossalter Model	26.70 m
INS		83 m
GPS		1.42 m
GPS/INS		0.51 m
INS/Odo		49.39 m

6. Our Proposal: Enhanced INS/Odometer Data Fusion

The trajectory reconstruction in the INS/GPS method is mainly based on two parameters, as illustrated in Equation (12): heading angle (yaw) and the traveled distance. In this work:

- in order to accurately estimate the traveled distance, the odometer measurements are improved using the wheel radius model that we proposed in Section 4.2; and
- in order to accurately estimate the heading angle, we propose to use the "INS" with the "Madgwick filter" algorithm [23].

The Madgwick filter is based on a quaternion representation, allowing the use of accelerometer and magnetometer measurements in an analytically-derived and optimized gradient descent algorithm to compute the direction of the gyroscope measurement errors as a quaternion derivative and to accurately estimate the attitude of the moving object.

In ideal conditions, i.e., absence of noise and magnetic deviation, the relation between the acceleration in the Earth frame "a^E" and the acceleration in the sensor frame "a^S" is given by Equation (4):

$$a_q^S = q^{-1} \otimes a_q^E \otimes q \tag{4}$$

where:

- $\otimes$: is the quaternion multiplication.
- a_q^S: is the quaternion form of "a^S", which can be written such as: $a_q^S = \begin{bmatrix} 0 & a_x^S & a_y^S & a_z^S \end{bmatrix}^T$
- a_q^E: is the quaternion form of "a^E". In static cases, $a_q^E = \begin{bmatrix} 0 & 0 & g \end{bmatrix}^T$ where g is the acceleration due to the gravity at the Earth's surface ($g \approx 9.8\ m.s^{-2}$).

The relation between "m^E" and "m^S" is as follows:

$$m_q^S = q^{-1} \otimes m_q^E \otimes q \tag{5}$$

where:

- m_q^S: is the quaternion form of "m^S", which can be written such as: $m_q^S = \begin{bmatrix} 0 & m_x^S & m_y^S & m_z^S \end{bmatrix}^T$
- m_q^E: is the quaternion form of "m^E".

The kinematic equation of a rigid body that describes the variation of the attitude in terms of the quaternion, defined from the angular rate measurements delivered by the gyroscope, is given by the following equation:

$$\dot{q} = \frac{1}{2} q \otimes \omega_q^S \tag{6}$$

where "ω_q^S" is the quaternion form of "ω^S".

We have chosen the Madgwick filter (Algorithm 1) rather than Kalman algorithm because their performances were compared in [23] and the results indicate that the Madgwick filter reaches levels of accuracy exceeding that of the Kalman algorithm; <0.6° static RMS error, <1.7° dynamic RMS error. Hence, the new system model of the INS/odometer integration is depicted in Figure 13.

Figure 14 represents the results of the INS/odometer data fusion method before and after the improvements that we propose. From this figure, we can clearly see that the approach we proposed greatly ameliorates the trajectory reconstruction, especially in bends.

Algorithm 1. Gradient descent-based orientation filter.

$$\hat{h}^{E}_{q,t} = \hat{q}_{t-1} \otimes m^{S}_{q,t} \otimes \hat{q}^{-1}_{t-1}$$

$$\hat{m}^{E}_{q,t} = \begin{bmatrix} 0 & 0 & \sqrt{(\hat{h}^{E}_{x,t})^2 + (\hat{h}^{E}_{y,t})^2} & \hat{h}^{E}_{z,t} \end{bmatrix}^T$$

$$F_t = \begin{bmatrix} \hat{q}^{-1}_{t-1} \otimes a^{E}_{q,t} \otimes \hat{q}_{t-1} - a^{S}_{q,t} \\ \hat{q}^{-1}_{t-1} \otimes m^{E}_{q,t} \otimes \hat{q}_{t-1} - m^{S}_{q,t} \end{bmatrix}$$

$\hat{q}_{e,t} = J_t^T F_t$, where J_t is the Jacobian matrix of F_t.

$$\hat{\omega}^{s}_{e,t} = 2\hat{q}_{t-1} \otimes \hat{q}_{e,t}$$

$$\dot{\hat{\omega}}^{S}_{b,t} = \omega^{S}_{e,t}$$

$\hat{\omega}^{S}_{t} = \omega^{S}_{t} - \zeta^{S}\omega_{b,t}$ where ζ^{S} is the integral gain

$$\dot{\hat{q}}_t = \frac{1}{2}\hat{q}_{t-1} \otimes \hat{\omega}^{S}_{q,t} - \beta\frac{\hat{q}_{e,t}}{\| \hat{q}_{e,t} \|}$$

β is the divergence rate of q_t expressed as the magnitude of quaternion derivative corresponding to the gyroscope measurement error.
q_e, ω_e are the quaternion and angular rate errors.

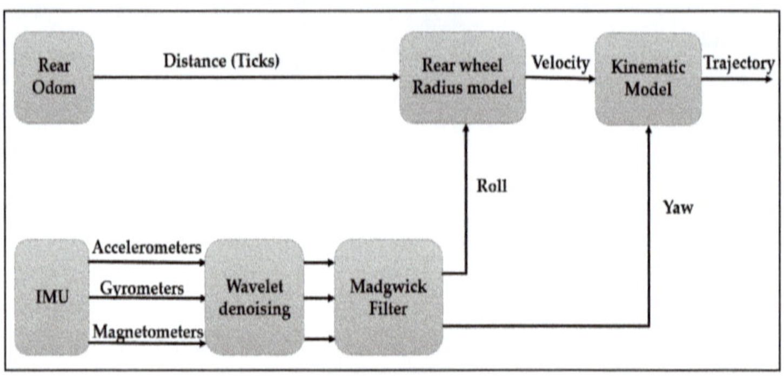

Figure 13. Schematic diagram of the new INS/odometer integration model.

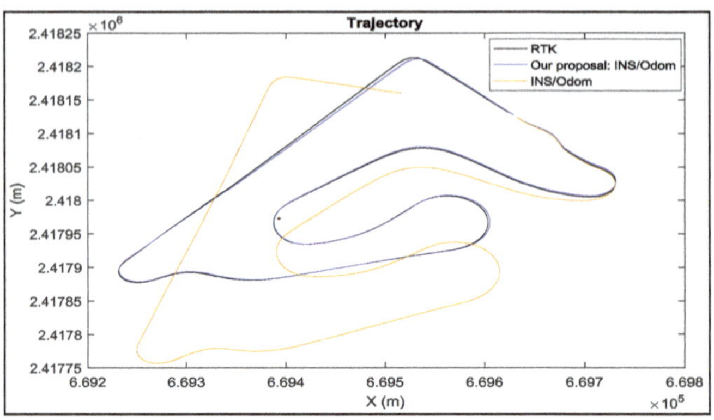

Figure 14. The results of our proposal compared to the traditional method INS/odo.

From Table 2 and Figure 14, it is clear that the integration of GPS/INS data and the proposed integration of odometer/INS data methods give the best reconstruction.

We are mainly interested in the reconstruction of bends instead of the whole trajectory in order to evaluate the behavior of the driver in bends and to compare it with a "safe trajectory". For this reason, the second step consisted of evaluating the accuracy of the motorcycle trajectory in bends achieved by the proposed method and the traditional "GPS/INS integration" method. The two algorithms were tested to reconstruct bends, as shown in Figure 3, with six drivers. Each driver realized three complete trajectories (in total, 18 are achieved). Three indices were used for the quantitative evaluation [5]: bias error, error variance, and maximum error.

The three indices are defined by the following equations:

$$\text{Bias error} = \sqrt{E_x^2 + E_y^2} \tag{7}$$

$$E_x = \frac{1}{t_{\max}} \sum_{t=1}^{t_{\max}} \Delta x(t) \text{ and } E_y = \frac{1}{t_{\max}} \sum_{t=1}^{t_{\max}} \Delta y(t) \tag{8}$$

where $\Delta x(t)$ and $\Delta y(t)$ are the positioning error in the east and north directions at epoch t ($t = 1, 2, \ldots, t_{\max}$) respectively and:

$$\text{Error variance} = \sqrt{V_x + V_y} \tag{9}$$

$$V_x = \frac{1}{t_{\max}} \sum_{t=1}^{t_{\max}} (\Delta x(t) - E_x)^2 \text{ and } V_y = \frac{1}{t_{\max}} \sum_{t=1}^{t_{\max}} (\Delta y(t) - E_y)^2 \tag{10}$$

$$\text{Maximum error} = \max\left[\sqrt{\Delta x(t)^2 + \Delta y(t)^2}\right] \tag{11}$$

According to Table 3 and Figure 15, the proposed method is more efficient than the traditional INS/GPS system. The accuracy of the proposed system is equal to the accuracy of a DGPS. The accuracy of the reconstruction is evaluated according to variance of the error, i.e., the precision varies between "Bias error ± Error variance". The accuracy obtained with the proposed method varies between 23 cm ± 25.20 cm and 36.32 cm ± 31.38 cm with a maximum error of 89.41 cm obtained in bend 2 while the accuracy obtained using the INS/GPS integration varies between 64.93 cm ± 34.56 cm and 83.15 cm ± 44.62 cm with a maximum error of 1.3085 m obtained in bend 4. An improvement of 61% in the accuracy is achieved compared to the INS/GPS method.

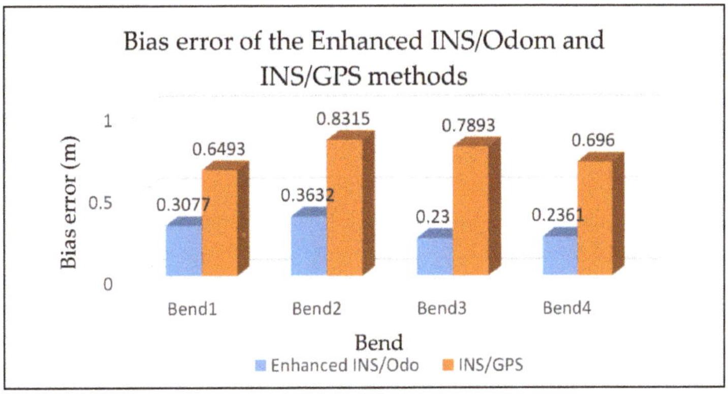

Figure 15. Bias error comparison.

Table 3. Experimental results of the enhanced INS/Odo and INS/GPS methods for four bends.

	Bend 1		Bend 2		Bend 3		Bend 4	
	Enhanced INS/Odom	INS/GPS	Enhanced INS/Odom	INS/GPS	Enhanced INS/Odom	INS/GPS	Enhanced INS/Odom	INS/GPS
Bias error	0.3077	0.6493	0.3632	0.8315	0.2300	0.7893	0.2361	0.6960
Error variance	0.1869	0.3456	0.3138	0.4462	0.2520	0.4194	0.2605	0.3908
Max error	0.5603	1.1426	0.8941	1.2192	0. 6927	1.1210	0.6927	1.3085
Improvement ratio	53%		56%		71%		66%	

Figure 16 shows bend measurement using the GPS/INS method (red curve) and its measurement using the proposed INS/odometer system (blue curve) in "Géoportail" compared to the reference bend given by "GPS RTK" (black curve). We can distinguish that the proposed method gives higher accuracy than the standard GPS/INS method.

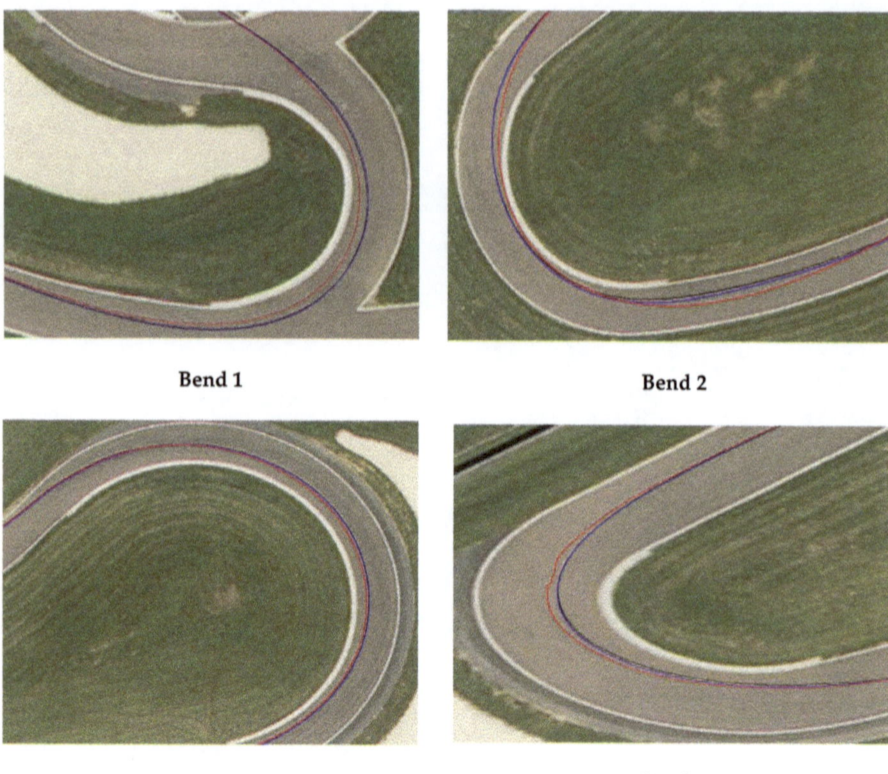

Figure 16. Comparison between the reference trajectory (black curve), GPS/ INS solution (red curve), and the proposed method (blue curve) in the four bends (Bend 1 to 4) showed in Figure 3.

7. Conclusions and Perspectives

In this paper, a comparative study of motorcycle trajectory reconstruction involving algorithms and ad hoc sensors was realized at a sampling time of 10 ms. Several methods and algorithms were evaluated in order to identify an optimal low-cost system further to a multi-sensor instrumentation for an accurate motorcycle bend reconstruction with high accuracy. We provide an experimental setup with a precise ground truth obtained through a GPS RTK.

An enhanced navigation system was proposed using INS/odometer data fusion combined with a Madgwick filter and a wheel radius calculation. The proposed approach represents a good alternative to the traditional INS/GPS system, especially during GPS outage where the accuracy of the GPS/INS solution is degraded. The results are given using a real dataset provided by different drivers.

The obtained accuracy is equal to that of a DGPS. However, the DGPS suffers from several problems:

- Poor dynamic characteristics: the GPS has a low frequency, thus, it provides the state information at low update rates.
- Low availability: accuracy is degraded for a small number of visible satellites (error can achieve 10 m).
- Data latency.
- Multipath errors: these errors occur when the GPS signal is reflected by objects such as large buildings or large areas of obstacles before it reaches the receiver antenna which increases the signal propagation time. This causes an overvaluation of the flight time and, therefore, generates positioning errors.

Our approach overcomes all of these problems and ensures accurate results at high frequency, availability, and solution continuity, which allows to objectively evaluate bend-taking maneuvers and to better train riders to adopt a secure bend.

The designed system will be used by the French Gendarmerie in order to objectively evaluate bend-taking practices. High precision is required to accurately reconstruct bends achieved by gendarmes and to compare them to safe bends. It can also be used for the initial training and retraining in order to better train riders to learn and estimate a safe trajectory.

As a perspective to this work, the proposed algorithm could be investigated to:

- Identify areas for the design and/or assessment of driving assistance devices dedicated to PTWs.
- Improve the positioning accuracy more by combining models (GPS-RTK, IMU/odo).
- Test and validate our proposed algorithm in M-ABS and MTC (motorcycle traction control) systems.

Author Contributions: This paper was a collaborative work among all authors who are part of VIROLO++ project. All authors participated in the design of the proposed model, experiments and writing the paper.

Funding: This research received grant from ANR funding agency in VIROLO++ project, grant number: ANR-15-CE22-0008-03.

Conflicts of Interest: The authors declare no conflict of interest.

References

1. NHTSA's National Center for Statistics and Analysis. NHTSA: Motorcycles Traffic Safety Fact Sheet. Available online: http://www-nrd.nhtsa.dot.gov/Pubs/810990.PDF (accessed on 13 July 2018).
2. Passion Moto Securite. Available online: http://moto-securite.fr/virage-2/ (accessed on 13 July 2018).
3. Espié, S.; Larnaudie, B.; Vincke, B.; Delgehier, F.; Sadoun, R.; Bouaziz, S.; Reynaud, J.P. In-depth study of bend taking practices, towards evaluation and (re)training tools : The VIROLO++ research project. In Proceedings of the 11th International Motorcycle Conference, Cologne, Germany, 3–4 October 2016.
4. ANR. Available online: http://www.agence-nationale-recherche.fr/Projet-ANR-15-CE22-0008 (accessed on 13 July 2018).
5. Koyama, Y.; Tanaka, T. High-Precision Motorcycle Trajectory Measurements Using GPS. *SICE J. Control Meas. Syst. Integr.* **2011**, *4*, 199–205. [CrossRef]
6. Gasbarro, L.; Beghi, A.; Frezza, R.; Nori, F.; Spagnol, C. Motorcycle Trajectory Reconstruction by Integration of Vision and MEMS Accelerometers. In Proceedings of the 43rd IEEE Conference on Decision and Control, Nassau, Bahamas, 14–17 December 2004.

7. Guarnieri, A.; Pirotti, F.; Vettore, A. Low-cost mems sensors and vision system for motion and position estimation of a scooter. *Sensors* **2013**, *13*, 1510–1522. [CrossRef] [PubMed]
8. Whipple, F.J. The stability of the motion of the bicycle. *Q. J. Pure Appl. Math.* **1899**, *30*, 312–321.
9. Waegli, A.; Schorderet, A.; Prongué, C.; Skaloud, J. Accurate Trajectory and Orientation of a Motorcycle derived from low-cost Satellite and Inertial Measurement Systems. In Proceedings of the 7th ISEA Conference, Biarritz, France, 2–6 June 2008.
10. Fiche Technique HONDA CBF 1000 (Donnees Constructeur). Available online: http://www.moto-net.com/article/fiche-technique-honda-cbf-1000-donnees-constructeur.html (accessed on 13 July 2018).
11. Maestro A2200-1. Available online: http://update.maestro-wireless.com/GNSS/A2200-A/Maestro_A2200_GPS_Module_Product_Brief_v14-2.pdf (accessed on 13 July 2018).
12. Drotek. IMU 10DOF-MPU9250 + MS5611. Available online: http://www.drotek.com/shop/fr/home/466-imu-10dof-mpu9250-ms5611.html (accessed on 13 July 2018).
13. AS5047P High-Resolution Position Sensors. Available online: http://ams.com/eng/Products/Magnetic-Position-Sensors/Angle-Position-On-Axis/AS5047P (accessed on 13 July 2018).
14. Septentrio.AltusAPS3G. Available online: http://www.septentrio.com/products/gnss-receivers/rover-base-receivers/smart-antennas/aps3g (accessed on 13 July 2018).
15. Xsens. MTi: Miniature MEMS Based AHRS. Available online: https://www.xsens.com/products/mti/ (accessed on 13 July 2018).
16. Cossalter, V. *Motorcycle Dynamics*, 2nd ed.; Lulu: Morrisville, NC, USA, 2016.
17. Larnaudie, B.; Bouaziz, S.; Maurin, T.; Espíe, S.; Reynaud, R. Experimental Motorcycle platform for Dynamic model extraction. In Proceedings of the IEEE Intelligent Vehicles Symposium (IV), Tokyo, Japan, 13–15 June 2006; pp. 492–497.
18. Gruyer, D.; Lambert, A.; Mourllion, B. Etat de L'art des Stratégies de Fusion de Données Pour La Localisation. *Livrable Arcos Thème* **2004**, *3*, t3.
19. Sakhi, S. Centrale d'Acquisition Temps Réel pour le Trace d'Engins Mobiles. Ph.D. Thesis, Ecole Militaire Polytechnique d'Alger, Bordj El Bahri, Algérie, 2014.
20. Eling, C.; Klingbeil, L.; Kuhlmann, H. Real-Time Single-Frequency GPS/MEMS-IMU Attitude Determination of Lightweight UAVs. *Sensors* **2015**, *15*, 26212–26235. [CrossRef] [PubMed]
21. North, E.; Georgy, J.; Iqbal, U.; Tarbochi, M.; Noureldin, A. *Improved Inertial/Odometry/GPS Positioning of Wheeled Robots Even in GPS-Denied Environments*; InTech Europe: Rijeka, Croatia, 2012.
22. Rogers-Marcovitz, F.; George, M.; Seegmiller, N.; Kelly, A. Aiding Off-Road Inertial Navigation with High Performance Models of Wheel Slip. In Proceedings of the International Conference on Intelligent Robots and Systems (IEEE/RSJ), Vilamoura, Portugal, 7–12 October 2012.
23. Madgwick, S.O.; Harrison, A.J.; Vaidyanathan, R. Estimation of IMU and MARG orientation using a gradient descent algorithm. In Proceedings of the IEEE International Conference on Rehabilitation Robotics Rehab Week Zurich, ETH Zurich Science City, Switzerland, 29 June–1 July 2011.

sensors

MDPI

Article

Comparison of Data Preprocessing Approaches for Applying Deep Learning to Human Activity Recognition in the Context of Industry 4.0

Xiaochen Zheng [1], Meiqing Wang [2] and Joaquín Ordieres-Meré [1,*]

1 Department of Industrial Engineering, ETSII, Universidad Politécnica de Madrid, 28006 Madrid, Spain; xiaochen.zheng@alumnos.upm.es

2 School of Mechanical Engineering and Automation, Beihang University (BUAA), Beijing 100083, China; sy1514206@buaa.edu.cn

* Correspondence: j.ordieres@upm.es; Tel.: +34-913-36-3144

Received: 24 May 2018; Accepted: 29 June 2018; Published: 3 July 2018

Abstract: According to the Industry 4.0 paradigm, all objects in a factory, including people, are equipped with communication capabilities and integrated into cyber-physical systems (CPS). Human activity recognition (HAR) based on wearable sensors provides a method to connect people to CPS. Deep learning has shown surpassing performance in HAR. Data preprocessing is an important part of deep learning projects and takes up a large part of the whole analytical pipeline. Data segmentation and data transformation are two critical steps of data preprocessing. This study analyzes the impact of segmentation methods on deep learning model performance, and compares four data transformation approaches. An experiment with HAR based on acceleration data from multiple wearable devices was conducted. The multichannel method, which treats the data for the three axes as three overlapped color channels, produced the best performance. The highest overall recognition accuracy achieved was 97.20% for eight daily activities, based on the data from seven wearable sensors, which outperformed most of the other machine learning techniques. Moreover, the multichannel approach was applied to three public datasets and produced satisfying results for multi-source acceleration data. The proposed method can help better analyze workers' activities and help to integrate people into CPS.

Keywords: deep learning; data preprocessing; Human Activity Recognition (HAR); Internet of things (IoT); Industry 4.0

1. Introduction

Recent advances in manufacturing industry and Internet of Things (IoT) technology have paved the way for a systematical deployment of cyber-physical systems (CPS), making networked machines perform more efficiently, collaboratively, and resiliently, and transforming manufacturing industries to the Industry 4.0 era [1,2]. According to the Industry 4.0 paradigm, all objects of the factory world are equipped with integrated processing and communication capabilities. This facilitates the vision of the "smart factory", which enables centralized decision-making while requiring distributed manufacturing equipment and resources [3,4]. More "things", even people, need to be connected to the system [5]. In contrast to computer-integrated manufacturing (CIM), the Industry 4.0 movement is not gravitating towards workerless production facilities. Instead, people should be integrated into the cyber-physical structure in such a way that their individual skills and talents can be fully realized [6,7].

The development of IoT technology has also promoted the improvement of Human Activity Recognition (HAR), which is based on copious sensors. HAR has been widely applied in surveillance-based security, context-aware computing, ambient assistive living, and assembly tasks analysis [8–14].

A variety of machine learning algorithms have been used to process human activity data in the big data environment [15–20]. In a recently reported study [8], the performance of several common classification methods were compared for recognizing eight daily activities, using the acceleration data collected from wearable sensors in seven different body positions. An overall accuracy of 89% was achieved using the random forest (RF) method, which outperformed artificial neural network (ANN), decision tree (DT), k-nearest neighbors (k-NN), naive Bayes (NB), and support vector machine (SVM) methods.

Deep learning is a paradigm of machine learning that enables computational models consisting of multiple processing layers to learn representations of data with multiple levels of abstraction [21]. Many studies have proven that the use of deep learning can improve the performance of many applications, especially speech and visual object recognition, in addition to many other domains [21,22]. As a powerful feature extraction mechanism, deep learning has also been used to perform HAR in recent years, and significant improvement has been achieved [23,24]. The convolutional neural network (CNN) is one of the most important deep learning approaches that has been used to perform HAR, and has produced satisfying results in a number of studies [25].

Data preprocessing plays an important role in machine learning and deep learning algorithms, and proper preprocessing of data is compulsory for achieving better HAR performance [26,27]. Kotsiantis et al. [26] defined data preprocessing as including data cleaning, normalization, transformation, feature extraction, and selection. Some of the most well-known algorithms for each step of data preprocessing are presented in their study. More specifically, when performing HAR tasks using inertial data from wearable devices, a segmentation operation is necessary, because raw inertial data fluctuate greatly over time. The segmented data should be transformed into proper formats as the inputs of the deep learning models. Spectrograms are a commonly used data preprocessing method for acceleration data. A spectrogram of an inertial signal is a new representation of the signal as a function of frequency and time. Previous studies [23] have shown that spectrogram representation is essential for extracting interpretable features that represent the intensity differences among nearest inertial data points. A method that combines shallow features and those obtained from deep learning models, in order to overcome the defects that resource limitations cause and the simple design of the deep learning models, was proposed in [23]. However, during our experiment it was found that the spectrogram representation of the acceleration signal does not always produce better classification results, and introducing shallow features does not always improve the overall performance, especially when the dataset is sufficiently large and contains multi-source sensor data.

The aim of this study is to compare different data preprocessing approaches for deep leaning supported HAR tasks in different scenarios, like single or multiple sensors, and provide references for future studies. In this paper, a deep learning algorithm was used to classify daily human activities on the basis of the acceleration data that has been provided by wearable devices in different body positions. The study focused on two important steps—data segmentation and data transformation—of preprocessing acceleration data for deep learning algorithms. A comparison among five data segmentation options was undertaken and the impact of segment length on activity recognition accuracy was analyzed. Four different data transformation methods were compared, including raw acceleration data, the multichannel method, the spectrogram method, and the spectrogram integrated with shallow features method. The highest overall recognition accuracy achieved in this study was 99.42% for eight daily activities, based on the data from seven wearable sensors, which outperformed most traditional machine learning techniques. Beside the above-mentioned dataset, the chosen multichannel method was also applied to three public HAR datasets, and the results were compared against existing studies.

2. Materials and Methods

The framework of the study is illustrated in Figure 1. The proposed method includes data segmentation, data transformation, deep learning model training, and testing. Human activities are time-dependent, and the raw acceleration data from wearable sensors fluctuates greatly over time, making classification impossible when using a single data point [28]. Most HAR methods are based on a series of data collected in a certain time interval. A segmentation operation is necessary before applying any classification method [8]. The data segments are then transformed into images with four different methods, in order to produce the inputs for the deep learning module. Each input corresponds to a specific deep CNN model and generates a specific classifier. The preprocessed data samples are separated into training and testing samples before the training process. The testing samples are selected randomly. Their quantity depends on the segment options and the total number of samples. More details of each step of the workflow are provided in the following sections.

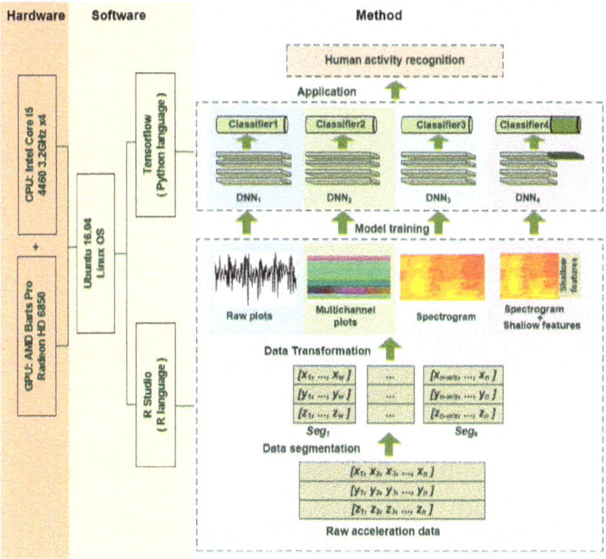

Figure 1. Framework of the proposed method.

2.1. Data Segmentation

The raw time-dependent acceleration dataset is separated into segments during the data segmentation process. All of the following HAR-related operations, including feature extraction, classification and validation, etc., are based on these segments. The length of the segments depends on the application context and sampling rate of the sensors. Increasing the length of the segments can improve recognition accuracy, but the training time will be increased and more time will be required to obtain sufficient data. This will cause a delay in response for real-time applications [23] and restrict the application scenarios. In most of the existing studies, segments of 1 to 10 s are considered for HAR [29].

2.2. Data Transformation

In order to generate the proper inputs for the deep learning models, four different data transformation methods were adopted in this study. These methods transform the raw data segments into different type of representations, from which the deep learning models can extract features automatically. The four methods are explained in detail below.

2.2.1. Raw Plot

The raw plot method transforms the acceleration data directly to time series images. The three axes are grouped by column, and the data collected from different positions are grouped by row on the same image, if applicable. Both the x-axis and y-axis resolution of the produced image are the same as the length of the segment, and the color is black and white. For example, Figure 2 shows the image that is generated from an acceleration data segment, which contains 21 separate sub-images that correspond to three axes (by column) and seven sensors (by row). In this plot, the length of the segments represents the number of values included in this segment. The image resolution (512 × 512 pixels) is not related to the lengths (512 × 3) of segments. Higher resolutions may produce better results, but the training time will also increase. This method can represent the temporal acceleration variance. The deep learning models are able to extract activity features based on the intensity and shape of the plot at different locations and on different levels.

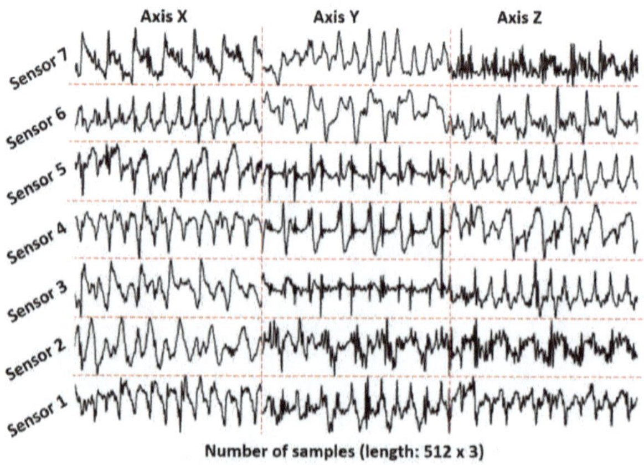

Figure 2. Raw acceleration plot of time domain: segment length 512 and sampling rate 50 Hz (the dotted lines are added manually for better clarification; image resolution is 512 × 512 pixels).

2.2.2. Multichannel Method

Unlike the raw plot method, the multichannel method treats the data for the three axes as three overlapped color channels that correspond to red, green, and blue components in the RGB color format. The amplitude of the acceleration signal, which is in the range (−20,20), is projected to a corresponding color value, which is in the range (0,1). In this case, the temporal variance of the acceleration data is transformed into color variance. The three acceleration values of each point are represented as one pixel in the image. The x-axis resolution of the image is the same as the length of the segment, and the y-axis resolution is the number of sensors. The data collected from different sensors are grouped by row. The advantage of this method is that it reduces the image size enormously and results in a much less training time than the raw plot method. Figure 3 illustrates the principle of this method and an example image produced with this method. The data segment used in this figure is the same as the one used in Figure 2.

2.2.3. Spectrogram

The spectrogram of an inertial signal represents the frequency features of the signal in the time domain. It is the magnitude squared of the short-time Fourier transform (STFT). STFT is used to determine the sinusoidal frequency and phase content of local sections of a signal that changes over time [23,30]. The procedure for computing the spectrogram is to divide a longer time signal into short windows of equal lengths, and then compute the Fourier transform separately for each shorter window. The study by Ravì et al. [23] proved that the spectrogram representation is essential for extracting interpretable features to capture the intensity differences between the nearest inertial data points. The spectrogram representation also provides a form of time and sampling rate invariance.

This enables the classification to be more robust against data shifting in time and against changes in the amplitude of the signal and sampling rate.

Figure 4 shows the spectrogram generated from the same data segment that is used in Figures 2 and 3. Here, the resolution of *y*-axis is 350 pixels and the resolution of *x*-axis is determined by the segment length (L), STFT window length (W), and overlap length (P) by the following equation:

$$\text{Res}_\text{x} = 3 \times (\text{L} - \text{W})/(\text{W} - \text{P}), \tag{1}$$

The spectrograms of different axes and sensors exhibit different patterns. There is also a difference between different activities.

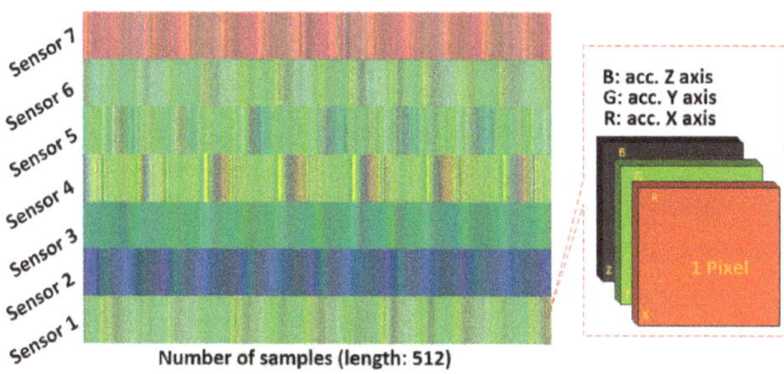

Figure 3. Multichannel RGB color plot on time domain. The segment length is 512 and the sampling rate is 50 Hz (image resolution 512 × 7 pixels).

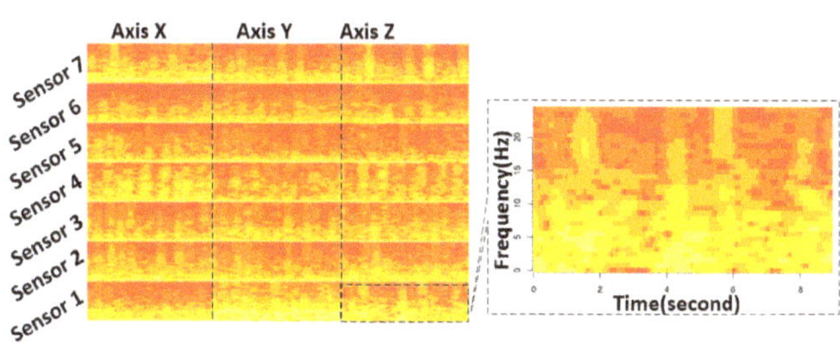

Figure 4. Spectrogram plot of the acceleration data. The segment length is 512 and the sampling rate is 50 Hz, with a short-time Fourier transform (STFT) window length of 64 STFT and an overlap length of 60 STFT (image resolution 336 × 350 pixels).

2.2.4. Spectrogram Combined with Shallow Features

Previous studies [23] have shown that when data resources are limited, the features that are derived from a deep learning method are sometimes less discriminating than a complete set of predefined shallow features. To overcome this problem, a method of combining both shallow and deep-learned features was proposed in [23], in order to provide complementary information for the classification. This method was also used in this paper, to compare the results against those of the other three methods. The aim is to determine if this method outperforms other methods for multi-source acceleration data. As suggested in [23], 15 shallow features are extracted from the raw acceleration data of each axis and each sensor, as shown in Table 1. These shallow features are combined with the deep-learned features to form the last layer of the deep CNN model.

Table 1. Shallow features extracted from acceleration data.

Data	Features
Raw signal	max, min, mean, median, variance, kurtosis, skewness, zero-cross, root mean square, standard deviation, interquartile range
First derivative	mean, variance, root mean square, standard deviation

2.3. Deep Learning Method

After preprocessing, the original acceleration data segments are transformed into different types of images, to which the deep learning methods are applied. In this study, the deep CNN algorithm [21–31] is used. Figure 5 shows the overall workflow of the proposed deep CNN method. Different models were built that correspond to the outputs of the four data transformation methods. Each model has its own parameters, such as the number of convolutional layers, the learning rate, pooling size, etc. Following the approach in [23], the shallow features are merged with the deep-learned features on the last fully connected layer, as shown in Figure 5. More details of the deep learning models are available online [32].

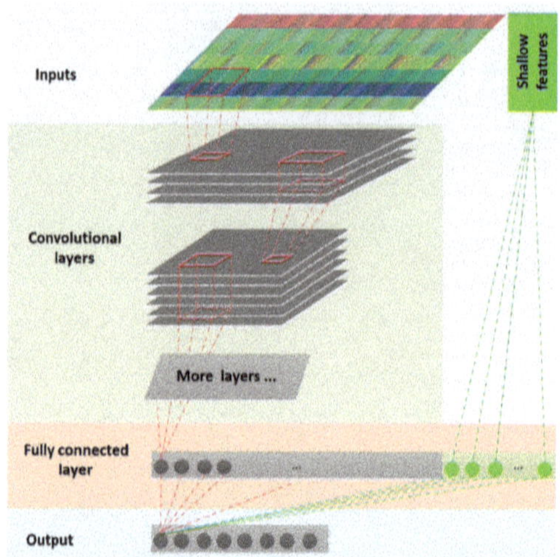

Figure 5. Workflow of deep convolutional neural network (CNN) models.

3. Results

3.1. Dataset and Experimental Setup

The dataset contributed by Sztyler et al. [8] was adopted to test the proposed methods. The reasons for this were that it is up-to-date and, according to the authors, is the most complete, realistic, and transparent dataset for on-body position detection that is currently available [8]. This dataset contains the acceleration data of eight activities—climbing stairs down (A1), climbing stairs up (A2), jumping (A3), lying (A4), jogging (A5), standing (A6), sitting (A7), and walking (A8)—of 15 subjects (age 31.9 $\pm$ 12.4, height 173.1 $\pm$ 6.9, and weight 74.1 $\pm$ 13.8, with eight males and seven females). For each activity, the acceleration of seven body positions—chest (P1), forearm (P2), head (P3), shin (P4), thigh (P5), upper arm (P6), and waist (P7)—were recorded simultaneously. The subjects performed each activity for roughly 10 min, except for jumping (about 1.7 min) due to the physical exertion. In total, the dataset covers 1065 min of acceleration data for each position and axes, with a sampling rate of 50 Hz. We filtered and reorganized the dataset to make it suitable for training deep learning models. The detailed processing method and the prepared datasets are available online [32].

As shown in Figure 1, the experiments in this study were implemented with a computer equipped with a four-core Intel Core i5-4460 3.2GHz CPU, an AMD Barts Pro Radeon HD 6850 Graphic Processing Unit (GPU) and a 12 GB of random-access memory (RAM). The operating system is Ubuntu Linux 16.04 64-bit version. Built on top of these is a software combination of RStudio and TensorFlow.

The data preprocessing was performed with RStudio, including data segmentation, data transformation, and shallow feature extracting. The details and complete code is also available in R-markdown format [32]. The deep learning model training and testing were conducted with TensorFlow (Version 1.0), and the model was built in Python (Version 2.7) language. TensorFlow is an interface for expressing machine learning algorithms, and an application for executing such algorithms, including training and inference algorithms for deep neural network models. More specifically, the TF.Learn module of TensorFlow was adopted for creating, configuring, training, and evaluating the deep learning model. TF.Learn is a high-level Python module for distributed machine learning inside TensorFlow. It integrates a wide range of state-of-art machine learning algorithms built on top of TensorFlow's low-level APIs for small- to large-scale supervised and unsupervised problems. The details of building deep learning models with TensorFlow are provided online, and some of the trained models are also available [32].

3.2. Results and Discussion

There are two evaluation schemes for the activity recognition model, which are a person-dependent method and a person-independent, leave-one-out method [17]. For person-dependent evaluation, the data from the same subject are separated to training samples and testing samples. For person-independent evaluation, the data of one or more subjects are excluded from the training process and used for testing. In our study, considering the small number of subjects we have, and in order to compare with a previous study [8], we used the person-dependent method. The classifiers were trained and evaluated for each subject individually. The data of each subject were segmented with a non-overlapping method to avoid over-fitting caused by data duplication in training and testing datasets. Ten percent of the segmented samples were used as testing data, and the remaining samples were used as training data. Sequential selection of samples in time was applied in order to avoid the over-fitting caused by predicting past based on future. All segment lengths were power values of 2 in order to better perform STFT when generating spectrogram images.

These segments were transformed into raw acceleration plots, multichannel plots and spectrogram images, according to the preprocessing methods that were introduced above. For each segment, the 15 shallow features that appear in Table 1 were extracted for each position and axis. Since each segment contains the acceleration data of three axis and seven positions, 315 shallow features were extracted for each segment. The details of data transformation and feature extracting are available [32].

Different deep learning models were built and trained for each combination of the five segmentation options and four data transformation methods. The introduced methods were evaluated for each individual subject. Table 2 presents the aggregated classification results of all 15 subjects, based on different segmentation and transformation combinations. The highest overall accuracy was 97.19%, using the multichannel method based on a segment length of 512 (10.24 s).

Table 2. Overall accuracy (%) of the four data transformation methods, based on five segmentation options.

Segment Length	Raw Plot	Multichannel	Spectrogram	Spectrogram and Shallow Features
64	92.44	94.60	92.86	90.39
128	93.05	96.14	93.37	90.42
256	93.45	96.58	93.94	92.02
512	94.97	97.19	95.56	93.58
1024	82.13	92.81	91.54	85.55

The results show that the multichannel method achieved the best performance for all segment lengths. For each of these four transformation methods, the performance improved with the increase of the segment length, from segment length 64 (2.56 s) to 512 (10.24 s). There is an accuracy decrease from segment length 512 (10.24 s) to 1024 (20.48 s). A possible explanation is the significant drop of training sample numbers. The accuracy of the multichannel method is more stable than other methods, among different segment lengths. This means that the performance variance of the multichannel method is less than that of others, and its classification accuracy is less dependent on segment lengths, which implies that this method is more suitable for short-time HAR tasks. Moreover, the introduction of shallow features did not increase performance as expected. In fact, it slightly decreased performance compared to the spectrogram method. One possible explanation is that the number of shallow features, which was 315, was too many, and they were confused with features extracted by the deep learning models.

With the same data preprocessing method, the classification accuracies of different individuals were different due to the variation of data quality, dataset size, and individual behaviors. Table 3 summarizes the overall classification accuracies of the 15 subjects, based on a segment length of 512 (10.24 s) with the four data preprocessing methods.

Table 3. Variation of overall classification accuracies (%) of 15 subjects based on a segment length of 512 (10.24 s) with four preprocessing methods.

Subject	Raw Plot	Multichannel	Spectrogram	Spectrogram & Shallow Features
Mean	95.25	97.58	95.81	93.92
Min.	92.42	93.91	91.61	88.46
Max.	97.22	99.56	98.57	97.18
Sd.	1.72	2.11	2.35	2.74

Leaving out the impact of the segment length, the four models that were based on the segment length of 512 (10.24 s) were compared in detail. Table 4 presents the classification accuracy of each of the eight activities that the four models produced.

Table 4. Performance of each model based on a segment length of 512 (10.24 s). A1: climbing down; A2: climbing up; A3: jumping; A4: lying; A5: running; A6: sitting; A7: standing; and A8: walking.

		A1	A2	A3	A4	A5	A6	A7	A8
Raw plot	Precision (%)	97.16	97.99	99.61	99.59	95.18	99.15	92.06	99.49
	Recall (%)	95.41	96.89	98.78	99.18	91.40	98.53	85.24	99.08
	Overall Acc. (%)		94.97			95% CI: (0.9434, 0.9556)			
Multichannel	Precision (%)	97.65	97.96	99.74	99.89	96.29	99.63	96.99	99.72
	Recall (%)	95.56	96.53	99.49	100.00	93.33	99.34	95.04	99.53
	Overall Acc. (%)		97.19			95% CI: (0.9670, 0.9763)			
Spectrogram	Precision (%)	97.65	97.23	99.92	98.60	98.84	97.47	91.18	97.76
	Recall (%)	95.65	96.05	100.00	97.56	98.96	96.55	82.73	96.08
	Overall Acc. (%)		94.56			95% CI: (0.9251, 0.9618)			
Spectrogram & Shallow features	Precision (%)	94.92	98.25	91.51	98.60	95.92	96.60	93.39	95.38
	Recall (%)	91.05	98.59	83.33	97.56	93.14	93.75	88.42	91.51
	Overall Acc. (%)		93.58			95% CI: (0.9157, 0.9512)			

Regarding to training time, the multichannel method also achieved outstanding performance. As shown in Figure 6, the multichannel model took only 40 min to reach the highest accuracy, whereas the other methods required at least 360 min. This proved that the multichannel method provided the best performance, in this case from both accuracy and training time points of view.

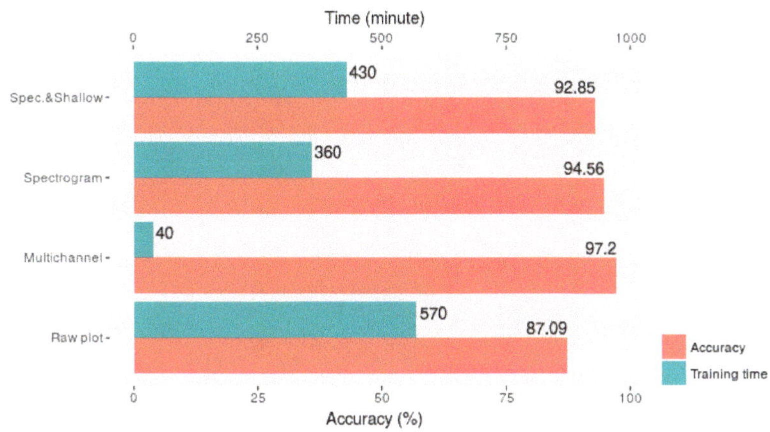

Figure 6. Classification accuracy and training time of the four data transformation methods.

Considering the classification accuracy of each activity, the multichannel method perfectly classified the 68 climbing down (A1) samples, as presented in Table 5. It produced a relatively lower accuracy for running activity (A5), where 5 out of 105 running samples were misclassified as standing activity (A7).

Table 5. Confusion matrix generated by the multichannel model based on a segment length of 512 (A1: climbing down; A2: climbing up; A3: jumping; A4: lying; A5: running; A6: sitting; A7: standing; and A8: walking).

Original	Prediction							
	A1	**A2**	**A3**	**A4**	**A5**	**A6**	**A7**	**A8**
A1	68	0	0	0	0	0	0	0
A2	0	78	2	0	1	0	0	1
A3	0	3	22	0	0	0	0	0
A4	0	0	0	81	1	0	0	0
A5	0	6	0	0	98	1	3	0
A6	0	0	0	0	0	92	1	0
A7	0	0	0	1	5	1	86	0
A8	0	1	0	0	0	0	2	100

The classification above is based on the acceleration data that were collected from seven body positions. In real life scenarios, it is difficult to obtain such a complete dataset. Therefore, activity classification using the data from each single position was also undertaken in this study. The combination of segment length 512 (10.24 s) and the multichannel method was used to better compare with the above-mentioned results. Figure 7 shows the overall classification accuracy for the eight activities. The data from the head produced the lowest accuracy (79.32%), whereas the data collected from the shin provided the highest accuracy (90.51%). This result agrees with practical experience that the movements of the head are more stable than other body positions, whereas the movements of the shin are more closely related to different activities, especially to such dynamic ones such as running, jumping, climbing up, and climbing down. By combining the data from the two positions with the data of highest accuracies, the shin and forearm, an overall accuracy of 93.00% was achieved. This is close to the result that was obtained based on the data from all of the seven positions, which was 97.20%.

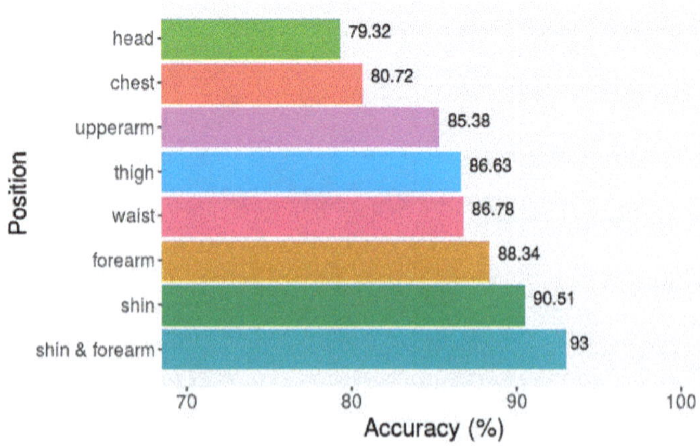

Figure 7. Overall classification accuracies of eight activities based on data from seven single positions and two combined positions.

Compared to other traditional classification techniques, such as ANN, DT, k-NN, NB, SVM, and RF, deep learning methods improved the classification accuracy significantly. Figure 8 shows a comparison of the results achieved by the proposed multichannel deep learning method (marked as DL) based on the segment length of 64 (1.28 s) and the results reported in [8], using the same dataset with a similar segment length of one second. It is shown that the deep learning method achieved an overall classification accuracy that was 7.22% higher than RF.

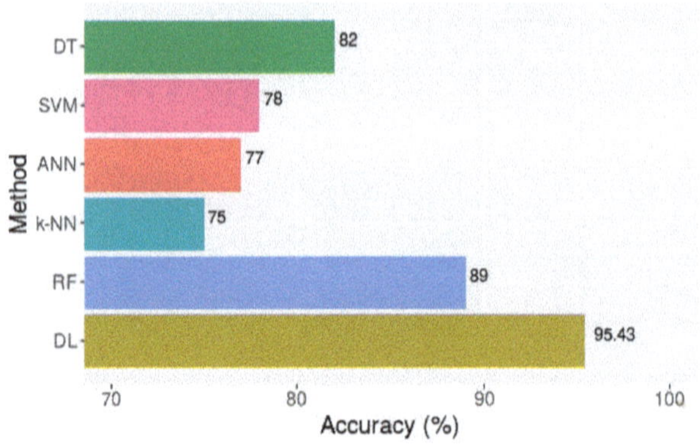

Figure 8. Accuracy of different classification methods [8].

Beside the dataset used above, in order to testify its feasibility, the proposed multichannel data preprocessing method was also applied to another three public HAR datasets, which are WISDM v1.1 (daily activity data collected by a smartphone in a laboratory, with a sampling rate of 20 Hz) [33], WISDM v2.0 (daily activity data collected by a smartphone in an uncontrolled environment, with a sampling rate of 20 Hz) [34,35], and Skoda (manipulative gestures performed in a car maintenance scenario, with sampling rate of 98 Hz) [36]. These datasets were used by Ravì et al. [23], and we used the same segment length as they did, which is a non-overlapping window size of 4 s (for the Skoda dataset) and 10 s (for the WISDM v1.1 and WISDM v2.0 datasets).

The comparison about the per-class precision and recall values obtained by the proposed multichannel transformation method (abbreviated as MCT in the tables) against the results produced by [23] is presented in Table 6. The result shows that the proposed method outperforms the spectrogram integrated with shallow features method in most activities, except the walking and jogging in the WISDM v1.1 dataset and walking in the WISDM v2.0 dataset. In terms of the multi-sensor Skoda dataset, the proposed method perfectly classified most activities, except the open and close left front door activities. This comparison result reveals that the proposed multichannel method is more suitable for multi-source data, although it can also achieve good results for singular sensor data.

Table 6. Precision (%) and recall (%) obtained by the proposed multichannel (MCT) method and existing study [23] in three public datasets.

				Dataset 1: WISDM v1.1			
		Walking	**Jogging**	**Sitting**	**Standing**	**Upstairs**	**Downstairs**
Ravì et al. [23]	Prec.	99.37	99.64	97.85	98.15	95.52	94.44
	Rec.	99.37	99.40	98.56	97.25	95.13	95.90
MCT	Prec.	98.34	98.11	100.00	100.00	96.14	98.44
	Rec.	97.31	97.53	100.00	100.00	93.10	97.67
				Dataset 2: WISDM v2.0			
		Jogging	**Lying Down**	**Sitting**	**Stairs**	**Standing**	**Walking**
Ravì et al. [23]	Prec.	98.01	88.65	87.32	85.00	82.05	97.17
	Rec.	97.73	85.85	89.28	76.98	82.11	97.19
MCT	Prec.	98.76	96.85	90.25	87.03	91.02	95.85
	Rec.	97.95	94.96	82.05	75.00	85.94	94.81
				Dataset 3: Skoda			
		Write on Notepad	**Open Hood**	**Close Hood**	**Check Gaps Front**	**Open Left Front Door**	
Ravì et al. [23]	Prec.	96.67	97.78	89.47	91.15	100.00	
	Rec.	91.34	97.78	94.44	92.79	100.00	
MCT	Prec.	100.00	99.54	100.00	100.00	80.00	
	Rec.	100.00	100.00	100.00	100.00	60.00	
		Close Left Front Door	**Close Both Left Door**	**Check Trunk Gaps**	**Open and Close Trunk**	**Check Steer Wheel**	
Ravì et al. [23]	Prec.	88.89	92.86	98.78	100.00	93.55	
	Rec.	80.00	94.20	97.59	98.04	100.00	
MCT	Prec.	99.18	100.00	100.00	100.00	94.44	
	Rec.	100.00	100.00	100.00	100.00	88.89	

4. Discussions and Conclusions

In this paper, preprocessing techniques in human activity recognition tasks by deep learning have been considered as a design parameter, and they were shown to be relevant. By comparing different data preprocessing approaches, we came to the following conclusions. Firstly, the length of data segment significantly impacts the final classification accuracy of the deep learning model. The accuracy improves with the increasing of the segment length, and the increasing rate is slower when the segment length is longer. This result agrees with the findings of previous studies that HAR are usually based on data segments of 1 to 10 s. Secondly, four different data transformation methods were compared, and the multichannel method achieved the best performance in both classification accuracy and training time. Unlike the reports of previous studies, we found that the introducing of shallow features did not increase the final accuracy when the experiments were based on a large and multisource dataset. By comparing the classification accuracy based on the data from seven different body positions, it was found that the acceleration data from the shin produced the highest accuracy of 90.51%. A satisfactory accuracy of 93.00% was achieved by combining the data from the shin and forearm. Moreover, we compared the proposed method against some of other common machine learning methods, based on the same dataset, and it was proven that the deep learning method outperforms others impressively. Finally, we applied the proposed multichannel method to three more public datasets, including the car maintenance activity data in a workshop. The results proved that our method can achieve satisfying recognition accuracy. It can help better analyze workers' activities in a factory environment and help integrate people into the cyber-physical structure in an Industry 4.0 context.

Author Contributions: Methodology: X.Z. and J.O.-M.; Formal analysis: X.Z.; Data curation: X.Z. and M.W.; Writing (original draft preparation): X.Z.; Writing (review and editing): J.O.-M.; Supervision: J.O.-M.

Funding: This research received no external funding.

Acknowledgments: The authors thank the support of the China Scholarship Council.

Conflicts of Interest: The authors declare no conflict of interest.

References

1. Chen, S.; Chen, Y.; Hsu, C. A New approach to integrate internet-of-things and software-as-a-service model for logistic systems: A case study. *Sensors* **2014**, *14*, 6144–6164. [CrossRef] [PubMed]
2. Lee, J.; Bagheri, B.; Kao, H. A cyber-physical systems architecture for industry 4.0-based manufacturing systems. *Manuf. Lett.* **2015**, *3*, 18–23. [CrossRef]
3. Xu, X. From cloud computing to cloud manufacturing. *Robot. Comput. Integr. Manuf.* **2012**, *28*, 75–86. [CrossRef]
4. Ooi, K.; Lee, V.; Tan, G.W.; Hew, T.; Hew, J. Cloud computing in manufacturing: The next industrial revolution in malaysia? *Expert Syst. Appl.* **2018**, *93*, 376–394. [CrossRef]
5. Hao, Y.; Helo, P. The role of wearable devices in meeting the needs of cloud manufacturing: A case study. *Robot. Comput. Integr. Manuf.* **2017**, *45*, 168–179. [CrossRef]
6. Putnik, G. Advanced manufacturing systems and enterprises: Cloud and ubiquitous manufacturing and an architecture. *J. Appl. Eng. Sci.* **2012**, *10*, 127–134.
7. Gorecky, D.; Schmitt, M.; Loskyll, M.; Zühlke, D. Human-machine-interaction in the industry 4.0 era. In Proceedings of the 12th IEEE International Conference on Industrial Informatics (INDIN), Porto Alegre, Brazil, 27–30 July 2014; pp. 289–294.
8. Sztyler, T.; Stuckenschmidt, H. On-body localization of wearable devices: An investigation of position-aware activity recognition. In Proceedings of the IEEE International Conference on Pervasive Computing and Communications (PerCom), Sydney, Australia, 14–18 March 2016; pp. 1–9.
9. Kwon, Y.; Kang, K.; Bae, C. Unsupervised learning for human activity recognition using smartphone sensors. *Expert Syst. Appl.* **2014**, *41*, 6067–6074. [CrossRef]
10. Ward, J.A.; Lukowicz, P.; Troster, G.; Starner, T.E. Activity recognition of assembly tasks using body-worn microphones and accelerometers. *IEEE Trans. Pattern Anal. Mach. Intell.* **2006**, *28*, 1553–1567. [CrossRef] [PubMed]
11. Shoaib, M.; Bosch, S.; Incel, O.D.; Scholten, H.; Havinga, P.J. Complex human activity recognition using smartphone and wrist-worn motion sensors. *Sensors* **2016**, *16*, 426. [CrossRef] [PubMed]
12. Attal, F.; Mohammed, S.; Dedabrishvili, M.; Chamroukhi, F.; Oukhellou, L.; Amirat, Y. Physical human activity recognition using wearable sensors. *Sensors* **2015**, *15*, 31314–31338. [CrossRef] [PubMed]
13. Nakai, D.; Maekawa, T.; Namioka, Y. Towards unsupervised measurement of assembly work cycle time by using wearable sensor. In Proceedings of the IEEE International Conference on Pervasive Computing and Communication Workshops (PerCom Workshops), Sydney, Australia, 14–18 March 2016; pp. 1–4.
14. Koskimaki, H.; Huikari, V.; Siirtola, P.; Laurinen, P.; Roning, J. Activity recognition using a wrist-worn inertial measurement unit: A case study for industrial assembly lines. In Proceedings of the 17th Mediterranean Conference on Control and Automation, Thessaloniki, Greece, 24–26 June 2009; pp. 401–405.
15. Kim, E.; Helal, S.; Cook, D. Human activity recognition and pattern discovery. *IEEE Pervasive Comput.* **2010**, *9*, 48–53. [CrossRef] [PubMed]
16. Lee, Y.; Cho, S. Activity recognition with android phone using mixture-of-experts co-trained with labeled and unlabeled data. *Neurocomputing* **2014**, *126*, 106–115. [CrossRef]
17. Bulling, A.; Blanke, U.; Schiele, B. A Tutorial on human activity recognition using body-worn inertial sensors. *ACM Comput. Surv.* **2014**, *46*, 33. [CrossRef]

18. Álvarez de la Concepción, M.A.; Soria Morillo, L.M.; Gonzalez-Abril, L.; Ortega Ramírez, J.A. Discrete techniques applied to low-energy mobile human activity recognition. *A New approach. Expert Syst. Appl.* **2014**, *41*, 6138–6146. [CrossRef]
19. Clark, C.C.; Barnes, C.M.; Stratton, G.; McNarry, M.A.; Mackintosh, K.A.; Summers, H.D. A review of emerging analytical techniques for objective physical activity measurement in humans. *Sports Med.* **2016**, *47*, 439–447. [CrossRef] [PubMed]
20. Hassan, M.M.; Uddin, M.Z.; Mohamed, A.; Almogren, A. A Robust human activity recognition system using smartphone sensors and deep learning. *Future Gener. Comput. Syst.* **2018**, *81*, 307–313. [CrossRef]
21. LeCun, Y.; Bengio, Y.; Hinton, G. Deep learning. *Nature* **2015**, *521*, 436–444. [CrossRef] [PubMed]
22. Krizhevsky, A.; Sutskever, I.; Hinton, G.E. Imagenet classification with deep convolutional neural networks. In Proceedings of the 25th International Conference on Neural Information Processing Systems, Lake Tahoe, NV, USA, 3–6 December 2012; pp. 1097–1105.
23. Ravì, D.; Wong, C.; Lo, B.; Yang, G. A Deep learning approach to on-node sensor data analytics for mobile or wearable devices. *IEEE J. Biomed. Health Inf.* **2017**, *21*, 56–64. [CrossRef] [PubMed]
24. Ronao, C.A.; Cho, S. Human activity recognition with smartphone sensors using deep learning neural networks. *Expert Syst. Appl.* **2016**, *59*, 235–244. [CrossRef]
25. Hammerla, N.Y.; Halloran, S.; Ploetz, T. Deep, Convolutional, and Recurrent Models for Human Activity Recognition Using Wearables. 2016. Available online: https://arxiv.org/abs/1604.08880 (accessed on 23 May 2018).
26. Kotsiantis, S.; Kanellopoulos, D.; Pintelas, P. Data preprocessing for supervised leaning. *Int. J. Comput. Sci.* **2006**, *1*, 111–117.
27. Zhang, L.; Zhang, L.; Du, B. Deep learning for remote sensing data: A technical tutorial on the state of the art. *IEEE Geosci. Remote Sens. Mag.* **2016**, *4*, 22–40. [CrossRef]
28. Lara, O.D.; Labrador, M.A. A Survey on human activity recognition using wearable sensors. *IEEE Commun. Surv. Tutor.* **2013**, *15*, 1192–1209. [CrossRef]
29. Preece, S.J.; Goulermas, J.Y.; Kenney, L.P.; Howard, D. A comparison of feature extraction methods for the classification of dynamic activities from accelerometer data. *IEEE Trans. Biomed. Eng.* **2009**, *56*, 871–879. [CrossRef] [PubMed]
30. Sejdić, E.; Djurović, I.; Jiang, J. Time–frequency feature representation using energy concentration: An overview of recent advances. *Digit. Signal Process.* **2009**, *19*, 153–183. [CrossRef]
31. LeCun, Y.; Bengio, Y. Convolutional networks for images, speech, and time series. In *The Handbook of Brain Theory and Neural Networks*; MIT Press: Cambridge, MA, USA, 1998; pp. 255–258.
32. Zheng, X.; Ordieres, J. Step-by-Step Introduction to Acceleration Data Classification Using Deep Learning Methods. 2017. Available online: https://www.researchgate.net/publication/317180890 (accessed on 23 May 2018).
33. Kwapisz, J.R.; Weiss, G.M.; Moore, S.A. Activity recognition using cell phone accelerometers. *ACM SigKDD Explor. Newslett.* **2011**, *12*, 74–82. [CrossRef]
34. Lockhart, J.W.; Weiss, G.M.; Xue, J.C.; Gallagher, S.T.; Grosner, A.B.; Pulickal, T.T. Design considerations for the WISDM smart phone-based sensor mining architecture. In Proceedings of the Fifth International Workshop on Knowledge Discovery from Sensor Data, San Diego, CA, USA, 21–24 August 2011; pp. 25–33.
35. Weiss, G.M.; Lockhart, J.W. *The Impact of Personalization on Smartphone-Based Activity Recognition*; AAAI Technical Report WS-12-05; Fordham University: New York, NY, USA, 2012; pp. 98–104.
36. Zappi, P.; Lombriser, C.; Stiefmeier, T.; Farella, E.; Roggen, D.; Benini, L.; Troster, G. Activity recognition from on-body sensors: Accuracy-power trade-off by dynamic sensor selection. *Lect. Notes Comput. Sci.* **2008**, *4913*, 17–33.

sensors

MDPI

Article

Fire Source Range Localization Based on the Dynamic Optimization Method for Large-Space Buildings

Guoyong Wang [1], Xiaoliang Feng [2,*] and Zhenzhong Zhang [3]

[1] Luoyang Institute of Science and Technology, Luoyang 471000, China; standardwgy@163.com
[2] College of Electrical Engineering, Henan University of Technology, Zhengzhou 450001, China
[3] Bureau of Geophysical Prospecting, China National Petroleum Corporation, Zhuozhou 072751, China; zhangzhenzhong@cnpc.com.cn
* Correspondence: fengxl2002@163.com

Received: 27 April 2018; Accepted: 11 June 2018; Published: 15 June 2018

Abstract: This paper is concerned to the fire localization problem for large-space buildings. Two kinds of circular fire source arrangement localization methods are proposed on the basis of the dynamic optimization technology. In the Range-Point-Range frame, a dynamic optimization localization is proposed to globally estimate the circle center of the circular arrangement to be determined based on all the point estimates of the fire source. In the Range-Range-Range frame, a dynamic optimization localization method is developed by solving a non-convex optimization problem. In this way, the circle center and the radius are obtained simultaneously. Additionally, the dynamic angle bisector method is evaluated. Finally, a simulation with three simulation scenes is provided to illustrate the effectiveness and availability of the proposed methods.

Keywords: fire source localization; dynamic optimization; global information; the Range-Point-Range frame; the Range-Range-Range frame; sensor array

1. Introduction

In recent decades, more and more large-space buildings have been utilized as product storage and manufacture places, which are usually designed with complicated structures and are crowded with many kinds of materials and products. For these places, fire safety has become one of the most important and difficult problems, and increasing attention has been by governments, factories, firemen, researchers, and engineers [1–6]. For fighting against fire disasters, a critical important issue is how to determine the location of a fire source, in order to provide accurate information and necessary guidance for firefighting. Therefore, it is necessary to develop fire location methods to determine the fire source location as quickly and accurately as possible, for intelligent firefighting, especially in large-space buildings.

In fact, the fire source location method is one kind of wireless indoor positioning technique [7–11]. The existing fire source location methods are mainly developed on the basis of image processing technology, fiber-optic temperature measurement, temperature sensor arrays, and smoke sensor arrays. The fire location systems based on image information are commonly used for the large open spaces, and are easy impacted by barriers [12–14]. The fiber-based temperature measurement methods utilize the impacts of temperature on the anti-Stokes spectral lines in the Raman scattering process in an optical fiber to determine the fire sources position, which is suitable for tunnels and other such scenes, however, the cost of installation and maintenance of optical fiber is too expensive [15–18]. Moreover, several temperature field-based positioning methods aiming at environments, such as mines and forests, are given in [19–22]. These methods usually require that the fire releases more heat. Recently, the fire monitoring and location methods based on temperature/smoke sensor arrays have been paid increasing attention, in which the measurements are sometimes non-line-of-sight propagated. The fire source burning process produces the hot gases or smoke, which rise up to the ceiling and spread in a nearly circular shape [23–28]. With two temperature sensor arrays, a prototype system for determining the fire location was proposed in [18], in which three temperature sensors of a four-sensor array were utilized to estimate the angle of the fire source to the sensor array, and the fire source location can be obtained with the help of two estimated angles and the distance between two sensor arrays. This method was identified as the far-field mode in [24], due to the assumption that the distance between the fire source and the sensor array is much large than the distance intervals in the sensor array. In [24], a far-field fire location approach and a near-field method were presented and compared. In the near-field fire location method, the distance and the angle of the fire source to the sensor array were estimated simultaneously, with signal source localization of the planar wave fronts, but not the planar wave fronts, which was utilized in the far-field approach. The position estimation of the smolder source was realized in [28], using the smoke sensor arrays arranged in the planar circular mode. In [25], a fire location position algorithm was also developed with the signal source localization of a bilinear sensor array. As shown in [27,29], the output results of the above fire source location methods were a number of dispersedly estimated point coordinates of the fire source. The fire source point coordinate estimates were solved locally and independently with the sensor data sampled in the current time, but not the global sensor data sampled in the time series. Therefore, the clustering algorithm was introduced in the fire source localization method in the Range-Point-Range (RPR) frame in [27]. Additionally, two effective fire range estimation approaches were proposed in the Range-Range-Range (RRR) frame in [29]. The circle range estimates were obtained globally in the form of the inscribed circle and the circum-circle of a quadrangle, which was solved recursively using a dynamic angular bisector method. However, it did not discuss the optimality of the circle ranges solved by the dynamic angular bisector method in [29].

Inspired by the above discussion, in this paper, the method of determining the circle range which covers the fire source is developed based on the dynamic optimization theory. Firstly, in the RPR frame, all the point estimates of the fire source obtained by the fire source point location method are considering to obtain a circle center of the circular fire source range to be determined, the sum distance from which each fire point estimate is minimizes. The radius is the standard deviation of the distances from the circle center to every fire point estimate. Actually, compared with to the clustering algorithm in [27], this method determines the circle center and radius based on all the fire point estimates. Therefore, the circular fire source range is determined globally, according to [29]. However, the dynamic optimization problem solved in this algorithm is not recursive. When several new fire point estimates are obtained by using the current measurements of sensor arrays, all the obtained fire point estimates in the time series should be considered to determine the circle center and radius of the circular fire source range to be determined. Although the location accuracy can be improved along with the fire source range determining process, the computing complexity is increasing poor. In order to reduce the computing complexity of the dynamic optimization problem, a circular fire source range location method is also developed in the RRR frame. In [29], the quadrangle is solved by using the dynamic angle bisector method with global information, which, in this paper, is utilized to determine a new circum-circle, the circle center and radius of which is solved by minimizing the distances of the circle center to the four vertices of the given quadrangle. An interesting discovery is that the circle center is similar to the one obtained by the dynamic angle bisector method. It seems that the circle center obtained by using the dynamic angle bisector method is optimal in the sense of minimizing the distances of the circle center to the four vertices of the quadrangle solved with the global information.

The major contributions of this paper are three-fold. First, with all the fire point estimates, a circular fire source range estimation method is proposed in the RPR frame. As shown in [27], the clustering algorithms used to obtain the circular fire source range inevitably discard some fire point estimates. However, in this paper, the circular fire source range is solved globally by using a dynamic optimization algorithm based on all the point estimates of the fire source. Secondly, the above circular fire source range location method is developed in the RRR frame to reduce the computational complexity. The global angle information in time series are recursively utilized to obtain a quadrangle by using the dynamic angle bisector method. What is different with the method in [29] is that the circle center and the radius are determined by solving a non-convex optimization problem. In fact, the non-convex optimal problem can be treated as a standard to evaluate the dynamic angle bisector method. Additionally, the optimality of the dynamic angle bisector method is analyzed by the numeric examples in the simulation section.

It should be noted that, for the case where the angle information are not uniform on the two sides of the bisectors, the proposed circular fire source arrangement methods have better location results than the dynamic angle bisector method in [29], as illustrated in the simulation.

The rest of this paper is organized as follows: In Section 2, the fire location problem is presented, and a fire source point localization principle is briefly introduced. On this basis, two fire source range localization methods based on the dynamic optimization algorithm are proposed in Section 3. The first one is designed with the help of the fire point estimates in the RPR frame in Section 3.1, and in Section 3.2, the second method is developed with the quadrangle solved by the dynamic angle bisector method, which is studied in the RRR frame. In Section 4, a simulation with three different simulation scenes are provided to prove the effectiveness of the proposed methods. Finally, the conclusion of this paper is given in Section 5.

2. Problem Formulation

The structured fire scene studied in this paper is similar to the ones in [6,12,29]. An ignition fire source is considered in a large-space building with static wind. The fire source burning process produces the hot gases, which rise up to the ceiling and spread in a nearly circular shape at a constant current velocity from, and around, the center and form a hot upper layer [6,12,29], as shown in Figure 1. In Figure 1, two temperature sensor arrays are placed in different locations on the ceiling to monitor the air temperature. In each sensor array, four homogeneous temperature sensors are arranged in a square shape with the side length d. The distance between the two sensor arrays is L. The distances from the fire source to the two sensor arrays are D_1 and D_2, respectively. Then the expectation of the temperature field around the fire source can be described as [12]:

$$E\{T(x,y,t)\}|_{(x-x_0)^2+(y-y_0)^2=c^2} = f(t) \tag{1}$$

where (x_0, y_0) is the point coordinate of fire source and $f(t)$ is the temperature function at location (x, y), the distance from which to the fire source is c.

For the structured fire scene described above, several effective fire source point location methods are proposed in [6]. Taking the famous far-field algorithm as an example, the main steps of the fire source point location scheme can be sketched as follows:

(1) *Delay Estimation*. As shown in Figure 1, there are different distances from the fire source to different temperature sensors. According to the temperature field expectation in Equation (1), the delay time of the same temperature time series spread to different sensors can be estimated by the correlation function method [12], or the gray relation analysis method [29].

(2) *Angle Estimation*. Denote $\tau_{i,j}$, $(i,j = 1,2,3,4)$ as the delay time from S_i to S_j, f_A as the sample rate of the temperature sensors, $\alpha(k)$ is an intersection angle crossed by the horizontal line and the line from the sensor array to the fire source, as shown in Figure 1. Based on the planar waves assumption of the far-field algorithm, for the sensor set $\{S_1, S_2, S_3\}$, one can obtain:

$$|v|\frac{\tau_{12}}{f_A} = d\cos(\alpha(k)) \tag{2}$$

$$|v|\frac{\tau_{13}}{f_A} = d\sin(\alpha(k)) \tag{3}$$

then:

$$\hat{\alpha}(k) = \arctan\left(\frac{\tau_{12}}{\tau_{13}}\right), \hat{v} = f_A \frac{d\cos(\hat{\alpha}(k))}{\tau_{12}} \tag{4}$$

Similarly, for different sensor sets in the same sensor array, such as $\{S_1, S_2, S_4\}$ and $\{S_1, S_3, S_4\}$, different estimates of $\alpha(k)$ can also be obtained. In the sensor array, the sensor at S_1 is the reference node. For the sake of convenience, the angle estimate is denoted as $\hat{\alpha}_{i,j}(k)$, in which k indicates the sampled times, $i \in \{A, B\}$ represents the different sensor arrays, and $j = 1, 2, 3$ signifies the different sensor sets.

(3) *Fire Source Point Estimation.* For every combination $\{\hat{\alpha}_{A,j}(k), \hat{\alpha}_{B,l}(k)\}, (j, l = 1, 2, 3)$, the fire source point can be estimated as follows [12]:

$$\hat{x}_{0,j,l}(k) = L\frac{\tan(\hat{\alpha}_{B,l}(k))}{\tan(\hat{\alpha}_{A,j}(k)) + \tan(\hat{\alpha}_{B,l}(k))} + x_1 \tag{5}$$

$$\hat{y}_{0,j,l}(k) = L\frac{\tan(\hat{\alpha}_{B,l}(k))}{\tan(\hat{\alpha}_{A,j}(k)) + \tan(\hat{\alpha}_{B,l}(k))} + y_1 \tag{6}$$

where (x_1, y_1) is the coordinate of S_1 in the sensor array A.

According to the fire source point estimation method given in Equations (5) and (6), nine estimates of the fire source can be solved at one time. Afterward, one can obtain a great deal of discretely ruleless estimate points. This kind of result limits the ability to guide firefighters. Therefore, some modified fire location methods are developed based on the fusion method [12,29]: the clustering algorithms [27]. In the next section, two kinds of location methods are designed to determine more accurate and compact ranges of the fire source, based on the above fire source point estimation method.

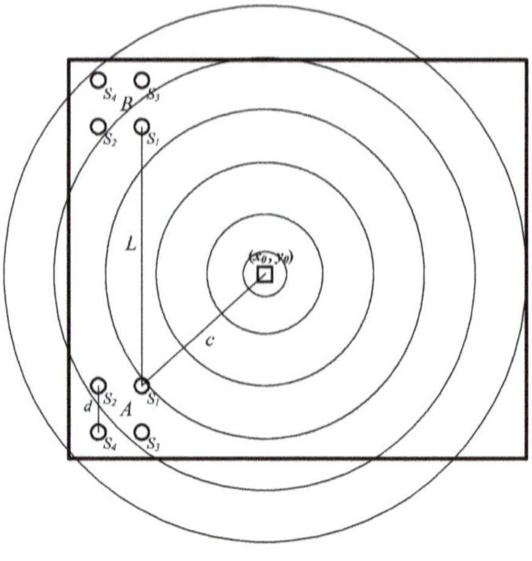

Figure 1. The fire source localization scene.

3. Main Results

3.1. Dynamic Optimization Localization Method in the RPR Frame

According to the fire source point estimation method mentioned in the above section, several ruleless discrete location points can be solved at one time. In order to fuse this location information into one estimate point, the mean of the three angle estimates for each sensor array is taken as the final angle estimate at one time, then the fusion fire source points are solved by Equations (5) and (6) in [12]. As noted in [29], this fusion result is solved only based on the sensor data at the current time, but not all of the sensor data in the time series. Therefore, in [27], a fire source range location method in the RPR frame is proposed based on an improved clustering algorithm. The fire source point estimates in the time series are utilized to determine a circular range which contains the real fire source with a large probability. However, this fire source circle range does not contain all the fire source point estimates in the clustering algorithm [29].

In this section, a fire source range location method is studied in the RPR frame, based on the dynamic optimization algorithm. All the estimates $\left(\hat{x}_{0,j,l}(t), \hat{y}_{0,j,l}(t)\right)$, $j, l = 1,2,3;\ t = 1,2,\cdots,k$ obtained by Equations (5) and (6) contain some information of the fire source. Thus, all of them should be considered to deal with the globally determined circular range. Therefore, in this section, the circle center of the circular sharp arrangement to be determined is considered as the solution of the following optimization problem:

$$\min sum\left((\hat{x}_0^k - x_0)^2 + (\hat{y}_0^k - y_0)^2 \right) \tag{7}$$

where $(\hat{x}_0^k, \hat{y}_0^k)$ is the matrix including all the fire source point estimates in the time series obtained by the fire source point estimation method mentioned in the above section.

It is indicated in Equation (7) that the sum of the distances between the circle center to be solved and every fire source point estimates in the time series should be the minimum.

Denote the solution of the optimization problem of Equation (1) as $(\hat{x}_0(k), \hat{y}_0(k))$. Take it as the circle center of the circular range to be determined, and take the standard deviation of the distances from $(\hat{x}_0(k), \hat{y}_0(k))$ to every fire source point estimate $\left(\hat{x}_{0,j,l}(t), \hat{y}_{0,j,l}(t)\right)$, $j, l = 1,2,3;\ t = 1,2,\cdots,k$ as the radius of the circular range to be determined. Namely:

$$r(k) = std\left\{ (\hat{x}_{0,j,l}(t) - \hat{x}_0(k))^2 + (\hat{y}_{0,j,l}(t) - \hat{y}_0(k))^2 \right\}, j, l = 1,2,3;\ t = 1,2,\cdots,k \tag{8}$$

where, $\left(\hat{x}_{0,j,l}(t), \hat{y}_{0,j,l}(t)\right)$, $j, l = 1,2,3;\ t = 1,2,\cdots,k$ are just the constituent elements of the matrix $(\hat{x}_0^k, \hat{y}_0^k)$.

In this way, a circular fire source range is determined by solving the optimization problem in Equation (7).

Remark 1. *All the fire source point estimates in the time series are utilized to solve the dynamic optimization problem in Equation (7). Therefore, the circle center is solved based on the global information.*

However, the dynamic optimization problem solved in this algorithm is not recursive. When several new fire point estimates are obtained by using the current measurements of sensor arrays, all the obtained fire point estimates in the time series should be considered to determine the circle center and radius of the circular fire source range to be determined. Although the location accuracy can be improved along with the fire source range determining process, the computing complexity is increasing poor. In order to reduce the computing complexity of the dynamic optimization problem in the RPR frame, a circular fire source range location method is also developed in the RRR frame in the next subsection.

3.2. Dynamic Optimization Localization Method in the RRR Frame

In the fire source localization scene shown in Figure 1, for each sensor array i, $i \in \{A, B\}$, three angle estimates can be obtained at each time k, according to Equations (2) and (3). In the RRR frame [29], the up and down bounds of the angle estimate range can be obtained by using the statistical mean and variance of the three angle estimates:

$$\hat{\alpha}_i(k) = \frac{1}{3}\sum_{j=1}^{3}\hat{\alpha}_{i,j}(k),\ \delta_i^2 = \frac{1}{2}\sum_{j=1}^{3}\left(\hat{\alpha}_{i,j}(k) - \hat{\alpha}_i(k)\right)^2 \tag{9}$$

$$\hat{\alpha}_i^H(k) = \hat{\alpha}_i(k) + \delta_i,\ \hat{\alpha}_i^L(k) = \hat{\alpha}_i(k) - \delta_i \tag{10}$$

For the global angle estimates in the time series, the corresponding statistical mean and variance can be recursively calculated by:

$$\hat{\alpha}_i^k = \frac{\hat{\alpha}_i(k) + (k-1)\hat{\alpha}_i^{k-1}}{k},\ \left(\delta_i^k\right)^2 = \frac{\left(\hat{\alpha}_i(k) - \hat{\alpha}_i^k\right)^2 + (k-1)\left(\delta_i^{k-1}\right)^2}{k-1} \tag{11}$$

Then, the up and down bounds of global angle estimate range can be given by:

$$\hat{\alpha}_i^{k,H}(k) = \hat{\alpha}_i^k(k) + \delta_i^k,\ \hat{\alpha}_i^{k,L}(k) = \hat{\alpha}_i^k(k) - \delta_i^k \tag{12}$$

The overlapped area of the global angle estimate ranges of the two sensor arrays contains the fire source with large probability, as shown in Figure 2. In [29], a dynamic angle bisector method was proposed to determine a circular fire source arrange, the circle center and radius of which were calculated on the basis of this overlapped area, which is a quadrangle. Actually, using the statistical features of the global data to obtain the quadrangle which contains the fire source with large probability is an excellent data compression method, which can also be used to reduce the computing complexity of the dynamic optimization problem in the proposed fire source range location method in Section 3.1.

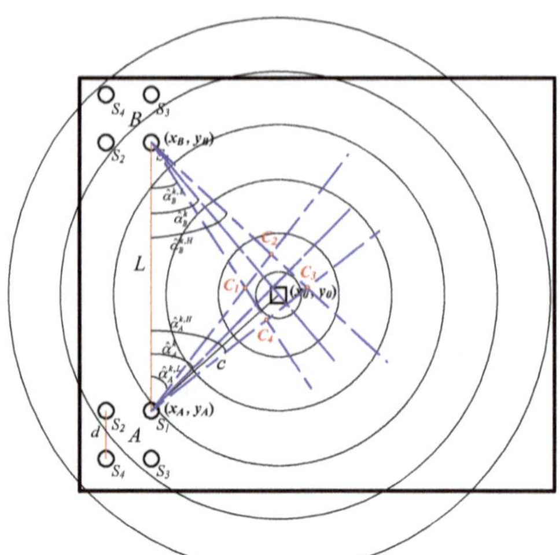

Figure 2. The fire source arrangement localization principle in the RRR frame.

As show in Figure 2, set the reference vertices' coordinates of two sensor arrays as (x_A, y_A), (x_B, y_B), then the four vertices $C_j(c_j^k, d_j^k)$, $(j = 1,2,3,4)$ of the quadrangle overlapped by the two global angle estimate ranges can be solved by [29]:

$$\begin{cases} c_1^k = \frac{\tan(\hat{\alpha}_A^{k,L})x_1 - \tan(\hat{\alpha}_B^{k,H})x_2 - y_1 + y_2}{\tan(\hat{\alpha}_A^{k,L}) - \tan(\hat{\alpha}_B^{k,H})}; d_1^k = \frac{y_2/\tan(\hat{\alpha}_B^{k,H}) - y_1/\tan(\hat{\alpha}_A^{k,L}) - x_2 + x_1}{1/\tan(\hat{\alpha}_B^{k,H}) - 1/\tan(\hat{\alpha}_A^{k,L})} \\ c_2^k = \frac{\tan(\hat{\alpha}_A^{k,H})x_1 - \tan(\hat{\alpha}_B^{k,H})x_2 - y_1 + y_2}{\tan(\hat{\alpha}_A^{k,H}) - \tan(\hat{\alpha}_B^{k,H})}; d_2^k = \frac{y_2/\tan(\hat{\alpha}_B^{k,H}) - y_1/\tan(\hat{\alpha}_A^{k,H}) - x_2 + x_1}{1/\tan(\hat{\alpha}_B^{k,H}) - 1/\tan(\hat{\alpha}_A^{k,H})} \\ c_3^k = \frac{\tan(\hat{\alpha}_A^{k,L})x_1 - \tan(\hat{\alpha}_B^{k,L})x_2 - y_1 + y_2}{\tan(\hat{\alpha}_A^{k,L}) - \tan(\hat{\alpha}_B^{k,L})}; d_3^k = \frac{y_2/\tan(\hat{\alpha}_B^{k,L}) - y_1/\tan(\hat{\alpha}_A^{k,L}) - x_2 + x_1}{1/\tan(\hat{\alpha}_B^{k,L}) - 1/\tan(\hat{\alpha}_A^{k,L})} \\ c_4^k = \frac{\tan(\hat{\alpha}_A^{k,H})x_1 - \tan(\hat{\alpha}_B^{k,L})x_2 - y_1 + y_2}{\tan(\hat{\alpha}_A^{k,H}) - \tan(\hat{\alpha}_B^{k,L})}; d_4^k = \frac{y_2/\tan(\hat{\alpha}_B^{k,L}) - y_1/\tan(\hat{\alpha}_A^{k,H}) - x_2 + x_1}{1/\tan(\hat{\alpha}_B^{k,H}) - 1/\tan(\hat{\alpha}_A^{k,L})} \end{cases} \quad (13)$$

On this basis, the fire localization problem can be formulated to the following dynamic optimization problem, in which the circle center and radius of the circular fire region to be determined can be solved simultaneously:

$$\begin{gathered} \min r^2 \\ s.t. \ (x_0 - c_j^k)^2 + (y_0 - d_j^k)^2 \le r^2 (j = 1,2,3,4) \end{gathered} \quad (14)$$

where (x_0, y_0) and r are the center's coordinate to be solved, respectively.

It should be noted that Equation (14) is only one of many optimization formulations and not the one and only. It is noted that Equation (14) is a non-convex optimization problem and it is difficult to solve. Thereby, in order to successfully solve this non-convex optimization, there is an available method to transform it to a set of convex optimization problems. In fact, the non-convex optimization problem given by Equation (14) can be expressed equivalently as the following set of four convex optimization problems, namely:

$$\begin{gathered} \min (x_0 - c_1^k)^2 + (y_0 - d_1^k)^2 \\ s.t. \ (x_0 - c_j^k)^2 + (y_0 - d_j^k)^2 \le (x_0 - c_1^k)^2 + (y_0 - d_1^k)^2 (j = 2,3,4) \end{gathered} \quad (15)$$

$$\begin{gathered} \min (x_0 - c_2^k)^2 + (y_0 - d_2^k)^2 \\ s.t. \ (x_0 - c_j^k)^2 + (y_0 - d_j^k)^2 \le (x_0 - c_2^k)^2 + (y_0 - d_2^k)^2 (j = 1,3,4) \end{gathered} \quad (16)$$

$$\begin{gathered} \min (x_0 - c_3^k)^2 + (y_0 - d_3^k)^2 \\ s.t. \ (x_0 - c_j^k)^2 + (y_0 - d_j^k)^2 \le (x_0 - c_3^k)^2 + (y_0 - d_3^k)^2 (j = 1,2,4) \end{gathered} \quad (17)$$

$$\begin{gathered} \min (x_0 - c_4^k)^2 + (y_0 - d_4^k)^2 \\ s.t. \ (x_0 - c_j^k)^2 + (y_0 - d_j^k)^2 \le (x_0 - c_4^k)^2 + (y_0 - d_4^k)^2 (j = 1,2,3) \end{gathered} \quad (18)$$

Denote the optimal solutions of Equations (15)–(18) as $(x_{0,j}^k, y_{0,j}^k)$, $(j = 1,2,3,4)$, and set:

$$r_{0,j}^k = \sqrt{(x_{0,j}^k - c_1^k)^2 + (y_{0,j}^k - d_1^k)^2}, j = 1,2,3,4 \quad (19)$$

Then, the optimal global solution on the radius of the circle to be determined is:

$$r_0^k = \min\left\{r_{0,1}^k, r_{0,2}^k, r_{0,3}^k, r_{0,4}^k\right\} = r_{0,p}^k$$

where $p \in \{1,2,3,4\}$. The corresponding estimate of center's coordinate of the expected circle is:

$$\hat{x}_0^k = x_{0,p}^k, \hat{y}_0^k = y_{0,p}^k \quad (20)$$

Therefore, in the RRR frame, a fire source localization arrangement covering the fire source is presented with the dynamic optimization theory and the data compression method based on the statistical features of the global data.

Remark 2. *Although the set of four convex optimization problems shown in Equations (15)–(18) can equivalently transform the non-convex optimization problem of Equation (14), but it cannot be directly solved with the CVX software package, which requires that constraints must be in the most simplified form. Thus, the optimization problems in (15)–(18) should be equivalently expressed as:*

$$\begin{gathered}\min\,(x_0-c_1^k)^2+(y_0-d_1^k)^2\\ s.t.\,2x_0(c_1^k-c_j^k)+2y_0(d_1^k-d_j^k)^2\le(c_1^k)^2+(d_1^k)^2-(c_j^k)^2-(d_j^k)^2(j=2,3,4)\end{gathered}\tag{21}$$

$$\begin{gathered}\min\,(x_0-c_2^k)^2+(y_0-d_2^k)^2\\ s.t.\,2x_0(c_2^k-c_j^k)+2y_0(d_2^k-d_j^k)^2\le(c_2^k)^2+(d_2^k)^2-(c_j^k)^2-(d_j^k)^2(j=1,3,4)\end{gathered}\tag{22}$$

$$\begin{gathered}\min\,(x_0-c_3^k)^2+(y_0-d_3^k)^2\\ s.t.\,2x_0(c_3^k-c_j^k)+2y_0(d_3^k-d_j^k)^2\le(c_3^k)^2+(d_3^k)^2-(c_j^k)^2+(d_j^k)^2(j=1,2,4)\end{gathered}\tag{23}$$

$$\begin{gathered}\min\,(x_0-c_4^k)^2+(y_0-d_4^k)^2\\ s.t.\,2x_0(c_4^k-c_j^k)+2y_0(d_4^k-d_j^k)^2\le(c_4^k)^2+(d_4^k)^2-(c_j^k)^2+(d_j^k)^2(j=1,2,3)\end{gathered}\tag{24}$$

which can be directly solved with the CVX software package.

Remark 3. *Theoretically speaking, the angle bisector method in [29] subjectively fixes the circle center and the radius of the circular fire source arrangement to be determined, while the dynamic optimization method looks more reasonable because it can obtain a dynamic optimal solution of the estimate of circle center and the radius which, in the meantime, is also global. Clearly, there is a question that the subjective circle center cannot theoretically be guaranteed to be reasonable and perfect. The optimization problem in this section can be taken as a standard to evaluate the effectiveness of the angle bisector method. It should be noted that the circle center and the radius are estimated, respectively, although they have a very close connection. In fact, the circle center can be considered as a solution of the optimization problem that the minimum distance from a point to be determined in the quadrangle to cover all the four vertices, and the circum-circle radius estimated by the angle bisector method is the shortest one of the four distances from the vertices to the determined circle center, but not the distance solved by the last optimization problem. Therefore, there are some areas in the quadrangle that cannot be covered by the circum-circle estimated by the angle bisector method. For the dynamic optimal method in the RRR frame, the circle center and radius are estimated simultaneously in a consistent optimization standard. Thus, the circum-circle determined by the estimated circle center and radius cover the whole quadrangle. Therefore, the output of the dynamic optimization localization method in the RRR frame is another kind of circum-circle, which covers the whole quadrangle. It differs from the output of the angle bisector method with the circum-circle.*

Remark 4. *Similar to the angle bisector method in [29], the dynamic optimal localization method in the RRR frame is sensitive to the statistical character of the measurement data on the fire position. For cases where the mean of the fire source point estimates are not zero, the location performances will be reduced to some extent. Nevertheless, the dynamic optimization localization method in the RPR frame has stronger robustness for the change of statistical characters of the measurement data.*

4. Simulation

In this section, a numerical simulation with three different simulation scenes are provided to prove the effectiveness and availability of the two proposed dynamic optimal localization methods. The simulation settings are similar to those set in [12,29], as shown in Tables 1 and 2.

No matter the localization methods designed in the RPR frame or the RRR frame, the processes of delay time estimation and angle estimation are coincident, even for the principle of fire source point estimation. Therefore, the simulation of angle estimates do not influence the validity of the simulation results. The angle estimate simulation method in [29] is adopted to simplify the simulation process.

Namely, several fire source point estimates are simulated firstly, and then the angle estimates can be computed according to the triangle principle.

Table 1. The simulation settings.

Simulation Background	Simulation Setting
The length of the buildings	11 m (meters)
The width of the buildings	11 m
the real point of fire source	(5 m, 5 m)
The distance between the two temperature sensor arrays	$L = 9$ m
The distance between the two sensors in a array	$d_0 = 0.1$ m
The coordinate of the reference node of sensor array A	(0.5 m, 1 m)
The coordinate of the reference node of sensor array B	(1 m, 10 m)
The sampling frequency	500 Hz

Table 2. The abbreviations of the algorithms used in this section.

The Algorithms	Abbreviations
The angle bisector method with the circum-circle	Algorithm A
The dynamic optimization localization method in the RPR frame	Algorithm B1
The dynamic optimization localization method in the RRR frame	Algorithm B2
The localization method based on VB-ASCKF in the RRR frame	Algorithm C
The localization method in the RRR frame with clustering technology	Algorithm D

In order to show the effectiveness and robustness of the proposed fire localization methods, they are compared with three other kinds of fire location methods in three different fire source localization simulation scenes.

Simulation Scene 1. The fire source point estimates are simulated based on the real fire coordinate by adding Gaussian noise. The estimation results of Algorithm A and Algorithm B1 are compared in Figure 3.

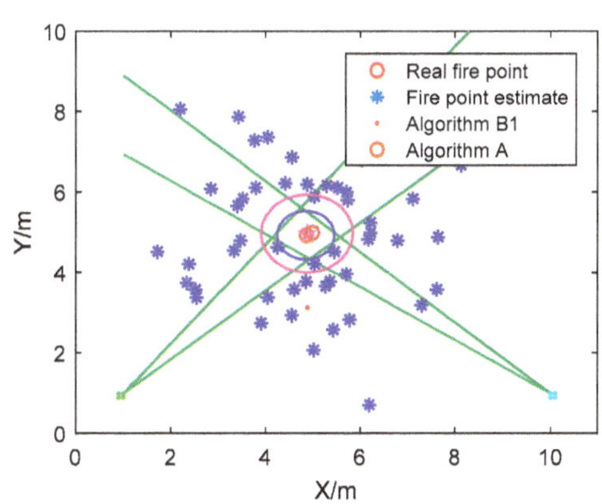

Figure 3. The circular fire source arrangement solved by Algorithm A and Algorithm B1.

Although Algorithm A and Algorithm B1 can both be implemented with global data to obtain the circular fire source arrangement estimation, the estimated fire source arrangements are different. This is because the standard to guide the algorithm is different, which has been analyzed with detail in Remark 3. In addition, in Algorithm B1, it is subjective to take the standard deviation of the distances from the circle center to every fire source point estimate as the radius of the circular range to be determined. Thus, the circular fire source arrangement determined by Algorithm B1 is greater than

Algorithm A, and the circle centers of the real fire point and its estimates by Algorithms A and B1 are shown in Table 3.

Table 3. The circle centers of real fire point and its estimates by Algorithms A and B1.

The Real Fire Point	Algorithm A	Algorithm B1
(5, 5)	(4.849, 4.935)	(4.873, 4.98)

Simulation Scene 2. The fire source point estimates are simulated based on the real fire coordinate by adding Gaussian noise. The estimation results of Algorithm A and Algorithm B2 are compared in Figure 4.

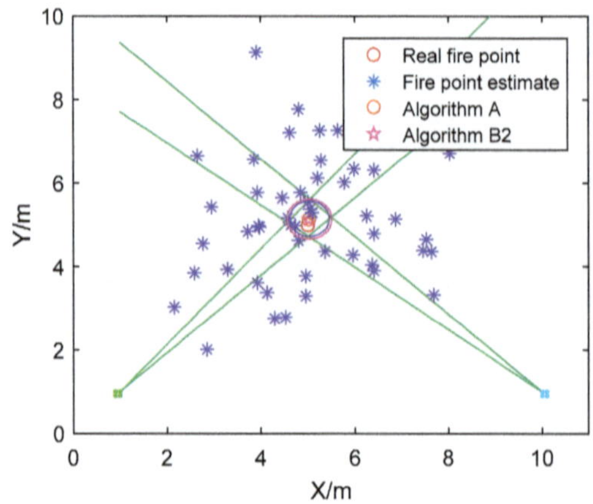

Figure 4. The circular fire source arrangement solved by Algorithm A and Algorithm B2.

As illustrated in Figure 4, the circle centers of the circular fire source arrangement estimated by Algorithm A and Algorithm B2 are very similar, as shown in Table 4, the relative error is $O(10^{-2})$. Although the circles estimated by Algorithm A and Algorithm B2 are both the circum-circle of the quadrangle, the circum-circle obtained by Algorithm B2 can cover the whole quadrangle, while the one solved by Algorithm B2 cannot. The reason for this result has been analyzed in Remark 3.

Table 4. The circle centers of real fire points and its estimates by Algorithms A and B2.

The Real Fire Point	Algorithm A	Algorithm B2
(5, 5)	(5.027, 5.144)	(5.027, 5.133)

Simulation Scene 3. The fire source point estimates are simulated based on the real fire coordinate by adding random mixed noise combined with Gaussian noise, Rayleigh noise, and sine wave signals. As shown in Remark 4, the dynamic optimal localization method in the RRR frame and the angle bisector method are sensitive to the statistical characters of the measurement data. Thus, the localization results are only compared by three robust localization methods: with the dynamic optimization localization method in the RPR frame, the localization method in the RPR frame with clustering technology, and the VB-ASCK method.

The simulation results given in Figure 5 are the average results of 100 Monte Carlo simulations. It is inevitable that the data from several measurements cannot be utilized to estimate the circle center and radius of the circular fire source arrangement. Therefore, the other two localization methods, with global measurement data, have better accuracy than Algorithm D. In Algorithm C, the mixed noise is taken as Gaussian noise with unknown covariance. Although the specific variational Bayesian technology is used to adaptively estimate the unknown covariance, the estimation accuracy of Algorithm C is worse than that of the dynamic optimization localization method in the RPR frame. As shown in Table 5, the mean estimation error of Algorithm D is 1.7, while that of Algorithm C is 1.2 and that of Algorithm B1 is 0.9.

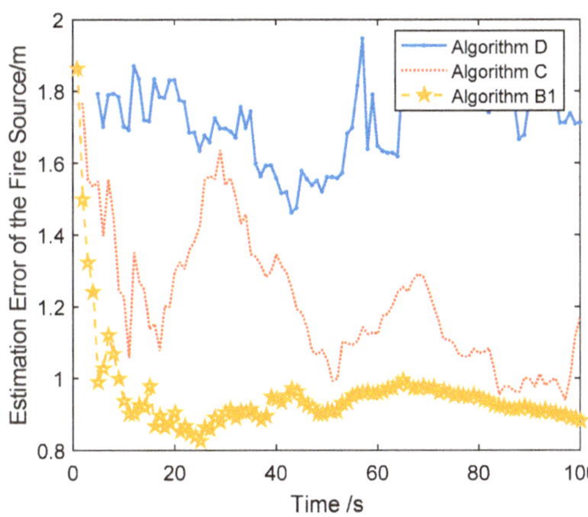

Figure 5. The estimation error curves of the fire source by the Algorithms B1, C, and D.

Table 5. Mean estimation error of the fire source by the Algorithms B1, C, and D.

The Algorithms	Algorithm B1	Algorithm C	Algorithm D
Mean estimation error	0.9	1.2	1.7

5. Conclusions

In this paper, two circular fire source arrangement localization methods are proposed on the basis of the dynamic optimization technology. In the RPR frame, the dynamic optimization localization method is global, but not recursive, so the computing complexity of the dynamic optimization process is increasingly poor. Inspired by the data compression technology in the angle bisector method, a dynamic optimization localization method is developed by solving a non-convex optimization problem. The non-convex optimization problem can be treated as a standard to evaluate the dynamic angle bisector method. It should be noted that the circle radius is subjectively obtained by the dynamic optimization localization method. How to build an optimal index to solve the circle radius and the circle center optimally and synchronously is still an open topic for future research.

Author Contributions: G.W. and X.F. designed the fire source location method; G.W. and Z.Z. implemented the experiments and analyzed the experimental data; and X.F. and Z.Z. wrote and modified the paper.

Funding: This work was supported by the National Natural Science Foundation of China (grant nos. 61503174, 61333005, and 61304258), the Natural Science Foundation of Henan Province of China (grant nos.162102210196 and 15A413011), and the Scientific Research Foundation of Henan University of Technology (grant no. 2015RCJH14).

Conflicts of Interest: The authors declare no conflict of interest.

References

1. Cram, D.; Hatch, C.; Tyler, S.; Ochoa, C. Use of Distributed Temperature Sensing Technology to Characterize Fire Behavior. *Sensors* **2016**, *16*, 1712. [CrossRef] [PubMed]

2. Lu, Y.; Wang, H.; Wang, K. Fire Location for High and Large-Span Space Buildings based on Binocular Stereo Vision. *Int. J. Smart Home* **2016**, *10*, 271–282.
3. Sun, Y.; Huang, H.; Li, F.; Yu, D.; Hao, C. Research on Zoning Control of Fire in Large Space Building. *J. Wuhan Univ. Technol.* **2017**, *2*, 144–147.
4. Mao, Y.; Wang, H.; Lu, Y.; Qin, L. Precise Localisation Method for Fire Sources in Large-space Buildings. *Comput. Appl. Softw.* **2016**, *2*, 169–172.
5. Ying, L.; Wang, H.; Qin, L. Accurate fire location method in high and large-span space buildings. *J. Jilin Univ.* **2016**, *6*, 2067–2073.
6. Zhang, Q.; Yang, X. Numerical Simulation and Detection Response Analysis of Fire in a Large Space Building. In Proceedings of the International Conference on Intelligent Computation Technology and Automation, Changsha, China, 25–26 October 2014; pp. 388–391.
7. Song, T.; Tang, B.; Zhao, M.; Deng, L. An accurate 3-D fire location method based on sub-pixel edge detection and non-parametric stereo matching. *Measurement* **2014**, *4*, 160–171. [CrossRef]
8. Hackner, A.; Oberpriller, H.; Ohnesorge, A.; Hechtenberg, V.; Müller, G. Heterogeneous sensor arrays: Merging cameras and gas sensors into innovative fire detection systems. *Sens. Actuators B Chem.* **2016**, *231*, 497–505. [CrossRef]
9. Song, H.; Lee, Y. A Development of Fire Detective Control System using the Fusion of a CCD Image and an Infrared Image. *J. Korean Inst. Illum. Elec. Install. Eng.* **2017**, *5*, 60–66. [CrossRef]
10. Jiang, M.; Yao, B.; Cheng, Y. Response characteristics of distributed optical fiber temperature testing system in highway tunnel fire detection. *Fire Saf. Sci.* **2013**, *4*, 194–200.
11. Feng, R. Optical Fiber Grating Temperature Sensing Tunnel Fire Alarm System. *Trans. Sci. Technol.* **2009**, *1*, 89–91.
12. Sun, M.; Tang, Y.; Yang, S.; Dong, F. Fiber optic distributed temperature sensing for fire source localization. *Meas. Sci. Technol.* **2017**, *8*, 085102. [CrossRef]
13. Sun, M.; Tang, Y.; Yang, S.; Li, J.; Sigrist, M.W.; Dong, F. Fire Source Localization Based on Distributed Temperature Sensing by a Dual-Line Optical Fiber System. *Sensors* **2016**, *16*, 829. [CrossRef] [PubMed]
14. Chen, X.; Li, K.; Li, J.; Hong, Y. Security technology research on high-temperature fire area blasting in surface mine. *Opencast Min. Technol.* **2010**, *4*, 73–75.
15. Martínez-de Dios, J.R.; Merino, L.; Caballero, F.; Ollero, A. Automatic Forest-Fire Measuring Using Ground Stations and Unmanned Aerial Systems. *Sensors* **2011**, *11*, 6328–6353. [CrossRef] [PubMed]
16. Shan, Y.; Ma, Y.; Fu, H.; Li, W.; Wang, C. Application of distributed optical fiber temperature measurement system in coal mine fire monitoring system. *Chin. J. Sens. Actuators* **2014**, *5*, 22–25.
17. Noordin, N.; Ney, H. Localization in wireless sensor network for forest fire detection. In Proceedings of the 3rd IEEE International Symposium on Telecommunication Technologies, Kuala Lumpur, Malaysia, 28–30 November 2016; pp. 87–90.
18. Berentsen, M.; Kaiser, T. Fire Location Estimation using Temperature Sensor Arrays. In Proceedings of the 12th International Conference on Automatic Fire Detection, Gaithersburg, MD, USA, 25 March 2001; pp. 432–443.
19. Wang, S.; Berentsen, M.; Kaiser, T. Signal processing algorithms for fire localization using temperature sensor arrays. *Fire Saf. J.* **2005**, *8*, 689–697. [CrossRef]
20. Stern-Gottfried, J.; Rein, G.; Bisby, L.; Torero, J.L. Experimental review of the homogeneous temperature assumption in post-flashover compartment fires. *Fire Saf. J.* **2010**, *4*, 249–261. [CrossRef]
21. Duan, S.; Cheng, K.; Ge, Q.; Wen, C. Wireless Intelligent Fire Fighting Systems Software Platform R&D. In Proceedings of the 2011 International Conference on Internet of Things and International Conference on Cyber, Physical and Social Computing, Dalian, China, 19–22 October 2011; pp. 94–99.
22. Duan, S.; Ge, Q.; Wen, C. New method for fire source clustering localization based on sensor arrays. *J. Nanjing Univ. Sci. Technol.* **2011**, *35*, 59–65.
23. Zhuang, Z.; Zhang, X.; Kalin, L. Method Research on Early Fire Source Localization Based on Planar Circular Sensor Arrays. *Chin. J. Sens. Actuators* **2009**, *8*, 1208–1212.
24. Ge, Q.; Wen, C.; Duan, S. Fire Localization Based on Range-Range-Range Model for Limited Interior Space. *IEEE Trans. Instrum. Meas.* **2014**, *9*, 2223–2237. [CrossRef]
25. Liu, H.; Darabi, H.; Banerjee, P.; Liu, J. Survey of Wireless Indoor Positioning Techniques and Systems. *IEEE Trans. Syst. Man Cybern. Part C* **2007**, *37*, 1067–1080. [CrossRef]

26. He, S.; Hu, T.; Chan, S.H.G. Contour-based Trilateration for Indoor Fingerprinting Localization. In Proceedings of the 2015 ACM Conference on Embedded Networked Sensor Systems, Seoul, Korea, 1–4 November 2015; pp. 225–238.
27. Sun, G.; Chen, J.; Guo, W.; Liu, K.R. Signal processing techniques in network-aided positioning: A survey of state-of-the-art positioning designs. *IEEE Signal Process. Mag.* **2005**, *22*, 12–23.
28. Fonollosa, J.; Solórzano, A.; Marco, S. Chemical Sensor Systems and Associated Algorithms for Fire Detection: A Review. *Sensors* **2018**, *18*, 553. [CrossRef] [PubMed]
29. Sawada, A.; Higashino, T.; Oyabu, T.; Takei, Y.; Nanto, H.; Toko, K. Gas sensor characteristics for smoldering fire caused by a cigarette smoke. *Sens. Actuators B Chem.* **2008**, *130*, 88–93. [CrossRef]

sensors

Article

Data Fusion Architectures for Orthogonal Redundant Inertial Measurement Units

Eric Gagnon [1,*], **Alexandre Vachon** [2] **and Yanick Beaudoin** [3]

1 Defence Research and Development Canada, Quebec, QC G3J 1X5, Canada
2 Numérica Technologies Inc., Quebec, QC G2E 4P8, Canada; Alexandre.Vachon.NUMERICA@drdc-rddc.gc.ca
3 Département de génie électrique et génie informatique, Université Laval, Quebec, QC G1V 0A6, Canada; yanick.beaudoin.1@ulaval.ca
* Correspondence: Eric.Gagnon@drdc-rddc.gc.ca; Tel.: +1-418-844-4000 (ext. 4908)

Received: 18 May 2018; Accepted: 8 June 2018; Published: 12 June 2018

Abstract: This work looks at the exploitation of large numbers of orthogonal redundant inertial measurement units. Specifically, the paper analyses centralized and distributed architectures in the context of data fusion algorithms for those sensors. For both architectures, data fusion algorithms based on Kalman filter are developed. Some of those algorithms consider sensors location, whereas the others do not, but all estimate the sensors bias. A fault detection algorithm, based on residual analysis, is also proposed. Monte-Carlo simulations show better performance for the centralized architecture with an algorithm considering sensors location. Due to a better estimation of the sensors bias, the latter provides the most precise and accurate estimates and the best fault detection. However, it requires a much longer computational time. An analysis of the sensors bias correlation is also done. Based on the simulations, the biases correlation has a small effect on the attitude rate estimation, but a very significant one on the acceleration estimation.

Keywords: orthogonal redundant inertial measurement units; data fusion architectures; sensors bias

1. Introduction

In the beginning of the 90s, a novel sensor design technique changed the manufacturing and operation of inertial navigation systems (INS). Micro-electro-mechanical systems (MEMS) integrate the classical mechanical design in the integrated circuit fabrication technology. This new class of small inertial measurement units (IMUs) can then be fabricated in large batches, drastically reducing the production costs. However, their smaller size is problematic; a smaller sensor is more sensitive to temperature and has a smaller scale factor, resulting in a lower signal-to-noise ratio than classical IMUs. On the other hand, the MEMS' small size and low cost make them suitable to build arrays of redundant sensors for fused measurements objective. The combination of inertial sensors in arrays has therefore become an important field of research lately [1].

The most common purpose for redundant inertial measurement units (RIMUs) in navigation systems is to facilitate the detection and isolation of faulty sensors [2]. The precision and accuracy improvement provided by using multiple measurements is often a secondary objective. In this optic, the studied sensors are, more than often, positioned in a non-orthogonal skewed configuration, known as SRIMU, or on regular polyhedrons. In a SRIMU, the sensors are distributed on a cone, in a way that there is a constant angle between two consecutive sensors [3]. This configuration provides the maximum redundancy and, hence, encapsulates the maximum amount of information for a specific number of IMUs [4]. The regular polyhedrons are, on their parts, the optimal configuration for navigation and fault detection and isolation (FDI) [5].

However, these configurations are interesting only for a relatively low number of redundancies and for costly sensors [6]. When dealing with a large number of IMUs, these optimal configurations are very complex and the interest of lowering the cost of the redundancy by using the minimal number of IMUs is overshadowed by the cost due to the complexity of the assembly. In this case, an orthogonal configuration, where triads of sensors are mounted parallel one to the others, can be interesting. This configuration is much easier to assemble, but, in comparison to the SRIMU and polyhedrons configurations, has poor FDI performance [6]. In order to provide the same level of robustness (same number of detected faults), a higher number of IMUs is required in the orthogonal configuration. However, for RIMU based on MEMS, which are relatively cheap and small, this is not an issue. As a result, the orthogonal configuration should be seriously considered for those sensors [6].

The handling of these redundant measurements can be categorised in two types: centralized or distributed. In the centralized architecture, all sensor measurements are directly processed by a single central data fusion filter in order to obtain the state estimates. This architecture is the optimal [7] and most common design in current INS [8], even though it is not necessarily the most efficient one. With multiple sensors of different types, the centralized filter computation might be time-consuming [8], and it is not the ideal method for a fault tolerant multi-sensors scheme [9]. However, when only IMUs are used, the centralized architecture remains time-consuming, but does not exhibit FDI sub-performance [2].

On its part, the distributed architecture is a multi-stage architecture where parallel local filters independently process the measurements of their corresponding sensors and their outputs are used by a master data fusion filter which produces the final state estimates. Also, the master filter can send back information to the local filters in order to reset some components of their internal state. As the centralized architecture, this architecture can offer optimal performance, but only when all filters model the full state and when they are all ran at the same rate [7]. The main advantage of this architecture is its better FDI performance [9]. This architecture has been applied to multiple types of multi-sensors navigation systems [9,10] and to multiple IMUs systems [2,4].

For vehicle navigation, the fusion of IMUs is mainly studied as part of navigation algorithm aided by Global Positioning System (GPS). However, some works [11–13] discard the GPS and rely on alternative (radar or vision) measurement devices, while others rely purely on inertial sensors, accelerometers only [14,15] or accelerometers and gyroscopes [16].

Considering the fictitious forces due to sensors location is not common in RIMU data fusion algorithms, but has been done [2,16]. Similarly, estimating the sensors bias within the algorithms has been considered [17–19]. It is also the case for the analysis of correlation between the sensor random walk noises [20,21], but not a correlation between the biases of the sensors.

This paper intends to simultaneously address those three aspects (sensors locations, biases estimation and biases correlation) within navigation based on IMUs only. Above all, the paper studies their effects on each type of data fusion architecture. The main objective is the identification of the ideal architecture for data fusion of a large number of orthogonal RIMU. For the two studied architectures, data fusion algorithms based on Kalman filter are used (Section 2). Those algorithms, notably, estimate the sensors bias. For both architectures, an algorithm which considers sensors location, and one which does not, are proposed. A baseline algorithm, averaging the measurements, is used as a comparison point. A fault detection and isolation algorithm, based on residual statistical analysis, is also presented (Section 3). Those algorithms are then compared using Monte-Carlo simulations (Section 4). The simulated application consists in an artillery shell (spin stabilized projectile) equipped with strapdown orthogonal RIMU having, as a first case, uncorrelated biases and, as a second case, correlated ones. A simulations-based analysis is mandatory for this application; it would have been impossible to test the same amount of sensors locations, noises, biases and faults in the field.

2. Data Fusion Algorithms

This section proposes data fusion algorithms based on Kalman filter for MEMS RIMU. MEMS sensors are characterized by large random walk and bias instability [22]. The random walk is, by nature, a white noise which can be efficiently processed by a Kalman filter, but the bias instability is a second order random walk [22] and requires a specific treatment in order to be estimated by a Kalman filter.

Obviously, it is impossible, even with MEMS, to have perfectly collocated sensors. The distance between the IMUs can be very small and much smaller than the sensitivity of the sensors, but this distance still influence the quality of the estimations. The Grubin transformation [23], which expresses the fictitious forces applied on a point not located at the centre of mass of a body, is used to include the effect of the angular motion on the accelerometer measurements by taking into account the sensors location in the models. A triad of accelerometers, located at the sensor position ($\mathbf{s}_{sb}$), then measures three accelerations: the nominal linear acceleration ($\mathbf{a}$), the angular acceleration ($\dot{\boldsymbol{\omega}} \times \mathbf{s}_{sb}$) and the centrifugal acceleration ($\boldsymbol{\omega} \times (\boldsymbol{\omega} \times \mathbf{s}_{sb})$). On its part, the gyroscopes triad is not affected by its location and still measures the angular velocity ($\boldsymbol{\omega}$) only.

For the centralized and distributed architectures, the Kalman filter prediction model required to consider the sensors location will be presented. Then, by eliminating the fictitious forces, a model for an algorithm not considering sensors location will be derived. Similarly, for the baseline architecture, the algorithm discarding sensors location is derived from the algorithm considering sensors location.

2.1. Centralized Architecture

In a typical GPS/INS navigation solution, the coupling between the acceleration and attitude rate measurements appears, through the rotation matrix, in the velocity and position vectors [24]. However, without GPS receiver, the velocity, position and attitude are unobservable, and can, without performance degradation, be eliminated from the data fusion algorithm in order to decrease its complexity. For the centralized architecture, the filter state is then composed of the acceleration estimates ($\mathbf{a}_e$), the attitude rate estimates ($\boldsymbol{\omega}_e$) and the biases of all accelerometers ($\mathbf{b}_a$) and of all gyroscopes ($\mathbf{b}_\omega$):

$$\mathbf{x} = \begin{bmatrix} \mathbf{a}_e & \mathbf{b}_{a1} & \cdots & \mathbf{b}_{an} & \boldsymbol{\omega}_e & \mathbf{b}_{\omega 1} & \cdots & \mathbf{b}_{\omega n} \end{bmatrix}^T \tag{1}$$

On its part, the state propagation equation, which is linear, is:

$$\mathbf{x}_{k+1} = \begin{bmatrix} \mathbf{a}_e \\ \mathbf{b}_{a1} \\ \vdots \\ \mathbf{b}_{an} \\ \boldsymbol{\omega}_e \\ \mathbf{b}_{\omega 1} \\ \vdots \\ \mathbf{b}_{\omega n} \end{bmatrix}_{k+1} = \begin{bmatrix} \mathbf{I}_3 & \mathbf{0}_3 & \cdots & \mathbf{0}_3 & \mathbf{0}_3 & \mathbf{0}_3 & \cdots & \mathbf{0}_3 \\ \mathbf{0}_3 & \mathbf{I}_3 & \cdots & \mathbf{0}_3 & \mathbf{0}_3 & \mathbf{0}_3 & \cdots & \mathbf{0}_3 \\ \vdots & \vdots & \ddots & \vdots & \vdots & \vdots & \ddots & \vdots \\ \mathbf{0}_3 & \mathbf{0}_3 & \cdots & \mathbf{I}_3 & \mathbf{0}_3 & \mathbf{0}_3 & \cdots & \mathbf{0}_3 \\ \mathbf{0}_3 & \mathbf{0}_3 & \cdots & \mathbf{0}_3 & \mathbf{I}_3 & \mathbf{0}_3 & \cdots & \mathbf{0}_3 \\ \mathbf{0}_3 & \mathbf{0}_3 & \cdots & \mathbf{0}_3 & \mathbf{0}_3 & \mathbf{I}_3 & \cdots & \mathbf{0}_3 \\ \vdots & \vdots & \ddots & \vdots & \vdots & \vdots & \ddots & \vdots \\ \mathbf{0}_3 & \mathbf{0}_3 & \cdots & \mathbf{0}_3 & \mathbf{0}_3 & \mathbf{0}_3 & \cdots & \mathbf{I}_3 \end{bmatrix} \mathbf{x}_k + \begin{bmatrix} \mathbf{v}_a \\ \mathbf{v}_{ba1} \\ \vdots \\ \mathbf{v}_{ban} \\ \mathbf{v}_\omega \\ \mathbf{v}_{b\omega 1} \\ \vdots \\ \mathbf{v}_{b\omega n} \end{bmatrix}_k \tag{2}$$

and the nonlinear output equation is:

$$y_k = \begin{bmatrix} a_{m1} \\ a_{m2} \\ \vdots \\ a_{mn} \\ \omega_{m1} \\ \omega_{m2} \\ \vdots \\ \omega_{mn} \end{bmatrix}_{k+1} = h(x_k) = \begin{bmatrix} a_e + \dot{\omega}_e \times \mathbf{s_{sb1}} + \omega_e \times (\omega_e \times \mathbf{s_{sb1}}) + \mathbf{b_{a1}} \\ a_e + \dot{\omega}_e \times \mathbf{s_{sb2}} + \omega_e \times (\omega_e \times \mathbf{s_{sb2}}) + \mathbf{b_{a2}} \\ \vdots \\ a_e + \dot{\omega}_e \times \mathbf{s_{sbn}} + \omega_e \times (\omega_e \times \mathbf{s_{sbn}}) + \mathbf{b_{an}} \\ \omega_e + \mathbf{b_{\omega 1}} \\ \omega_e + \mathbf{b_{\omega 2}} \\ \vdots \\ \omega_e + \mathbf{b_{\omega n}} \end{bmatrix} + \begin{bmatrix} w_{a1} \\ w_{a2} \\ \vdots \\ w_{an} \\ w_{\omega 1} \\ w_{\omega 2} \\ \vdots \\ w_{\omega n} \end{bmatrix}_k \tag{3}$$

The Kalman filter state propagation matrix is the identity matrix of Equation (2), while its output matrix is the Jacobian of Equation (3):

$$\mathbf{C}_k = \left.\frac{\partial h}{\partial x}\right|_k = \begin{bmatrix} \mathbf{I}_3 & \mathbf{I}_3 & \mathbf{0}_3 & \cdots & \mathbf{0}_3 & \mathbf{a}_{\omega_1} & \mathbf{0}_3 & \mathbf{0}_3 & \cdots & \mathbf{0}_3 \\ \mathbf{I}_3 & \mathbf{0}_3 & \mathbf{I}_3 & \cdots & \mathbf{0}_3 & \mathbf{a}_{\omega_2} & \mathbf{0}_3 & \mathbf{0}_3 & \cdots & \mathbf{0}_3 \\ \vdots & \vdots & \vdots & \ddots & \vdots & \vdots & \vdots & \vdots & \ddots & \vdots \\ \mathbf{I}_3 & \mathbf{0}_3 & \mathbf{0}_3 & \cdots & \mathbf{I}_3 & \mathbf{a}_{\omega_n} & I_3 & \mathbf{0}_3 & \cdots & \mathbf{0}_3 \\ \mathbf{0}_3 & \mathbf{0}_3 & \mathbf{0}_3 & \cdots & \mathbf{0}_3 & \mathbf{I}_3 & \mathbf{I}_3 & \mathbf{0}_3 & \cdots & \mathbf{0}_3 \\ \mathbf{0}_3 & \mathbf{0}_3 & \mathbf{0}_3 & \cdots & \mathbf{0}_3 & \mathbf{I}_3 & \mathbf{0}_3 & \mathbf{I}_3 & \cdots & \mathbf{0}_3 \\ \vdots & \vdots & \vdots & \ddots & \vdots & \vdots & \vdots & \cdots & \ddots & \mathbf{0}_3 \\ \mathbf{0}_3 & \mathbf{0}_3 & \mathbf{0}_3 & \cdots & \mathbf{0}_3 & \mathbf{I}_3 & \mathbf{0}_3 & \mathbf{0}_3 & \cdots & \mathbf{I}_3 \end{bmatrix} \tag{4}$$

where the $\mathbf{a}_{\omega_i}$ matrix is the Jacobian of the ith accelerometer measurements with respect to the attitude rate estimates:

$$\mathbf{a}_{\omega\mathbf{i}} = \begin{bmatrix} q_e s_{y_i} + r_e s_{zi} & p_e s_{y_i} - 2q_e s_{xi} & p_e s_{zi} - 2r_e s_{xi} \\ q_e s_{xi} - 2p_e s_{y_i} & p_e s_{xi} + r_e s_{zi} & q_e s_{zi} - 2r_e s_{y_i} \\ r_e s_{xi} - 2p_e s_{zi} & r_e s_{y_i} - 2q_e s_{zi} & p_e s_{xi} + q_e s_{y_i} \end{bmatrix} \tag{5}$$

where p_e, q_e and r_e, the components of ω_e, are the estimates of, respectively, the roll, pitch and yaw rates, and s_{xi}, s_{y_i} and s_{zi}, the components of $\mathbf{s_{sbi}}$, are the locations of the ith accelerometer along, respectively, the body frame x, y and z axes, with respect to the body centre of mass.

Assuming that all sensors have the same specifications, the state noise covariance matrix ($\mathbf{Q}$) and the measurement noise matrix ($\mathbf{R}$) of the Kalman filter are:

$$\mathbf{Q} = \begin{bmatrix} \boldsymbol{\sigma}^2_{\mathbf{ae}} & \mathbf{0}_3 & \cdots & \mathbf{0}_3 & \mathbf{0}_3 & \mathbf{0}_3 & \cdots & \mathbf{0}_3 \\ \mathbf{0}_3 & \sigma^2_{ba1}\mathbf{I}_3 & \cdots & \mathbf{0}_3 & \mathbf{0}_3 & \mathbf{0}_3 & \cdots & \mathbf{0}_3 \\ \vdots & \vdots & \ddots & \vdots & \vdots & \vdots & \ddots & \vdots \\ \mathbf{0}_3 & \mathbf{0}_3 & \cdots & \sigma^2_{ban}\mathbf{I}_3 & \mathbf{0}_3 & \mathbf{0}_3 & \cdots & \mathbf{0}_3 \\ \mathbf{0}_3 & \mathbf{0}_3 & \cdots & \mathbf{0}_3 & \boldsymbol{\sigma}^2_{\boldsymbol{\omega}\mathbf{e}} & \mathbf{0}_3 & \cdots & \mathbf{0}_3 \\ \mathbf{0}_3 & \mathbf{0}_3 & \cdots & \mathbf{0}_3 & \mathbf{0}_3 & \sigma^2_{b\omega 1}\mathbf{I}_3 & \cdots & \mathbf{0}_3 \\ \vdots & \vdots & \ddots & \vdots & \vdots & \vdots & \ddots & \vdots \\ \mathbf{0}_3 & \mathbf{0}_3 & \cdots & \mathbf{0}_3 & \mathbf{0}_3 & \mathbf{0}_3 & \cdots & \sigma^2_{b\omega n}\mathbf{I}_3 \end{bmatrix} \tag{6}$$

$$\mathbf{R} = \begin{bmatrix} \sigma^2_{wa1}\mathbf{I}_3 & \cdots & \mathbf{0}_3 & \mathbf{0}_3 & \cdots & \mathbf{0}_3 \\ \vdots & \ddots & \vdots & \vdots & \ddots & \vdots \\ \mathbf{0}_3 & \cdots & \sigma^2_{wan}\mathbf{I}_3 & \mathbf{0}_3 & \cdots & \mathbf{0}_3 \\ \mathbf{0}_3 & \cdots & \mathbf{0}_3 & \sigma^2_{w\omega 1}\mathbf{I}_3 & \cdots & \mathbf{0}_3 \\ \vdots & \ddots & \vdots & \vdots & \ddots & \vdots \\ \mathbf{0}_3 & \cdots & \mathbf{0}_3 & \mathbf{0}_3 & \cdots & \sigma^2_{w\omega n}\mathbf{I}_3 \end{bmatrix} \tag{7}$$

where σ^2_{bai} is the variance of the ith accelerometer bias, $\sigma^2_{b\omega i}$ is the variance of the ith gyroscope bias, σ^2_{wai} is the variance of the ith accelerometer random walk and $\sigma^2_{w\omega i}$ is the variance of the ith gyroscope random walk. The matrices $\boldsymbol{\sigma}^2_{\mathbf{ae}}$ and $\boldsymbol{\sigma}^2_{\boldsymbol{\omega}\mathbf{e}}$ contain the variances of the jerk and angular acceleration, respectively. The variances of the three axes are expressed, individually, along the diagonals of the matrices. The variances of the jerk and of the angular acceleration are estimated based on the expected movements of the sensor, by computing the maximum variation in the expected consecutive measurements. Artillery munitions trajectory is predictable and a good estimation of this variance is achievable. However, for vehicles having unpredictable trajectory (e.g., missiles), the estimation would not be as good.

On their parts, the bias and random walk variances are set to mimic the sensor specifications. As the same sensors are used in all IMUs, all the accelerometers random walk variances $\left(\sigma^2_{wa1\ldots n}\right)$ are equal. This is also the case for the accelerometers bias variances $\left(\sigma^2_{ba1\ldots n}\right)$, for the gyroscopes random walk variances $\left(\sigma^2_{w\omega 1\ldots n}\right)$ and for the gyroscopes bias variances $\left(\sigma^2_{b\omega 1\ldots n}\right)$.

As they cannot be measured, the angular accelerations $(\dot{\omega})$ must be estimated. This can be done by additional components in the Kalman filter state, or by backward finite differences (Equation (8)). The latter is selected in order to decrease the size of the filter state, hence the computational time:

$$\dot{\omega}_{ek} = \begin{bmatrix} \dot{p}_e \\ \dot{q}_e \\ \dot{r}_e \end{bmatrix} = \frac{1}{\delta_t} \begin{bmatrix} p_{ek-1} - p_{ek-2} \\ q_{ek-1} - q_{ek-2} \\ r_{ek-1} - r_{ek-2} \end{bmatrix} \tag{8}$$

For algorithm not considering sensors location, the fictitious forces are null. The term $\dot{\omega}_e \times \mathbf{s}_{\mathbf{s}\mathbf{b}i} + \omega_e \times (\omega_e \times \mathbf{s}_{\mathbf{s}\mathbf{b}i})$ in Equation (3) and the $\mathbf{a}_{\omega_i}$ matrix in Equation (4) are then null. There is therefore no relation between the accelerometer and gyroscope measurements, and no coupling between the axes. Then, in order to diminish the computational time, all axes and sensor types are processed separately. The algorithm not considering sensors location is therefore split in three data fusion filters for the acceleration estimation and three others for the attitude rate estimation. The model of each filter is a subset of the complete model (Equations (2) and (3)); it corresponds to the rows and lines of the estimated parameter and the related sensors bias. The model is then:

$$x_{k+1} = \begin{bmatrix} a_e \\ b_{a1} \\ b_{a2} \\ \vdots \\ b_{an} \end{bmatrix}_{k+1} = \begin{bmatrix} 1 & 0 & 0 & \cdots & 0 \\ 0 & 1 & 0 & \cdots & 0 \\ 0 & 0 & 1 & \cdots & 0 \\ \vdots & \vdots & \vdots & \ddots & \vdots \\ 0 & 0 & 0 & \cdots & 1 \end{bmatrix} x_k + \begin{bmatrix} v_a \\ v_{ba1} \\ v_{ba2} \\ \vdots \\ v_{ban} \end{bmatrix}_k \tag{9}$$

$$y_k = \begin{bmatrix} a_{m1} \\ a_{m2} \\ \vdots \\ a_{mn} \end{bmatrix}_k = \begin{bmatrix} 1 & 1 & 0 & \cdots & 0 \\ 1 & 0 & 1 & \cdots & 0 \\ \vdots & \vdots & \vdots & \ddots & \vdots \\ 1 & 0 & 0 & \cdots & 1 \end{bmatrix} x_k + \begin{bmatrix} w_{a1} \\ w_{a2} \\ \vdots \\ w_{an} \end{bmatrix}_k . \tag{10}$$

and the covariance matrices are:

$$\mathbf{Q} = \begin{bmatrix} \sigma_{ae}^2 & 0 & 0 & \cdots & 0 \\ 0 & \sigma_{ba_1}^2 & 0 & \cdots & 0 \\ 0 & 0 & \sigma_{ba_2}^2 & \cdots & 0 \\ \vdots & \vdots & \vdots & \ddots & \vdots \\ 0 & 0 & 0 & \cdots & \sigma_{ba_n}^2 \end{bmatrix} \tag{11}$$

$$\mathbf{R} = \begin{bmatrix} \sigma_{wa_1}^2 & 0 & \cdots & 0 \\ 0 & \sigma_{wa_2}^2 & \cdots & 0 \\ \vdots & \vdots & \ddots & \vdots \\ 0 & 0 & \cdots & \sigma_{wa_n}^2 \end{bmatrix} \tag{12}$$

The previous equations are for the accelerometers. The processing of the gyroscopes is identical, but with the accelerations being replaced by the attitude rates and their corresponding standard deviations. The models of the six filters are then identical. The only difference resides in the values within the covariance matrices, which are related to the estimated parameter.

In the filters initial state, the acceleration and attitude rate estimates are set based on the expected launching state, and the sensors bias are set at 0. On its part, the initial error covariance matrix is set equal to the state noise matrix.

2.2. Distributed Architecture

For the distributed architecture, the state of each local filter is composed of 12 elements:

$$x = \begin{bmatrix} \boldsymbol{a}_e & \boldsymbol{\omega}_e & \boldsymbol{b}_a & \boldsymbol{b}_\omega \end{bmatrix}^T \tag{13}$$

The state equation is therefore:

$$x_{k+1} = \begin{bmatrix} \boldsymbol{a}_e \\ \boldsymbol{\omega}_e \\ \boldsymbol{b}_a \\ \boldsymbol{b}_\omega \end{bmatrix}_{k+1} = \begin{bmatrix} \mathbf{I}_3 & \mathbf{0}_3 & \mathbf{0}_3 & \mathbf{0}_3 \\ \mathbf{0}_3 & \mathbf{I}_3 & \mathbf{0}_3 & \mathbf{0}_3 \\ \mathbf{0}_3 & \mathbf{0}_3 & \mathbf{I}_3 & \mathbf{0}_3 \\ \mathbf{0}_3 & \mathbf{0}_3 & \mathbf{0}_3 & \mathbf{I}_3 \end{bmatrix} x_k + \begin{bmatrix} \boldsymbol{v}_a \\ \boldsymbol{v}_\omega \\ \boldsymbol{v}_{ba} \\ \boldsymbol{v}_{b\omega} \end{bmatrix}_k \tag{14}$$

while the output equation is:

$$y_k = \begin{bmatrix} \boldsymbol{a}_m \\ \boldsymbol{\omega}_m \end{bmatrix}_{k+1} = h(x_k) = \begin{bmatrix} \boldsymbol{a}_e + \dot{\boldsymbol{\omega}}_e \times \boldsymbol{s}_{sb} + \boldsymbol{\omega}_e \times (\boldsymbol{\omega}_e \times \boldsymbol{s}_{sb}) + \boldsymbol{b}_a \\ \boldsymbol{\omega}_e + \boldsymbol{b}_\omega \end{bmatrix} + \begin{bmatrix} \boldsymbol{w}_a \\ \boldsymbol{w}_\omega \end{bmatrix}_k \tag{15}$$

A Kalman filter is implemented with its output matrix being the Jacobian of Equation (15):

$$\mathbf{C}_k = \left.\frac{\partial h}{\partial x}\right|_k = \begin{bmatrix} \mathbf{I}_3 & \mathbf{a}_\omega & \mathbf{I}_3 & \mathbf{0}_3 \\ \mathbf{0}_3 & \mathbf{I}_3 & \mathbf{0}_3 & \mathbf{I}_3 \end{bmatrix} \tag{16}$$

Assuming that all sensors have the same specifications, its covariance matrices are:

$$\mathbf{Q} = \begin{bmatrix} \boldsymbol{\sigma}_{ae}^2 & \mathbf{0}_3 & \mathbf{0}_3 & \mathbf{0}_3 \\ \mathbf{0}_3 & \boldsymbol{\sigma}_{\omega e}^2 & \mathbf{0}_3 & \mathbf{0}_3 \\ \mathbf{0}_3 & \mathbf{0}_3 & \sigma_{ba}^2 \mathbf{I}_3 & \mathbf{0}_3 \\ \mathbf{0}_3 & \mathbf{0}_3 & \mathbf{0}_3 & \sigma_{b\omega}^2 \mathbf{I}_3 \end{bmatrix} \tag{17}$$

$$\mathbf{R} = \begin{bmatrix} \sigma_{wa}^2 \mathbf{I}_3 & \mathbf{0}_3 \\ \mathbf{0}_3 & \sigma_{w\omega}^2 \mathbf{I}_3 \end{bmatrix} \tag{18}$$

As for the centralized architecture, the angular accelerations cannot be measured and are estimated using backward finite differences (Equation (8)).

For the algorithm not considering sensors location, as for the centralized architecture, the fictitious forces are null and, for each sensor, the data fusion process can be separated in six filters handling the sensor types and axes individually. When the sensors location is not considered, the subset model used in each filter is composed of the rows and lines of Equations (14)–(16) corresponding to the estimated parameter and its related sensor bias:

$$x_{k+1} = \begin{bmatrix} a_e \\ b_a \end{bmatrix}_{k+1} = \begin{bmatrix} 1 & 0 \\ 0 & 1 \end{bmatrix} x_k + \begin{bmatrix} v_a \\ v_{ba} \end{bmatrix}_k \tag{19}$$

$$y_k = a_{mk} = \begin{bmatrix} 1 & 1 \end{bmatrix} x_k + w_{ak} \tag{20}$$

and the covariance matrices are:

$$\mathbf{Q} = \begin{bmatrix} \sigma_{ae}^2 & 0 \\ 0 & \sigma_{ba}^2 \end{bmatrix} \tag{21}$$

$$\mathbf{R} = \sigma_{wa}^2 \tag{22}$$

The previous equations are for the accelerometers. The processing of the gyroscopes is identical, but with the accelerations being replaced by the attitude rates and their corresponding standard deviations.

The state of each local filter is initialized based on the expected launching state. The sensors bias are set at 0, and each local filter error covariance matrix is set identical to the state noise matrix.

For both cases, considering or not the sensors location, following the local filters computation, a master data fusion algorithm computes the mean of all estimations for each parameter. Those means are the final estimates, which are sent to the local filters to reset their estimates before the next time step.

2.3. Baseline Architecture

In order to better assess the performance of two previous architectures, a baseline architecture is proposed. The baseline architecture computes for each axis (x, y and z) and sensor type (gyroscope and accelerometer), the measurements average. However, because of the centrifugal and angular accelerations, pre-computations are required on the accelerometer measurements before computing their means.

First, the angular velocity is obtained by directly computing, on each axis separately, the mean of the gyroscope measurements.

The acceleration and attitude rate estimations are then used to remove the angular and centrifugal accelerations from the accelerometer measurements:

$$\boldsymbol{a}_i = \boldsymbol{a}_{mi} - \left(\dot{\boldsymbol{\omega}}_e \times \boldsymbol{s}_{sb}\right) - \left(\boldsymbol{\omega}_e \times \left(\boldsymbol{\omega}_e \times \boldsymbol{s}_{sb}\right)\right) \tag{23}$$

As previously, the angular accelerations, required in the linear accelerations computation, are estimated by backward finite differences (Equation (8)).

The acceleration estimates are the means of the Equation (23) subtractions:

$$a_{xe} = \frac{\sum_{i=1}^{n} \left(a_{mxi} - \left(p_e q_e s_{y_i} - q_e^2 s_{xi} + p_e r_e s_{zi} - r_e^2 s_{xi}\right) - \left(\dot{q}_e s_{zi} - \dot{r}_e s_{y_i}\right)\right)}{n} \tag{24}$$

$$a_{y_e} = \frac{\sum_{i=1}^{n} \left(a_{my_i} - \left(p_e q_e s_{xi} - p_e^2 s_{y_i} + q_e r_e s_{zi} - r_e^2 s_{y_i}\right) - \left(\dot{r}_e s_{xi} - \dot{p}_e s_{zi}\right)\right)}{n} \tag{25}$$

$$a_{ze} = \frac{\sum_{i=1}^{n}\left(a_{mzi} - \left(p_e r_e s_{xi} - p_e^2 s_{zi} + q_e r_e s_{y_i} - q_e^2 s_{zi}\right) - \left(\dot{p}_e s_{y_i} - \dot{q}_e s_{xi}\right)\right)}{n} \tag{26}$$

For the algorithm not considering sensors location, the fictitious forces terms can be removed from the previous equations. The acceleration estimates then become the average, on each axis separately, of the accelerometers measurements only.

3. Fault Detection and Isolation Algorithm

The FDI algorithm, identical for the three architectures, is based on a statistical analysis of the residual. There are systematic ways to deal with faulty sensors [6,25], but as this work realizes relative comparisons of data fusion architectures, a simple FDI method is used.

The residual is the difference between the sensor measurement, and the parameter and sensor bias estimates:

$$r_{ax_1} = a_{x1} - (a_{xe} + b_{ax_1}) \tag{27}$$

For the baseline architecture, where the bias is not estimated, the latter is considered null in the residual computation.

If this residual is larger than four times the standard deviation of the sensor random walk for two consecutive measurements, the sensor is considered faulty. The two consecutive 4σ threshold is selected in order to limit false warnings, while ensuring detection of faulty sensors. The sensor random walk being considered as a normal distribution, there is 0.0063% (1 in 15,787) chance that a healthy sensor gives a value larger than the threshold. With a sample time of 0.001 s, this is a false warning every 16 s. For two consecutive measurements, this percentage drops to 4.01×10^{-9}% (1 in 249,229,369), which is approximately a false warning every 29 days. This percentage is equivalent to a single value larger than a threshold of 5.9σ, approximately. However, because of the exponential tail of the normal distribution, a faulty sensor, which is a sensor with a random walk of standard deviation larger than the nominal sensor, has more chance of producing two consecutive measurements larger than 4σ than a single measurement larger than 5.9σ.

4. Performance Analysis

The architectures and algorithms are tested, in simulation, on a spin-stabilized projectile. A typical low quadrant elevation, low muzzle velocity and Northward launch trajectory is obtained from a projectile simulator developed in a non-spinning body frame [23]. The sensors measurements are simulated by adding random noises and bias instabilities to the nominal accelerations and attitude rates. Except when otherwise mentioned, the noises and biases are completely uncorrelated.

Two series of tests are done. First, the quality of the estimation of each architecture and algorithm, without faulty sensors, is tested (Section 4.1). Then, the performance of the fault detection algorithm is analysed (Section 4.2). For both series, the sensors are distributed around the projectile centre of mass. Hence, each accelerometer measures a different acceleration based on the sensed fictitious forces at its location.

4.1. Estimation Precision and Accuracy

As non-collocated sensors are used, the distance between the IMUs, their relative locations and their numbers influence the quality of the estimation of each algorithm. These three parameters are separately analysed.

4.1.1. Relative Locations of the IMUs

The effects of the relative locations of the IMUs are studied by comparing a symmetric 27 IMUs configuration to a configuration where the 27 IMUs are randomly located. In the random configuration, the sensors are all located within the cube of the symmetric configuration (Figure 1) and the randomly chosen positions are kept for all simulations. The random configuration is physically impracticable, but it is interesting to evaluate the sensors dissymmetry effect on the algorithms.

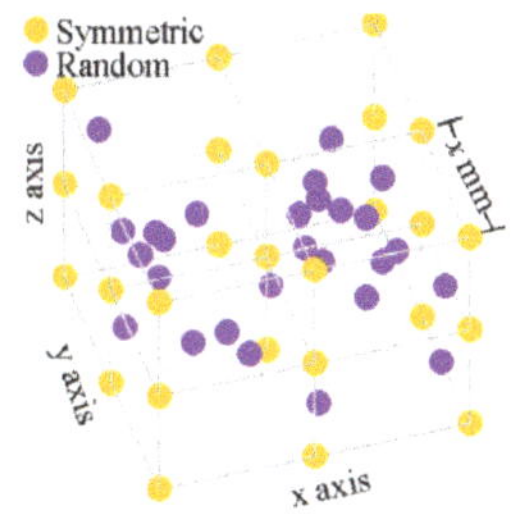

Figure 1. Symmetric and random 27 IMUs configurations.

Monte-Carlo simulations, composed of 350 runs each, are executed for 10 and 100 mm separations between the symmetric IMUs (20 and 200 mm edges cubes), and for their corresponding random locations. As the gyroscope measurements are not affected by their locations, the estimation of the attitude rate is not affected by the random configuration. However, the quality of the acceleration estimation decreases with the random configuration. Tables 1 and 2 show the standard deviation of the acceleration estimation errors.

Table 1. Standard deviation of the acceleration estimation errors, for the symmetric and random 27 IMUs configurations within the 20 mm edges cube.

IMUs Distribution	Axis	Locations Considered (m/s^2)			Locations Unconsidered (m/s^2)		
		Centralized	**Distributed**	**Baseline**	**Centralized**	**Distributed**	**Baseline**
Symmetric	*x* axis	0.0136	0.0146	0.0150	0.0136	0.0094	0.0150
	y axis	0.0139	0.0158	0.0149	0.0139	0.0098	0.0149
	z axis	0.0139	0.0160	0.0149	0.0139	0.0097	0.0149
Random	*x* axis	0.0142	0.0150	0.0171	0.0158	0.0124	0.0169
	y axis	0.0142	0.0174	0.0173	16.01	16.11	16.01
	z axis	0.0142	0.0173	0.0174	22.93	23.08	22.93

Table 2. Standard deviation of the acceleration estimation errors, for the symmetric and random 27 IMUs configurations within the 200 mm edges cube.

IMUs Distribution	Axis	Locations Considered (m/s^2)			Locations Unconsidered (m/s^2)		
		Centralized	**Distributed**	**Baseline**	**Centralized**	**Distributed**	**Baseline**
Symmetric	*x* axis	0.0135	0.0279	0.0148	0.0135	0.0092	0.0148
	y axis	0.0137	0.0227	0.0148	0.0137	0.0096	0.0148
	z axis	0.0138	0.0231	0.0148	0.0138	0.0096	0.0148
Random	*x* axis	0.0151	0.0377	0.0876	0.0839	0.0833	0.0842
	y axis	0.0144	0.0416	0.0910	160.14	161.12	160.16
	z axis	0.0143	0.0469	0.0914	229.29	230.75	229.32

As expected, without symmetry in the measured fictitious accelerations, the three algorithms not considering sensors location provide completely unusable results. On their part, the algorithms considering sensors location are differently affected by the random configuration. For the three architectures, the performance degradation is larger for the 200 mm edges cube than for the 20 mm one. Also, for the 200 mm edges cube, the degradation is much larger for the baseline architecture than for the two others. The knowledge of the sensor noises characteristics, which is introduced into the Kalman filters, helps at conserving the quality of the estimation with the centralized and distributed architectures. When the locations are considered, the main reason to explain the performances degradation due to random configuration is the sensors bias estimation. The centralized architecture provides a better estimation of the biases (Section 4.3) than the distributed architecture. It is, therefore, less affected by the random configuration. Both architectures use the redundancy and apparent observability differently; the centralized architecture uses it to improve the estimation of the biases, while the distributed architecture uses it to improve the acceleration and attitude rate estimations.

4.1.2. Relative Distance between the IMUs

The distance between 27 IMUs is varied from 1 mm to 100 mm. A symmetric 27 IMUs configuration, a cube with nine IMUs evenly distributed on each face and one IMU in the middle of the cube, at the centre of mass, is selected (Figure 1). This configuration creates symmetries in the sensed fictitious accelerations.

Figure 2 presents the resulting standard deviation of the acceleration estimation errors and Figure 3 shows the standard deviation of the attitude rate estimation errors. Those results are obtained with a Monte-Carlo simulation composed of 350 runs for each sensor separation. For each run covering the full projectile trajectory, the standard deviation of the estimation errors is computed. The resulting standard deviations are then averaged over the 350 runs, and the result presented as a function of the sensors separation.

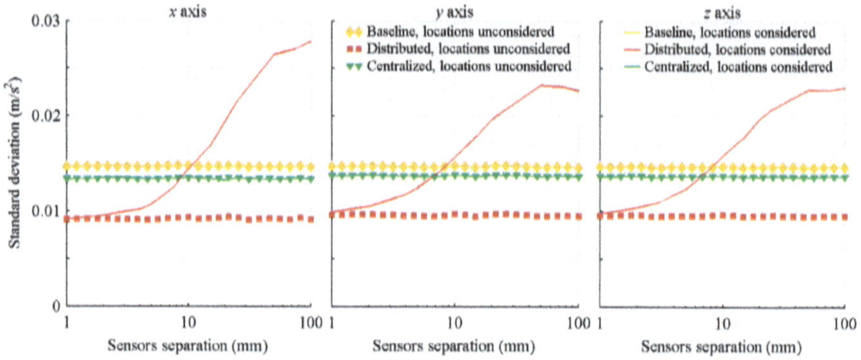

Figure 2. Standard deviation of the acceleration estimation errors, with respect to the distance between the IMUs.

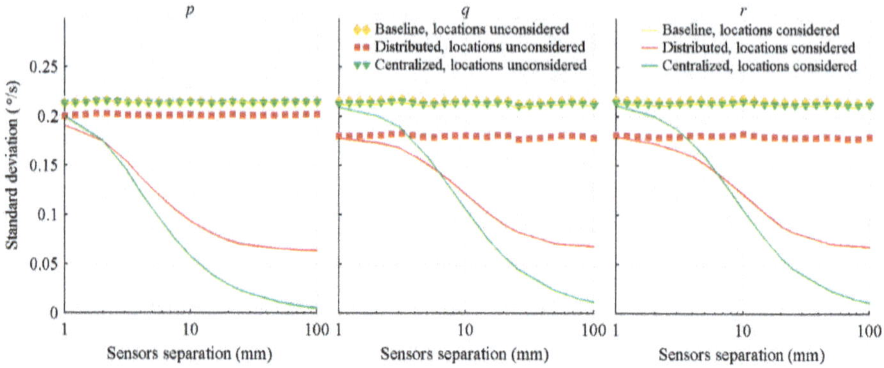

Figure 3. Standard deviation of the attitude rate estimation errors, with respect to the distance between the IMUs.

As a symmetric sensors configuration is used, the algorithms not considering sensors location are still very efficient. For the acceleration estimation with the centralized and baseline architectures, there is no gain at considering the relative locations of the sensors in the algorithms. The fictitious accelerations sensed by one sensor are always cancelled by the corresponding symmetrical sensor. For the distributed architecture, there is even a loss in efficiency when the sensors location is considered in the algorithm. In this architecture, the acceleration of each IMU is estimated prior to the averaging. Therefore, portions of the fictitious accelerations are considered as real accelerations and are not canceled by the sensors symmetry.

Considering the sensors location in the algorithms significantly improves the precision of the attitude rate estimation. By taking into account the sensors location, the accelerometers are combined with the gyroscopes in the attitude rate estimation. When the sensors separation grows, the fictitious accelerations increase and are more easily discriminated over the accelerations of the sensor centre of mass and noises. The performance improvement is therefore more important for larger separation. Also, the centralized architecture better exploits those fictitious accelerations than the distributed architecture. The first one considers the sensors configuration as a whole, and therefore, simultaneously estimates the fictitious accelerations sensed by all accelerometers. On its part, the distributed architecture treats each IMU separately and does not exploit the sensors configuration as a whole. It only uses the location of each IMU with respect to the centre of mass. Therefore, portions of the fictitious accelerations are considered as real accelerations. The worse results of the distributed architecture, in comparison to the centralized architecture ones, are emphasized by its poor biases estimation (Section 4.3).

Also, due to the nature of the projectile trajectory and to the Kalman filters tuning, the three projectile axes show similar performance. The motions of the projectile along its y and z axes are not significantly different, and are in the same order of magnitude as its motion along its x axis (linear deceleration). The only parameter which is significantly different than the others is the projectile x axis spin rate which decreases rapidly. Therefore, a larger value is given to the variance of the projectile spin rate $\left(\sigma_p^2\right)$. For the algorithms not considering sensors location, the improvements of the centralized and distributed architectures in comparison to the baseline one are slightly less important for this parameter (left graphic of Figure 3). However, for the algorithms considering sensors location, the improvements are more significant for this parameter than for the others. The nominal spin rate of the projectile being much larger than the two other rates, the fictitious accelerations due to this rate are larger, and therefore, are more easily distinguishable.

4.1.3. Number of Near Symmetrically Located IMUs

As demonstrated in Section 4.1.2, the symmetry of the sensors significantly influences the estimation performance. Therefore, in this analysis, the locations are selected in order to have a close to symmetric distribution for all numbers of IMUs. In this way, the algorithms not considering sensors location can be applied and provide relatively good performance.

Also, a 10 mm separation between the sensors is selected. For close to symmetric distribution, this distance is the larger one allowing the packaging of all sensors within a projectile fuze. Furthermore, based on Section 4.1.2, this distance still provides a significant gain in the attitude rate estimation.

A Monte-Carlo simulation composed of 350 runs is then executed for each IMU number, which is varied from 2 to 60 by steps of 2. For each run, covering the full projectile trajectory, the standard deviation of the estimation errors is computed. The resulting standard deviations are then averaged over the 350 runs, and the result presented as a function of the number of IMUs in Figures 4 and 5.

As expected, a trend coherent with the signal averaging theory, a diminution of the standard deviation proportional to $\frac{1}{\sqrt{n}}$, is obtained for all cases. For the algorithms not considering sensors location, the distributed architecture provides the best performance. Because of the local filtering, the mean of the master filter is computed on data having a smaller standard deviation than the raw measurements, resulting in a less noisy final estimation for the distributed architecture than for the baseline architecture. On its part, because of the covariance matrices tuning which gives equal weight to all measurements, the centralized architecture is, when the sensors location is not considered, similar to directly computing the mean of the measurements. The small performance improvement, in comparison to the baseline architecture, comes from the a priori knowledge of the trajectory and noises introduced into the Kalman filter. The two Kalman filterbased architectures, therefore, use the redundancy differently: the centralized one improves the biases estimation (Section 4.3), while the distributed one improves the parameter estimations.

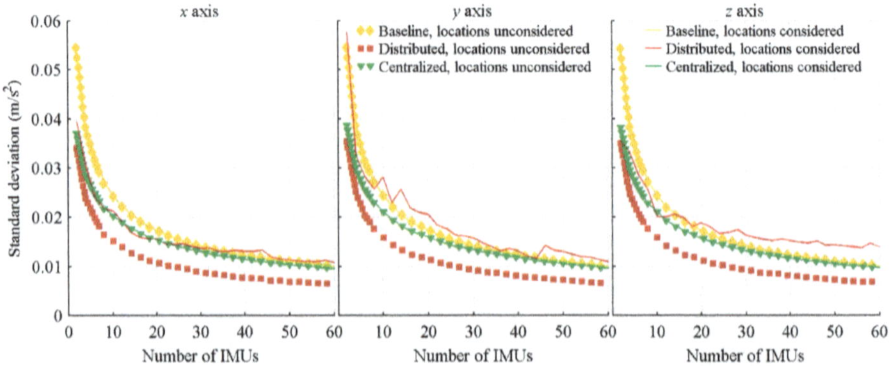

Figure 4. Standard deviation of the acceleration estimation errors, with respect to the number of IMUs.

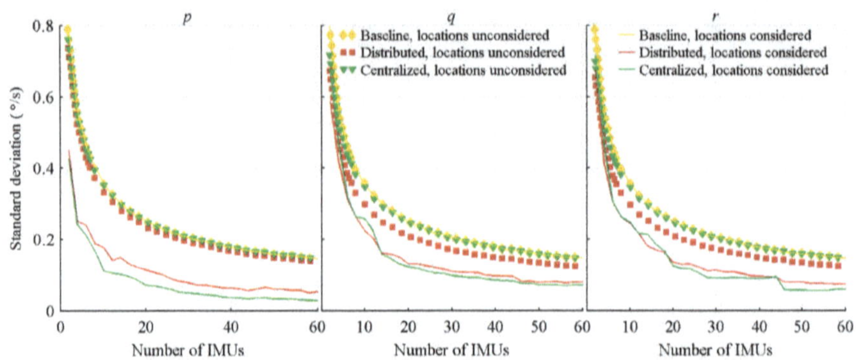

Figure 5. Standard deviation of the attitude rate estimation errors, with respect to the number of IMUs.

However, as in Section 4.1.1, due to how it handles the fictitious accelerations, the centralized architecture with its algorithm considering sensors location is the best option for nearly all tested number of IMUs. The light geometric dissymmetry for some numbers of sensors, and the non-linear variation, with respect to the number of IMUs, of the distance between the centre of mass of the IMUs configuration and the farthest IMU, explain the non-smooth behaviour of the distributed architecture acceleration estimation and of the centralized and distributed architectures attitude rate estimation.

As in Section 4.1.1, the algorithms considering sensors location are, for the attitude rate estimation, better than those not considering them, but the distributed architecture acceleration estimation is worse when the sensors location is considered. Also, the acceleration estimation errors of the centralized and baseline architectures are not significantly modified by the inclusion of the sensors location within the algorithms. As in Section 4.1.1, because of the large nominal value of the spin rate, the estimation of this parameter differs from the others; the centralized and distributed architectures with their algorithms not considering sensors location are slightly less efficient for the spin rate axis than for the other axes, the opposite being observed for those architectures when the sensors location is considered in the algorithms.

All previous results were obtained with fully uncorrelated bias instabilities. However, MEMS are expected to have some correlations between them. Therefore, a second series of tests, with fully correlated biases is done. For these tests, the same bias is applied to all sensors generating measurements for the same parameter. There are no correlations between the axes and, as it is nearly impossible to know the sensors' correlation in a real system, the knowledge of the correlation is not included in the Kalman filter covariance matrices.

The following graphics compare the attitude rate (Figure 6) and acceleration (Figure 7) estimation errors of the correlated biases tests to the previous uncorrelated ones. Obviously, the biases of the sensors measuring the same parameter are not expected to be fully correlated nor fully uncorrelated. However, comparing the two sets of results provides bounds for the achievable performance. In order to keep the graphics readable, the results of the algorithms considering sensors location were included only. For correlated biases, the three algorithms not considering sensors location provide estimations nearly identical to those of the baseline algorithm considering sensors location.

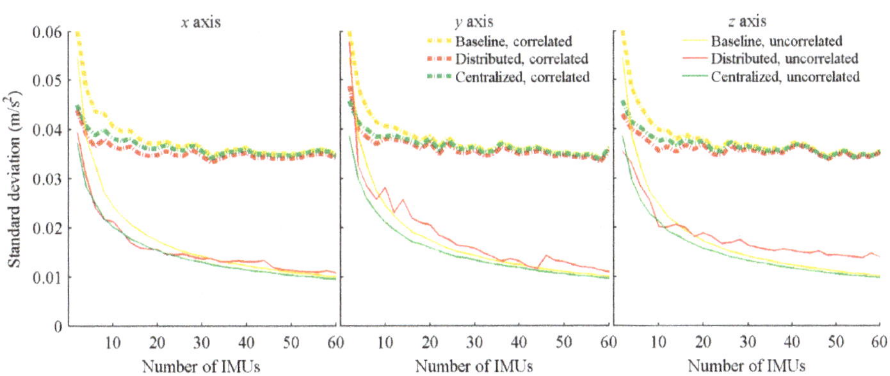

Figure 6. Standard deviation of the acceleration estimation errors, with correlated and uncorrelated sensors bias, and function of the number of IMUs.

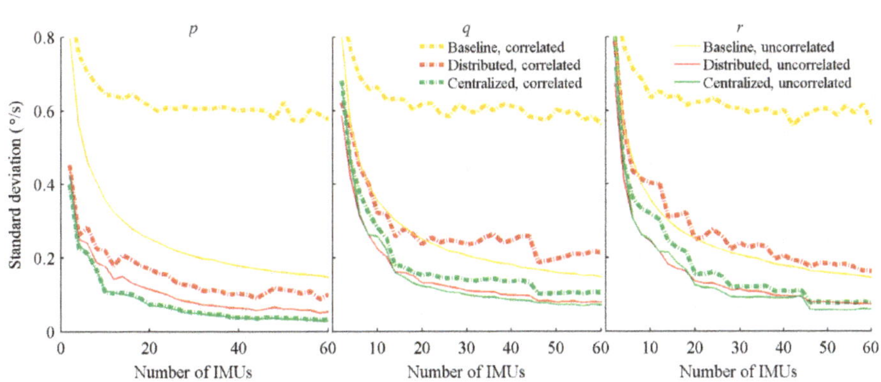

Figure 7. Standard deviation of the attitude rate estimation errors, with correlated and uncorrelated sensors bias, and function of the number of IMUs.

When the sensors location is considered, the accelerometers provide supplementary information on the attitude rate, allowing the distinction between the biases and vehicle motions. This is the main explanation of the improvements, with respect to the number of sensors, obtained with fully correlated biases. On their parts, the three architectures' acceleration estimation and the baseline attitude rate estimate provide, as mentioned in the literature [20], small improvements with respect to the number of considered IMUs. The biases cannot be clearly distinguished from the vehicle motions, and increasing the number of IMUs does not fix the issue.

Up to this point, the analysis is focusing on the precision of the estimation. The second important characteristic of the quality of an estimation is its accuracy, characterized by the mean of the estimation errors. For the tested cases, the mean is always near 0 and does not show a specific trend with respect to the number of IMUs.

The third studied characteristic of the algorithms is the relative computational time, which is presented in Figure 8 as a function of the number of IMUs. As the sensors' bias correlation does not affect the computational time, only the times of the fully uncorrelated biases tests are presented. The reference computational time is the computational time of the baseline architecture with its algorithm not considering sensors location in the case of two IMUs. Table 3 summarizes the operation that requires the most computational time for each architecture and the number of times this operation must be performed.

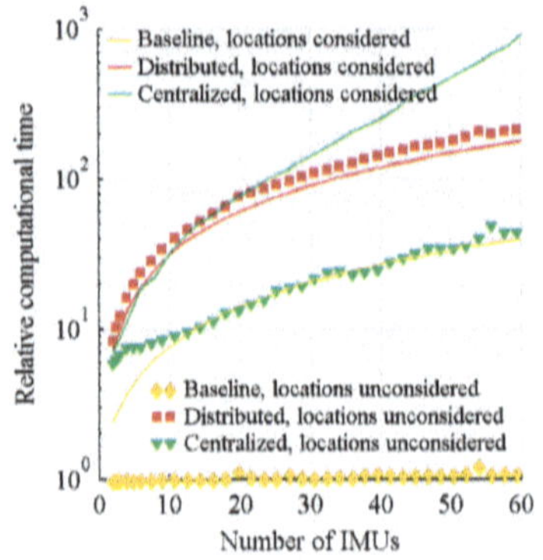

Figure 8. Relative computational time.

Table 3. Most time consuming operation for each architecture.

Architecture	Locations Considered	Locations Unconsidered
Centralized	Invert a $6n \times 6n$ matrix	Invert 6 $n \times n$ matrices
Distributed	Invert n 6×6 matrices	Compute $6n$ divisions
Baseline	Compute $3n$ cross-products	Mean computations

For the baseline architecture, the data fusion is implemented through, quickly performed, basic mathematical operations and the computation time variation as a function of the number of IMUs is marginal for the algorithm not considering sensors location. However, for the algorithm considering sensors location, the estimation of the angular acceleration which requires n cross-products is more time consuming and produces an increase of the computational time proportional to the number of sensors.

For the distributed architecture, the computational time increments come from the time required to compute the added local filters. For the algorithm not considering sensors location, each local filter has to compute six divisions, while for the algorithm considering sensors location, it must invert a 6×6 square matrix. Both operations are similar in term of computational time, the small difference being rather due to the larger amount of data handling operations for the algorithm not considering sensors location.

For the centralized architecture, the computational time is driven by the inversion of the Kalman filter covariance matrix. When the sensors location is considered, the addition of an IMU adds six rows and six columns to the matrix to be inverted. However, when the sensors location is not considered, it rather adds one row and one column to each of the six covariance matrices. For a relatively small symmetric matrix, the inversion is quickly performed. Therefore, among the algorithms not considering sensors location, the centralized architecture, which requires less handling operations, is faster than the distributed one. However, for large matrix, typical of the centralized architecture with its algorithm considering sensors location, the computational time is much longer.

The presented data fusion algorithms do not provide attitude, velocity and position estimations, which are provided by a standard INS. Also, without GPS receiver, there is no corrections possibility on these estimations. The estimated attitude rate is therefore directly integrated to estimate the attitude. On its part, the estimated acceleration, which is measured in the body frame, is transposed, by a rotation matrix, in the North-East-Down (NED) frame and integrated, in this frame, to estimate the position and velocity.

The standard deviation of the position errors is shown in Figure 9. As a comparison point, the typical precision of a military grade GPS receiver is included in the graphics.

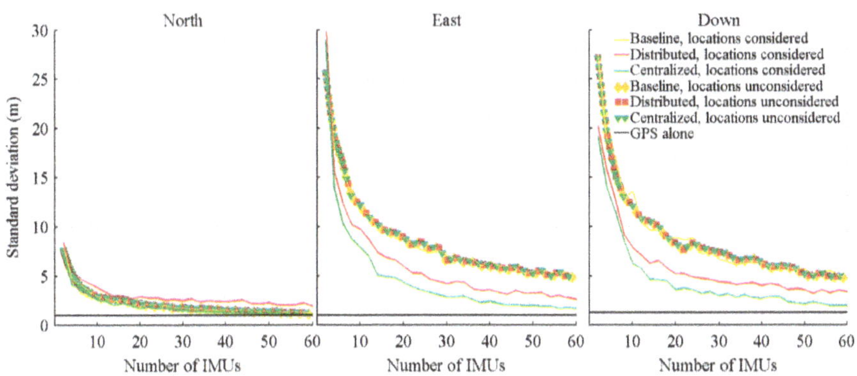

Figure 9. Standard deviation of the position errors when the projectile hits the ground, with respect to the number of IMUs.

Even if the precision of the acceleration and attitude rate estimations differ between the algorithms, the standard deviation of the position errors when the projectile hits the ground is not significantly different between the algorithms not considering sensors location, the gaps being not even visible in the graphics. The distributed architecture is better than the two others by, approximately, 0.01 m.

However, for the algorithms considering sensors location, the centralized and distributed architectures offer a significant gain over the baseline architecture, the improvements coming from the better attitude rate estimation. The accelerations are measured in the body frame, while the positions are expressed in the NED coordinates. To migrate from the body frame to the NED frame, the accelerations must be rotated. The rotation is performed by a rotation matrix which is directly affected by the attitude rate estimation.

4.2. Fault Detection

The efficiency of the algorithm proposed for FDI is tested on a specific number of IMUs and a fixed number of faults. The symmetric 27 IMUs configuration of Section 4.1.2, with three faults for the accelerometers and three faults for the gyroscopes, is used to assess the FDI performance of the three architectures equipped with the proposed FDI algorithm. A single form of faulty sensor, a sensor producing a measurement related to the vehicle motion, but with a larger random walk than the sensor typical one, is studied. The faulty sensor, the time of the fault and the amplitude of the faulty random walk are randomly set prior to each simulation. The amplitude of the random walk of the faulty sensor is, at least, 1.75 times that of a healthy sensor, and the fault occurs, at least, 1 s prior to the end of the simulation.

In this section, only the algorithms considering sensors location are studied. The algorithms not considering sensors location quickly consider all accelerometers as faulty, because they assimilate the fictitious accelerations to the sensors bias. Figures 10 and 11, respectively, show the mean numbers of correctly detected faulty sensors and of false warnings, based on Monte-Carlo simulations of 100 runs each.

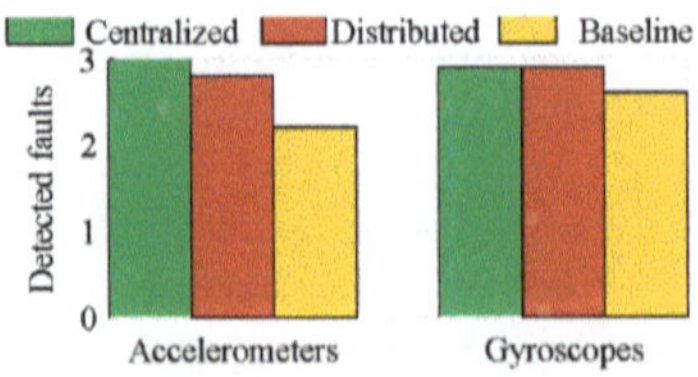

Figure 10. Detected faults for the algorithms considering sensors location.

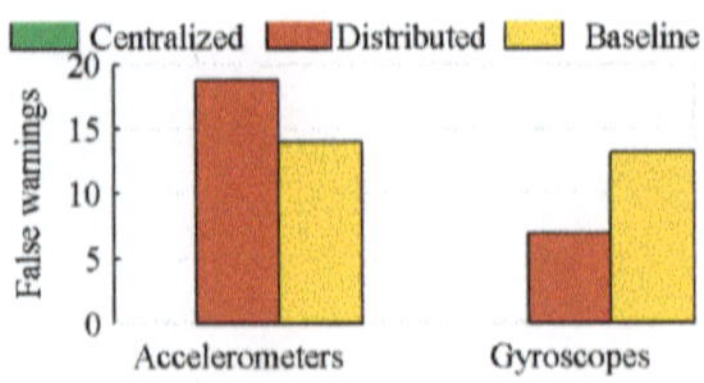

Figure 11. False warnings for the algorithms considering sensors location.

Because of its inability to correctly estimate the sensors bias, the distributed architecture generates a lot of false warnings, while the centralized architecture correctly identifies nearly all faulty sensors. The effect of this poor biases estimation is shown in Figure 12. This figure presents the 4σ thresholds of the centralized and distributed architectures over the measurements of a healthy sensor, when 27 IMUs are used. Because of its drift, at 8.9 s, this healthy sensor is considered faulty by the distributed architecture. The ideal threshold, added to the graphic, is the threshold computed on the nominal biased value. The latter is the measured value without noise.

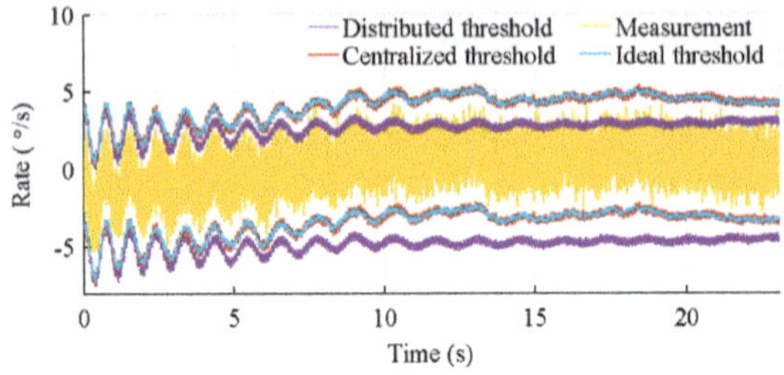

Figure 12. Typical 4σ thresholds of the centralized and distributed architectures with the algorithms considering sensors location, for 27 IMUs.

The bias estimation also explains why there are more false warnings for the accelerometers than for the gyroscopes. Because of the fictitious forces, the measured acceleration magnitude is much larger than the measured attitude rate one. The distributed architecture has difficulty to estimate the sensors bias in presence large measured values (Section 4.3).

The proposed FDI algorithm is therefore dependent of the number of IMUs. With more IMUs, the standard deviation of the residual decreases and the bias estimation improves. A smaller standard deviation smoothes the threshold bounds, while a bias estimation improvement tightens them. Figure 13 shows, for 8 and 64 IMUs, the threshold of the centralized architecture and the detection time of a faulty sensor. Because of a better threshold estimation, the fault is detected nearly 6 s earlier with 64 IMUs.

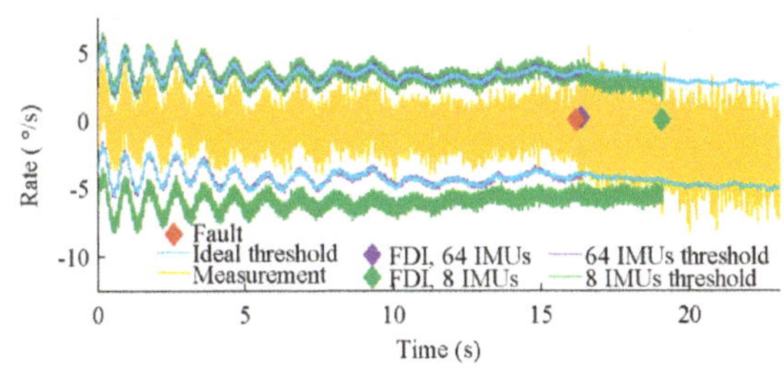

Figure 13. Typical 4σ threshold of the centralized architecture with the algorithm considering sensors location, for 8 and 64 IMUs.

4.3. Bias Estimation

Most of the results obtained previously are highly influenced by the inclusion of the sensors bias estimation within the centralized and distributed architectures. An analysis of this bias estimation is therefore done.

Both, the distributed and centralized architectures, have an unobservable subspace. However, they both show the typical behaviour of a fully observable system. Notably, they are able to come back to the true value after a significant outlier. This apparent observability comes from the correlation between the measurements, which is used to extract supplementary information for the sensors bias estimation. The centralized architecture is however better at estimating those biases, as shown in Figure 14, which shows the estimation of a bias on a gyroscope measuring the spin rate in a 8 and a 64 sensors configuration, the sensors location being considered. Furthermore, also visible on Figure 14, with more sensors, which favour the extraction of supplementary information, the biases estimation is improved.

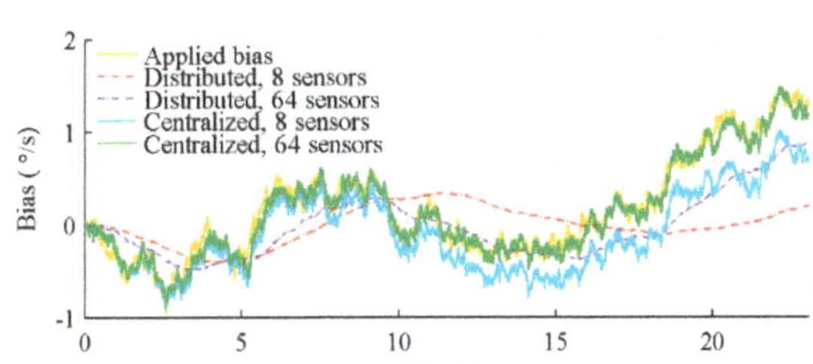

Figure 14. Bias estimation of gyroscopes measuring p, for 8 and 64 IMUs, with algorithms considering sensors location.

Figure 15 illustrates the difficulty of the distributed architecture, due to the a priori knowledge of the trajectory (sensor movements) introduced into the Kalman filter state noise covariance matrix, to estimate the sensors bias when the expected measurement variation is large. In this figure, the same bias is applied on gyroscopes measuring p (large value) and q (small value). This figure shows that the centralized architecture bias estimation is marginally affected by the amplitude of the measurements, while the distributed architecture one is significantly better for q.

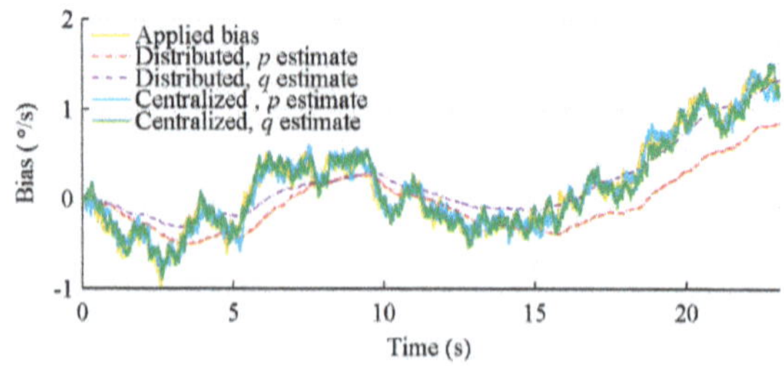

Figure 15. Bias estimation of gyroscopes measuring p and q, for 64 IMUs, with algorithms considering sensors location.

5. Conclusions

This work has analysed two data fusion architectures, a centralized architecture and a distributed architecture, to treat the measurements of a large amount of orthogonal redundant inertial measurement units. For each architecture, an algorithm considering the location of the sensors and another one not considering it have been proposed. In the centralized architecture, all sensor measurements are simultaneously processed by a single Kalman filter. In the distributed architecture, the measurements of each sensor are filtered by a local Kalman filter before being sent to a master data fusion filter which computes the final estimate. A baseline architecture, which directly computes the mean of the measurements, is used as a point of comparison. As this work focused on Micro-Electronic-Mechanical Systems, whose one of their main weaknesses is their large bias instability, the implemented Kalman filters were designed to estimate the sensors bias.

The Monte-Carlo simulations demonstrated that the centralized architecture with its algorithm considering sensors location provides the best performance. Its estimations of the accelerations, attitude rates and sensors bias are the most precise and accurate. It is also less affected when the IMUs are randomly located and its acceleration estimation does not degrade when the distance between the IMUs is increased. Additionally, it provides better fault detection capabilities, detecting nearly all faults and generating few false warnings. However, its better performance comes at the expense of a much longer computational time. The distributed architecture with its algorithm considering sensors location provides, at a much lower computational time, good estimations of the accelerations and attitude rates. However, the quality of these estimations degrades when the IMUs are located randomly, and its fault detection capabilities are very limited. The centralized and distributed architectures with their algorithms not considering sensors location remain viable for nearly symmetric locations, but they are completely unusable when the IMUs are located randomly, and for fault detection. With correlated sensors bias, only the centralized and distributed architectures with their algorithms considering sensors location are able to obtain a significant improvement, with respect to the number of IMUs, of the attitude rate estimation. Also, in presence of sensors bias correlation, no matter which architecture is used, the quality of the acceleration estimation is barely modified by the number of sensors.

The results and analyses of this paper were obtained under perfect conditions for the filtering algorithms, which means that the latter were always set with perfect knowledge of the sensors' characteristics and their locations. This study could be extended to cases considering an approximate knowledge of this information. Also, a more in-depth analysis, involving real flight data with a specific number of IMUs, would be done to validate the simulation results.

Author Contributions: E.G. and A.V. conceived and designed the experiments. A.V. and Y.B. developed the algorithms. A.V. performed the tests. E.G., A.V., and Y.B. analyzed the results and wrote the paper.

Conflicts of Interest: The authors declare no conflict of interest.

References

1. Nilsson, J.-O.; Skog, I. Inertial sensors arrays—A literature review. In Proceedings of the 2016 European Navigation Conference, Helsinki, Finland, 30 May–2 June 2016. [CrossRef]
2. Bancroft, J.; Lachapelle, G. Data Fusion Algorithms for Multiple Inertial Measurement Units. *Sensors* **2011**, *11*, 6771–6798. [CrossRef] [PubMed]
3. Pesja, A. Optimum Skewed Redundant Inertial Navigators. *AIAA J.* **1974**, *12*, 899–902. [CrossRef]
4. Waegli, A.; Guerrier, S.; Skaloud, J. Redundant MEMS-IMU integrated with GPS for Performance Assessment in Sports. In Proceedings of the IEEE/ION PLANS 2008, Monterey, CA, USA, 5–8 May 2008. [CrossRef]
5. Shim, D.-S.; Yang, C.-K. Optimal Configuration of Redundant Inertial Sensors for Navigation and FDI Performance. *Sensors* **2010**, *10*, 6497–6512. [CrossRef] [PubMed]
6. Giroux, R. Capteurs bas de Gamme et Systèmes de Navigation Inertielle: Nouveaux Paradigmes D'application, Ph.D. Thesis, École de Technologie Supérieure, Montréal, QC, Canada, 17 September 2004.
7. Levy, L.J. Suboptimality of Cascaded and Federated Kalman Filters. In Proceedings of the 52nd Annual Meeting of the Institute of Navigation, Cambridge, MA, USA, 19–21 June 1996.
8. Allerton, D.; Jia, H. A Review of Multisensor Fusion Methodologies for Aircraft Navigation Systems. *J. Navig.* **2005**, *58*, 405–417. [CrossRef]
9. Lawrence, P.; Berarducci, M. Comparison of federated and centralized Kalman filters with fault detection considerations. In Proceedings of the IEEE PLANS 1994, Las Vegas, NV, USA, 11–15 April 1994. [CrossRef]
10. Sukkarieh, S.; Gibbens, P.; Groscholsky, B.; Willis, K.; Durrant-Whyte, H.F. A Low-Cost, Redundant Inertial Measurement Unit for Unmanned Air Vehicles. *Int. J. Robot. Res.* **2000**, *19*, 1089–1103. [CrossRef]
11. Miller, M.M.; Soloviev, A.; Uijt de Haag, M.; Veth, M.; Raquet, J.; Klausutis, T.J.; Touma, J.E. *Navigation in GPS Denied Environments: Feature-Aided Inertial Systems*; Air Force Research Laboratory: Eglin AFB, FL, USA, 2010.
12. Bachrach, A.; Prentice, S.; He, R.; Roy, N. RANGE—Robust Autonomous Navigation in GPS-denied. *J. Field Robot.* **2011**, *28*, 644–666. [CrossRef]
13. Wu, A.D.; Johnson, E.N.; Kaess, M.; Dellaert, F.; Chowdhary, G. Autonomous Flight in GPS-Denied Environmnets Using Monocular Vision and Inertial Sensors. *J. Aerosp. Inf. Syst.* **2013**, *10*, 172–186. [CrossRef]
14. Grigorie, T.L.; Botez, R.M.; Sandu, D.G.; Grigorie, O. Experimental Testing of Data Fusion Algorithm for Miniaturized Inertial Sensors in Redundant Configurations. In Proceedings of the 2014 International Conference on Mathematical Methods, Mathematical Models and Simulation in Science and Engineering, Interlaken, Switzerland, 22–24 February 2014.
15. Chen, T. Design and Analysis of a Fault-Tolerant Coplanar Gyro-Free Inertial Measurement Unit. *J. Microelectromechan. Syst.* **2008**, *17*, 201–212. [CrossRef]
16. Skog, I.; Nilsson, J.-O.; Händel, P.; Nehorai, A. Inertial Sensor Arrays, Maximum Likelihood, and Cramér–Rao Bound. *IEEE Trans. Signal Process.* **2016**, *64*, 4218–4227. [CrossRef]
17. Sabatini, A. Kalman-Filter-Based Orientation Determination Using Inertial/Magnetic Sensors: Observabiliy Analysis and Performance Evaluation. *Sensors* **2011**, *11*, 9182–9206. [CrossRef] [PubMed]
18. Tereshkov, V. A Simple Observer for Gyro and Accelerometer Biases in Land Navigation Systems. *J. Navig.* **2015**, *68*, 635–645. [CrossRef]
19. Noureldin, A.; Karamat, T.; Eberts, M.; El-Sahfie, A. Performance Enhancement of MEMS-Based INS/GPS Integration for Low-Cost Navigation Applications. *IEEE Trans. Veh. Technol.* **2009**, *58*, 1077–1096. [CrossRef]
20. Jiang, C.; Xue, L.; Chang, H.; Yuan, G.; Yuan, W. Signal Processing of MEMS Gyroscope Arrays to Improve Accuracy Using a 1st Order Markov for Rate Signal Modeling. *Sensors* **2012**, *12*, 1720–1737. [CrossRef] [PubMed]
21. Schwaab, M.; Reginya, S.; Sikora, A.; Abramov, E. Measurement analysis of multiple MEMS sensor array. In Proceedings of the 24th Saint Petersburg International Conference on Integrated Navigation Systems, St. Petersburg, Russia, 29–31 May 2017. [CrossRef]
22. Woodman, O. *An. Introduction to Inertial Navigation*; University of Cambridge: Cambridge, MA, USA, 2007.

23. Zipfel, P.H. *Modeling and Simulation of Aerospace Vehicle Dynamics*, 2nd ed.; AIAA Inc.: Reston, VA, USA, 2000; ISBN 1-56347-875-7.
24. Crassidis, J.L.; Junkins, J.L. *Optimal Estimation of Dynamic Systems*; Chapman & Hall/CRC: Boca Raton, FL, USA, 2011; ISBN 1-58488-391-X.
25. Ray, A.; Phoba, S. Calibration and estimation of redundant signals. In Proceedings of the 2002 American Control Conference, Anchorage, AK, USA, 8–10 May 2002. [CrossRef]

sensors

Article

An Extension to Deng's Entropy in the Open World Assumption with an Application in Sensor Data Fusion

Yongchuan Tang [1,2,*], Deyun Zhou [1] and Felix T. S. Chan [2,*]

[1] School of Electronics and Information, Northwestern Polytechnical University, Xi'an 710072, China; dyzhou@nwpu.edu.cn

[2] Department of Industrial and Systems Engineering, The Hong Kong Polytechnic University, Hong Kong, China

* Correspondence: tangyongchuan@mail.nwpu.edu.cn (Y.T.); f.chan@polyu.edu.hk (F.T.S.C.)

Received: 6 May 2018; Accepted: 8 June 2018; Published: 11 June 2018

Abstract: Quantification of uncertain degree in the Dempster-Shafer evidence theory (DST) framework with belief entropy is still an open issue, even a blank field for the open world assumption. Currently, the existed uncertainty measures in the DST framework are limited to the closed world where the frame of discernment (FOD) is assumed to be complete. To address this issue, this paper focuses on extending a belief entropy to the open world by considering the uncertain information represented as the FOD and the nonzero mass function of the empty set simultaneously. An extension to Deng's entropy in the open world assumption (EDEOW) is proposed as a generalization of the Deng's entropy and it can be degenerated to the Deng entropy in the closed world wherever necessary. In order to test the reasonability and effectiveness of the extended belief entropy, an EDEOW-based information fusion approach is proposed and applied to sensor data fusion under uncertainty circumstance. The experimental results verify the usefulness and applicability of the extended measure as well as the modified sensor data fusion method. In addition, a few open issues still exist in the current work: the necessary properties for a belief entropy in the open world assumption, whether there exists a belief entropy that satisfies all the existed properties, and what is the most proper fusion frame for sensor data fusion under uncertainty.

Keywords: Dempster-Shafer evidence theory (DST); uncertainty measure; open world; closed world; Deng entropy; extended belief entropy; sensor data fusion

1. Introduction

Uncertain information processing plays a key role in complex systems of many fields such as sensor networks [1,2], pattern recognition [3,4], decision-making [5,6], supply chain network management [7,8], complex network [9] and target tracking [10,11]. Uncertain information may come from sensors with different credibilities and experts's subjective judgement. The heterogeneous sources and reliable degree increase the complexity and uncertainty of information process. The Dempster-Shafer evidence theory (DST) [12,13] has a promising efficiency in uncertain information processing such as information fusion [14,15]. However, there are still a few open issues in the DST framework that need further study. Firstly, the approaches of managing the conflicting belief masses still needs further refining [16,17]. Secondly, the reasonable ways of generating the mass functions for the practical applications [18,19]. Thirdly, uncertainty quantification with the possible measures in the DST framework [20,21], and the necessary properties a new belief entropy should obey [22–24]. Fourthly, rules of combining the body of evidence vary under different circumstances [25–27]. Inspired by the open world assumption in [28–30], this paper focus on designing an uncertainty measure for the open world in the DST framework.

Uncertainty measure for belief structures is a hot topic for uncertain information processing in DST framework [31]. Many uncertainty measures are derived from Shannon entropy, including Hohle's confusion measure [32] which is based on the mass function and the belief function of a proposition, Yager's dissonance measure [33] which is based on the mass function and the plausibility function of a proposition, Dubois and Prade's weighted Hartley entropy [34] which is based on the mass function of a proposition and the corresponding element number. Both Klir & Ramer's discord measure [35] and Klir & Parviz's strife measure [36] are based on two mass functions of different propositions as well as their element number. Recently, some new uncertainty measures emerge, including the general formulation for second-order uncertainties proposed by Yager [37], the maximum likelihood estimation proposed by Denoeux [38], the non–conflicting parts-based conflict measure in belief functions proposed by Daniel [39], Deng entropy proposed by Deng [21] as well as its modification by Tang et al. [40] and uncertainty measure based on interval probabilities [41,42]. According to [21], Deng entropy shows some advantages in some cases in comparison with some other uncertainty measures, and it has been applied in real applications such as the fault diagnosis [43], decision making [44] and sensor data fusion [45]. However, we noticed that all the aforementioned research works handle the uncertainty measure of uncertain information in the closed world, the uncertainty environment in the open world, where there are more sources of uncertainty, has been ignored. Inspired by some research works, especially for the concept of the open world in [28–30], we argue that the uncertainty measure in the open world should be different from that in the closed world because the circumstance of uncertainty is different. In the open world assumption of the DST framework, the uncertainty exists in the information expressed by (1) the mass functions of focal elements; (2) the nonzero mass function of the empty set and (3) the possible incompleteness of the frame of discernment (FOD). In the previous uncertainty measures, the uncertain information expressed by the mass function of empty set and the possible incompleteness of the FOD is ignored; which is the reason of this work.

In order to measure the uncertainty of belief structures in the open world, an extended uncertainty measure named the extension to Deng's entropy in the open world assumption (EDEOW) is proposed. EDEOW is based on the Deng entropy and it can be regarded as a generalization of the Deng entropy. By handling the uncertain information represented by the mass function of the empty set and the uncertain FOD of DST in the open world, EDEOW has the capability of measuring uncertain degree of belief structures for the open world assumption. To verify the usefulness and applicability of the extended belief entropy, some numerical examples and an application on sensor data fusion are presented and discussed in this paper. It should be noticed that a few open issues still exist in the current work, for example, (1) what are the necessary properties for a belief entropy in the open world assumption; (2) whether there exists a belief entropy that satisfies all the existingproperties in the closed world; and (3) what is the proper fusion frame for sensor data fusion under uncertainty.

This rest of this paper is organized as follows. The preliminaries are introduced in Section 2. In Section 3, the EDEOW for uncertainty measure of belief functions is proposed, as well as some illustrative numerical examples. The EDEOW-based information fusion approach and its application in sensor data fusion is presented in Section 4. In Section 5, some open issues for future research related to the work are discussed. Section 6 draws the conclusion of this paper.

2. Preliminaries

The background of Dempster-Shafer evidence theory (DST), Shannon entropy and Deng entropy are introduced.

2.1. Dempster-Shafer Evidence Theory

Some basic definitions in DST are presented as follows [12,13].

Definition 1. *Assume that* $\Omega = \{\theta_1, \theta_2, \ldots, \theta_i, \ldots, \theta_N\}$ *is a nonempty set with N mutually exclusive and exhaustive events,* Ω *is the frame of discernment (FOD). The power set of* Ω *consists of* 2^N *elements denoted as follows:*

$$2^\Omega = \left\{ \begin{array}{l} \emptyset, \{\theta_1\}, \{\theta_2\}, \ldots, \{\theta_N\}, \{\theta_1, \theta_2\}, \\ \ldots, \{\theta_1, \theta_2, \ldots, \theta_i\}, \ldots, \Omega \end{array} \right\}. \tag{1}$$

Definition 2. *A mass function m is a mapping from the power set* 2^Ω *to the interval [0,1]. m satisfies:*

$$m(\emptyset) = 0, \sum_{A \in \Omega} m(A) = 1. \tag{2}$$

If $m(A) > 0$*, then A is called a focal element.* $m(A)$ *indicates the support degree of the evidence on the proposition A.*

Definition 3. *A body of evidence (BOE), also known as a basic probability assignment (BPA) or basic belief assignment (BBA), is defined as the focal sets and the corresponding mass functions:*

$$(\Re, m) = \left\{ \langle A, m(A) \rangle : A \in 2^\Omega, m(A) > 0 \right\}, \tag{3}$$

where $\Re$ *is a subset of the power set* 2^Ω.

Definition 4. *A BPA m can also be represented by the belief function Bel or the plausibility function Pl, defined as follows:*

$$Bel(A) = \sum_{\emptyset \neq B \subseteq A} m(B) \quad , \quad Pl(A) = \sum_{B \cap A \neq \emptyset} m(B). \tag{4}$$

Definition 5. *In Dempster-Shafer evidence theory (DST), two independent mass functions* m_1 *and* m_2 *can be fused with Dempster's rule of combination:*

$$m(A) = (m_1 \oplus m_2)(A) = \frac{1}{1-k} \sum_{B \cap C = A} m_1(B) m_2(C), \tag{5}$$

where k is a normalization factor defined as follows:

$$k = \sum_{B \cap C = \emptyset} m_1(B) m_2(C). \tag{6}$$

It should be noted that the classical definitions of DST are defined in the closed world. In the open world assumption, Dempster's rule of combination is extended and named as the generalized combination rule (GCR) by Deng in [30].

Definition 6. *In [30], the fusion result of two empty sets is defined as* $\emptyset_1 \cap \emptyset_2 = \emptyset$*, which means that the intersection between two empty sets is still an empty set. Given two BPAs (*m_1 *and* m_2*), the generalized combination rule (GCR) is defined as follows:*

$$\begin{array}{l} m(A) = \dfrac{(1 - m(\emptyset)) \sum_{B \cap C = A} m_1(B) \cdot m_2(C)}{1 - K}, \\ K = \sum_{B \cap C = \emptyset} m_1(B) \cdot m_2(C), \\ m(\emptyset) = m_1(\emptyset) \cdot m_2(\emptyset), \\ m(\emptyset) = 1 \quad iff \quad K = 1. \end{array} \tag{7}$$

2.2. Shannon Entropy

As the information entropy for uncertainty measure, Shannon entropy has been applied and generalized in many areas such as complexity network [46–48].

Definition 7. *Shannon entropy is defined as [49]:*

$$H = -\sum_{i=1}^{N} p_i \log_b p_i, \tag{8}$$

where N is the number of basic states, p_i is the probability of state i, p_i satisfies $\sum_{i=1}^{N} p_i = 1$.

If the unit of information is bit, then $b = 2$. In this case, Shannon entropy is:

$$H = -\sum_{i=1}^{N} p_i \log_2 p_i. \tag{9}$$

2.3. Deng Entropy

As an extension of Shannon entropy in the framework of DST, Deng entropy is proposed in [21]. Some properties and behaviors are discussed in [21,24]. The application of Deng entropy can be found in [45,50].

Definition 8. *In FOD X, Deng entropy, denoted as E_d, is defined as:*

$$E_d(m) = -\sum_{A \subseteq X} m(A) \log_2 \frac{m(A)}{2^{|A|} - 1}, \tag{10}$$

where $|A|$ denotes the cardinality of the proposition A.

According to [21], the Deng entropy has some advantages in some cases in comparison with some other uncertainty measures in Table 1.

Table 1. Uncertainty measures in DST framework.

Uncertainty Measure	Definition
Hohle's confusion measure [32]	$C_H(m) = -\sum_{A \subseteq X} m(A) \log_2 Bel(A)$
Yager's dissonance measure [33]	$E_Y(m) = -\sum_{A \subseteq X} m(A) \log_2 Pl(A)$
Dubois &Prade's weighted Hartley entropy [34]	$E_{DP}(m) = \sum_{A \subseteq X} m(A) \log_2 \lvert A \rvert$
Klir & Ramer's discord measure [35]	$D_{KR}(m) = -\sum_{A \subseteq X} m(A) \log_2 \sum_{B \subseteq X} m(B) \frac{\lvert A \cap B \rvert}{\lvert B \rvert}$
Klir & Parviz's strife measure [36]	$S_{KP}(m) = -\sum_{A \subseteq X} m(A) \log_2 \sum_{B \subseteq X} m(B) \frac{\lvert A \cap B \rvert}{\lvert A \rvert}$
George & Pal's total conflict measure [51]	$TC_{GP}(m) = \sum_{A \subseteq X} m(A) \sum_{B \subseteq X} m(B) \left(1 - \frac{\lvert A \cap B \rvert}{\lvert A \cup B \rvert}\right)$

However, Equation (10) will be unavailable if $|A| = 0$. Thus, the uncertainty measure in the closed world of the DST framework should be extended. In the open world assumption [19,30,52,53], the uncertain information represented by the nonzero mass function of the empty set and the incomplete FOD should be handled properly and cautiously.

3. New Uncertainty Measure in the Open World

In the DST framework, the uncertain information is modelled not only by mass functions, the FOD is also an important source of uncertainty [40]. In addition, in the open world assumption, the mass value of the empty set may not be zero, which also indicates the incompleteness of the FOD [30]. With this background, how to measure the uncertain degree in the open world assumption of the DST framework is a new perspective and an important issue. According to literature review, no existing uncertainty measure addresses this problem, which is the reason for this work.

Example 1. *Consider a set of BPAs with the FOD $X = \{a, b\}$, the mass functions with nonzero mass value of the empty set:*

$$m(\{a\}) = 0.5, m(\{a,b\}) = 0.3, m(\emptyset) = 0.2. \tag{11}$$

It is obvious that the Deng entropy E_d in Equation (10) is not available for the uncertainty measure of BPAs in this case. The denominator of the *log* function with respect to $m(\emptyset) = 0.2$ will be $(2^0 - 1 = 0)$, which is illegal. This is because the Deng entropy is only based on the mass function of the focal element and the cardinality of the corresponding proposition. In the open world assumption, the mass value of empty set may not be zero. In addition, how to define the element number in incomplete FOD is also an open issue. The same question also exists in other uncertainty measures listed in Table 1. The works in [37,38,40,42] also pay no attention to the possible nonzero mass function of the empty set as well as the possible incomplete element number in the FOD. A new uncertainty measure which is extended from the Deng entropy in the closed world, named the extension to Deng's entropy in the open world assumption, is proposed especially for the problems mentioned above.

3.1. An Extension to Deng's Entropy in the Open World Assumption

Definition 9. *The extension to Deng's entropy in the open world assumption is defined as follows:*

$$E_{edeow}(m) = -\sum_{A \subseteq X} m(A) \log_2 \frac{m(A)}{\left(2^{(|A| + \lceil m(\emptyset)|X| \rceil)} - 1\right)}, \tag{12}$$

where $|A|$ is the cardinality of the proposition A, X is the FOD, $|X|$ denotes the certain element number in the FOD, $\lceil m(\emptyset)|X| \rceil$ is proposed to denote the uncertain element number in the FOD with respect to the corresponding proposition (A). '$\lceil \quad \rceil$' is the symbol of the ceiling function, which means the smallest integer that is no smaller than the independent variable, e.g., $\lceil 0.3 \rceil = 1$.

The extended measure addresses three parts of uncertainty in the DST framework, including the uncertain information expressed by the mass functions of focal elements, the mass function of the empty set and the possible incompleteness of the FOD. In detail, inspired by the existed uncertainty measures and the Deng entropy, the EDEOW handles two aspects of uncertainty according to the following methods:

- In the closed world where $m(\emptyset) = 0$, the uncertainty represented by the mass function $m(A)$ of the focal element as well as the corresponding cardinality $|A|$.
- In the open world where $m(\emptyset) \neq 0$, the nonzero mass function $m(\emptyset)$ of the empty set can be an indicator of the completeness or incompleteness of the FOD; currently, $\lceil m(\emptyset)|X| \rceil$ is chosen to express this uncertainty.

It should be noted that, in the EDEOW defined in Equation (12), the proposition A is no longer limited as a traditional focal element, it can also be an empty set $\emptyset$ which means uncertainty in the FOD [30]. In addition, apart from the $\lceil m(\emptyset)|X| \rceil$, there must exist many types of expressions to express the incompleteness of the FOD.

Recall the BPAs in Equation (11), with the EDEOW, the uncertainty degree of the BPAs can be calculated as follows:

$$\begin{aligned} E_{edeow}(m) &= -0.5\log_2 \frac{0.5}{2^{1+\lceil 0.2\times 2\rceil}-1} - 0.3\log_2 \frac{0.3}{2^{2+\lceil 0.2\times 2\rceil}-1} \\ &- 0.2\log_2 \frac{0.2}{2^{0+\lceil 0.2\times 2\rceil}-1} = 3.1202. \end{aligned} \tag{13}$$

With the proposed EDEOW, the problem in Example 1 can be handled. The BPAs with a nonzero mass function of the empty set can be handled now with the extended measure.

3.2. Numerical Example and Discussion

Example 2. *In FOD* $X = \{a\}$*, the mass functions are:*

$$m(\{a\}) = 1, m(\emptyset) = 0. \tag{14}$$

According to the BPAs in Equation (14), the mass value of the empty set is 0, which indicates the BPAs are assigned in the closed world. The uncertain degree with Shannon entropy H, Deng entropy E_d and the EDEOW E_{edeow} can be calculated respectively as follows:

$$\begin{aligned} H(m) &= -1 \times \log_2 1 = 0, \\ E_d(m) &= -1 \times \log_2 \frac{1}{2^1-1} = 0, \\ E_{edeow}(m) &= -1 \times \log_2 \frac{1}{2^{(1+0)}-1} = 0. \end{aligned} \tag{15}$$

Obviously, the mass function $m(\{a\}) = 1$ assigns a belief of 100% on the proposition $\{a\}$, which means the uncertain degree of the proposition is 0. In this case, the measuring result of the EDEOW is consistent with that of Shannon entropy and Deng entropy.

Example 3. *In FOD* $X = \{a, b, c, d\}$*, the mass functions are:*

$$m(\{a\}) = m(\{b\}) = m(\{c\}) = m(\{d\}) = 0.25, m(\emptyset) = 0. \tag{16}$$

The mass value of the empty set is 0, the BPAs are assigned in the closed world. The uncertain degree measured by H, E_d and E_{edeow} can be calculated respectively as follows:

$$\begin{aligned} H(m) &= (-0.25 \times \log_2 0.25) \times 4 = 2.0, \\ E_d(m) &= \left(-0.25 \times \log_2 \frac{0.25}{2^1-1}\right) \times 4 = 2.0, \\ E_{edeow}(m) &= \left(-0.25 \times \log_2 \left(\frac{0.25}{2^{(1+0)}-1}\right)\right) \times 4 = 2.0. \end{aligned} \tag{17}$$

According to the measuring results shown in Equations (15) and (17), if a mass function is assigned on the single subset, then the EDEOW can be degenerated to Deng entropy in the closed world. More importantly, the EDEOW satisfies the property of probabilistic consistency if the BPAs are only assigned on the single subset in the closed world. It should be noted that Shannon entropy and Deng entropy are not available if the BPAs are assigned in the open world where the mass value of the empty set is nonzero; as is shown in Example 1 and the following Example 4.

Example 4. *In a changing FOD* $|X|$*, consider the mass functions given as follows:*

$$m(\{1\}) = 0.2, m(\{2\}) = 0.3, m(\emptyset) = 0.5. \tag{18}$$

The mass value of the empty set is 0.5, the BPAs are assigned in the open world assumption. The uncertain degree measured by H, E_d and E_{edeow} are presented in Table 2. Calculation results show that Shannon entropy E_d cannot reflect the changes of the cardinality in the FOD $|X|$ (even if we treat the empty set $\emptyset$ as an uncertain proposition with nonzero set to make this function applicable in this case), while the Deng entropy is not applicable in this case for the reason that $m(\emptyset) \neq 0$. Only the EDEOW can successfully express the enlarging in the FOD as the value of $E_{edeow}(m)$ increases with the increasing of the $|X|$.

Table 2. Uncertain degree with different measures in Example 4.

Uncertainty Measure	$\|X\| = 2$	$\|X\| = 3$	$\|X\| = 5$	$\|X\| = 7$	$\|X\| = 9$
$E_d(m)$	-	-	-	-	-
$H(m)$	1.4855	1.4855	1.4855	1.4855	1.4855
$E_{edeow}(m)$	2.2780	3.6816	4.8426	5.9160	6.9512

In the following example adopted from [21], the EDEOW is compared with some other uncertainty measures in the DST framework including Deng entropy E_d, Yager's dissonance measure E_Y, Dubois &Prade's weighted Hartley entropy E_{DP}, Hohle's confusion measure C_H, Klir & Ramer's discord measure D_{KR}, Klir & Parviz's strife measure S_{KP} and George & Pal's total conflict measure TC_{GP}.

Example 5. *In the FOD $X = \{1, 2, ..., 14, 15\}$, 15 certain elements are denoted as element 1, 2, ..., 14, and 15. The mass functions are as follows:*

$$m(\{6\}) = 0.05, m(\{3,4,5\}) = 0.05, m(Y) = 0.8, m(X) = 0.1. \tag{19}$$

The element number in the proposition Y changes from 0 to 14, as is shown in Table 3. If the element number of Y is 0, which means Y is an empty set and the FOD may be incomplete, the BPAs are assigned in the open world assumption. In this case, the uncertainty measures E_d, E_Y, E_{DP}, C_H, D_{KR}, S_{KP} and TC_{GP}, which are defined in the closed world is not applicable (N/A). Mathematically, E_Y and C_H can be applied to calculate the uncertain degree if and only if the constraint of "BPAs are for focal element" is ignored which means a possible modification of the definition of Yager's dissonance measure E_Y and Hohle's confusion measure C_H. If the element number of Y changes from 1 to 14, then all the uncertainty measures presented in the Section Preliminaries are available for measuring the uncertain degree. The uncertain degree of the BPAs with different uncertainty measures are presented in Table 3, where there is a large discrepancy among the values of the uncertainty measures especially for the proposition $Y = \emptyset$. Compared with the analysis in [21], the new changes exist in the nonzero mass value of the empty set. The E_{edeow} is the only proper measure in this case compared with other measures listed in Table 3. Of course, we also believe that there are new proper measures for this case since a new measure is always being proposed, e.g., the new entropy in [23].

Figure 1 presents the uncertain degree of different uncertainty measures visually. Intuitively, if a big mass value is assigned on the empty set, which means a big uncertain degree in the FOD, in this case, the EDEOW can measure the uncertain degree. It seems that Yager's dissonance measure E_Y and Hohle's confusion measure C_H can be generalized to measure the uncertain degree in the open world assumption where the mass value of the empty set is nonzero. However, Figure 1 shows that the uncertain degree measured by E_Y and C_H does not increase along with the increasing element number in the proposition Y. The E_d, E_{DP}, D_{KR}, S_{KP} and TC_{GP} are all not available for uncertainty measure in the open world assumption because of the limitation in the *log* function of the definitions. Above all, the other uncertainty measures in Table 1 can only be applied in the closed world. Only the EDEOW can successfully measure the uncertainty degree of belief functions in this case. In addition, the EDEOW is identical to Deng entropy in the closed world, which ensures a successful possible extension of the Deng entropy.

Table 3. The EDEOW E_{edeow}, Deng entropy E_d, Yager's dissonance measure E_Y, Dubois & Prade's weighted Hartley entropy E_{DP}, Hohle's confusion measure C_H, Klir & Ramer's discord measure D_{KR}, Klir & Parviz's strife measure S_{KP} and George & Pal's total conflict measure TC_{GP} with the variable proposition Y. (For computing the E_Y and C_H, we treat '$Y = \emptyset$' as a special proposition in this case to compute the corresponding values.)

Proposition	E_{edeow}	E_d	E_Y	E_{DP}	C_H	D_{KR}	S_{KP}	TC_{GP}
$Y=\emptyset$	14.7216	N/A	(0.5312)	N/A	(1.0219)	N/A	N/A	N/A
$Y=\{1\}$	2.6623	2.6623	0.3952	0.4699	1.0219	6.4419	3.3804	0.3317
$Y=\{1,2\}$	3.9303	3.9303	0.3952	1.2699	1.0219	5.6419	3.2956	0.3210
$Y=\{1,2,3\}$	4.9082	4.9082	0.1997	1.7379	1.0219	4.2823	2.9709	0.2943
$Y=\{1,...,4\}$	5.7878	5.7878	0.1997	2.0699	1.0219	3.6863	2.8132	0.2677
$Y=\{1,...,5\}$	6.6256	6.6256	0.1997	2.3274	1.0219	3.2946	2.7121	0.2410
$Y=\{1,...,6\}$	7.4441	7.4441	0.0074	2.5379	1.0219	2.4888	2.4992	0.2250
$Y=\{1,...,7\}$	8.2532	8.2532	0.0074	2.7158	1.0219	2.4562	2.5198	0.2219
$Y=\{1,...,8\}$	9.0578	9.0578	0.0074	2.8699	1.0219	2.4230	2.5336	0.2170
$Y=\{1,...,9\}$	9.8600	9.8600	0.0074	3.0059	1.0219	2.3898	2.5431	0.2108
$Y=\{1,...,10\}$	10.6612	10.6612	0.0074	3.1275	1.0219	2.3568	2.5494	0.2037
$Y=\{1,...,11\}$	11.4617	11.4617	0.0074	3.2375	1.0219	2.3241	2.5536	0.1959
$Y=\{1,...,12\}$	12.2620	12.2620	0.0074	3.3379	1.0219	2.2920	2.5562	0.1877
$Y=\{1,...,13\}$	13.0622	13.0622	0.0074	3.4303	1.0219	2.2605	2.5577	0.1791
$Y=\{1,...,14\}$	13.8622	13.8622	0.0074	3.5158	1.0219	2.2296	2.5582	0.1701

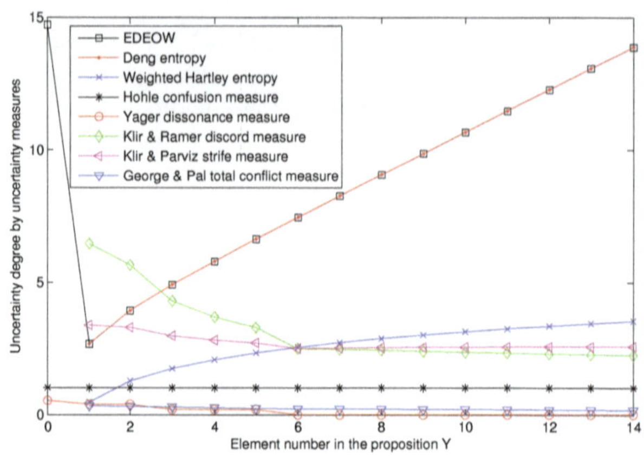

Figure 1. Comparison among different uncertainty measures.

3.3. A Discussion on the Properties of the Extended Measure

As is discussed in [24], the Deng entropy does not match some of the essential properties for a uncertainty measure in the DST framework. In detail, the Deng entropy satisfies the property of 'probabilistic consistency', but the properties such as the 'set consistency', the 'subadditivity', the 'additivity' and the 'monotonicity' are all broken by the Deng entropy. In addition, the range of the Deng entropy is greater than $[0, \log_2 |X|]$. Since the EDEOW is just a simple extension of the Deng entropy, the EDEOW inherits the shortcomings of the Deng entropy with respect to these properties; this should be addressed in the following work.

We noticed that there are new rules of properties defined in a recent research [23], which should be taken into consideration in the ongoing work. Although the extension to Deng's entropy in the open world assumption only satisfies the property of 'Probabilistic consistency', we noticed that the newly defined measure in [23] does not satisfy the 'subadditivity' property, and the distance-based measure in [20] does not satisfy the properties of 'probability consistency' and 'set consistency'. In short, the property of the belief entropy is still an open issue in the closed world as well as the open world assumption in the DST framework.

4. EDEOW-Based Uncertain Information Fusion Approach

An uncertain information fusion approach based on the EDEOW is proposed to illustrate the usefulness and applicability of the extended measure. The framework of the new approach based on the EDEOW is presented in Figure 2, which is a modification of the methods in [44,54]. Firstly, the uncertain information in the closed world and the open world assumption are modelled as BPAs in DST framework. Then, the EDEOW is adopted to measure the uncertain degree of the BPAs without distinguishing the difference of belief functions in the closed world or the open world assumption, which is accomplished by the advantages that the EDEOW is the extension of an uncertainty measure from the closed world. After that, the uncertain degree measured by the EDEOW is used as the weight of each BPA for modification of the BPAs. Finally, the generalized combination rule in [30] is adopted to combine the BPAs. As a result, applications will be based on the fusion results, such as decision making and fault diagnosis.

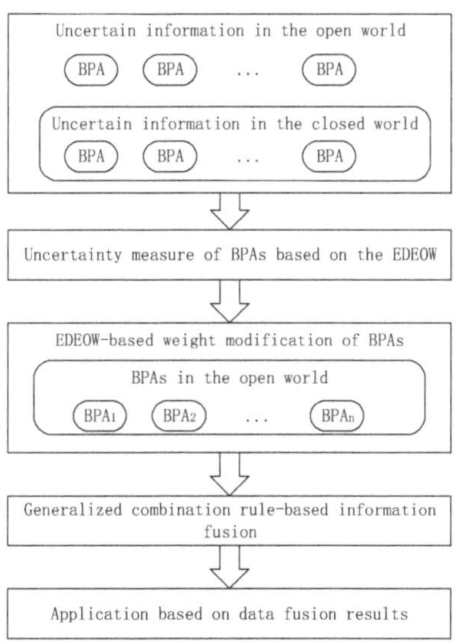

Figure 2. Framework of EDEOW-based uncertain information fusion approach in the open world.

The case study in [55] is adopted and modified for verifying the effectiveness of the extended measure, as well as illustrating the EDEOW-based information fusion approach in Figure 2. According to experience and historical data, there are three types of identified fault types in the motor rotor denoted as $F_1 = \{Rotor\ \ unbalance\}$, $F_2 = \{Rotor\ \ misalignment\}$ and $F_3 = \{Pedestal\ \ looseness\}$ respectively. The vibration signal is collected by three acceleration sensors placed in different positions. The acceleration sensors can collect the signals at different frequencies denoted as $Freq1$, $Freq2$ and $Freq3$, the signals will be used as the judgement variables of fault types. The monitoring results of sensors are modelled as BPAs in Table 4 adopted from [55].

Table 4. Data for fault diagnosis modelled as BPAs [55].

	Freq1				*Freq2*		*Freq3*			
	$\{F2\}$	$\{F3\}$	$\{F1,F2\}$	$\emptyset$	$\{F2\}$	$\emptyset$	$\{F1\}$	$\{F2\}$	$\{F1,F2\}$	$\emptyset$
$m_{s_1}(\cdot)$	0.8176	0.0003	0.1553	0.0268	0.6229	0.3771	0.3666	0.4563	0.1185	0.0586
$m_{s_2}(\cdot)$	0.5658	0.0009	0.0646	0.3687	0.7660	0.2341	0.2793	0.4151	0.2652	0.0404
$m_{s_3}(\cdot)$	0.2403	0.0004	0.0141	0.7452	0.8598	0.1402	0.2897	0.4331	0.2470	0.0302

For each frequency, the BPAs reported by three sensors are denoted as $m_{s_1}(\cdot)$, $m_{s_2}(\cdot)$ and $m_{s_3}(\cdot)$. $\{F1,F2,F3\}$ is the FOD of this application in the closed world. Here, in this paper, in order to adapt the experiment data for the application of the extended measure in the open world, the belief functions of $\{F1,F2,F3\}$ are assumed to be assigned to the empty set $\emptyset$, which extends the uncertainty of the FOD from the closed world to the open world. This is reasonable, because there may exist unknown fault types.

4.1. Uncertainty Measure of BPAs with EDEOW

In real applications, the reliability of each sensor is unknown. Thus, the uncertain degree of sensor reports should be measured properly. In the DST framework, the belief entropy is proposed for measuring the uncertainty of BPAs. Once the sensor reports are modelled as BPAs, the uncertain degree of sensor reports can be measured based on the EDEOW in Equation (12). For example, for the BPAs of *Freq*1, the uncertain degree with the EDEOW is calculated as follows:

$$\begin{aligned}
E_{edeow}(m_{s_1}) &= -0.8176\log_2\frac{0.8176}{2^{1+\lceil 0.0268\times 3\rceil}-1} \\
&\quad -0.0003\log_2\frac{0.0003}{2^{1+\lceil 0.0268\times 3\rceil}-1} \\
&\quad -0.1553\log_2\frac{0.1553}{2^{2+\lceil 0.0268\times 3\rceil}-1} \\
&\quad -0.0268\log_2\frac{0.0268}{2^{0+\lceil 0.0268\times 3\rceil}-1} \\
&= 2.5306, \\
E_{edeow}(m_{s_2}) &= -0.5658\log_2\frac{0.5658}{2^{2+\lceil 0.3687\times 3\rceil}-1} \\
&\quad -0.0009\log_2\frac{0.0009}{2^{2+\lceil 0.3687\times 3\rceil}-1} \\
&\quad -0.0646\log_2\frac{0.0646}{2^{2+\lceil 0.3687\times 3\rceil}-1} \\
&\quad -0.3687\log_2\frac{0.3687}{2^{0+\lceil 0.3687\times 3\rceil}-1} \\
&= 3.6877, \\
E_{edeow}(m_{s_3}) &= -0.2403\log_2\frac{0.2403}{2^{2+\lceil 0.7452\times 3\rceil}-1} \\
&\quad -0.0004\log_2\frac{0.0004}{2^{2+\lceil 0.7452\times 3\rceil}-1} \\
&\quad -0.0141\log_2\frac{0.0141}{2^{2+\lceil 0.7452\times 3\rceil}-1} \\
&\quad -0.7452\log_2\frac{0.7452}{2^{0+\lceil 0.7452\times 3\rceil}-1} \\
&= 4.0040.
\end{aligned} \tag{20}$$

The uncertain degree of *Freq*2 and *Freq*3 can also be calculated by Equation (12). The results are presented in Table 5.

Table 5. Uncertainty measure results of sensor reports based on EDEOW.

$E_{edeow}(\cdot)$	*Freq1*	*Freq2*	*Freq3*
$E_{edeow}(m_{s_1})$	2.5306	3.3024	3.2887
$E_{edeow}(m_{s_2})$	3.6877	1.9991	3.5804
$E_{edeow}(m_{s_3})$	4.0040	1.9475	3.5305

4.2. EDEOW-Based Modification of BPAs

The EDEOW of each BPA is used as the weight factor of each sensor report. With a process of normalization, the weight of each BPA in *Freq*1 is calculated as follows:

$$
\begin{aligned}
w_{S_1} &= E_{edeow}(m_{S_1}) \;/\; \sum_{i=1}^{3} E_{edeow}(m_{S_i}) = 0.2476, \\
w_{S_2} &= E_{edeow}(m_{S_2}) \;/\; \sum_{i=1}^{3} E_{edeow}(m_{S_i}) = 0.3608, \\
w_{S_3} &= E_{edeow}(m_{S_3}) \;/\; \sum_{i=1}^{3} E_{edeow}(m_{S_i}) = 0.3917.
\end{aligned} \tag{21}
$$

Similarly, the weight of BPAs in *Freq*2 and *Freq*3 can be calculated. After normalization, the weight of each BPA for *Freq*1, *Freq*2 and *Freq*3 is listed in Table 6.

Table 6. EDEOW-based weight factor of BPAs after normalization.

w_{S_i}	*Freq*1	*Freq*2	*Freq*3
w_{S_1}	0.2476	0.4556	0.3162
w_{S_2}	0.3608	0.2758	0.3443
w_{S_3}	0.3917	0.2687	0.3395

The modification of BPAs for each frequency can be calculated with the following equation:

$$m_w(\cdot) = w_{S_i} m_{S_i}. \tag{22}$$

Based on the normalized weight factor in Table 6, with Equation (22), the modified BPAs of *Freq*1 is calculated as follows:

$$
\begin{aligned}
& m_w(\{F2\}) = 0.5006, \\
& m_w(\{F3\}) = 0.0005, \\
& m_w(\{F1, F2\}) = 0.0673, \\
& m_w(\emptyset) = 0.4315.
\end{aligned} \tag{23}
$$

The modification of BPAs for *Freq*2 and *Freq*3 can be calculated with Equation (22). The BPAs after modification of each frequency is shown in Table 7.

Table 7. Modified mass function based on EDEOW.

	*Freq*1				*Freq*2		*Freq*3			
	$\{F2\}$	$\{F3\}$	$\{F1, F2\}$	$\emptyset$	$\{F2\}$	$\emptyset$	$\{F1\}$	$\{F2\}$	$\{F1, F2\}$	$\emptyset$
$m_w(\cdot)$	0.5006	0.0005	0.0673	0.4315	0.7260	0.2740	0.3104	0.4342	0.2126	0.0427

4.3. Generalized Combination Rule-Based Data Fusion

In the open world assumption, classical Dempster's rule of combination is not applicable [30]. In this paper, the generalized combination rule in [30] is chosen for data fusion in the proposed approach. Since the original n sets of BPAs have been modified as one set of BPAs by EDEOW-based weight factors, the modified BPAs should be fused $(n-1)$ times according to the chosen combination rule.

There are three sets of BPAs before modification. Thus, the modified BPAs should be combined two times with generalized combination rule in Equation (7). For frequency *Freq*1, the fusion results are shown as follows:

$$\begin{aligned} &m(\{F2\}) = ((m_w \oplus m_w)_1 \oplus m_w)_2 (\{F2\}) = 0.9181, \\ &m(\{F3\}) = ((m_w \oplus m_w)_1 \oplus m_w)_2 (\{F3\}) = 0.0000, \\ &m(\{F1, F2\}) = ((m_w \oplus m_w)_1 \oplus m_w)_2 (\{F1, F2\}) = 0.0015, \\ &m(\emptyset) = ((m_w \oplus m_w)_1 \oplus m_w)_2 (\emptyset) = 0.0803. \end{aligned} \tag{24}$$

The BPAs of *Freq*2 and *Freq*3 are also fused three times with the generalized combination rule, the results are shown in Table 8.

Table 8. Sensor data fusion results with different methods.

	*Freq*1				*Freq*2		*Freq*3			
	$\{F2\}$	$\{F3\}$	$\{F1, F2\}$	$\emptyset$	$\{F2\}$	$\emptyset$	$\{F1\}$	$\{F2\}$	$\{F1, F2\}$	$\emptyset$
Jiang et al.'s method [55]	0.8861	0.0002	0.0582	-	0.9621	-	0.3384	0.5904	0.0651	-
Tang et al.'s method [54]	0.8891	0.0003	0.0427	-	0.9784	-	0.3318	0.6332	0.0349	-
The proposed method	0.9181	0.0000	0.0015	0.0803	0.9796	0.0206	0.3303	0.6459	0.0238	0.0001

With the fusion results shown in Table 8, *F*2 significantly has the highest support degree among all the frequencies, therefore, we can judge that the fault type is *F*2. The experiment results are consistent with [54,55], which verifies the effectiveness of the EDEOW. In addition, the proposed method has a higher support degree on the recognized fault type *F*2 than that in [54,55], which is good for decision-making by engineers in real applications.

5. Open Issues for Future Work

There is no universally accepted measure for uncertainty quantification in the DST framework. Many new measures are still being proposed within one year [23,56]. To match the open world assumption [19,30,52,53], an extended measure for quantification of uncertain degree in the DST framework is proposed in this paper. It should be noted that the extended measure is a simple extension of the Deng entropy. A lot of open issues exist in the extended measure as well as the other measures for the open world assumption in the DST framework.

The first one exists in the scope of the uncertainty measures in the DST framework. According to the current research, we find that the theory of belief entropy or uncertainty measures in the DST is still not solid and needs further deep research. We suggest that the following research work on this topic should take into consideration the open world assumption.

The second open issue exists in the properties of the extended measure, which is a shortcoming inherited from the Deng entropy. According to the research work in [24], the Deng entropy only satisfies the property of probabilistic consistency with respect to the five requirements for a total uncertainty measure. The following work should focus on improving the measure or developing a totally new uncertainty measure for the open world assumption by taking into consideration all of the properties discussed in [22–24,57].

Thirdly, the following work needs to investigate what happens if the mass on the empty set is not null with different size of the universe because the new measure in the open world must address these two parameters. In addition, what is the meaning of having an entropy measure that changes in accordance with the cardinality of the universe? For instance, for a FOD $|X|$, we will have the same measure result $E_{edeow}(m) = 0$ for the mass function $m(\emptyset) = 1$ and $m(\{a\}) = 1$. Currently, we have difficulty answering all of these questions in this simple extended measure.

Fourthly, there are still no universally accepted properties for a belief entropy or uncertainty measure in the closed world and for the open world assumption, which is a big problem for developing a new belief entropy. For example, even the newly defined measure in [23] does not satisfy the 'subadditivity' property. Another example is that the measure in [20] does not satisfy the properties of 'probability consistency' and 'set consistency'. We believe that there are new properties that should be obeyed by the measures in the open world assumption.

Finally, in the application of sensor data fusion, fusion frame and combination rule need further study. There are more fusion methods and the combination rules in the research works [25,26,53,58] that need to be investigated cautiously.

6. Conclusions

An extended uncertainty measure for belief structures in the open world assumption, named the EDEOW, is proposed in this paper. The extended measure can successfully quantify the uncertain degree of belief structures not only in the closed world, but also in the open world. With the extended measure, more uncertain information in DST framework is taken into consideration while applying information processing, including the possible incomplete FOD and the nonzero mass function of the empty set, of which both are sources of uncertainty in the DST framework in the open world assumption. To verify the usefulness and applicability of the extended measure, the EDEOW is adopted to design a new information fusion approach in the open world circumstance. Numerical examples and the application on the sensor data fusion-based fault diagnosis verify the effectiveness of the proposed method. The limitations and open issues for possible future research are also discussed in this paper.

Author Contributions: Y.T., D.Z. and F.T.S.C. designed and performed the research. Y.T. wrote and revised the manuscript. Y.T. performed the computation. D.Z. and F.T.S.C. analyzed the data. All authors discussed the results and commented on the manuscript.

Funding: The work is partially supported by the Innovation Foundation for Doctor Dissertation of Northwestern Polytechnical University (Grant No. CX201705).

Acknowledgments: The authors greatly appreciate the editor's encouragements and the three anonymous reviewers' valuable and constructive comments which have led to improve the manuscript significantly.

Conflicts of Interest: The authors declare no conflict of interest. The founding sponsors had no role in the design of the study; in the collection, analyses or interpretation of data; in the writing of the manuscript; nor in the decision to publish the results.

References

1. Deng, Z.; Zhang, P.; Qi, W.; Liu, J.; Gao, Y. Sequential covariance intersection fusion Kalman filter. *Inf. Sci.* **2012**, *189*, 293–309. [CrossRef]
2. Dou, Y.; Ran, C.; Gao, Y. Weighted measurement fusion Kalman estimator for multisensor descriptor system. *Int. J. Syst. Sci.* **2016**, *47*, 2722–2732. [CrossRef]
3. Zhou, K.; Martin, A.; Pan, Q.; Liu, Z. SELP: Semi–supervised evidential label propagation algorithm for graph data clustering. *Int. J. Approx. Reason.* **2018**, *92*, 139–154. [CrossRef]
4. Liu, Z.G.; Pan, Q.; Dezert, J.; Mercier, G. Hybrid classification system for uncertain data. *IEEE Trans. Syst. Man Cybern. Syst.* **2017**, *47*, 2783–2790. [CrossRef]
5. Sabahi, F.; Akbarzadeh-T, M.R. Introducing validity in fuzzy probability for judicial decision-making. *Int. J. Approx. Reason.* **2014**, *55*, 1383–1403. [CrossRef]
6. Chan, F.T.; Bhagwat, R.; Chan, H.K. The effect of responsiveness of the control-decision system to the performance of FMS. *Comput. Ind. Eng.* **2014**, *72*, 32–42. [CrossRef]
7. Tang, Y.; Zhou, D.; Xu, S. An improved physarum solver for supply chain network with fuzzy numbers. *Int. J. Unconv. Comput.* **2017**, *13*, 161–184.
8. Chan, F.T.S.; Chan, H.K. An AHP model for selection of suppliers in the fast changing fashion market. *Int. J. Adv. Manuf. Technol.* **2010**, *51*, 1195–1207. [CrossRef]
9. Zhang, X.; Mahadevan, S. A game theoretic approach to network reliability assessment. *IEEE Trans. Reliab.* **2017**, *66*, 875–892. [CrossRef]
10. Gao, Y.; Ran, C.J.; Sun, X.J.; Deng, Z.L. Optimal and self-tuning weighted measurement fusion Kalman filters and their asymptotic global optimality. *Int. J. Adapt. Control Signal Process.* **2010**, *24*, 982–1004. [CrossRef]

11. Jin, X.; Du, J.; Bao, J. Target tracking of a linear time invariant system under irregular sampling regular paper. *Int. J. Adv. Robot. Syst.* **2012**, *9*, 219.
12. Dempster, A.P. Upper and lower probabilities induced by a multi-valued mapping. *Ann. Math. Stat.* **1967**, *38*, 325–339. [CrossRef]
13. Shafer, G. *A Mathematical Theory of Evidence*; Princeton University Press: Princeton, NJ, USA, 1976.
14. Su, X.; Mahadevan, S.; Xu, P.; Deng, Y. Dependence assessment in human reliability analysis using evidence theory and AHP. *Risk Anal.* **2015**, *35*, 1296–1316. [CrossRef] [PubMed]
15. Dubois, D.; Liu, W.; Ma, J.; Prade, H. The basic principles of uncertain information fusion. An organized review of merging rules in different representation frameworks. *Inf. Fusion* **2016**, *32*, 12–39. [CrossRef]
16. Jøsang, A.; Daniel, M.; Vannoorenberghe, P. Strategies for combining conflicting dogmatic beliefs. In Proceedings of the 6th International Conference on Information Fusion, Cairns, Australia, 8–11 July 2003; pp. 1133–1140.
17. Schubert, J. Conflict management in Dempster-Shafer theory using the degree of falsity. *Int. J. Approx. Reason.* **2011**, *52*, 449–460. [CrossRef]
18. Tang, Y.; Zhou, D.; Jiang, W. A new fuzzy-evidential controller for stabilization of the planar inverted pendulum system. *PLoS ONE* **2016**, *11*, e0160416. [CrossRef] [PubMed]
19. Jiang, W.; Zhan, J.; Zhou, D.; Li, X. A method to determine generalized basic probability assignment in the open world. *Math. Probl. Eng.* **2016**, *2016*, 3878634. [CrossRef]
20. Yang, Y.; Han, D. A new distance-based total uncertainty measure in the theory of belief functions. *Knowl. Based Syst.* **2016**, *94*, 114–123. [CrossRef]
21. Deng, Y. Deng entropy. *Chaos Solitons Fractals* **2016**, *91*, 549–553. [CrossRef]
22. Harmanec, D.; Klir, G.J. Measuring total uncertainty in Dempster-Shafer theory: A novel approach. *Int. J. Gen. Syst.* **1994**, *22*, 405–419. [CrossRef]
23. Jirousek, R.; Shenoy, P.P. A new definition of entropy of belief functions in the Dempster-Shafer theory. *Int. J. Approx. Reason.* **2018**, *92*, 49–65. [CrossRef]
24. Abellan, J. Analyzing properties of Deng entropy in the theory of evidence. *Chaos Solitons Fractals* **2017**, *95*, 195–199. [CrossRef]
25. Smets, P. Belief functions: The disjunctive rule of combination and the generalized Bayesian theorem. *Int. J. Approx. Reason.* **1993**, *9*, 1–35. [CrossRef]
26. Murphy, C.K. Combining belief functions when evidence conflicts. *Decis. Support Syst.* **2000**, *29*, 1–9. [CrossRef]
27. Pichon, F.; Destercke, S.; Burger, T. A consistency-specificity trade-off to select source behavior in information fusion. *IEEE Trans. Cybern.* **2015**, *45*, 598–609. [CrossRef] [PubMed]
28. Smets, P. The combination of evidence in the transferable belief model. *IEEE Trans. Pattern Anal. Mach. Intell.* **1990**, *12*, 447–458. [CrossRef]
29. Daniel, M. A relationship of conflicting belief masses to open world assumption. In Proceedings of the International Conference on Belief Functions, Prague, Czech Republic, 21–23 September 2016; pp. 146–155.
30. Deng, Y. Generalized evidence theory. *Appl. Intell.* **2015**, *43*, 530–543. [CrossRef]
31. Yager, R.R. On the fusion of imprecise uncertainty measures using belief structures. *Inf. Sci.* **2011**, *181*, 3199–3209. [CrossRef]
32. Hohle, U. Entropy with respect to plausibility measures. In Proceedings of the 12th IEEE International Symposium on Multiple-Valued Logic, Paris, France, 25–26 May 1982; pp. 167–169.
33. Yager, R.R. Entropy and specificity in a mathematical theory of evidence. *Int. J. Gen. Syst.* **1983**, *9*, 249–260. [CrossRef]
34. Dubois, D.; Prade, H. A note on measures of specificity for fuzzy sets. *Int. J. Gen. Syst.* **1985**, *10*, 279–283. [CrossRef]
35. Klir, G.J.; Ramer, A. Uncertainty in Dempster–Shafer theory: A critical re-examination. *Int. J. Gen. Syst.* **1991**, *18*, 155–166. [CrossRef]
36. Klir, G.J.; Parviz, B. A note on the measure of discord. In Proceedings of the Eighth International Conference on Uncertainty in Artificial Intelligence, Stanford, CA, USA, 17–19 July 1992; pp. 138–141.
37. Yager, R.R. Dempster-Shafer structures with general measures. *Int. J. Gen. Syst.* **2012**, *41*, 395–408. [CrossRef]
38. Denoeux, T. Maximum likelihood estimation from uncertain data in the belief function framework. *IEEE Trans. Knowl. Data Eng.* **2013**, *25*, 119–130. [CrossRef]

39. Daniel, M. Conflict between belief functions: A new measure based on their non-conflicting parts. In Proceedings of the Third International Conference on Belief Functions: Theory and Applications, Oxford, UK, 26–28 September 2014; pp. 321–330.
40. Zhou, D.; Tang, Y.; Jiang, W. A modified belief entropy in Dempster-Shafer framework. *PLoS ONE* **2017**, *12*, e0176832. [CrossRef] [PubMed]
41. Wang, X.; Song, Y. Uncertainty measure in evidence theory with its applications. *Appl. Intell.* **2017**, 1–7. [CrossRef]
42. Song, Y.; Wang, X.; Lei, L.; Yue, S. Uncertainty measure for interval-valued belief structures. *Measurement* **2016**, *80*, 241–250. [CrossRef]
43. Yuan, K.; Xiao, F.; Fei, L.; Kang, B.; Deng, Y. Conflict management based on belief function entropy in sensor fusion. *SpringerPlus* **2016**, *5*, 638. [CrossRef] [PubMed]
44. Zhou, D.; Tang, Y.; Jiang, W. An improved belief entropy and its application in decision-making. *Complexity* **2017**, *2017*, 4359195. [CrossRef]
45. Tang, Y.; Zhou, D.; Zhuang, M.; Fang, X.; Xie, C. An improved evidential-IOWA sensor data fusion approach in fault diagnosis. *Sensors* **2017**, *17*, 2143. [CrossRef] [PubMed]
46. Cao, S.; Dehmer, M.; Shi, Y. Extremality of degree-based graph entropies. *Inf. Sci.* **2014**, *278*, 22–33. [CrossRef]
47. Chen, Z.; Dehmer, M.; Shi, Y. A note on distance-based graph entropies. *Entropy* **2014**, *16*, 5416–5427. [CrossRef]
48. Cao, S.; Dehmer, M. Degree-based entropies of networks revisited. *Appl. Math. Comput.* **2015**, *261*, 141–147. [CrossRef]
49. Shannon, C.E. A mathematical theory of communication. *ACM Sigmob. Mob. Comput. Commun. Rev.* **2001**, *5*, 3–55. [CrossRef]
50. Tang, Y.; Zhou, D.; He, Z.; Xu, S. An improved belief entropy-based uncertainty management approach for sensor data fusion. *Int. J. Distrib. Sens. Netw.* **2017**, *13*, 1550147717718497. [CrossRef]
51. George, T.; Pal, N.R. Quantification of conflict in Dempster-Shafer framework: A new approach. *Int. J. Gen. Syst.* **1996**, *24*, 407–423. [CrossRef]
52. Smets, P.; Kennes, R. The transferable belief model. *Artif. Intell.* **1994**, *66*, 191–234. [CrossRef]
53. Smets, P. Analyzing the combination of conflicting belief functions. *Inf. Fusion* **2007**, *8*, 387–412. [CrossRef]
54. Tang, Y.; Zhou, D.; Xu, S.; He, Z. A weighted belief entropy-based uncertainty measure for multi-sensor data fusion. *Sensors* **2017**, *17*, 928. [CrossRef] [PubMed]
55. Jiang, W.; Xie, C.; Zhuang, M.; Shou, Y.; Tang, Y. Sensor data fusion with Z-numbers and its application in fault diagnosis. *Sensors* **2016**, *16*, 1509. [CrossRef] [PubMed]
56. Schubert, J. Counter-deception in information fusion. *Int. J. Approx. Reason.* **2017**, *91*, 152–159. [CrossRef]
57. Destercke, S.; Burger, T. Toward an axiomatic definition of conflict between belief functions. *IEEE Trans. Cybern.* **2013**, *43*, 585–596. [CrossRef] [PubMed]
58. Su, X.; Mahadevan, S.; Han, W.; Deng, Y. Combining dependent bodies of evidence. *Appl. Intell.* **2015**, *44*, 634–644. [CrossRef]

sensors

MDPI

Article

Multi-Focus Fusion Technique on Low-Cost Camera Images for Canola Phenotyping

Thang Cao [1], Anh Dinh [1,*], Khan A. Wahid [1], Karim Panjvani [1] and Sally Vail [2]

[1] Department of Electrical and Computer Engineering, University of Saskatchewan, Saskatoon, SK S7N 5A9, Canada; thang.cao@usask.ca (T.C.); khan.wahid@usask.ca (K.A.W.); karim.panjvani@usask.ca (K.P.)

[2] Agriculture and Agri-Food Canada, Ottawa, ON K1A 0C5, Canada; Sally.Vail@agr.gc.ca

* Correspondence: anh.dinh@usask.ca; Tel.: +1-306-966-5344

Received: 19 March 2018; Accepted: 5 June 2018; Published: 8 June 2018

Abstract: To meet the high demand for supporting and accelerating progress in the breeding of novel traits, plant scientists and breeders have to measure a large number of plants and their characteristics accurately. Imaging methodologies are being deployed to acquire data for quantitative studies of complex traits. Images are not always good quality, in particular, they are obtained from the field. Image fusion techniques can be helpful for plant breeders with more comfortable access plant characteristics by improving the definition and resolution of color images. In this work, the multi-focus images were loaded and then the similarity of visual saliency, gradient, and color distortion were measured to obtain weight maps. The maps were refined by a modified guided filter before the images were reconstructed. Canola images were obtained by a custom built mobile platform for field phenotyping and were used for testing in public databases. The proposed method was also tested against the five common image fusion methods in terms of quality and speed. Experimental results show good re-constructed images subjectively and objectively performed by the proposed technique. The findings contribute to a new multi-focus image fusion that exhibits a competitive performance and outperforms some other state-of-the-art methods based on the visual saliency maps and gradient domain fast guided filter. The proposed fusing technique can be extended to other fields, such as remote sensing and medical image fusion applications.

Keywords: image fusion; multi-focus; weight maps; gradient domain; fast guided filter.

1. Introduction

The sharp increase in demand for global food raises the awareness of the public, especially agricultural scientists, to global food security. To meet the high demand for food in 2050, agriculture will need to produce almost 50 percent more food than was produced in 2012 [1]. There are many ways to improve yields for canola and other crops. One of the solutions is to increase breeding efficiency. In the past decade, advances in genetic technologies, such as next generation DNA sequencing, have provided new methods to improve plant breeding techniques. However, the lack of knowledge of phenotyping capabilities limits the ability to analyze the genetics of quantitative traits related to plant growth, crop yield, and adaptation to stress [2]. Phenotyping creates opportunities not only for functional research on genes, but also for the development of new crops with beneficial features. Image-based phenotyping methods are those integrated approaches that enable the potential to greatly enhance plant researchers' ability to characterize many different traits of plants. Modern advanced imaging methods provide high-resolution images and enable the visualization of multi-dimensional data. The basics of image processing have been thoroughly studied and published. Readers can find useful information on image fusion in the textbooks by Starck or Florack [3,4]. These methods allow plant breeders and researchers to obtain exact data, speed up image analysis, bring high throughput and high dimensional phenotype data for modeling, and estimate plant growth and structural development during the plant life cycle. However, with low-cost and low-resolution sensors, phenotyping would meet some obstacles to solve low-resolution images. To cope with this issue, image fusion is a valuable choice.

Image fusion is a technique that combines many different images to generate a fused image with highly informative and reliable information. There are several image fusion types, such as multi-modal and multi-sensor image fusion. In multi-modal image fusion, two different kinds of images are fused, for example, combining a high-resolution image with a high-color image. In multi-sensor image fusion, images from different types of sensors are combined, for example, combining an image from a depth sensor with an image from a digital sensor. Image fusion methods can be divided into three levels depending on the processing: pixel level, feature level, and decision level [5,6]. Image fusion at the pixel level refers to an imaging process that occurs in the pixel-by-pixel manner in which each new pixel of the fused image obtains a new value. At a higher level than the pixel level, feature-level image fusion first extracts the relevant key features from each of the source images and then combines them for image-classification purposes such as edge detection. Decision-level image fusion (also named as interpretation-level or symbol-level image fusion) is the highest level of image fusion. Decision-level image fusion refers to a type of fusion in which the decision is made based on the information separately extracted from several image sources.

Over the last two decades, image fusion techniques have been widely applied in many areas: medicine, mathematics, physics, and engineering. In plant science, many image fusion techniques are being deployed to improve the classification accuracy for determining plant features, detecting plant diseases, and measuring crop diversification. Fan et al. [7] well implemented a Kalman filtering fusion to improve the accuracy of the prediction on the citrus maturity. In a related work, a feature-level fusion technique [8] was successfully developed to detect some types of leaf disease with excellent results. In other similar research, apple fruit diseases were detected by using feature-level fusion in which two or more color and feature textures were combined [9]. Decision-level fusion techniques have been deployed to detect crop contamination and plague [10]. Dimov et al. [11] have also implemented the Ehler's fusion algorithm (decision level) to measure the diversification of the three critical crop systems with the highest classification accuracy. These findings suggest that image-fusion techniques at many levels are broadly applied in the plant science sector because they offer the highest classification accuracy.

Currently, many multi-focus image fusion techniques have been developed [12,13]. The techniques can be categorized into two classes: spatial domain methods and frequency domain methods. In the spatial domain methods, source images are directly fused, in which the information of image pixels are directly used without any pre-processing or post-processing. In the frequency domain methods, source images are transformed into frequency domain, and then combined. Therefore, frequency domain methods are more complicated and time-consuming than spatial domain methods. Many studies investigated multi-focus image fusion techniques in spatial and frequency domains to improve the outcomes. Wan et al. [14] proposed a method based on the robust principal component analysis in the spatial domain. They developed this method to form a robust fusion technique to distinguish focused and defocused areas. The method outperforms wavelet-based fusion methods and provides better visual perception; however, it has a high computation cost. In the similar spatial domain, a multi-focus image fusion method based on region [15] was developed, in which, their algorithm offers smaller distortion and a better reflection of the edge information and importance of the source image. Similarly, Liu et al. [16] investigated a fusion technique based on dense scale invariant feature transform (SIFT) in the spatial domain. The method performs better than other techniques in terms of visual perception and performance evaluation, but it requires a high amount of memory. In the frequency domain, Phamila and Amutha [17] conducted a method based on Discrete Cosine Transform. The process computes and obtains the highest variance of the 8×8 DCT coefficients to reconstruct the fused image. However, the method suffers from undesirable side effects such as blurring and artifact. In a recently published article, the authors reviewed the works using sparse representation (SR)-based methods on multi-sensor systems [18]. Based on sparse representation, the same authors also developed the image fusing method for multi-focus and multi-modality images [19]. This SR method learns an over-complete dictionary from a set of training images for image fusion, it may result in a huge increment of computational complexity.

To deal with these obstacles mentioned above, a new multi-focus image fusion based on the image quality assessment (IQA) metrics is proposed in this paper. The proposed fusion method is developed based on crucial IQA metrics and a gradient domain fast guided image filter (GDFGIF). This approach is motivated by the fact that visual saliency maps, including visual saliency, gradient similarity, and chrominance similarity maps, outperform most of the state-of-the-art IQA metrics in terms of the prediction accuracy [20]. According to Reference [20], visual saliency similarity, gradient similarity, and chrominance maps are vital metrics in accounting for the visual quality of image fusion techniques. In most cases, changes of visual saliency (VS) map can be a good indicator of distortion degrees and thus, VS map is used as a local weight map. However, VS map does not work well for the distortion type of contrast change. Fortunately, the image gradient can be used as an additional feature to compensate for the lack of contrast sensitivity of the VS map. In addition, VS map does not work well for the distortion type change of color saturation. This color distortion cannot be well represented by gradient either since usually the gradient is computed from the luminance channel of images. To deal with this color distortion, two chrominance channels are used as features to represent the quality degradation caused by color distortion. These IQA metrics have been proved to be stable and have the best performance [20]. In addition, gradient domain guided filter (GDGIF) [21] and fast guided filter (FGF) [22] are adopted in this work as the combination of GDGIF and FGF and can offer fast and better fused results, especially near the edges, where halo artifacts appear in the original guided image filter. This study focuses on how to fuse multi-focus color images to enhance the resolution and quality of the fused image using a low-cost camera. The proposed image fusion method was developed and compared with other state-of-the-art image fusion methods. In the proposed multi-focus image fusion, two or more images captured by the same sensor from the same visual angle but with a different focus are combined to obtain a more informative image. For example, a fused image with clearer canola seedpods can be produced by fusing many different images of a canola plant acquired by the same Pi camera at the same angle with many different focus lengths.

2. Methodology

2.1. Data Acquisition System

This image fusion work is part of the development of a low-cost, high throughput phenotyping mobile system for canola in which a low-cost Raspberry Pi camera is used as a source of image acquisition. This system includes a 3D Time-of-Flight camera, a Pi camera, a Raspberry Pi3 (RP3), and appropriate power supplies for the cameras and the mini computer (Raspberry Pi3). A built-in remote control allows the user to start and stop image recording as desired. Figure 1 shows various components of the data acquisition system. Data are recorded in the SD card of the RP3 and retrieved using USB connection to a laptop before the images are processed. The Kuman for Raspberry Pi 3 Camera Module with adjustable focus is used in this system. This camera is connected to the Raspberry Pi using the dedicated CSI interface. The Pi camera equips to the 5 megapixels OV5647 sensor. It is capable of capturing 2592 × 1944 pixels static images; it also supports video capturing of 1080 p at 30 fps, 720 p at 60 fps, and 640 × 480 p at 60/90 formats.

The testing subjects were the canola plants at different growing stages. The plants were growing in a controlled environment and also in the field. To capture images of the canola, the plants were directly placed underneath the Pi camera that fixed on the tripod at a distance of 1000 mm (Figure 1). Each canola plant was recorded at 10 fps for 3 s. The time between each change of the focal length is 10 s. Only frame number 20 of each video stream acquired from the Pi camera was extracted to store in the database for later use. The reason for selecting the 20th frame is that the plants and the camera are required to be stable before the images are being captured and processed. Only the regions containing the plant in the selected images were cropped and used for multi-focus image fusion methods.

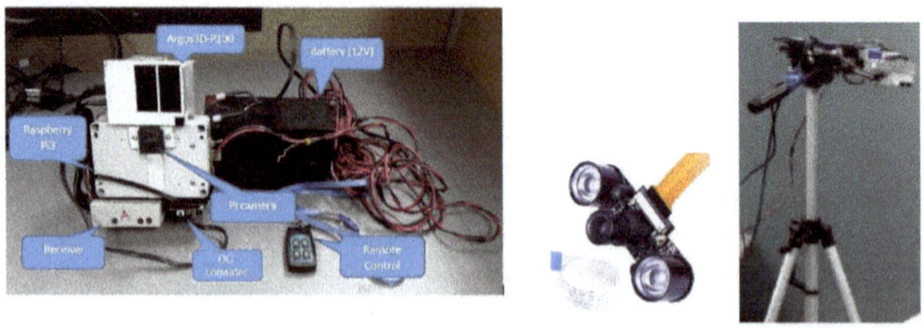

(a) (b) (c)

Figure 1. (**a**) Low-cost mobile phenotyping system; (**b**) Adjustable focus Pi camera; (**c**) System mounted on a tripod looking down to the canola plants.

2.2. Image Fusion Algorithm

In the proposed fusion approach, three image quality assessment (IQA) metrics: visual saliency similarity, gradient similarity, and chrominance similarity (or color distortion) are measured to obtain their weight maps. Then these weight maps are refined by a gradient domain fast guided filter in which, a gradient domain guided filter proposed by Reference [21] and a fast guided filter proposed by Reference [22] are combined. The workflow of the proposed multi-focus image fusion algorithm is illustrated in Figure 2. The detail of the proposed algorithm is described as follows.

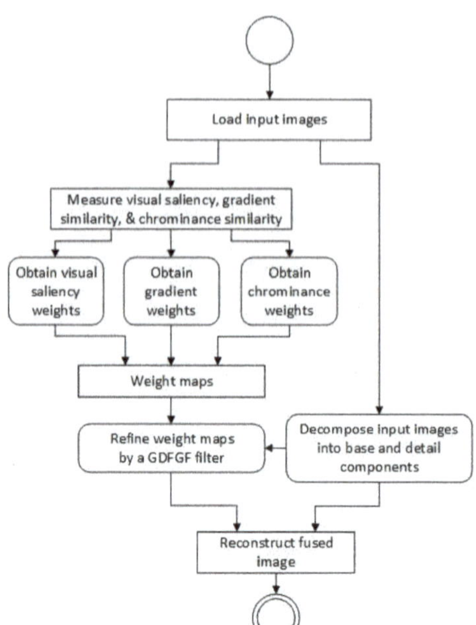

Figure 2. Workflow of the proposed multi-focus image fusion algorithm.

First, each input image is decomposed into a base and detailed component, which contain the large-scale and small-scale variations in intensity. A Gaussian filter is used for each source image to obtain its base component, and the detailed component can be easily obtained by subtracting the base component from the input image, as given by:

$$B_n = I_n * G_{r,\sigma} \tag{1}$$

$$D_n = I_n - B_n \tag{2}$$

where B_n and D_n are the base and detail component of the nth input image, respectively. $*$ denotes convolution operator, and $G_{r,\sigma}$ is a 2-D Gaussian smoothing filter.

Several measures were used to obtain weight maps for image fusing. Visual saliency similarity, gradient similarity, and chrominance maps are vital metrics in accounting for the visual quality of image fusion techniques [20]. In most cases, changes of visual saliency (VS) map can be a good indicator of distortion degrees and thus, VS map is used as a local weight map. However, VS map does not work very well for the distortion type of contrast change. Fortunately, the gradient modulus can be used as an additional feature to compensate for the lack of contrast sensitivity of the VS map. In addition, VS map does not work well for the distortion type change of color saturation. This color distortion cannot be well represented by gradient either since usually gradient is computed from the luminance channel of images. To deal with this color distortion, two chrominance channels are used as features to represent the quality degradation caused by color distortion. Motivated by these metrics, an image fusion method is designed based on the measurement of the three key visual features of input images.

2.2.1. Visual Saliency Similarity Maps

A saliency similarity detection algorithm proposed by [23] is adopted to calculate visual saliency similarity due to its higher accuracy and low computational complexity. This algorithm is constructed by combining three simple priors: frequency, color, and location. The visual saliency similarity maps are calculated as

$$VS_n^k = SF_n^k \cdot SC_n^k \cdot SD_n^k \tag{3}$$

where SF_n^k, SC_n^k, SD_n^k are the saliency at pixel k under frequency, color and location priors. SF_n^k is calculated by

$$SF_n^k = (IL_n^k * g)^2 + (Ia_n^k * g)^2 + (Ib_n^k * g)^2)^{1/2} \tag{4}$$

where IL_n^k, Ia_n^k, Ib_n^k are three resulting channels transformed from the given RGB input image, I_n to CIEL*a*b* space. * denotes the convolution operation. CIEL*a*b* is an opponent color system that a* channel represents green-red information while b* channel represents blue-yellow information. If a pixel has a smaller (greater) a* value, it would seem greenish (reddish). If a pixel has a smaller (greater) b* value, it would seem blueish (yellowish). Then, if a pixel has a higher a* or b* value, it would seem warmer; otherwise, colder. The color saliency SC_n at pixel k is calculated using

$$SC_n^k = 1 - \exp\left(-\frac{(Ia_n^k)^2 + (Ib_n^k)^2}{\sigma_C^2}\right) \tag{5}$$

where σ_C is a parameter. $Ia_n^k = \frac{Ia_n^k - mina}{maxa - mina}$, $Ib_n^k = \frac{Ib_n^k - minb}{maxb - minb}$, $mina(maxa)$ is the minimum (maximum) value of the Ia and $minb$ ($maxb$) is the minimum (maximum) value of the Ib.

Many studies found that the regions near the image center are more attractive to human visual perception [23]. It can thus be suggested that regions near the center of the image will be more likely to be "salient" than the ones far away from the center. The location saliency at pixel k under the location prior can be formulated by

$$SD_n^k = \exp\left(-\frac{||k-c||^2}{\sigma_D^2}\right) \quad (6)$$

where σ_D is a parameter. c is the center of the input image I_n. Then, the visual saliency is used to construct the visual saliency (VS) maps, given by

$$VSm = VS * G_{r,\sigma} \quad (7)$$

where $G_{r,\sigma}$ is a Gaussian filter.

2.2.2. Gradient Magnitude Similarity

According to Zhang et al. [24], the gradient magnitude is calculated as the root mean square of image directional gradients along two orthogonal directions. The gradient is usually computed by convolving an image with a linear filter such as the classic Sobel, Prewitt and Scharr filters. The gradient magnitude similarity algorithm proposed by Reference [24] is adopted in this study. This algorithm uses a Scharr gradient operator, which could achieve slightly better performance than Sobel and Prewitt operators [25]. With the Scharr gradient operator, the partial derivatives GMx_n^k and GMy_n^k of an input image I_n are calculated as:

$$\begin{aligned} GMx_n^k &= \frac{1}{16}\begin{bmatrix} 3 & 0 & -3 \\ 10 & 0 & -10 \\ 3 & 0 & -3 \end{bmatrix} * I_n^k \\ GMy_n^k &= \frac{1}{16}\begin{bmatrix} 3 & 0 & -3 \\ 10 & 0 & -10 \\ 3 & 0 & -3 \end{bmatrix} * I_n^k \end{aligned} \quad (8)$$

The gradient modulus of the image I_n is calculated by

$$GM_n = \sqrt{GMx^2 + GMy^2} \quad (9)$$

The gradient is computed from the luminance channel of input images that will be introduced in the next section. Similar to the visual saliency maps, the gradient magnitude (GM) maps is constructed as

$$GMm = GM * G_{r,\sigma} \quad (10)$$

2.2.3. Chrominance Similarity

The RGB input images are transformed into an opponent color space, given by

$$\begin{bmatrix} L \\ M \\ N \end{bmatrix} = \begin{bmatrix} 0.06 & 0.63 & 0.27 \\ 0.30 & 0.04 & -0.35 \\ 0.34 & -0.6 & 0.17 \end{bmatrix} \begin{bmatrix} R \\ G \\ B \end{bmatrix} \quad (11)$$

The L channel is used to compute the gradients introduced in the previous section. The M and N (chrominance) channels are used to calculate the color distortion saliency, given by

$$M_n = 0.30 * R + 0.04 * G - 0.35 * B \quad (12)$$

$$N_n = 0.34 * R - 0.6 * G + 0.17 * B \quad (13)$$

$$C_n = M_n \cdot N_n \tag{14}$$

Finally, the chrominance similarity or color distortion saliency (CD) maps are calculated by

$$CDm = C * G_{r,\sigma} \tag{15}$$

2.2.4. Weight Maps

Using three measured metrics above, the weight maps are computed as given by

$$W_n = (VSm)^{\alpha} \cdot (GMm)^{\beta} \cdot (CDm)^{\gamma} \tag{16}$$

where α, β, and γ are parameters used to control the relative importance of visual saliency (VS), gradient saliency (GM), and color distortion saliency (CD). From these weight maps, W at each location k, the overall weight maps of each input image can be obtained.

$$W_n^k = \begin{cases} 1, & if\ W_n^k = \max\left(W_1^k,\ W_2^k, \ldots, W_N^k\right), \\ 0, & otherwise, \end{cases} \tag{17}$$

where N is the number of input images, W_n^k is the weight value of the pixel k in the nth image. Then proposed weight maps are determined by normalizing the saliency maps as follows:

$$W_n^k = \frac{W_n^k}{\sum_{n=1}^{N} W_n^k}, \ \forall \mathrm{n} = 1, 2, \ldots, \mathrm{N} \tag{18}$$

These weight maps are then refined by a gradient domain guided filter described in the next section.

2.2.5. Gradient Domain Fast Guided Filter

The gradient domain guided filter proposed by Reference [21] is adopted to optimize the initial weight maps. By using this filter, the halo artifacts can be more effectively suppressed. It is also less sensitive to its parameters but still has the same complexity as the guided filter. The gradient domain guided filter has good edge-preserving smoothing properties as the bilateral filter, but it does not suffer from the gradient reversal artifacts. The filtering output is a local linear model of the guidance image. This is one of the fastest edge-preserving filters. Therefore, the gradient domain guided filter can apply in image smoothing to avoid ringing artifacts.

It is assumed that the filtering output Q is a linear transform of the guidance image G in a local window w_k centered at pixel k.

$$Q_i = a_k G_i + b_k, \ \forall i \in w_k \tag{19}$$

where $(a_k,\ b_k)$ are some linear coefficients assumed to be constant in the local window w_k with the size of $(2\zeta 1 + 1) \times (2\zeta 1 + 1)$. The linear coefficients $(a_k,\ b_k)$ can be estimated by minimizing the cost function in the window w_k between the output image Q and the input image P

$$E_{(a_k, b_k)} = \sum_{i \in w_k} \left[(Q_i - P_i)^2 + \frac{\lambda}{\hat{\Gamma}_G(k)}(a_k - \gamma_k)^2\right] \tag{20}$$

where γ_k is defined as

$$\gamma_k = 1 - \frac{1}{1 + e^{\eta(\chi(k)) - \mu_{\chi,\infty}}} \tag{21}$$

$\mu_{\chi,\infty}$ is the mean value of all $\chi(k)$. η is calculated as $4/(\mu_{\chi,\infty} - \min(\chi(k)))$.

$\hat{\Gamma}_G(k)$ is a new edge-aware weighting used to measure the importance of pixel k with respect to the whole guidance image. It is defined by using a local variance of 3×3 windows and $(2\zeta 1+1) \times (2\zeta 1+1)$ windows of all pixels by

$$\hat{\Gamma}_G(k) = \frac{1}{N}\sum_{i=1}^{N} \frac{\chi(k)+\varepsilon}{\chi(i)+\varepsilon} \tag{22}$$

where $\chi(k) = \sigma_{G,1}(k)\sigma_{G,\zeta 1}(k)$. ζ_1 is the window size of the filter.

The optimal values of a_k and b_k are computed by

$$a_k = \frac{\mu_{G\odot X,\zeta 1}(k) - \mu_{G,\zeta 1}(k)\mu_{X,\zeta 1}(k) + \frac{\lambda}{\hat{\Gamma}_G(k)}\gamma_k}{\sigma^2_{G,\zeta 1}(k) + \frac{\lambda}{\hat{\Gamma}_G(k)}} \tag{23}$$

$$b_k = \mu_{X,\zeta 1}(k) - a_k\,\mu_{G,\zeta 1}(k) \tag{24}$$

The final value of $\hat{Q}_i$ is calculated by

$$\hat{Q}_i = \overline{a}_k G_i + \overline{b}_k \tag{25}$$

where $\overline{a}_k$ and $\overline{b}_k$ are the mean values of a_k and b_k in the window, respectively. $\overline{a}_k$ and $\overline{b}_k$ are computed by

$$\overline{a}_k = \frac{1}{|w_{\zeta 1}(k)|}\sum_{i \in w_{\zeta 1}(k)} a_k \tag{26}$$

$$\overline{b}_k = \frac{1}{|w_{\zeta 1}(k)|}\sum_{i \in w_{\zeta 1}(k)} b_k \tag{27}$$

where $|w_{\zeta 1}(k)|$ is the cardinality of $w_{\zeta 1}(k)$.

2.2.6. Refining Weight Maps by Gradient Domain Guided Filter

Due to these weight maps being noisy and not well aligned with the object boundaries, the proposed approach deploys a gradient domain guided filter to refine the weight maps. The gradient domain guided filter is used at each weight map W_n with the corresponding input image I_n. However, the weigh map W_D_n used W_B_n as the guidance image to improve the W_D_n, it is calculated by

$$W\_B_n = G_{r1,\varepsilon 1}(W_n, I_n) \tag{28}$$

$$W\_D_n = G_{r2,\ \varepsilon 2}(W\_B_n, I_n) \tag{29}$$

where $r1$, $\varepsilon 1$ and $r2$, and $\varepsilon 2$ are the parameters of the guided filter. W_B_n and W_D_n are the refined weight maps of the base and detail layers, respectively. Both weight maps W_B_n and W_D_n are deployed using mathematical morphology techniques to remove small holes and unwanted regions in the focus and defocus regions. The morphology techniques are described as bellow,

$$\begin{aligned} mask\ &= W_n < threshold \\ temp1 &= imfill(mask,'holes') \\ temp2 &= 1 - temp1 \\ temp3 &= imfill(temp2,'holes') \\ W_n(refined)\ &= bwareaopen(temp3, threshold) \end{aligned} \tag{30}$$

Then, the values of the N refined weight maps are normalized such that they sum to one at each pixel k. Finally, the fused base and detail layer images are calculated and blended to fuse the input images, as given by

$$\overline{B}_n = W\_B_n * B_n \quad (31)$$

$$\overline{D}_n = W\_D_n * D_n \quad (32)$$

$$Fused_n = \overline{B}_n + \overline{D}_n \quad (33)$$

The fast-guided filter is improved by the guided filter proposed by Reference [22]. This algorithm is adopted for reducing the processing of gradient domain guided filter time complexity. Before processing the gradient domain guided filter, the rough transmission map and the guidance image employ nearest the neighbor interpolation down-sampling. After gradient domain guided filter processing, the gradient domain guided filter output image uses bilinear interpolation for up-sampling and obtains the refining transmission map. Using this fast-guided filter, the gradient domain guided filter performs better than the original one. Therefore, the proposed filter was named as the gradient domain fast guided filter.

3. Results and Discussion

3.1. Multi-Focus Image Fusion

This section describes the comprehensive experiments conducted to evaluate and verify the performance of the proposed approach. The proposed algorithm was developed to fit many types of multi-focus images that are captured by any digital camera or Pi camera. The proposed method was also compared with five multi-focus image fusion techniques: the multi-scale weighted gradient based method (MWGF) [26], the DCT based Laplacian pyramid fusion technique (DCTLP) [27], the image fusion with guided filtering (GFF) [28], the gradient domain-based fusion combined with a pixel-based fusion (GDPB) [29], and the image matting (IM)-based fusion algorithm [30]. The codes of these methods were downloaded and run on the same computer to compare to the proposed method.

The MWGF method is based on the image structure saliency and two scales to solve the fusion problems raised by anisotropic blur and miss-registration. The image structure saliency is used because it reflects the saliency of local edge and corner structures. The large-scale measure is used to reduce the impacts of anisotropic blur and miss-registration on the focused region detection, while the small-scale measure is used to determine the boundaries of the focused regions. The DCTLP presents an image fusion method using Discrete Cosine Transform based Laplacian pyramid in the frequency domain. The higher level of pyramidal decomposition, the better quality of the fused image. The GFF method is based on fusing two-scale layers by using a guided filter-based weighted average method. This method measures pixel saliency and spatial consistency at two scales to construct weight maps for the fusion process. The GDPB method fuses luminance and chrominance channels separately. The luminance channel is fused by using a wavelet-based gradient integration algorithm coupled with a Poisson Solver at each resolution to attenuate the artifacts. The chrominance channels are fused based on a weighted sum of the chrominance channels of the input images. The image mating fusion (IM) method is based on three steps: obtaining the focus information of each source image by morphological filtering, applying an image matting technique to achieve accurate focused regions of each source image, and combining these fused regions to construct the fused image.

All methods used the same input images as the ones applied in the proposed technique. Ten multi-focused image sequences were used in the experiments. Four of them were canola images captured by setting well-focused and manual changing focal length of the Pi camera; the others were selected from the general public datasets used for many image fusion techniques. These general datasets are available in Reference [31,32]. In the first four canola database sets, three of them were artificial multi-focus images obtained by using LunaPic tool [33], one of them was a multi-focus image acquired directly from the Pi camera after cropping the region of interest as described in Section 2.1.

The empirical parameters of the gradient domain fast guided filter and VS metrics were adjusted to obtain the best outputs. The parameters of the gradient domain fast guided filter (see Equation (22)) consisted of a window size filter ($\zeta 1$), a small positive constant (ε), subsampling of the fast-guided filter (s), and a dynamic range of input images (L). The parameters of VS maps (Equation (16)), including α, β, and γ, were used to control visual saliency, gradient similarity, and color distortion measures, respectively. These empirical parameters of the gradient domain fast guided filters were experimentally set as s = 4, L = 9, and two pairs of $\zeta 1(1) = 4$, $\varepsilon(1) = 1.0e-6$ and $\zeta 1(2) = 4$, $\varepsilon(2) = 1.0e-6$ for optimizing base and detail weight maps. Other empirical parameters of VS maps were set as $\alpha = 1$, $\beta = 0.89$, and $\gamma = 0.31$.

Surprisingly, when changing these parameters of the VS maps, such as, $\alpha = 0.31$, $\beta = 1$, and $\gamma = 0.31$, the fused results had a similar quality to the first parameter settings. It can be thus concluded that to obtain focused regions, both visual saliency and gradient magnitude similarity can be used as the main saliencies. In addition, the chrominance colors (M and N) also contributed to the quality of the fused results. For example, when increasing the parameters of M and N, the blurred regions appeared in the fused results. Figure 3 shows the outputs of the proposed algorithm, including visual saliency, gradient magnitude similarity, and chrominance colors. The red oval denotes the defocused region of the input image (Figure 3a).

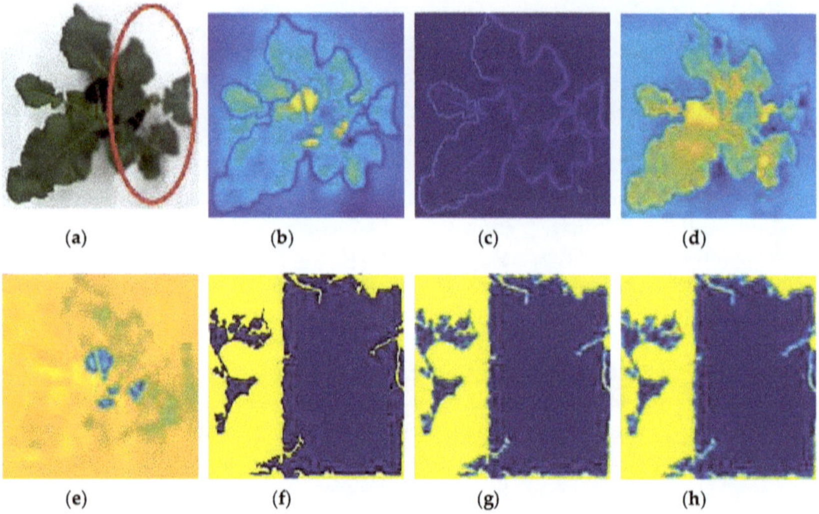

Figure 3. An example of a source image and its saliencies and weight maps: (**a**) A source image; (**b**) visual saliency; (**c**) gradient saliency; (**d**) chrominance color (M); (**e**) chrominance color (N); (**f**) weight maps; (**g**) refined base weight map; (**h**) refined detail weight map.

3.2. Comparison with Other Multi-Fusion Methods

In this section, a comprehensive assessment, including both subjective and objective assessment, is used to evaluate the quality of fused images obtained from the proposed and other methods. Subjective assessments are the methods used to evaluate the quality of an image through many factors, including viewing distance, display device, lighting condition, vision ability, etc. However, subjective assessments are expensive and time consuming. Therefore, objective assessments—mathematical models—are designed to predict the quality of an image accurately and automatically.

For subjective or perceptual assessment, the comparisons of these fused images are shown from Figures 4–7. The figures show the fused results of the "Canola 1", "Canola 2", "Canola 4" and "Rose flower" image sets. In these examples, (a) and (b) are two source multi-focus images, and (c), (d), (e), (f), (g), and (h) are the fused images obtained with the MWGF, DCTLP, GFF, GDPB, IM, and the proposed methods, respectively. In almost all the cases, the MWGF method offers quite good fused images; however, sometimes it fails to deal with the focused regions. For example, the blurred regions remain in the fused image as marked by the red circle in Figure 4c. The DCTLP method offers fused images as good as the MWGF but causes blurring of the fused images in all examples. The IM method also provides quite good results; however, ghost artifacts remain in the fused images, as shown in Figure 4g, Figure 6g, and Figure 7g. Although the fused results of the GFF method reveal good visual effects at first glance, small blurred regions are still remained at the edge regions (the boundary between focused and defocused regions) of the fused results. This blurring of edge regions can be seen in the "Rose flower" fused images in Figure 7e. The fused images of the GDPB method have unnatural colors and too much brightness. The fused results of the GDPB are also suffered from the ghost artifacts on the edge regions and on the boundary between the focused and defocused regions. It can be clearly seen that the proposed algorithm can obtain clearer fused images and better visual quality and contrast than other algorithms due to its combination of the gradient domain fast-guided filter and VS maps. The proposed algorithm offers fused images with fewer block artifacts and blurred edges.

In addition to subjective assessments, an objective assessment without the reference image was also conducted. Three objective metrics, including mutual information (MI) [34], structural similarity (QY) [35], and the edge information-based metric Q(AB/F) [36] were used to evaluate the fusion performance of different multi-focus fusion methods.

The mutual information (MI) measures the amount of information transferred from both source images into the resulting fused image. It is calculated by

$$MI = 2\left(\frac{I(X,F)}{H(F)+H(X)} + \frac{I(Y,F)}{H(F)+H(Y)}\right) \quad (34)$$

where $I(X,F)$ is the mutual information of the input image X and fused image F. $I(Y,F)$ is the mutual information of the input image Y and fused image F. $H(X)$, $H(Y)$, and $H(F)$ denotes the entropies of the input image X, Y, and used image F, respectively.

Figure 4. Results: (**a**) Source image 1 of Canola 1; (**b**) Source image 2 of canola 1; (**c**) Reconstructed image using MWGF; (**d**) DCTLP; (**e**) GFF; (**f**) GDPB; (**g**) IM; (**h**) proposed method.

The structural similarity (QY) measures the corresponding regions in the reference original image x and the test image y. It is defined as

$$Q(x,y,f|w) = \begin{cases} \lambda(w)SSIM(x,f|w) + (1-\lambda(w))SSIM(y,f|w), & for\ SSIM(x,y|w) \geq 0.75 \\ max\{SSIM(x,f|w),\ SSIM(y,f|w)\}, & for\ SSIM(x,y|w) < 0.75 \end{cases} \tag{35}$$

where $\lambda(w) = \frac{s(x|w)}{s(x|w)+s(y|w)}$ is the local weight, and $s(x|w)$ and $s(y|w)$ are the variances of w_x and w_y, respectively.

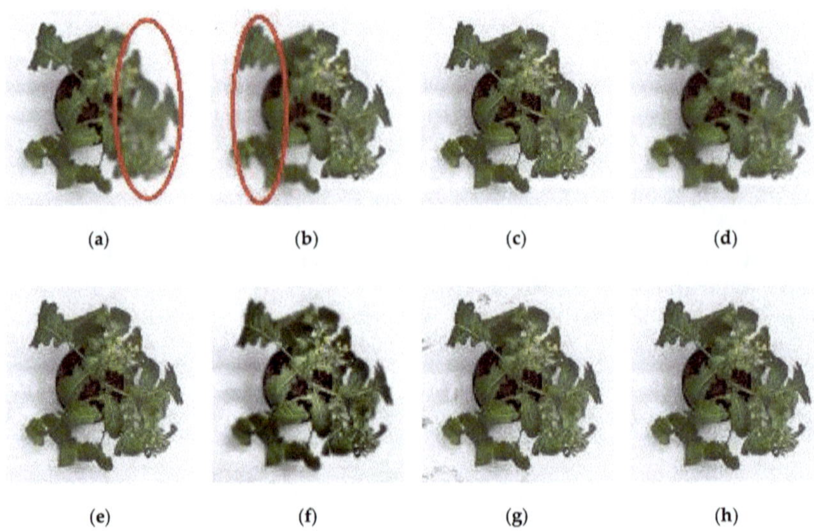

Figure 5. (**a**) Source image 1 of Canola 2; (**b**) Source image 2 of Canola 2; (**c**) MWGF; (**d**) DCTLP; (**e**) GFF; (**f**) GDPB; (**g**) IM; (**h**) proposed method.

Figure 6. (**a**) Source image 1 of Canola 4; (**b**) Source image 2 of Canola 4; (**c**) MWGF; (**d**) DCTLP; (**e**) GFF; (**f**) GDPB; (**g**) IM; (**h**) proposed method.

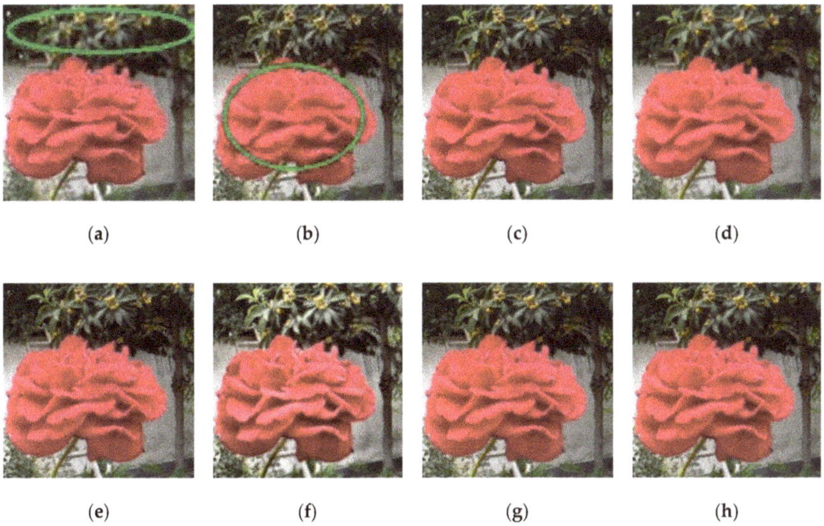

Figure 7. Results: (a) Source image 1 of Rose; (b) Source image 2 of Rose; (c) MWGF; (d) DCTLP; (e) GFF; (f) GDPB; (g) IM; (h) proposed method.

The edge information-based metric $Q^{AB/F}$ measures the amount of edge information that is transferred from input images to the fused image. For the fusion of source images A and B resulting in a fused image F, gradient strength $g(n,m)$ and orientation $\alpha(n,m)$ are extracted at each pixel (n, m) from an input image, as given by

$$g_A(n,m) = \sqrt{s_A^x(n,m)^2 + s_A^y(n,m)^2} \tag{36}$$

$$\alpha_A = tan^{-1}(\frac{s_A^y(n,m)}{s_A^x(n,m)}) \tag{37}$$

where $s_A^x(n,m)$ and $s_A^y(n,m)$ are the output of the horizontal and vertical Sobel templates centered on pixel $p_A(n,m)$ and convolved with the corresponding pixels of input image A. The relative strength and orientation values of $G^{AF}(n,m)$ and $A^{AF}(n,m)$ of the input image A with respect to the fused image F are calculated by

$$G^{AF}(n,m) = \begin{cases} \frac{g_{F(n,m)}}{g_{A(n,m)}} & if \; g_{A(n,m)} > g_{F(n,m)} \\ \frac{g_{A(n,m)}}{g_{F(n,m)}}, & otherwise \end{cases} \tag{38}$$

$$A^{AF}(n,m) = 1 - \frac{|\alpha_A(n,m) - \alpha_F(n,m)|}{\pi/2} \tag{39}$$

From these values, the edge strength and orientation values are derived, as given by

$$Q_g^{AF}(n,m) = \frac{\Gamma_g}{1 + e^{K_g(G^{AF}(n,m)-\sigma_g)}} \tag{40}$$

$$Q_\alpha^{AF}(n,m) = \frac{\Gamma_\alpha}{1 + e^{K_\alpha(A^{AF}(n,m)-\sigma_\alpha)}} \tag{41}$$

$Q_g^{AF}(n,m)$ and $Q_\alpha^{AF}(n,m)$ model information loss between the input image A and the fused image F. The constants Γ_g, K_g, σ_g and Γ_α, K_α, σ_α determine the exact shape of the sigmoid functions used to form the edge strength and orientation preservation values (Equations (40) and (41)). Edge information preservation values are formed by

$$Q^{AF}(n,m) = Q_g^{AF}(n,m) Q_\alpha^{AF}(n,m) \tag{42}$$

with $0 \leq Q^{AF}(n,m) \leq 1$. The higher value of $Q^{AF}(n,m)$, the less loss of information of the fused image.

The fusion performance $Q^{AB/F}$ is evaluated as a sum of local information preservation estimates between each of the input images and fused image, it is defined as

$$Q^{AB/F} = \frac{\sum_{n=1}^{N} \sum_{m=1}^{M} Q^{AF}(n,m) w^A(n,m) + Q^{BF}(n,m) w^B(n,m))}{\sum_{j=1}^{N} \sum_{j=1}^{M} (w^A(i,j) + w^B(i,j))} \tag{43}$$

where $Q^{AF}(n,m)$ and $Q^{BF}(n,m)$ are edge information preservation values, weighted by $w^A(n,m)$ and $w^B(n,m)$, respectively.

Table 1 illustrates the quantitative assessment values of five different multi-focus fusion methods and the proposed method. The larger the value of these metrics, the better image quality is. The values shown in bold represent the highest performance. From Table 1, it can be seen that the proposed method produces the highest quality scores for all three objectives metrics except for QY with "Canola 2" datasets and QAB/F with "Book" (extra images were also run to test the performance). These largest quality scores imply that the proposed method performed well, stably, and reliably. Overall, it can be concluded that the proposed method reveals the competitive performance when compared with previous multi-focus fusion methods both in visual perception and objective metrics. Table 2 describes the ranking of the proposed method with others based on the quality of fused images. The performance (including quality of the images and the processing time) is scaled from 1 to 6. The results show the outperformance of the proposed technique with other techniques previously published.

4. Summary and Conclusions

To improve the description and quality images, especially images acquired from the digital camera or the Pi camera for canola phenotyping, an image fusion method is necessary. A new multi-focus image fusion method was proposed with the combination of the VS maps and gradient domain fast guided filters. In the proposed algorithm, the VS maps were first deployed to obtain visual saliency, gradient magnitude similarity saliency, and chrominance saliency (or color distortions), then the initial weight map was constructed with a mix of three metrics. Next, the final decision weight maps were obtained by optimizing the initial weight map with a gradient domain fast guided filter at two components. Finally, the fused results were retrieved by the combination of two-component weight maps and two-component source images that present large-scale and small-scale variations in intensity. The proposed method was compared with five proper representative fusion methods both in subjective and objective evaluations. Based on the experiment's results, the proposed fusion method presents a competitive performance with or outperforms some state-of-the-art methods based on the VS maps measure and gradient domain fast guided filter. The proposed method can use digital images which are captured by either a high-end or low-end camera, especially the low-cost Pi camera. This fusion method can be used to improve the images for trait identification in phenotyping of canola or other species.

On the other hand, some limitations of the proposed multi-focus image fusion, such as small-blurred regions in the boundaries between the focused and defocused regions and computational cost, are worthwhile to investigate. Morphological techniques and optimizing the multi-focus fusion algorithm are also recommended for further study.

Furthermore, 3D modeling from enhancing depth images and image fusion techniques should be investigated. The proposed fusion technique can be implemented in the phenotyping system which has multiple sensors, such as thermal, LiDAR, or high-resolution sensors to acquire multi-dimensional images to improve the quality or resolution of the 2D and 3D images. The proposed system and fusion techniques can be applied in plant phenotyping, remote sensing, robotics, surveillance, and medical applications.

Table 1. Comparison of the proposed method with other methods.

Index	Source Images	Methods					
		MWGF	DCTLP	IM	GFF	GDPB	Proposed Algorithm
QMI	Canola 1	1.224	1.042	1.124	1.190	0.656	1.288
	Canola 2	1.220	0.946	1.164	1.147	0.611	1.230
	Canola 3	1.165	0.981	1.043	1.148	0.573	1.212
	Canola 4	1.320	0.943	1.400	1.060	0.570	1.400
	Doll	0.664	0.918	0.881	0.310	0.732	1.011
	Rose	1.049	1.133	1.002	0.440	0.736	1.147
	Jug	1.065	1.085	0.974	0.347	0.742	1.094
	Diver	1.168	1.207	1.190	0.515	0.910	1.210
	Book	0.957	1.188	1.152	0.487	0.900	1.234
	Notebook	1.118	1.181	1.141	0.463	0.745	1.190
QY	Canola 1	0.958	0.851	0.812	0.948	0.755	0.970
	Canola 2	0.981	0.859	0.901	0.967	0.762	0.980
	Canola 3	0.961	0.856	0.752	0.955	0.737	0.970
	Canola 4	0.777	0.799	0.980	0.913	0.700	0.980
	Doll	0.902	0.950	0.960	0.800	0.872	0.980
	Rose	0.973	0.979	0.973	0.829	0.901	0.980
	Jug	0.995	0.990	0.970	0.970	0.779	0.995
	Diver	0.975	0.971	0.976	0.744	0.918	0.976
	Book	0.952	0.956	0.959	0.647	0.850	0.977
	Notebook	0.987	0.983	0.991	0.844	0.816	0.992
QAB/F	Canola 1	0.958	0.885	0.938	0.930	0.883	0.974
	Canola 2	0.987	0.987	0.981	0.987	0.987	0.987
	Canola 3	0.955	0.621	0.937	0.841	0.607	0.970
	Canola 4	0.906	0.492	0.915	0.529	0.481	0.915
	Doll	0.987	0.986	0.987	0.986	0.987	0.987
	Rose	0.987	0.987	0.987	0.986	0.987	0.987
	Jug	0.987	0.987	0.987	0.986	0.987	0.987
	Diver	0.986	0.986	0.986	0.986	0.986	0.986
	Book	0.984	0.980	0.984	0.984	0.983	0.984
	Notebook	0.986	0.987	0.987	0.986	0.987	0.987

Table 2. Ranking the performance of fused images of the proposed method with other methods based on the results from Table 1.

Source Images	Methods					
	MWGF	DCTLP	IM	GFF	GDPB	Proposed Algorithm
Canola 1	2	5	4	3	6	1
Canola 2	2	5	4	3	6	1
Canola 3	2	5	4	3	6	1
Canola 4	2	4	1	3	5	1
Doll	5	2	3	6	4	1
Rose	3	2	4	5	6	1
Jug	3	2	4	6	5	1
Diver	4	2	3	6	5	1
Book	4	2	3	6	5	1
Notebook	4	3	2	6	5	1

Author Contributions: T.C. and K.P. conceived, designed, and performed the experiments; T.C. also analyzed the data; A.D., K.W., S.V. contributed revising, reagents, materials, analysis tools; T.C., A.D., and K.W. wrote the paper.

Funding: The work is supported by The Canada First Research Excellence Fund through the Global Institute for Food Security, University of Saskatchewan, Canada.

Conflicts of Interest: The authors declare no conflict of interest.

References

1. Food and Agriculture Organization of the United Nations. The Furture of Food and Agriculture—Trends and Challenges. Available online: http://www.fao.org/3/a-i6583e.pdf/ (accessed on 28 February 2018).
2. Li, L.; Zhang, Q.; Huang, D. A review of imaging techniques for plant phenotyping. *Sensors* **2014**, *14*, 20078–20111. [CrossRef] [PubMed]
3. Starck, J.L.; Fionn, S.; Bijaoui, A. *Image Processing and Data Analysis: The Multiscale Approach*; Cambridge University Press: Cambridge, UK, 1998.
4. Florack, L. *Image Structure*; Springer: Dordrecht, Netherlands, 1997.
5. Nirmala, D.E.; Vaidehi, V. Comparison of Pixel-level and feature level image fusion methods. In Proceedings of the 2015 2nd International Conference on Computing for Sustainable Global Development (INDIACom), New Delhi, India, 11–13 March 2015; pp. 743–748.
6. Olding, W.C.; Olivier, J.C.; Salmon, B.P. A Markov Random Field model for decision level fusion of multi-source image segments. In Proceedings of the 2015 IEEE International Geoscience and Remote Sensing Symposium (IGARSS), Milan, Italy, 26–31 July 2015; pp. 2385–2388.
7. Liang, F.; Xie, J.; Zou, C. Method of Prediction on Vegetables Maturity Based on Kalman Filtering Fusion and Improved Neural Network. In Proceedings of the 2015 8th International Conference on Intelligent Networks and Intelligent Systems (ICINIS), Tianjin, China, 1–3 November 2015. [CrossRef]
8. Padol, P.B.; Sawant, S.D. Fusion classification technique used to detect downy and Powdery Mildew grape leaf diseases. In Proceedings of the 2016 International Conference on Global Trends in Signal Processing, Information Computing and Communication (ICGTSPICC), Jalgaon, India, 22–24 December 2016. [CrossRef]
9. Samajpati, B.J.; Degadwala, S.D. Hybrid approach for apple fruit diseases detection and classification using random forest classifier. In Proceedings of the 2016 International Conference on Communication and Signal Processing (ICCSP), Melmaruvathur, India, 6–8 April 2016. [CrossRef]
10. West, T.; Prasad, S.; Bruce, L.M.; Reynolds, D.; Irby, T. Rapid detection of agricultural food crop contamination via hyperspectral remote sensing. In Proceedings of the 2009 IEEE International Geoscience and Remote Sensing Symposium (IGARSS 2009), Cape Town, South Africa, 12–17 July 2009. [CrossRef]
11. Dimov, D.; Kuhn, J.; Conrad, C. Assessment of cropping system diversity in the fergana valley through image fusion of landsat 8 and sentinel-1. *ISPRS Ann. Photogramm. Remote Sens. Spat. Inf. Sci.* **2016**, *III-7*, 173–180. [CrossRef]
12. Kaur, G.; Kaur, P. Survey on multifocus image fusion techniques. In Proceedings of the International Conference on Electrical, Electronics, and Optimization Techniques (ICEEOT), Chennai, India, 3–5 March 2016. [CrossRef]
13. Sun, J.; Han, Q.; Kou, L.; Zhang, L.; Zhang, K.; Jin, Z. Multi-focus image fusion algorithm based on Laplacian pyramids. *J. Opt. Soc. Am.* **2018**, *35*, 480–490. [CrossRef] [PubMed]
14. Wan, T.; Zhu, C.; Qin, Z. Multifocus image fusion based on robust principal component analysis. *Pattern Recognit. Lett.* **2013**, *34*, 1001–1008. [CrossRef]
15. Li, X.; Wang, M. Research of multi-focus image fusion algorithm based on Gabor filter bank. In Proceedings of the 2014 12th International Conference on Signal Processing (ICSP), Hangzhou, China, 19–23 October 2014. [CrossRef]
16. Liu, Y.; Liu, S.; Wang, Z. Multi-focus image fusion with dense SIFT. *Inf. Fusion* **2015**, *23*, 139–155. [CrossRef]
17. Phamila, Y.A.V.; Amutha, R. Discrete Cosine Transform based fusion of multi-focus images for visual sensor networks. *Signal Process.* **2014**, *95*, 161–170. [CrossRef]
18. Zhang, Q.; Liu, Y.; Blum, R.S.; Han, J.; Tao, D. Sparse Representation based Multi-sensor Image Fusion: A Review. *arXiv*, 2017.
19. Zhang, Q.; Liu, Y.; Blum, R.S.; Han, J.; Tao, D. Sparse representation based multi-sensor image fusion for multi-focus and multi-modality images. *J. Inf. Fusion* **2018**, *40*, 57–75. [CrossRef]

20. Zhou, Q.; Liu, X.; Zhang, L.; Zhao, W.; Chen, Y. Saliency-based image quality assessment metric. In Proceedings of the 2016 3rd International Conference on Systems and Informatics (ICSAI), Shanghai, China, 19–21 November 2016. [CrossRef]
21. Kou, F.; Chen, W.; Wen, C.; Li, Z. Gradient Domain Guided Image Filtering. *IEEE Trans. Image Process.* **2015**, *24*, 4528–4539. [CrossRef] [PubMed]
22. Zhang, Q.; Li, X. Fast image dehazing using guided filter. In Proceedings of the 2015 IEEE 16th International Conference on Communication Technology (ICCT), Hangzhou, China, 18–20 October 2015. [CrossRef]
23. Zhang, L.; Gu, Z.; Li, H. SDSP: A novel saliency detection method by combining simple priors. In Proceedings of the 2013 20th IEEE International Conference on Image Processing (ICIP), Melbourne, VIC, Australia, 15–18 September 2013. [CrossRef]
24. Zhang, L.; Shen, Y.; Li, H. VSI: A Visual Saliency-Induced Index for Perceptual Image Quality Assessment. *IEEE Trans. Image Process.* **2014**, *23*, 4270–4281. [CrossRef] [PubMed]
25. Zhang, L.; Zhang, L.; Mou, X.; Zhang, D. FSIM: A Feature Similarity Index for Image Quality Assessment. *IEEE Trans.Image Process.* **2011**, *20*, 2378–2386. [CrossRef] [PubMed]
26. Zhou, Z.; Li, S.; Wang, B. Multi-scale weighted gradient-based fusion for multi-focus images. *Inf. Fusion* **2014**, *20*, 60–72. [CrossRef]
27. Naidu, V.; Elias, B. A novel image fusion technique using dct based laplacian pyramid. *Int. J. Inven. Eng. Sci.* **2013**, *1*, 1–9.
28. Li, S.; Kang, X.; Hu, J. Image Fusion with Guided Filtering. *IEEE Trans. Image Process.* **2013**, *22*, 2864–2875. [CrossRef] [PubMed]
29. Paul, S.; Sevcenco, I.S.; Agathoklis, P. Multi-Exposure and Multi-Focus Image Fusion in Gradient Domain. *J. Circuits Syst. Comput.* **2016**, *25*, 1650123. [CrossRef]
30. Li, S.; Kang, X.; Hu, J.; Yang, B. Image matting for fusion of multi-focus images in dynamic scenes. *Inf. Fusion* **2013**, *14*, 147–162. [CrossRef]
31. Saeedi, J.; Faez, K. Multi Focus Image Dataset. Available online: https://www.researchgate.net/profile/Jamal_Saeedi/publication/273000238_multifocus_image_dataset/data/54f489b80cf2ba6150635697/multi-focus-image-dataset.rar (accessed on 5 February 2018).
32. Nejati, M.; Samavi, S.; Shiraniv, S. Lytro Multi Focus Dataset. Available online: http://mansournejati.ece.iut.ac.ir/content/lytro-multi-focus-dataset (accessed on 5 February 2018).
33. LunaPic Tool. Available online: https://www.lunapic.com (accessed on 5 February 2018).
34. Hossny, M.; Nahavandi, S.; Creighton, D. Comments on 'Information measure for performance of image fusion'. *Electron. Lett.* **2008**, *44*, 1066. [CrossRef]
35. Yang, C.; Zhang, J.; Wang, X.; Liu, X. A novel similarity based quality metric for image fusion. *Inf. Fusion* **2008**, *9*, 156–160. [CrossRef]
36. Xydeas, C.; Petrović, V. Objective image fusion performance measure. *Electron. Lett.* **2000**, *36*, 308–309. [CrossRef]

sensors

MDPI

Article

A Multimodal Deep Log-Based User Experience (UX) Platform for UX Evaluation

Jamil Hussain [1], Wajahat Ali Khan [1], Taeho Hur [1], Hafiz Syed Muhammad Bilal [1], Jaehun Bang [1], Anees Ul Hassan [1], Muhammad Afzal [2] and Sungyoung Lee [1,*]

1 Ubiquitous Computing Lab, Department of Computer Science and Engineering, Kyung Hee University, Giheung-gu, Yongin-si, Gyeonggi-do, Seoul 446-701, Korea; jamil@oslab.khu.ac.kr (J.H.); wajahat.alikhan@oslab.khu.ac.kr (W.A.K.); hth@oslab.khu.ac.kr (T.H.); bilalrizvi@oslab.khu.ac.kr (H.S.M.B.); jhb@oslab.khu.ac.kr (J.B.); anees@oslab.khu.ac.kr (A.U.H.)

2 Department of Software, College of Electronics and Information Engineering, Sejong University, Seoul 05006, Korea; mafzal@sejong.ac.kr

* Correspondence: sylee@oslab.khu.ac.kr; Tel.: +82-31-201-2514

Received: 16 March 2018; Accepted: 15 May 2018; Published: 18 May 2018

check for updates

Abstract: The user experience (UX) is an emerging field in user research and design, and the development of UX evaluation methods presents a challenge for both researchers and practitioners. Different UX evaluation methods have been developed to extract accurate UX data. Among UX evaluation methods, the mixed-method approach of triangulation has gained importance. It provides more accurate and precise information about the user while interacting with the product. However, this approach requires skilled UX researchers and developers to integrate multiple devices, synchronize them, analyze the data, and ultimately produce an informed decision. In this paper, a method and system for measuring the overall UX over time using a triangulation method are proposed. The proposed platform incorporates observational and physiological measurements in addition to traditional ones. The platform reduces the subjective bias and validates the user's perceptions, which are measured by different sensors through objectification of the subjective nature of the user in the UX assessment. The platform additionally offers plug-and-play support for different devices and powerful analytics for obtaining insight on the UX in terms of multiple participants.

Keywords: user experience evaluation; user experience measurement; eye-tracking; facial expression; galvanic skin response; EEG; interaction tracker; self-reporting; user experience platform; mix-method approach

1. Introduction

The user experience (UX) is a multi-faceted research area that includes diverse aspects of the experiential and affective use of a product, system, or service [1,2]. A UX assessment helps uncover the important aspects of designing high-quality interactive products and providing an overall positive UX [3]. The UX involves the user beliefs, preferences, thoughts, feelings, and behaviors when interacting with the product, system, or service [1]. It is thus subjective by nature, highly dependent on the use context [4], and linked to the potential benefit obtained from the product, system, or service [5]. The UX is measured using different constructs related to usability (perspicuity, efficiency, etc.), user perception (stimulation, dependability, novelty, etc.), and human emotional reaction [6] using various methods [7]. For example, a user's feelings can be captured if the user "thinks aloud" while performing tasks. Similarly, the UX can also be interpreted by means of a daily diary over a certain period, such as a long-term diary study [8], day reconstruction method [9], repertory grid technique (RGT) [10], and experience sampling method (ESM) [11]. Additionally, the user can be observed by various means, such as a camera, sensor, user interaction tracker, and screen capture devices [7].

The subjective aspect of the UX, however, can make UX assessment difficult. Traditional methods of UX assessment rely on self-reported measurements, usability studies (performance), and observations [6,12], which may be unable to uncover the true user emotional experience [3]. A common method of expressing emotional and cognitive aspects is via retrospective self-reported verbal or written questionnaires [13–15], whereby the user is asked questions relating to their experience. However, this conventional method is highly subjective in nature and thus dependent on user interpretation, recollection, and bias [3]. Even when the questionnaire items are clear, most participants have difficulty engaging in honest and accurate introspection [3]; hence, they do not faithfully articulate their true emotions, abilities, and experiences. Meanwhile, open-ended interview methods [12] may avoid the confusion engendered by the specific-questioning process and can thus enhance the quality of user responses. Nevertheless, this method cannot completely solve the issues relating to self-reporting and self-discourse [16].

Observational methods can resolve the latter two problems, primarily when the user is unaware of the observation [17,18]. The user observation can reveal information relating to the task/user performance, efficiency, and errors while the user interacts with the system. However, both self-reporting and observation methods are unable to determine the psychological states of the user while employing the system. Both of these methods require skilled researchers for data recording, analysis, and interpretation, the latter of which is another source of subjectivity.

In short, determining how the user feels while employing the product, system, or service presents a significant hurdle for the UX evaluation. In reality, the user may have difficulty identifying, interpreting, and reporting their feelings and thoughts during or after use of the product, system, or service. Hence, in addition to self-reporting and observational measurements, UX researchers currently use physiological measurements to assess the user experience based on quantitative metrics.

In general, biometric sensors can detect emotional arousal and stress, motivation, and visual attention, states that have a direct relationship with user cognitive and affective conditions [3]. For instance, an eye tracker can detect visual attention [19–21], electroencephalography (EEG) can detect user motivations and emotional responses [22–24], the galvanic skin response (GSR) can measure stress and arousal through skin conductivity [25,26], and electrocardiogram (ECG) and electromyogram (EMG) can measure stress levels and muscle-arousing activities [27,28]. However, existing research has focused on limited methods and techniques to uncover the true experience of a user employing a product.

To address the limitations of the above individualistic methods and approaches, we propose an innovative "lean UX platform", which employs a mix-method approach by combining observational, self-reported, and physiological measurements. It can evaluate the overall user experience over time by acquiring and synchronizing multimodal data while the user interacts with a product, system, or service.

The rest of this paper is structured as follows: in Section 2, UX evaluation methods of related work are described. In Section 3, the proposed lean UX platform architecture is generally described. In Section 4, the overall proposed platform is presented with respect to its architecture and implementation. In Section 5, execution scenarios are presented as case studies of "mind-mining" evaluations. Section 6 presents the evaluation and discussion, and Section 7 concludes the work.

2. Related Work

Many approaches have been proposed to acquire the user experience in various ways, including the questionnaire, facial analysis, vocal analysis, biometrics, and others. We classify these user experience evaluation methods (UXEMs) into three categories: (i) self-reported measurement, whereby the participant reports their feelings and thoughts in the form of a questionnaire, survey, or poll without expert intervention; (ii) observational measurement, a non-intrusive means of observing the user while interacting with the product, system, or service; and (iii) physiological measurement, whereby sensors are mounted on the user's body for collecting physical information as quantifiable data. The following subsections detail the above categories.

2.1. Self-Reported Measurement

The self-reported approach has been used for a long time as a UXEM. Different tools have been developed to gather the self-reported data from users who express their feelings about the given product, system, or service [12]. No comprehensive solutions exist for extracting the holistic UX, and every method has its positive and negative aspects [5].

For emotion measurement via self-reporting in response to a stimulus, numerous methods have been used, such as the two-dimensional (2D) emotion space (ES) [13,29], to gather data by moving a mouse in the 2D space in response to valence and arousal. However, it cannot be applied to low-fidelity prototypes. Similarly, expressing experiences and emotions ("3E") [14] uses a semi-structured method by providing a predefined template in which the user experience and sentiment data are entered as a daily diary. In addition, the day reconstruction method [9] is a well-known approach for capturing the user's daily experience through their reporting of three important experiences or encounters each day. However, these methods are laborious and require researchers to analyze the gathered data [3,5,7,12].

Furthermore, the affect grid [15] provides a simple and easy scale for measuring affects in a 2D form, while the differential emotions scale (DES) [30] provides diverse categories of emotion to evaluate the user emotions. In addition, the Geneva emotion wheel [31] provides a wheel-shaped emotion scale through which a participant expresses their emotions, and PrEmo [32] uses cartoon animation to obtain the user's emotional responses in the form of dynamic facial, body, and vocal expressions. However, the scale is subjective. The EMO2 [33] tool provides a rating scale in one and two dimensions for emotion measurement while using the product. Emocards and Emofaces [34] use a non-verbal, quick, and easy method that employs emotion cards (cartoon faces) indicating the user emotions while using the product. However, these approaches are intrusive during the given task.

Different questionnaires have been referenced in the literature for measuring various UX aspects, such as affect, aesthetics, attractiveness, pragmatics, hedonics, mental efforts, and satisfaction levels [6,27,35–38]. Lavie and Tractinsky [39] developed an aesthetics scale for website perceived aesthetics in terms of classic and expressive aesthetics. AttrackDiff [35] and User Experience Questionnaire (UEQ) [6] facilitate a rapid assessment of the user experience by obtaining the user's expressed feelings, impressions, and attitudes after using the respective product. However, these assessments only indirectly reflect the experience, and do not focus on the actual experience. The mental effort scale [40] is an easy means of assessing how much effort is needed to complete a task; nevertheless, it requires other tools to obtain the holistic perspective.

2.2. Observational Measurement

Observational measurement is an alternative approach to self-reporting or other methods of measuring user behavior. Situations exist in which the observational measurement method may be more scientifically valid than other methods when the participant is nonverbal or limited in his/her verbal or cognitive ability and is thus unaware and unable to report the behavior. Observational measurement enables detailed descriptions of behavior and its social and non-social contexts. Different methods and techniques, such as video-based facial expression analysis (FEA) [41], emotion from human voice [42], and tracking user interaction [43,44] by logging user actions have been employed for user experience assessment.

Humans communicate considerable emotional information, both voluntarily and involuntary, through the movement of facial muscles. Facial expressions can be used in methods to understand a person's emotional response and valence. Facial expression analysis detects muscle groups in action during different emotional responses, such as smiling, crying, and moving the inner and outer brows. Facial response provides a passive means of measuring a person's experience. For example, the Facereader [45,46] software analyzes real-time videos for facial expression analysis by tracking the user emotional state during interactions with products or software. It also calculates the gaze direction, head orientation, and person characteristics. However, generate Facereader data are limited to six basic emotions: joy, anger, sadness, surprise, fear, and disgust. The relationship between the learning performance and user emotions expressed through the face was examined by Whitehill et al. [47], who found that a user smiles less when they learn more. Their findings show that a user smiles more when they feel embarrassed. In sum, FEA provides a useful approach to assessing affective responses of emotion valences. However, it is unable to identify emotional arousal.

Emotions can be recognized in the human voice using different statistical methods and voice features [42,48]. For example, anger can be detected from a high-pitched voice and faster speech rate. Numerous previous studies mentioned the most significant features for audio-based emotion recognition, such as intensity, duration, pitch, and spectral energy distribution [48].

Furthermore, analytical trackers are software systems that determine how the user interacts with the system. Several tracker systems can trace common user interactions, such as page tracking, event tracking, app/screen tracking, user time, exception tracking, custom dimensions, and metrics. For example, Google Analytics [43,49], Piwik [50], Appsee, and UXCam [44,51] are systems that can track common user interactions to assess the performance of a product by focusing on key performance indicators (KPIs) [52], such as daily active users (DAU), monthly active users (MAU), page views (PV), and unique visitors (UV). It can decipher the user context to better evaluate the performance of the use context, system, or product. However, these KPIs do not reflect the reason and emotion behind the user behavior.

Nevertheless, observational measurements have several challenges relating to the experience over time, and the measurement may vary on account of systematic or stochastic methods [53]. Additionally, observational methods are unable to determine the user's psychological state while employing the system.

2.3. Physiological Measurement

In this section, we explore different biometric sensors, which obtain physical information as quantifiable data for the UX assessment. These tools can be used to validate the traditional measures or add extra information to the conventionally obtained data to extract the actual user perception of the product, system, or service. Herein, we briefly explore this specific research and related technologies.

Eye tracking is a powerful technology that tracks light corneal reflection and pupil dilation [54] for the identification of eye and gaze moments [19–21]. These data can be used to provide important insight that is unachievable from other techniques, such as user visual attention (locating a user's eye positions) and distraction [19–21,55]. Eye tracking data reveal important information relating to arousal, engagement, fatigue, and interest because the eye is unable to deceive [3]. Thus, the issues relating to traditional measurement are avoided. User perception relating to tasks was investigated by Tzafilkou et al. [56] by using eye tracking data, such as eye fixation. These data were linked with user events while the user interacted with an interface; fixation duration, which was associated with user attention; and pupil size. Those authors used the gaze data to assess the self-efficacy and ease of use along with the questionnaire data. For example, a person gazing at the same point in the user interface felt more comfortable.

Similarly, Zheng [22] used eye tracking data with EEG signals to extract user emotions by fusing both the feature and decision levels to improve the emotion recognition model accuracy. The author used the pupil diameter as a metric for emotion classification, such as the pupil diameter changing in accordance with different emotional states. Stable patterns were also extracted for emotion recognition over time in both EEG and eye-tracking data. Meanwhile, Sanfilipp [57] used eye tracking data for tracking the user eye position and eye movements, while employing other biometric sensors for personnel training in situational awareness.

In addition to validating traditional methods, eye-tracking data provides information, such as how motion and background complexity can influence a player's performance in a game environment. Eye fixation data are obtained while the user shoots in a game and the background complexity is measured. Moreover, eye tracking data can eliminate the obstacles of language or culture in UX assessments. For example, Sivajii [58] combined "think-aloud" data with eye tracking data (from multilingual country users) for website usability testing. Their findings showed that the results differed across different cultures on account of the "high-power distance", that is, the unequal power distribution. In high-power distance cultures, feelings and thoughts are more likely to be less expressed, whereas low-power distance cultures are more open and more readily reveal their feelings and thoughts. Thus, eye tracking data remove these obstacles in true UX extraction while interacting with a website. In short, eye tracking technology assists the traditional UX assessment methods by adding validating and complementary data in the form of visual attention.

Similarly, facial EMG [59] is used for the measurement of emotional states (e.g., arousal and valence) during gaming for positive improvement. However, facial EMG requires a proper laboratory setting and technical knowledge for handling artifacts, while engendering obtrusiveness and intimacy issues. Facial coding is another observational method for capturing behavior from facial expressions.

In the UX domain, multiple biometric sensors are used to detect affective information that can validate and complement the traditional methods. Each biometric sensor can detect a portion of the person's behaviors. For example, eye tracking can detect visual attention. However, it does not provide adequate information on the user's emotional states. Similarly, EEG and GSR [26,60] are effective at extracting the user emotional state in terms of arousal; however, they do not provide particular data relating to the emotional valence. GSR is a less effective method for measuring emotions.

Thus, we conclude that one method's weakness is the strength of another method. Consequently, for effective UX measurement, a mix-method approach is the best solution for extracting the true emotional experience. The mix-method approach provides more accurate and precise information about the user while the user interacts with the product, system, or service for the UX assessment. Nevertheless, this approach requires skilled UX researchers and developers to integrate multiple devices, synchronize data, analyze them, and produce informed decisions relating to the UX. Therefore, a single platform is required that can provide an integrated environment in a seamless manner with real-time synchronization and powerful visualizations for measuring the UX of any product, system, or service.

3. Lean UX Platform Architecture

Our proposed platform is based on the "lean UX" [61], which incorporates product development through the continuous measurement of a so-called "learning loop" (build—measure—learn), as shown in Figure 1. The main focus of the proposed platform is measuring and learning (inferencing), both implicitly and explicitly, from the subject's usage behaviors and emotional responses. Consequently, UX research can be simplified by incorporating the human behavior research [5,62].

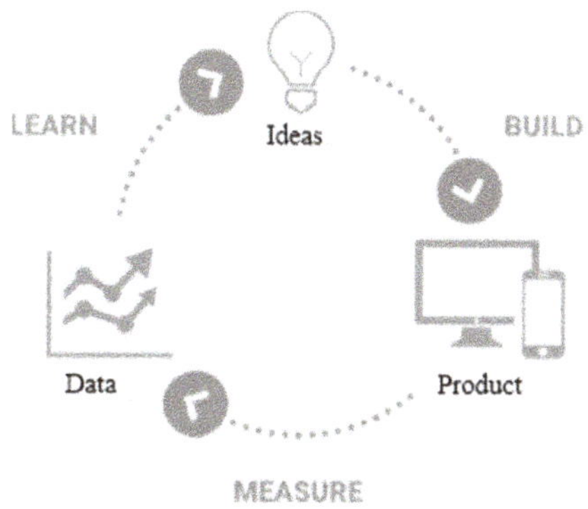

Figure 1. Lean UX learning loop.

The proposed platform collects the user data through different methods and sensors, such as audio, video, and biometrics, as well as user interaction data and surveys, such as self-reported data use for the UX evaluation [7]. The abstract view of the proposed platform is shown in Figure 2.

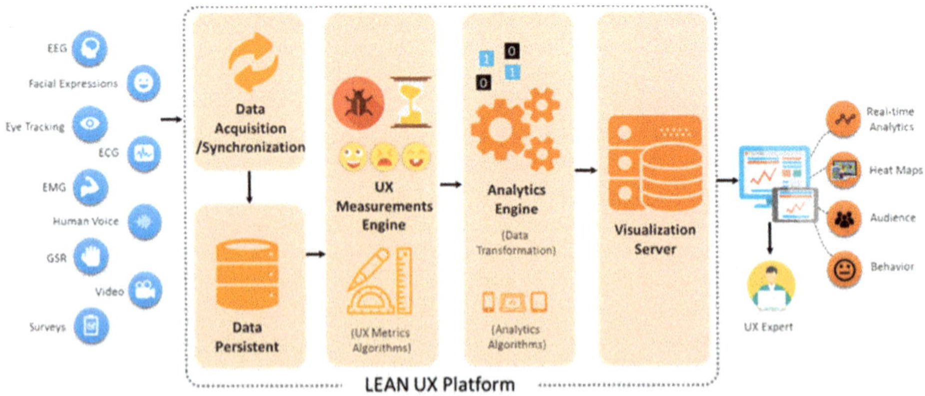

Figure 2. Proposed platform overview.

The platform is composed of four layers: the data layer (DL), UX measurement layer (UXML), analytics layer (AL), and visualization server (VS). The detailed architecture of the proposed platform is shown in Figure 3.

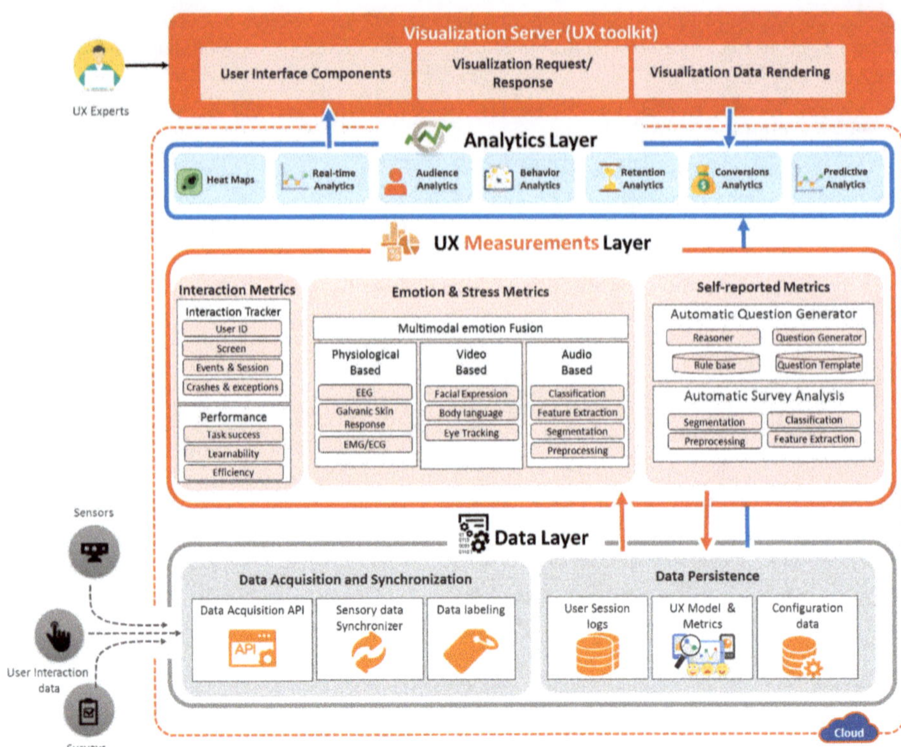

Figure 3. Lean UX platform architecture.

In short, the DL acquires and stores the data acquired from the multiple data sources, including audio devices, video equipment, biometric devices, surveys, and user interaction logs. The data acquired by DL is mainly employed by UXML to deduce the user's emotion, perception, and usage experience. UXML deals with UX metric extractions of a particular phenomenon or object that will help quantify the UX of a person with respect to the product. The extracted information is then used by the AL upon the UX expert request to enable different types of analytics to infer the informed decision. The final layer is the visualization server, which serves as a toolkit for the UX expert to evaluate the digital product. It is a web-based application that is used to realize the different features, analytics, and visualizations based on UX measurement metrics of collected data.

4. Lean UX Platform Implementation Details

4.1. Data Layer

In the development of the "mining minds" platform, we created the data curation layer (DCL) [63,64] task to acquire, curate, and persist the data acquired from multimodal sources. We adopted the same DCL implementation for the Lean UX platform that collects multimodal data to determine the UX [65]. Data acquisition deals with the real-time data acquisition and synchronization obtained from heterogeneous data sources. A label is assigned according to the nature of the data and the data persists as a user session log for use by the UX measurement layer, which can then define the UX corresponding metrics. Upon receiving the UX metrics, they are calculated by the UXML and then stored in the data layer.

4.1.1. Multimodal Data Acquisition and Synchronization

Data Acquisition and Synchronization (DAS) is a Representational State Transfer (REST; RESTful) web service that acquires real-time data from multiple data sources. After acquiring the data, synchronization is performed based on the time stamp of the device and queued based on the event for identification of the context. In the UX domain, every event is linked with a timestamp and heavily depends on the context [62]. All attached devices and sensors send data independently (they each have an independent clock); thus, a logical clock is needed to synchronize all attached devices to the Lean UX platform. Therefore, a synchronization mechanism based on a time frame—the so-called "complete and incomplete sync" [64]—is implemented. The details of the implemented algorithm are presented in [64]; it was developed using the Node.JS platform [64,66] to more efficiently handle events and non-blocking communication. After synchronization, each received data packet is labeled according to the data nature.

4.1.2. Data Persistence

Data persistence relates to two kinds of data persistence: relational database (RDBMS) and big data. RDBMS stores data related to the user model, which stores information related to user cognition, physical characteristics, sensory input, and UX. The context model stores information about the contextual factors; the device model stores information about different characteristics of the devices, such as screen resolution and their abilities of displaying content [67]; the UX model [68] stores the UX structural model [4]; and the configuration settings store information about the connected devices and experiment setting. An overview of the object model is shown in Figure 4. All the sensory data are directly collected from the devices and stored as big data for further analysis.

4.2. UX Measurements Layer

The UX measurements layer is the core of the lean UX platform for inference and modeling of the UX evaluation. It is composed of three main modules that deal with interaction metrics, emotion and stress metrics, and self-reported metrics.

4.2.1. User Interaction Metrics

This module handles the collection of the user interactions and calculating the system performance. It monitors each user's actions by determining how they use the application, problems they are experiencing, and how to resolve them. It views the application from the user perspective by pinpointing the performance, usability, and UX issues [69]. This module deals with real-time qualitative analytics along with traditional quantitative analytics (with numbers) by adding qualitative data on top of the quantitative data, thereby enabling UX experts to transform data into information, and information into insights.

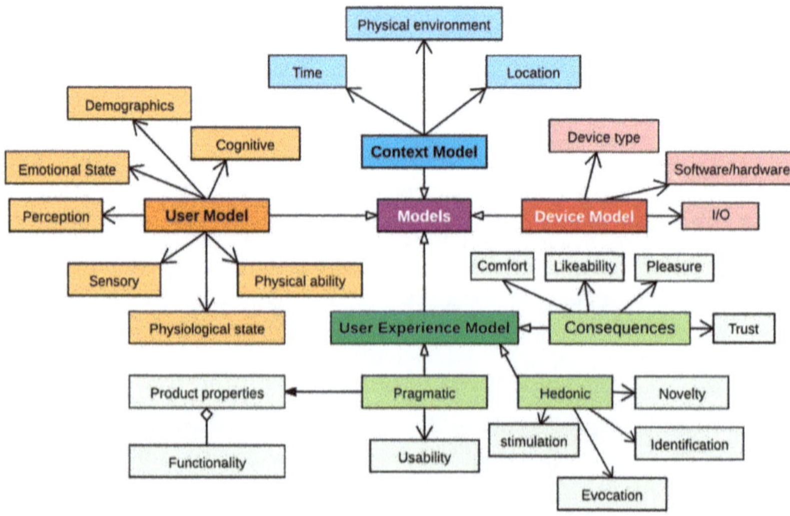

Figure 4. Lean UX Model.

The module consists of two main sub-modules: user behavior metrics and performance metrics. The user behavior metrics track common user interactions, such as page/screen, event, user timing, cross-domain tracking, tasks, crashes, exceptions, and custom dimensions [43,49]. The performance metric reveals how well users are using the product. It is also valuable in estimating the degree of a particular usability issue [70]. For example, if users are making several errors during a task, this means that there is room for enhancement. This module deals with the user interaction data collected by the analytics tracker during usage of the system, such as the task success, time spent on the task, errors, efficiency, and learnability.

4.2.2. Emotion and Stress Metrics

- *Physiological-based Emotion Recognition*: We use the biometric measurement to understand the emotional engagement of user while the user interacts with the system. We use multimodal data from various sensors, such as eye tracking, for visual attention and EEG for quick detection of emotions, motivations, engagement (arousal) in the cognitive workload and frustration level. We will add more modules that can measure emotional arousal and stress by the galvanic skin response (GSR) via measuring changes in the conductivity of the skin, and we will use EMG/ECG for the detection of muscle activity, motion, stress, and arousal. In this study, we implemented the eye tracking and EEG modules.
- *Video-based Emotion Recognition*: The video-based emotion recognition metric consists of two methods and sub-modules: facial expression analysis [41] and body language analysis. Automatic facial expression analysis (AFEA) plays an important role in producing deeper insights in human emotional reactions (valence), such as fear, happiness, sadness, surprise, anger, disgust, or neutrality. For AFEA, we used an inexpensive webcam to capture video of a participant in order to reduce the overall financial cost. Our developed AFEA first detects the face in a given

video frame or image by applying the Viola Jones cascaded classifier algorithm. Second, different facial landmarks features are detected (e.g., eyes, brows, mouth, nose) as the face model. Finally, the face model is fed into the classifier to provide emotions and facial expression metrics as labels [41]. Non-verbal gestures (i.e., body language) play a significant part in the communication process and can yield critical insight into one's experience while interacting with any computing system. We will use a depth camera to recognize emotions through user body language in upcoming version of lean UX platform release.

- *Audio-based Emotion Recognition*: We used an automatic method of measuring human emotions by analyzing the human voice collected through a microphone while using the system [71], such as anger, sadness, and happiness. The trained model is built on the emotion audio data collected from lab students using a microphone recording by manually labeling each audio clip, Berlin Emotional Speech (EMO-DB) [72], and SEMAINE corpus [73]. The model classifies incoming audio to the platform as seven basic emotions: fear, happiness, sadness, surprise, anger, disgust, or neutrality. A Voice Activity Detection (VAD) VAD technique that consists of short-time energy (STE) and short-time zero-crossing rate (STZCR) [74,75] is used to remove the background noise and eliminate the silent parts from audio signals. The speech signals are divided into frames, then STE detects the energy within each frame for voice segmentation. Afterward, STZCR is calculated from the rate of change of speech signal within a particular time window. These two features are used to extract the speech segment for emotion recognition and removed the unwanted frames from signals. The output of VAD is used by the audio feature extraction to extract the audio features such as pitch, log-energy, teager energy operator (TEO), and zero ZCR. Subsequently, we have employed the feature level fusion using a set of rules to choose the right emotions as a previous study [75].
- *Multimodal Data Fusion*: The primary goal of multimodal fusion is to enhance the accuracy of prediction classifiers [76]. It shows the importance of making a multimodal fusion framework that could effectively extract emotions from different modalities in human-centric environment. The benefit of using multimodal data from different devices is to get deep insights of human emotions and motivations. The platform fuses the different emotional measurements, such as audio, video, physiological, and eye tracking to acquire the complete overview of the user's emotional experience by using the mixed method approach to measure the actual user's emotional experience, as shown in Figure 5.

There are three different types of fusion level [76–78]: feature level, decision level, and hybrid level. Feature-level [79–84] is also known as early fusion, that fuses the features extracted from different modalities (e.g., audio, textual, EEG, and eye tracking) for prediction. Decision-level fusion [79,85] is called as late fusion, where the individual's modalities classifier examined the features, gives the results, and then fuses the results to give a final decision. In feature-level, we combined the features of EEG and eye tracking (pupil size) for user sentiment recognition either positive, negative, or neutral. The combined feature vector has been used for the prediction classification. In decision-level fusion, we have employed the feature vector from each input modality and fed into individual classifier as shown in Figure 5. We have adopted the 10-fold cross-validation to estimate the performance of each recognizer. We have used the mean values of all prediction confidences score for prediction fusion. In our case, the rule-based approach has calculated the final label of the prediction as shown in the formula given below:

$$\begin{aligned} \mathrm{K} &= \operatorname{argmax} w_1 C_i^a + w_2 C_i^v + w_3 C_i^p + w_4 C_i^t),\ \mathrm{i} = 1,2,3,\ldots, C \\ &= \{\text{fear, happiness, sadness, surprise, anger, disgust, neutrality}\} \end{aligned}$$

where w_1, w_2, w_3, and w_4 represent the weights of each prediction classifier. We have assigned the equal weights (0.1) to each classifier. C represents the classifier classes such as fear, happiness, sadness, surprise, anger, disgust, and neutrality, and C_i^a, C_i^v, C_i^p, and C_i^t represent the confidences score for audio, video, physiological, and textual modalities respectively. The textual modality has used the same prediction model discussed in Section 4.2.3, after speech to text conversion.

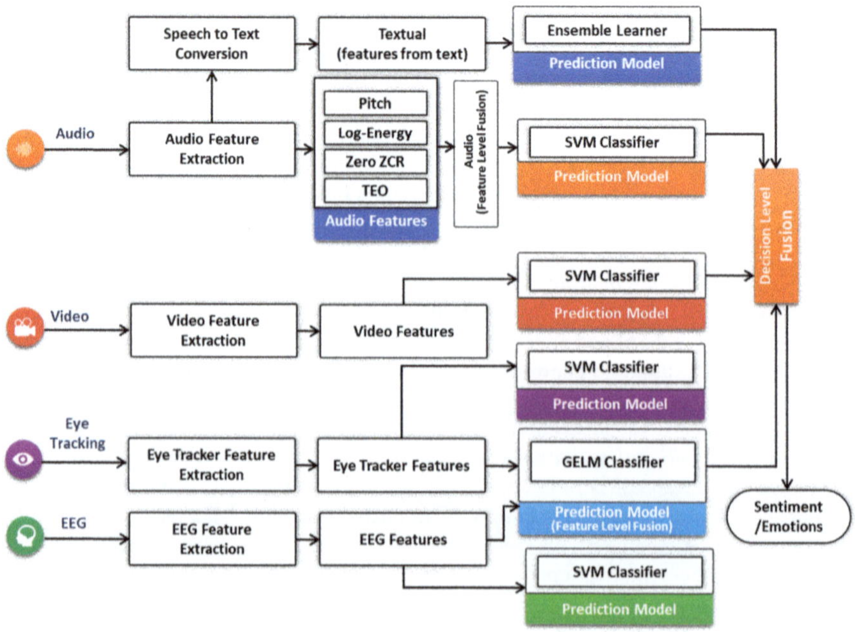

Figure 5. Hybrid level fusion for affect computing.

4.2.3. Self-Reported Metrics

Self-reported metrics [35] deal with post-tasks that explicitly ask questions about the participant for information about their opinion and their interaction with the system, for example, overall interaction, ease of use, satisfaction, effectiveness, and efficacy. It consists of two main modules: automatic question generation and automatic survey analysis. Automatic question generation asks questions based on UX measurement information that triangulates [86] stated answers with biometric unconscious responses. The reasoner [87] component uses the UX measurement information as input data, which are quantified by emotion and stress metrics and interaction metric modules. Based on input facts, the reasoner fires the rules. The fired rules are passed to the question generator, which uses the predefined question templates to ask selective questions against the post-task performed by the participant.

The rule base was constructed from the existing standardized usability and UX questionnaires, including AttrakDiff [35], User Experience Questionnaire (UEQ) [6], Questionnaire for User Interaction Satisfaction (QUIS) [36], Single Ease Question [88], Software Usability Measurement Inventory (SUMI) [37], and Software Usability Scale (SUS) [38]. The production rules "IF-THEN" was used to associate the selected questionnaires with post-task UX measurements from user observational data. First, we extracted all questions of bipolar words and merged the duplicate one, arranged it as an LTR (negative to positive), and assigned an ID to each bipolar word that uses an index, as shown in Appendix A Table A1, to load the bipolar word based on the reasoner action. Accordingly, the question template is filled by the question generator module. The partial list of candidate rules is presented in Table 1.

Table 1. A partial list of candidate rules.

Rule ID	Condition (IF)	Action (THEN)
R1	**IF** emotional_state = "anger" **AND** congnitive_state = "stress" **AND** usability.tasksuccess = "failure"	T1, WL1, WL13
R2	**IF** emotional_state = "anger" **AND** congnitive_state = "confuse" **AND** usability.tasksuccess = "failure"	T1, WL1, WL21
R3	**IF** emotional_state = "disgust" **AND** congnitive_state = "confuse" **AND** usability.tasksuccess = "failure"	T1, WL19, WL21
⋮	⋮	⋮
Rn	**IF** emotional_state = "happy" **AND** usability.tasksuccess = "complate"	T1, WR14, WR9

We created predefined templates that store the question template repertory by ID, such as T1. One sample question template structure that uses the question generator component is the following:

I was __________________ with the ______________ complete the task.

The question generator selects and completes the template based on the resultant fired rules, e.g., R1, R3, and R4 based on the UX measurements facts.

Example 1. *I was feeling annoyed with the confusing UI to complete the task.*

Example 2. *I was feeling unfriendly with the unpleasant UI to complete the task.*

Example 3. *I was pleased with the time taken to complete the task.*

Additionally, the question generator adds a free text field, user emotions Likert scale emoticons as shown in Figure 6 and then sends it to the participants for obtaining the response. The obtained user's response is persisted in the database for analysis.

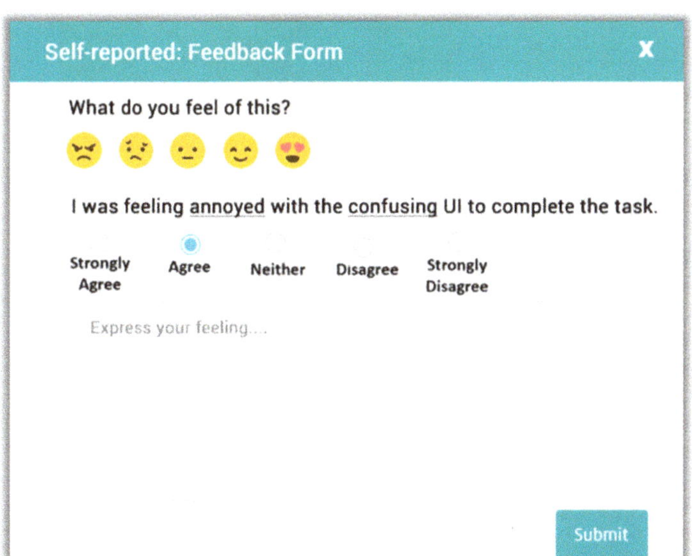

Figure 6. Self-reported feedback form.

The automatic survey analysis deals with the analysis of closed-ended and open-ended questionnaires. Analysis of the former deals with the response transformation, measurement of central tendency, variance, confidence interval, and scale consistency by assigning the questions items to UX model. For example, word annoying belongs to the "attractiveness", and "confusing" belongs to the "perspicuity" of UX scale. Based on that UX scale, UX moderator evaluates the UX of the project.

The latter analysis deals with the free text user responses. First, it loads all user free text responses, which are preprocessed before applying the topic modeling using Latent Dirichlet Allocation (LDA). LDA is an unsupervised generative statistical model, which assumes that each document may be consisted of different topics and words distribution over each topic. We implemented the pyLDAvis (https://github.com/bmabey/pyLDAvis), a python library for interactive topic model visualization for the extraction of topics from the collected user's feedback. All the collected user's textual feedbacks are processed by preprocessor module to remove the numeric data, erase punctuation, remove stop words, convert text into lower case, and stemming. We set number of topics to 30, 50 number of words per topic, and 1000 number of iterations to interpret the results. LDA extracts the topics and assigns a topic name based on dictionary words. Based on the collected topics on a different project, we will extract the important UX constructs/dimensions, for the inclusion in the UX model.

We have built the classifiers to classify the user textual feedback either as positive or negative along with emotions using automatic survey analysis module. Both positives and negatives user feelings are related to the post-task for determining the UX consequences. The overall workflow of the aforementioned process of the self-reporting metric is shown in Figure 7.

The workflow of sentiment and emotion analyzer is shown in Figure 8. The workflow consists of three main steps (a) Feature construction (b) Feature Extraction and Selection (c) Learning of prediction model. The details of these steps are described in the subsequent sections.

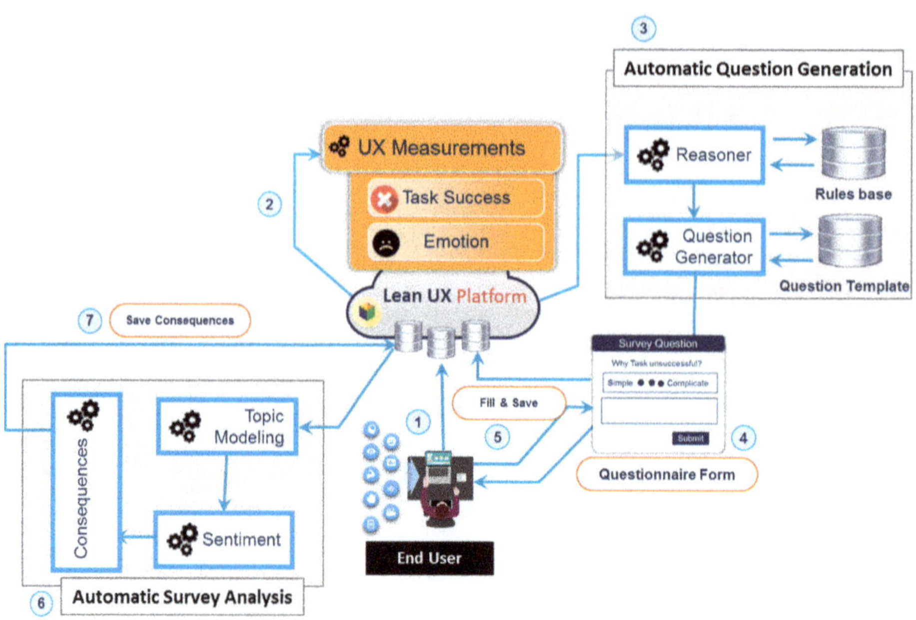

Figure 7. Survey workflow: triangulation of UX metric with self-reporting.

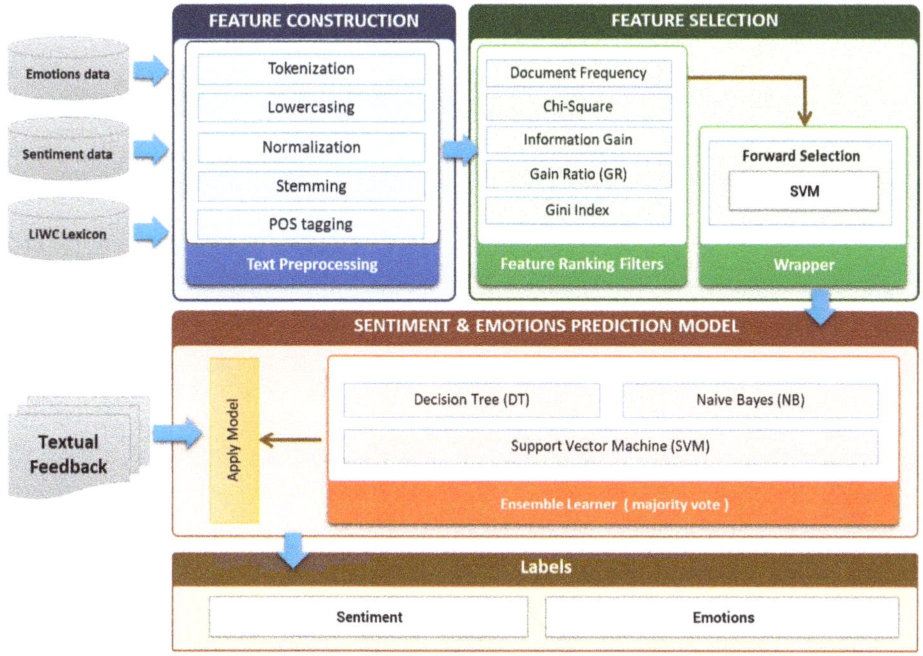

Figure 8. The workflow of sentiment and emotion analyzer.

(a) Feature construction

In text classification, conversion of text into feature vector is an essential task. The construction of an adequate feature space from the raw and unstructured text for better learning performance is necessary for text classification. It is essential to include only relevant/appropriate features for text representation. In the recent literature, different features representation methods have been used to represent text, for textual classification. These are bag-of-words (BOW), linguistic patterns using part-of-speech (POS) tags, high order n-gram features (character n-grams and word n-grams), dependency parsing tree, semantic features (lexicons and dictionaries) and structural features [89,90]. In this study, we used BOW, POS tags, semantic features (lexicons and dictionaries). For feature construction, we have applied preprocessing step to make the initial feature vectors which are suitable for further feature extraction and selection process. The preprocessing step contains tokenization, stop-word removal, and stemming (Porter algorithm). We used PENN Treebank scheme [91] for POS tagging pattern. For example, the feature "excellent interface" filtered by the POS tag pattern "JJ NN" and "was disappointed" feature is filtered out by the pattern "VBD VBN". TF-IDF term weight scheme have been applied for word vector creation.

(b) Feature Selection

Feature selection is the way to extract and select the most important and relevant features. It reduces the dimensionality feature space without losing too much information for an accurate prediction. The selected features are used to train the predictive model. We have employed filter method and wrapper method for effective features selection. In the filtering method, the subset of important features/relevant features is selected by ranking them according to specific scoring schemes based on the intrinsic properties of the features. The low scoring features are removed while highest scoring features are selected. The filter uses a fast evaluation function and is independent of the classifier. In the filter based method, we have used the filters like chi-square, Gini index, gain ratio, and information gain as shown in Figure 9.

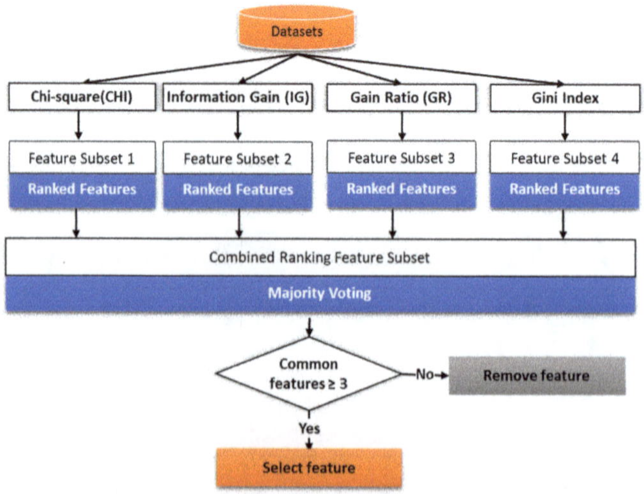

Figure 9. Filter base feature selection process.

The word-vector is input in feature selection module. The individual filter assigns weight to each feature using their internal logic and select the initial subset features. We apply the majority voting method for the final feature selection. We set the threshold value to 3 that checks for common features selected by at least three filters. Then we have applied the wrapper method (forward selection process), in subset feature selection. In wrapper method, various subsets of features are generated and evaluated. The forward selection starts with an empty selection of features/attributes and, in each iteration, it adds new attribute of the given recordset. We have applied 10-fold cross-validation using SVM learner to estimate the performance, if the added attribute gives the higher performance then is added to the selection. Then a new round is started with the modified selection. We have added the stopping behavior to stop the iteration if no significant increase in performance.

(c) Learning prediction model (Ensemble Learner)

We have employed the ensemble learning method for sentiment and emotion classification. Ensemble learning combines the predictions of multiple base learners to improve performance over a single learner. In this work, we have employed majority voting technique in conjunction with three base learners namely, Support Vector Machine (SVM), Naïve Bayes (NB) and Decision Tree. Based on the majority voting of base learners, the user textual feedback is classified into either positive or negative class along with basic emotions (Joy, anger, fear, sadness, and surprise).

4.3. Analytics Layer

The analytics layer is responsible for providing different analytics based on the UX expert query. It is composed of a heat map, real-time, audience, behavior, retention, conversion, and predictive analytics. Heat-map analytics provide a comprehensive solution to present user interaction data in a more intuitive way. We used the Heatmap.js library to create a heat map from the eye tracker fixation metric and analytical tracker using click data, which helped us determine the obstructions and ignored parts in the user interface.

The audience analytics module uses the subject's data to understand the audience habits and determine what makes them more or less likely to take the action in a system. The behavioral analytics module deals with how and why the user acts based on the retrospective analysis. The retention analytics module checks how often the user returns to the product/application in a specific time frame and to check if either the user increases the frequency with changes in the user interface.

Conversion analytics module measures the user state of change in terms of the conversion rate, checks the cause of each process of success or failure, and determines why a user failed to complete certain tasks. Based on this information, we can identify the main hurdles that the user faces while completing some specific steps in the application and how to overcome those hurdles. Predictive analytics make predictions about user's next move by using supervised machine learning to forecast the next move based on user observational data.

4.4. Visualization Server (UX Toolkit)

The visualization server is a client application that is used by the UX expert to evaluate the product, system, or service. It is a web application for realizing the different features, analytics, and visualizations based on UX measurement metrics and collected data. The UX toolkit is designed as responsive and adaptive so that it can operate on any device and operating system. The toolkit user interface is shown in Figure 10. We developed the toolkit using the Django platform. For markup language, HTML 5 along with JavaScript libraries, such as D3.js, were used. For API design, the Django rest platform was used. The Lean UX toolkit evaluates the product with respect to momentary, episodic, and cumulative UX based on the study design. It provides plug and plays support to attach sensors and devices according to the design study. Before collecting the multimodal user interaction data, the application must be registered to the Lean UX platform through the UX toolkit, and SDK code should be added to the application with assigned registered code. From that point, the UX expert can check the real-time visualization that is generated by analytics based on collected data to evaluate the momentary UX. The UX expert can also evaluate the episodic and cumulative UX in a retrospective manner. It also provides access to all the question templates and rules to modify according to the application. The rest of Lean UX toolkit workflow and screenshots are presented in Appendix B.

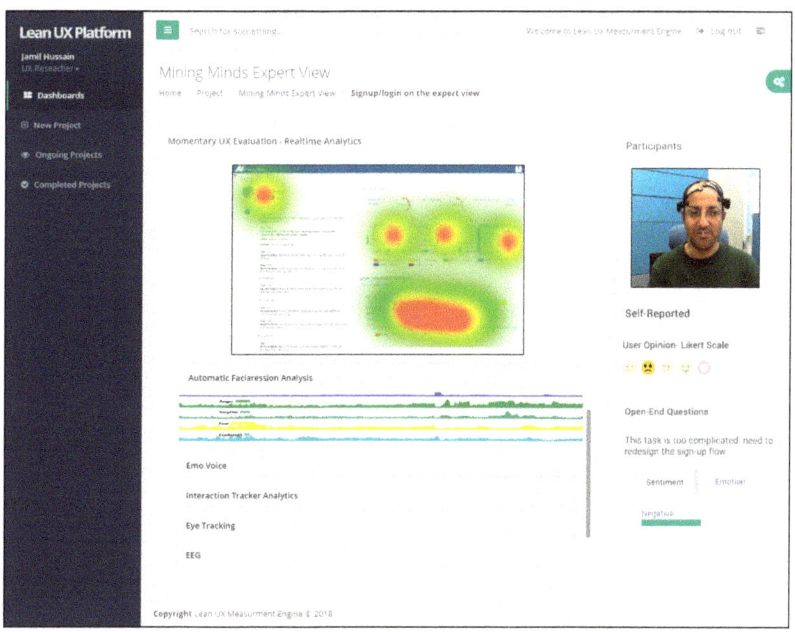

Figure 10. User Interface of UX toolkit.

5. Execution Scenarios as Case Studies of Mining Minds Evaluation

In this section, a conceptual case study is discussed. The Ubiquitous Computing Laboratory (UCLab) development team developed a platform related to health and wellness named "Mining Minds" [63]. The product is concerned with the well-being and interaction of the users. We evaluated the Mining Minds expert view through the LEAN UX platform.

First, we set up an experiment through the Lean UX toolkit. Multiple sensors were connected to obtain a response in the form of user interactions, video, audio, EEG, and eye tracking. These sensors sent data to the lean UX platform whenever participants used the Mining Minds application. The multimodal sensory data were gathered, synchronized, and labeled before persistence and routing to different UX measuring modules. Depending on the data size and nature, the data persisted either as big data or in a relational database. Each measuring module analyzed the user stimulus and measured the metrics of the UX. The self-reporting module automatically generated the UX questionnaire. The results of the measurement confirmed the user responses. The overall workflow of the proposed platform is shown in Figure 11.

After an expert-defined duration, we investigated the user experience in three modes: momentary, episodic, and cumulative, through the Lean UX toolkit. We evaluated different UX results based on queries. All concerning analytical modules of the analytics layer generated interactive representations of reports in the form of graphs based on UX measured metrics. Furthermore, UX experts can evaluate and decide the area of improvement of that product and then indicate it to the development team. An abstract view of the case study is shown in Figure 12.

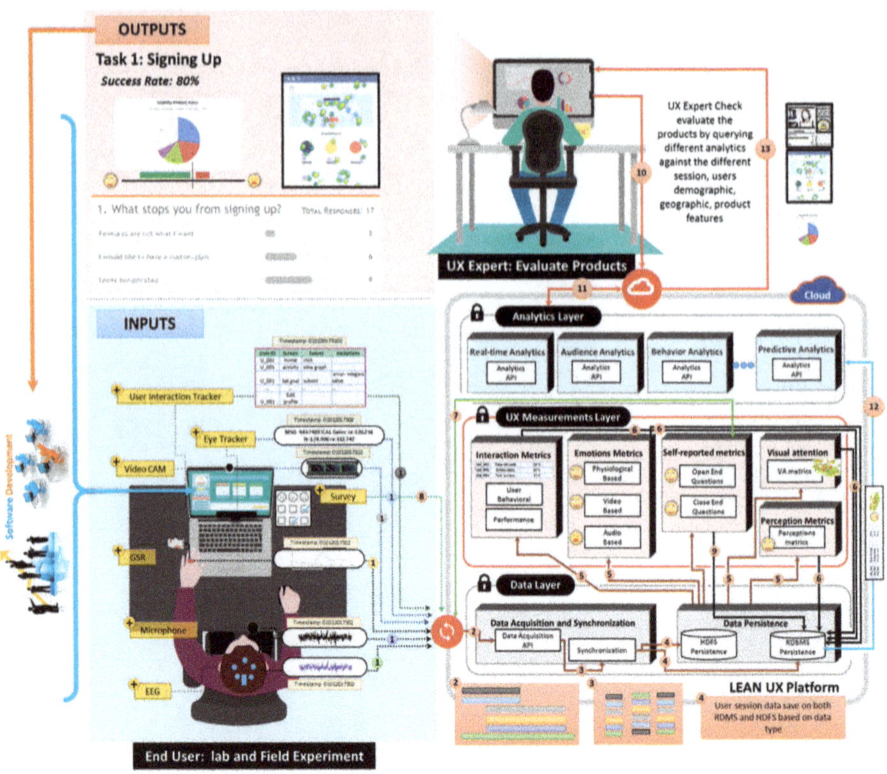

Figure 11. Overall workflow of the proposed platform.

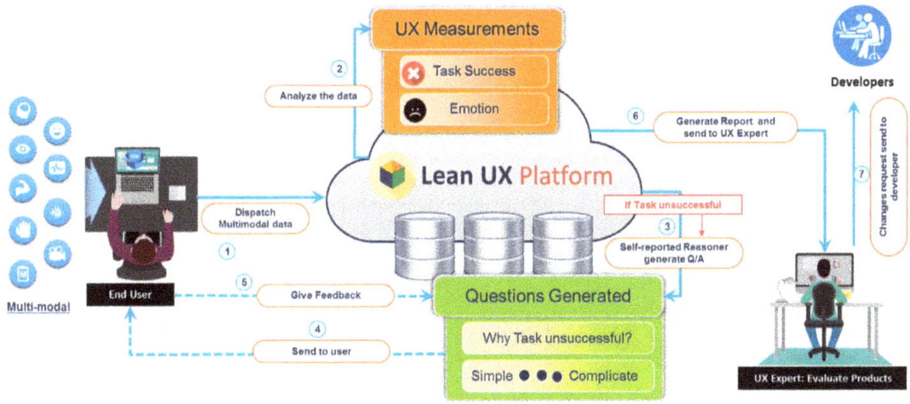

Figure 12. How it works.

6. Results and Evaluation

The proposed platform was evaluated from different aspects, such as multimodal data acquisition error rate, synchronization accuracy, individual UX measurements metrics ranging from interactions, multimodal emotions recognizers, and self-reported assessments.

Sample: The 10 participants were university students (70% male and 30% female), and their ages ranges from 19 to 44 (29 mean). They had a mixed race and were used for evaluation of the Lean UX platform. Each session was 20 min on average. The data were collected from different devices, e.g., EEG, camera, eye tracking, interaction tracker, text analyzer, and microphone. The results and discussion of evaluations are presented further below.

6.1. Multimodal Data Acquisition and Data Synchronization Process

The accuracy of the data acquisition and synchronization process was validated by connected different devices, such as EEG, camera, second-generation Kinect, eye-tracker, and PC with the Lean UX platform cloud. All data streams from multimodal data sources were acquired, synchronized at server endpoints, and checked for data accuracy using a three-second window size. The rate of missing data packets was used to measure the accuracy of data acquisition module shown in Table 2. The results show a 0.03% average error rate, which is very low, meaning that the platform acquired and processed multimodal data safely.

Table 2. The data acquisition process accuracy.

No. of API Calls	Missing Data Packets	Error Rate
20,000	2	0.010
40,000	5	0.012
60,000	9	0.015
80,000	12	0.015
120,000	21	0.017
Average		0.03

The results are shown in Figure 13. The multimodal data from all devices were effectively synchronized at cloud endpoints in milliseconds. The synchronization module synchronized all sensors, stimuli and API data streams in real-time without manual post-synchronization of data. For example, the eye-tracker, Kinect, camera, microphone, EEG, and interaction tracker communicated at 300 ms, 200 ms, 450 ms, 562 ms, 860 ms, and 1318 ms, respectively, at the first window frame. The synchronization module recognized that all incoming data streams belonged to the single event. The results depicted that all data streams were well synchronized in a real-time manner, showing the perfection of the synchronization module.

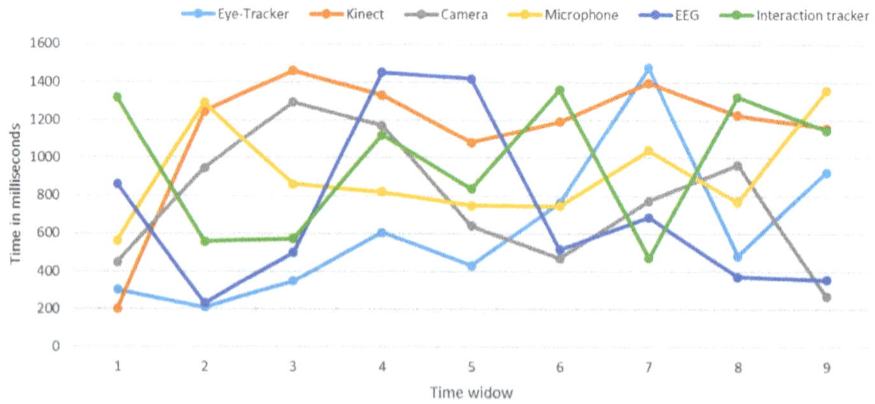

Figure 13. Multi-modal data sync testing per time-window.

6.2. Emotion and Stress Metrics

Video-based emotion recognition: We evaluated the video-based emotion recognition component by using five datasets: Cohn-Kanade dataset [92], JAFEE dataset [93], USTC-NVIE dataset [94], Yale B face dataset [95], and FEI face dataset [96]. Table 3 shows the confusion matrix of automatic facial expressions for Cohn-Kanade dataset. Figure 14 shows the average accuracy for each dataset. The results show a high accuracy for the happy, anger, sadness and surprise, while relatively low accuracy for the fear and disgust. Additionally, fear and disgust were mixed with sadness and anger owing to the subject's expressions. However, generally, the model accuracy was quite reasonable compared to the other video-based emotion recognizers. There were some challenges for effective emotion recognition for heterogeneous populations with respect to demographic, cultural, and impairment aspects, which can be resolved by improving the landmarking techniques to classify the emotions for face impairment.

Table 3. Facial Expression confusion matrix using Cohn-Kanade dataset (unit %).

Expression	Happy	Anger	Sad	Surprise	Fear	Disgust	Neutral
Happy	99	0	0	1	0	0	0
Anger	0	98	0	1	0	1	0
Sad	0	0	98	0	1	0	1
Surprise	0	1	1	96	0	2	0
Fear	0	1	1	1	95	1	1
Disgust	0	1	1	0	1	97	0
Neutral	0	0	1	0	0	0	99
Overall Accuracy				97.429%			

Audio-based emotion recognition: The result of audio-based emotion metric extraction is shown in Table 4 for Emo-DB dataset. The results show a high accuracy for anger and surprise, while a relatively low accuracy is shown for happy and disgust. Additionally, happy and anger were mixed owing to the high sound pitch, while sadness and neutral were mixed owing to the soft voice. However, generally, the model accuracy was quite reasonable compared to the other audio-based emotion recognizers. There were some challenges, such as tone differences and voice pitch, which made the audio-based emotion recognition difficult.

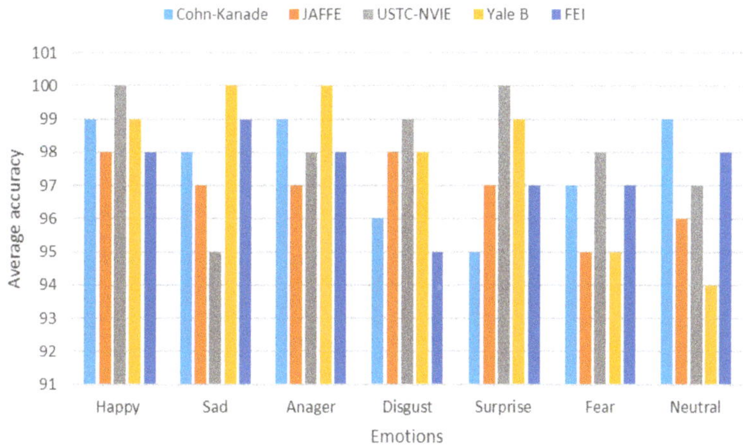

Figure 14. Recognition average accuracy for each dataset.

Table 4. Audio base emotion recognition confusion matrix using Emo-DB dataset (unit %).

Expression	Happy	Anger	Sad	Surprise	Fear	Disgust	Neutral
Happy	83	10	0	7	0	0	0
Anger	2	92	0	1	0	5	0
Sad	0	0	87	0	2	0	11
Surprise	6	3	0	89	0	2	0
Fear	0	1	1	8	87	1	2
Disgust	0	7	2	6	2	80	3
Neutral	0	0	10	0	2	0	88
Overall Accuracy			86.571%				

EEG-based emotion recognition: For the EEG-base emotion recognition, we used four features—differential entropy (DE), power spectral density (PSD), rational asymmetry (RASM), and differential asymmetry (DASM)—to extract the most stable pattern for emotion recognition either positive, negative, or neutral. The results are shown in Figure 15, where DE achieves a higher accuracy for all frequency bands compared to the other features. From the experiment and results, we identified that the DE feature is more suitable to fuse with other features of emotion recognizers, such pupil size of the eye tracking data.

Pupil Diameter: We performed different experiments based on the pupil size metric using an eye tracker to observe how the pupil size changed in accordance with different emotional states. From the experiments, we found that the pupil size increased (dilated) in a sorrowful state, and was smallest in a calm state, as shown in Figure 16.

For both positive and negative emotions, the pupil size was larger compared with neutral, which showed a correlation with different emotions. We extracted different features: power spectral density (PSD) and differential entropy (DE) from the pupil size to measure the emotional arousal, and the DE feature outperformed PSD.

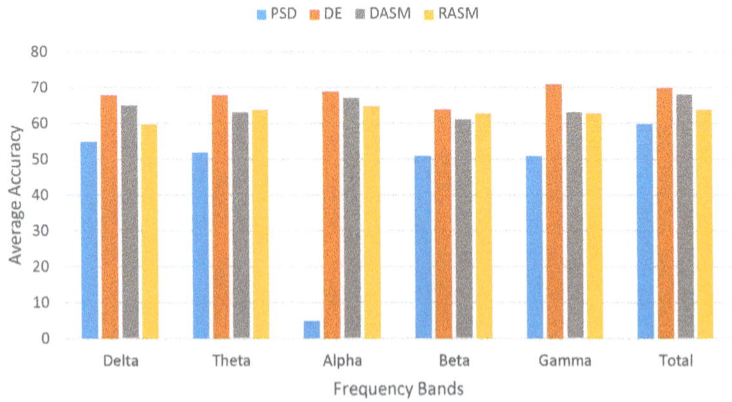

Figure 15. Average accuracy of the classifier using different features on different frequency bands.

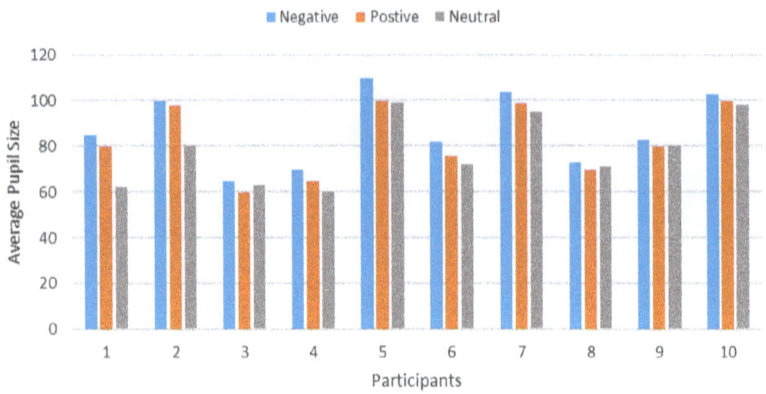

Figure 16. The average pupil size of each trail.

Emotion fusion: Table 5 shows the average accuracy of emotion fusion. The results show that fusion accuracy is higher than the individual classifier accuracy. Furthermore, we used the paired t-test ($p < 0.05$) to evaluate the accuracies of all the methods. The t-test analysis showed no significant differences between the feature level and decision level fusion.

Table 5. The average accuracies of each classifier and fusion method.

Subject	Facial Expression	Audio Base	Textual	EEG (DE)	Eye Tracking	Fusion	
						Feature Level	Decision Level
1	95	84	91	68	80	96	96
2	92	82	89	63	82	97	98
3	100	80	94	64	83	98	99
4	98	83	89	62	89	93	98
5	98	84	93	76	90	92	93
6	90	83	94	70	81	97	98
7	94	84	94	72	87	91	93
8	93	83	91	69	85	94	94
9	93	80	92	64	80	95	93
10	98	82	92	70	87	98	96
Average	**95.1**	**82.5**	**91.9**	**67.8**	**84.4**	**95.1**	**95.8**

6.3. Self-Reported Metric

The open-ended question analyzer module assessed the affective content (sentiment and emotions) by using lexicon-based dictionaries; POS-tagging; bag-of-words; and in combination with classifiers, such as SVM or NB. We used multiple lexicon dictionaries (e.g., LIWIC, and Custom), and annotated a training dataset at different levels—document, paragraph, sentence, and word level—to help extract the true emotions from the user textual response. For feature selection, filter and wrapper approaches were used for the selection of optimal features that improved the classification accuracy. For the experiment, we used five datasets that are widely used for text-based sentiment analysis. The results of experiments shown in Table 6 reveal that the ensemble method with minimal feature selection strategies can effectively increase the accuracy of classification compared with the baseline classifier.

Table 6. Average accuracies of each classifier for each dataset.

Dataset	Classifier	# of Features	Accuracy
Movie	SVM	3625 ± 1209	93
	NB	2400 ± 1375	92
	DT	3816 ± 1254	88
	Ensemble	3779 ± 1314	**94**
	Average	**3405**	**91.75**
Book	SVM	2199 ± 1066	87
	NB	2612 ± 1074	86
	DT	2031 ± 1048	83
	Ensemble	2956 ± 1021	**89**
	Average	**2449**	**86.25**
Electronic	SVM	1323 ± 474	85
	NB	1002 ± 1090	**89**
	DT	1938 ± 625	87
	Ensemble	1760 ± 855	86
	Average	**1505**	**86.75**
Kitchen	SVM	1843 ± 770	89
	NB	1566 ± 470	86
	DT	1600 ± 787	89
	Ensemble	1969 ± 877	**90**
	Average	**1744**	**88.5**
Music	SVM	642 ± 296	**89**
	NB	819 ± 276	87
	DT	855 ± 267	86
	Ensemble	362 ± 155	88
	Average	**669**	**87.5**

7. Conclusions

Understanding user feelings, thoughts, and needs are very important to engaging, sustaining, and increasing the purchase of a product, system, or service. The UX assessment reveals the user feeling about the product, system, or service and their functionalities. The user may have difficulty expressing their feelings and thoughts about a product, system, or service through traditional methods. Sometimes they may be unable to interpret their own feelings in order to describe them. The physiological measurements in assessing UX can detect emotional arousal and stress, motivation, and visual attention that have direct relationships with user cognitive and affective states in a non-intrusive way. The mixed-method approach showed importance in the UX evaluation methods by providing more accurate and precise information about the user while interacting with the product.

However, this approach requires a skilled researcher to integrate multiple devices, synchronize data, analyze data, and make informed decisions. Thus, we developed the Lean UX platform to provide an integrated environment in a seamless manner with real-time synchronization and powerful visualizations. The platform offers plug-and-play support for data collection from different devices and powerful real-time analytics visualization to enable insights of time spans of the user experience with multiple participants. Further, it helps identify the areas of improvement after assessment of any product, system, or service to improve the overall UX.

However, improvements can be made in terms of the classifier performance. We will add more classifiers and biometric sensors, such as GSR and ECG/EMG, to the Lean UX platform to extract the true user emotional experience. Finally, we will increase the datasets for effective emotional state recognition.

Author Contributions: J.H. proposed and formulated the idea, conceived, designed and performed all the experiments and wrote the paper; T.H. worked on the video-based emotion recognition; J.B. worked on audio-base emotion recognition; H.S.M.B. wrote the paper, and designed and formatted the contents; A.U.H. worked on sentiment analysis; W.A.K. proofread the paper; M.A. revised the manuscript critically for important contents and contributed to scientific discussion; S.L. provided advisory comments, remarks and financial support for the paper.

Funding: This research was supported by an Institute for Information & Communications Technology Promotion (IITP) grant funded by the Korean government (MSIT) (No. 2017-0-00655). This work was supported by the MSIT (Ministry of Science and ICT), Korea, under the ITRC (Information Technology Research Center) support program (IITP-2017-0-01629) supervised by the IITP (Institute for Information & communications Technology Promotion) and NRF- 2016K1A3A7A03951968.

Conflicts of Interest: The authors declare no conflict of interest.

Appendix A

Table A1 shows the selected items from existing UX questionnaires.

Table A1. The average accuracies of each classifier and fusion method.

Question ID	Bipolar Words	
	WL	**WR**
1	annoying	enjoyable
2	not understandable	understandable
3	dull	Creative
4	difficult to learn	easy to learn
5	inferior	valuable
6	boring	exciting
7	not interesting	interesting
8	unpredictable	predictable
9	slow	fast
10	inventive	conventional
11	obstructive	supportive
12	bad	good
13	complicated	easy
14	unlikable	pleasing
15	usual	leading edge
16	unpleasant	pleasant
17	not secure	secure
18	motivating	demotivating
19	Does not meets expectations	meet expectations
20	inefficient	effient
21	confusing	clear
22	impractical	practical
23	cluttered	organized
24	unattractive	attractive
25	unfriendly	friendly
26	conservative	innovative

Table A1. *Cont.*

Question ID	Bipolar Words	
	WL	WR
27	technical	human
28	isolating	connective
29	unprofessional	professional
30	cheap	premium
31	alienating	integrating
32	separates me	brings me closer
33	unpresentable	presentable
34	cautious	bold
35	undemanding	challenging
36	ordinary	novel
37	rejecting	inviting
38	repelling	appealing
39	disagreeable	likeable

Appendix B

Appendix B depicts the lean UX platform toolkit. The moderator first login into the Lean UX platform toolkit, to create a new project by clicking on "Add new project" button. The details of project should be entering in step-wise form such as project information, UX evaluation type (anticipated UX, momentary UX, episodic UX, and cumulative UX), and input modalities/stimuli (video cam, MIC, screen recording, interaction tracker, EEG, Eye tracking, and survey) shown in Figure A1. The input modalities are dependent on the UX evaluation type. For example, if moderator selects only anticipated UX, then the evaluation will be performed using "survey". In the survey, we are using User Experience Questionnaire (UEQ) scale to collect the user experience for measuring the UX, contains six dimension scales such as novelty, stimulation, attractiveness, dependability, and efficiency. While for the other type of UX evaluation, all types of input modalities will be available. The moderator can select any type of input modalities, depend on their study.

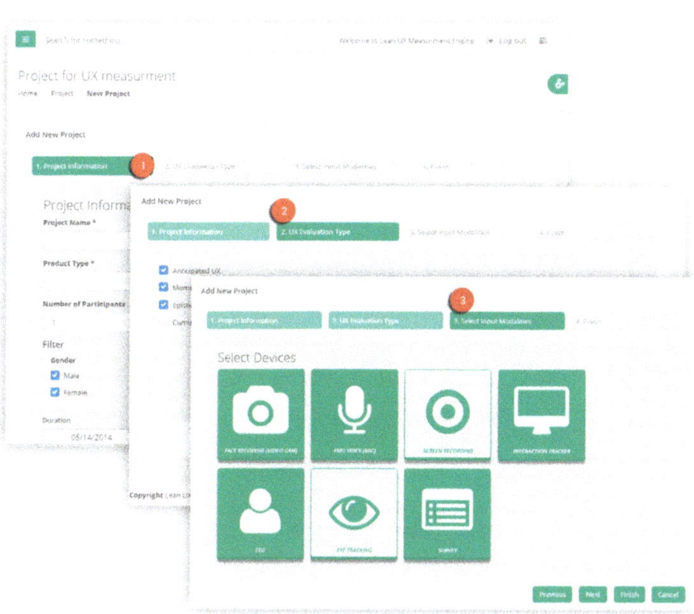

Figure A1. Project creation step-wise process.

After successful creation of the project, the moderator can add tasks to project as shown in Figures A2 and A3. The system will generate automatically the project Id, which is used by the interaction tracker module, to track the user interaction as discussed in Section 4.2.1. The moderator first adds the JavaScript code in the header of each page by assigning the project id. The JavaScript code is also responsible to display the feedback form on the completion of the task or error situation.

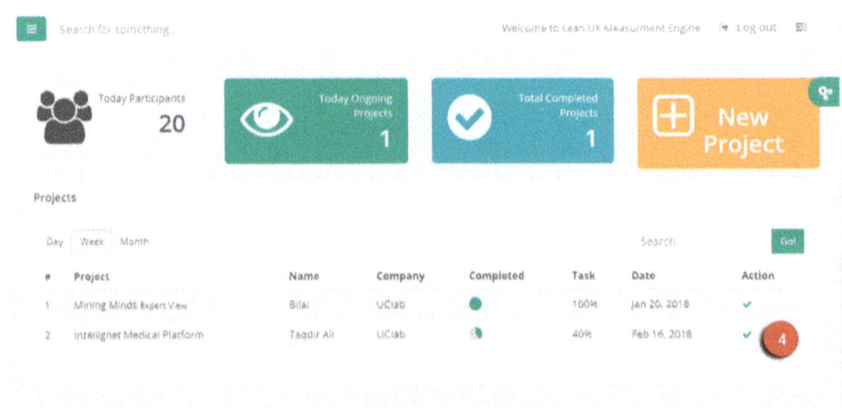

Figure A2. Dashboard of Lean UX- List of created projects for UX evaluation.

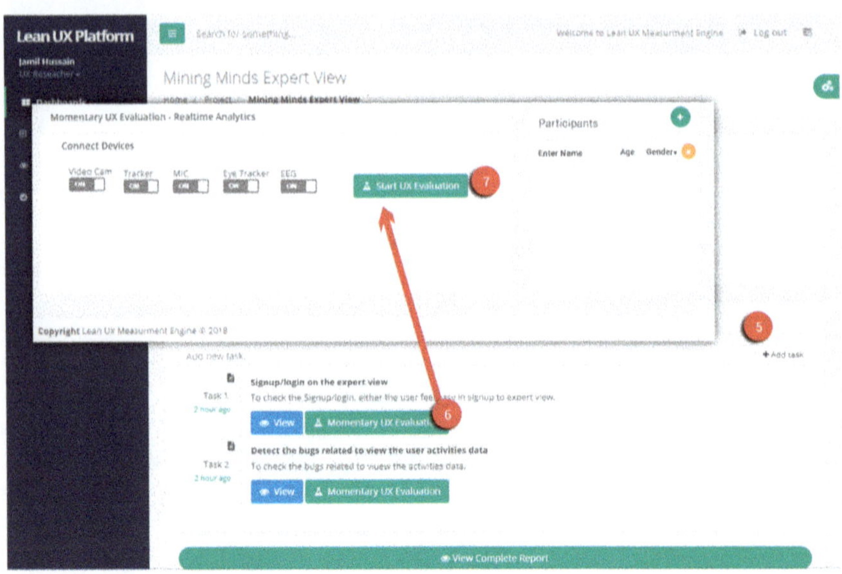

Figure A3. Task creation and evaluation process by task wise.

The moderator can collect the UX measurement data by connected the sensors, sensors connectivity is auto checked by the system. The moderator should add the participant information by adding their name, age, and gender. By clicking on "Start UX evaluation" button, all measurement modules will start collecting the data and perform real-time UX measures related to emotions, user interaction, and self-reported as shown in Figure A4.

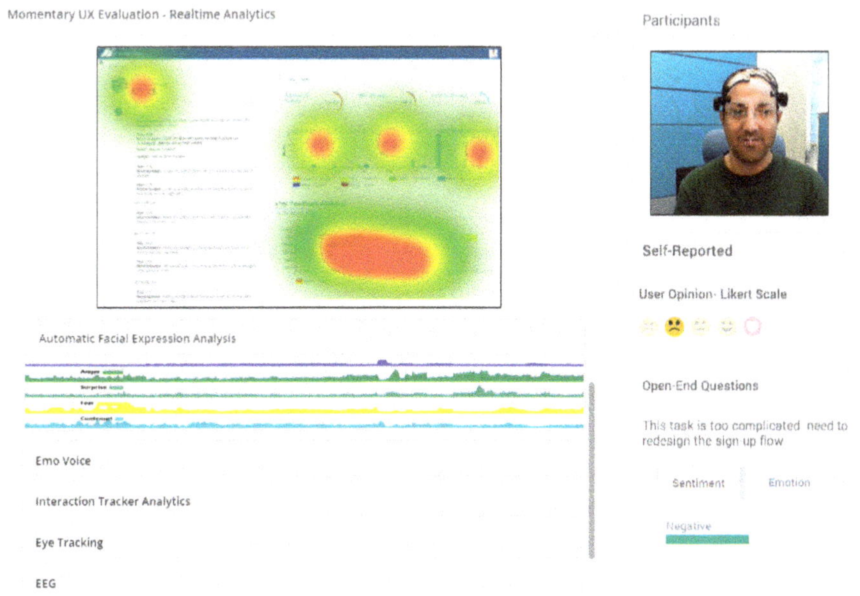

Figure A4. Momentary UX evaluation: real-time data collection and UX metric measurement.

The moderator can check the different modalities measures such as automatic facial expression analysis, emo voice, interaction tracker analytics (e.g., the heatmap of user click and mouse move data), eye tracking, and EEG. At the successful/unsuccessful of task, the self-reported feedback form will be appeared on the participant screen to collect the self-reported feedback. The participants can express their feeling in both Likert scale and free text format. The self-reported data will be available on the submission of self-reported form by the participant to the moderator. The open end question analyzer will analyze the free text self-reported feedback to extract the user sentiment and emotions related to that task shown in Figure A4. This evaluation process will repeat for all participants who will participate in the study for each task. The moderator can check the results of UX evaluation at the task level and project level as shown in Figures A5 and A6.

The moderator can check the results by applying the participant's filter such as emotions by numbers (fusing the different modalities emotions), overall emotions, self-reported sentiment, and task completion rates.

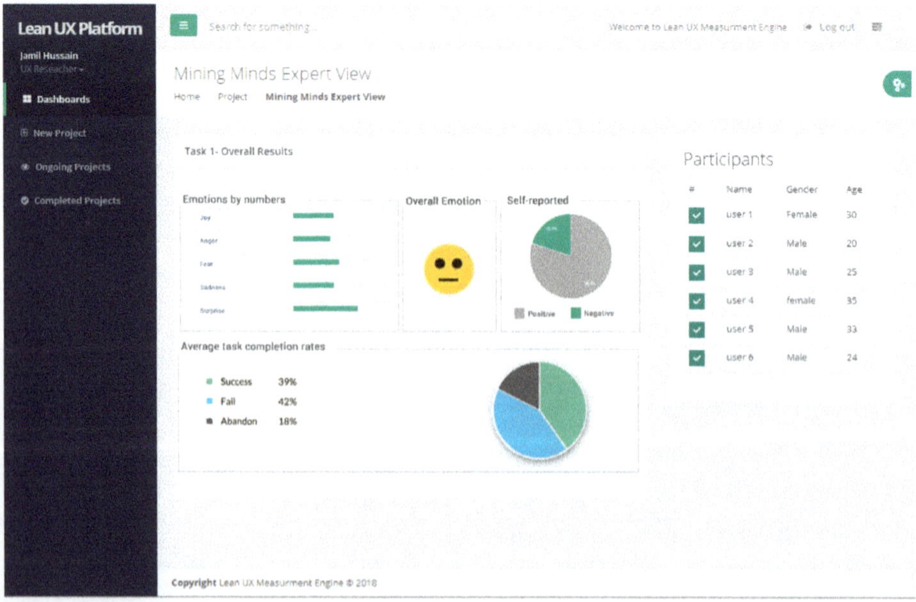

Figure A5. Task wise result view.

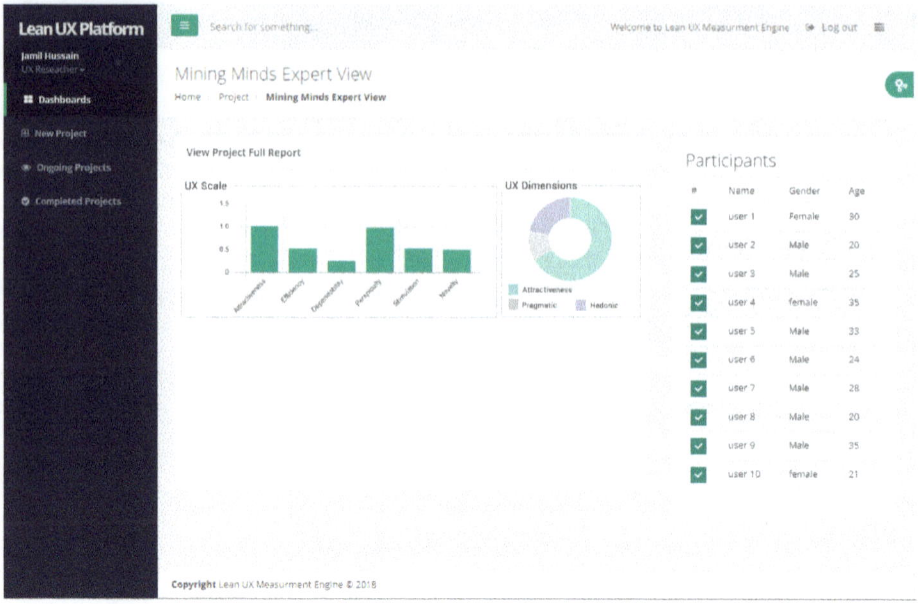

Figure A6. Project wise result view.

References

1. Hassenzahl, M.; Tractinsky, N. User experience—A research agenda. *Behav. Inf. Technol.* **2006**, *25*, 91–97. [CrossRef]
2. Liang, Y.; Liu, Y.; Loh, H.T. Exploring Online Reviews for User Experience Modeling. In *DS 75–7: Proceedings of the 19th International Conference on Engineering Design (ICED13), Design for Harmonies, Vol. 7: Human Behaviour in Design, Seoul, Korea, 19–22.08. 2013*; Sungkyunkwan University: Seoul, Korea, 2013.
3. Kula, I.; Atkinson, R.K.; Branaghan, R.J.; Roscoe, R.D. Assessing User Experience via Biometric Sensor Affect Detection. In *End-User Considerations in Educational Technology Design*; IGI Global: Hershey, PA, USA, 2017; p. 123.
4. Law, E.L.-C.; van Schaik, P. *Modelling User Experience–An Agenda for Research and Practice*; Oxford University Press: Oxford, UK, 2010; ISBN 0953-5438.
5. Roto, V.; Law, E.; Vermeeren, A.; Hoonhout, J. User experience white paper. Bringing clarity to the concept of user experience. Result from Dagstuhl Seminar on Demarcating User Experience, September 15–18 (2010). *Disponible en ligne le* **2011**, *22*, 6–15.
6. Laugwitz, B.; Held, T.; Schrepp, M. Construction and Evaluation of a User Experience Questionnaire. In *Symposium of the Austrian HCI and Usability Engineering Group*; Springer: Berlin/Heidelberg, Germany, 2008; pp. 63–76.
7. All About, U.X. Available online: http://www.allaboutux.org/all-methods (accessed on 29 March 2007).
8. Bolger, N.; Davis, A.; Rafaeli, E. Diary methods: Capturing life as it is lived. *Annu. Rev. Psychol.* **2003**, *54*, 579–616. [CrossRef] [PubMed]
9. Karapanos, E.; Zimmerman, J.; Forlizzi, J.; Martens, J.-B. User Experience over Time: An Initial Framework. In Proceedings of the SIGCHI Conference on Human Factors in Computing Systems, Boston, MA, USA, 4–9 April 2009; ACM: New York, NY, USA, 2009; pp. 729–738.
10. Fallman, D.; Waterworth, J. Dealing with User Experience and Affective Evaluation in HCI Design: A Repertory Grid Approach. In Proceedings of the Conference on Human Factors in Computing Systems, Portland, OR, USA, 2–7 April 2005; pp. 2–7.
11. Scollon, C.N.; Prieto, C.-K.; Diener, E. Experience Sampling: Promises and Pitfalls, Strength and Weaknesses. In *Assessing Well-Being*; Springer: Dordrecht, The Netherlands, 2009; pp. 157–180.
12. Vermeeren, A.P.; Law, E.L.-C.; Roto, V.; Obrist, M.; Hoonhout, J.; Väänänen-Vainio-Mattila, K. User Experience Evaluation Methods: Current State and Development Needs. In Proceedings of the 6th Nordic Conference on Human-Computer Interaction: Extending Boundaries, Reykjavik, Iceland, 16–20 October 2010; ACM: New York, NY, USA, 2010; pp. 521–530.
13. Schubert, E. *Continuous Measurement of Self-Report Emotional Response to Music*; Oxford University Press: Oxford, UK, 2001.
14. Tähti, M.; Arhippainen, L. A Proposal of collecting Emotions and Experiences. *Interact. Exp. HCI* **2004**, *2*, 195–198.
15. Russel, J.A.; Weiss, A.; Mendelsohn, G.A. Affect grid: A single-item scale of pleasure and arousal. *J. Personal. Soc. Psychol.* **1989**, *57*, 493–502. [CrossRef]
16. Van Gog, T.; Paas, F.; Van Merriënboer, J.J.; Witte, P. Uncovering the problem-solving process: Cued retrospective reporting versus concurrent and retrospective reporting. *J. Exp. Psychol. Appl.* **2005**, *11*, 237. [CrossRef] [PubMed]
17. Goodman, E.; Kuniavsky, M.; Moed, A. Observing the User Experience: A Practitioner's Guide to User Research. *IEEE Trans. Prof. Commun.* **2013**, *56*, 260–261. [CrossRef]
18. Kuniavsky, M. *Observing the User Experience: A Practitioner's Guide to User Research*; Morgan Kaufmann: Burlington, MA, USA, 2003; ISBN 0-08-049756-X.
19. Fu, B.; Noy, N.F.; Storey, M.-A. Eye tracking the user experience–An evaluation of ontology visualization techniques. *Semant. Web J.* **2017**, *8*, 23–41. [CrossRef]
20. Qu, Q.-X.; Zhang, L.; Chao, W.-Y.; Duffy, V. User Experience Design Based on Eye-Tracking Technology: A Case Study on Smartphone APPs. In *Advances in Applied Digital Human Modeling and Simulation*; Springer: Cham, Switzerland, 2017; pp. 303–315.
21. Bojko, A. *Eye Tracking the User Experience: A Practical Guide to Research*; Rosenfeld Media: New York, NY, USA, 2013; ISBN 1-933820-91-8.
22. Zheng, W.-L.; Zhu, J.-Y.; Lu, B.-L. Identifying stable patterns over time for emotion recognition from EEG. *IEEE Trans. Affect. Comput.* **2017**. [CrossRef]
23. Li, X.; Yan, J.-Z.; Chen, J.-H. Channel Division Based Multiple Classifiers Fusion for Emotion Recognition Using EEG signals. In *ITM Web of Conferences*; EDP Sciences: Les Ulis, France, 2017; Volume 11, p. 07006.

24. Liu, Y.-J.; Yu, M.; Zhao, G.; Song, J.; Ge, Y.; Shi, Y. Real-time movie-induced discrete emotion recognition from EEG signals. *IEEE Trans. Affect. Comput.* **2017**. [CrossRef]
25. Mundell, C.; Vielma, J.P.; Zaman, T. Predicting Performance Under Stressful Conditions Using Galvanic Skin Response. *arXiv*, 2016.
26. Nourbakhsh, N.; Chen, F.; Wang, Y.; Calvo, R.A. Detecting Users' Cognitive Load by Galvanic Skin Response with Affective Interference. *ACM Trans. Interact. Intell. Syst.* **2017**, *7*, 12. [CrossRef]
27. Greene, S.; Thapliyal, H.; Caban-Holt, A. A survey of affective computing for stress detection: Evaluating technologies in stress detection for better health. *IEEE Consum. Electron. Mag.* **2016**, *5*, 44–56. [CrossRef]
28. Basu, S.; Bag, A.; Aftabuddin, M.; Mahadevappa, M.; Mukherjee, J.; Guha, R. Effects of Emotion on Physiological Signals. In Proceedings of the 2016 IEEE Annual India Conference (INDICON), Bangalore, India, 16–18 December 2016; pp. 1–6.
29. Schubert, E. Measuring emotion continuously: Validity and reliability of the two-dimensional emotion-space. *Aust. J. Psychol.* **1999**, *51*, 154–165. [CrossRef]
30. Izard, C.E. *The Differential Emotions Scale: DES IV-A; [A Method of Measuring the Meaning of Subjective Experience of Discrete Emotions]*; University of Delaware: Newark, DE, USA, 1993.
31. Sacharin, V.; Schlegel, K.; Scherer, K.R. Geneva Emotion Wheel Rating Study. Available online: https://archive-ouverte.unige.ch/unige:97849 (accessed on 29 March 2017).
32. Desmet, P. Measuring emotion: Development and Application of an Instrument to Measure Emotional Responses to Products. In *Funology*; Springer: Dordrecht, The Netherlands, 2003; pp. 111–123.
33. Laurans, G.; Desmet, P.M.A.; Karlsson, M.A.; van Erp, J. Using Self-Confrontation to Study User Experience: A New Approach to the Dynamic Measurement of Emotions while Interacting with Products. In *Design & Emotion*; Chalmers University of Technology: Gothenburg, Sweden, 2006; Volume 2006.
34. Desmet, P.; Overbeeke, K.; Tax, S. Designing products with added emotional value: Development and appllcation of an approach for research through design. *Des. J.* **2001**, *4*, 32–47. [CrossRef]
35. Hassenzahl, M.; Burmester, M.; Koller, F. AttrakDiff: A Questionnaire to Measure Perceived Hedonic and Pragmatic Quality. In *Mensch & Computer*; Springer: Berlin, Germany, 2003; pp. 187–196.
36. Norman, K.L.; Shneiderman, B.; Harper, B.; Slaughter, L. *Questionnaire for User Interaction Satisfaction*; University of Maryland: College Park, MD, USA, 1998.
37. Kirakowski, J.; Corbett, M. SUMI: The software usability measurement inventory. *Br. J. Educ. Technol.* **1993**, *24*, 210–212. [CrossRef]
38. Brooke, J. SUS-A quick and dirty usability scale. *Usability Eval. Ind.* **1996**, *189*, 4–7.
39. Lavie, T.; Tractinsky, N. Assessing dimensions of perceived visual aesthetics of web sites. *Int. J. Hum. Comput. Stud.* **2004**, *60*, 269–298. [CrossRef]
40. Paas, F.G.; Van Merriënboer, J.J. The efficiency of instructional conditions: An approach to combine mental effort and performance measures. *Hum. Factors* **1993**, *35*, 737–743. [CrossRef]
41. Siddiqi, M.H.; Alam, M.G.R.; Hong, C.S.; Khan, A.M.; Choo, H. A Novel Maximum Entropy Markov Model for Human Facial Expression Recognition. *PLoS ONE* **2016**, *11*, e0162702. [CrossRef] [PubMed]
42. El Ayadi, M.; Kamel, M.S.; Karray, F. Survey on speech emotion recognition: Features, classification schemes, and databases. *Pattern Recognit.* **2011**, *44*, 572–587. [CrossRef]
43. Plaza, B. Google Analytics for measuring website performance. *Tour. Manag.* **2011**, *32*, 477–481. [CrossRef]
44. Scherr, S.A.; Elberzhager, F.; Holl, K. An Automated Feedback-Based Approach to Support Mobile App Development. In Proceedings of the 2017 43rd Euromicro Conference on Software Engineering and Advanced Applications (SEAA), Vienna, Austria, 30 August–1 September 2017; pp. 44–51.
45. Den Uyl, M.J.; Van Kuilenburg, H. The FaceReader: Online Facial Expression Recognition. In Proceedings of Measuring Behavior 2005, 5th Internaltional Conference on Methods and Techniques in Behavioral Research, Wageningen, The Netherlands, 30 August–2 September 2005; Volume 30, pp. 589–590.
46. Zaman, B.; Shrimpton-Smith, T. The FaceReader: Measuring Instant Fun of Use. In Proceedings of the 4th Nordic Conference on Human-Computer Interaction: Changing Roles, Oslo, Norway; ACM: New York, NY, USA, 2006; pp. 457–460.
47. Whitehill, J.; Bartlett, M.; Movellan, J. Automatic Facial Expression Recognition for Intelligent Tutoring Systems. In Proceedings of the 2008 IEEE Computer Society Conference on Computer Vision and Pattern Recognition Workshops, Anchorage, AK, USA, 23–28 June 2008; pp. 1–6.

48. Noroozi, F.; Marjanovic, M.; Njegus, A.; Escalera, S.; Anbarjafari, G. Audio-visual emotion recognition in video clips. *IEEE Trans. Affect. Comput.* **2017**. [CrossRef]
49. Clifton, B. *Advanced Web Metrics with Google Analytics*; John Wiley & Sons: Hoboken, NJ, USA, 2012; ISBN 1-118-23958-X.
50. Miller, S.A. *Piwik Web Analytics Essentials*; Packt Publishing Ltd.: Birmingham, UK, 2012; ISBN 1-84951-849-1.
51. Liu, X.; Zhu, S.; Wang, W.; Liu, J. Alde: Privacy Risk Analysis of Analytics Libraries in the Android Ecosystem. In *International Conference on Security and Privacy in Communication Systems, Guangzhou, China, 10–12 October 2016*; Springer: Cham, Switzerland, 2016; pp. 655–672.
52. Alepuz, I.; Cabrejas, J.; Monserrat, J.F.; Perez, A.G.; Pajares, G.; Gimenez, R. Use of Mobile Network Analytics for Application Performance Design. In Proceedings of the 2007 Network Traffic Measurement and Analysis Conference (TMA), Dublin, Ireland, 21–23 June 2017; pp. 1–6.
53. Girard, J.M.; Cohn, J.F. A primer on observational measurement. *Assessment* **2016**, *23*, 404–413. [CrossRef] [PubMed]
54. Zheng, W.-L.; Dong, B.-N.; Lu, B.-L. Multimodal Emotion Recognition Using EEG and Eye Tracking Data. In Proceedings of the 2014 36th Annual International Conference of the IEEE Engineering in Medicine and Biology Society (EMBC), Chicago, IL, USA, 26–30 August 2014; pp. 5040–5043.
55. Bergstrom, J.R.; Schall, A. *Eye Tracking in User Experience Design*; Elsevier: Walthan, MA, USA, 2014; ISBN 0-12-416709-8.
56. Tzafilkou, K.; Protogeros, N. Diagnosing user perception and acceptance using eye tracking in web-based end-user development. *Comput. Hum. Behav.* **2017**, *72*, 23–37. [CrossRef]
57. Sanfilippo, F. A multi-sensor fusion framework for improving situational awareness in demanding maritime training. *Reliab. Eng. Syst. Saf.* **2017**, *161*, 12–24. [CrossRef]
58. Sivaji, A.; Ahmad, W.F.W. Benefits of Complementing Eye-Tracking Analysis with Think-Aloud Protocol in a Multilingual Country with High Power Distance. In *Current Trends in Eye Tracking Research*; Springer: Cham, Switzerland, 2014; pp. 267–278.
59. Vrana, S.R. The psychophysiology of disgust: Differentiating negative emotional contexts with facial EMG. *Psychophysiology* **1993**, *30*, 279–286. [CrossRef] [PubMed]
60. Bacic, D. Understanding Business Dashboard Design User Impact: Triangulation Approach Using Eye-Tracking, Facial Expression, Galvanic Skin Response and EEG Sensors. Available online: http://aisel.aisnet.org/amcis2017/HumanCI/Presentations/21/ (accessed on 15 May 2018).
61. Klein, L. *UX for Lean Startups: Faster, Smarter User Experience Research and Design*; O'Reilly Media, Inc.: Sebastopol, CA, USA, 2013; ISBN 1-4493-3504-7.
62. Meneweger, T.; Wurhofer, D.; Obrist, M.; Beck, E.; Tscheligi, M. Characteristics of Narrative Textual Data Linked to User Experiences. In Proceedings of the CHI'14 Extended Abstracts on Human Factors in Computing Systems, Toronto, ON, Canada, 26 April–1 May 2014; ACM: New York, NY, USA, 2014; pp. 2605–2610.
63. Banos, O.; Amin, M.B.; Khan, W.A.; Afzal, M.; Hussain, M.; Kang, B.H.; Lee, S. The Mining Minds digital health and wellness framework. *Biomed. Eng. Online* **2016**, *15*, 76. [CrossRef] [PubMed]
64. Amin, M.B.; Banos, O.; Khan, W.A.; Muhammad Bilal, H.S.; Gong, J.; Bui, D.-M.; Cho, S.H.; Hussain, S.; Ali, T.; Akhtar, U. On curating multimodal sensory data for health and wellness platforms. *Sensors* **2016**, *16*, 980. [CrossRef] [PubMed]
65. Lin, K.-Y.; Chien, C.-F.; Kerh, R. UNISON framework of data-driven innovation for extracting user experience of product design of wearable devices. *Comput. Ind. Eng.* **2016**, *99*, 487–502. [CrossRef]
66. Node.js. Available online: https://nodejs.org/en/ (accessed on 29 March 2017).
67. Hussain, J.; Khan, W.A.; Afzal, M.; Hussain, M.; Kang, B.H.; Lee, S. Adaptive User Interface and User Experience Based Authoring Tool for Recommendation Systems. In *International Conference on Ubiquitous Computing and Ambient Intelligence*; Springer: Cham, Switzerland, 2014; pp. 136–142.
68. Hussain, J.; Lee, S. Identifying User Experience (UX) Dimensions from UX Literature Reviews. Available online: http://www.riss.kr/search/detail/DetailView.do?p_mat_type=1a0202e37d52c72d&control_no=f631e21b1c0c2bd1b36097776a77e665 (accessed on 15 May 2018).
69. Hussain, J.; Hassan, A.U.; Bilal, H.S.M.; Ali, R.; Afzal, M.; Hussain, S.; Bang, J.; Banos, O.; Lee, S. Model-based adaptive user interface based on context and user experience evaluation. *J. Multimodal User Interfaces* **2018**, *12*, 1–16. [CrossRef]
70. Albert, W.; Tullis, T. *Measuring the User Experience: Collecting, Analyzing, and Presenting Usability Metrics*; Newnes: Oxford, UK, 2013; ISBN 0-12-415792-0.

71. Banos, O.; Villalonga, C.; Bang, J.; Hur, T.; Kang, D.; Park, S.; Le-Ba, V.; Amin, M.B.; Razzaq, M.A.; Khan, W.A. Human Behavior Analysis by Means of Multimodal Context Mining. *Sensors* **2016**, *16*, 1264. [CrossRef] [PubMed]
72. Ververidis, D.; Kotropoulos, C. A State of the Art Review on Emotional Speech Databases. Available online: http://citeseerx.ist.psu.edu/viewdoc/download?doi=10.1.1.420.6988&rep=rep1&type=pdf (accessed on 15 May 2018).
73. McKeown, G.; Valstar, M.F.; Cowie, R.; Pantic, M. The SEMAINE Corpus of Emotionally Coloured Character Interactions. In Proceedings of the 2010 IEEE International Conference on Multimedia and Expo (ICME), Suntec City, Singapore, 19–23 July 2010; pp. 1079–1084.
74. Yang, X.; Tan, B.; Ding, J.; Zhang, J.; Gong, J. Comparative Study on Voice Activity Detection Algorithm. In Proceedings of the 2010 International Conference on Electrical and Control Engineering (ICECE), Wuhan, China, 25–27 June 2010; pp. 599–602.
75. Ooi, C.S.; Seng, K.P.; Ang, L.-M.; Chew, L.W. A new approach of audio emotion recognition. *Expert Syst. Appl.* **2014**, *41*, 5858–5869. [CrossRef]
76. Zeng, Z.; Pantic, M.; Roisman, G.I.; Huang, T.S. A survey of affect recognition methods: Audio, visual, and spontaneous expressions. *IEEE Trans. Pattern Anal. Mach. Intell.* **2009**, *31*, 39–58. [CrossRef] [PubMed]
77. D'mello, S.K.; Kory, J. A review and meta-analysis of multimodal affect detection systems. *ACM Comput. Surv. CSUR* **2015**, *47*, 43. [CrossRef]
78. Patwardhan, A.S. Multimodal Mixed Emotion Detection. In Proceedings of the 2017 2nd International Conference on Communication and Electronics Systems (ICCES), Suntec City, Singapore, 19–23 July 2010; pp. 139–143.
79. Poria, S.; Cambria, E.; Howard, N.; Huang, G.-B.; Hussain, A. Fusing audio, visual and textual clues for sentiment analysis from multimodal content. *Neurocomputing* **2016**, *174*, 50–59. [CrossRef]
80. Wöllmer, M.; Weninger, F.; Knaup, T.; Schuller, B.; Sun, C.; Sagae, K.; Morency, L.-P. Youtube movie reviews: Sentiment analysis in an audio-visual context. *IEEE Intell. Syst.* **2013**, *28*, 46–53. [CrossRef]
81. Mansoorizadeh, M.; Charkari, N.M. Multimodal information fusion application to human emotion recognition from face and speech. *Multimed. Tools Appl.* **2010**, *49*, 277–297. [CrossRef]
82. Sarkar, C.; Bhatia, S.; Agarwal, A.; Li, J. Feature Analysis for Computational Personality Recognition Using Youtube Personality Data Set. In Proceedings of the 2014 ACM Multi Media on Workshop on Computational Personality Recognition, Orlando, FL, USA, 7 November 2014; ACM: New York, NY, USA, 2014; pp. 11–14.
83. Poria, S.; Cambria, E.; Hussain, A.; Huang, G.-B. Towards an intelligent framework for multimodal affective data analysis. *Neural Netw.* **2015**, *63*, 104–116. [CrossRef] [PubMed]
84. Wang, S.; Zhu, Y.; Wu, G.; Ji, Q. Hybrid video emotional tagging using users' EEG and video content. *Multimed. Tools Appl.* **2014**, *72*, 1257–1283. [CrossRef]
85. Dobrišek, S.; Gajšek, R.; Mihelič, F.; Pavešić, N.; Štruc, V. Towards efficient multi-modal emotion recognition. *Int. J. Adv. Robot. Syst.* **2013**, *10*, 53. [CrossRef]
86. Jick, T.D. Mixing qualitative and quantitative methods: Triangulation in action. *Adm. Sci. Q.* **1979**, *24*, 602–611. [CrossRef]
87. Ali, R.; Afzal, M.; Hussain, M.; Ali, M.; Siddiqi, M.H.; Lee, S.; Kang, B.H. Multimodal hybrid reasoning methodology for personalized wellbeing services. *Comput. Biol. Med.* **2016**, *69*, 10–28. [CrossRef] [PubMed]
88. Sauro, J.; Dumas, J.S. Comparison of Three One-Question, Post-Task Usability Questionnaires. In Proceedings of the SIGCHI Conference on Human Factors in Computing Systems, Boston, MA, USA, 4–9 April 2009; ACM: New York, NY, USA, 2009; pp. 1599–1608.
89. Yousefpour, A.; Ibrahim, R.; Hamed, H.N.A. Ordinal-based and frequency-based integration of feature selection methods for sentiment analysis. *Expert Syst. Appl.* **2017**, *75*, 80–93. [CrossRef]
90. Xia, R.; Zong, C.; Li, S. Ensemble of feature sets and classification algorithms for sentiment classification. *Inf. Sci.* **2011**, *181*, 1138–1152. [CrossRef]
91. Taylor, A.; Marcus, M.; Santorini, B. The Penn Treebank: An Overview. In *Treebanks*; Springer: Dordrecht, The Netherlands, 2003; pp. 5–22.
92. Lucey, P.; Cohn, J.F.; Kanade, T.; Saragih, J.; Ambadar, Z.; Matthews, I. The Extended Cohn-Kanade Dataset (ck+): A Complete Dataset for Action Unit and Emotion-Specified Expression. In Proceedings of the 2010 IEEE Computer Society Conference on Computer Vision and Pattern Recognition Workshops (CVPRW), San Francisco, CA, USA, 13–18 June 2010; pp. 94–101.

93. Lyons, M.J.; Akamatsu, S.; Kamachi, M.; Gyoba, J.; Budynek, J. The Japanese Female Facial Expression (JAFFE) Database. In Proceedings of the Third International Conference on Automatic Face and Gesture Recognition, Nara, Japan, 14–16 April 1998; pp. 14–16.
94. Krumhuber, E.G.; Manstead, A.S. Can Duchenne smiles be feigned? New evidence on felt and false smiles. *Emotion* **2009**, *9*, 807. [CrossRef] [PubMed]
95. Lee, K.-C.; Ho, J.; Kriegman, D.J. Acquiring linear subspaces for face recognition under variable lighting. *IEEE Trans. Pattern Anal. Mach. Intell.* **2005**, *27*, 684–698. [PubMed]
96. Thomaz, C.E. FEI Face Database. Available online: http://fei.edu.br/~cetfacedatabase.html (accessed on 2 October 2012).

sensors

MDPI

Article

A Framework of Covariance Projection on Constraint Manifold for Data Fusion

Muhammad Abu Bakr and Sukhan Lee *

Intelligent Systems Research Institute, Sungkyunkwan University, Suwon, Gyeonggi-do 440-746, Korea; abubakr@skku.edu

* Correspondence: lsh1@skku.edu; Tel.: +82-31-299-6471

† This paper is an extended version of the paper entitled "A general framework for data fusion and outlier removal in distributed sensor networks", presented at IEEE International Conference on Multisensor Fusion and Integration for Intelligent Systems, Daegu, Korea, 16–18 November 2017.

Received: 3 May 2018; Accepted: 15 May 2018; Published: 17 May 2018

Abstract: A general framework of data fusion is presented based on projecting the probability distribution of true states and measurements around the predicted states and actual measurements onto the constraint manifold. The constraint manifold represents the constraints to be satisfied among true states and measurements, which is defined in the extended space with all the redundant sources of data such as state predictions and measurements considered as independent variables. By the general framework, we mean that it is able to fuse any correlated data sources while directly incorporating constraints and identifying inconsistent data without any prior information. The proposed method, referred to here as the Covariance Projection (CP) method, provides an unbiased and optimal solution in the sense of minimum mean square error (MMSE), if the projection is based on the minimum weighted distance on the constraint manifold. The proposed method not only offers a generalization of the conventional formula for handling constraints and data inconsistency, but also provides a new insight into data fusion in terms of a geometric-algebraic point of view. Simulation results are provided to show the effectiveness of the proposed method in handling constraints and data inconsistency.

Keywords: Bar-Shalom Campo; Covariance Projection method; data fusion; distributed architecture; Kalman filter; linear constraints; inconsistent data

1. Introduction

Data fusion is the process of obtaining a more meaningful and precise estimate of a state by combining data from multiple sources. The architecture of multisensor data fusion can be broadly categorized into two, depending on the way raw data are processed: (1) Centralized fusion architecture [1], where raw data from multiple sources are directly sent to and fused in the central node for state estimation and (2) Distributed fusion architecture [1,2], where data measured at multiple sources are processed independently at individual nodes to obtain local estimates before they are sent to the central node for fusion. In the centralized architecture, it is possible to apply data fusion methodology such as the Kalman filter [3] to all raw data received to yield optimal estimates in the sense of minimum variance. However, the centralized architecture can be costly especially for a large system in terms of infrastructure and communication overheads at the central node, let alone the issues of reliability and scalability. On the other hand, the distributed architecture is advantageous in reliability and scalability, with lower infrastructure and communication costs. Although advantageous, the distributed architecture needs to address statistical dependency among the local state estimates received from multiple nodes for fusion. This is due to the fact that local state estimates at individual nodes can be subject to the same process noise [4] and to double counting, i.e., sharing the same data sources among them [5]. Ignoring such statistical dependency or cross-correlation among multiple nodes leads to inconsistent results, causing divergence in data fusion [6]. In the case of known cross-correlation, the Bar-Shalom Campo (BC) formula [7] provides a consistent fusion result for a pair of redundant data sources, where the fused estimate is based on maximum likelihood [8]. A generalization to more than two data sources with known cross-correlations is given by weighted fusion algorithms of the generalized Millman's formula (GMF) [9] and weighted Kalman filter (WKF) [10].

Sensors often provide spurious and inconsistent data due to unexpected situations such as short duration spike faults, sensor glitches, a permanent failure or slowly developing failure due to sensor elements [5,11]. Since these types of uncertainties are not attributable to the inherent noise, they are difficult to predict and model. The fusion of inconsistent sensor data with correct data can lead to severely inaccurate results [12]. For example, when exposed to abnormalities and outliers, a Kalman filter would easily diverge [13]. Hence, a data validation scheme is required to identify and eliminate the sensor faults/outliers/inconsistencies before fusion.

The detection of inconsistency needs either a priori information often in the form of specific failure model(s) or data redundancy [5]. Model-based approaches use the generated residuals between the model outputs and actual measurements to detect and remove faults. For instance, in [14], the Nadaraya–Watson estimator and a priori observations are used to validate sensor measurements. Similarly, a priori system model information as a reference is used to detect failures in filtered state estimates [15–17]. However, requirement of the prior information restricts the usage of these methods in the general case where prior information is not available or unmodeled failure occurs. A method to detect spurious data based on the Bayesian framework is proposed in [18,19]. The method adds a term to the Bayesian formulation which has the effect of increasing the covariance of the fused probability distribution when measurement from one of the sensor is inconsistent with respect to the other. However, the method is based on heuristics and assumes independence of sensor estimates in its analysis. In [20], the Covariance Union (CU) method is proposed where the fused covariance is enlarged to cover all local means and covariances in such a way that the fused estimate is consistent under spurious data. However, the method incurs high computational cost and results in an inappropriately large conservative fused result.

In some applications, the state variables observed in a multisensory system may be subject to additional constraints. These constraints can arise due to the basic laws of physics, kinematics or geometry consideration of a system or due to the mathematical relations to satisfy among states. For instance, the energy conservation laws in an undamped mechanical system; Kirchhoff's laws in electric circuits; a road constraint in a vehicle-tracking scenario [21]; an orthonormal constraint in quaternion-based estimation [22] etc. These constraints if properly included can lead to improvement in state estimation and data fusion.

Various methods have been proposed to incorporate linear constraints among the state variables of dynamic systems [23–28]. For instance, the dimensionality reduction method converts a constrained estimation problem to an unconstrained one of lower dimension by eliminating some state variables using the constraints [25]. However, the state variables in a reduced dimension model may become difficult to interpret and their physical meaning may be lost [23]. The pseudo-measurement method satisfies the linear constraints among state variables by treating the state constraints as additional perfect/noise-free measurements [26–28]. However, this method increases the computational complexity of state estimation due to an increase in the dimension of augmented measurement. Furthermore, due to the singularity of augmented measurement covariance, the method may cause numerical problems [23,29]. A popular approach, the estimate projection method, projects the unconstrained estimate obtained from conventional Kalman filtering onto the constraint subspace using classical optimization methods [23,24]. Unfortunately, the method may not lead to the true constraint optimum since the projection method merely gives the solution as a feasible point that is closest to the unconstrained minimum.

This paper presents a unified and general data fusion framework, referred to as the Covariance Projection (CP) method to fuse multiple data sources under arbitrary correlations and linear constraints as well as data inconsistency. The method projects the probability distribution of true states and measurements around the predicted states and actual measurements onto the constraint manifold representing the constraints to be satisfied among true states and measurements. The proposed method also provides a framework for identifying and removing outliers in a fusion architecture where only sensor estimates may be available at the fusion center. This paper is an extended and improved version of the conference paper [30]. What was presented in [30] is a preliminary new framework of data fusion that we proposed. On the other hand, what is presented here represents a much more detailed implementation and refinement of the concept proposed in the conference paper. Specifically, this paper includes the following additions: (1) a detailed analysis of the equivalence of the proposed method to conventional methods for fusing redundant data sources; (2) handling linear constraints simultaneously under the proposed data fusion framework; (3) refining the mathematical formula and technical descriptions associated with them; and (4) detailed analysis of the method with additional simulations that deal with state estimation and data fusion in the presence of correlations, outliers and constraints.

2. Problem Statement

Consider a distributed sensor architecture [1], where each sensor system is equipped with a tracking system to provide local estimates of some quantity of interest in the form of mean and covariance. Assume the following linear dynamic system model for each local sensor system,

$$x_k = A_{k-1}\, x_{k-1} + B_{k-1}u_{k-1} + w_{k-1} \tag{1}$$

where k is the discrete time, A_k is the system matrix, B_k is the input matrix, u_k is the input vector and x_k is the state vector. The system process is affected by zero mean Gaussian noise w_k with covariance matrix Q. The sensor measurements are approximated as,

$$z_{k_i} = H_{k_i}x_k + v_{k_i},\ i = 1, \ldots, n \tag{2}$$

where H_k is the observation matrix and n represents the number of sensors. v_{k_i} is Gaussian noise with covariance matrices R_i, $i = 1, 2, \ldots, n$. Each sensor systems employs a Kalman filter to provide local state mean estimate $\hat{x}_k$ and its covariance P_k [31]. A prediction of the state estimate $\hat{x}_k^-$ and its estimation error covariance P_k^- can be computed based on process model (1),

$$\hat{x}_k^- = A_{k-1}\, x_{k-1} + B_{k-1}u_{k-1} \tag{3}$$

$$P_k^- = A_{k-1}P_{k-1}A_{k-1}^T + Q_{k-1} \tag{4}$$

The Kalman filter then provides the state estimate $\hat{x}_k$ and its covariance P_k as,

$$\hat{x}_k = \hat{x}_k^- + K_k\left(z_k - H_k\hat{x}_k^-\right) \tag{5}$$

$$P_k = (I - K_kH_k)P_k^-(I - K_kH_k)^T + K_kR_kK_k^T \tag{6}$$

with the Kalman gain, $K_k = P_k^- H_k^T\left(H_kP_k^- H_k^T + R_k\right)^{-1}$. The estimates provided by sensor systems are assumed to be correlated due to the common process noise or double counting, that is, $P_{ij} = cov\left(\hat{x}_i, \hat{x}_j\right) \neq 0$. To ensure optimality of the fused results, the cross-covariance P_{ij} should be properly incorporated.

Due to the inherent nature of the sensor and environmental factors [32], the sensor measurements can also be perturbed by unmodel random faults e_{k_i},

$$z_{k_i} = H_i x_k + v_{k_i} + e_{k_i},\ i = 1, \ldots, n \tag{7}$$

Subsequently, the estimates computed by local sensor systems may be spurious and inconsistent. Therefore, validation of the sensor estimates is required to remove inconsistencies before the fusion process.

In addition, the states x_k of the sensor systems may subject to linear constraints due to the geometry of the system environment or the mathematical description of the system [23], such that,

$$Cx_k = c \tag{8}$$

where $C \in \mathbb{R}^{n \times m}$ and $c \in \mathbb{R}^n$ are both known. C is assumed to have a full row rank. The constraints provide deterministic information about the state variables and can be used to improve the fusion accuracy.

These issues of correlations, data inconsistency and state constraints motivate the development of the Covariance Projection method, which is described next.

3. Proposed Approach

The proposed method first represents the probability of true states and measurements in the extended space around the data from state predictions and sensor measurements, where the extended space is formed by taking states and measurements as independent variables. Any constraints among true states and measurements that should be satisfied are then represented as a constraint manifold in the extended space. This is shown schematically in Figure 1a for filtering as an example (refer to Equations (1)–(6)). Data fusion is accomplished by projecting the probability distribution of true states and measurements onto the constraint manifold.

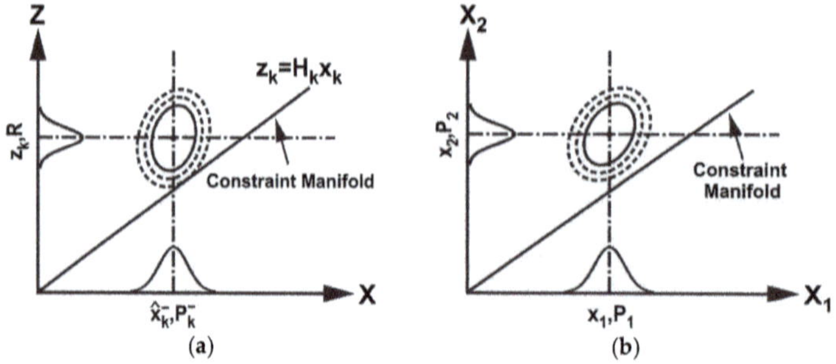

Figure 1. (**a**) Probability of true states and measurements in the extended space around the data from state predictions and sensor measurements and constraint manifold (**b**) Extended space representation of two data sources with constraint manifold.

More specifically, consider two mean estimates, $\hat{x}_1$ and $\hat{x}_2$, of the state $x \in \mathbb{R}^N$, with their respective covariances as $P_1, P_2 \in \mathbb{R}^{N \times N}$. Furthermore, the estimates are assumed to be correlated with cross-covariance P_{12}. The mean estimates and their covariances together with their cross-covariance in $\mathbb{R}^N$ are then transformed to an extended space of $\mathbb{R}^{2N}$ along with the linear constraint between the two estimates:

$$\hat{x} = \begin{bmatrix} \hat{x}_1 \\ \hat{x}_2 \end{bmatrix}, P = \begin{bmatrix} P_1 & P_{12} \\ P_{12}^T & P_2 \end{bmatrix}, C_1\hat{x}_1 = C_2\hat{x}_2 \tag{9}$$

where C_1 and C_2 are constant matrices of compatible dimensions. In the case where $\hat{x}_1$ and $\hat{x}_2$ estimate the same entity, C_1 and C_2 become identity matrix I. Figure 1b illustrates schematically the fusion of $\hat{x}_1$ and $\hat{x}_2$ in the extended space based on the proposed CP method. Fusion takes place by finding the point on the constraint manifold that represents the minimum weighted distance from $\hat{x}$ in $\mathbb{R}^{2N}$, where the weight is given by P. As seen later, the proposed CP method with the minimum weighted distance is shown to be equivalent to the minimum variance estimates but advantageous for dealing with additional constraints and data inconsistency.

To find a point on the constraint manifold with minimum weighted distance, we apply the whitening transform (WT) defined as, $W = D^{-1/2}E^T$, where D and E are the eigenvalue and eigenvector matrices of P. Applying WT,

$$\hat{x}^W = W\hat{x},\ P^W = WPW^T,\ M^W = WM$$

where the matrix $M = [C_1\ C_2]^T$ is the subspace of the constraint manifold. Figure 2 illustrates the transformation of the probability distribution as an ellipsoid into a unit circle after WT. The probability distribution is then orthogonally projected on the transformed manifold M^W to satisfy the constraints between the data sources in the transformed space as illustrated in Figure 2. Inverse WT is applied to obtain the fused mean estimate and covariance in the original space,

$$\tilde{x} = W^{-1}P_rW\hat{x} \tag{10}$$

$$\tilde{P} = W^{-1}P_rP_r{}^TW^{-T} \tag{11}$$

where $P_r = M^W\left(M^{W^T}M^W\right)^{-1}M^{W^T}$ is the orthogonal projection matrix. Using the definition of various components in (10) and (11), a close form simplification can be obtained as,

$$\tilde{x} = M\left(M^TP^{-1}M\right)^{-1}M^TP^{-1}\hat{x} \tag{12}$$

$$\tilde{P} = M\left(M^TP^{-1}M\right)^{-1}M^T \tag{13}$$

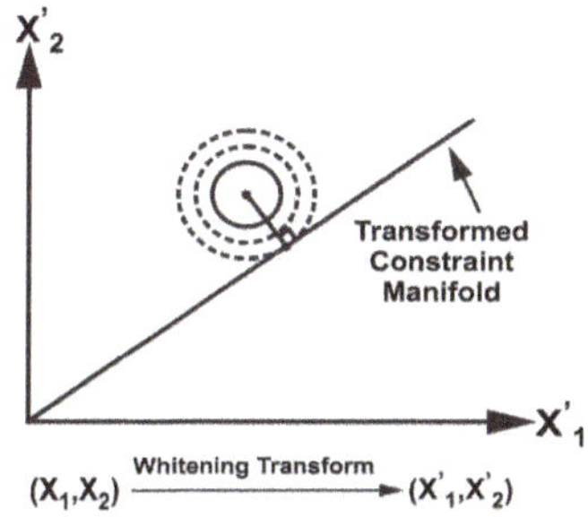

Figure 2. Whitening transform and projection.

The details of the simplification are provided in Appendix A. Due to the projection in extended space of $\mathbb{R}^{2N}$, (12) and (13) provide a fused result with respect to each data source. In the case where $\hat{x}_1$ and $\hat{x}_2$ estimate the same entity, that is, $M = [I_N\ I_N]^T$, the fused result will be same for the two data sources. As such, a close form equation for fusing redundant data sources in $\mathbb{R}^N$ can be obtained from (12) and (13) as,

$$\tilde{x} = \left(M^TP^{-1}M\right)^{-1}M^TP^{-1}\hat{x} \tag{14}$$

$$\tilde{P} = \left(M^TP^{-1}M\right)^{-1} \tag{15}$$

Given n mean estimates $\hat{x}_1, \hat{x}_2, \ldots, \hat{x}_n$ of a state $x \in \mathbb{R}^N$ with their respective covariances $P_1, P_2, \ldots, P_n \in \mathbb{R}^{N\times N}$ and known cross-covariances P_{ij}, $i,j = 1,\ldots,n$, (14) and (15) can be used to obtain the optimal fused mean estimate and covariance with $M = [I_{N1}, I_{N2}, \ldots, I_{Nn}]^T$.

For fusing correlated estimates from n redundant sources, the CP method is equivalent to the weighted fusion algorithms [9,10], which compute the fused mean estimate and covariance as a summation of weighted individual estimates as,

$$\tilde{x} = \sum_{i=1}^{n} c_i \hat{x}_i \tag{16}$$

$$\tilde{P} = \sum_{i,j=1}^{n} c_i P_{ij} c_j^T \tag{17}$$

with $\sum_{i=1}^{n} c_i = I$. Where the weights c_i are determined by solving some cost function of (16) and (17) such that, $\sum_{i=1}^{n} c_i = I$. Equivalently, the CP fused mean and covariance can be written as,

$$\tilde{x} = L\hat{x} \tag{18}$$

$$\tilde{P} = LPL^T \tag{19}$$

where $L = [L_1, L_2, \ldots, L_n] = \left(M^T P^{-1} M\right)^{-1} M^T P^{-1}$ and $\sum_{i=1}^{n} L_i = I$. In the particular case of two data sources, the CP fused solution reduces to the well-known Bar-Shalom Campo formula [7],

$$\tilde{x} = (P_2 - P_{21})(P_1 + P_2 - P_{12} - P_{21})^{-1}\hat{x}_1 + (P_1 - P_{12})(P_1 + P_2 - P_{12} - P_{21})^{-1}\hat{x}_2 \tag{20}$$

$$\tilde{P} = P_1 - (P_1 - P_{12})(P_1 + P_2 - P_{12} - P_{21})^{-1}(P_1 - P_{21}) \tag{21}$$

Although equivalent to the traditional approaches in fusing redundant data sources, the proposed method offers a generalized framework not only for fusing correlated data sources but also for handling linear constraints and data inconsistency simultaneously within the framework.

The proposed method provides an unbiased and optimal fused estimate in the sense of minimum mean square error (MMSE).

Theorem 1. *For n unbiased mean estimates $\hat{x}_1, \hat{x}_2, \ldots, \hat{x}_n$, the fused mean estimate $\tilde{x}$ provided by the CP method is an unbiased estimator of x, that is, $E(\tilde{x}) = E(x)$.*

Proof. From (18), we can write,

$$\tilde{x} = [L_1, L_2, \ldots, L_n]\begin{bmatrix} \hat{x}_1 \\ \hat{x}_2 \\ \vdots \\ \hat{x}_n \end{bmatrix}$$

$$\tilde{x} = L_1\hat{x}_1 + L_2\hat{x}_2 + \ldots + L_n\hat{x}_n$$

$$\tilde{x} = \sum_{i=1}^{n} L_i\hat{x}_i$$

Taking the expectation on both sides, we get,

$$E(\tilde{x}) = E(\sum_{i=1}^{n} L_i\hat{x}_i)$$

$$E(\tilde{x}) = \sum_{i=1}^{n} L_i E(\hat{x}_i)$$

Since the sensor estimates $\hat{x}_1, \hat{x}_2, \ldots, \hat{x}_n$ are unbiased, we have $E(\hat{x}_1) = E(\hat{x}_2) = \cdots E(\hat{x}_n) = E(x)$,

$$E(\widetilde{x}) = E(x)$$

where $\sum_{i=1}^{n} L_i = I$. This concludes that the fused state estimate $\widetilde{x}$ is an unbiased estimate of x. □

Theorem 2. *The fused covariance $\widetilde{P}$ of the CP method is smaller than the individual covariances, that is, $\widetilde{P} \leq P_i$, $i = 1, 2, \ldots, n$.*

Proof. From equation (15), we can write,

$$\widetilde{P} = \left(M^T P^{-1} M\right)^{-1}$$

By Schwartz matrix inequality, we have,

$$\widetilde{P} = \left[(P^{-\frac{1}{2}}M)^T (P^{\frac{1}{2}}M_i)\right]^T \times \left[(P^{-\frac{1}{2}}M)^T (P^{-\frac{1}{2}}M)\right]^{-1} \times \left[(P^{-\frac{1}{2}}M)^T (P^{\frac{1}{2}}M_i)\right] \leq \left(P^{\frac{1}{2}}M_i\right)^T \left(P^{\frac{1}{2}}M_i\right) = P_i$$

where M is the constraint between data sources and $M_i = [I_{Ni}, 0, \ldots, 0]^T$ is the constraint matrix for P_i. The equality holds for $P_i = P_{ij}$, that is, $\widetilde{P} = P_i$, when $P_i = P_{ij}$, $j = 1, 2, \ldots, n$.

It can be observed from (14) and (15) that computation of cross-covariance P_{ij} is needed to compute the fused mean and covariance. Cross-covariance among the local estimates can be computed as [9,10,33],

$$P_{ij} = [I - K_i H_i]\left[A P_{ij}^{k-1} A^T + BQB^T\right][I - K_j H_j]^T \tag{22}$$

where K_i and K_j are the Kalman gain of source i and j respectively for $i, j = 1, \ldots, n$ and P_{ij}^{k-1} represent the cross covariance of the previous cycle between source i and j. □

4. Fusion in the Presence of Spurious Data

Due to the inherent nature of sensor devices and the real-world environment, the sensor observations may also be affected by random faults. Subsequently, the local estimates provided by sensor systems in a distributed architecture may be spurious and inconsistent. This may cause the fusion methodologies to fail since they are based on the assumption of consistent input sensor estimates. Therefore, a validation scheme is required to detect and remove the spurious estimates from the fusion pool. The proposed approach exploits the constraint manifold among sensor estimates to identify any data inconsistency. The identification of inconsistent data is based on the distance from the constraint manifold to the mean of redundant data sources in the extended space that provides a confidence measure with the relative disparity among data sources. Assuming a joint multivariate normal distribution for the data sources, the data confidence can be measured by computing the distance from the constraint manifold as illustrated in Figure 3.

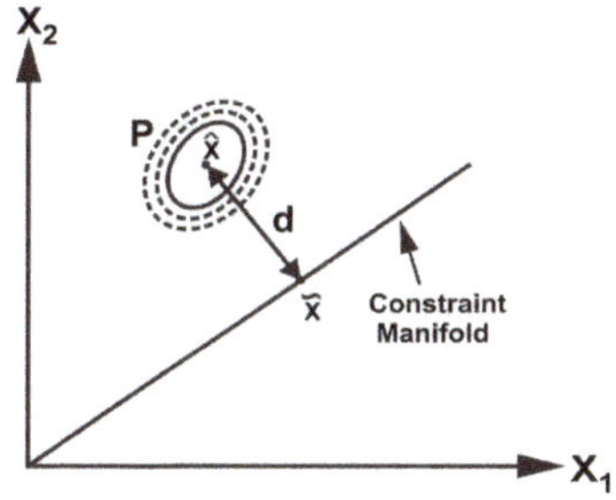

Figure 3. The distance of the multi-variate distribution from the constraint manifold.

Consider the joint space representation of n sensor estimates $(\hat{x}_{N1}, P_1), (\hat{x}_{N2}, P_n), \ldots, (\hat{x}_{Nn}, P_1)$,

$$\hat{x} = \begin{bmatrix} P_1 & P_{12} & \ldots & P_{1n} \\ P_{12}^T & P_2 & \ldots & \vdots \\ \vdots & \vdots & \ddots & \vdots \\ P_{1n}^T & \ldots & \ldots & P_{nn} \end{bmatrix}$$

where N is the dimension of the state vector. The distance d from the constraint manifold can be calculated as,

$$d = (\hat{x} - \tilde{x})^T P^{-1} (\hat{x} - \tilde{x}) \tag{23}$$

where $\tilde{x}$ is the point on the manifold and can be obtain by using (12). In the case of two data sources with mean $\hat{x}_1, \hat{x}_2 \in \mathbb{R}^N$ and respective covariance matrices P_1 and $P_2 \in \mathbb{R}^{N\times N}$. The distance d can be obtained as,

$$d = \begin{bmatrix} (\hat{x}_1 - \tilde{x})^T & (\hat{x}_2 - \tilde{x})^T \end{bmatrix} \begin{bmatrix} P_1 & 0 \\ 0 & P_2 \end{bmatrix}^{-1} \begin{bmatrix} \hat{x}_1 - \tilde{x} \\ \hat{x}_2 - \tilde{x} \end{bmatrix}$$

The point on the manifold is given as,

$$\tilde{x} = P_2(P_1 + P_2)^{-1}\hat{x}_1 + P_1(P_1 + P_2)^{-1}\hat{x}_2$$

Simplifying, we get,

$$d = [\hat{x}_1 - \hat{x}_2]^T (P_1 + P_2)^{-1} [\hat{x}_1 - \hat{x}_2] \tag{24}$$

The details of simplifications are provided in Appendix B. From (24), it can be observed that distance d is a weighted distance between the two data sources and it can provide a measure of nearness or farness of the two data sources to each other. A large value of d implies a large separation while a small d signifies closeness of the data sources. In other words, the distance from the manifold provides an indication of the relative disparity among data sources.

Theorem 3. *For n data sources of N dimension, the d distance (23) follow a chi-squared distribution with nN degrees of freedom (DOF), that is, $d \sim \chi^2(Nn)$.*

Proof. From (23) we have,

$$d = (\hat{x} - \tilde{x})^T P^{-1} (\hat{x} - \tilde{x}) \tag{25}$$

Under the whitening transformation, $WP^{-1}W = I$, $W\hat{x} = \hat{x}^W$ and $W\tilde{x} = \tilde{x}^W$. Thus, we can write,

$$(\hat{x} - \tilde{x})^T P^{-1} (\hat{x} - \tilde{x}) = (\hat{x}^W - \tilde{x}^W)^T (\hat{x}^W - \tilde{x}^W) \Rightarrow (W(\hat{x} - \tilde{x}))^T (W(\hat{x} - \tilde{x})) = y^T y \tag{26}$$

where for normal distribution $(\hat{x} - \tilde{x})$, $y = W(\hat{x} - \tilde{x})$ is an independent standard normal distribution $\mathcal{N}(0,1)$. For N dimensions of the state vector, the right-hand side of (26) is $\sum_{i=1}^N y_i^2$, thus distance d follows a chi-square distribution with N DOF, that is, $d \sim \chi^2(N)$. For n data sources with N states,

$$d \sim \chi^2(Nn)$$

Since d is a chi-square distribution with Nn DOF, then for any significance level $\alpha \in (0,1)$, $\chi^2_\alpha(Nn)$ is defined such that the probability,

$$P\left\{d \geq \chi^2_\alpha(Nn)\right\} = \alpha$$

This is depicted in Figure 4. Hence, to have a confidence of $100 \times (1 - \alpha)$ percent, d should be less than respective critical value as illustrated in Figure 4. A chi-square table [34] can be used to obtain the critical value for the confidence distance with a particular significance level and DOF. A value of $\alpha = 0.05$ is assumed in this paper unless specified. □

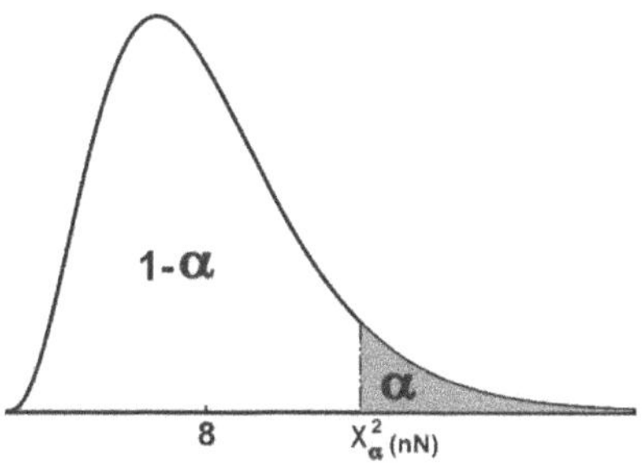

Figure 4. Chi-square distribution with 8 degreed of freedom. The unshaded area represents a cumulative probability associated with chi-square statistics $\chi^2_\alpha(Nn)$.

4.1. Inconsistency Detection and Exclusion

To obtain reliable and consistent fusion results, it is important that the inconsistent estimates be identified and excluded before fusion. For this reason, at each time step when the fusion center receives computed estimates from sensor nodes, distance d is calculated. A chi-square table is then used to obtain the critical value for a particular significance level and DOF. A computed distance d less than the critical value mean that we are confident about the closeness of sensor estimates and that they can be fused together to provide a better estimate of the underlying states. On the other hand, a distance d greater than or equal to the critical value indicate spuriousness of the sensor estimates. At least one of the sensor estimate is significantly different than the other sensor estimates. To exclude the outliers, a distance from the manifold is computed for every estimate and compared with the respective critical values. For n mean estimates $\hat{x}_1, \hat{x}_2, \ldots, \hat{x}_n$ with respective covariances $P_1, P_2, \ldots, P_n$ and cross-covariances P_{ij} for $i, j = 1, \ldots, n$, the hypothesis and decision rule are summarized as follows,

Hypotheses:

$$H_0 : \hat{x}_1 = \hat{x}_2 = \cdots = \hat{x}_n$$

$$H_1 : \hat{x}_1 \neq \hat{x}_2 \neq \cdots \neq \hat{x}_n$$

Compute: $d = (\hat{x} - \tilde{x})^T P^{-1} (\hat{x} - \tilde{x})$

Decision Rule: Accept H_0 if $d < \chi^2_\alpha(Nn)$
Reject H_0 if $d \geq \chi^2_\alpha(Nn)$

If the hypothesis H_0 is accepted, then the estimates are optimally fused using (14) and (15). On the other hand, rejection of null hypothesis means that at least one of the sensor estimates is significantly different than the other estimates. Then, a distance from the manifold is computed for each of the estimates as,

$$d_i = (\hat{x}_i - \tilde{x})^T P_i^{-1} (\hat{x}_i - \tilde{x}),\ i = 1, 2, \ldots, n$$

with $\tilde{x}$ computed using (14). The outliers are identified and eliminated based on the respective critical value, that is, if $d_i \geq \chi^2_\alpha(N)$, they are rejected, where N is the dimension of an individual data source.

4.2. Effect of Correlation on d Distance

Since the estimates provided by multiple data sources are correlated, it is important to consider the effect of cross-correlation in the calculation of confidence distance. Consider two sensor estimates $\hat{x}_1 \in \mathbb{R}^1$ and $\hat{x}_2 \in \mathbb{R}^1$ with respective variances σ_1^2 and σ_2^2 and cross-covariance $\sigma_{12}^2 = \rho\sqrt{\sigma_1^2 \sigma_2^2}$, where

$\rho \in [-1,1]$ is the correlation coefficient. The d distance for the pair of multivariate Gaussian estimates $(\hat{x}_1, \sigma_1^2)$ and $(\hat{x}_2, \sigma_2^2)$, with cross-covariance σ_{12}^2 can be written as,

$$d = \frac{[\hat{x}_1 - \hat{x}_2]^2}{\sigma_1^2 + \sigma_2^2 - \sigma_{12}^2 - \sigma_{21}^2} \tag{27}$$

It is apparent that the distance between the mean values is affected by the correlation between the data sources. Figure 5 illustrates the dependency of confidence distance d on the correlation coefficient. Figure 5a shows the distance d with changing correlation coefficient from -1 to 1. It can be observed that a positive correlation between the data sources results in a large d distance. This means that a slight separation between the positively correlated data sources indicates spuriousness with high significance as compared to negatively correlated and uncorrelated data sources. Figure 5b shows the scenario in which a data source (with changing mean and constant variance) is moving away from another data source (with constant mean and constant variance). The distance d is plotted for various values of correlation coefficients. The y-axis shows the percentage of rejection of the null hypothesis H_0. It can be noted that ignoring the cross-correlation in distance d results in underestimated or overestimated confidence and may lead to an incorrect rejection of the true null hypothesis (Type I error) or incorrect retaining of false null hypothesis (Type II error). The proposed framework inherently takes care of any cross-correlation among multiple data sources in the computation of distance d.

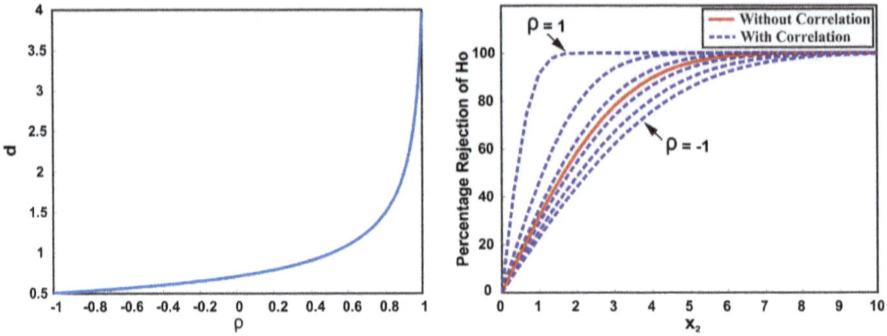

(**a**) (**b**)

Figure 5. Effect of correlation on d distance (**a**) d distance with correlation $\rho \in [-1,1]$; (**b**) percentage of rejecting the null hypothesis H_0 with different correlation values.

Example: *Consider a numerical simulation with the constant state,*

$$x_k = 10$$

Three sensors are used to estimate the state x_k, where the measurements of the sensors are affected with respective variances of R_1, R_2 and R_3. The values for the parameters assumed in the simulation are,

$$Q = 2,\ R_1 = 0.5,\ R_2 = 1,\ R_3 = 0.9$$

The sensor measurements are assumed to be cross-correlated. It is also assumed that the sensor 1, sensor 2 and sensor 3 measurements are independently corrupted by unmodeled noise and produce inconsistent data for 33%, 33% and 34% of the time respectively. The sensors compute local estimates of the state and send it to the fusion center. Three strategies for combining the local sensor estimates are compared: (1) CP, which fuses the three sensor estimates using (14) and (15) without removing outliers; (2) CP WO-d means the outliers were identified and rejected before fusion based on (27) with $\sigma_{12}^2 = 0$, that is, correlation in computation of d is ignored and; (3) CP WO-dC, reject the outliers based on (27) with taking into account the cross-correlation. Figure 6 shows the fused solution of three sensors when the estimate provided by sensor 2 is in disagreement with sensor 1 and 3. It can be observed from Figure 6 that neglecting the cross-correlation in CP WO-d results in Type II error, that is, all the three estimates are fused despite the fact that estimate 2 is inconsistent. CP WO-dC correctly identifies and eliminates the spurious estimate before the fusion process. Figure 7 shows the estimated position after the fusion of the three sensors' estimates for 100 samples. It can be seen that the presence of outliers greatly affects the outcome of multisensor data fusion. As depicted in Figure 7, eliminating outliers before fusion improves the estimation performance. The fused samples of CP WO-d and CP WO-dC on average lies closer to the actual state. Figure 7 also shows the difference in fusion performance when outliers are identified with and without considering cross-correlation. It can be noted that neglecting the correlation affects the estimation quality because of Type I and Type II errors.

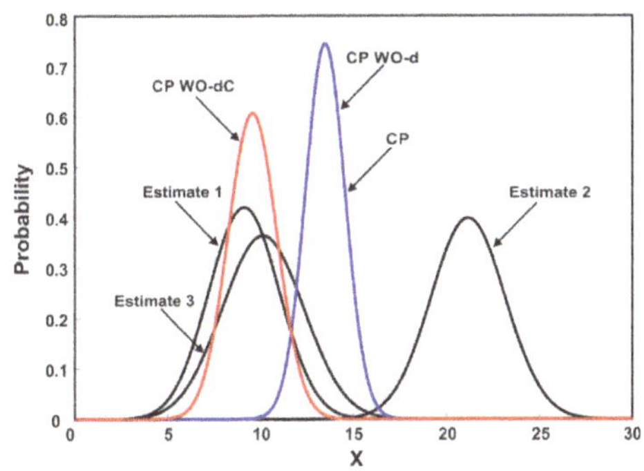

Figure 6. Three-sensor fusion when the estimate of sensor 2 is inconsistent. Neglecting the cross-correlation results in Type II error.

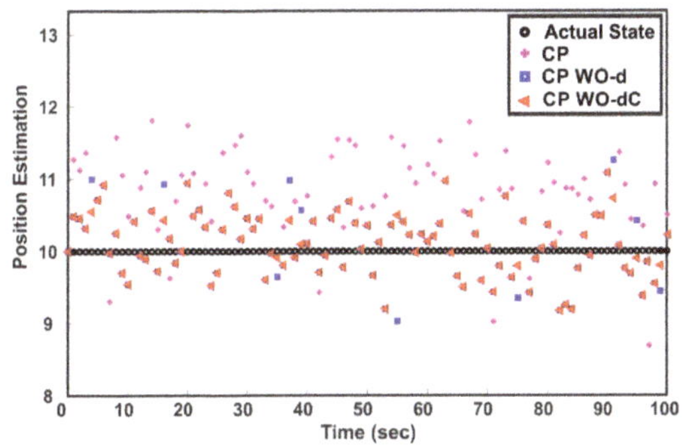

Figure 7. Estimated position after three-sensor fusion in presence of inconsistent estimates.

5. Fusion under Linear Constraints

The system model of a linear dynamic system takes into account the relation and dependencies among components of the state vector. In some applications, however, the state variables may be subject to additional constraints due to the basic laws of physics, geometry of the system environment or due to the mathematical description of the state vector. Imposing such certain information in an otherwise probabilistic setting should yield a more accurate estimate that is guaranteed to be feasible.

Consider a linear dynamic system model,

$$x_k = A_{k-1}\, x_{k-1} + B_{k-1}u_{k-1} + w_{k-1} \tag{28}$$

$$z_{k_i} = H_{k_i}x_k + v_{k_i},\ i = 1, \ldots, n \tag{29}$$

where k represents the discrete-time index, A_k is the system matrix, B_k is the input matrix, u_k is the input vector and x_k is the state vector. The system process noise w_k with covariance matrix Q and measurement noise v_k with covariance R are assumed to be correlated, that is,

$$E\left\{\begin{bmatrix} w_k \\ v_k \end{bmatrix}\begin{bmatrix} w_k \\ v_k \end{bmatrix}^T\right\} = \begin{bmatrix} Q & P_{QR} \\ P_{QR}^T & R \end{bmatrix}$$

The state $x_k \in \mathbb{R}^N$ is known to be constrained as,

$$Cx_k = c = 0 \tag{30}$$

For $c \neq 0$, the state space can be translated by a factor c such that $C\overline{x}_k = 0$. After constrained state estimation, the state space can be translated back by the factor c to satisfy $Cx_k =$ c. Hence, without loss of generality, the $c = 0$ case is considered for analysis here. The matrix $C \in \mathbb{R}^{n\times m}$ is assumed to have a full row rank. A row deficient matrix C signifies the presence of redundant constraints. In such a case, we can simply remove the linearly dependent rows from C. In the following, the estimate projection (EP) method is briefly reviewed which is followed by the Covariance Projection (CP) method for linear constraints among state variables.

5.1. Estimate Projection Method

The estimate projection (EP) method [23,24] projects the unconstrained estimate obtained from Kalman filtering onto the constraint subspace to satisfy the linear constraints among state variables. Let us denote the unconstrained filtered estimate and constrained estimate as $(\hat{x}^u, P^u)$ and $(\hat{x}^p, P^p)$, respectively. Then the following optimization problem is solved to obtain the constrained estimate,

$$\min_{\hat{x}^p}(\hat{x}^p - \hat{x}^u)^T U(\hat{x}^p - \hat{x}^u) \text{ such that } C\hat{x}^p = 0 \tag{31}$$

where U is any symmetric positive definite weighting matrix. Solving (31) using Lagrange multipliers results in a constrained mean and covariance,

$$\hat{x}^p = J\hat{x}^u \tag{32}$$

$$P^p = JP^uJ^T \tag{33}$$

where J is the projector on the null space of constrained matrix C, defined as,

$$J = I - U^{-1}C^T\left(CU^{-1}C^T\right)^{-1}C$$

Any symmetrical positive definite matrix can be used as a weighting matrix U to obtain the constrained estimate but the two most common choices are identity matrix I and inverse of unconstrained covariance $P^{u^{-1}}$.

5.2. Covariance Projection Method for Linear Constraints

The CP framework incorporates any linear constraints among states without any additional processing. Let us denote the constrained filtered estimate of the CP method as $(\hat{x}^c, P^c)$. The extended space representation of the state predictions and measurements of multiple sensors can be written as,

$$\hat{x} = \begin{bmatrix} \hat{x}_k^- \\ z_{k_1} \\ \vdots \\ z_{k_n} \end{bmatrix}, P = \begin{bmatrix} P_k^- & P_{P_k^- R_1} & \cdots & P_{P_k^- R_n} \\ P_{P_k^- R_1}^T & R_1 & \cdots & \vdots \\ \vdots & \vdots & \ddots & \vdots \\ P_{P_k^- R_n}^T & \cdots & \cdots & R_n \end{bmatrix}$$

Then the CP estimate in the presence of linear constraints among states can be obtained using (12) and (13) as,

$$\hat{x}^c = M_c \left(M_c^T P^{-1} M_c \right)^{-1} M_c^T P^{-1} \hat{x} \tag{34}$$

$$P^c = M_c \left(M_c^T P^{-1} M_c \right)^{-1} M_c^T \tag{35}$$

where the M_c matrix is the subspace of the constraint among the state prediction x_k^- and sensor measurements z_{k_i} as well as linear constraints C among state variables. The subspace of the linear constraint among state prediction and sensor measurements can be written as,

$$M = [I_N, H_1, H_2, \ldots, H_n]^T$$

Then, M_c is a combination of M and C, that is,

$$M_c \in (M, C)$$

The projection of the probability distribution of true states and measurements around the predicted states and actual measurements onto the constraint manifold M_c in the extended space provide the filtered or fused estimate of state prediction and sensor measurements as well as completely satisfying the linear constraints among the states directly in one step.

It can be observed from (31) that the EP method forces the unconstrained estimate on the linear constraint under some norm to satisfy the linear constraints among state variables. Consequently, the true optimality cannot be guaranteed due to the fact that the projected point close to the unconstrained estimate does not imply that it is close to the true constrained state [29]. On the other hand, the unified constraint matrix M_c of the CP method ensures that the linear constraints among state variables are exactly satisfied. Furthermore, as compared to the EP method that needs the online projection steps (32) and (33) after filtering, the M_c matrix for the CP method can be computed offline. This means that the EP method is computationally less efficient. Additionally, the proposed CP method is inherently suitable for taking care of any cross-correlation in the constrained estimation process.

6. Simulation Results

In this section, illustrative examples are provided to demonstrate the effectiveness of the theoretical results derived in the previous sections. We use the Monte Carlo technique [35], a method extensively used in a wide variety of fields such as physical science, computational biology, statistics, computational geometry, artificial intelligence, engineering, decision theory, and quantitative finance (see, e.g., the recent works [36–40]).

6.1. Tracking in the Presence of Correlations and Outliers

Consider a target tracking scenario characterized by the following dynamic system model,

$$x_k = \begin{bmatrix} 1 & T \\ 0 & 1 \end{bmatrix} x_{k-1} + \begin{bmatrix} \frac{T^2}{2} \\ T \end{bmatrix} w_{k-1} \tag{36}$$

with the state vector $x_k = [x, \dot{x}]^T$. Where x and $\dot{x}$ are the position and velocity of the target at time k, respectively. $T = 0.5$ s is the sampling period. The system process is affected by zero mean Gaussian noise with covariance matrix Q. Four sensors are employed to track the movement of the target, where the sensor measurements are approximated by the following equation,

$$z_{k_i} = \begin{bmatrix} 1 & 0 \\ 0 & 1 \end{bmatrix} x_k + v_{k_i},\ i = 1,2,3,4 \tag{37}$$

The measurements of the sensors are affected by noise v_{k_i} with respective covariances of R_1, R_2, R_3 and R_4. The covariances of the process noise and sensor measurement noises used in the simulation are,

$$Q = 3.5,\ R_1 = diag(5, 3.5),\ R_2 = diag(2, 8),$$
$$R_3 = diag(7, 2.1),\ R_4 = diag(2.5, 5)$$

Starting from an initial value of [100, 3], in each time step the individual sensor uses (36) to predict the state of the target and then update the state prediction by its own sensor measurements. The local estimates are assumed to be correlated and (22) is used to compute the cross-correlation among local estimates. The estimated mean and covariances of the states by each sensor are sent to the fusion center, where they are fused by the CP method. Table 1 summarizes the trace of covariance of local sensors along with the trace of fused covariance provided by the CP method. As seen from Table 1, the trace of covariance provided by the CP method is less than the individual local state estimates, that is, $trace\widetilde{P} \leq traceP_i, i = 1, \ldots, 4$. This means that the fused result is better than each of the local state estimates.

Table 1. The accuracy comparison in terms of trace of matrices.

$traceP_1$	$traceP_2$	$traceP_3$	$traceP_4$	$trace\widetilde{P}$
4.3271	2.9109	3.9656	3.4321	1.2674

In order to further verify this theoretical result, we compute the mean square error (MSE) as,

$$S_{MSE}(k) = \frac{1}{V} \sum_{i=1}^{V} [\hat{x}_i(k) - x_i(k)]^T [\hat{x}_i(k) - x_i(k)]$$

where V is the number of Monte Carlo trials and $\hat{x}_i(k)$ and $x_i(k)$ are the estimated and actual state vector respectively. Since,

$$trace(P_i) = E[(\hat{x}_i - x)^T (\hat{x}_i - x)]$$

Then we have [41],

$$S_{MSE}(k) = trace(P_i),\ k \to \infty,\ V \to \infty \tag{38}$$

The simulation is carried out for 1000 Monte Carlo runs and the local estimates provided by four sensor nodes along with the fused result of the CP method are shown in Figure 8. The straight lines in Figure 8 denote the trace of error covariance matrices and the solid curve represents the MSE of local and fused estimates. It can be observed from Figure 8 that the MSE of the individual sensor node fluctuates around the trace, which is consistent with (38). Furthermore, the accuracy relation of local sensor estimates and fused estimates in terms of MSE in Figure 8 is coincident with the theoretical results in Table 1.

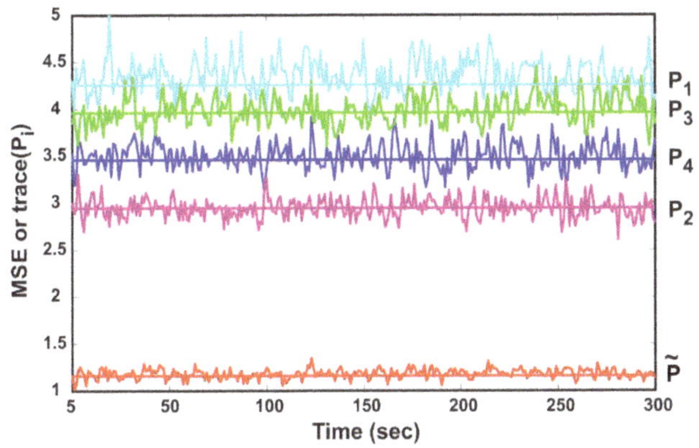

Figure 8. The mean square error (MSE) and trace (P_i) of local and fused estimates.

Consider the same dynamic example of four sensors with the following system and measurement equations,

$$x_k = \begin{bmatrix} 1 & T \\ 0 & 1 \end{bmatrix} x_{k-1} + \begin{bmatrix} \frac{T^2}{2} \\ T \end{bmatrix} u_{k-1} + w_{k-1} \tag{39}$$

$$z_{k_i} = \begin{bmatrix} 1 & 0 \\ 0 & 1 \end{bmatrix} x_k + v_{k_i} + e_{k_i},\ i = 1,2,3,4 \tag{40}$$

where the process and measurement noise parameters are the same. Now, it is also assumed that the sensor 1, sensor 2, sensor 3 and sensor 4 measurements are independently affected by unmodeled random noise e_{k_i} for 5%, 15%, 20% and 10% of the time, respectively, and thus the estimates provided by the sensors are spurious. The control input alternate between 1 and -1 and set to a value of 1 if $\dot{x} < 30$ otherwise it is changed to -1 until $\dot{x} < 5$. Starting from an initial value of [100, 3], in each time step the individual sensor node compute local filtered estimates. The estimated mean and covariances by each sensor are sent to the fusion center, where they are fused. The three fusion strategies of CP (fusion without outlier removal), CP WO-d (outlier removal without considering cross-correlation) and CP WO-dC (taking care of correlation in outlier removal) are compared based on root mean squared error (RMSE) between the actual state value and fused estimate of the state. The inconsistency is detected with significance level $\alpha = 0.05$. Figure 9a,b illustrate the RMSE of the target position and velocity respectively versus time for 1000 Monte Carlo runs. Table 2 summarizes the average RMSE of position and velocity. It can be observed from Figure 9 and Table 2 that the presence of outliers deteriorates the performance of multisensor data fusion. Eliminating the outliers before fusion greatly improves the estimation quality. Figure 9 and Table 2 also shows the difference in fusion performance of CP WO-d and CP WO-dC, when outliers are identified with and without consideration of cross-correlation in distance d respectively. It can be noted that consideration of correlation in distance d improves the estimation quality in presence of outliers by avoiding Type I and Type II errors.

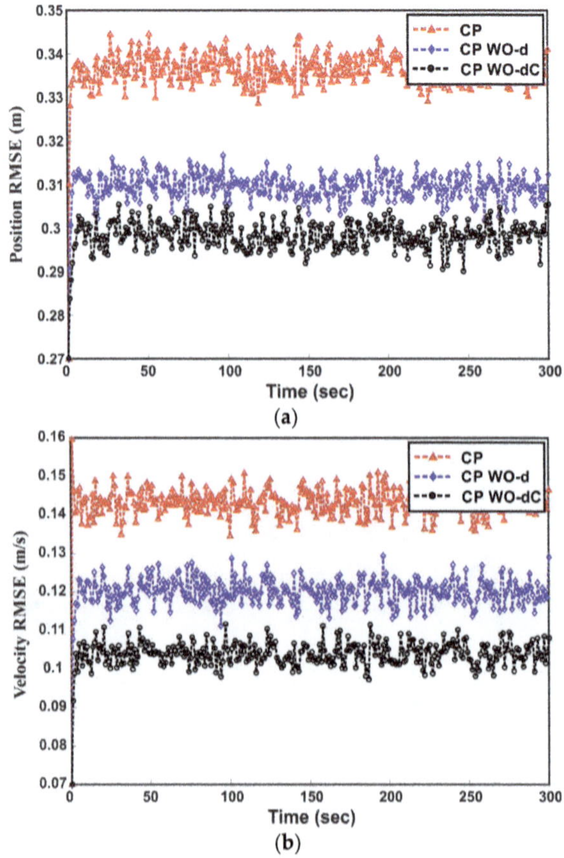

Figure 9. Illustration of multisensor data fusion in the presence of inconsistent estimates. (**a**) Position root mean squared error (RMSE); (**b**) velocity RMSE.

Table 2. Average RMSE for 1000 Monte Carlo Runs.

Average RMSE	CP	CP WO-d	CP WO-dC
Position (m)	101.1605	93.0824	89.6491
Velocity (m/s)	43.0339	36.0899	31.2033

6.2. *Target Tracking in the Presence of Linear Constraints*

Consider a 2D target tracking problem [24] with the following system equation,

$$x_k = \begin{bmatrix} 1 & 0 & T & 0 \\ 0 & 1 & 0 & T \\ 0 & 0 & 1 & 0 \\ 0 & 0 & 0 & 1 \end{bmatrix} x_{k-1} + \begin{bmatrix} 0 \\ 0 \\ T\sin\theta \\ T\cos\theta \end{bmatrix} u_{k-1} + w_{k-1} \tag{41}$$

where $x_k = \left[x_1, x_2, \dot{x}_1, \dot{x}_2\right]^T$ is the state vector, with the first two states as the north and east position with the last two states as the north and east velocity of the target respectively. A sensor measures the position and velocity of the system as,

$$z_k = Hx_k + v_k \tag{42}$$

with $H = \begin{bmatrix} 1 & 3 & 2 & 1 \\ 3 & 1 & 0 & 1 \\ 2 & 1 & 3 & 3 \\ 2 & 2 & 0 & 1 \end{bmatrix}$. w_k and v_k are the Gaussian process and measurement noise, respectively. Suppose that we have additional information that the vehicle is moving on the road with a heading angle of θ from the east x_2, then we can write,

$$\tan\theta = \tfrac{x_1}{x_2} \Rightarrow x_1 - x_2 \tan\theta = 0$$

$$\tan\theta = \tfrac{\dot{x}_1}{\dot{x}_2} \Rightarrow \dot{x}_1 - \dot{x}_2 \tan\theta = 0$$

Due to the heading of the vehicle, the states are dependent on each other, thus providing us additional constraints,

$$\begin{bmatrix} 1 & -\tan\theta & 0 & 0 \\ 0 & 0 & 1 & -\tan\theta \end{bmatrix} x_k = \begin{bmatrix} 0 \\ 0 \end{bmatrix} \tag{43}$$

Based on the constraint matrix $M = \begin{bmatrix} I_4 & H \end{bmatrix}^T$ between the state prediction and sensor measurement and linear constraints (43) among the state variables, the unified constraint matrix M_c can be obtained as,

$$M_c = \begin{bmatrix} \tan\theta & 1 & 0 & 0 & 3+\tan\theta & 1+3\tan\theta & 1+2\tan\theta & 2+2\tan\theta \\ 0 & 0 & \tan\theta & 1 & 1+2\tan\theta & 1 & 3+3\tan\theta & 1 \end{bmatrix}^T \tag{44}$$

The covariance of the process and measurement noise are assumed to be,

$$Q = \begin{bmatrix} 100 & 20 & 0 & 0 \\ 20 & 100 & 0 & 0 \\ 0 & 0 & 30 & 10 \\ 0 & 0 & 10 & 30 \end{bmatrix}, \; R = \begin{bmatrix} 500 & 10 & 20 & 30 \\ 10 & 500 & 15 & 10 \\ 20 & 15 & 100 & 20 \\ 30 & 10 & 20 & 100 \end{bmatrix}$$

The target is set to an initial state of $\begin{bmatrix} 0 & 0 & 15\tan\theta & 15 \end{bmatrix}$. The sampling time T is set to 1 s, heading angle $\theta = \pi/3$ and control input $u_k = 1$ if $\dot{x}_1\ or\ \dot{x}_2 < 30$ otherwise it is changed to -1 until $\dot{x}_1\ or\ \dot{x}_2 < 5$. The process and measurement noise are assumed to be correlated with covariance,

$$P_{QR} = \rho F_1 F_2^T$$

where F_1 and F_2 are the cholesky decomposition of Q and R respectively. ρ is the correlation coefficient and assumed as 0.4 in the simulation.

Starting from the initial value, in each time step (41) is used to predict the states of the vehicle. The states are then updated with the sensor measurements (42). Using M_c from (44) in (34) and (35), the proposed method directly satisfies the constraint between the perdition and measurement as well as the linear constraints among states due to the heading of the vehicle. On the other hand, the EP method first employs a Kalman filter to obtain the unconstrained state estimates and then project the unconstrained state estimates on the linear constraints subspace to satisfy the constraints among states. The performance of the proposed method is compared with the unconstrained estimate and estimate provided by the EP method in terms of RMSE. Table 3 summarizes the average RMSE of the individual states for 1000 Monte Carlo runs. It can be observed from Table 3 that the RMSE of the CP method is lower than the other estimators for all states. For the EP method, the use of $P^{u^{-1}}$ as a weighting parameter provides better results than using I. The comparative results of different methods can be also seen in Figure 10a,b, where Figure 10a is the RMSE of the northerly position and Figure 10b is the RMSE of northerly velocity. It can be seen that the proposed method performs better as compared to the unconstrained state estimation and EP method. Figure 11a,b show the variance of northerly position and northerly velocity of the vehicle, respectively. It is proven that the EP method with weighting parameter $P^{u^{-1}}$ has the smallest covariance [24]. This can also be observed from Figure 11a,b where the EP method with weight $P^{u^{-1}}$ provides smaller variance than the unconstrained estimate and EP method by identitying matrix I as the weighting parameter. However, the CP method results in even smaller state variance than all the competing methods as seen in Figure 11a,b.

Table 3. Average RMSE for 1000 Monte Carlo Runs.

Methods	Average RMSE			
	x_1 (m)	x_2 (m)	$\dot{x}_1$ (m/s)	$\dot{x}_2$ (m/s)
Unconstrained	95.479	87.496	52.413	49.972
CP	50.39	29.093	36.624	21.145
EP (I)	55.823	32.229	39.648	22.891
EP ($P^{u^{-1}}$)	53.023	30.613	37.186	21.469

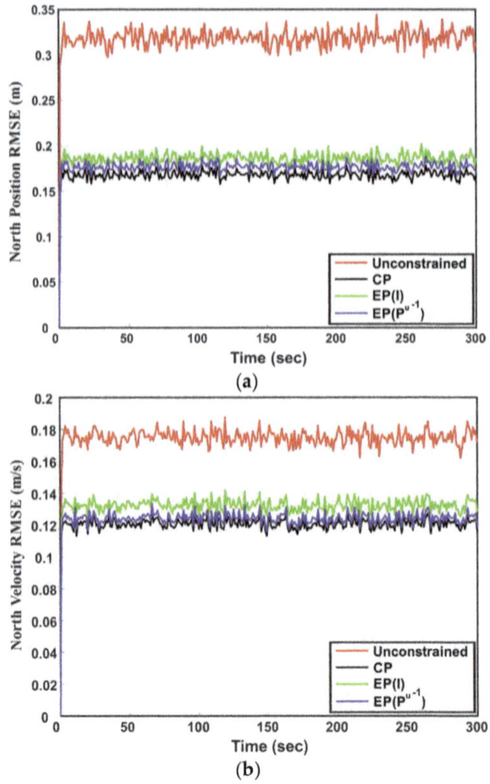

Figure 10. Simulation results for the constrained and unconstrained dynamic system. The covariance projection (CP) method is compared with the unconstrained state estimate and estimate projection (EP) method. (**a**) RMSE of northerly position of vehicle over 1000 runs; (**b**) RMSE of northerly velocity of vehicle over 1000 runs.

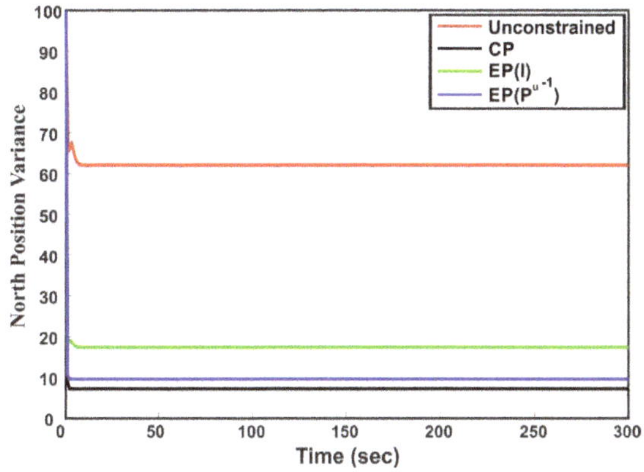

(a)

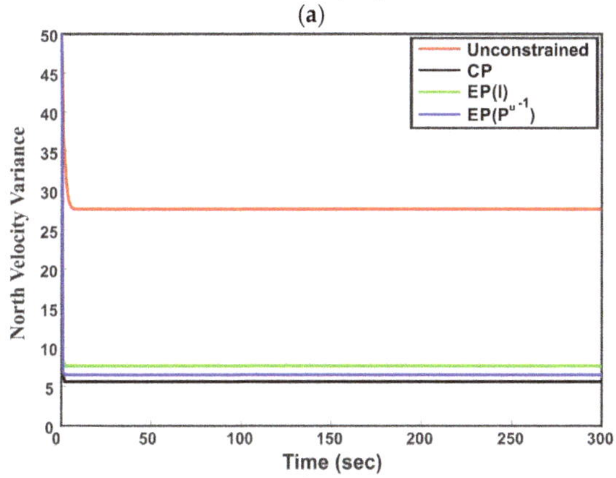

(b)

Figure 11. Comparative results of different methods in terms of states variance. (**a**) Variance of the northerly position of vehicle; (**b**) Variance of the northerly velocity of vehicle.

7. Conclusions

In this paper, we propose a general approach to data fusion under arbitrary correlations and linear constraints as well as data inconsistency. The proposed method provides an unbiased and optimal solution in the sense of MMSE for fusing data from multiple sources. The method improves the fusion accuracy by automatically detecting and removing inconsistent estimates from the fusion pool by a statistical confidence measure. Moreover, it is shown that by considering the cross-correlation among local estimates, the proposed method avoids the deterioration of the fusion accuracy due to Type I and Type II errors. Without any additional manipulation, the proposed method completely satisfies any linear constraints among state variables. The method improves the accuracy of constrained state estimation by satisfying the linear constraints among state variables and provides better results than the unconstrained state estimation and constrained state estimation using the estimate projection method.

Future work includes the extension of the proposed method for state estimation of non-linear dynamic systems. Another avenue is examining the proposed method for incorporating non-linear constraints among state variables.

Author Contributions: Sukhan Lee proposed the concept of data fusion based on covariance projection in the extended space as well as its application to outlier removal, while Muhammad Abu Bakr carried out the mathematical implementation and simulation analysis. Abu Bakr drafted the paper. Lee supervised Abu Bakr with critical assessment of the draft for quality revision.

Funding: This research was supported, in part, by the "Space Initiative Program" of the National Research Foundation (NRF) of Korea (NRF-2013M1A3A3A02042335), sponsored by the Korean Ministry of Science, ICT and Planning (MSIP), and, in part, by the "3D Visual Recognition Project" of the Korea Evaluation Institute of Industrial Technology (KEIT) (10060160), and, in part, by the "Project of e-Drive Train Platform Development for small and medium Commercial Electric Vehicles based on IoT Technology" of Korea Institute of Energy Technology Evaluation and Planning (KETEP) (20172010000420), sponsored by the Korea Ministry of Trade, Industry and Energy (MOTIE).

Conflicts of Interest: The authors declare no conflict of interest.

Appendix A

The fused mean estimate and covariance of the covariance projection (CP) method are given as,

$$\widetilde{x} = W^{-1}P_r W\hat{x} \tag{A1}$$

$$\widetilde{P} = W^{-1}P_r WPW^T {P_r}^T W^{-T} \tag{A2}$$

Putting $W = D^{-1/2}E^T$, $P_r = M^W\left(M^{W^T}M^W\right)^{-1}M^{W^T}$ and $M^W = WM$ in (A2), we get,

$$\widetilde{P} = W^{-1}\left(WM\left(M^T W^T WM\right)^{-1}M^T W^T\right) \times \left(WM\left(M^T W^T WM\right)^{-1}M^T W^T\right)^T W^{-T}$$

Let $\alpha = M^T W^T WM$, then,

$$\widetilde{P} = W^{-1}WM\alpha^{-1}M^T W^T WM\alpha^{-T}M^T W^T W^{-T}$$

$$\widetilde{P} = M\alpha^{-T}M^T \tag{A3}$$

Putting the value of α in (A3) and simplifying, we get,

$$\widetilde{P} = M\left(M^T ED^{-1}E^T M\right)^{-1}M^T \tag{A4}$$

By eigenvalue decomposition, we know that,

$$P = EDE^T \Rightarrow P^{-1} = ED^{-1}E^T$$

So,

$$\widetilde{P} = M\left(M^T P^{-1}M\right)^{-1}M^T \tag{A5}$$

Similarly, using definitions of various components in CP fused mean (A1), we have,

$$\widetilde{x} = W^{-1}\left(WM\left(M^T W^T WM\right)^{-1}M^T W^T\right)W\hat{x}$$

$$\widetilde{x} = M\left(M^T W^T WM\right)^{-1}M^T W^T W\hat{x} \tag{A6}$$

Since $W^T W = P^{-1}$, (A6) can be simplified as,

$$\widetilde{x} = M\left(M^T P^{-1}M\right)^{-1}M^T P^{-1}\hat{x} \tag{A7}$$

Appendix B

The weighted distance from the joint mean of two data sources to the point on the manifold can be calculated as,

$$d = \begin{bmatrix} (\hat{x}_1 - \widetilde{x})^T & (\hat{x}_2 - \widetilde{x})^T \end{bmatrix} \begin{bmatrix} P_1 & 0 \\ 0 & P_2 \end{bmatrix}^{-1} \begin{bmatrix} (\hat{x}_1 - \widetilde{x}) \\ (\hat{x}_2 - \widetilde{x}) \end{bmatrix}$$

$$d = (\hat{x}_1 - \widetilde{x})^T P_1^{-1} (\hat{x}_1 - \widetilde{x}) + (\hat{x}_2 - \widetilde{x})^T P_2^{-1} (\hat{x}_2 - \widetilde{x}) \quad \text{(A8)}$$

$$\hat{x}_1 - \widetilde{x} = \hat{x}_1 - P_2(P_1 + P_2)^{-1}\hat{x}_1 + P_1(P_1 + P_2)^{-1}\hat{x}_2$$

$$\hat{x}_1 - \widetilde{x} = \left[I - P_2(P_1 + P_2)^{-1}\right]\hat{x}_1 - \left[P_1(P_1 + P_2)^{-1}\right]\hat{x}_2$$

Since $P_1(P_1 + P_2)^{-1} + P_2(P_1 + P_2)^{-1} = I$

$$\hat{x}_1 - \widetilde{x} = P_1(P_1 + P_2)^{-1}[\hat{x}_1 - \hat{x}_2] \quad \text{(A9)}$$

Similarly,

$$\hat{x}_2 - \widetilde{x} = \left[I - P_1(P_1 + P_2)^{-1}\right]\hat{x}_2 - \left[P_2(P_1 + P_2)^{-1}\right]\hat{x}_1$$

$$\hat{x}_2 - \widetilde{x} = P_2(P_1 + P_2)^{-1}[\hat{x}_2 - \hat{x}_1]$$

$$\hat{x}_2 - \widetilde{x} = -P_2(P_1 + P_2)^{-1}[\hat{x}_1 - \hat{x}_2] \quad \text{(A10)}$$

Putting (A9) and (A10) in (A8) we get,

$$d = \left(P_1(P_1 + P_2)^{-1}[\hat{x}_1 - \hat{x}_2]\right)^T P_1^{-1}\left(P_1(P_1 + P_2)^{-1}[\hat{x}_1 - \hat{x}_2]\right) + \left(-P_2(P_1 + P_2)^{-1}[\hat{x}_1 - \hat{x}_2]\right)^T P_2^{-1}\left(-P_2(P_1 + P_2)^{-1}[\hat{x}_1 - \hat{x}_2]\right)$$

$$d = \left([\hat{x}_1 - \hat{x}_2]^T (P_1 + P_2)^{-1} P_1\right) P_1^{-1}\left(P_1(P_1 + P_2)^{-1}[\hat{x}_1 - \hat{x}_2]\right) + \left([\hat{x}_1 - \hat{x}_2]^T (P_1 + P_2)^{-1} P_2\right) P_2^{-1}\left(P_2(P_1 + P_2)^{-1}[\hat{x}_1 - \hat{x}_2]\right)$$

$$d = [\hat{x}_1 - \hat{x}_2]^T \left[(P_1 + P_2)^{-1}(P_1 + P_2)(P_1 + P_2)^{-1}\right][\hat{x}_1 - \hat{x}_2]$$

$$d = [\hat{x}_1 - \hat{x}_2]^T (P_1 + P_2)^{-1} \quad \text{(A11)}$$

References

1. Liggins, M., II; Hall, D.; Llinas, J. *Handbook of Multisensor Data Fusion: Theory and Practice*; CRC Press: Boca Raton, FL, USA, 2017.
2. Hall, D.; Chong, C.; Llinas, J.; Liggins, M., II. *Distributed Data Fusion for Network-Centric Operations*; CRC Press: Boca Raton, FL, USA, 2012.
3. Kalman, R. A new approach to linear filtering and prediction problems. *J. Basic Eng.* **1960**, *82*, 35–45. [CrossRef]
4. Bar-Shalom, Y. On the track-to-track correlation problem. *IEEE Trans. Automat. Contr.* **1981**, *26*, 571–572. [CrossRef]
5. Bakr, M.A.; Lee, S. Distributed Multisensor Data Fusion under Unknown Correlation and Data Inconsistency. *Sensors* **2017**, *17*, 2472. [CrossRef] [PubMed]
6. Maybeck, P. *Stochastic Models, Estimation, and Control*; Academic Press: Cambridge, MA, USA, 1982.
7. Bar-Shalom, Y.; Campo, L. The effect of the common process noise on the two-sensor fused-track covariance. *IEEE Trans. Aerosp.* **1986**, *AES-22*, 803–805. [CrossRef]
8. Chang, K.C.; Saha, R.K.; Bar-Shalom, Y. On optimal track-to-track fusion. *IEEE Trans. Aerosp. Electron. Syst.* **1997**, *33*, 1271–1276. [CrossRef]

9. Shin, V.; Lee, Y.; Choi, T. Generalized Millman's formula and its application for estimation problems. *Signal Process.* **2006**, *86*, 257–266. [CrossRef]
10. Sun, S.; Deng, Z. Multi-sensor optimal information fusion Kalman filter. *Automatica* **2004**, *40*, 1017–1023. [CrossRef]
11. Khaleghi, B.; Khamis, A.; Karray, F.; Razavi, S. Multisensor data fusion: A review of the state-of-the-art. *Inf. Fusion* **2013**, *14*, 28–44. [CrossRef]
12. Abdulhafiz, W.; Khamis, A. Handling data uncertainty and inconsistency using multisensor data fusion. *Adv. Artif. Intell.* **2013**, *2013*, 11. [CrossRef]
13. Durovic, Z.; Kovacevic, B. QQ-plot approach to robust Kalman filtering. *Int. J. Control* **1995**, *61*, 837–857. [CrossRef]
14. Wellington, S.; Atkinson, J.; Sion, R. Sensor validation and fusion using the Nadaraya-Watson statistical estimator. In Proceedings of the Fifth International Conference on Information Fusion, Annapolis, MD, USA, 8–11 July 2002.
15. Hage, J.A.; Najjar, M.E.; Pomorski, D. Multi-sensor fusion approach with fault detection and exclusion based on the Kullback–Leibler Divergence: Application on collaborative multi-robot system. *Inf. Fusion* **2017**, *37*, 61–76. [CrossRef]
16. Del Gobbo, D.; Napolitano, M.; Famouri, P.; Innocenti, M. Experimental application of extended Kalman filtering for sensor validation. *IEEE Trans. Control Syst. Technol.* **2001**, *9*, 376–380. [CrossRef]
17. Brumback, B.; Srinath, M. A fault-tolerant multisensor navigation system design. *IEEE Trans. Aerosp. Electron. Syst.* **1987**, *AES-23*, 738–756. [CrossRef]
18. Kumar, M.; Garg, D.; Zachery, R. A method for judicious fusion of inconsistent multiple sensor data. *IEEE Sens. J.* **2007**, *7*, 723–733. [CrossRef]
19. Kumar, M.; Garg, D.; Zachery, R. A generalized approach for inconsistency detection in data fusion from multiple sensors. In Proceedings of the 2006 American Control Conference, Minneapolis, MN, USA, 14–16 June 2006.
20. Uhlmann, J. Covariance consistency methods for fault-tolerant distributed data fusion. *Inf. Fusion* **2003**, *4*, 201–215. [CrossRef]
21. Kirubarajan, T.; Bar-Shalom, Y.; Pattipati, K.; Kadar, I. Ground target tracking with variable structure IMM estimator. *IEEE Trans. Aerosp. Electron. Syst.* **2000**, *36*, 26–46. [CrossRef]
22. Bernstein, D.S.; Hyland, D.C. Compartmental Modeling and Second-Moment Analysis of State Space Systems. *SIAM J. Matrix Anal. Appl.* **1993**, *14*, 880–901. [CrossRef]
23. Simon, D. Kalman filtering with state constraints: a survey of linear and nonlinear algorithms. *IET Control Theory Appl.* **2010**, *4*, 1303–1318. [CrossRef]
24. Simon, D.; Chia, T. Kalman filtering with state equality constraints. *IEEE Trans. Aerosp. Electron. Syst.* **2002**, *38*, 128–136. [CrossRef]
25. Wen, W.; Durrant-Whyte, H. Model-based multi-sensor data fusion. In Proceedings of the 1992 IEEE International Conference on Robotics Automation, Nice, France, 12–14 May 1992.
26. Tahk, M.; Speyer, J. Target tracking problems subject to kinematic constraints. *IEEE Trans. Automat. Contr.* **1990**, *35*, 324–326. [CrossRef]
27. Porrill, J. Optimal Combination and Constraints for Geometrical Sensor Data. *Int. J. Robot. Res.* **1988**, *7*, 66–77. [CrossRef]
28. De Geeter, J.; Van Brussel, H.; De Schutter, J.; Decreton, M. A smoothly constrained Kalman filter. *IEEE Trans. Pattern Anal. Mach. Intell.* **1997**, *19*, 1171–1177. [CrossRef]
29. Hewett, R.J.; Heath, M.T.; Butala, M.D.; Kamalabadi, F. A Robust Null Space Method for Linear Equality Constrained State Estimation. *IEEE Trans. Signal Process.* **2010**, *58*, 3961–3971. [CrossRef]
30. Bakr, M.A.; Lee, S. A general framework for data fusion and outlier removal in distributed sensor networks. In Proceedings of the 2017 IEEE International Conference on Multisensor Fusion and Integration for Intelligent Systems (MFI), Daegu, Korea, 16–18 November 2017.
31. Simon, D. *Optimal State Estimation: Kalman, H [Infinity] and Nonlinear Approaches*; Wiley-Interscience: Hoboken, NJ, USA, 2006; ISBN 0470045337.
32. Jiang, L. Sensor Fault Detection and Isolation Using System Dynamics Identification Techniques. Ph.D. Thesis, University of Michigan, Ann Arbor, MI, USA, 2011.
33. Bar-Shalom, Y.; Willett, P.; Tian, X. *Tracking and Data Fusion*; YBS publishing: Storrs, CT, USA, 2011.

34. Walpole, R.; Myers, R.; Myers, S.; Ye, K. *Probability and Statistics for Engineers and Scientists*; Macmillan: Basingstoke, UK, 1993.
35. Fishman, G.S. *Monte Carlo*; Springer: New York, NY, USA, 1996; ISBN 978-1-4419-2847-4.
36. Rillo, G.; Morales, M.A.; Ceperley, D.M.; Pierleoni, C. Coupled electron-ion Monte Carlo simulation of hydrogen molecular crystals. *J. Chem. Phys.* **2018**, *148*, 102314. [CrossRef] [PubMed]
37. Baudry, G.; Macharis, C.; Vallée, T. Range-based Multi-Actor Multi-Criteria Analysis: A combined method of Multi-Actor Multi-Criteria Analysis and Monte Carlo simulation to support participatory decision making under uncertainty. *Eur. J. Oper. Res.* **2017**, *264*, 257–269. [CrossRef]
38. Ma, Y.; Chen, X.; Biegler, L.T. Monte-Carlo-simulation-based optimization for copolymerization processes with embedded chemical composition distribution. *Comput. Chem. Eng.* **2018**, *109*, 261–275. [CrossRef]
39. Vîlcu, A.-D.; Vîlcu, G.-E. An algorithm to estimate the vertices of a tetrahedron from uniform random points inside. *Ann. Mat. Pura Appl.* **2018**, *197*, 487–500. [CrossRef]
40. Hua, X.; Cheng, Y.; Wang, H.; Qin, Y. Robust Covariance Estimators Based on Information Divergences and Riemannian Manifold. *Entropy* **2018**, *20*, 219. [CrossRef]
41. Ljung, L. System Identification. In *Signal Analysis and Prediction*; Prochazka, A., Kinsbury, N., Payner, P.J.W., Uhlir, L., Eds.; Birkhäuser: Boston, MA, USA, 1998.

sensors

MDPI

Article

A Weighted Combination Method for Conflicting Evidence in Multi-Sensor Data Fusion

Fuyuan Xiao * and Bowen Qin

School of Computer and Information Science, Southwest University, No.2 Tiansheng Road, BeiBei District, Chongqing 400715, China; qinbowen_swu@163.com

* Correspondence: xiaofuyuan@swu.edu.cn

Received: 30 March 2018; Accepted: 1 May 2018; Published: 9 May 2018

check for updates

Abstract: Dempster–Shafer evidence theory is widely applied in various fields related to information fusion. However, how to avoid the counter-intuitive results is an open issue when combining highly conflicting pieces of evidence. In order to handle such a problem, a weighted combination method for conflicting pieces of evidence in multi-sensor data fusion is proposed by considering both the interplay between the pieces of evidence and the impacts of the pieces of evidence themselves. First, the degree of credibility of the evidence is determined on the basis of the modified cosine similarity measure of basic probability assignment. Then, the degree of credibility of the evidence is adjusted by leveraging the belief entropy function to measure the information volume of the evidence. Finally, the final weight of each piece of evidence generated from the above steps is obtained and adopted to modify the bodies of evidence before using Dempster's combination rule. A numerical example is provided to illustrate that the proposed method is reasonable and efficient in handling the conflicting pieces of evidence. In addition, applications in data classification and motor rotor fault diagnosis validate the practicability of the proposed method with better accuracy.

Keywords: multi-sensor data fusion; conflicting evidence; Dempster–Shafer evidence theory; belief entropy; similarity measure; data classification; fault diagnosis

1. Introduction

Multi-sensor data fusion technology has received significant attention in a variety of fields, as it combines the collected information from multi-sensors, which can enhance the robustness and safety of a system. In wireless sensor networks applications, however, the data that are collected from the sensors are often imprecise and uncertain [1]. How to model and handle the uncertainty information is still an open issue. To address this problem, many mathematical approaches have been presented, such as the fuzzy sets theory [2,3], that focuses on the intuitive reasoning by taking into account human subjectivity and imprecision; the intuitionistic fuzzy sets theory [4] which generalizes fuzzy sets by considering the uncertainty in the assignment of membership degree known as the hesitation degree; evidence theory [5–7], as a general framework for reasoning with uncertainty, with understood connections to other frameworks such as probability, possibility, and imprecise probability theories; rough sets theory [8,9] where its methodology is concerned with the classification and analysis of imprecise, uncertain, or incomplete information and knowledge, which is considered one of the first non-statistical approaches in data analysis; evidential reasoning [10,11] which is a generic evidence-based multi-criteria decision analysis (MCDA) approach for dealing with problems having both quantitative and qualitative criteria under various uncertainties including ignorance and randomness; Z numbers [12,13], that intend to provide a basis for computation with numbers which are not totally reliable; D numbers theory [14–17] which is a generalization of Dempster–Shafer theory, but does not follow the commutative law; and so on [18–21]. In addition, mixed intelligent methods have been applied in decision making [22], risk analysis [23], supplier selection [24], pattern recognition [25], classification [26], human reliability analysis [27], and fault diagnosis [28], etc. In this paper, we focus on evidence theory to deal with the uncertain problem of multi-sensor data fusion.

Dempster–Shafer evidence theory was firstly presented by Dempster [5] in 1967; later, it was extended by Shafer [6] in 1976. Dempster–Shafer evidence theory is effective to model both of the uncertainty and imprecision without prior information, so it is widely applied in various fields for information fusion [29–32]. Nevertheless, it may result in counter-intuitive results when combining highly conflicting pieces of evidence [33]. To address this issue, many methods have been presented in recent years [34–36]. On the one hand, some researchers focused on amending Dempster's combination rule. On the other hand, some researchers tried to pretreat the bodies of evidence before using Dempster's combination rule. In terms of of amending Dempster's combination rule, the major works contain Smets's unnormalized combination rule [37], Dubois and Prade's disjunctive combination rule [38], and Yager's combination rule [39]. However, the modification of combination rule often breaks the good properties, like commutativity and associativity. Furthermore, if the sensor failure gives rise to the counter-intuitive results, the modification of combination rule is considered to be unreasonable. Therefore, in order to resolve the fusion problem of highly conflicting pieces of evidence, researchers prefer to pretreat the bodies of evidence. With respect to pretreating the bodies of evidence, the main works contain Murphy's simple average approach of the bodies of evidence [40], and Deng et al.'s weighted average of the masses based on distance of evidence [41]. Deng et al.'s method [41] conquered the deficiency of the method in [40]. However, the impact of evidence itself was neglected in the decision-making process.

Hence, in this paper, a weighted combination method for conflicting pieces of evidence in multi-sensor data fusion is proposed to resolve fusion problem of highly conflicting evidence. First, the credibility degree of each piece of evidence is determined on the basis of the modified cosine similarity measure of basic probability assignment [42]. Then, credibility degree of each piece of evidence is modified by adopting the belief entropy function [43] to measure the information volume of the evidence. Finally, the modified credibility degree of each piece of evidence is used to adjust its corresponding body of evidence to obtain the weighted averaging evidence before using Dempster's combination rule. A numerical example is given to illustrate the feasibility and effectiveness of the proposed method. Additionally, the proposed method is applied in data classification and motor rotor fault diagnosis, which validates the practicability of it.

The rest of this paper is organized as follows. Section 2 briefly introduces the preliminaries of this paper. After that, Section 3 proposes the novel method, which is based on the similarity measure of evidence and belief function entropy. Then, Section 4 gives a numerical example to show the effectiveness of the proposed method. A statistical experiment is carried out in Section 5. Afterwards, the proposed method is applied to *Iris* data set classification, and motor rotor fault diagnosis is performed in Section 6. Finally, Section 7 gives the conclusions.

2. Preliminaries

2.1. Data Fusion

Data fusion can be identified as a combination of multiple sources to obtain improved information with less expensive, higher quality, or more relevant information [44]. General data fusion structure can be classified into three types based on the different stages: data-level, feature-level, and decision-level, as referred in [45].

In the data-level fusion, all raw data from sensors for a measured object are combined directly. Then, a feature vector is extracted from the fused data. Fusion of data at this level consists of the maximum information so that it can generate good results. However, sensors used in the data-level fusion, such as the sensors reporting vibration signals, must be homogeneous. As a consequence, the data-level fusion is limited in the actual application environment, because many physical quantities can be measured for a more comprehensive analysis. In the feature-level fusion, heterogeneous sensors can be used to report the data. According to the types of collected raw data, the features are extracted from the sensors. Then, these heterogeneous sensor data are combined at the feature-level stage. All of the feature vectors are combined into a single feature vector, which is then utilized in a special classification model for decision-making. In the decision-level fusion, the processes of feature extraction and pattern recognition are sequentially conducted for the data collected from each sensor. Then, the produced decision vectors are combined by using decision-level fusion techniques such as the Bayesian method, Dempster–Shafer evidence theory, or behavior knowledge space.

Because of the advantages of multi-sensor data fusion technology, it has been widely applied in various fields, such as in fault diagnosis [46–48], target tracking [49,50], health care analysis [51,52], image processing [53], attack detection [54], estimation of ship dynamics [55], and characterization of built environments [56].

In this paper, we focus on decision-level fusion, and try to improve the performance of the system based on Dempster–Shafer evidence theory.

2.2. Dempster-Shafer Evidence Theory

Dempster–Shafer evidence theory was firstly proposed by Dempster [5] and was then further developed by Shafer [6]. Dempster–Shafer evidence theory, as a generalization of Bayesian inference, asks for weaker conditions, which makes it more flexible and effective to model both the uncertainty and imprecision. The basic concepts are introduced as below.

Definition 1. *Let U be a set of mutually exclusive and collectively exhaustive events, indicated by*

$$U = \{C_1, C_2, \ldots, C_i, \ldots, C_N\}. \tag{1}$$

The set U is called frame of discernment. The power set of U is indicated by 2^U, where

$$2^U = \{\emptyset, \{C_1\}, \{C_2\}, \ldots, \{C_N\}, \{C_1, C_2\}, \ldots, \{C_1, C_2, \ldots, C_i\}, \ldots, U\}, \tag{2}$$

and $\emptyset$ is an empty set. If $A \in 2^U$, A is called a proposition or hypothesis.

Definition 2. *For a frame of discernment U, a mass function is a mapping m from 2^U to [0, 1], formally defined by*

$$m : 2^U \rightarrow [0, 1], \tag{3}$$

which satisfies the following condition:

$$m(\emptyset) = 0 \text{ and } \sum_{A \in 2^U} m(A) = 1. \tag{4}$$

In Dempster–Shafer evidence theory, a mass function can be also called as a basic probability assignment (BPA). If $m(A)$ is greater than 0, A will be called as a focal element, and the union of all of the focal elements is known as the core of the mass function.

Definition 3. *For a proposition $A \subseteq U$, the belief function $Bel : 2^U \rightarrow [0, 1]$ is defined as*

$$Bel(A) = \sum_{B \subseteq A} m(B). \tag{5}$$

The plausibility function $Pl : 2^U \rightarrow [0, 1]$ is defined as

$$Pl(A) = 1 - Bel(\bar{A}) = \sum_{B \cap A \neq \emptyset} m(B), \tag{6}$$

where $\bar{A} = U - A$.

Apparently, $Pl(A)$ is equal or greater than $Bel(A)$, where the function Bel is the lower limit function of proposition A and the function Pl is the upper limit function of proposition A.

Definition 4. *Let the two BPAs be* m_1 *and* m_2 *on the frame of discernment* U. *Assuming that these BPAs are independent, Dempster's rule of combination, denoted by* $m = m_1 \oplus m_2$, *known as the orthogonal sum, is defined as below:*

$$m(A) = \begin{cases} \frac{1}{1-K} \sum\limits_{B \cap D = A} m_1(B) m_2(D), & A \neq \emptyset, \\ 0, & A = \emptyset, \end{cases} \tag{7}$$

with

$$K = \sum_{B \cap D = \emptyset} m_1(B) m_2(D), \tag{8}$$

where B and D are also the elements of 2^U, *and K is a constant that presents the conflict between the two BPAs.*

Note that Dempster's combination rule is only practicable for the two BPAs with the condition $K < 1$.

2.3. Modified Cosine Similarity Measure of BPAs

A modified cosine similarity measure is proposed by Jiang [42]. Because it considers three important factors, namely, angle, distance, and vector norm, the modified cosine similarity measure is an efficient approach to measure the similarity between vectors more precisely. The modified cosine similarity measure among the BPAs can determine whether the pieces of evidence conflict with each other. A large similarity indicates that this piece of evidence has more support from another piece of evidence, while a small similarity indicates that this piece of evidence has less support from another piece of evidence.

Definition 5. *Let* $E = [e_1, e_2, \ldots, e_n]$ *and* $F = [f_1, f_2, \ldots, f_n]$ *be two vectors of* R^n. *The modified cosine similarity between vectors E and F is defined as*

$$SI(E,F) = \begin{cases} \frac{1}{2}\{\alpha^{-P} + min(\frac{|E|}{|F|}, \frac{|F|}{|E|})\} si_{cos}(E,F), & E \neq 0, F \neq 0, \\ 0, & E = 0 \text{ or } F = 0, \end{cases} \tag{9}$$

where α *is a constant whose value is greater than 1, P is the Euclidean distance between the two vectors E and F,* α^{-P} *is the distance-based similarity measure,* $min(\frac{|E|}{|F|}, \frac{|F|}{|E|})$ *is the minimum of* $\frac{|E|}{|F|}$ *and* $\frac{|F|}{|E|}$, *and* $si_{cos}(E,F)$ *is the cosine similarity. The larger the* α *is, the greater the distance impact on vector similarity will be.*

Definition 6. *Let* m_1 *and* m_2 *be the BPAs in the frame of discernment* $U = \{C_1, C_2, \ldots, C_N\}$. *The two vectors are expressed as*

$$\begin{aligned} Bel_i &= [Bel_i(C_1), Bel_i(C_2), \ldots, Bel_i(C_N)], \quad & i = 1,2, \\ Pl_i &= [Pl_i(C_1), Pl_i(C_2), \ldots, Pl_i(C_N)], \quad & i = 1,2. \end{aligned} \tag{10}$$

Then, the belief function vector similarity $SI(Bel_1, Bel_2)$ *and the plausibility function vector similarity* $SI(Pl_1, Pl_2)$ *can be calculated. The new similarity of BPAs is defined as*

$$SI_{BPA} = (1 - \lambda) * SI(Bel_1, Bel_2) + \lambda * SI(Pl_1, Pl_2), \tag{11}$$

with

$$0 \leq \lambda \leq 1, \tag{12}$$

where λ is the total uncertainty of BPAs, which is defined as

$$\lambda = \frac{\sum_{i=1}^{2}\sum_{j=1}^{N}(Pl_i(C_j) - Bel_i(C_j))}{\sum_{i=1}^{2}\sum_{j=1}^{N}(Pl_i(C_j))}. \tag{13}$$

Because $Pl_i(C_j) \geq Bel_i(C_j)$ and $Bel \geq 0$, if $Pl_i(C_j) = Bel_i(C_j)$, then $\lambda = 0$. Otherwise, if $Bel_i(C_j) = 0$, then $\lambda = 1$. The larger the uncertainty λ is, the greater the influence on the similarity of BPA will be.

2.4. Belief Entropy

A novel type of belief entropy, known as the Deng entropy, was first proposed by Deng [43]. When the uncertain information is expressed by probability, the Deng entropy degenerates to the Shannon entropy. Hence, the Deng entropy is regarded as a generalization of the Shannon entropy. It is an efficient mathematical tool to measure the uncertain information, especially when the uncertain information is expressed by the BPA. Because of its advantage in measuring the uncertain information, the Deng entropy is applied in a variety of areas [57,58]. The basic concepts are introduced below.

Definition 7. *Let B be a hypothesis or proposition of the BPA m in the frame of discernment U and $|B|$ be the cardinality of B. The Deng entropy of the BPA m is defined as follows:*

$$E_d(m) = -\sum_{B \subseteq U} m(B)\, log\, \frac{m(B)}{2^{|B|} - 1}. \tag{14}$$

When the belief value is only allocated to the singleton, the Deng entropy degenerates to the Shannon entropy, i.e.,

$$E_d(m) = -\sum_{B \in U} m(B)\, log\, \frac{m(B)}{2^{|B|} - 1} = -\sum_{B \in U} m(B)\, log\, m(B). \tag{15}$$

The larger the value of the cardinality of the hypothesis or proposition, the larger the value the Deng entropy of evidence, which means that the piece of evidence involves more information. Therefore, if a piece of evidence has a large Deng entropy value, it has more support from other pieces of evidence, indicating that this piece of evidence plays an important role in the evidence combination.

3. The Proposed Method

In this paper, a weighted combination method for conflicting pieces of evidence multi-sensor data fusion is proposed by combining the modified cosine similarity measure of evidence with the belief entropy function. In contrast to the method of Jiang et al. [42], in the proposed method, the impact of evidence itself is considered in the process of fusion of multiple pieces of evidence by leveraging the belief entropy [43], i.e., a useful uncertainty measure tool, to measure the information volume of each piece of evidence, so that the proposed method can combine multiple pieces of evidence with greater accuracy. This will be discussed further in the next section.

3.1. Process Steps

The proposed method is composed of the following procedures. The credibility degree of the pieces of evidence is first determined on the basis of the similarity measure among the BPAs. Then, the credibility degree is modified by leveraging the belief entropy function to measure the information volume of the evidence. Afterwards, the final weight of each piece of evidence is obtained and adopted to adjust the body of evidence before using Dempster's combination rule. The specific calculation processes are listed as follows. The flowchart of the proposed method is shown in Figure 1.

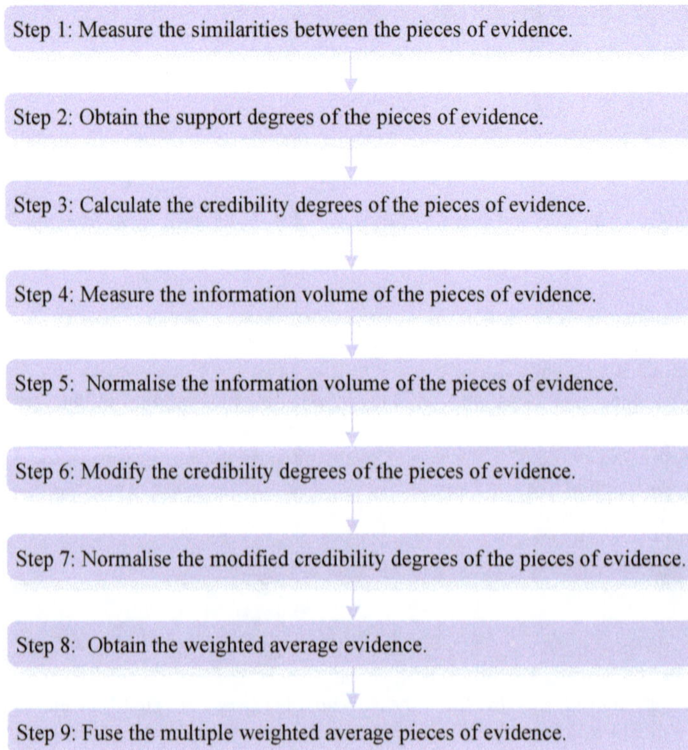

Figure 1. The flowchart of the proposed method.

Step 1: Measure the similarities between the pieces of evidence.

The similarity measure $SI_{BPA}(ij)$ between the BPAs m_i and m_j can be obtained by Equations (11)–(13). Then, a similarity measure matrix (SMM) can be constructed as follows:

$$SMM = \begin{bmatrix} SI_{BPA}(11) & \cdots & SI_{BPA}(1i) & \cdots & SI_{BPA}(1k) \\ \vdots & \vdots & \vdots & \vdots & \vdots \\ SI_{BPA}(i1) & \cdots & SI_{BPA}(ii) & \cdots & SI_{BPA}(ik) \\ \vdots & \vdots & \vdots & \vdots & \vdots \\ SI_{BPA}(k1) & \cdots & SI_{BPA}(ki) & \cdots & SI_{BPA}(kk) \end{bmatrix}. \quad (16)$$

Step 2: Obtain the support degrees of the pieces of evidence.

The support degree of the BPA m_i $(i = 1, \ldots, k)$, denoted as $SD(m_i)$, is defined as follows:

$$SD(m_i) = \sum_{j=1, j \neq i}^{k} SI_{BPA}(ij). \quad (17)$$

Step 3: Calculate the credibility degrees of the pieces of evidence.

The credibility degree of the BPA m_i $(i = 1, \ldots, k)$, denoted as $CD(m_i)$, is defined as follows:

$$CD(m_i) = \frac{SD(m_i)}{\sum_{l=1}^{k} SD(m_l)}. \quad (18)$$

Step 4: Measure the information volume of the pieces of evidence.

According to Equation (14), the belief entropy $E_d(m_i)$ of the BPA m_i $(i = 1, \ldots, k)$ can be calculated. To avoid assigning zero weight to the evidence, the information volume $IV(m_i)$ is used for measuring the uncertain information of m_i. It is defined as follows:

$$IV(m_i) = e^{E_d(m_i)} = e^{-\sum_{B \subseteq U} m(B)\ log\ \frac{m(B)}{2^{|B|}-1}}. \tag{19}$$

Step 5: Normalize the information volume of the pieces of evidence.

The information volume of the BPA m_i $(i = 1, \ldots, k)$ will be normalized as below:

$$\overline{IV}(m_i) = \frac{IV(m_i)}{\sum_{l=1}^{k} IV(m_l)}. \tag{20}$$

Step 6: Modify the credibility degrees of the pieces of evidence.

Based on the normalized information volume, the credibility degree of the BPA m_i $(i = 1, \ldots, k)$ will be modified, denoted as $MCD(m_i)$:

$$MCD(m_i) = CD(m_i) \times \overline{IV}(m_i)^{(\frac{\sum_{l=1}^{k} CD(m_l)}{k} - CD(m_i))}. \tag{21}$$

Step 7: Normalize the modified credibility degrees of the pieces of evidence.

The modified credibility degree $MCD(m_i)$ of the BPA m_i $(i = 1, \ldots, k)$ will be normalized as below, and is considered as the final weight to adjust the bodies of evidence.

$$\overline{M}CD(m_i) = \frac{MCD(m_i)}{\sum_{l=1}^{k} MCD(m_l)}. \tag{22}$$

Step 8: Obtain the weighted average evidence.

Based on the modified credibility degree of the BPA m_i $(i = 1, \ldots, k)$, the weighted average evidence $WAE(m)$ is defined as follows:

$$WAE(m) = \sum_{i=1}^{k} (\overline{M}CD(m_i) \times m_i). \tag{23}$$

Step 9: Fuse multiple weighted average pieces of evidence.

When k number of pieces of evidence exist, the weighted average evidence will be fused through Dempster's combination rule Equation (7) via $k - 1$ times as below,

$$Fus(m) = (((WAE(m) \oplus WAE(m))_1 \oplus \cdots)_h \oplus WAE(m))_{(k-1)}. \tag{24}$$

Ultimately, we can obtain the final fusion result of the evidence.

3.2. Algorithm

Let $m = \{m_1, \ldots, m_i, \ldots, m_k\}$ be a set of multiple pieces of evidence. After receiving k pieces of evidence, a fusion result is expected to be generated for decision-making support. The weighted fusion method for multiple pieces of evidence is outlined in Algorithm 1.

As shown in Algorithm 1, it provides a formal expression in terms of the specific calculation processes of the proposed method listed in Section 3.1. To be specific, Lines 2–7 explain how to measure the similarities between the pieces of evidence and construct the similarity measure matrix for k pieces of evidence. Lines 9–11 show how to obtain the support degrees for k pieces of evidence. Lines 13–15 represent how to calculate the credibility degrees for k pieces of evidence. Lines 17–19 explain how to measure the information volumes for k pieces of evidence. Lines 21–23 express how to normalize the information volumes for k pieces of evidence. Lines 25–27 state how to modify the credibility degrees for k pieces of evidence. Lines 29–31 show how to normalize the modified credibility degrees for k pieces of evidence. Line 33 describes how to obtain the weighted average evidence based on k pieces of evidence. Lines 35–37 depict how to generate the fusion result.

Algorithm 1: A weighted fusion method for multiple pieces of evidence.

Input: A set of multiple pieces of evidence $m = \{m_1, \ldots, m_i, \ldots, m_k\}$;
Output: Fusion result $Fus(m)$;
1 /* Step 1 */
2 **for** $i = 1; i \leq k$ **do**
3   **for** $j = 1; j \leq k$ **do**
4   Calculate $SI_{BPA}(ij)$ with Equations (11)–(13);
5   **end**
6 **end**
7 Construct the similarity measure matrix SMM;
8 /* Step 2 */
9 **for** $i = 1; i \leq k$ **do**
10   Obtain the support degree $SD(m_i)$ with Equation (17);
11 **end**
12 /* Step 3 */
13 **for** $i = 1; i \leq k$ **do**
14   Calculate the credibility degree $CD(m_i)$ with Equation (18);
15 **end**
16 /* Step 4 */
17 **for** $i = 1; i \leq k$ **do**
18   Measure the information volume $IV(m_i)$ with Equation (19);
19 **end**
20 /* Step 5 */
21 **for** $i = 1; i \leq k$ **do**
22   Normalise the information volume $\overline{IV}(m_i)$ with Equation (20);
23 **end**
24 /* Step 6 */
25 **for** $i = 1; i \leq k$ **do**
26   Obtain the modified credibility degree $MCD(m_i)$ with Equation (21);
27 **end**
28 /* Step 7 */
29 **for** $i = 1; i \leq k$ **do**
30   Normalise the modified credibility degree $\overline{MCD}(m_i)$ with Equation (22)
31 **end**
32 /* Step 8 */
33 Obtain the weighted average evidence $WAE(m)$ with Equation (23);
34 /* Step 9 */
35 **for** $h = 1; h \leq k - 1$ **do**
36   Calculate the fusion result $Fus(m)$ by combining $WAE(m)$ with Equation (7);
37 **end**

---

4. Numerical Example

In this section, in order to demonstrate the feasibility and effectiveness of the proposed method, a numerical example is illustrated.

Example 1. *Consider the decision-making problem of the multi-sensor-based target recognition system from [59] associated with five different kinds of sensors to observe objects, where $U = \{a, b, c\}$. Here, a, b, and c are the three objects in the frame of discernment U. The five BPAs that are collected by the system are listed as shown in Table 1.*

Table 1. The basic probability assignments (BPAs) for the example.

Pieces of Evidence	BPAs			
	$\{a\}$	$\{b\}$	$\{c\}$	$\{a,b,c\}$
$m_1(\cdot)$	0.30	0.20	0.10	0.40
$m_2(\cdot)$	0.00	0.90	0.10	0.00
$m_3(\cdot)$	0.60	0.10	0.10	0.20
$m_4(\cdot)$	0.70	0.10	0.10	0.10
$m_5(\cdot)$	0.70	0.10	0.10	0.10

Step 1: The similarity measure $SI_{BPA}(ij)$ $(i, j = 1, 2, 3, 4, 5)$ between the BPAs m_i and m_j can be constructed as below:

$$SMM = \begin{pmatrix} 1.0000 & 0.3730 & 0.8144 & 0.7478 & 0.7478 \\ 0.3730 & 1.0000 & 0.1958 & 0.1568 & 0.1568 \\ 0.8144 & 0.1958 & 1.0000 & 0.9340 & 0.9340 \\ 0.7478 & 0.1568 & 0.9340 & 1.0000 & 1.0000 \\ 0.7478 & 0.1568 & 0.9340 & 1.0000 & 1.0000 \end{pmatrix}.$$

Step 2: The support degree $SD(m)$ of the BPA m_i $(i = 1, 2, 3, 4, 5)$ is calculated as shown in Table 2.

Table 2. The calculated results in terms of support degree, credibility degree, information volume, normalized information volume, credibility degree, and modified credibility degree of BPAs.

Items	Pieces of Evidence				
	m_1	m_2	m_3	m_4	m_5
$SD(m)$	2.6830	0.8824	2.8782	2.8386	2.8386
$CD(m)$	0.2214	0.0728	0.2375	0.2342	0.2342
$IV(m)$	19.480	1.5984	8.4351	5.1423	5.1423
$\overline{IV}(m)$	0.4895	0.0402	0.2119	0.1292	0.1292
$MCD(m)$	0.2248	0.0484	0.2517	0.2512	0.2512
$\overline{MCD}(m)$	0.2188	0.0471	0.2450	0.2445	0.2445

Step 3: The credibility degree $CD(m)$ of the BPA m_i $(i = 1, 2, 3, 4, 5)$ is obtained as shown in Table 2.

Step 4: The information volume $IV(m)$ of the BPA m_i $(i = 1, 2, 3, 4, 5)$ is measured as shown in Table 2.

Step 5: The information volume of the BPA m_i $(i = 1, 2, 3, 4, 5)$ is normalized as shown in Table 2, denoted by $\overline{IV}(m)$.

Step 6: The credibility degree $MCD(m)$ of the BPA m_i $(i = 1, 2, 3, 4, 5)$ is modified as shown in Table 2.

Step 7: The modified credibility degree $\overline{MCD}(m)$ of the BPA m_i $(i = 1, 2, 3, 4, 5)$ is normalized as shown in Table 2.

Step 8: The weighted average evidence $WAE(m)$ is computed as shown in Table 3.

Table 3. The weighted average evidence ($WAE(m)$) and final fusion result ($Fus(m)$).

Items	BPAs			
	$\{a\}$	$\{b\}$	$\{c\}$	$\{a,b,c\}$
$WAE(m)$	0.5550	0.1596	0.1000	0.1854
$Fus(m)$	0.9713	0.0204	0.0073	0.0010

Step 9: By fusing the weighted average evidence via Dempster's combination rule four times, the final fusion result $Fus(m)$ of evidence can be produced as shown in Table 3.

From Example 1, it is obvious that m_2 highly conflicts with other pieces of evidence. The fusing results that are obtained by different combination approaches are presented in Table 4. In addition, the comparisons of target a's BPA in terms of different combination rules are shown in Figure 2.

Table 4. Evidence fusion results based on different combination rules.

Evidences	Methods	BPAs				Target
		$\{a\}$	$\{b\}$	$\{c\}$	$\{a,b,c\}$	
m_1, m_2	Dempster [5]	0.0000	0.9153	0.0847	0.0000	b
	Murphy [40]	0.1187	0.7518	0.0719	0.0576	b
	Deng et al. [41]	0.1187	0.7518	0.0719	0.0576	b
	Qian et al. [59]	0.1187	0.7518	0.0719	0.0576	b
	Proposed method	0.1187	0.7518	0.0719	0.0576	b
m_1, m_2, m_3	Dempster [5]	0.0000	0.9153	0.0847	0.0000	b
	Murphy [40]	0.3324	0.5909	0.0540	0.0227	b
	Deng et al. [41]	0.4477	0.4546	0.0644	0.0333	-
	Qian et al. [59]	0.6110	0.2861	0.0659	0.0370	a
	Proposed method	0.5779	0.3070	0.0714	0.0438	a
m_1, m_2, m_3, m_4	Dempster [5]	0.0000	0.9153	0.0847	0.0000	b
	Murphy [40]	0.6170	0.3505	0.0272	0.0053	a
	Deng et al. [41]	0.8007	0.1640	0.0283	0.0070	a
	Qian et al. [59]	0.8472	0.1221	0.0249	0.0058	a
	Proposed method	0.8785	0.0857	0.0271	0.0076	a
m_1, m_2, m_3, m_4, m_5	Dempster [5]	0.0000	0.9153	0.0847	0.0000	b
	Murphy [40]	0.8389	0.1502	0.0099	0.0010	a
	Deng et al. [41]	0.9499	0.0411	0.0080	0.0010	a
	Qian et al. [59]	0.9525	0.0393	0.0074	0.0008	a
	Proposed method	0.9713	0.0204	0.0073	0.0010	a

As shown in Table 4, no matter how many pieces of evidence support target a, Dempster's combination method [5] always generates a counterintuitive result. As the number of pieces of evidence increases to three, Murphy's combination method [40] and Deng et al.'s combination method [41] cannot deal with the highly conflicting pieces of evidence very well, because the BPA values of object a generated by Murphy's method [40] and Deng et al.'s method [41] are 33.24% and 44.77%, respectively, which are smaller than 50%. When the number of pieces of evidence increases from four to five, Murphy's combination method [40] and Deng et al.'s combination method [41] work well, and the BPA values of object a generated by Murphy's method [40] and Deng et al.'s method [41] increase up to 83.89% and 94.99%, respectively.

On the other hand, as shown in Table 4, Qian et al.'s combination method [59] and the proposed method show reasonable results and can efficiently deal with the highly conflicting pieces of evidence as the number of pieces of evidence increases from three to five. In the face of five pieces of evidence, the BPA value of object a generated by the proposed method increases to 97.13% which is much higher than for other combination approaches, as shown in Figure 2. Therefore, it is concluded that the proposed method is as feasible and effective as related approaches.

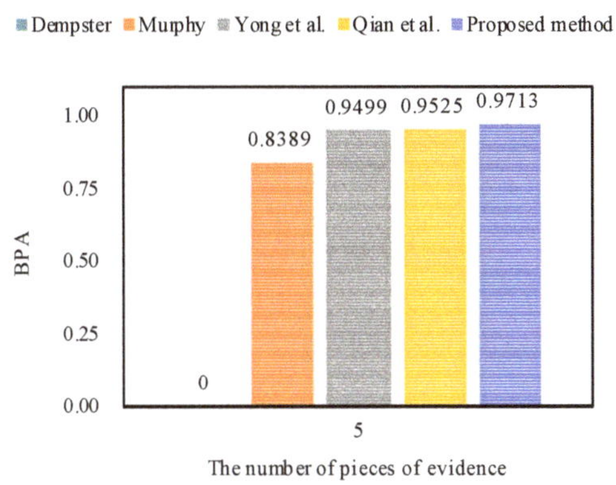

Figure 2. The comparisons of target a's BPA in terms of different methods.

5. Statistical Experiment

In this section, in order to make a sound comparison, a statistical experiment is carried out with multiple pieces of initial data for the comparison of the proposed method with other related methods.

This statistical experiment is implemented based on Example 1. In the experimental setting, for generating multiple initial data 100 times, we provide a variation range $[-0.1, 0.1]$ for each BPA of m_1, and vary the values of BPAs of m_1 randomly.

Then, the generated multiple pieces of initial data are fused by utilizing the different methods, namely, Dempster's combination method [5], Murphy's combination method [40], Deng et al.'s combination method [41], Jiang et al.'s combination method [42], and the proposed method.

The experimental results of target a's BPA generated by different combination methods are shown in Figure 3. From the comparison results, it is obvious that Murphy's combination method [40], Deng et al.'s combination method [41], Jiang et al.'s combination method [42], and the proposed method are more efficient than Dempster's combination method [5], because Dempster's combination method cannot effectively deal with the conflicting pieces of evidence, and thus always generates counterintuitive results where target a's BPA value is 0 (under 0.5). In contrast, the other methods can effectively cope with the conflicting evidence and recognize the target a, where its corresponding BPA value is always larger than 0.5 under multiple experiments. On the other hand, because Murphy's combination method is a simply average-weighted approach to the bodies of evidence, its overall performance is poorer than that of Deng et al.'s combination method, Jiang et al.'s combination method, and the proposed method to a certain extent.

Furthermore, as shown in Figure 3a, Jiang et al.'s combination method [42] which is based on the modified cosine similarity measure, is more effective than Deng et al.'s combination method [41] that is based on the Jousselme distance as a whole. This is the reason that the modified cosine similarity measure is considered in this study.

In order to improve the performance of Jiang et al.'s combination method, we investigate and find that in the process of fusion of multiple pieces of evidence, the impact of the evidence itself is overlooked in their method. Hence, we also take the belief entropy into consideration to measure the information volume of each piece of evidence in the course of fusion and design the proposed method. Consequently, as shown in Figure 3b, it can be noted that the proposed method is superior to Jiang et al.'s combination method [42] with a higher target a BPA value.

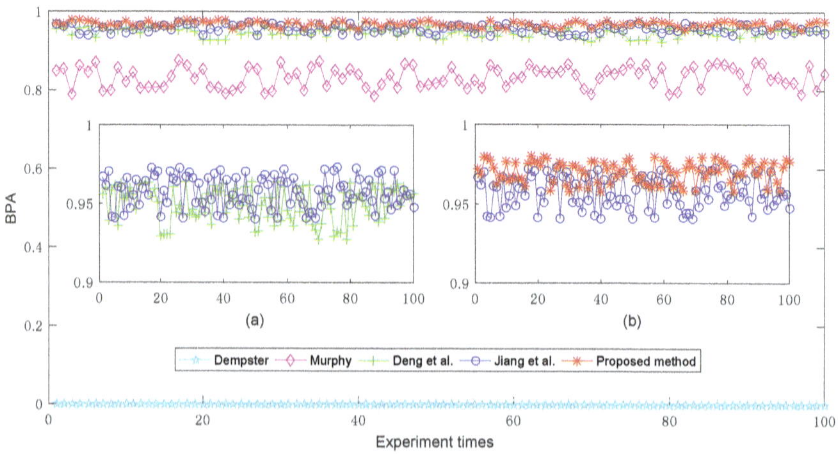

Figure 3. The comparisons of target a's BPAs obtained by different combination methods where the multiple BPAs are generated randomly 100 times. (**a**) The comparisons of Deng et al.'s combination method and Jiang et al.'s combination method; (**b**) The comparisons of Jiang et al.'s combination method and the proposed method.

6. Applications

In this section, the proposed approach is applied to *Iris* data set classification and motor rotor fault diagnosis, respectively, to validate its practicability, in which the experimental data in [48,59] are leveraged for the comparison among different approaches.

6.1. Iris Data Set Classification

Consider the *Iris* data set classification problem associated with a frame of discernment U consisting of three species of *Iris* flowers given by $U = \{setosa, versicolor, virginica\} = \{Se, Ve, Vi\}$ in terms of four numerical attributes of *Iris* flowers given by $\{sepal\ length\ (SL), sepal\ width\ (SW), petal\ length\ (PL), petal\ width\ (PW)\}$, where the BPAs of *Iris* instances are modeled with noisy data and given in Table 5 from [59].

Table 5. The BPAs of *Iris* flower instances.

BPAs	Attributes			
	$\{SL\}$	$\{SW\}$	$\{PL\}$	$\{PW\}$
$m\{Se\}$	0.3337	0.0000	0.6699	0.6996
$m\{Ve\}$	0.3165	0.9900	0.2374	0.2120
$m\{Vi\}$	0.2816	0.0100	0.0884	0.0658
$m\{Se, Ve\}$	0.0307	0.0000	0.0000	0.0000
$m\{Se, Vi\}$	0.0052	0.0000	0.0000	0.0000
$m\{Ve, Vi\}$	0.0272	0.0000	0.0043	0.0226
$m\{Se, Ve, Vi\}$	0.0052	0.0000	0.0000	0.0000

Step 1: The similarity measure $SI_{BPA}(ij)$ $(i, j = SL, SW, PL, PW)$ between the BPAs m_i and m_j can be constructed as below:

$$SMM = \begin{pmatrix} 1.0000 & 0.3324 & 0.7965 & 0.7750 \\ 0.3324 & 1.0000 & 0.2056 & 0.1794 \\ 0.7965 & 0.2056 & 1.0000 & 0.9867 \\ 0.7750 & 0.1794 & 0.9867 & 1.0000 \end{pmatrix}.$$

Step 2: The support degree of the BPA m_i $(i = SL, SW, PL, PW)$ is calculated as follows:

$$SD(m_{SL}) = 1.9039,$$
$$SD(m_{SW}) = 0.7174,$$
$$SD(m_{PL}) = 1.9888,$$
$$SD(m_{PW}) = 1.9411.$$

Step 3: The credibility degree of the BPA m_i $(i = SL, SW, PL, PW)$ is obtained as below:

$$CD(m_{SL}) = 0.2906,$$
$$CD(m_{SW}) = 0.1095,$$
$$CD(m_{PL}) = 0.3036,$$
$$CD(m_{PW}) = 0.2963.$$

Step 4: The information volume of the BPA m_i $(i = SL, SW, PL, PW)$ is measured as follows:

$$IV(m_{SL}) = 7.8287,$$
$$IV(m_{SW}) = 1.0842,$$
$$IV(m_{PL}) = 3.4202,$$
$$IV(m_{PW}) = 3.4998.$$

Step 5: The information volume of the BPA m_i $(i = SL, SW, PL, PW)$ is normalised as follows:

$$\overline{IV}(m_{SL}) = 0.4945,$$
$$\overline{IV}(m_{SW}) = 0.0685,$$
$$\overline{IV}(m_{PL}) = 0.2160,$$
$$\overline{IV}(m_{PW}) = 0.2210.$$

Step 6: The credibility degree of the BPA m_i $(i = SL, SW, PL, PW)$ is modified as below:

$$MCD(m_{SL}) = 0.2991,$$
$$MCD(m_{SW}) = 0.0751,$$
$$MCD(m_{PL}) = 0.3296,$$
$$MCD(m_{PW}) = 0.3177.$$

Step 7: The modified credibility degree of the BPA m_i $(i = SL, SW, PL, PW)$ is normalized as follows:

$$\overline{MCD}(m_{SL}) = 0.2928,$$
$$\overline{MCD}(m_{SW}) = 0.0736,$$
$$\overline{MCD}(m_{PL}) = 0.3226,$$
$$\overline{MCD}(m_{PW}) = 0.3111.$$

Step 8: The weighted average evidence is computed as below:

$$m(\{Se\}) = 0.5314,$$
$$m(\{Ve\}) = 0.3080,$$
$$m(\{Vi\}) = 0.1322,$$
$$m(\{Se, Ve\}) = 0.0090,$$
$$m(\{Se, Vi\}) = 0.0015,$$
$$m(\{Ve, Vi\}) = 0.0164,$$
$$m(\{Se, Ve, Vi\}) = 0.0015.$$

Step 9: By fusing the weighted average evidence via Dempster's combination rule four times, the final fusion result of the evidence can be produced as follows:

$$m(\{Se\}) = 0.8693,$$
$$m(\{Ve\}) = 0.1254,$$
$$m(\{Vi\}) = 0.0053,$$
$$m(\{Se, Ve\}) = 1 \times 10^{-7},$$
$$m(\{Se, Vi\}) = 7 \times 10^{-10},$$
$$m(\{Ve, Vi\}) = 1 \times 10^{-6},$$
$$m(\{Se, Ve, Vi\}) = 5 \times 10^{-11}.$$

The fusion results based on different combination approaches that were applied on the *Iris* data set are presented in Table 6. From the experimental results, it can be seen that Dempster's combination method [5] and Murphy's combination method [40] always generate counterintuitive results and classify the species of *Iris* flower as *versicolor*, even when the number of pieces of evidence increases from two (m_{SL}, m_{SW}) to four ($m_{SL}, m_{SW}, m_{PL}, m_{PW}$). By contrast, Deng et al.'s combination method [41] works well when the number of pieces of evidence is increased up to four ($m_{SL}, m_{SW}, m_{PL}, m_{PW}$), because it can classify the species of *Iris* flower as the target *setosa* with a belief value of 73.01%.

Table 6. The comparison of different methods applied in the *Iris* data set classification.

Evidence	Methods	BPAs							Target
		$\{Se\}$	$\{Ve\}$	$\{Vi\}$	$\{Se, Ve\}$	$\{Se, Vi\}$	$\{Ve, Vi\}$	$\{Se, Ve, Vi\}$	
m_{SL}, m_{SW}	Dempster [5]	0.0000	0.9916	0.0084	0.0000	0.0000	0.0000	0.0000	*Ve*
	Murphy [40]	0.0655	0.8828	0.0505	6×10^{-4}	4×10^{-5}	5×10^{-4}	1×10^{-5}	*Ve*
	Deng et al. [41]	0.0655	0.8828	0.0505	6×10^{-4}	4×10^{-5}	5×10^{-4}	1×10^{-5}	*Ve*
	Qian et al. [59]	0.0655	0.8828	0.0505	6×10^{-4}	4×10^{-5}	5×10^{-4}	1×10^{-5}	*Ve*
	Proposed method	0.0655	0.8828	0.0505	6×10^{-4}	4×10^{-5}	5×10^{-4}	1×10^{-5}	*Ve*
m_{SL}, m_{SW}, m_{PL}	Dempster [5]	0.0000	0.9968	0.0032	0.0000	0.0000	0.0000	0.0000	*Ve*
	Murphy [40]	0.2112	0.7749	0.0139	8×10^{-6}	2×10^{-7}	9×10^{-6}	3×10^{-8}	*Ve*
	Deng et al. [41]	0.3219	0.6534	0.0247	2×10^{-5}	4×10^{-7}	2×10^{-5}	5×10^{-8}	*Ve*
	Qian et al. [59]	0.5678	0.4036	0.0287	2×10^{-5}	4×10^{-7}	2×10^{-5}	5×10^{-8}	*Se*
	Proposed method	0.5206	0.4421	0.0372	2×10^{-5}	5×10^{-7}	2×10^{-5}	7×10^{-8}	*Se*
$m_{SL}, m_{SW}, m_{PL}, m_{PW}$	Dempster [5]	0.0000	0.9988	0.0012	0.0000	0.0000	0.0000	0.0000	*Ve*
	Murphy [40]	0.4422	0.5546	0.0032	8×10^{-8}	5×10^{-10}	6×10^{-7}	3×10^{-11}	*Ve*
	Deng et al. [41]	0.7301	0.2652	0.0047	1×10^{-7}	7×10^{-10}	9×10^{-7}	5×10^{-11}	*Se*
	Qian et al. [59]	0.8338	0.1617	0.0045	9×10^{-8}	6×10^{-10}	9×10^{-7}	4×10^{-11}	*Se*
	Proposed method	0.8693	0.1254	0.0053	1×10^{-7}	7×10^{-10}	1×10^{-6}	5×10^{-11}	*Se*

Obviously, Qian et al.'s combination method [59] and the proposed method show reasonable results and classify the species of *Iris* flower as the target *setosa* with 83.38% and 86.93% belief values, respectively. Therefore, we can conclude that the proposed method is more efficient than other related methods with better accuracy of data classification, as shown in Figure 4. The reason is that the proposed method not only takes the interplay between the pieces of evidence into account, but also considers the impacts of the pieces of evidence themselves.

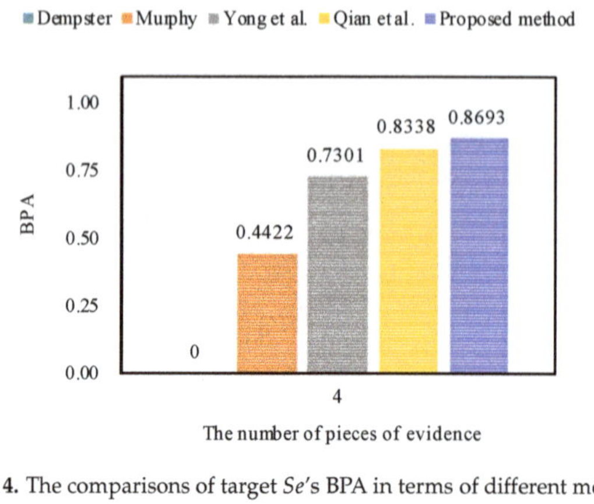

Figure 4. The comparisons of target *Se*'s BPA in terms of different methods.

6.2. Motor Rotor Fault Diagnosis

Supposing there are three types of faults for a motor rotor given by $\{F_1, F_2, F_3\}$ = {*rotor unbalance, rotor misalignment, pedestal looseness*} in the frame of discernment U. We place a set of vibration acceleration sensors at different places for gathering the vibration signals given by $S = \{S_1, S_2, S_3\}$.

The acceleration vibration frequency amplitudes at $1X$, $2X$, and $3X$ frequencies are considered as the fault feature variables. The collected sensor reports at $1X$, $2X$, and $3X$ frequencies modeled as BPAs are shown in Tables 7–9, respectively, in which $m_1(\cdot)$, $m_2(\cdot)$, and $m_3(\cdot)$ represent the BPAs modeled from the three vibration acceleration sensors S_1, S_2, and S_3.

Table 7. The collected sensor reports at the frequency of $1X$ modeled as BPAs.

BPA	$\{F_2\}$	$\{F_3\}$	$\{F_1,F_2\}$	$\{F_1,F_2,F_3\}$
$m_1(\cdot)$	0.8176	0.0003	0.1553	0.0268
$m_2(\cdot)$	0.5658	0.0009	0.0646	0.3687
$m_3(\cdot)$	0.2403	0.0004	0.0141	0.7452

Table 8. The collected sensor reports at the frequency of $2X$ modeled as BPAs.

BPA	$\{F_2\}$	$\{F_1,F_2,F_3\}$
$m_1(\cdot)$	0.6229	0.3771
$m_2(\cdot)$	0.7660	0.2341
$m_3(\cdot)$	0.8598	0.1402

Table 9. The collected sensor reports at the frequency of $3X$ modeled as BPAs.

BPA	$\{F_1\}$	$\{F_2\}$	$\{F_1,F_2\}$	$\{F_1,F_2,F_3\}$
$m_1(\cdot)$	0.3666	0.4563	0.1185	0.0586
$m_2(\cdot)$	0.2793	0.4151	0.2652	0.0404
$m_3(\cdot)$	0.2897	0.4331	0.2470	0.0302

6.2.1. Motor Rotor Fault Diagnosis at $1X$ Frequency

By conducting the steps in Section 3, the weighted average evidence with regard to motor rotor fault diagnosis at $1X$ frequency is obtained as below:

$$m(\{F_2\}) = 0.5442,$$
$$m(\{F_3\}) = 0.0006,$$
$$m(\{F_1,F_2\}) = 0.0773,$$
$$m(\{F_1,F_2,F_3\}) = 0.3780.$$

Then, the final fusion results for motor rotor fault diagnosis at $1X$ frequency are computed as follows:

$$m(\{F_2\}) = 0.9055,$$
$$m(\{F_3\}) = 0.0002,$$
$$m(\{F_1,F_2\}) = 0.0404,$$
$$m(\{F_1,F_2,F_3\}) = 0.0541.$$

6.2.2. Motor Rotor Fault Diagnosis at $2X$ Frequency

By carrying out the steps in Section 3, the weighted average evidence with respect to motor rotor fault diagnosis at $2X$ frequency is obtained as follows:

$$m(\{F_2\}) = 0.7387,$$
$$m(\{F_1,F_2,F_3\}) = 0.2613.$$

Afterwards, the final fusion results in terms of motor rotor fault diagnosis at $2X$ frequency are generated as below:

$$m(\{F_2\}) = 0.9822,$$
$$m(\{F_1,F_2,F_3\}) = 0.0178.$$

6.2.3. Motor Rotor Fault Diagnosis at 3X Frequency

By applying the steps in Section 3, the weighted average evidence with respect to motor rotor fault diagnosis at 3X frequency is obtained as follows:

$$m(\{F_1\}) = 0.3111,$$
$$m(\{F_2\}) = 0.4346,$$
$$m(\{F_1, F_2\}) = 0.2115,$$
$$m(\{F_1, F_2, F_3\}) = 0.0428.$$

Then, the final combination results for motor rotor fault diagnosis at 3X frequency are shown below:

$$m(\{F_1\}) = 0.3345,$$
$$m(\{F_2\}) = 0.6321,$$
$$m(\{F_1, F_2\}) = 0.0333,$$
$$m(\{F_1, F_2, F_3\}) = 0.0001.$$

From the experimental results as shown in Tables 10–12, it can be seen that the proposed method diagnoses the fault type as F_2, in accordance with Jiang et al.'s method [48].

Furthermore, the proposed method outperforms Jiang et al.'s method [48] in dealing with the uncertainty as shown in Figures 5–7, because by utilizing the proposed method, the belief degrees allocated to the target fault type F_2 at 1X frequency, 2X frequency and 3X frequency increase up to 90.55%, 98.22%, and 63.21%, respectively; however, by using Jiang et al.'s method [48], the belief degrees allocated to the target F_2 at 1X frequency, 2X frequency and 3X frequency are 88.61%, 96.21%, and 59.04%, respectively.

Additionally, by utilizing the proposed method, the uncertainty $\{F_1, F_2\}$ falls from 0.0582 to 0.0541, and the uncertainty $\{F_1, F_2, F_3\}$ falls from 0.0555 to 0.0404 at 1X frequency; the uncertainty $\{F_1, F_2, F_3\}$ decreased from 0.0371 to 0.0178 at 2X frequency; the uncertainty $\{F_1, F_2\}$ falls from 0.0651 to 0.0333, and the uncertainty $\{F_1, F_2, F_3\}$ drops from 0.0061 to 0.0001 at 3X frequency. As a result, the proposed method can diagnose motor rotor faults more accurately than the related work.

Table 10. Fusion results by using different combination methods at 1X frequency.

Method	$\{F_2\}$	$\{F_3\}$	$\{F_1, F_2\}$	$\{F_1, F_2, F_3\}$	Target
Jiang et al. [48]	0.8861	0.0002	0.0582	0.0555	F_2
Proposed method	0.9055	0.0002	0.0404	0.0541	F_2

Table 11. Fusion results by using different combination methods at 2X frequency.

Method	$\{F_2\}$	$\{F_1, F_2, F_3\}$	Target
Jiang et al. [48]	0.9621	0.0371	F_2
Proposed method	0.9822	0.0178	F_2

Table 12. Fusion results by using different combination methods at 3X frequency.

Method	$\{F_1\}$	$\{F_2\}$	$\{F_1, F_2\}$	$\{F_1, F_2, F_3\}$	Target
Jiang et al. [48]	0.3384	0.5904	0.0651	0.0061	F_2
Proposed method	0.3345	0.6321	0.0333	0.0001	F_2

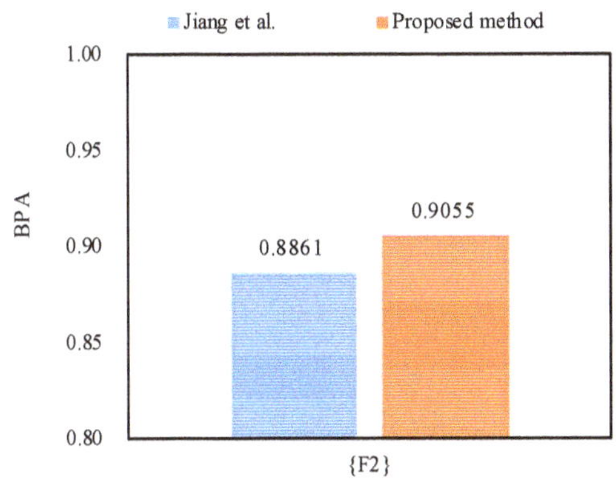

Figure 5. The comparison of the BPA of the target F_2 at 1X frequency.

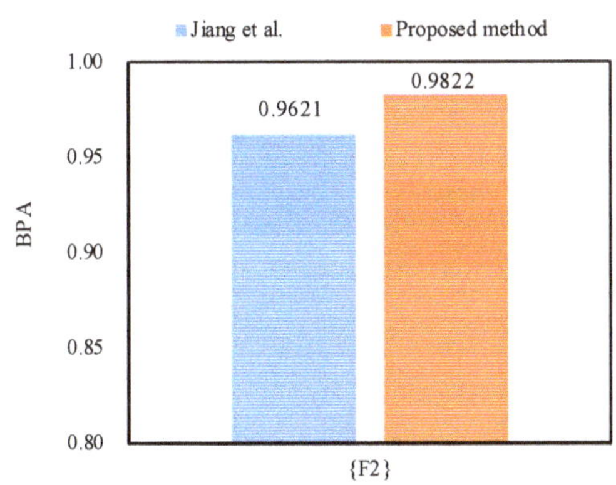

Figure 6. The comparison of the BPA of the target F_2 at 2X frequency.

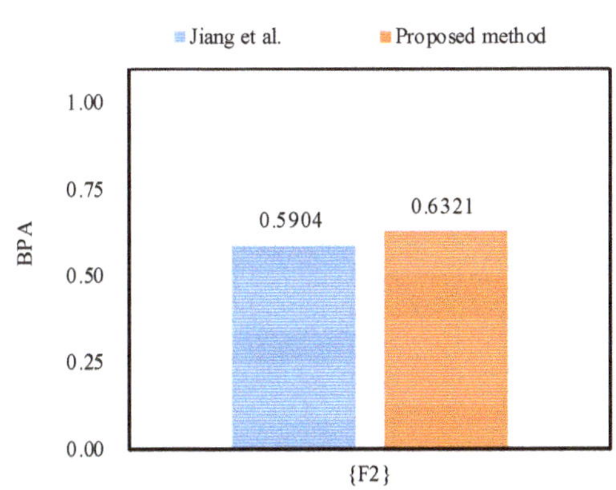

Figure 7. The comparison of the BPA of the target F_2 at 3X frequency.

7. Conclusions

In this paper, a weighted combination method for conflicting evidence in multi-sensor data fusion was proposed by combining the modified cosine similarity measure of the pieces of evidence with the belief entropy function. The proposed method was a kind of pretreatment of the bodies of evidence, which was effective to handle the conflicting pieces of evidence in a multi-sensor environment. A numerical example was illustrated to show the feasibility and effectiveness of the proposal. In addition, applications in data classification and motor rotor fault diagnosis were presented to validate the practicability of the proposed method, where it outperformed the related methods with better accuracy.

Author Contributions: F.X. contributed most of the work in this paper. B.Q contributed the experiments in this paper.

Funding: "This research was funded by the National Natural Science Foundation of China grant numbers 61672435, 61702427, 61702426, and the 1000-Plan of Chongqing by Southwest University grant number SWU116007."

Acknowledgments: The authors greatly appreciate the reviews' suggestions and the editor's encouragement.

Conflicts of Interest: The authors declare no conflict of interest.

References

1. Jin, X.B.; Sun, Y.X. Pei-Radman fusion estimation algorithm for multisensor system applied in state monitoring. *Lect. Notes Control Inf. Sci.* **2006**, *344*, 963–968.
2. Zadeh, L.A. Fuzzy sets. *Inf. Control* **1965**, *8*, 338–353. [CrossRef]
3. Mardani, A.; Jusoh, A.; Zavadskas, E.K. Fuzzy multiple criteria decision-making techniques and applications–Two decades review from 1994 to 2014. *Expert Syst. Appl.* **2015**, *42*, 4126–4148. [CrossRef]
4. Jiang, W.; Wei, B.; Liu, X.; Li, X.; Zheng, H. Intuitionistic fuzzy evidential power aggregation operator and its application in multiple criteria decision-making. *Int. J. Syst. Sci.* **2018**, *49*, 582–594. [CrossRef]
5. Dempster, A.P. Upper and lower probabilities induced by a multivalued mapping. *Ann. Math. Stat.* **1967**, *38*, 325–339. [CrossRef]
6. Shafer, G. A mathematical theory of evidence. *Technometrics* **1978**, *20*, 242. [CrossRef]
7. Jiang, W.; Chang, Y.; Wang, S. A method to identify the incomplete framework of discernment in evidence theory. *Math. Prob. Eng.* **2017**, *2017*, doi:10.1155/2017/7635972. [CrossRef]
8. Walczak, B.; Massart, D. Rough sets theory. *Chem. Intell. Lab. Syst.* **1999**, *47*, 1–16. [CrossRef]
9. Greco, S.; Matarazzo, B.; Slowinski, R. Rough sets theory for multicriteria decision analysis. *Eur. J. Oper. Res.* **2001**, *129*, 1–47. [CrossRef]
10. Yang, J.B.; Xu, D.L. Evidential reasoning rule for evidence combination. *Artif. Intell.* **2013**, *205*, 1–29. [CrossRef]
11. Fu, C.; Xu, D.L. Determining attribute weights to improve solution reliability and its application to selecting leading industries. *Ann. Oper. Res.* **2014**, *245*, 401–426. [CrossRef]
12. Zadeh, L.A. A note on Z-numbers. *Inf. Sci.* **2011**, *181*, 2923–2932. [CrossRef]
13. Kang, B.; Chhipi-Shrestha, G.; Deng, Y.; Hewage, K.; Sadiq, R. Stable Strategies Analysis Based on the Utility of Z-number in the Evolutionary Games. *Appl. Math. Comput.* **2018**, *324*, 202–217. [CrossRef]
14. Bian, T.; Zheng, H.; Yin, L.; Deng, Y. Failure mode and effects analysis based on D numbers and TOPSIS. *Qual. Reliab. Eng. Int.* **2018**, doi:10.1002/qre.2268. [CrossRef]
15. Xiao, F. A novel multi-criteria decision making method for assessing health-care waste treatment technologies based on D numbers. *Eng. Appl. Artif. Intell.* **2018**, *71*, 216–225. [CrossRef]
16. Xiao, F. An intelligent complex event processing with D numbers under fuzzy environment. *Math. Prob. Eng.* **2016**, *2016*. [CrossRef]
17. Deng, X.; Deng, Y. D-AHP method with different credibility of information. *Soft Comput.* **2018**. [CrossRef]
18. Gao, Y.; Ran, C.J.; Sun, X.J.; Deng, Z.L. Optimal and self-tuning weighted measurement fusion Kalman filters and their asymptotic global optimality. *Int. J. Adapt. Control Signal Process.* **2010**, *24*, 982–1004. [CrossRef]
19. Gao, Y.; Jia, W.J.; Sun, X.J.; Deng, Z.L. Self-tuning multisensor weighted measurement fusion Kalman filter. *IEEE Trans. Aerosp. Electron. Syst.* **2009**, *45*, 179–191.
20. Jin, X.B.; Dou, C.; Su, T.L.; Lian, X.F.; Shi, Y. Parallel irregular fusion estimation based on nonlinear filter for indoor RFID tracking system. *Int. J. Distrib. Sens. Netw.* **2016**, *2016*, 1–11. [CrossRef]
21. Zhou, X.; Hu, Y.; Deng, Y.; Chan, F.T.S.; Ishizaka, A. A DEMATEL-based completion method for incomplete pairwise comparison matrix in AHP. *Ann. Oper. Res.* **2018**. [CrossRef]

22. Xu, H.; Deng, Y. Dependent evidence combination based on Shearman coefficient and Pearson coefficient. *IEEE Access* **2018**, *6*, 11634–11640. [CrossRef]
23. Dutta, P. Uncertainty modeling in risk assessment based on Dempster–Shafer theory of evidence with generalized fuzzy focal elements. *Fuzzy Inf. Eng.* **2015**, *7*, 15–30. [CrossRef]
24. Liu, T.; Deng, Y.; Chan, F. Evidential supplier selection based on DEMATEL and game theory. *Int. J. Fuzzy Syst.* **2018**, *20*, 1321–1333. [CrossRef]
25. Denoeux, T. A k-nearest neighbor classification rule based on Dempster–Shafer theory. *IEEE Trans. Syst. Man Cybern.* **1995**, *25*, 804–813. [CrossRef]
26. Liu, Z.; Quan, P.; Dezert, J.; Han, J.W.; You, H. Classifier fusion with contextual reliability evaluation. *IEEE Trans. Cybern.* **2017**, *PP*, 1–14. [CrossRef] [PubMed]
27. Zheng, X.; Deng, Y. Dependence assessment in human reliability analysis based on evidence credibility decay model and IOWA operator. *Ann. Nuclear Energy* **2018**, *112*, 673–684. [CrossRef]
28. Xiao, F. A novel evidence theory and fuzzy preference approach-based multi-sensor data fusion technique for fault diagnosis. *Sensors* **2017**, *17*, 2504. [CrossRef] [PubMed]
29. Jiang, W.; Wang, S. An uncertainty measure for interval-valued evidences. *Int. J. Comput. Commun. Control* **2017**, *12*, 631–644. [CrossRef]
30. Xiao, F. An improved method for combining conflicting evidences Based on the similarity measure and belief function entropy. *Int. J. Fuzzy Syst.* **2017**, 1–11. [CrossRef]
31. Zheng, H.; Deng, Y. Evaluation method based on fuzzy relations between Dempster-Shafer belief structure. *Int. J. Intell. Syst.* **2017**, doi:10.1002/int.21956. [CrossRef]
32. Jiang, W.; Yang, T.; Shou, Y.; Tang, Y.; Hu, W. Improved evidential fuzzy c-means method. *J. Syst. Eng. Electron.* **2018**, *29*, 187–195.
33. Zadeh, L.A. A simple view of the Dempster–Shafer theory of evidence and its implication for the rule of combination. *AI Mag.* **1986**, *7*, 85–90.
34. Lefevre, E.; Colot, O.; Vannoorenberghe, P. Belief function combination and conflict management. *Inf. Fusion* **2002**, *3*, 149–162. [CrossRef]
35. Deng, X.; Jiang, W. An evidential axiomatic design approach for decision making using the evaluation of belief structure satisfaction to uncertain target values. *Int. J. Intell. Syst.* **2018**, *33*, 15–32. [CrossRef]
36. Jiang, W.; Hu, W. An improved soft likelihood function for Dempster-Shafer belief structures. *Int. J. Intell. Syst.* **2018**. [CrossRef]
37. Smets, P. The combination of evidence in the transferable belief model. *IEEE Trans. Pattern Anal. Mach. Intell.* **1990**, *12*, 447–458. [CrossRef]
38. Dubois, D.; Prade, H. Representation and combination of uncertainty with belief functions and possibility measures. *Comput. Intell.* **1988**, *4*, 244–264. [CrossRef]
39. Yager, R.R. On the Dempster–Shafer framework and new combination rules. *Inf. Sci.* **1987**, *41*, 93–137. [CrossRef]
40. Murphy, C.K. Combining belief functions when evidence conflicts. *Decis. Support Syst.* **2000**, *29*, 1–9. [CrossRef]
41. Deng, Y.; Shi, W.; Zhu, Z.; Liu, Q. Combining belief functions based on distance of evidence. *Decis. Support Syst.* **2004**, *38*, 489–493.
42. Jiang, W.; Wei, B.; Qin, X.; Zhan, J.; Tang, Y. Sensor data fusion based on a new conflict measure. *Math. Prob. Eng.* **2016**, *2016*. [CrossRef]
43. Deng, Y. Deng entropy. *Chaos Solitons Fractals* **2016**, *91*, 549–553. [CrossRef]
44. Khaleghi, B.; Khamis, A.; Karray, F.O.; Razavi, S.N. Multisensor data fusion: A review of the state-of-the-art. *Inf. Fusion* **2013**, *14*, 28–44. [CrossRef]
45. Niu, G.; Yang, B.S.; Pecht, M. Development of an optimized condition-based maintenance system by data fusion and reliability-centered maintenance. *Reliab. Eng. Syst. Saf.* **2010**, *95*, 786–796. [CrossRef]
46. Yunusa-Kaltungo, A.; Sinha, J.K. Sensitivity analysis of higher order coherent spectra in machine faults diagnosis. *Struct. Health Monit.* **2016**, *15*, 555–567. [CrossRef]
47. Yunusa-Kaltungo, A.; Sinha, J.K.; Nembhard, A.D. A novel fault diagnosis technique for enhancing maintenance and reliability of rotating machines. *Struct. Health Monit.* **2015**, *14*, 231–262. [CrossRef]
48. Jiang, W.; Xie, C.; Zhuang, M.; Shou, Y.; Tang, Y. Sensor data fusion with Z-numbers and its application in fault diagnosis. *Sensors* **2016**, *16*, 1509. [CrossRef] [PubMed]

49. Akselrod, D.; Sinha, A.; Kirubarajan, T. Information flow control for collaborative distributed data fusion and multisensor multitarget tracking. *IEEE Trans. Syst. Man Cybern. Part C* **2012**, *42*, 501–517. [CrossRef]
50. Dallil, A.; Oussalah, M.; Ouldali, A. Sensor fusion and target tracking using evidential data association. *IEEE Sens. J.* **2013**, *13*, 285–293. [CrossRef]
51. Kashanian, H.; Dabaghi, E. Feature dimension reduction of multisensor data fusion using principal component fuzzy analysis. *Int. J. Eng.* **2017**, *30*, 493–499.
52. Hernandez-Penaloza, G.; Belmonte-Hernandez, A.; Quintana, M.; Alvarez, F. A Multi-sensor Fusion Scheme to Increase Life Autonomy of Elderly People with Cognitive Problems. *IEEE Access* **2018**, *6*, 12775–12789. [CrossRef]
53. Santos, E.N.D.; Silva, M.J.D. Advanced image processing of wire-mesh sensor data for two-phase flow investigation. *IEEE Latin Am. Trans.* **2015**, *13*, 2269–2277. [CrossRef]
54. Mohammadi, A.; Yang, C.; Chen, Q.W. Attack detection/isolation via a secure multisensor fusion framework for cyberphysical systems. *Complexity* **2018**, *2018*, 1–8. [CrossRef]
55. Santi, F.; Pastina, D.; Bucciarelli, M. Estimation of ship dynamics with a multi-platform Radar imaging system. *IEEE Trans. Aerosp. Electron. Syst.* **2017**, *53*, 2769–2788. [CrossRef]
56. Geiß, C.; Thoma, M.; Pittore, M.; Wieland, M.; Dech, S.W.; Taubenbock, H. Multitask active learning for characterization of built environments with multisensor earth observation data. *IEEE J. Sel. Top. Appl. Earth Obs. Remote Sens.* **2017**, *PP*, 1–15.
57. Zhang, Q.; Li, M.; Deng, Y. Measure the structure similarity of nodes in complex networks based on relative entropy. *Phys. A Stat. Mech. Appl.* **2018**, *491*, 749–763. [CrossRef]
58. Jiang, W.; Wei, B.; Liu, X.; Li, X.; Zheng, H. Intuitionistic fuzzy power aggregation operator based on entropy and its application in decision making. *Int. J. Intell. Syst.* **2018**, *33*, 49–67. [CrossRef]
59. Qian, J.; Guo, X.; Deng, Y. A novel method for combining conflicting evidences based on information entropy. *Appl. Intell.* **2017**, *46*, 876–888. [CrossRef]

sensors

MDPI

Article

A Closed-Form Error Model of Straight Lines for Improved Data Association and Sensor Fusing

Volker Sommer

Department of Computer Science and Media, Beuth University of Applied Sciences, Luxemburger Str. 10, D-13353 Berlin, Germany; sommer@beuth-hochschule.de; Tel.: +49-30-4504-5154

Received: 12 March 2018; Accepted: 13 April 2018; Published: 17 April 2018

Abstract: Linear regression is a basic tool in mobile robotics, since it enables accurate estimation of straight lines from range-bearing scans or in digital images, which is a prerequisite for reliable data association and sensor fusing in the context of feature-based SLAM. This paper discusses, extends and compares existing algorithms for line fitting applicable also in the case of strong covariances between the coordinates at each single data point, which must not be neglected if range-bearing sensors are used. Besides, in particular, the determination of the covariance matrix is considered, which is required for stochastic modeling. The main contribution is a new error model of straight lines in closed form for calculating quickly and reliably the covariance matrix dependent on just a few comprehensible and easily-obtainable parameters. The model can be applied widely in any case when a line is fitted from a number of distinct points also without a priori knowledge of the specific measurement noise. By means of extensive simulations, the performance and robustness of the new model in comparison to existing approaches is shown.

Keywords: linear regression; covariance matrix; data association; sensor fusing; SLAM

1. Introduction

Contour points acquired by active sensors using sonar, radar or LiDAR [1], or extracted from image data [2,3], are a key source of information for mobile robots in order to detect obstacles or to localize themselves in known or unknown environments [4,5]. For this purpose, often, geometric features are extracted from raw data since in contrast to detailed contours, features are uniquely described just by a limited set of parameters, and their extraction works as additional filtering in order to improve reliability when dealing with sensor noise and masking [6]. However, the performance of feature-based localization or SLAM strongly depends on exact the determination of a feature vector y from measured raw data. Moreover, especially for data association, as well as for sensor fusing, not only the feature parameters are needed, but also a reliable estimation of their uncertainty is required. Generally, in the case of non-linear multi-sensor fusing, likelihood-based models can be applied (see [7]), which employ Bayesian filtering [8] or the calculation of entropies [9] to quantify uncertain information. Alternatively, especially for localization and map building, sensor fusing often is achieved with a Kalman filter (EKF). For this purpose, the covariance matrix R is required, which encapsulates the variances of the single elements in y and their dependencies.

This will be obvious if one looks at the standard algorithm for updating an estimated system state $\hat{x}$ typically by means of EKF; compare [10–12]: new measurements y are plausible if their deviations from expected measurements $\hat{y} = h(\hat{x})$ dependent on the in general non-linear measurement model $h(\hat{x})$ are within a limited range. For exact calculation of this limit, usually the Mahalanobis metric is applied (see [11,13]), which considers the covariance matrix S of the innovation $\nu = y - \hat{y}$ with $S = R + H \cdot \hat{P} \cdot H^T$ dependent on R, the covariance matrix $\hat{P}$ of the system state and using $H = \nabla h(\hat{x})$. A new measurement y will be considered to relate to an already known feature vector $\hat{y}$ if its distance is below a given threshold r_{th} with $\nu^T S^{-1} \nu < r_{th}^2$. Only in this case, the system state vector $\hat{x}$ can be updated by means of $\Delta\hat{x} = K \cdot \nu$ using the Kalman gain $K = \hat{P} \cdot H^T \cdot S$, again depending on the covariance matrix R of the measurements, while otherwise $\hat{x}$ and $\hat{P}$ are expanded by the new feature.

Thus, for reliable map building, errors in the step of data association should be strictly avoided by means of exact knowledge of the covariance matrix at each measurement, since otherwise, multiple versions of certain features would be included in the map, while other features erroneously are ignored.

Particularly in artificial environments, straight lines in a plane are frequently used as features, since these are defined by just two parameters and can be clearly and uniquely determined. In contrast to point features, lines in images are almost independent of illumination and perspective, and a number of measurements can be taken along their length to localize them accurately and to distinguish them from artifacts [14]. Moreover, already, a single line enables a robot to determine its orientation and perpendicular distance, which clearly improves localization accuracy. Thus, many tracking systems have been proposed based on line features, either using range-bearing scans [15,16] or applying visual servoing (see [17,18]), and also, recently, this approach has been successfully implemented [19–21]. However, due to missing knowledge of the covariance matrix, for data association, often, suboptimal solutions like the Euclidean distance in Hough space [15] or other heuristics are used [22].

Obviously, fitting data to a straight line is a well-known technique, addressed in a large number of papers [23–25] and textbooks [26–28]. In [29], a recent overview of algorithms in this field is outlined. As shown in [30,31], if linear regression is applied to data with uncertainties in the x- and y-direction, always both coordinates must be considered as random variables. In [32], Arras and Siegwart suggest an error model for range-bearing sensors including a covariance matrix, affected exclusively by noise in the radial direction. Pfister et al. introduce weights into the regression algorithm in order to determine the planar displacement of a robot from range-bearing scans [33]. In [34], a maximum likelihood approach is used to formulate a general strategy for estimating the best fitted line from a set of non-uniformly-weighted range measurements. Furthermore, merging of lines and approximating the covariance matrix from an iterative approach is considered. In [30], Krystek and Anton point out that the weighting factors of the single measurements depend on the orientation of a line, which therefore can only be determined numerically. This concept has been later extended to the general case with covariances existing between the coordinates of each data point [35].

Since linear regression is sensitive with respect to outliers, split-and-merge algorithms must be applied in advance, if a contour consists of several parts; see [36,37]. In cases of strong interference, straight lines can still be identified by Hough-transformation (compare [38–40]), or alternatively, RANSAC algorithms can be applied; see [41,42]. Although these algorithms work reliably, exact determination of line parameters and estimating their uncertainties still requires linear regression [43].

In spite of a variety of contributions in this field, a straightforward, yet accurate algorithm for determining the covariance matrix of lines reliably, quickly and independently of the a priori mostly unknown measurement noise is missing. In Section 4, such a model in closed-form is proposed depending on just a few clearly-interpretable and easily-obtainable parameters. Besides its low complexity and great clarity, the main advantage of the covariance matrix in closed form results from the fact that it can be calculated from the same data points as used for line fitting without the need to provide additional reliability information of the measurements, which in many cases is not available.

Beforehand, in the next two paragraphs, existing methods for the linear regression and calculation of the covariance matrix are reviewed with certain extensions focusing on the usage of range-bearing sensors, which cause strong covariances between the x- and y-coordinates. Based on these theoretical foundations, Section 5 exhibits detailed simulation results in order to compare the precision and robustness of the presented algorithms.

2. Determination of the Accurate Line Parameters

In 2D-space, each straight line is uniquely described by its perpendicular distance d from the origin and by the angle ϕ between the positive x-axis and this normal line; see Figure 1. In order to determine these two parameters, the mean squared error *MSE* considering the perpendicular distances of N measurement points from the fitted line needs to be minimized. For this purpose, each perpendicular distance ρ_i of point i is calculated either from polar or with $x_i = r_i \cos\theta_i$ and $y_i = r_i \sin\theta_i$ alternatively in Cartesian coordinates as:

$$\rho_i = d_i - d = r_i \cos(\theta_i - \phi) - d = x_i \cos\phi + y_i \sin\phi - d \quad (1)$$

Then, *MSE* is defined as follows dependent on ϕ and d:

$$MSE(\phi, d) = \sum_{i=1}^{N} (s_i \rho_i)^2 \quad (2)$$

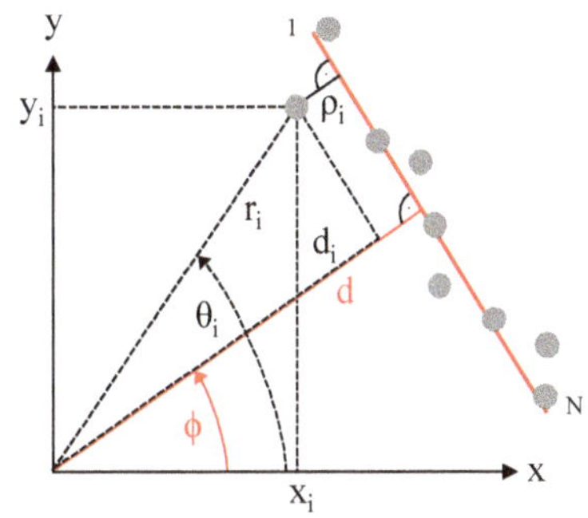

Figure 1. Parameters of measured raw data and a fitted straight line.

In (2), optional scaling values s_i are included in order to consider the individual reliability of each measurement point. By calculating the derivatives of (2) with respect to ϕ and d and setting both to zero, the optimum values of these parameters can be analytically derived assuming all s_i to be constant, i.e., independent of ϕ and d. The solution has been published elsewhere (compare [32]), and in the Appendix of this paper, a straightforward derivation is sketched, yielding for ϕ and d:

$$\phi = \frac{1}{2} \cdot \text{atan2}\left(-2\sigma_{xy}, \sigma_y^2 - \sigma_x^2\right) \quad (3)$$

$$d = \bar{x}\cos\phi + \bar{y}\sin\phi \quad (4)$$

The function atan2() means the four quadrant arc tangent, which calculates ϕ always in the correct range. If d becomes negative, its modulus must be taken, and the corresponding ϕ has to be altered by plus or minus π. In these equations, $\bar{x}$ and $\bar{y}$ denote the mean values of all N measurements x_i and y_i, while σ_x^2, σ_y^2 and σ_{xy} denote the variances and the covariance:

$$\sigma_x^2 = \frac{1}{N}\sum_{i=1}^{N} w_i\,(x_i - \bar{x})^2 \quad (5)$$

$$\sigma_y^2 = \frac{1}{N}\sum_{i=1}^{N} w_i\,(y_i - \bar{y})^2 \quad (6)$$

$$\sigma_{xy} = \frac{1}{N}\sum_{i=1}^{N} w_i\,(x_i - \bar{x})\,(y_i - \bar{y}) \quad (7)$$

$$\bar{x} = \frac{1}{N}\sum_{i=1}^{N} w_i x_i \tag{8}$$

$$\bar{y} = \frac{1}{N}\sum_{i=1}^{N} w_i y_i \tag{9}$$

In (5)–(9), normalized weighting factors w_i are used with $\frac{1}{N}\sum_{i=1}^{N} w_i = 1$ and $0 \leq w_i \leq 1$, calculated dependent on the chosen scaling values s_i:

$$w_i = \frac{s_i^2}{\frac{1}{N}\sum_{i=1}^{N} s_i^2} \tag{10}$$

As pointed out in [35], for accurate line matching, the scaling values s_i must not be assumed to be constant since in general, they depend on ϕ. This can be understood from Figure 2, which shows for one measurement point i the error ellipse spanned by the standard deviations $\sigma_{x,i}$ and $\sigma_{y,i}$, while the rotation of the ellipse is caused by the covariance $\sigma_{xy,i}$.

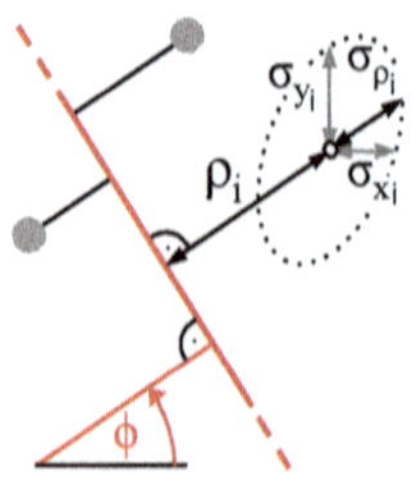

Figure 2. Optimum setting of weighting parameter for each data point.

Apparently, as a measure of confidence, only the deviation $\sigma_{\rho,i}$ perpendicular to the line is relevant, while the variance of any data point parallel to the fitted line does not influence its reliability. Thus, the angle ϕ given in (3) will only be exact, if the error ellipse equals a circle, which means that all measurements exhibit the same standard deviations in the x- as in the y-direction, and no covariance exists. Generally, in order to determine optimum line parameters with arbitrary variances and covariance of each measurement i, in Equation (2) the inverse of $\sigma_{\rho,i}$ dependent on ϕ has to be used as scaling factor s_i, yielding:

$$MSE(\phi) = \sum_{i=1}^{N} \frac{\rho_i^2(\phi)}{\sigma_{\rho,i}^2(\phi)} \tag{11}$$

In this formula, which can only be solved numerically, the variance $\sigma_{\rho,i}^2$ needs to be calculated dependent on the covariance matrix of each measurement point i. In the case of line fitting from range-bearing scans, the covariance matrix $\underline{R}_{r\theta,i}$ can be modeled as a diagonal matrix since both parameters r_i and θ_i are measured independently, and thus, their covariance $\sigma_{r\theta,i}$ equals zero:

$$\underline{R}_{r\theta,i} = \begin{pmatrix} \sigma_{r,i}^2 & 0 \\ 0 & \sigma_{\theta,i}^2 \end{pmatrix} \tag{12}$$

Typically, this matrix may also be considered as constant, thus independent of index i, assuming that all measured radii and angles are affected by the same noise, i.e., $\underline{R}_{r\theta,i} \approx \underline{R}_{r\theta}$.

With known variances $\sigma_{r,i}^2$ and $\sigma_{\theta,i}^2$ and for a certain ϕ, now $\sigma_{\rho,i}^2$ is determined by evaluating the relation between ρ_i and the distances d_i of each data point with $1 \leq i \leq N$. According to (1) and with the distance d written as the mean of all d_i, it follows:

$$\rho_i = d_i - \frac{1}{N}\sum_{j=1}^{N} d_j = \left(\frac{N-1}{N}\right) d_i - \frac{1}{N} \sum_{\substack{j=1 \\ (j \neq i)}}^{N} d_j \tag{13}$$

Since noise-induced variations of all distances d_i are uncorrelated with each other, now the variance $\sigma^2_{\rho,i}$ is calculated by means of summing over all variances $\sigma^2_{d,i}$:

$$\sigma^2_{\rho,i} = \left(\frac{N-1}{N}\right)^2 \sigma^2_{d,i} + \frac{1}{N^2} \sum_{\substack{j=1 \\ (j \neq i)}}^{N} \sigma^2_{d,j} \tag{14}$$

In order to derive $\sigma^2_{d,i}$, changes of d_i with respect to small deviations of r_i and θ_i from their expected values $\bar{r}_i$ and $\bar{\theta}_i$ are considered with $d_i = \bar{d}_i + \Delta d_i$, $r_i = \bar{r}_i + \Delta r_i$ and with $\theta_i = \bar{\theta}_i + \Delta\theta_i$:

$$\Delta d_i = \Delta d_i^r + \Delta d_i^\theta \tag{15}$$

The terms on the right side of (15) can be determined independently of each other, since Δr_i and $\Delta\theta_i$ are assumed to be uncorrelated. With $d_i = r_i \cdot \cos(\theta_i - \phi)$, it follows:

$$\Delta d_i^r = \Delta r_i \cdot \cos(\bar{\theta}_i - \phi) \tag{16}$$

and:

$$\Delta d_i^\theta = \bar{r}_i \left[\cos(\bar{\theta}_i - \phi + \Delta\theta_i) - \cos(\bar{\theta}_i - \phi)\right] \approx -\bar{r}_i \left[\frac{\Delta\theta_i^2}{2} \cos(\bar{\theta}_i - \phi) + \Delta\theta_i \sin(\bar{\theta}_i - \phi)\right] \tag{17}$$

In the last line, the addition theorem was applied for $\cos(\bar{\theta}_i - \phi + \Delta\theta_i)$, and for small variations, the approximations $\cos(\Delta\theta_i) \approx 1 - \frac{\Delta\theta_i^2}{2}$ and $\sin(\Delta\theta_i) \approx \Delta\theta_i$ are valid.

The random variables Δr_i and $\Delta\theta_i$ are assumed to be normally distributed with variances $\sigma^2_{r,i}$ and $\sigma^2_{\theta,i}$. Thus, the random variable $\Delta\theta_i^2$ exhibits a χ^2-distribution with variance $2(\sigma^2_{\theta,i})^2$ (see [44]), and the variance of d_i is calculated from (15)–(17) as the weighted sum with $\bar{r}_i$ and $\bar{\theta}_i$ approximately replaced by r_i and θ_i, respectively:

$$\sigma^2_{d,i} = \left(\sigma^2_{r,i} + \frac{(\sigma^2_{\theta,i})^2}{2}\right) \cos^2(\theta_i - \phi) + \sigma^2_{\theta,i} \sin^2(\theta_i - \phi) \tag{18}$$

When applying this algorithm, a one-dimensional minimum search of *MSE* according to (11) needs to be executed, yielding the optimum ϕ of the straight line. For this purpose, $\sigma^2_{\rho,i}$ is inserted from (14) considering (18) and ρ_i is determined according to (1) by calculating d from (4) and (8)–(10) with $s_i = 1/\sigma_{\rho,i}$.

Obviously, numerical line fitting can also be accomplished if measurements are available in Cartesian coordinates x_i and y_i. In this case, the covariance matrix $\underline{R}_{xy,i}$ of each measurement point must be known, defined as:

$$\underline{R}_{xy,i} = \begin{pmatrix} \sigma^2_{x,i} & \sigma_{xy,i} \\ \sigma_{xy,i} & \sigma^2_{y,i} \end{pmatrix} \tag{19}$$

Furthermore, the partial derivatives of d_i according to (1) with respect to x_i and y_i need to be calculated:

$$\underline{J}_{d,i} = \begin{pmatrix} \frac{\partial d_i}{\partial x_i} & \frac{\partial d_i}{\partial y_i} \end{pmatrix} = \begin{pmatrix} \cos\phi & \sin\phi \end{pmatrix} \tag{20}$$

Then, $\sigma^2_{d,i}$ follows dependent on $\underline{R}_{xy,i}$ and $\underline{J}_{d,i}$:

$$\sigma^2_{d,i} = \underline{J}_{d,i} \cdot \underline{R}_{xy,i} \cdot (\underline{J}_{d,i})^T = \sigma^2_{x,i} \cos^2\phi + \sigma_{xy,i} \sin\phi\cos\phi + \sigma^2_{y,i} \sin^2\phi \tag{21}$$

If raw data stem from a range-bearing scan, $\underline{R}_{xy,i}$ can be calculated from $\underline{R}_{r\theta,i}$ by exploiting the known dependencies between the polar and Cartesian plane. For this purpose, the Jacobian matrix $\underline{J}_{xy,i}$ is determined:

$$\underline{J}_{xy,i} = \begin{pmatrix} \frac{\partial x_i}{\partial r_i} & \frac{\partial x_i}{\partial \theta_i} \\ \frac{\partial y_i}{\partial r_i} & \frac{\partial y_i}{\partial \theta_i} \end{pmatrix} = \begin{pmatrix} \cos\theta_i & -r_i \sin\theta_i \\ \sin\theta_i & r_i \cos\theta_i \end{pmatrix} \tag{22}$$

Then, the covariance matrix $\underline{R}_{xy,i}$ will depend on $\underline{R}_{r\theta,i}$, if small deviations from the mean value of the random variables r_i and θ_i and a linear model are assumed:

$$\underline{R}_{xy,i} = \underline{J}_{xy,i} \cdot \underline{R}_{r\theta,i} \cdot (\underline{J}_{xy,i})^T \tag{23}$$

According to (23), generally a strong covariance $\sigma_{xy,i}$ in $\underline{R}_{xy,i}$ must be considered, if measurements are taken by range-bearing sensors.

By means of applying (21)–(23) instead of (18) for searching the minimum of *MSE* dependent on ϕ, the second order effect regarding $\Delta\theta_i$ is neglected. This yields almost the same formula as given in [35], though the derivation differs, and in [35], additionally, the variance of d is ignored assuming $\sigma^2_{\rho,i} = \sigma^2_{d,i}$, which according to (14) is only asymptotically correct for large N.

Finally, it should be noted that the numerical determination of ϕ according to (11) means clearly more complexity compared to the straightforward solution according to Equation (3). Later, in Section 5, it will be analyzed under which conditions this additional computational effort actually is required.

3. Analytic Error Models of Straight Lines

In the literature, several methods are described to estimate errors of ϕ and d and their mutual dependency. Thus, the covariance matrix $R_{d\phi}$ must be known, defined as:

$$\underline{R}_{d\phi} = \begin{pmatrix} \sigma^2_d & \sigma_{d\phi} \\ \sigma_{d\phi} & \sigma^2_\phi \end{pmatrix} \tag{24}$$

For this purpose, a general method in nonlinear parameter estimation is the calculation of the inverse Hessian matrix at the minimum of *MSE*. Details can be found in [30,35], while in [45], it is shown that this procedure may exhibit numerical instability. In Section 5, results using this method are compared with other approaches.

Alternatively, in [32,46], an analytic error model is proposed based on fault analysis of the line parameters. In this approach, the effect of variations of each single measurement point defined by $\underline{R}_{xy,i}$ with respect to the covariance matrix of the line parameters $\underline{R}_{d\phi}$ is considered, based on (3) and (4). Thereto, the Jacobian matrix $\underline{J}_{d\phi,i}$ with respect to x_i and y_i is determined, defined as:

$$\underline{J}_{d\phi,i} = \begin{pmatrix} \frac{\partial d}{\partial x_i} & \frac{\partial d}{\partial y_i} \\ \frac{\partial \phi}{\partial x_i} & \frac{\partial \phi}{\partial y_i} \end{pmatrix} \tag{25}$$

With this matrix, the contribution of a single data point i to the covariance matrix between d and ϕ can be written as:

$$\underline{R}_{d\phi,i} = \underline{J}_{d\phi,i} \cdot \underline{R}_{xy,i} \cdot \underline{J}^T_{d\phi,i} \tag{26}$$

For determining the partial derivatives of d in (25), Equation (4) is differentiated after expanding it by (8) and (9), yielding:

$$\frac{\partial d}{\partial x_i} = w_i \frac{\cos\phi}{N} + (\bar{y}\cos\phi - \bar{x}\sin\phi)\frac{\partial\phi}{\partial x_i} \tag{27}$$

$$\frac{\partial d}{\partial y_i} = w_i \frac{\sin\phi}{N} + (\bar{y}\cos\phi - \bar{x}\sin\phi)\frac{\partial\phi}{\partial y_i} \tag{28}$$

Differentiating ϕ according to (3) with respect to x_i gives the following expression with $u = -2\sigma_{xy}$ and $v = \sigma_y^2 - \sigma_x^2$:

$$\frac{\partial\phi}{\partial x_i} = \frac{1}{2(u^2+v^2)}\left(\frac{\partial u}{\partial x_i}v - \frac{\partial v}{\partial x_i}u\right) \tag{29}$$

The partial derivation of u in (29) is calculated after expanding it with (7) and (8) as:

$$\frac{\partial u}{\partial x_i} = -\frac{2}{N}\cdot\frac{\partial}{\partial x_i}\left(\sum_{i=1}^{N} w_i x_i y_i - \bar{y}\sum_{i=1}^{N} w_i x_i\right) = -\frac{2w_i}{N}(y_i - \bar{y}) \tag{30}$$

while partial derivation of v with (5), (6) and (8) yields:

$$\frac{\partial v}{\partial x_i} = -\frac{1}{N}\cdot\frac{\partial}{\partial x_i}\left(\sum_{i=1}^{N} w_i x_i^2 - \frac{1}{N}\left(\sum_{i=1}^{N} w_i x_i\right)^2\right) = -\frac{2w_i}{N}(x_i - \bar{x}) \tag{31}$$

Finally, after substituting all terms with u and v in (29), it follows:

$$\frac{\partial\phi}{\partial x_i} = w_i \frac{\left(\sigma_x^2 - \sigma_y^2\right)(y_i - \bar{y}) - 2\sigma_{xy}(x_i - \bar{x})}{N\left(\left(\sigma_x^2 - \sigma_y^2\right)^2 + 4\sigma_{xy}^2\right)} \tag{32}$$

Correspondingly, for the partial derivative of ϕ with respect to y_i, the following result is obtained:

$$\frac{\partial\phi}{\partial y_i} = w_i \frac{\left(\sigma_x^2 - \sigma_y^2\right)(x_i - \bar{x}) + 2\sigma_{xy}(y_i - \bar{y})}{N\left(\left(\sigma_x^2 - \sigma_y^2\right)^2 + 4\sigma_{xy}^2\right)} \tag{33}$$

Now, after inserting (27), (28), (32) and (33) into (25), the covariance matrix of d and ϕ (24) is calculated by summing over all N data points since the noise contributions of the single measurements can be assumed to be stochastically independent of each other:

$$\underline{R}_{d\phi} = \sum_{i=1}^{N} \underline{R}_{d\phi,i} = \sum_{i=1}^{N} \underline{J}_{d\phi,i}\cdot\underline{R}_{xy,i}\cdot\underline{J}_{d\phi,i}^{T} \tag{34}$$

Equation (34) enables an exact calculation of the variances σ_d^2, σ_ϕ^2 and of the covariance $\sigma_{d\phi}$ as long as the deviations of the measurements stay within the range of a linear approach, and as long as Equations (3) and (4) are valid. In contrast to the method proposed in [35], no second derivatives and no inversion of the Hessian matrix are needed, and thus, more stable results can be expected.

However, both algorithms need some computational effort especially for a large number of measurement points. Moreover, they do not allow one to understand the effect of changing parameters on $R_{d\phi}$, and these models can only be applied, if for each data point, the covariance matrix $\underline{R}_{xy,i}$ is available. Unfortunately, for lines extracted from images, this information is unknown, and also, in the case of using range-bearing sensors, only a worst case estimate of σ_r is given in the data sheet, while σ_θ is ignored.

4. Closed-Form Error Model of a Straight Line

In this section, a simplified error model in closed form is deduced, which enables a fast, clear and yet, for most applications, sufficiently accurate calculation of the covariance matrix $\underline{R}_{d\phi}$ in any case when line parameters d and ϕ have been determined from a number of discrete data points.

Thereto, first, the expected values of the line parameters d and ϕ, denoted as $\bar{d}$ and $\bar{\phi}$, are assumed to be known according to the methods proposed in Section 2 with $\bar{d} \approx d$ and $\bar{\phi} \approx \phi$. Besides, for modeling the small deviation of d and ϕ, the random variables Δd and $\Delta\phi$ are introduced. Thus, with $d = \bar{d} + \Delta d$ and $\phi = \bar{\phi} + \Delta\phi$, it follows for the variances and the covariance:

$$\sigma_d^2 = \sigma_{\Delta d}^2 \qquad \sigma_\phi^2 = \sigma_{\Delta\phi}^2 \qquad \sigma_{d\phi} = \sigma_{\Delta d\Delta\phi} \tag{35}$$

Next, Δd and $\Delta\phi$ shall be determined dependent on a random variation of any of the N measured data points. For this purpose, Figure 3 is considered, which shows the expected line parameters and the random variables Δd and $\Delta\phi$.

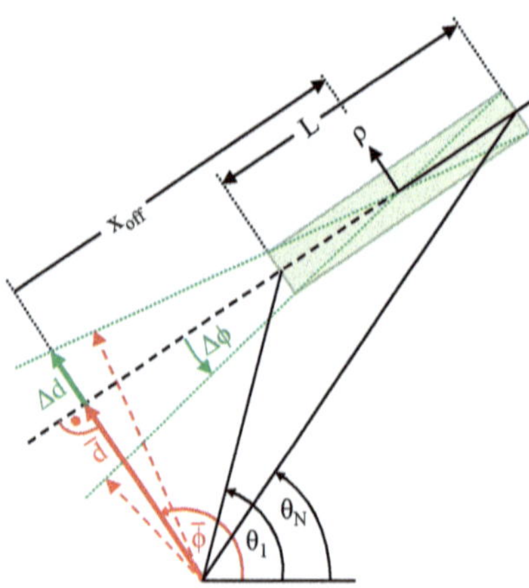

Figure 3. Dependency between Δd, $\Delta\phi$ and geometric parameters.

In order to derive expressions for Δd and $\Delta\phi$ depending on the random variables ρ_i, Figure 4 shows an enlargement of the rectangular box depicted in Figure 3 along the direction of the line $\tilde{x}$.

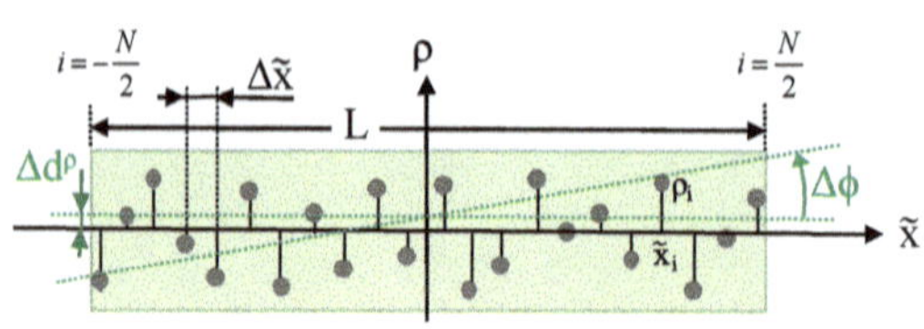

Figure 4. Details of Figure 3 with the deviation of data points along the axis $\tilde{x}$.

First, the effect of variations of any ρ_i on $\Delta\phi$ is considered. Since $\Delta\phi$ is very small, this angle may be replaced by its tangent, which defines the slope Δm of the line with respect to the direction $\tilde{x}$. Here, only ρ_i is considered as random variable, but not $\tilde{x}_i$. Thus, the standard formula for the slope of a regression line can be applied (see, e.g., [26] Chapter 2), which will minimize the mean squared distance in the direction of ρ, if all $\tilde{x}_i$ are assumed to be exactly known:

$$\Delta\phi \approx \tan(\Delta\phi) = \Delta m = \frac{\sigma_{\rho\tilde{x}}}{\sigma_{\tilde{x}}^2} = \frac{\sum_i \rho_i \cdot \tilde{x}_i}{\sum_i \tilde{x}_i^2} \tag{36}$$

Now, in order to calculate the variance of $\Delta\phi$, a linear relation between $\Delta\phi$ and each ρ_i is required, which is provided by the first derivation of (36) with respect to ρ_i:

$$\frac{\partial\Delta\phi}{\partial\rho_i} = \frac{\tilde{x}_i}{\sum_i \tilde{x}_i^2} \tag{37}$$

Then, the variance of $\Delta\phi$ dependent on the variance of ρ_i can be specified. From (37), it follows:

$$\sigma_{\Delta\phi,i}^2 = \sigma_{\rho,i}^2 \cdot \left(\frac{\partial\Delta\phi}{\partial\rho_i}\right)^2 = \sigma_{\rho,i}^2 \cdot \frac{\tilde{x}_i^2}{\left(\sum_i \tilde{x}_i^2\right)^2} \tag{38}$$

If $\sigma_{\rho,i}^2$ is assumed to be approximately independent of i, it may be replaced by σ_ρ^2 and can be estimated from (2) with ρ_i taken from (1) and setting all s_i to $1/N$:

$$\sigma_{\rho,i}^2 \approx \sigma_\rho^2 = \frac{1}{N}\sum_{i=1}^{N} \rho_i(\phi, d)^2 \tag{39}$$

It should be noted that for a bias-free estimation of σ_ρ^2 with (39), the exact line parameters ϕ and d must be used in (1), which obviously are not available. If instead, estimated line parameters according to Section 2 are taken, e.g., by applying (3) and (4), calculated from the same data as used in (39), an underestimation of σ_ρ^2 especially for small N can be expected, since ϕ and d are determined by minimizing the variance of ρ of these N data points. This is referred to later.

Next, from (38), the variance of $\Delta\phi$ results as the sum over all N data points, since all ρ_i are independent of each other:

$$\sigma_{\Delta\phi}^2 = \sum_i \sigma_{\Delta\phi,i}^2 \approx \sigma_\rho^2 \cdot \frac{\sum_i \tilde{x}_i^2}{\left(\sum_i \tilde{x}_i^2\right)^2} = \sigma_\rho^2 \cdot \frac{1}{\sum_i \tilde{x}_i^2} \tag{40}$$

Equations (40) with (35) and (39) enables an exact calculation of σ_ϕ^2 dependent on the N data points of the line.

However, from (40), a straightforward expression can be derived, which is sufficiently accurate in most cases and enables a clear understanding of the influencing parameters on σ_ϕ^2; compare Section 5. For this purpose, according to Figure 3, the length L of a line segment is determined from the perpendicular distance d and from the angles θ_1 and θ_N of the first and N-th data point, respectively:

$$L = d \cdot |\tan(\phi - \theta_N) - \tan(\phi - \theta_1)| \tag{41}$$

Furthermore, a constant spacing $\Delta\tilde{x}$ between adjacent data points is assumed:

$$\Delta\tilde{x} \approx \frac{L}{N-1}. \tag{42}$$

Applying this approximation, the sum over all squared $\tilde{x}_i$ can be rewritten, yielding for even N as depicted in Figure 4:

$$\sum_i \tilde{x}_i^2 \approx 2 \cdot \sum_{i=1}^{N/2} \left[\frac{\Delta \tilde{x}}{2} (2i-1) \right]^2 = \Delta \tilde{x}^2 \cdot \sum_{i=1}^{N/2} \frac{(2i-1)^2}{2} \tag{43}$$

The last sum can be transformed into closed form as:

$$\sum_{i=1}^{N/2} \frac{(2i-1)^2}{2} = \frac{\frac{N}{2}\left(4\left(\frac{N}{2}\right)^2 - 1\right)}{6} = \frac{N(N^2-1)}{12} \tag{44}$$

With N odd, the sum must be taken twice from 1–$\frac{N-1}{2}$, since in this case, the central measurement point has no effect on $\sigma^2_{\Delta\phi,i}$, yielding:

$$\sum_i \tilde{x}_i^2 \approx 2 \cdot \sum_{i=1}^{\frac{(N-1)}{2}} [\Delta \tilde{x} \cdot i]^2 = \Delta \tilde{x}^2 \cdot \sum_{i=1}^{\frac{(N-1)}{2}} 2 \cdot i^2 \tag{45}$$

Again, the last sum can be written in closed form, which gives the same result as in (44):

$$\sum_{i=1}^{\frac{(N-1)}{2}} 2 \cdot i^2 = \frac{\frac{N-1}{2}\left(\frac{N-1}{2}+1\right)\left(2 \cdot \frac{N-1}{2}+1\right)}{3} = \frac{N(N^2-1)}{12} \tag{46}$$

Finally, by substituting (43) with (44) or (45) with (46) into (40) and regarding (35), as well as (42), a simple analytic formula for calculating the variance of ϕ is obtained, just depending on L, N and the variance of ρ:

$$\sigma_\phi^2 \approx \sigma_\rho^2 \cdot \frac{12}{L^2 \cdot N} \cdot \frac{N-1}{N+1} \overset{N \gg 1}{\approx} \sigma_\rho^2 \cdot \frac{12}{L^2 \cdot N} \tag{47}$$

The last simplification in (47) overestimates σ_ϕ^2 a little bit for small N. Interestingly, this error compensates quite well for a certain underestimation of σ_ρ^2 according to (39), assuming that the line parameters ϕ and d are determined from the same data as σ_ρ^2; see Section 5.

Next, in order to deduce the variance σ_d^2, again, Figure 3 is considered. Apparently, the first part of the random variable Δd is strongly correlated with $\Delta\phi$ since any mismatch in ϕ is transformed into a deviation Δd by means of the geometric offset x_{off} with:

$$\Delta d^\phi = -x_{off} \cdot \Delta\phi \tag{48}$$

Actually, with a positive value for x_{off}, as depicted in Figure 3 the correlation between Δd and $\Delta\phi$ becomes negative, since positive values of $\Delta\phi$ correspond to negative values of Δd. According to Figure 3, x_{off} is determined from ϕ and d, as well as from θ_1 and θ_N:

$$x_{off} = \frac{d}{2} \cdot [\tan(\phi - \theta_N) + \tan(\phi - \theta_1)] \tag{49}$$

Alternatively, x_{off} can be taken as the mean value from all N data points of the line segment:

$$x_{off} = \frac{d}{N} \cdot \sum_{i=1}^{N} \tan(\phi - \theta_i) \tag{50}$$

Nevertheless, it should be noted that Δd is not completely correlated with $\Delta\phi$, since also in the case $x_{off} = 0$, the error Δd will not be zero.

Indeed, as a second effect, each single ρ_i has a direct linear impact on the variable Δd. For this purpose, in Figure 4, the random variable Δd^{ρ} is depicted, which describes a parallel shift of the regression line due to variation in ρ_i, calculated as the mean value over all ρ_i:

$$\Delta d^{\rho} = \frac{1}{N} \cdot \sum_i \rho_i \tag{51}$$

Combining both effects, variations in d can be described as the sum of two uncorrelated terms, Δd^{ϕ} and Δd^{ρ}:

$$\Delta d = \Delta d^{\phi} + \Delta d^{\rho} = -x_{off} \cdot \Delta\phi + \frac{1}{N} \cdot \sum_i \rho_i \tag{52}$$

This missing correlation between $\Delta\phi$ and the sum over all ρ_i is also intuitively accessible: if the latter takes a positive number, it will not be possible to deduce the sign or the modulus of $\Delta\phi$. From (52) and with $E(\Delta d^{\phi} \cdot \Delta d^{\rho}) = 0$, $E(\Delta d^{\phi}) = 0$ and $E(\Delta d^{\rho}) = 0$, the variance σ_d^2 can be calculated as:

$$\sigma_d^2 = E([\Delta d]^2) = E([\Delta d^{\phi}]^2) + E([\Delta d^{\rho}]^2) = x_{off}^2 \cdot E([\Delta\phi]^2) + \frac{1}{N^2} \cdot E\left(\left[\sum_i \rho_i\right]^2\right) \tag{53}$$

$$\approx x_{off}^2 \cdot \sigma_\phi^2 + \frac{1}{N} \cdot \sigma_\rho^2 \tag{54}$$

In the last step from (53) to (54), again, the independence of the single measurements from each other is used; thus, the variance of the sum of the N data points approximates N-times the variance σ_ρ^2. Finally, the covariance between ϕ and d needs to be determined. Based on the definition, it follows with $\sigma_{d\phi} = \sigma_{\Delta d \Delta\phi}$:

$$\sigma_{d\phi} = E(\Delta d \cdot \Delta\phi) = E(\Delta d^{\phi} \cdot \Delta\phi) + E(\Delta d^{\rho} \cdot \Delta\phi) = -x_{off} \cdot E([\Delta\phi]^2) = -x_{off} \cdot \sigma_\phi^2 \tag{55}$$

By means of (47), (54) and (55), now, the complete error model in closed form is known, represented by the covariance matrix $\underline{R}_{d\phi}$ given as:

$$\underline{R}_{d\phi} \approx \sigma_\rho^2 \cdot \begin{pmatrix} \frac{12 \cdot x_{off}^2}{L^2 \cdot N} + \frac{1}{N} & \frac{-12 \cdot x_{off}}{L^2 \cdot N} \\ \frac{-12 \cdot x_{off}}{L^2 \cdot N} & \frac{12}{L^2 \cdot N} \end{pmatrix} \tag{56}$$

Applying this error model is easy since no knowledge of the variances and covariance for each single measurement is needed, which in practice is difficult to acquire. Instead, just the number N of preferably equally-spaced points used for line fitting, the variance σ_ρ^2 according to (39), the length L of the line segment calculated with (41) and its offset x_{off} according to (49) or (50) must be inserted.

5. Simulation Results

The scope of this section is to compare the presented algorithms for linear regression and error modeling based on statistical evaluation of the results. Segmentation of raw data is not considered; if necessary, this must be performed beforehand by means of well-known methods like Hough transformation or RANSAC; compare Section 1. Thus, for studying the performance reliably and repeatably, a large number of computer simulations was performed, applying a systematic variation of parameters within a wide range, which would not be feasible if real measurements are used.

For this purpose, straight lines with a certain perpendicular distance d from the origin and within a varying range of normal angles ϕ have been specified. Each of these lines is numerically described by a number of N points either given in Cartesian (x_i, y_i) or in polar $(r_i\ \theta_i)$ coordinates. In order to simulate the outcome of a real range-bearing sensor as much as possible, the angular coordinate was varied between θ_1 and θ_N. To each measurement, a certain amount of normally-distributed noise with σ_x, σ_y and σ_{xy} or alternatively with σ_r and σ_θ was added. Further, for each ϕ, a number of $N_s = 1000$ sets of samples was generated, in order to allow statistical evaluation of the results. The first simulation was performed with $N = 40$ equally-spaced points affected each by uncorrelated noise in the x- and y-direction with standard deviations $\sigma_x = \sigma_y = 5$ cm. This is a typical situation when a line is calculated from binary pixels, and in Figure 5a, a bundle of the simulated line segments is shown. The deviations Δ_ϕ and Δ_d taken as the mean value over all N_s samples of the estimated ϕ and d from their true values, respectively, are depicted in Figure 5b,c, comparing four algorithms as presented in Section 2: The triangles mark the outcome of Equations (3) and (4) with all weights set to one, whereas the squares are calculated according to the same analytic formulas, but using individual weighting factors applying (10) with $s_i = 1/\sigma_{\rho,i}$. The perpendicular deviations $\sigma_{\rho,i}$ are determined according to (14) and (21) with ϕ taken from (3) without weights. Obviously, in this example, all triangles coincide with the squares since each measurement i is affected by the same noise and thus for any ϕ, all weighting factors are always identical. The blue lines in Figure 5b,c show the results when applying the iterative method according to (11) with the minimum of *MSE* found numerically. For this purpose, $\sigma^2_{\rho,i}$ is inserted from (14) considering (21) and ρ_i is taken from (1) and d is calculated from (4), (8)–(10) with $s_i = 1/\sigma_{\rho,i}$. The black lines (KA) depict the deviations of d and ϕ obtained according to Krystek and Anton in [35]. Both numerical algorithm yield the same results, which is not surprising, since the variances $\sigma^2_{\rho,i}$ used as weighting factors are all identical. Further, here, the analytical algorithms provide exactly the same performance as the numerical ones, since for $\sigma_x = \sigma_y$. the weighting factors show no dependency on ϕ, and for that case, the analytical formulas are optimal.

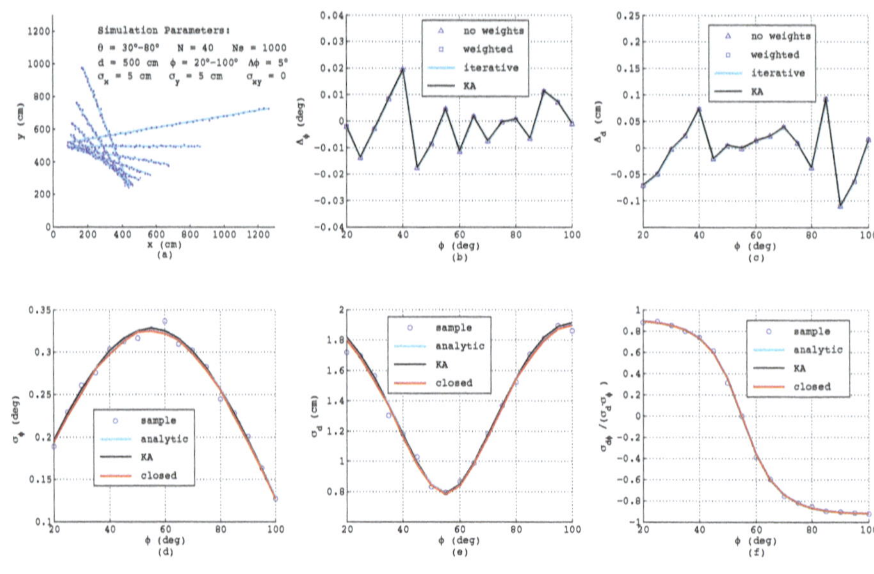

Figure 5. Simulation results for equidistant measurement points superimposing normally-distributed and uncorrelated noise in the x- and y-direction.

The lower subfigures depict the parameters of the covariance matrix $\underline{R}_{d\phi}$, again as a function of ϕ comparing different methods. Here, the circles represent numerical results obtained from the definitions of variance and covariance by summing over all N_s passes with $1 \leq k \leq N_s$, yielding d_k and ϕ_k, respectively:

$$\sigma_d^2 = \frac{1}{N_s} \sum_{k=1}^{N_s} (d_k - d)^2 \tag{57}$$

$$\sigma_{\phi}^{2} = \frac{1}{N_s} \sum_{k=1}^{N_s} (\phi_k - \phi)^2 \tag{58}$$

$$\sigma_{d\phi} = \frac{1}{N_s} \sum_{k=1}^{N_s} (d_k - d)\,(\phi_k - \phi) \tag{59}$$

Since these numerical results serve just as a reference for judging the accuracy of the error models, in the formulas above, the true values for d and ϕ have been used. The required line parameters d_k and ϕ_k in (57)–(59) can be estimated with any of the four described methods, since minor differences in d_k and ϕ_k have almost no effect on the resulting variances and the covariance. The blue lines in Figure 5d–f show the results of the analytic error model as described in Section 3, and the black lines represent the outcomes of the algorithm from Krystek and Anton [35], while the red lines corresponds to the model in closed-form according to (56) in Section 4 with L and x_{off} taken from (41) and (49), respectively. Interestingly, although the theoretical derivations differ substantially, the results match very well, which especially proves the correctness of the simplified model in closed-form. Since this model explicitly considers the effect of the line length L and of the geometric offset x_{off}, the behavior of the curves can be clearly understood: The minimum of L will occur if ϕ equals the mean value of θ_{min} and θ_{max}, i.e., at $\phi = 55^\circ$, and exactly at this angle, the maximum standard deviation σ_ϕ occurs. Further, since L linearly depends on ϕ, a quadratic dependence of σ_ϕ on ϕ according to (47) can be observed. With respect to Figure 5e, the minimum of σ_d also appears at $\phi = 55^\circ$ corresponding to $x_{off} = 0$. At this angle, according to (54), the standard deviation of d is given as $\sigma_d \approx \sigma_\rho/\sqrt{N} = 5/\sqrt{40} = 0.79$, while the covariance $\sigma_{\rho d}$ calculated according to (55) and with it the correlation coefficient shown in Figure 5f vanish.

When comparing the results, one should be aware that in the simulations of the analytic error models, the exact variances $\sigma_{x_i}^2$, $\sigma_{y_i}^2$ and σ_{xy_i} are used; thus, in practice, the achievable accuracies will be worse. On the other hand, when applying the new error model in closed-form, the variance σ_ρ^2 is calculated as the mean value of all ρ_i^2 from the actual set of N data points according to (39), and hence, is always available.

Nevertheless, if in this equation, the estimated line parameters ϕ and d are used, which are calculated, e.g., according to (3) and (4) using the same measurements as in (39), no unbiased σ_ρ^2 can be expected. This is reasoned from the fact that for each set of N data points, the mean quadratic distance over all ρ_i^2 is minimized in order to estimate ϕ and d. Thus, the numeric value of σ_ρ^2 will always be smaller than its correct value calculated with the exact line parameters. This effect can be clearly observed from Figure 6, which shows for the same simulation parameters as depicted in Figure 5a the dependency of σ_ρ^2 on the number of points on the line N, averaged over N_s sets of samples: only in the case of using the exact line parameters in (39), which obviously are only available in a simulation, actually the correct $\sigma_\rho^2 = 25\ \text{cm}^2$ is obtained as shown by the triangles. If however, in each run, σ_ρ^2 is calculated with the estimated ϕ and d as indicated by the squares, a clear deviation especially at low N occurs. Only asymptotically for large N when ϕ converges to its exact value, the correct σ_ρ^2 is reached. Fortunately, this error can be compensated quite well by means of multiplying σ_ρ^2 with a correction factor $c = \frac{N+1}{N-1}$ as shown by the dashed line in Figure 6. Due to the strongly non-linear relation between ϕ and any ρ_i, this correction works much better than simply exchanging in (39) the divisor N by $N-1$ as often used in statistics. Since c is the inverse of the term neglected in the approximation of σ_ϕ^2 in (47), the closed-form of the covariance matrix $\underline{R}_{d\phi}$ according to (56) yields almost unbiased results also for small N if σ_ρ^2 is calculated according to (39) with estimated line parameters ϕ and d. Although not shown here, the proposed bias compensation works well for a large range of measurement parameters. For a reliable determination of σ_ρ^2 from N data points of a line segment, N should be at least in the order of 10.

Figure 7 shows the results when simulating a range-bearing scan with a constant angular offset $\Delta\theta = (\theta_{max} - \theta_{min})/(N-1)$ between adjacent measurements. Each measurement is distorted by adding normally-distributed noise with standard deviations $\sigma_r = 5$ cm and $\sigma_\theta = 0.1^\circ$. This is a more challenging situation, since now that the measurements are not equispaced, each data point exhibits individual variances $\sigma_{x,i}$, $\sigma_{y,i}$ dependent on ϕ, and moreover, a covariance $\sigma_{xy,i}$ exists. As can be seen, the errors of the estimated ϕ and d as depicted in Figure 7b,c exhibit the same order of magnitude as before; yet, both analytic results differ slightly from each other and are less accurate compared to the numerical solutions. Both numerical methods yield quasi-identical results, since for the chosen small noise amplitudes, the differences between both algorithms have no impact on the resulting accuracy.

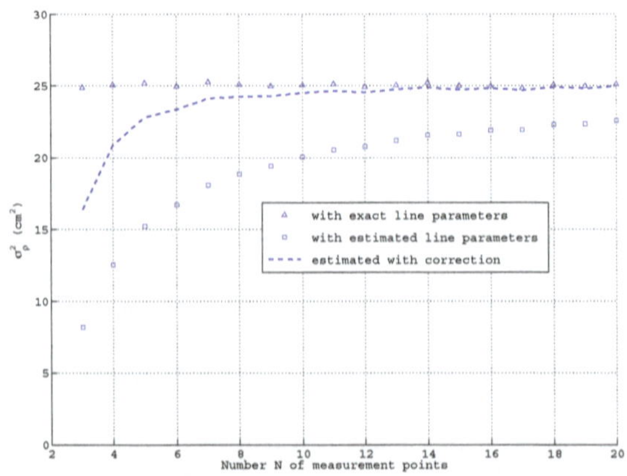

Figure 6. Variance of ρ dependent on the number N of measured data points, using the same simulation parameters as indicated in Figure 5a.

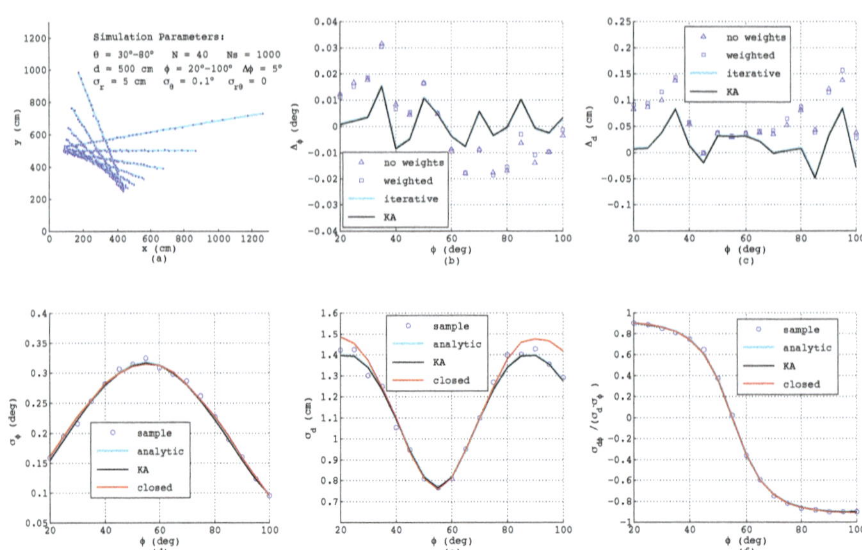

Figure 7. Results from simulated range-bearing scans superimposing low noise in the r- and θ-direction.

Regarding the error models, Figure 7d–f reveal that in spite of unequal distances between the measurement points and varying $\sigma_{\rho,i}$, the results of the closed-form model match well with the analytic and numeric results. Only σ_d shows a certain deviation at steep and flat lines with ϕ below 30° or above 80°. This is related to errors in x_{off}, since in this range of ϕ, the points on the lines measured with constant $\Delta\theta$ have clearly varying distances, and thus, (49) yields just an approximation of the effective offset of the straight line.

The next Figure 8 shows the results with the models applied to short lines measured in the angular range of $30° \leq \theta \leq 40°$ with $N = 20$, while all other parameters are identical to those depicted in Figure 7a. As can be seen from Figure 8b,c, now, the analytical algorithms based on (3) and (4) are no longer adequate since these, independent of applying weights or not, yield much higher errors than the numerical approaches. All error models however still provide accurate results. Actually, the closed-form model even yields better accuracy than before, since the distances of the data points on the line between adjacent measurement and also $\sigma_{\rho,i}$ are more uniform compared to the simulations with long lines.

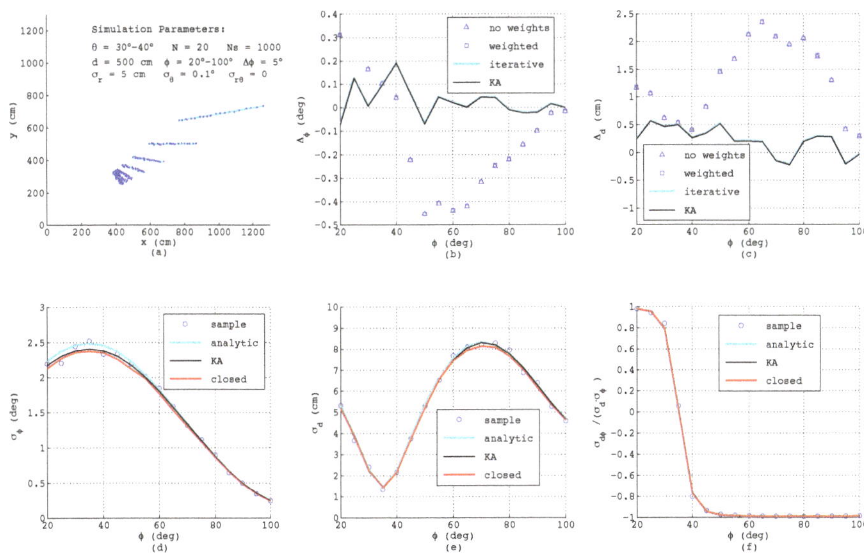

Figure 8. Results from simulated range-bearing scans of short lines superimposing low noise in the r- and θ-direction.

In order to check the limits of the models, Figure 9 depicts the results when applying large angular noise with $\sigma_\theta = 2°$. In this extreme case, also the numerical algorithms show systematic errors dependent on ϕ since the noise of ρ_i can no longer be assumed to be normally distributed. However, according to Figure 8b,c the iterative method as presented in Section 2 shows clear benefits in comparison to the KA algorithm proposed in [35], caused by the more accurate modeling of σ_{ρ_i}.

With respect to the outcome of the noise models in Figure 9d–f, now, only the analytic algorithm as presented in Section 3 still yields reliable results, while the KA-method based on matrix inversion reveals numerical instability. Due to the clear uneven distribution of measurements along the line, also the simplified error model in this case shows clear deviations, although at least the order of magnitude is yet correct.

Finally, Figure 10 shows typical results, if the sensor noise is not exactly known. In this example, the radial standard deviation was assumed to be 10 cm, whereas the exact value, applied when generating the measurements, was only 5 cm. The simulation parameters correspond to those in Figure 7, only the number of data points has been reduced to $N = 10$. According to Figure 10b,c, now, for calculating ϕ and d, the numerical methods yield no benefit over the analytical formulas with or without weights. Due to the only approximately known variance, the analytic error model, as well as the KA-method in Figure 10d–f reveal clear deviations from the reference results. Only the model in closed-form is still accurate, since it does not require any a priori information regarding sensor noise. In addition, these results prove the bias-free estimation of σ_ρ^2 with (39) also if N is low, as depicted in Figure 6.

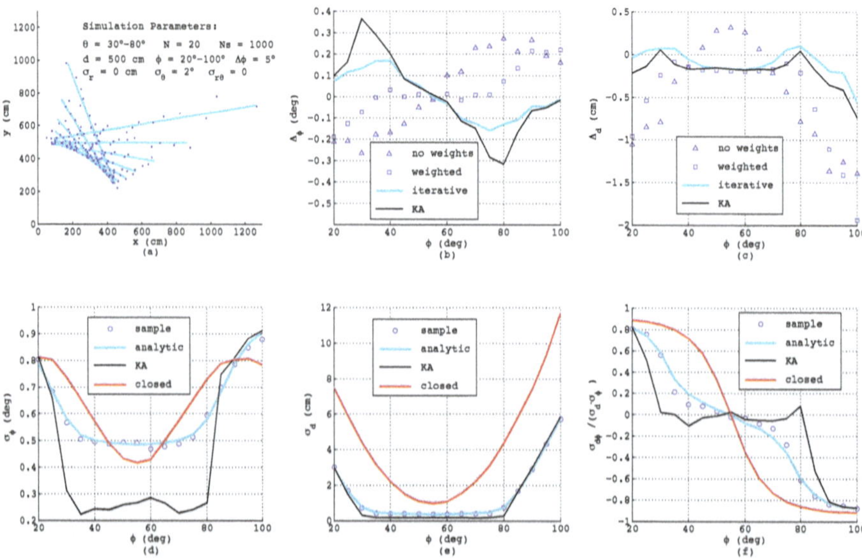

Figure 9. Results from simulated range-bearing scans superimposing high noise only in the θ-direction.

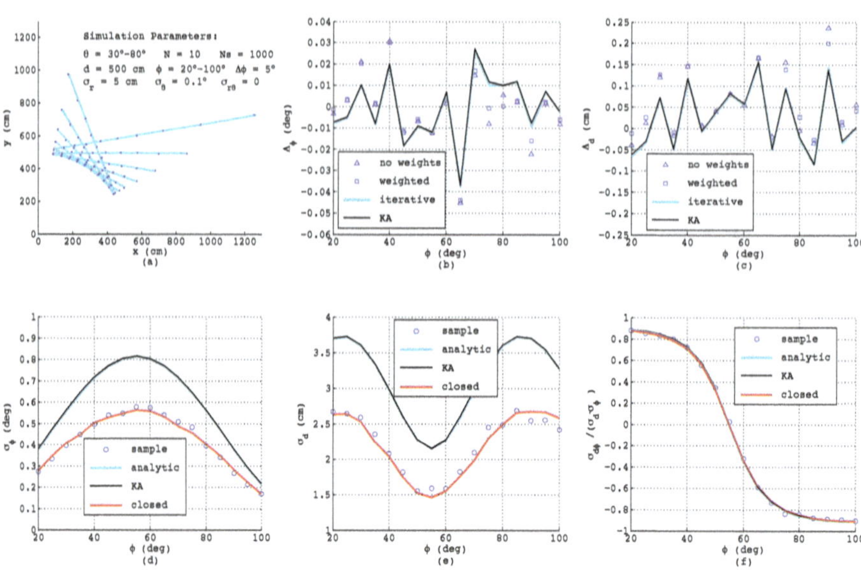

Figure 10. Results from simulated range-bearing scans with a low number of data points and only an approximately known noise level of the sensor.

6. Conclusions

In this study, the performance of linear regression is evaluated, assuming both coordinates as random variables. It is shown that, especially with range-bearing sensors, frequently used in mobile robotics, a distinct covariance of the noise in the x- and y-direction at each measurement point exists. In this case, analytical formulas assuming identical and uncorrelated noise will only provide accurate line parameters ϕ and d if the detected line segments are sufficiently long and the noise level stays below a certain limit. If these prerequisites are not fulfilled and if the sensor noise is known, numerical algorithms should be applied, which consider the reliability of each measurement point as a function of ϕ. For this, the performance of prior work can be improved by means of modeling the independence of the single data points exactly and by paying attention also to second order effects of the angular noise.

The main focus of this paper is on the derivation of the covariance matrix $\underline{R}_{d\phi}$ of straight lines. This information has a crucial impact on the performance of SLAM with line features, since for both, data association and sensor fusing, $\underline{R}_{d\phi}$ must be estimated precisely. For this purpose, the first analytical error models are reviewed, which however need exact knowledge of the measurement noise, although in many applications, this is not available. In addition, these approaches require high computational effort and do not allow one to comprehend the effect of measurement parameters on the resulting accuracy of an estimated straight line. Thus, a new error model in closed form is proposed, depending only on two geometric parameters, as well as on the number of points of a line segment. Besides, a single variance must be known, which is determined easily and reliably from the same measurements as used for line fitting. By means of this model, the covariance matrix can be estimated quickly and exactly. Moreover, it allows one to adapt measurement conditions in order to achieve the maximum accuracy of detected line features.

Acknowledgments: The author greatly appreciates the editor's encouragement and the valuable comments of the reviewers. This work has been partially supported by the German Federal Ministry of Education and Research (Grant No. BMBF 17N2008) and funded by the department of computer science and media at Beuth university.

Conflicts of Interest: The author declares no conflict of interest.

Appendix A. Analytic Derivation of Straight Line Parameters With Errors in Both Coordinates

For the derivation of the perpendicular distance d, the partial derivative of Equation (2) with respect to d is taken and set to zero, which directly gives Equation (4) using (8)–(10). In order to calculate ϕ, first the partial derivation of (2) with respect to ϕ must be calculated and set to zero, yielding:

$$\frac{1}{N}\sum_{i=1}^{N} s_i\left[x_i y_i\left(\cos^2\phi-\sin^2\phi\right)+\left(y_i^2-x_i^2\right)\sin\phi\,\cos\phi\right]+\frac{1}{N}\sum_{i=1}^{N} s_i d\left(x_i\sin\phi-y_i\cos\phi\right)=0 \quad \text{(A1)}$$

Now, the distance d can be replaced by (4), and after inserting the definitions of $\bar{x}$, $\bar{y}$, σ_x^2, σ_y^2 and σ_{xy} according to (5)–(9) considering (10), it follows from (A1) after reordering:

$$\sigma_{xy}\left(\cos^2\phi-\sin^2\phi\right)+\sin\phi\cos\phi\left(\sigma_y^2-\sigma_x^2\right)=0 \quad \text{(A2)}$$

Applying the theorem of Pythagoras and the addition theorems of angles, the terms with the sine and cosine can be rewritten:

$$\cos^2\phi-\sin^2\phi=2\cos^2\phi-1=\cos 2\phi \quad \text{(A3)}$$

$$\sin\phi\cos\phi=\frac{1}{2}\sin 2\phi \quad \text{(A4)}$$

Inserting these formulas into (A2) finally yields for ϕ:

$$\phi = \frac{1}{2}\arctan\left(\frac{-2\sigma_{xy}}{\sigma_y^2 - \sigma_x^2}\right) \tag{A5}$$

Equation (A5) calculates ϕ always in the range $-\pi/4 < \phi < \pi/4$, although according to Figure 1, this is only correct if $\sigma_y^2 > \sigma_x^2$, while in the case $\sigma_y^2 < \sigma_x^2$, an angle $\pi/2$ must be added to ϕ. Thus, as general solution (3) should be taken also avoiding a special consideration if σ_y^2 equals σ_x^2.

References

1. Everett, H.R. *Sensors for Mobile Robots*, 1st ed.; A. K. Peters Ltd.: New York, NY, USA, **1995**.
2. Canny, J. A Computational Approach to Edge Detection. *IEEE Trans. Pattern Anal. Mach. Intell.* **1986**, *8*, 679–698.
3. Guse, W.; Sommer, V. A New Method for Edge Oriented Image Segmentation. In Proceedings of the Picture Coding Symposium, Tokio, Japan, 2–4 September 1991.
4. Gao, Y.; Liu, S.; Atia, M.M.; Noureldin, A. INS/GPS/LiDAR Integrated Navigation System for Urban and Indoor Environments Using Hybrid Scan Matching Algorithm. *Sensors* **2015**, *15*, 23286–23302.
5. Lu, F.; Milios, E. Robot pose estimation in unknown environments by matching 2D range scans. *J. Intell. Robot. Syst.* **1997**, *18*, 249–275.
6. Arras, K.O.; Castellanos, J.A.; Schilt, M.; Siegwart, R. Feature-based multi-hypothesis localization and tracking using geometric constraints. *Robot. Auton. Syst.* **2003**, *44*, 41–53.
7. Rodriguez, S.; Paz, J.F.D.; Villarrubia, G.; Zato, C.; Bajo, J.; Corchado, J.M. Multi-Agent Information Fusion System to mange data from a WSN in a residential home. *Inf. Fusion* **2016**, *23*, 43–57.
8. Li, T.; Corchado, J.M.; Bajo, J.; Sun, S.; Paz, J.F.D. Effectiveness of Bayesian filters: An information fusion perspective. *Inf. Sci.* **2016**, *329*, 670–689.
9. Tang, Y.; Zhou, D.; Xu, S.; He, Z. A Weighted Belief Entropy-Based Uncertainty Measure for Multi-Sensor Data Fusion. *Sensors* **2017**, *17*, 928, doi:10.3390/s17040928..
10. Borenstein, J.; Everett, H.R.; Feng, L. *Navigating Mobile Robots, Systems and Techniques*, 1st ed.; A. K. Peters: Natick, MA, USA, 1996.
11. Neira, J.; Tardos, J.D. Data association in stochastic mapping using the joint compatibility test. *IEEE Trans. Robot. Autom.* **2001**, *17*, 890–897.
12. Durrant-Whyte, H.; Bailey, T. Simultaneous localization and mapping: Part I. *IEEE Robot. Autom. Mag.* **2006**, *13*, 99–110.
13. Blanco, J.L.; Gonzalez-Jimenez, J.; Fernandez-Madrigal, J.A. An Alternative to the Mahalanobis Distance for Determining Optimal Correspondences in Data Association. *Trans. Robot.* **2012**, *28*, 980–986.
14. Wang, H.; Liu, Y.H.; Zhou, D. Adaptive Visual Servoing Using Point and Line Features With an Uncalibrated Eye-in-Hand Camera. *IEEE Trans. Robot.* **2008**, *24*, 843–857.
15. Choi, Y.H.; Lee, T.K.; Oh, S.Y. A line feature based SLAM with low grade range sensors using geometric constraints and active exploration for mobile robot. *Auton. Robot.* **2008**, *24*, 13–27.
16. Yin, J.; Carlone, L.; Rosa, S.; Anjum, M.L.; Bona, B. Scan Matching for Graph SLAM in Indoor Dynamic Scenarios. In Proceedings of the Twenty-Seventh International Florida Artificial Intelligence Research Society Conference, Pensacola Beach, FL, USA, 21–23 May 2014; pp. 418–423.
17. Pasteau, F.; Narayanan, V.K.; Babel, M.; Chaumette, F. A visual servoing approach for autonomous corridor following and doorway passing in a wheelchair. *Robot. Auton. Syst.* **2016**, *75*, 28–40.
18. David, P.; DeMenthon, D.; Duraiswami, R.; Samet, H. Simultaneous pose and correspondence determination using line features. In Proceedings of the 2003 IEEE Computer Society Conference on Computer Vision and Pattern Recognition, Madison, WI, USA, 18–20 June 2003; Volume 2, pp. 424–431.
19. Marchand, É.; Fasquelle, B. Visual Servoing from lines using a planar catadioptric system. In Proceedings of the 2017 IEEE/RSJ International Conference on Intelligent Robots and Systems (IROS), Vancouver, BC, Canada, 24–28 September 2017; pp. 2935–2940.

20. Bista, S.R.; Giordano, P.R.; Chaumette, F. Combining Line Segments and Points for Appearance- based Indoor Navigation by Image Based Visual Servoing. In Proceedings of the IEEE/RSJ International Conference on Intelligent Robots and Systems (IROS), Vancouver, BC, Canada, 24–28 September 2017; pp. 2960–2967.
21. Xu, D.; Lu, J.; Wang, P.; Zhang, Z.; Liang, Z. Partially Decoupled Image-Based Visual Servoing Using Different Sensitive Features. *IEEE Trans. Syst. Man Cybern. Syst.* **2017**, *47*, 2233–2243.
22. Jeong, W.Y.; Lee, K.M. Visual SLAM with Line and Corner Features. In Proceedings of the 2006 IEEE/RSJ International Conference on Intelligent Robots and System (IROS), Beijing, China, 9–15 October 2006.
23. York, D. Least-squares fitting og a straight line. *Can. J. Phys.* **1966**, *44*, 1079–1086.
24. Krane, K.S.; Schecter, L. Regression line analysis. *Am. J. Phys.* **1982**, *50*, 82–84.
25. Golub, G.; van Loan, C. Ana analysis of the total least squares problem. *SIAM J. Numer. Anal.* **1980**, *17*, 883–893.
26. Weisberg, S. *Applied Linear Regression*, 3rd ed.; John Wiley & Sons: Hoboken, NJ, USA, 2005; ISBN 0-471-66379-4.
27. Draper, N.R.; Smith, H. *Applied Regression Analysis*, 3rd ed.; John Wiley & Sons: Hoboken, NJ, USA, 1988.
28. Seber, G.A.F.; Lee, A.J. *Linear Regression Analysis*, 2nd ed.; John Wiley & Sons: Hoboken, NJ, USA, 2003.
29. Amiri-Simkooeiab, A.R.; Zangeneh-Nejadac, F.; Asgaria, J.; Jazaeri, S. Estimation of straight line parameters with fully correlated coordinates. *J. Int. Meas. Confed.* **2014**, *48*, 378–386.
30. Krystek, M.; Anton, M. A weighted total least-squares algorithm for fitting a straight line. *Meas. Sci. Technol.* **2007**, *18*, 3438–3442.
31. Cecchi, G.C. Error analysis of the parameters of a least-squares determined curve when both variables have uncertainties. *Meas. Sci. Technol.* **1991**, *2*, 1127–1129.
32. Arras, K.O.; Siegwart, R.Y. Feature Extraction and Scene Interpretation for Map-Based Navigation and Map Building. In *Proceedings of SPIE: Mobile Robotics XII*; SPIE: Pittsburgh, PA, USA, 1997; pp. 42–53.
33. Pfister, S.T.; Kriechbaum, K.L.; Roumeliotis, S.I.; Burdick, J.W. A Weighted range sensor matching algorithms for mobile robot displacement estimation. In Proceedings of the IEEE International Conference on Robotics and Automation, Washington, DC, USA, 11–15 May 2002; Volume 4.
34. Pfister, A.T.; Roumeliotis, S.I.; Burdick, W. Weighted line fitting algorithms for mobile robot map building and efficient data representation. In Proceedings of the 2003 IEEE International Conference on Robotics and Automation, Taipei, Taiwan, 14–19 September 2003; pp. 1304–1311.
35. Krystek, M.; Anton, M. A least-squares algorithm for fitting data points with mutually correlated coordinates to a straight line. *Meas. Sci. Technol.* **2011**, *22*, 035101, doi:10.1088/0957-0233/22/3/035101.
36. Borges, G.A.; Aldon, M.J. A Split-and-Merge Segmentation Algorithm for Line Extraction in 2-D Range Images. In Proceedings of the 15th International Conference on Pattern Recognition, Barcelona, Spain, 3–7 September 2000; Volume 4.
37. Jian, M.; Zhang, C.F.; Yan, F.; Tang, M.Z. A global line extraction algorithm for indoor robot mapping based on noise eliminating via similar triangles rule. In Proceedings of the 35th Chinese Control Conference (CCC), Chengdu, China, 27–29 July 2016; pp. 6133–6138.
38. Illingworth, J.; Kittler, J. A survey of the hough transform. *Comput. Vis. Graph. Image Process.* **1988**, *44*, 87–116.
39. Kim, J.; Krishnapuram, R. A Robust Hough Transform Based on Validity. In Proceedings of the International Conference on Computational Intelligence, Anchorage, AK, USA, 4–9 May 1998; Volume 2, pp. 1530–1535.
40. Banjanovic-Mehmedovic, L.; Petrovic, I.; Ivanjko, E. Hough Transform based Correction of Mobile Robot Orientation. In Proceedings of the International Conference on Industrial Technology, Hammamet, Tunisia, 8–10 December 2004; Volume 3, pp. 1573–1578.
41. Fischler, M.A.; Bolles, R.C. Random Sample Consensus: A Paradigm for Model Fitting with Applications to Image Analysis and Automated Cartography. *Commun. ACM* **1981**, *25*, 381–395.
42. Liu, Y.; Gu, Y.; Li, J.; Zhang, X. Robust Stereo Visual Odometry Using Improved RANSAC-Based Methods for Mobile Robot Localization. *Sensors* **2017**, *17*, 2339.
43. Nguyen, V.; Martinelli, A.; Tomatis, N.; Siegwart, R. A Comparison of Line Extraction Algorithms using 2D Laser Rangefinder for Indoor Mobile Robotics. In Proceedings of the IEEE/RSJ International Conference on Intelligent Robots and System (IROS), Edmonton, AB, Canada, 2–6 August 2005.
44. Westfall, P.H. *Understanding Advanced Statistical Methods*; CRC Press: Boca Raton, FL, USA, 2013; Chapter 16.

45. Dovì, V.G.; Paladino, O.; Reverberi, A.P. Some remarks on the use of the inverse hessian matrix of the likelihood function in the estimation of statistical properties of parameters. *Appl. Math. Lett.* **1991**, *4*, 87–90.
46. Garulli, A.; Giannitrapani, A.; Rossi, A.; Vicino, A. Mobile robot SLAM for line-based environment representation. In Proceedings of the 44th IEEE Conference on Decision and Control, Seville, Spain, 15 December 2005; pp. 2041–2046.

sensors

MDPI

Article

Hybrid Adaptive Cubature Kalman Filter with Unknown Variance of Measurement Noise

Yuepeng Shi [1], Xianfeng Tang [2,*], Xiaoliang Feng [3,*], Dingjun Bian [4] and Xizhao Zhou [5]

1 College of Intelligent Manufacturing and Automation, Henan University of Animal Husbandry and Economy, Zhengzhou 450011, China; syp@hnuahe.edu.cn
2 Information Technology Center, Zhejiang University, Hangzhou 310027, China
3 College of Electrical Engineering, Henan University of Technology, Zhengzhou 450001, China
4 School of Information Engineering, Zhengzhou University, Zhengzhou 450000, China; bdjbray@163.com
5 School of Business, University of Shanghai for Science and Technology, Shanghai 200093, China; xizhaozhou@163.com
* Correspondence: txf1213@zju.edu.cn (X.T.); fxl@haut.edu.cn (X.F.)

Received: 23 September 2018; Accepted: 27 November 2018; Published: 7 December 2018

Abstract: This paper is concerned with the filtering problem caused by the inaccuracy variance of measurement noise in real nonlinear systems. A novel weighted fusion estimation method of multiple different variance estimators is presented to estimate the variance of the measurement noise. On this basis, a hybrid adaptive cubature Kalman filtering structure is proposed. Furthermore, the information filter of the hybrid adaptive cubature Kalman filter is also studied, and the stability and filtering accuracy of the filter are theoretically discussed. The final simulation examples verify the validity and effectiveness of the hybrid adaptive cubature Kalman filtering methods proposed in this paper.

Keywords: nonlinear system; hybrid adaptive filtering; weighted fusion estimation; square-root cubature Kalman filter; information filter

1. Introduction

In recent decades, nonlinear filtering has been widely used in military and civil fields such as target tracking, navigation, positioning, and intelligent manufacturing [1,2]. The theory and method of nonlinear filtering has became one of the most important research issues in the signal processing field, and has attracted increasing attention from researchers.

There are two main kinds of nonlinear filtering methods. The representative of the first kind of nonlinear filter is the extended Kalman filter (EKF), which linearizes the system model by Taylor expansion, holds the first order term, and ignores the second- and higher-order terms. The second kind of nonlinear filtering approximates the statistics of the system state, with examples being the unscented Kalman filter (UKF) and the cubature Kalman filter (CKF). Due to the model error of the linearization of the nonlinear system, the accuracy of the EKF is slightly lower, even leading to filtering divergence. Based on the unscented transformation to approximate the statistics of the system state, UKF was presented in [3]. Further, the cubature Kalman filter algorithm was proposed by Ienkaran in [4]. The CKF algorithm uses a third-degree spherical-radial cubature rule based on a Gaussian filtering framework. The algorithm has higher numerical stability and a smaller amount of calculation. Its excellent performance has made it widely used in various nonlinear system scenarios. Many advantages of CKF have attracted scholars to conduct in-depth research on it, considering that the traditional nonlinear filters often need to overcome the filtering divergence caused by high-dimensional operational errors. Drawing on the idea of square root filtering in the Kalman filter, Ienkaram and Haykin proposed the square-root cubature Kalman filter (SCKF) [5], which further improved the accuracy and stability of filtering. However, the traditional nonlinear filtering method requires knowledge of the mathematical model and the prior statistical information of noise when in practical application. Additionally, the statistical characteristics of noise in actual systems are usually indeterminate, which leads to a decline in the filtering accuracy.

For the problem of unknown statistical characteristics of measurement noise in real applications [6,7], Sage and Husa [8,9] proposed an excellent Sage–Husa suboptimal unbiased maximum a posteriori (MAP) estimator. Many scholars also adopted adaptive filtering techniques to improve the performance of the estimation algorithm, such as sliding window method [10], fading factor adjustment (FFA) [11–13], maximum a posteriori (MAP) estimator [14,15], and the variational Bayesian (VB) method [16–19], etc. Different methods for estimating the statistical characteristics of system noise are usually designed under different estimation criterions. How to use these methods to estimate the statistical characteristics of system noise is still an open issue.

For a class of nonlinear stochastic systems with inaccurate or unknown measurement noise variance (i.e., the priori measurement noise variance is not a precise value), an adaptive filtering algorithm based on SCKF is designed in this paper. Firstly, a novel fusion approach is proposed to estimate the measurement noise variance on the basis of the MAP and VB methods. Then, we use FFA to adjust the part of the variance matrix of the SCKF algorithm to obtain a hybrid adaptive SCKF algorithm (HASCKF) and the corresponding information filter (recorded as HASCIF). This is beneficial to reduce the effects on the adaptive filtering algorithm performance that may be caused by the estimation deviation of noise fusion. At the same time, the performance of the adaptive filtering algorithm is analyzed from two aspects based on the established HASCKF algorithm: the stability of the adaptive filtering algorithm and the filtering accuracy.

This paper is organized as follows: Section 2 formulates the nonlinear stochastic system and describes the problem of the inaccuracy of the the measurement noise variance. In Section 3, a novel noise variance fusion estimation algorithm HASCKF and the corresponding information filter are proposed based on the idea of weighted fusion. In Section 4, two simulation examples are utilized to display and verify the performance of the proposed algorithms. Section 5 provides the conclusions of this work.

2. Problem Description

Considering a class of discrete nonlinear stochastic systems, the state space model is described as follows [1]:

$$\mathbf{x}(k+1) = \boldsymbol{f}(\mathbf{x}(k)) + \mathbf{w}(k), \tag{1}$$

$$\mathbf{z}(k) = \boldsymbol{h}(\mathbf{x}(k)) + \mathbf{v}(k), \tag{2}$$

where $\mathbf{x}(k) \in R^n$ is the state of the target, $\mathbf{z}(k) \in R^m$ is the measurement, $\boldsymbol{f} : R^n \to R^n$ is the evolution process of the nonlinear state, and $\boldsymbol{h} : R^n \to R^m$ is the corresponding nonlinear measurement mapping.

The process noise $\mathbf{w}(k) \in R^n$ is a Gaussian white noise with zero means and variance $\mathbf{Q}(k)$.

The measurement noise $\mathbf{v}(k) \in R^m$ is a Gaussian white noise with zero means and variance $\mathbf{R}(k)$.

Hypothesis 1. *The process noise* $\mathbf{w}(k)$ *and measurement noise* $\mathbf{v}(k)$ *in the model are mutually statistically independent.*

Hypothesis 2. *The initial state of the system is* $\mathbf{x}(0)$*, with mean* $\mathbf{x}_0$ *and variance* $\mathbf{P}_0$*, and it is uncorrelated with* $\mathbf{v}(k)$ *and* $\mathbf{w}(k)$*.*

Hypothesis 3. *The process noise variance* $\mathbf{Q}(k)$ *is known, but the measurement noise variance is only with an inaccuracy prior state* $\mathbf{R}_0$*.*

For the nonlinear system described in (1) and (2), assuming that at the time k, we have the optimal estimation $\hat{\mathbf{x}}(k-1|k-1)$ and the square root matrix of the error variance matrix $\mathbf{S}(k-1|k-1)$. Then, the state estimate $\hat{\mathbf{x}}(k|k)$ and the square root of the variance matrix $\mathbf{S}(k|k)$ can be calculated according to the standard SCKF algorithm.

For nonlinear systems with determined noise variance, the SCKF algorithm has better estimation performance. However, when the priori value of the measurement noise variance is inaccurate, the final estimation $\hat{\mathbf{x}}(k|k)$ and $\mathbf{S}(k|k)$ will have large errors.

3. Hybrid Adaptive SCKF Algorithm (HASCKF)

In order to improve the adaptive filtering accuracy for an inaccurate modeling system with unknown measurement noise variance, and to let the estimated noise variance be closer to the true noise variance, in this paper, a novel noise variance fusion estimation algorithm HASCKF is proposed based on the idea of weighted fusion.

3.1. Estimation Method of Measurement Noise

Theorem 1. *Assume that the measurement noise variance estimated by MAP and VB methods at time k are denoted as $\hat{\mathbf{R}}^1(k)$ and $\hat{\mathbf{R}}^2(k)$, respectively. Then, the weighted fusion estimation of the measurement noise variance $\hat{\mathbf{R}}^g(k)$ is*

$$\hat{\mathbf{R}}^g(k) = \hat{\mathbf{R}}^1(k) + [T_1(k) - T_{12}(k)][T_1(k) + T_2(k) - 2T_{12}(k)]^{-1}[\hat{\mathbf{R}}^2(k) - \hat{\mathbf{R}}^1(k)], \tag{3}$$

where

$$\begin{cases} T_1(k) = \|\hat{\mathbf{R}}^g(k-1) - \hat{\mathbf{R}}^1(k)\|_F^2, \\ T_{12}(k) = tr[(\hat{\mathbf{R}}^g(k-1) - \hat{\mathbf{R}}^1(k))(\hat{\mathbf{R}}^g(k-1) - \hat{\mathbf{R}}^2(k))^T], \\ T_2(k) = \|\hat{\mathbf{R}}^g(k-1) - \hat{\mathbf{R}}^2(k)\|_F^2. \end{cases} \tag{4}$$

In the above equation, $\|\cdot\|_F$ represents the Frobenius norm of the matrix, and the initial value of the fusion estimation $\hat{\mathbf{R}}^g(0) = \mathbf{R}_0$.

Proof. The weighted fusion estimation of the measurement noise variance $\hat{\mathbf{R}}^g(k)$ can be expressed as the following linear combination:

$$\hat{\mathbf{R}}^g(k) = a_1(k)\hat{\mathbf{R}}^1(k) + a_2(k)\hat{\mathbf{R}}^2(k). \tag{5}$$

Under the condition $a_1(k) + a_2(k) = 1$, we minimize the performance criterion:

$$J(k) = tr\left\{ [(\mathbf{R}(k) - \hat{\mathbf{R}}^g(k)][(\mathbf{R}(k) - \hat{\mathbf{R}}^g(k)]^T \right\}$$

From $a_1(k) + a_2(k) = 1$, we can get $a_1(k) = (1 - a_2(k))$, then substituting it into formula (5), we obtain

$$\hat{\mathbf{R}}^g(k) = \hat{\mathbf{R}}^1(k) + a_2(k)[\hat{\mathbf{R}}^2(k) - \hat{\mathbf{R}}^1(k)]. \tag{6}$$

Then, the overall estimation error is

$$\begin{aligned} \tilde{\mathbf{R}}^g(k) &= \mathbf{R}(k) - \hat{\mathbf{R}}^g(k) \\ &= a_1(k)\tilde{\mathbf{R}}^1(k) + a_2(k)\tilde{\mathbf{R}}^2(k) \\ &= [1 - a_2(k)]\tilde{\mathbf{R}}^1(k) + a_2(k)\tilde{\mathbf{R}}^2(k). \end{aligned}$$

So, we can get

$$\begin{aligned} J(k) &= tr[\tilde{\mathbf{R}}^g(k)(\tilde{\mathbf{R}}^g(k))^T] \\ &= [1-a_2(k)]^2 tr[\tilde{\mathbf{R}}^1(k)(\tilde{\mathbf{R}}^1(k))^T] + [1-a_2(k)]tr[\tilde{\mathbf{R}}^1(k)(\tilde{\mathbf{R}}^2(k))^T]a_2(k) \\ &\quad + a_2(k)tr[\tilde{\mathbf{R}}^2(k)(\tilde{\mathbf{R}}^1(k))^T][1-a_2(k)] + a_2^2(k)tr[\tilde{\mathbf{R}}^2(k)(\tilde{\mathbf{R}}^2(k))^T]. \end{aligned} \tag{7}$$

Let $\dfrac{\partial J(k)}{\partial a_2(k)} = 0$, and after simplification, we can get

$$a_2(k) = [T_1(k) - T_{12}(k)][T_1(k) + T_2(k) - T_{12}(k) - T_{21}(k)]^{-1}, \tag{8}$$

where

$$\begin{cases} T_1(k) = tr[\tilde{\mathbf{R}}^1(k)(\tilde{\mathbf{R}}^1(k))^T] \\ \qquad = tr[(\mathbf{R}(k) - \hat{\mathbf{R}}^1(k))(\mathbf{R}(k) - \hat{\mathbf{R}}^1(k))^T] \\ \qquad = \|\mathbf{R}(k) - \hat{\mathbf{R}}^1(k)\|_F^2, \\ T_{12}(k) = tr[\tilde{\mathbf{R}}^1(k)(\tilde{\mathbf{R}}^2(k))^T] \\ \qquad = tr[(\mathbf{R}(k) - \hat{\mathbf{R}}^1(k))(\mathbf{R}(k) - \hat{\mathbf{R}}^2(k))^T], \\ T_{21}(k) = tr[\tilde{\mathbf{R}}^2(k)(\tilde{\mathbf{R}}^1(k))^T] \\ \qquad = tr[(\mathbf{R}(k) - \hat{\mathbf{R}}^2(k))(\mathbf{R}(k) - \hat{\mathbf{R}}^1(k))^T] \\ \qquad = T_{12}(k), \\ T_2(k) = tr[\tilde{\mathbf{R}}^2(k)(\tilde{\mathbf{R}}^2(k))^T] \\ \qquad = tr[(\mathbf{R}(k) - \hat{\mathbf{R}}^2(k))(\mathbf{R}(k) - \hat{\mathbf{R}}^2(k))^T] \\ \qquad = \|\mathbf{R}(k) - \hat{\mathbf{R}}^2(k)\|_F^2. \end{cases} \tag{9}$$

Since the measurement noise variance $\mathbf{R}(k)$ is uncertain, Equation (9) cannot be directly calculated. For this reason, we replace $\mathbf{R}(k)$ with fusion estimated value $\hat{\mathbf{R}}^g(k-1)$ of the measurement noise variance at time $(k-1)$. Obviously, the initial value satisfies $\hat{\mathbf{R}}^g(0) = \mathbf{R}_0$, then substituting it to Equations (9), (8), and (6), respectively, we can get Equations (3) and (4). □

Note 1 In the above theorem, since the measurement noise variance $\mathbf{R}(k)$ is uncertain, we replace $\mathbf{R}(k)$ with fusion estimate $\hat{\mathbf{R}}^g(k-1)$ of the measurement noise variance. This approximate substitution has certain rationality, especially for the case of constant noise variance and slowly varying noise variance. From Theorem 1, the physical meanings of $T_1(k)$ and $T_2(k)$ are equivalent to the estimated error variance of the two noise variance estimation methods, and $T_{12}(k)$ and $T_{21}(k)$ are similar to their cross-variance.

Inference 1. *If we do not consider the correlation between the estimation error of noise variance, that is $T_{12}(k) = T_{12}(k) = 0$, then the weighted fusion estimation of measurement noise variance $\hat{\mathbf{R}}^g(k)$ is*

$$\begin{cases} \hat{\mathbf{R}}^g(k) = [T_1^{-1}(k) + T_2^{-1}(k)]^{-1} T_1^{-1}(k)\hat{\mathbf{R}}^1(k) \\ \qquad\qquad + [T_1^{-1}(k) + T_2^{-1}(k)]^{-1} T_2^{-1}(k)\hat{\mathbf{R}}^2(k), \\ T^{-1}(k) = T_1^{-1}(k) + T_2^{-1}(k). \end{cases} \tag{10}$$

It can be directly derived according to the principle of simple convex combination fusion [1].

Note 2 Obviously, the result of inference 1 is easy to generalize to the case where the number of noise variance estimators $N_e \geq 3$. Assume that the error variance of the ith estimator $T_i(k) =$

$\left\|\hat{\mathbf{R}}^g(k-1)-\hat{\mathbf{R}}^i(k)\right\|_F^2$. Assuming that the estimation errors of various estimation methods are not related to each other, the fusion estimation of noise variance is

$$\begin{cases} \hat{\mathbf{R}}^g(k)=\left[\sum\limits_j^{N_e} T_j^{-1}(k)\right]^{-1}\sum\limits_j^{N_e} T_j^{-1}(k)\hat{\mathbf{R}}^j(k), \\ T_g^{-1}(k)=\sum\limits_j^{N_e} T_j^{-1}(k), \end{cases} \tag{11}$$

where $T_g(k)$ can be regarded as the error variance of noise variance fusion estimation. Since $T_i(k)\geq 0$, $T_g(k)\leq T_i(k)$, which indicates that the estimated noise variance after fusion is superior to that of any single noise variance estimator.

3.2. HASCKF Algorithm

Based on Theorem 1, combined with the fading factor adjustment technique [11], we propose the hybrid adaptive SCKF algorithm (HASCKF). The principle block diagram is shown in Figure 1.

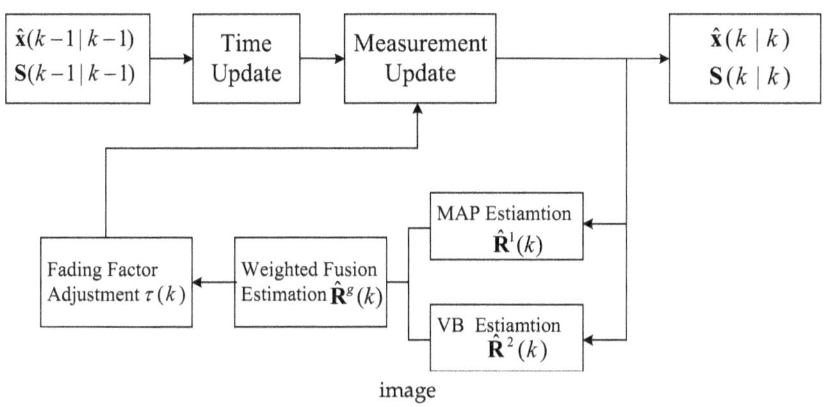

Figure 1. Principle block diagram of the HASCKF algorithm.

Firstly, the variance of measurement noise is estimated by the MAP estimator and VB method respectively. Then, the weighted fusion technique is introduced to fuse the two noise variance estimators. Finally, the variance matrix of the SCKF measurement update is adjusted by the fading factor adjustment technique to obtain the final state estimation and the root-mean square error (RMSE) variance matrix. The detailed implementation of HASCKF is described in the following Theorem 2.

Theorem 2. *Consider a class of nonlinear system as described in (1) and (2). Under the condition of Hypotheses 1–3, if the optimal estimate $\hat{\mathbf{x}}(k-1|k-1)$ and the square-root matrix of the estimation error variance $\mathbf{S}(k-1|k-1)$ have been obtained, the state estimate $\hat{\mathbf{x}}(k|k)$ and the square-root matrix of the estimation error variance $\mathbf{S}(k|k)$ can be calculated according to the following steps:*

Step 1: Time Update

$$\mathbf{x}_i(k-1|k-1)=\mathbf{S}(k-1|k-1)\boldsymbol{\xi}_i+\hat{\mathbf{x}}(k-1|k-1), \tag{12}$$

$$\mathbf{x}_i^*(k|k-1)=f\left(\mathbf{x}_i(k-1|k-1)\right), \tag{13}$$

$$\hat{\mathbf{x}}(k|k-1)=\frac{1}{N_x}\sum_{i=1}^{N_x}\mathbf{x}_i^*(k|k-1). \tag{14}$$

In (12)–(14), $i = 1, 2, \cdots, N_x$, $N_x = 2n$. The parameters ξ_i are given below (ε_i is an n-order unit vector):

$$\xi_i = \begin{cases} \sqrt{N_x/2} \cdot \varepsilon_i & , i = 1, 2, \cdots, n, \\ -\sqrt{N_x/2} \cdot \varepsilon_{i-n_x} & , i = n+1, n+2, \cdots, N_x. \end{cases}$$

Calculate the square root of the variance matrix:

$$\mathbf{S}(k|k-1) = Tria([\mathbf{X}^*(k|k-1) \quad \mathbf{S}_Q(k)]), \tag{15}$$

where $Tria(\cdot)$ represents a triangular operation. $\mathbf{S}_Q(k)$ represents the square-root of the new process noise variance $\mathbf{Q}(k)$, that is, $\mathbf{Q}(k) = \mathbf{S}_Q(k)\mathbf{S}_Q^T(k)$, and

$$\mathbf{X}^*(k|k-1) = \frac{1}{\sqrt{N_x}} \left[\mathbf{x}_1^*(k|k-1) - \hat{\mathbf{x}}(k|k-1) \ , \ \mathbf{x}_2^*(k|k-1) - \hat{\mathbf{x}}(k|k-1), \ \cdots, \mathbf{x}_{N_x}^*(k|k-1) - \hat{\mathbf{x}}(k|k-1) \right].$$

Step 2: Measurement Update

(1) According to Equations (16)–(18), we can calculate the predicted value $\hat{\mathbf{z}}(k|k-1)$.

$$\mathbf{x}_i(k|k-1) = \mathbf{S}(k|k-1)\xi_i + \hat{\mathbf{x}}(k|k-1), \tag{16}$$

$$\mathbf{z}_i(k|k-1) = \mathbf{h}(\mathbf{x}_i(k|k-1)), \tag{17}$$

$$\hat{\mathbf{z}}(k|k-1) = \frac{1}{N_x} \sum_{i=1}^{N_x} \mathbf{z}_i(k|k-1). \tag{18}$$

(2) The measurement noise variance $\hat{\mathbf{R}}^1(k)$ of the MAP estimator is calculated by using Lemma 1 (19) or Lemma 2 (20).

Lemma 1 ([14]). *When the measurement noise variance is constant, the suboptimal MAP estimate $\hat{\mathbf{R}}(k)$ of noise variance $\mathbf{R}(k)$ can be obtained by the recursive calculation:*

$$\begin{aligned} \hat{\mathbf{R}}(k) = \frac{1}{k}[(k-1)\hat{\mathbf{R}}(k-1) + \tilde{\mathbf{z}}(k)\tilde{\mathbf{z}}^T(k) \\ -\mathbf{z}(k|k-1)\mathbf{z}^T(k|k-1)], \end{aligned} \tag{19}$$

where $\tilde{\mathbf{z}}(k) = \mathbf{z}(k) - \hat{\mathbf{z}}(k|k-1)$ is the residual vector of measurement. The initial value $\hat{\mathbf{R}}(0) = \mathbf{R}_0$.

Lemma 2 ([15]). *When the measurement noise variance is time-varying, the suboptimal MAP estimate $\hat{\mathbf{R}}(k)$ of noise variance $\mathbf{R}(k)$ can be obtained by the recursive calculation:*

$$\begin{aligned} \hat{\mathbf{R}}(k) = [1 - d(k-1)]\hat{\mathbf{R}}(k-1) + d(k-1)[\tilde{\mathbf{z}}(k)\tilde{\mathbf{z}}^T(k) \\ -\mathbf{z}(k|k-1)\mathbf{z}^T(k|k-1)], \end{aligned} \tag{20}$$

where $d(k) = (1-b)/(1-b^{k+1})$. b is the forgetting factor, and its value range is usually between 0.95 and 0.99.

(3) Using the following Equation (21), combined with Equations (16)–(18), (22), (25), (26), and (23)–(32), we can iteratively compute the measurement noise variance estimation $\hat{\mathbf{R}}^2(k)$ of the VB method.

Estimation of prediction parameters of measurement noise variance by VB method:

$$\begin{cases} \zeta(k|k-1) = \rho \cdot \zeta(k-1), \\ \eta(k|k-1) = \rho \cdot \eta(k-1), \end{cases} \tag{21}$$

where "·" represents the point operation in Matlab. $\boldsymbol{\rho} = [\rho_1, \cdots, \rho_m]^T$, $\boldsymbol{\zeta}(k) = [\zeta_1(k), \cdots, \zeta_m(k)]^T$, $\boldsymbol{\eta}(k) = [\eta_1(k), \cdots, \eta_m(k)]^T$. $\zeta_i(k)$ and $\eta_i(k)$ are two parameters of the inverse gamma distribution. $\rho_i \subset (0,1)$ is the predictive weighting factor. It reflects the degree of correlation between the noise at the last moment and the noise at the current moment. When the difference between the measurement noise variance at the last moment and the measurement noise variance at the current moment is small, a larger ρ_i value should be used. Conversely, ρ_i should take a smaller value.

The square-root of the variance matrix:

$$\mathbf{S}_{zz}(k|k-1) = Tria([\mathbf{z}(k|k-1) \;\; \mathbf{S}_R(k)]), \tag{22}$$

where $\mathbf{S}_R(k)$ represents the square root of measurement noise variance $\mathbf{R}(k)$ (namely, $\mathbf{R}(k) = \mathbf{S}_R(k)\mathbf{S}_R^T(k)$) and

$$\begin{aligned} \mathbf{z}(k|k-1) &= \frac{1}{\sqrt{N_x}}[\mathbf{z}_1(k|k-1) - \hat{\mathbf{z}}(k|k-1) \\ &\quad \mathbf{z}_2(k|k-1) - \hat{\mathbf{z}}(k|k-1) \\ &\quad \cdots \quad \mathbf{z}_{N_x}(k|k-1) - \hat{\mathbf{z}}(k|k-1)], \\ \mathbf{P}_{xz}(k|k-1) &= \mathbf{x}(k|k-1)\mathbf{z}^T(k|k-1), \end{aligned} \tag{23}$$

where

$$\begin{aligned} \mathbf{x}(k|k-1) &= \frac{1}{\sqrt{N_x}}[\mathbf{x}_1(k|k-1) - \hat{\mathbf{x}}(k|k-1) \\ &\quad \mathbf{x}_2(k|k-1) - \hat{\mathbf{x}}(k|k-1) \\ &\quad \cdots \quad \mathbf{x}_{N_x}(k|k-1) - \hat{\mathbf{x}}(k|k-1)]. \end{aligned}$$

Then, the VB method gets $\hat{R}(k)$ through M iterations:

Iterative initialization: let $t = 1$, for a given number of iterations M, we have

$$\begin{cases} \boldsymbol{\zeta}(k) = [1/2, 1/2, \cdots, 1/2]^T + \boldsymbol{\zeta}(k|k-1), \\ \hat{\mathbf{x}}^1(k|k-1) = \hat{\mathbf{x}}(k|k-1). \end{cases} \tag{24}$$

Calculating the estimate of the measurement noise variance:

$$\hat{\mathbf{R}}^t(k) = diag(\boldsymbol{\eta}^t(k) \cdot / \boldsymbol{\zeta}(k)), \tag{25}$$

where $diag(A)$ represents a diagonal matrix composed of matrix A diagonal elements.

Use Equations (16) and (32) to calculate the t-th iteration state estimate $\hat{\mathbf{x}}^t(k|k)$ and the root of its mean square error matrix $\mathbf{S}^t(k|k)$. If $t < M$, update the parameter $\boldsymbol{\eta}^t(k)$.

$$\boldsymbol{\eta}^t(k) = \boldsymbol{\eta}(k|k-1) + (\mathbf{z}(k) - \hat{\mathbf{z}}^t(k|k-1))^{\cdot 2}/2 + diag\{P_{zz}^t(k|k)\}/2. \tag{26}$$

Let $t = t+1$, $\hat{\mathbf{x}}^t(k|k-1) = \hat{\mathbf{x}}^{t-1}(k|k)$, return to the beginning of the iterative.

When $t = M$, end the iteration, we can get

$$\begin{cases} \boldsymbol{\eta}(k) = \boldsymbol{\eta}^M(k) \\ \hat{\mathbf{R}}(k) = \hat{\mathbf{R}}^M(k) \end{cases}, \quad \begin{cases} \hat{\mathbf{x}}(k|k) = \hat{\mathbf{x}}^M(k|k) \\ \mathbf{S}(k|k) = \mathbf{S}^M(k|k) \end{cases}. \tag{27}$$

(4) Calculate the fusion estimation $\hat{\mathbf{R}}^g(k)$ of the measurement noise variance according to Equation (3).

(5) According to Equations (28) and (29), $\mathbf{S}_{zz}(k|k-1)$ and $\mathbf{P}_{xz}(k|k-1)$ are adaptively adjusted using the FFA fading factor.

$$\mathbf{S}_{zz}(k|k-1) = Tria\left([\mathbf{z}(k|k-1) \Big/ \cdot \sqrt{\tau(k)} \;\; \mathbf{S}_R(k)]\right) \tag{28}$$

$$\mathbf{P}_{xz}(k|k-1) = \frac{1}{\tau(k)}\mathbf{x}(k|k-1)\mathbf{z}^T(k|k-1), \tag{29}$$

where $\tau(k)$ is the adaptive factor, $0 < \tau(k) \le 1$, and it is calculated by the following Lemma 3.

Lemma 3 ([11]). *For nonlinear systems with unknown measurement noise variance, the adaptive fading factor is determined by the following equation:*

$$\tau(k) = \begin{cases} 1, & tr[\tilde{\mathbf{z}}(k|k-1)\tilde{\mathbf{z}}^T(k|k-1)] \le tr[\mathbf{S}_{zz}(k|k-1)\mathbf{S}_{zz}^T(k|k-1)], \\ \dfrac{tr[\mathbf{S}_{zz}(k|k-1)\mathbf{S}^T(k|k-1)]}{tr[\tilde{\mathbf{z}}(k|k-1)\tilde{\mathbf{z}}^T(k|k-1)]}, & tr[\tilde{\mathbf{z}}(k|k-1)\tilde{\mathbf{z}}^T(k|k-1)] > tr[\mathbf{S}_{zz}(k|k-1)\mathbf{S}_{zz}^T(k|k-1)]. \end{cases}$$

In the above equation, tr *denotes the trace of the matrix,* $\tilde{\mathbf{z}}(k|k-1) = \mathbf{z}(k) - \hat{\mathbf{z}}(k|k-1)$ *is the measurement residual error vector.*

(**6**) The following Equations (30)–(32) are used to obtain the updated estimation $\hat{\mathbf{x}}(k|k)$ and the root of its mean square error variance $\mathbf{S}(k|k)$.

$$\mathbf{K}(k) = \left(\mathbf{P}_{xz}(k|k-1)/\mathbf{S}_{zz}^T(k|k-1)\right)/\mathbf{S}_{zz}(k|k-1), \tag{30}$$

where the symbol "/" indicates the matrix right divide operation (e.g., $A/B = AB^{-1}$)

$$\hat{\mathbf{x}}(k|k) = \hat{\mathbf{x}}(k|k-1) + \mathbf{K}(k)[\mathbf{z}(k) - \hat{\mathbf{z}}(k|k-1)], \tag{31}$$

$$\mathbf{S}(k|k) = Tria\left([\mathbf{x}(k|k-1) - \mathbf{K}(k)\mathbf{z}(k|k-1)\ \mathbf{K}(k)\mathbf{S}_R(k)]\right). \tag{32}$$

Proof: We can directly derive Theorem 1 and Lemma 3 by Lemma 1, omitted here.

Note 3 Theorem 2 only shows the hybrid adaptive filtering algorithm when the number of noise variance estimators N_e is 2. Obviously, when $N_e > 2$, we first use various noise variance estimators to estimate $\hat{\mathbf{R}}^j(k)$ and then calculate the fused estimate $\hat{\mathbf{R}}^g(k)$ of the measurement noise variance according to Equation (11). In other words, only steps (2)–(4) in Theorem 2 need to be adjusted.

3.3. HASCIF Information Filter

Compared with traditional filtering, the information filter may not require prior information when it is initialized, and thus has better numerical performance. In addition, the use of an information filter to design a fusion algorithm is also simpler. In the information filter, the state estimate and its estimation error variance matrix are replaced by information vector and information matrix, respectively. Subsequently, we give its corresponding information filtering form (HASCIF) on the basis of HASCKF. According to the literature [20,21], we can obtain the implementation process of HASCIF as follows:

Step 1: Time update [20,21]

$$\hat{\mathbf{y}}(k|k-1) = \mathbf{Y}(k|k-1)\hat{\mathbf{x}}(k|k-1), \tag{33}$$

$$\mathbf{Y}(k|k-1) = \mathbf{P}^{-1}(k|k-1) = [\mathbf{S}(k|k-1)\mathbf{S}^T(k|k-1)]^{-1}, \tag{34}$$

where $\hat{\mathbf{y}}(k|k-1)$ and $\mathbf{Y}(k|k-1)$ are the predicted information vector and the predicted information matrix, respectively. $\hat{\mathbf{x}}(k|k-1)$ and $\mathbf{S}(k|k-1)$ can be calculated according to Equations (14) and (15).

Step 2: Measurement update

$$\begin{aligned}\hat{\mathbf{y}}(k|k) &= \mathbf{Y}(k|k)\hat{\mathbf{x}}(k|k) \\ &= \hat{\mathbf{y}}(k|k-1) + \mathbf{Y}(k|k-1)\mathbf{P}_{xz}(k|k)[\hat{\mathbf{R}}^{g}(k)]^{-1}\{\tilde{\mathbf{z}}(k|k-1) + \mathbf{P}_{xz}^{T}(k|k)\mathbf{Y}^{T}(k|k-1)\hat{\mathbf{x}}(k|k-1)\} \\ &\triangleq \hat{\mathbf{y}}(k|k-1) + \boldsymbol{\theta}(k),\end{aligned} \tag{35}$$

$$\begin{aligned}\mathbf{Y}(k|k) &= \mathbf{Y}(k|k-1) + \mathbf{Y}(k|k-1)\mathbf{P}_{xz}(k|k)[\hat{\mathbf{R}}^{g}(k)]^{-1}(k)\mathbf{P}_{xz}^{T}(k|k)\mathbf{Y}^{T}(k|k-1) \\ &\triangleq \mathbf{Y}(k|k-1) + \boldsymbol{\Theta}(k).\end{aligned} \tag{36}$$

In Equations (35) and (36), the information state vectors $\hat{\mathbf{y}}(k|k)$ and $\mathbf{Y}(k|k)$ are the information vector and information matrix of the state estimate, respectively. $\boldsymbol{\theta}(k)$ and $\boldsymbol{\Theta}(k)$ are the information contribution vector and the information contribution matrix. $\hat{\mathbf{R}}^{g}(k)$ is determined by formula (3), and the cross-variance matrix $\mathbf{P}_{xz}(k|k)$ can be calculated by Formula (29).

3.4. Performance Analysis of HASCKF

References [22–24] proposed the bounded convergence theorem of the UKF algorithm, and Reference [14] extended its theorem to the adaptive cubature Kalman filter (ACKF). In this section, the CKF bounded convergence theorem proposed in Reference [14] and the Cramer–Rao lower bound (CRLB) [25] are used to analyze the convergence of the HASCKF algorithm.

3.4.1. Stability Analysis

Assuming that the noise variances $\mathbf{Q}(k)$ and $\mathbf{R}(k)$ are known accurately, and considering the nonlinear systems (1), (2), and the standard CKF algorithm [4], the corresponding state error variance matrix can be written as [14,22,24]:

$$\begin{aligned}\mathbf{P}(k|k-1) &= \mathbf{x}(k|k-1)\mathbf{x}^{T}(k|k-1) + \mathbf{Q}(k-1) \\ &= \mathbf{A}(k)\mathbf{P}(k-1|k-1)\mathbf{A}^{T}(k) + \delta\mathbf{P}(k|k-1) + \mathbf{Q}(k-1) \\ &= \mathbf{A}(k)\mathbf{P}(k-1|k-1)\mathbf{A}^{T}(k) + \boldsymbol{\Xi}(k),\end{aligned} \tag{37}$$

$$\begin{aligned}\mathbf{P}_{zz}(k|k-1) &= \mathbf{z}(k|k-1)\mathbf{z}^{T}(k|k-1) + \mathbf{R}(k) \\ &= \mathbf{G}(k)\mathbf{P}(k|k-1)\mathbf{G}^{T}(k) + \delta\mathbf{P}_{zz}(k|k-1) + \mathbf{R}(k) \\ &= \mathbf{G}(k)\mathbf{P}(k|k-1)\mathbf{G}^{T}(k) + \boldsymbol{\Sigma}(k),\end{aligned} \tag{38}$$

$$\mathbf{K}(k) = \mathbf{P}(k|k-1)\mathbf{G}^{T}(k)[\mathbf{G}(k)\mathbf{P}(k|k-1)\mathbf{G}^{T}(k)+\boldsymbol{\Sigma}(k)]^{-1}, \tag{39}$$

where

$$\delta\mathbf{P}(k|k-1) = \mathbf{x}(k|k-1)\mathbf{x}^{T}(k|k-1) - \mathbf{A}(k)\mathbf{P}(k-1|k-1)\mathbf{A}^{T}(k), \tag{40}$$

$$\delta\mathbf{P}_{zz}(k|k-1) = \mathbf{z}(k|k-1)\mathbf{z}^{T}(k|k-1) - \mathbf{G}(k)\mathbf{P}(k|k-1)\mathbf{G}^{T}(k). \tag{41}$$

Each correlation matrix is defined as follows:

$$\begin{cases}\mathbf{A}(k) = \beta(k)\mathbf{F}(k), \\ \mathbf{B}(k) = \alpha(k)\mathbf{H}(k), \\ \mathbf{C}(k) = \mathbf{I} - \mathbf{K}(k)\alpha(k)\mathbf{H}(k),\end{cases} \tag{42}$$

$$\mathbf{G}(k) = \begin{cases}\alpha(k)\mathbf{H}(k)\gamma^{T}(k), & n \geq m, \\ \gamma^{T}(k)\alpha(k)\mathbf{H}(k), & n < m,\end{cases} \tag{43}$$

$$\begin{cases}\boldsymbol{\Xi}(k) = \delta\mathbf{P}(k|k-1) + \mathbf{Q}(k-1), \\ \boldsymbol{\Sigma}(k) = \delta\mathbf{P}_{zz}(k|k-1) + \mathbf{R}(k),\end{cases} \tag{44}$$

where

$$\mathbf{F}(k) = \frac{\partial \mathbf{f}}{\partial \mathbf{x}}\bigg|_{\mathbf{x}=\hat{\mathbf{x}}(k-1|k-1)}, \quad \mathbf{H}(k) = \frac{\partial \mathbf{h}}{\partial \mathbf{x}}\bigg|_{\mathbf{x}=\hat{\mathbf{x}}(k|k-1)}.$$

$\alpha(k)$, $\beta(k)$, and $\gamma(k)$ are auxiliary diagonal matrixes. Refer to Reference [22] for specific meanings.

Lemma 4 ([14]). *Consider the nonlinear systems (1), (2), and the standard CKF algorithm. If $\forall k \geq 0$, both satisfy the following two assumptions:*

(1) There are non-zero real numbers $a_{\min}$, $a_{\max}$, $b_{\max}$, $c_{\max}$, $g_{\min}$, and $g_{\max}$ existing to let the following formulas be established:

$$\begin{cases} a_{\min}^2\mathbf{I} \leq \mathbf{A}(k)\mathbf{A}^T(k) \leq a_{\max}^2\mathbf{I}, \quad \mathbf{B}(k)\mathbf{B}^T(k) \leq b_{\max}^2\mathbf{I}, \\ g_{\min}^2\mathbf{I} \leq \mathbf{G}(k)\mathbf{G}^T(k) \leq g_{\max}^2\mathbf{I}, \quad \mathbf{C}(k)\mathbf{C}^T(k) \leq c_{\max}^2\mathbf{I}, \\ [\mathbf{G}(k) - \mathbf{B}(k)][\mathbf{G}(k) - \mathbf{B}(k)]^T \leq (g_{\max} - b_{\max})^2\mathbf{I}. \end{cases} \tag{45}$$

(2) There are positive real numbers $p_{\min}$, $p_{\max}$, $q_{\max}$, $r_{\max}$, $\Xi_{\min}$, $\Xi_{\max}$, and $\Sigma_{\min}$ existing to let the following forms be established:

$$\begin{cases} p_{\min}\mathbf{I} \leq \mathbf{P}(k|k) \leq p_{\max}\mathbf{I}, \quad \mathbf{Q}(k) \leq q_{\max}\mathbf{I}, \\ \mathbf{R}(k) \leq r_{\max}\mathbf{I}, \quad \Xi(k) \leq \Xi_{\max}\mathbf{I}, \\ \Xi(k) > \Xi_{\min}\mathbf{I}, \quad \mathbf{\Sigma}(k) > \Sigma_{\min}\mathbf{I}, \end{cases} \tag{46}$$

where

$$\begin{cases} \Sigma_{\min} = \max(\Sigma_1, \Sigma_2), \\ \Sigma_1 = a_{\max}^2(g_{\max} - b_{\max})^2(p_{\max} + p_{\max}^2 a_{\max}^2 \Xi_{\min}^{-1}), \\ \Sigma_2 = b_{\max}^2(a_{\max}^2 p_{\max} + \Xi_{\max}) - g_{\max}^2(a_{\min}^2 p_{\min}\Xi_{\min}). \end{cases}$$

Then, the standard CKF state estimation error will be mean-square bounded, that is, the algorithm is stable and convergent.

Note 4 Lemma 4 shows that the statistical characteristics of noise are closely related to the stability of the CKF algorithm. In addition, if Lemma 4 is established, the SCKF with known noise statistics is also stable and convergent. This is because the theoretical framework of SCKF is consistent with that of CKF. The mean square root matrix $\mathbf{S}(k|k-1)$ and $\mathbf{S}(k|k)$ in CKF are only used when transferring the error variance. The root mean square matrix in the SCKF algorithm is obtained through triangulation of the matrix, thus avoiding the filter divergence caused by the non-positive definite variance matrix in the numerical calculation, so it has better stability while ensuring the accuracy of estimation with CKF.

Theorem 3. *If the standard SCKF algorithm is stable and convergent when the statistical characteristics of noise are known accurately, the introduction of the noise variance fusion estimator and the adaptive fade factor can ensure the stable convergence of the HASCKF algorithm.*

Proof. When the measured noise variance matrix $\mathbf{R}(k)$ is inaccurate, other sufficient conditions in the bounded convergence theorem can be satisfied, but the condition (46) will be affected.

(a) First only consider the influence of the weighted fusion noise estimator. Let $\Delta\mathbf{R}(k) = \hat{\mathbf{R}}^g(k) - \mathbf{R}(k)$, then formula (40) can be rewritten as follows:

$$\begin{aligned} \breve{\mathbf{\Sigma}}(k) &= \delta\mathbf{P}_{zz}(k|k-1) + \hat{\mathbf{R}}^g(k) \\ &= \delta\mathbf{P}_{zz}(k|k-1) + \mathbf{R}(k) + \Delta\mathbf{R}(k). \end{aligned} \tag{47}$$

When $\hat{\mathbf{R}}^g(k) \geq \mathbf{R}(k)$, then $\Delta\mathbf{R}(k) \geq 0$, and $\breve{\mathbf{\Sigma}}(k)$ will become larger. Obviously, condition (46) still holds and the HASCKF algorithm still converges steadily.

When $\hat{\mathbf{R}}^g(k) < \mathbf{R}(k)$, then $\Delta\mathbf{R}(k) < 0$, and $\breve{\mathbf{\Sigma}}(k)$ will become smaller. Now, condition (46) may not be satisfied. However, the noise variance weighted fusion estimator $\hat{\mathbf{R}}^g(k)$ estimates and corrects the inaccurate noise variance matrix in real time so that it gradually tracks the real value $\mathbf{R}(k)$, thus making $\Delta\mathbf{R}(k) \to 0$, $\breve{\mathbf{\Sigma}}(k) \to \mathbf{\Sigma}(k)$. In this way, $\Delta\mathbf{R}(k) \geq 0$ is gradually satisfied to Equation (42) to ensure stable convergence of the HASCKF algorithm.

(b) Further consider the effect of fading factors. According to Theorem 2, the variance matrix for measuring residuals is

$$\begin{aligned}\breve{\mathbf{P}}_{zz}(k|k-1) &= \mathbf{S}_{zz}(k|k-1)\mathbf{S}_{zz}^T(k|k-1)\\ &= \frac{1}{\tau(k)}\mathbf{z}(k|k-1)\mathbf{z}^T(k|k-1) + \hat{\mathbf{R}}^g(k).\end{aligned} \tag{48}$$

Then, Equation (41) is rewritten as:

$$\begin{aligned}\delta\breve{\mathbf{P}}_{zz}(k|k-1) &= \frac{1}{\tau(k)}\mathbf{z}(k|k-1)\mathbf{z}^T(k|k-1) - \mathbf{G}(k)\mathbf{P}(k|k-1)\mathbf{G}^T(k)\\ &= [\frac{1}{\tau(k)} - 1]\mathbf{z}(k|k-1)\mathbf{z}^T(k|k-1) + \mathbf{z}(k|k-1)\mathbf{z}^T(k|k-1) - \mathbf{G}(k)\mathbf{P}(k|k-1)\mathbf{G}^T(k)\\ &= [\frac{1}{\tau(k)} - 1]\mathbf{z}(k|k-1)\mathbf{z}^T(k|k-1) + \delta\mathbf{P}_{zz}(k|k-1)\\ &\triangleq \Delta\mathbf{P}_{zz}(k|k-1) + \delta\mathbf{P}_{zz}(k|k-1).\end{aligned} \tag{49}$$

Therefore, Equation (44) can be written as

$$\begin{aligned}\breve{\mathbf{\Sigma}}(k) &= \delta\breve{\mathbf{P}}_{zz}(k|k-1) + \hat{\mathbf{R}}^g(k)\\ &= \delta\mathbf{P}_{zz}(k|k-1) + \mathbf{R}(k) + \Delta\mathbf{P}_{zz}(k|k-1) + \Delta\mathbf{R}(k)\\ &\triangleq \mathbf{\Sigma}(k) + \Delta\mathbf{\Sigma}(k).\end{aligned} \tag{50}$$

It is known from the definition of fading factor $0 < \tau(k) \leq 1$, so we can get

$$\Delta\mathbf{P}_{zz}(k|k-1) = [1/\tau(k) - 1]\mathbf{z}(k|k-1)\mathbf{z}^T(k|k-1) \geq 0. \tag{51}$$

Similar to the analysis in (a), when $\Delta\mathbf{R}(k) \geq 0$ and $\Delta\mathbf{\Sigma}(k) \geq 0$, it is obvious that condition (46) holds and the stability of the HASCKF algorithm remains. When $\Delta\mathbf{R}(k) < 0$, if $\Delta\mathbf{P}_{zz}(k|k-1)$ is large enough, $\Delta\mathbf{\Sigma}(k) \geq 0$ can still be satisfied, and condition (46) still holds, so the HASCKF algorithm converges steadily. If $\Delta\mathbf{P}_{zz}(k|k-1)$ is not enough to guarantee $\Delta\mathbf{\Sigma}(k) \geq 0$, the introduction of the weighted fusion noise variance estimator can also make $\Delta\mathbf{R}(k) \to 0$, so that $\Delta\mathbf{\Sigma}(k) \geq 0$ is established stepwise to ensure stable convergence of the HASCKF algorithm.

In summary, the introduction of the noise variance fusion estimator and adaptive fading factor in the HASCKF algorithm can ensure the stable convergence of the algorithm. □

Note 5 This theorem combined with Note 2 shows that the noise variance estimation based on weighted fusion is superior to the estimate of any single noise variance estimator. Therefore, the hybrid adaptive HASCKF estimation algorithm has better stability than the adaptive SCKF using a single noise variance estimation algorithm.

3.4.2. Filtering Accuracy Analysis

There is a lower bound on the minimum variance unbiased estimator of the state of the nonlinear filtering algorithm. It is widely used to evaluate the performance of nonlinear estimation. In practice, the lower limit of Cramer–Rao Lower Bound (CRLB) is commonly used. Denote

$\mathbf{X}_{0:k} = \{\mathbf{x}(0), \mathbf{x}(1), \cdots, \mathbf{x}(k)\}$, $\mathbf{Z}_{0:k} = \{\mathbf{z}(0), \mathbf{z}(1), \cdots, \mathbf{z}(k)\}$, and $p(\mathbf{Z}_{0:k}, \mathbf{X}_{0:k})$ is the joint probability density of $(\mathbf{Z}_{0:k}, \mathbf{X}_{0:k})$. $\hat{\mathbf{x}}(k)$ is the unbiased estimation of $\mathbf{x}(k)$. Then, CRLB is defined as [25]:

$$\mathbf{P}(k|k) \triangleq E\{[\mathbf{x}(k) - \hat{\mathbf{x}}(k)][\mathbf{x}(k) - \hat{\mathbf{x}}(k)]^T\} \geq \mathbf{J}^{-1}(k), \tag{52}$$

where $\mathbf{J}(k)$ is the Fisher information matrix:

$$\mathbf{J}(k) = E\left[-\frac{\partial^2 \ln p(\mathbf{Z}_{0:k}, \mathbf{X}_{0:k})}{\partial^2 \mathbf{x}^2(k)}\right]. \tag{53}$$

Theorem 4 Assume that the nonlinear filter is applied to system (1), (2). Then

$$\mathbf{P}(k|k) \geq \mathbf{J}^{-1}(k), \tag{54}$$

where

$$\mathbf{J}(k) = [\mathbf{Q}(k-1) + \mathbf{F}(k)\mathbf{J}^{-1}(k-1)\mathbf{F}^T(k)]^{-1} + \mathbf{H}^T(k)\mathbf{R}^{-1}(k)\mathbf{H}(k). \tag{55}$$

Proof. We use $\mathbf{J}(k)$ to denote the information matrix of state $\mathbf{x}(k)$. Then, the information matrix $\mathbf{J}(k)$ can be recursively calculated according to the following formula [26]:

$$\mathbf{J}(k) = \mathbf{D}_{22}(k) - \mathbf{D}_{21}(k)[\mathbf{J}(k-1) + \mathbf{D}_{11}(k)]^{-1}\mathbf{D}_{12}(k), \tag{56}$$

where

$$\begin{cases} \mathbf{D}_{11}(k) = E\{-\Delta_{\mathbf{x}(k-1)}^{\mathbf{x}(k-1)} \ln p[\mathbf{x}(k)|\mathbf{x}(k-1)]\}, \\ \mathbf{D}_{12}(k) = E\{-\Delta_{\mathbf{x}(k-1)}^{\mathbf{x}(k)} \ln p[\mathbf{x}(k)|\mathbf{x}(k-1)]\}, \\ \mathbf{D}_{21}(k) = E\{-\Delta_{\mathbf{x}(k)}^{\mathbf{x}(k-1)} \ln p[\mathbf{x}(k)|\mathbf{x}(k-1)]\} \\ \qquad = \mathbf{D}_{12}^T(k), \\ \mathbf{D}_{22}(k) = E\{-\Delta_{\mathbf{x}(k)}^{\mathbf{x}(k)} \ln p[\mathbf{x}(k)|\mathbf{x}(k-1)]\} \\ \qquad\qquad + E\{-\Delta_{\mathbf{x}(k)}^{\mathbf{x}(k)} \ln p[\mathbf{z}(k)|\mathbf{x}(k)]\}, \end{cases} \tag{57}$$

where $\Delta_{\mathbf{x}}^{\mathbf{y}} = \nabla_{\mathbf{x}} \nabla_{y}^{T}$ means the second-order operator. $\nabla_{\mathbf{x}} = \frac{\partial}{\partial \mathbf{x}}$ is the first-order operator. Because process noise and measurement noise are Gaussian white noise, we have

$$\begin{aligned} &-\ln p[\mathbf{x}(k)|\mathbf{x}(k-1)] \\ &= -\ln\left\{\frac{e^{\{-\frac{1}{2}[\mathbf{x}(k)-\mathbf{f}(\mathbf{x}(k-1))]^T\mathbf{Q}^{-1}(k-1)[\mathbf{x}(k)-\mathbf{f}(\mathbf{x}(k-1))]\}}}{\sqrt{2\pi}\,|\mathbf{Q}(k-1)|}\right\} \\ &= c_1 + \frac{1}{2}[\mathbf{x}(k) - \mathbf{f}(\mathbf{x}(k-1))]^T\mathbf{Q}^{-1}(k-1)[\mathbf{x}(k) - \mathbf{f}(\mathbf{x}(k-1))] \end{aligned} \tag{58}$$

$$\begin{aligned} &-\ln p[\mathbf{z}(k)|\mathbf{x}(k)] \\ &= -\ln\left\{\frac{e^{\{-\frac{1}{2}[\mathbf{z}(k)-\mathbf{h}(\mathbf{x}(k))]^T\mathbf{R}^{-1}(k)[\mathbf{z}(k)-\mathbf{h}(\mathbf{x}(k))]\}}}{\sqrt{2\pi}\,|\mathbf{R}(k)|}\right\} \\ &= c_2 + \frac{1}{2}[\mathbf{z}(k) - \mathbf{h}(\mathbf{x}(k))]^T\mathbf{R}^{-1}(k)[\mathbf{z}(k) - \mathbf{h}(\mathbf{x}(k))], \end{aligned} \tag{59}$$

where constant $c_1 = \ln[\sqrt{2\pi}\,|\mathbf{Q}(k-1)|]$, $c_2 = \ln[\sqrt{2\pi}\,|\mathbf{R}(k)|]$.

Substituting Equations (58) and (59) into Equation (57), then we can get

$$\begin{cases} \mathbf{D}_{11}(k) = E\{[\nabla_{\mathbf{x}(k-1)}\mathbf{f}^T(\mathbf{x}(k-1))]\mathbf{Q}^{-1}(k-1)[\nabla_{\mathbf{x}(k-1)}\mathbf{f}^T(\mathbf{x}(k-1))]^T\} \\ \qquad = \mathbf{F}^T(k)\mathbf{Q}^{-1}(k-1)\mathbf{F}(k), \\ \mathbf{D}_{12}(k) = E\{[\nabla_{\mathbf{x}(k-1)}\mathbf{f}^T(\mathbf{x}(k-1))]\mathbf{Q}^{-1}(k-1) \\ \qquad = \mathbf{F}^T(k)\mathbf{Q}^{-1}(k-1) = \mathbf{D}_{21}^T(k), \\ \mathbf{D}_{22}(k) = \mathbf{Q}^{-1}(k-1) + E\{[\nabla_{\mathbf{x}(k)}\mathbf{h}^T(\mathbf{x}(k))]\mathbf{R}^{-1}(k)[\nabla_{\mathbf{x}(k)}\mathbf{h}^T(\mathbf{x}(k))]^T\} \\ \qquad = \mathbf{Q}^{-1}(k-1) + \mathbf{H}^T(k)\mathbf{R}^{-1}(k)\mathbf{H}(k). \end{cases} \tag{60}$$

$$\begin{aligned} \mathbf{J}(k) = {} & \mathbf{Q}^{-1}(k-1) + \mathbf{H}^T(k)\mathbf{R}^{-1}(k)\mathbf{H}(k) \\ & -\mathbf{Q}^{-1}(k-1)\mathbf{F}(k) \times [\mathbf{J}(k-1) + \mathbf{F}^T(k)\mathbf{Q}^{-1}(k-1)\mathbf{F}(k)]^{-1}\mathbf{F}^T(k)\mathbf{Q}^{-1}(k-1). \end{aligned} \tag{61}$$

Equation (56) can be obtained by applying a matrix inversion lemma to Equation (61). □

4. Simulation

In this section, two numerical simulation examples are provided to display and verify the performance of the SACKF algorithm proposed in this paper, mainly including the following contents:

(1) For a class of inaccurate modeling with unknown measurement noise variance, we compare the performance difference between the weighted fusion estimator (referred to as WF-NE) and the single noise estimator (e.g., MAP estimation or VB estimation, respectively denoted as MAP-NE and VB-NE).

(2) For a class of inaccurate modeling with unknown measurement noise variance, we study the advantages and disadvantages of the HASCKF algorithm and the standard SCKF algorithm, and the equivalence between the HASCKF and HASCIF algorithms.

Example 1. *This example is used to evaluate the performance of three kinds of noise estimators, MAP-NE, VB-NE, and WF-NE. Considering the following first-order nonlinear discrete dynamic system:*

$$x(k) = 0.5x(k-1) + \frac{0.2x(k-1)}{1 + x^2(k-1)} + w(k), \tag{62}$$

$$z(k) = \frac{x^2(k)}{20} + v(k), \tag{63}$$

where $w(k)$ and $v(k)$ are mutually independent Gaussian white noise sequences. Assume that the initial value of the system state and the error variance matrix are

$$x_0 = 2 \ , \ P_0 = 0.01, \tag{64}$$

and the system state initial value x_0 is independent of the two noises. The process noise statistic $Q(k) = 0.001$. In the following, simulation experiments are performed for two cases where the measurement noise variance $R(k)$ is a constant and piecewise continuous function.

(1) Situation 1: If the measurement noise variance is constant and $R(k) = 0.012$. Assume that the imprecise measurement noise variance of the initial value $\hat{R}_0 = 0.04$. In the simulation, MAP-NE adopted the estimator described in Equation (19). The parameters of VB-NE were selected as follows: $\rho = 1 - e^{-5}$, $\zeta(0) = 1$, $\eta(0) = 0.04$, $M = 1$.

In order to compare the performance of various algorithms, we adopted the absolute error (AE) and the mean absolute error (MAE), which are calculated as follows:

$$AE(k) = |y(k) - \hat{y}(k)| \quad , \quad MAE(k) = \frac{1}{N_s}\sum_{k=1}^{N_s} AE(k), \tag{65}$$

where $y(k)$ and $\hat{y}(k)$ are the value to be estimated and the estimated value respectively. N_s is the number of simulation steps. In this case, $N_s = 1000$.

The estimated noise variance of the three kinds of noise estimators are shown in Figures 2 and 3, and the estimation error is given in Table 1.

(2) Situation 2: If the measurement noise value is time varying, the true statistical characteristics meet the following formula:

$$R(k) = \begin{cases} R & 0 \leq k \leq 500 \\ 2R & 501 \leq k \leq 1000 \end{cases}, \; and \; R = 0.012. \tag{66}$$

Assume that the known inaccurate initial value of the measurement noise variance is $\hat{R}_0 = 0.08$. In the simulation, MAP-NE adopted the estimator described in Equation (20), and the forgetting factor b was 0.98. The parameters of VB-NE were selected as $\rho = 1 - e^{-5}$, $\zeta(0) = 1$, $\eta(0) = 0.08$, $M = 1$. The estimation results of the three kinds of noise estimators on the measured noise variance are shown in Figures 4 and 5, and the mean absolute errors of several algorithms are given in Table 2.

Table 1. The mean absolute error of three noise estimation algorithms in Situation 1.

Algorithm	MAP-NE	VB-NE	WF-NE
Mean absolute error	0.0010	0.0015	0.0005
CPU time cost	0.2188	0.2188	0.3725

Table 2. The mean absolute error of three noise estimation algorithms in Situation 2.

Algorithm	MAP-NE	VB-NE	WF-NE
Mean absolute error	0.0024	0.0022	0.0019

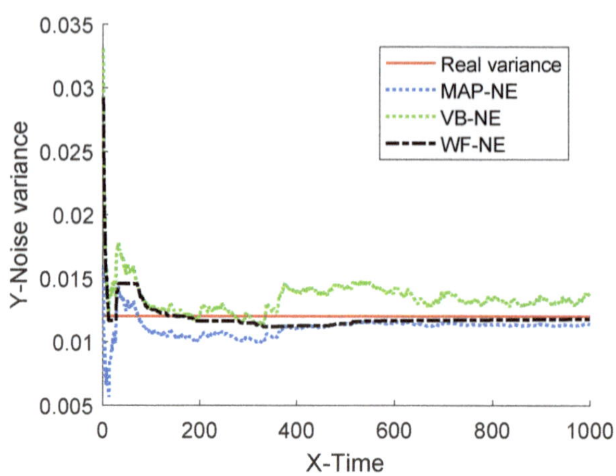

Figure 2. Estimation of measurement noise variance.

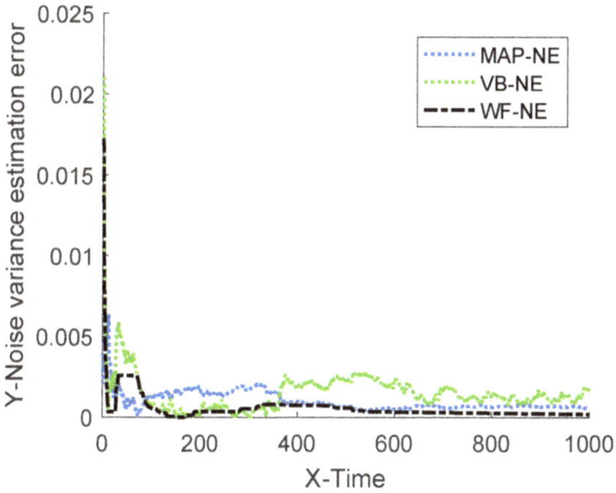

Figure 3. Absolute estimation error of measurement noise variance.

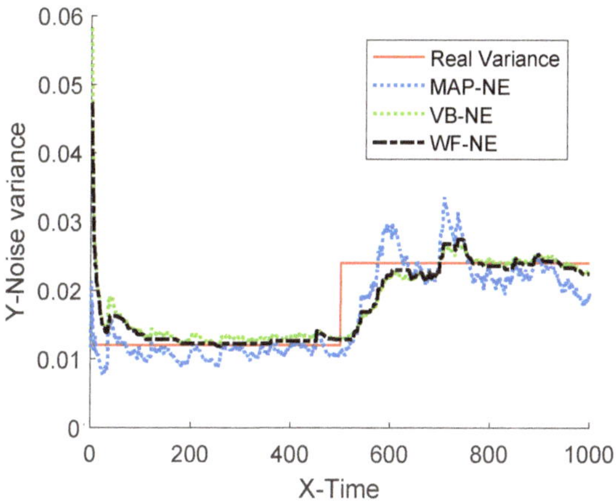

Figure 4. Estimation of measurement noise variance.

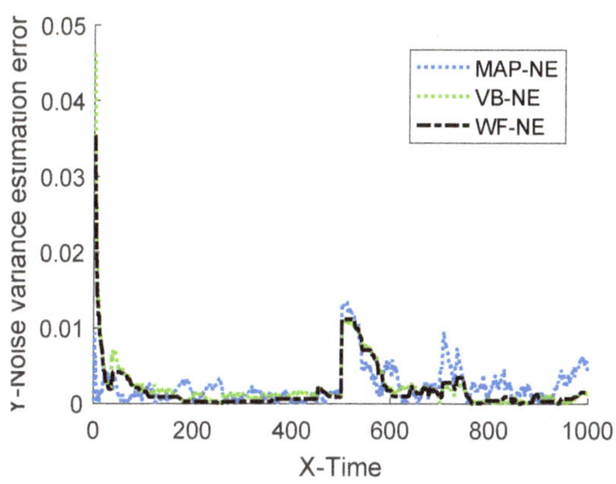

Figure 5. Absolute estimation error of measurement noise variance.

The simulation results of Example 1 show that the fusion estimator WF-NE proposed in Theorem 1 could obtain the best estimated result for either the constant unknown measurement noise variance or for the time-varying one, compared with the MAP-NE and VB-NE. It had the same result with the analysis conclusion in Note 2. In Situation 1, as shown in Figures 2 and 3, compared with the fusion results of MAP-NE and VB-NE, the WF-NE had better estimates in most simulation steps. Due to the randomness of noise, not all the noise variance estimates obtained by WF-NE were better than the ones solved by other methods, especially for the estimation of the time-varying variance in Situation 2. As shown in Figure 3, the MAP-NE method had relatively large estimation errors in the time interval [160–360], and the VB-NE method had relatively large estimation errors in the time interval [390–800]. Neththerless, the WF-NE could avoid large variation of the noise variance estimation error. Similarly, the general trend of the WF-NE method was best in Figure 4, although the MAP-NE method or the VB-NE method were best in some small time intervals. Although the CPU time cost of WF-NE (0.3725 for 1000 simulation steps) was larger than the other two methods (0.2188 for 1000 simulation steps), it was still acceptable. Meanwhile, the mean absolute errors of WF-NE were smaller than the other two methods, for boththe constant noise variance (in Situation 1) and for the time-varying variance (in Situation 2).

Example 2. *This example is used to verify the pros and cons of the hybrid adaptive SCKF estimation algorithm HASCKF proposed by Theorem 2 and the standard SCKF algorithm described in Reference [4]. Assume that the target is moving in a uniform linear motion on a two-dimensional plane. The system state* $x(k) = [x(k), \dot{x}(k), y(k), \dot{y}(k)]^T$*, where* $x(k)$ *and* $y(k)$ *are the position components in the east–north coordinate system.* $\dot{x}(k)$ *and* $\dot{y}(k)$ *are the corresponding velocity components, respectively. Then, the state equation can be described as:*

$$x(k) = \begin{bmatrix} 1 & T & 0 & 0 \\ 0 & 1 & 0 & 0 \\ 0 & 0 & 1 & T \\ 0 & 0 & 0 & 1 \end{bmatrix} \cdot x(k-1) + w(k), \tag{67}$$

where the sampling period $T = 0.5$ *s, the process noise* $w(k)$ *is zero mean Gaussian white noise. Its statistical characteristic is* $Q(k) = 0.1 \times diag(Q1, Q2)$*, and* $Q1 = \begin{bmatrix} T^3/3 & T^2/2 \\ T^2/2 & T \end{bmatrix}$.

Consider a radar to track the target. The nonlinear measurement equation can be expressed as

$$z(k) = \begin{bmatrix} \sqrt{x^2(k) + y^2(k)} \\ \arctan\left(\dfrac{y(k)}{x(k)}\right) \end{bmatrix} + v(k). \tag{68}$$

The real statistical characteristic of the measuring noise $R(k) = diag\{(4m)^2, (0.1^\circ)^2\}$*. During the simulation, the simulation time was set as* $N_s = 100$ *s. Suppose that the inaccuracy initial value of the measurement noise variance* $\hat{R}(0) = diag\{(81m)^2, (0.3^\circ)^2\}$*. MAP-NE adopts the estimator described in Equation (19), and the parameters of VB-NE were selected as* $\rho = [1-e^{-5}, 1-e^{-5}]^T$*,* $\zeta(0) = [1,1]^T$*,* $\eta(0) = [100, 0.02]^T$*,* $M = 1$*. The initial state estimate and the estimation error variance matrix are*

$x_0 = [10,000\,\text{m}, 150\,\text{m/s}, 15,000\,\text{m}, 200\,\text{m/s}]^T$,

$P_0 = diag\{(100\,\text{m})^2, (14\,\text{m/s})^2, (100\,\text{m})^2, (15\,\text{m/s})^2\}$.

The estimated results of SCKF and HASCKF algorithms are shown in Figures 6–10. The estimation error is given in Table 3.

From Figure 6 to Figure 10, it can be clearly seen that the standard SCKF estimation showed a large deviation after 40 s. However, HASCKF could still better estimate the state of the target. This is because the standard SCKF adopts an inaccurate prior noise variance $\hat{R}(0)$, while the HASCKF proposed in this paper estimates and corrects the inaccurate measurement noise variance, thus ensuring the accuracy and stability of the algorithm. As analyzed in Example 1, due to the randomness of noise, not all the noise variance estimates obtained by WF-NE were better. Therefore, the estimates of some components of the state obtained by SCKF were better than the one solved by HASCKF at some time in the interval [10 s, 40 s]. However, the general tend of the proposed HASCKF was better than SCKF, as shown in Figures 6–10.

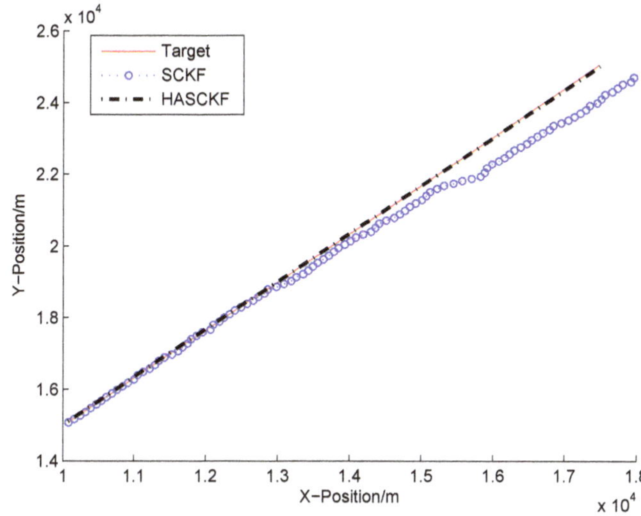

Figure 6. Target's trajectory and tracking result of SCKF and HASCKF.

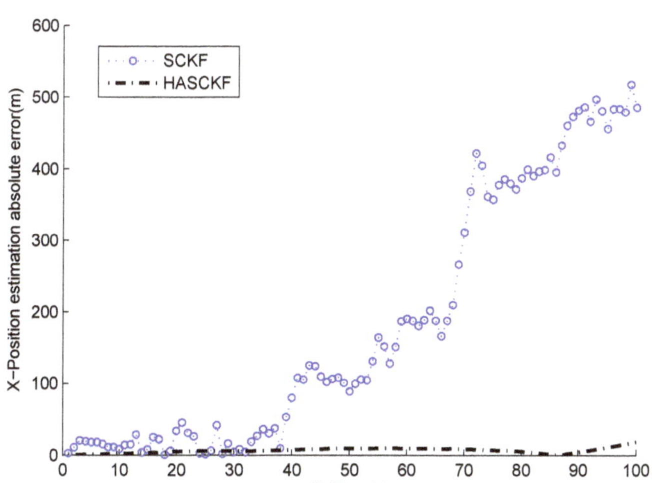

Figure 7. Absolute error curves of X-displacement.

Table 3. The mean absolute error of two algorithms.

Mean Absolute Error	Algorithms	
	SCKF	HASCKF
X-Position (m)	182.9477	6.4784
X-Velocity (m/s)	5.2113	0.7898
Y-Position (m)	129.5228	24.6147
Y-Velocity (m/s)	3.6766	1.0834

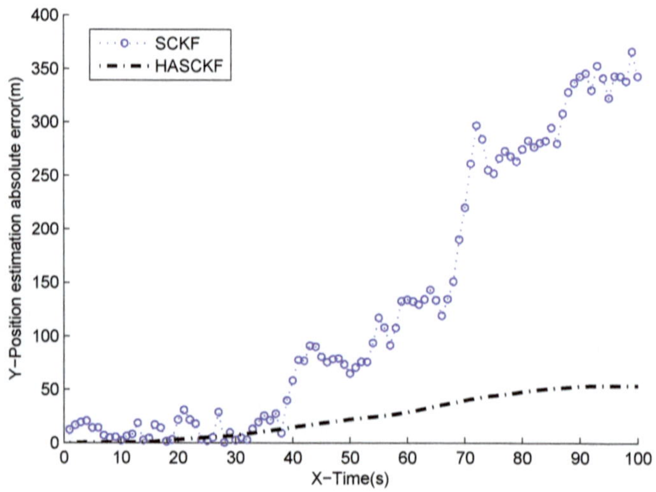

Figure 8. Absolute error curves of Y-displacement.

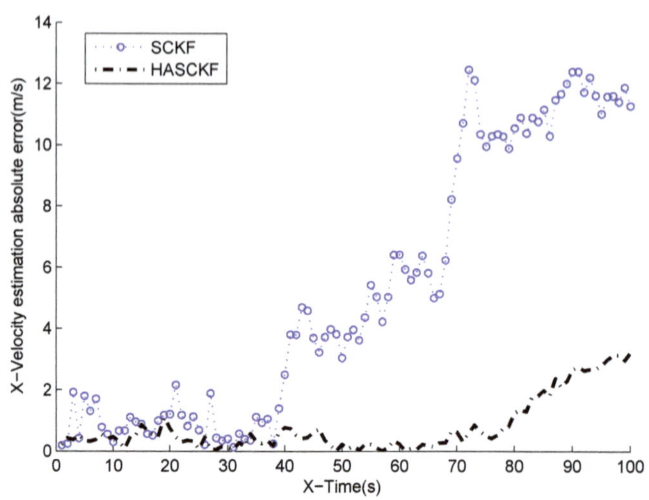

Figure 9. Absolute error curves of X-velocity.

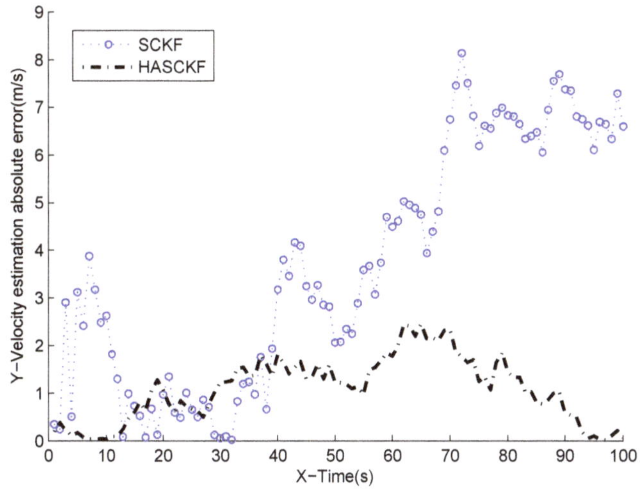

Figure 10. Absolute error curves of Y-velocity.

5. Conclusions

In this paper, aiming at the filtering problem caused by the inaccurate measurement noise variance in real systems, a weighted fusion estimation method is designed and a class of hybrid adaptive filtering structures is proposed, based on different measurement noise variance estimators. The stability and the filtering accuracy of the hybrid adaptive filter are analyzed and discussed theoretically. The simulation results showed that the algorithm had excellent accuracy and stability.

The work that can continue to be studied in the future includes further study on the research and design of fusion methods based on hybrid adaptive filtering in nonlinear centralized fusion framework and distributed fusion framework, research on functional equivalence among various fusion algorithms, and so on.

Author Contributions: Conceptualization, Y.S.; Methodology, Y.S.; Software, D.B.; Validation, X.Z.; Formal Analysis, X.F. and X.T.; Writing—Original Draft Preparation, Y.S. and X.F.; Writing—Review & Editing, X.Z. and X.T.

Funding: This research was supported in part by the National Natural Science Foundation of China (Grant Nos. 61503174, 61503336, 61503136, and 61602141), Henan higher education school young backbone teacher training program (2017GGJS160), Henan University of Animal Husbandry and Economy key scientific research innovation fund (XKYCXJJ2017009).

Conflicts of Interest: The authors declare no conflict of interest.

References

1. Han, C.Z.; Zhu, H.Y.; Duan, Z.S. *Multi-source Information Fusion*, 2nd ed.; Tsinghua University Press: Beijing, China, 2010.
2. Ge, Q.; Shao, T.; Yang, Q.; Shen, X.; Wen, C. Multisensor Nonlinear Fusion Methods Based on Adaptive Ensemble Fifth-Degree Iterated Cubature Information Filter for Biomechatronics. *IEEE Trans. Syst. Man Cybern. Syst.* **2017**, *46*, 912–925. [CrossRef]
3. Julier, S.J.; Uhlmann, J.K. New extension of the Kalman filter to nonlinear systems. In Proceedings of the Signal Processing, Sensor Fusion, and Target Recognition VI, Orlando, FL, USA, 28 July 1997; pp. 182–193.
4. Arasaratnam, I.; Haykin, S. Cubature kalman filter. *IEEE Trans. Autom. Control* **2009**, *54*, 1254–1269. [CrossRef]

5. Arasaratnam, I.; Haykin, S.; Thomas, R.H. Cubature kalman filtering forcontinuous-discrete systms: Theory and simulations. *IEEE Trans. Signal Process.* **2010**, *58*, 4977–4993. [CrossRef]
6. Hernandez, W.; De Vicente, J.; Sergiyenko, O.Y.; Fernández, E. Improving the Response of Accelerometers for Automotive Applications by Using LMS Adaptive Filters: Part II. *Sensors* **2010**, *10*, 952–962. [CrossRef] [PubMed]
7. Yang, Y.; Gao, W. A new learning statistic for adaptive filter based on predicted residuals. *Prog. Nat. Sci. Mater. Int.* **2006**, *16*, 833–837.
8. Sage, A.; Husa, G.W. Adaptive filtering with unknown prior statistics. In Proceedings of the Joint Automatic Control Conference, Boulder, CO, USA, 5–7 August 1969; pp. 760–769.
9. Sage, A.P.; Husa, G.W. Algorithms for sequential adaptive estimation of prior statistics. In Proceedings of the 8th Symp on Adaptive Processes Decision and Control, Session 6a, University Park, PA, USA, 17–19 November 1969; pp. 61–70.
10. Gao, S.; Hu, G.; Zhong, Y. Windowing and random weighting-based adaptive unscented Kalman filter. *Int. J. Adapt. Control Signal Process.* **2015**, *29*, 201–223. [CrossRef]
11. Gao, W.G.; Yang, Y.X.; Cuix, Q. Application of adaptive kalman filtering algorithm in imu/GPS integrated navigation system. *Geo-Spat. Inf. Sci.* **2007**, *10*, 22–26. [CrossRef]
12. Li, W.; Sun, S.; Jia, Y.; Du, J. Robust unscented Kalman filter with adaptation of process and measurement noise variances. *Digit. Signal Process.* **2016**, *48*, 93–103. [CrossRef]
13. Zheng, B.; Fu, P.; Li, B.; Yuan, X. A Robust Adaptive Unscented Kalman Filter for Nonlinear Estimation with Uncertain Noise variance. *Sensors* **2018**, *18*, 808. [CrossRef]
14. Ding, J.L.; Xiao, J. Design of adaptive cubature Kalman filter based on maximum a posteriori estimation. *Control Decis.* **2014**, *29*, 327–334.
15. Zhao, L.; Wang, X.X.; Sun, M.; Ding, J.C.; Yan, C. Adaptive UKF Filtering Algorithm Based on Maximum a Posterior Estimation and Exponential Weighting. *Acta Autom. Sin.* **2010**, *36*, 1007–1019. [CrossRef]
16. Sarkka, S.; Nummenmaa, A. Recursive noise adaptive kalman filtering by variational bayesian approximations. *IEEE Trans. Autom. Control* **2009**, *54*, 596–600. [CrossRef]
17. Ge, Q.; Wei, Z.; Liu, M.; Yu, J.; Wen, C. Adaptive Quantized Estimation Fusion Using Strong Tracking Filtering and Variational Bayesian. *IEEE Trans. Syst. Man Cybern. Syst.* **2017**, 1–12. [CrossRef]
18. Sun, C.; Zhang, Y.; Wang, G.; Gao, W. A New Variational Bayesian Adaptive Extended Kalman Filter for Cooperative Navigation. *Sensors* **2018**, *18*, 2538. [CrossRef] [PubMed]
19. Li, J.; Deng, F.; Chen, J. A Fast Distributed Variational Bayesian Filtering for Multisensor LTV System With Non-Gaussian Noise. *IEEE Trans. Cybern.* **2018**, 1–13. [CrossRef] [PubMed]
20. Chandra, K.P.B.; Gu, D.W.; Postlethwaite, I. Square root cubature information filter. *IEEE Sens. J.* **2013**, *13*, 750–758. [CrossRef]
21. Ge, Q.B.; Xu, D.X.; Wen, C.L. Cubature information filters with correlated noises and their applications in decentralized fusion. *Signal Process.* **2014**, *94*, 434–444. [CrossRef]
22. Xiong, K.; Zhang, H.Y.; Chan, C.W. Performance evaluation of ukf-based nonlinear fltering. *Automatica* **2006**, *42*, 261–270. [CrossRef]
23. Wu, Y.X.; Hu, D.W.; Hu, X.P. Comments on "performance evaluation of ukf-based nonlinear filtering". *Automatica* **2007**, *43*, 567–568. [CrossRef]
24. Xiong, K.; Zhang, H.Y.; Chan, C.W. Author's reply to "comments on 'performance evaluation of ukf-based nonlinear fltering"'. *Automatica* **2007**, *43*, 569–570. [CrossRef]
25. Farina, A.; Ristic, B.; Benvenuti, D. Tracking a ballistic target: Comparison of several nonlinear filters. *IEEE Trans. Aerosp. Electron. Syst.* **2002**, *38*, 854–867. [CrossRef]
26. Tichavsky, P.; Muravchik, C.H.; Nehorai, A. Posterior cramér-Rao bounds for discrete-time nonlinear filtering. *IEEE Trans. Signal Process.* **1998**, *46*, 1386–1396. [CrossRef]

sensors

Article

Gaussian Process Based Bayesian Inference System for Intelligent Surface Measurement

Ming Jun Ren [1], Chi Fai Cheung [2] and Gao Bo Xiao [1,*]

[1] State Key Laboratory of Mechanical System and Vibration, School of Mechanical Engineering, Shanghai Jiao Tong University, Shanghai 200245, China; renmj@sjtu.edu.cn

[2] State Key Laboratory of Ultra-Precision Machining Technology, The Hong Kong Polytechnic University, Hong Kong, China; mfbenny@inet.polyu.edu.hk

* Correspondence: gaobo.xiao@sjtu.edu.cn; Tel.: +86-21-3420-4616

Received: 29 October 2018; Accepted: 19 November 2018; Published: 21 November 2018

Abstract: This paper presents a Gaussian process based Bayesian inference system for the realization of intelligent surface measurement on multi-sensor instruments. The system considers the surface measurement as a time series data collection process, and the Gaussian process is used as mathematical foundation to establish an inferring plausible model to aid the measurement process via multi-feature classification and multi-dataset regression. Multi-feature classification extracts and classifies the geometric features of the measured surfaces at different scales to design an appropriate composite covariance kernel and corresponding initial sampling strategy. Multi-dataset regression takes the designed covariance kernel as input to fuse the multi-sensor measured datasets with Gaussian process model, which is further used to adaptively refine the initial sampling strategy by taking the credibility of the fused model as the critical sampling criteria. Hence, intelligent sampling can be realized with consecutive learning process with full Bayesian treatment. The statistical nature of the Gaussian process model combined with various powerful covariance kernel functions offer the system great flexibility for different kinds of complex surfaces.

Keywords: Surface measurement; multi-sensor measurement; surface modelling; data fusion; Gaussian process

1. Introduction

Surface size, geometry and texture are some of the most influential subjects in the field of precision engineering [1]. The development of advanced machining technologies allows vast application of complex surfaces superimposing multiple scales of feature in mechanical and optical engineering for their superior performance in terms size reduction and versatile functions [1–3]. There is a growing awareness of the importance of these new types of surfaces in modern science and technologies [4]. To ensure the functionality of the components, these surfaces are required to be fabricated with high precision in terms of form accuracy in sub-micron range and surface finishing at nanometric level. However, the geometric complexity of these advanced surfaces requires multi-scale measurement and characterization, which imposes a lot challenges for current precision surface metrology [5,6].

Extensive research has been conducted on developing various measurement instruments to fulfill a wide range of metrological needs [7–10], such as high precision coordinate measuring machines [7,8], micro topographical instruments [11], electron microscopy [12], optical interferometry [13–15], etc. Although these instruments are capable of performing accurate and efficient measurement at specific measurement range, few of them could realize high dynamic range multi-scale measurement with high efficiency and accuracy. Integrating several complementary sensors into an instrument therefore becomes a promising solution to address complicated measurement tasks. For instance, integrating tactile or scanning probe with imaging sensors is an attractive solution for precision measurement of large scale complex surfaces with high efficiency [16,17]. However, multi-sensor measurement also brings many challenges, such as more sophisticated measurement strategy, effective multi-sensor data fusion, and complex uncertainty propagation. The effectiveness of multi-sensor measurement largely depends on appropriate measurement strategy so as to guide the process to perform cooperative measurement [5]. Model-based methods have been considered as the effective solution for multi-sensor measurement [18–21]. Unlike the conventional sampling strategy which is normally designed prior the actual measurement, the model-based methods consider the measurement as a time-series optimization process such that sampling plan is adaptively determined based on the model which are established using prior measured dataset in each iteration. Obviously, the surface modelling forms the beating heart of the model-based sampling strategy. Although there are a large amount of literature on this topic [22–24], the surface modelling in multi-sensor system is still a challenging task since the measured datasets are normally embedded in different coordinate frames with different resolutions and different levels of uncertainties. This requires the surface modelling methods be capable of efficiently fusing multi-sensor measured datasets and hopefully be capable of performing self-assessment so as to give some hints to the sample distribution [25].

Considering the characteristics of the multi-sensor measurement, the modelling process can generally be divided into two steps, including data registration and fusion [26]. In the first step, data registration algorithm is used to register all the measured dataset to a common coordinate frame via matching process. Although there are several mature data registration algorithms, such as iterative closest point algorithm [27], the results would inevitably include systematic errors due to the existence of measurement errors, which is seldom considered in the measurement of complex surfaces. In the second step, the overlapped datasets are fused to construct a unique surface model. Advanced fusion algorithm should be capable of using the redundant data to deduce a fused result with reduced measurement uncertainty. Data fusion have widely been studied in fields like signal processing, computer vision, and control, and some of the fusion techniques have started to be used in coordinate metrology. A review of current data fusion in coordinate metrology have been presented by Wang et al. and categorized the existing methods into four groups, including repeated measurements, stitching, range image fusion, and 3D data fusion [28]. Repeated measurement and stitching techniques are widely used in optical interferometry for enlarging the measurement range. The fusion is relatively simple since the datasets are obtained from same sensors and should have save level of uncertainties. Range image and 3D data fusion is however difficult, which normally involves six degree of registration and sophisticated data fusion process. Several data fusion methods have been presented, such as weighted least square based fusion and scale-decomposition based fusion [28]. However, these methods are either suspicious to the geometry being modelled or difficult to perform self-assessment of the fused result, which makes it ineffective to be served as a modelling algorithms to perform automatic and intelligent sampling of the surface during the measurement.

Recently, Gaussian process (GP) has been used to address the multi-sensor data fusion problem [29]. GP is kind of a non-parametric Bayesian inference model which has been proved a powerful mathematical tool to model various complex surfaces in terms of spatial covariance functions [30–32]. The statistic nature of GP makes it capable of incorporating the measurement errors into the modelling process and assigning credibility to the constructed model. Taking these advantages, this paper presents a GP based Bayesian inference system for the realization of intelligent surface measurement on multi-sensor system. The system is responsible for multi-feature classification and multi-dataset regression. Multi-feature classification is capable of performing automatic extraction and classification of the features of the measured surfaces so as to design appropriate covariance kernel functions and corresponding initial sampling strategy. Multi-dataset regression is capable of optimizing the sampling strategy via multi-sensor data fusion and on-line sampling adaptation by taking the uncertainty of the Gaussian process model as critical sampling criteria. Hence, automatic measurement setup and intelligent sampling strategy can be realized with consecutive learning process with full Bayesian treatment. Experiments are conducted to verify the effectiveness of the proposed study.

The remaining sections are outlined as follows. Section 2 states the idea that how a surface measurement can be treated as a regression problem by taking GP as mathematical foundation. The GP-based Bayesian inference system is then presented in Section 3 with detailed introduction of the two main parts of the inference system, i.e., the multi-feature classification and multi-dataset regression. Experimental work is given in Section 4 to verify the effectiveness of the proposed method, and a conclusion is made in Section 5.

2. Preliminary

Due to the limited resolution and the error of a measurement instrument, surface measurement can generally be formulated as a regression problem of the form, where $S(x)$ and ε_x are the measured data and the associated measurement error at x, $f(x)$ is the evaluated surface model which is used to give prediction to the unmeasured area of the surface. Since $f(x)$ is generally an unknown function for a manufactured surface, its prediction is inevitably uncertain. Considering the statistic nature of the surface measurement, Gaussian process (GP) [31], a non-parametric Bayesian inference model, is used to address this problem.

GP is basically a collection of random variables, any finite subset of which obeys joint Gaussian distribution. It is generally be defined by a mean function $\mu(X)$ and a covariance matrix $K(X,X)$ [31]

$$GP(X) \sim N(\mu(X), K(X,X)) \tag{1}$$

where $X = [x_1, x_2, \ldots, x_n]$ is the locations of the measured data S. In present study, the mean function is set to be zero-offset by scaling the data appropriately such that it has an empirical mean of zero. The covariance matrix $K(X,X)$ is generally formulated by covariance kernel function $k(x_i, x_j)$ in the form as given by Equation (2) where (x_i, x_j) are the locations of arbitrary two points on the GP model.

$$K(\mathbf{X},\mathbf{X}) = \begin{bmatrix} k(\mathbf{x}_1,\mathbf{x}_1) & k(\mathbf{x}_1,\mathbf{x}_2) & k(\mathbf{x}_1,\mathbf{x}_3) & \cdots & k(\mathbf{x}_1,\mathbf{x}_n) \\ k(\mathbf{x}_2,\mathbf{x}_1) & k(\mathbf{x}_2,\mathbf{x}_2) & k(\mathbf{x}_2,\mathbf{x}_3) & \cdots & k(\mathbf{x}_2,\mathbf{x}_n) \\ k(\mathbf{x}_3,\mathbf{x}_1) & k(\mathbf{x}_3,\mathbf{x}_2) & k(\mathbf{x}_3,\mathbf{x}_3) & \cdots & k(\mathbf{x}_3,\mathbf{x}_n) \\ \vdots & \vdots & \vdots & \ddots & \vdots \\ k(\mathbf{x}_n,\mathbf{x}_1) & k(\mathbf{x}_n,\mathbf{x}_2) & k(\mathbf{x}_n,\mathbf{x}_3) & \cdots & k(\mathbf{x}_n,\mathbf{x}_n) \end{bmatrix} \tag{2}$$

Hence, according to the Bayesian theory, a prediction of f^* at an arbitrary location x^* on unmeasured region of the model can be evaluated by the joint distribution of f^* with measured data S as follows:

$$\begin{bmatrix} S \\ f^* \end{bmatrix} \sim N\left(0, \begin{bmatrix} K(\mathbf{X},\mathbf{X}) + \sigma_\varepsilon^2 \mathbf{I} & K(\mathbf{X},x^*) \\ K(\mathbf{X},x^*) & K(x^*,x^*) \end{bmatrix}\right) \tag{3}$$

where $\mathbf{I}$ is the identity matrix; σ_ε is a hyperparameter representing the variance of the random part of the measurement error. The covariance cov and the mean m^* of the f^* can then be obtained by the marginal distribution of f^* as given by Equations (4) and (5).

$$m^* = K(\mathbf{x}^*,\mathbf{X})\left(K(\mathbf{X},\mathbf{X}) + \sigma_\varepsilon^2 I\right)^{-1}\mathbf{Z} \tag{4}$$

$$\operatorname{cov}(f^*) = K(\mathbf{x}^*,\mathbf{x}^*) - K(\mathbf{x}^*,\mathbf{X})\left(K(\mathbf{X},\mathbf{X}) + \sigma_\varepsilon^2 I\right)^{-1}K(\mathbf{X},\mathbf{x}^*) \tag{5}$$

It is seen that the GP inference is determined by the covariance forms which represents the covariance of the points on the model. Hence the construction of appropriate covariance kernel function is the core of GP modelling. Figure 1 shows an example of GP modelling and prediction. The crosses are the measured points, in which x is the locations and y is the corresponding measured value. The line is the established GP model by taking the measured point as training data. Two predictions are made at unmeasured location x1 and x2. μ_1 and μ_2 are the predictions, and the σ_1 and σ_2 are their evaluated uncertainties.

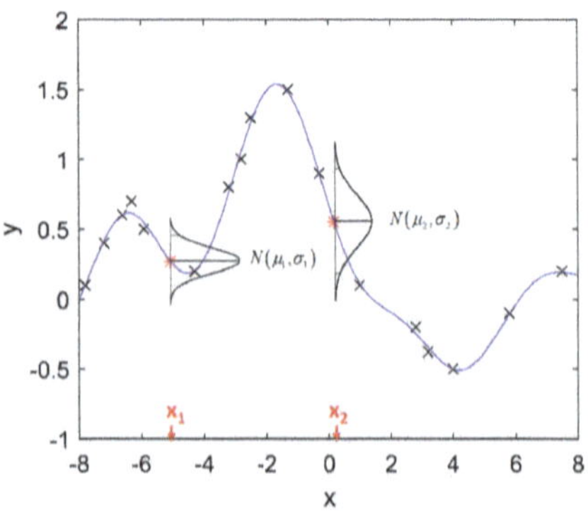

Figure 1. An example of GP modelling and prediction.

3. Gaussian Process Based Bayesian Inference System

3.1. System Configuration

Figure 2 shows the schematic diagram of the GP based Bayesian inference system (GP-BIS), which is mainly composed of four modules, including preprocessing, multi-feature classification, multi-dataset regression, and surface characterization. In preprocessing, the characteristics of the involved measurement sensors are analyzed and calibrated, including measurement range, resolution, speed, errors, etc. The designed models of the measured surfaces are transformed to a unified format and the properties of the measured parts such as hard and soft materials, and the required specifications, e.g., form accuracy and roughness, are loaded.

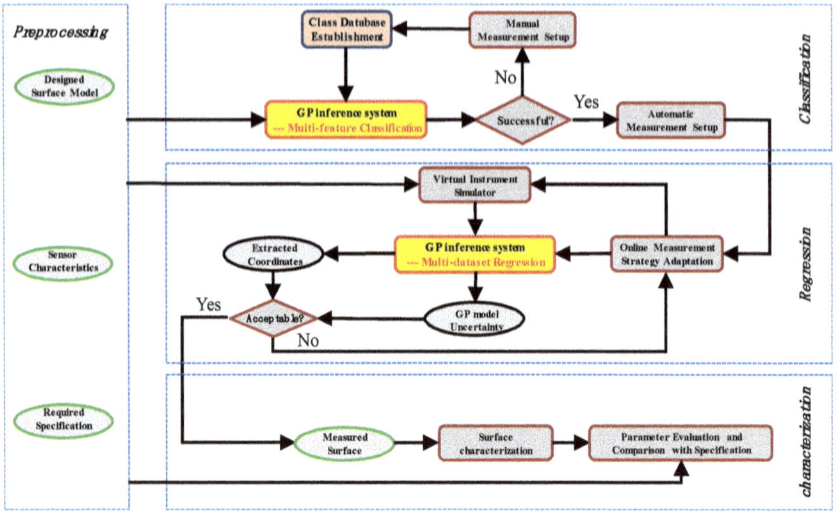

Figure 2. Schematic diagram of GP-BIS.

Taking the information of the preprocessing as input, multi-feature classification module extracts and classifies the geometric features of the designed models so as to design an appropriate covariance kernel functions for GP modelling. An initial measurement strategy is also loaded from a pre-established class database. The class database is responsible for collecting appropriate class of measurement strategies for specific tasks with respect to certain types of surface features based on the prior knowledge. If the contained surface features and the required specifications are successfully classified, the measurement strategy will be automatically loaded. Otherwise, manual design of the sampling strategy is required and a new class will be uploaded to the database so that they can be successfully classified in future. A certain class in the database would contain necessary information for the setup of the measurement instruments, including the types of the sensors and the associated parameters which will be used during measurement. For contact types of sensors, such as trigger probe, the initial sampling patterns may be also included for further optimization in GP modelling process. It is emphasized that the sampling strategies produced in this stage are only initial plans which require further optimization. Multi-dataset regression module is responsible further optimizing the initial measurement strategy. This is carried in two steps. Firstly, taking the designed covariance kernel function as input, GP model is established to fuse the datasets obtained from different sensors to obtain a unique measured surface. Secondly, both the bias and the uncertainty of the fused surface model is used to qualify the reliability of the measured surface, and is served as critical sampling criteria to perform adaptive sampling until reaching desired reliability. During the process, a virtual instrument simulator [33] can also be developed to analyze the error distribution of each measured data by taking the sensor characteristics and the adopted sampling strategy as input. The analyzed error distribution of each measured data can be used to aid the surface modelling and uncertainty analysis. At the end, the accepted surface model is then used to perform scale-decomposition, feature segmentation, or pattern analysis to fully characterize the quality of the machined surface in different scales [2].

3.2. Design of Covariance Kernel via Multi-Feature Classification

Covariance kernel is a function used to describe the spatial correlation of a geometric feature in GP modelling, and hence the design of a covariance kernel should be in accordance to the characteristics of the measured surface. For a machined surface with known CAD model, the characteristics of the surface topography, such as smoothness, roughness, and periodicity can be specified and classified easily. In GP-BIS, such knowledge is the key to design covariance kernel functions and would significantly improve the measurement efficiency and accuracy. There already exists various covariance kernel functions which possess variety of geometric characteristics [31,34]. Table 1 summarizes currently widely used kernel functions and their characteristics, including white noise (WN), linear (LIN), square exponential (SE), periodic (PER), Matérn class (MC), rational quadratic (RQ), neural network (NN), and piecewise polynomial (PP). Different kernel functions can be selected to study different kinds of surfaces in accordance with the characteristics. Currently, the basic kernel functions are pre-assigned to different types of topographies based on the characteristics as listed in Table 1. For instance, SE is infinitely differentiable and hence is suitable in studying smooth surface, while PER would be a good choice for the surfaces possessing periodic features.

For some surfaces containing multiple features at different scales, a single covariance kernel may not be flexible enough to describe the spatial correlation. In such case, the listed kernel functions can be served as base functions to form various composite kernel functions by combining them together. Figure 3 shows the design of the composite kernel functions. It starts from decomposing the geometric feature of a surface in different scales. Appropriate kernels are selected for the decomposed features in each scale and then a composite kernel function can be formed by simply combining the selected kernels. For instance, if a smooth surface is superimposed by periodic micro-structures, a combination of SE and PER can be designed to model two types of features simultaneously. For some scenario, even the decomposed feature in a scale may be complex that cannot be appropriately described by a single kernel. In such case, besides summation, multiplication among the kernels can also be performed to further enhance the flexibility of the model. Kernel summation can be used to simultaneously model different types of spatial correlation while kernel multiplication can be used enhance the flexibility of a kernel for modelling a complex spatial correlation. The fitness of the designed kernel function should well balance the model fitness and complexity which can be qualified by the Bayesian information criterion [34]. This offers GP a great flexibility in modelling different types of complex surfaces by combining kernel summation and multiplication.

Table 1. Geometric characteristics of 8 base kernel functions.

	Geometric Characteristic	Base Function	Gp Prior
WN	White noise	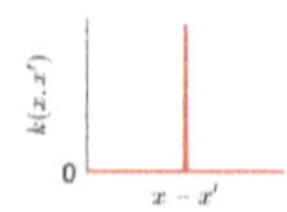	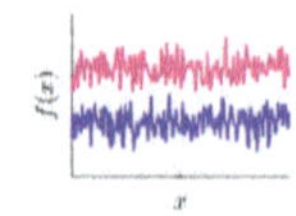
LIN	linearly varying amplitude	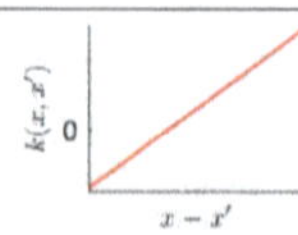	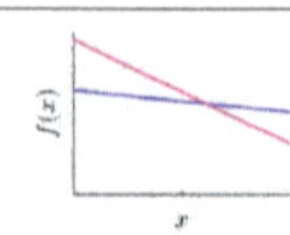
SE	Infinitely differentiable, offering smooth variations with a typical length scale	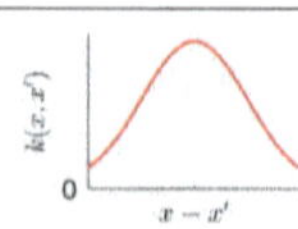	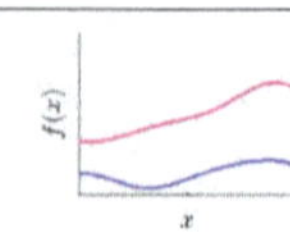
PER	With arbitrary roughness and period, suitable for periodic shape	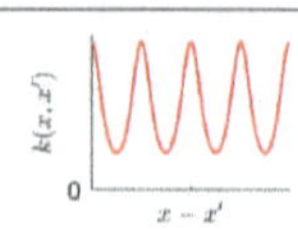	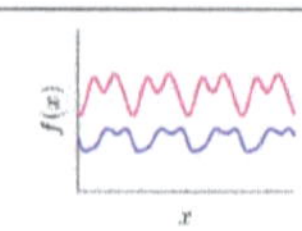
MC	Finite times differentiable, suitable for different roughness with appropriate parameters	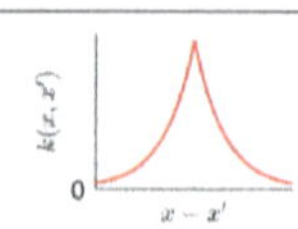	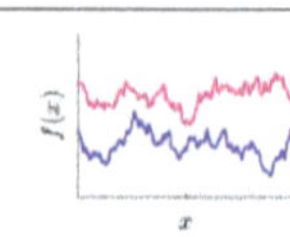
RQ	A mixture of SE with different length scales, more flexible with relatively more hyperparameters, suitable for smooth and multi-scaled shape	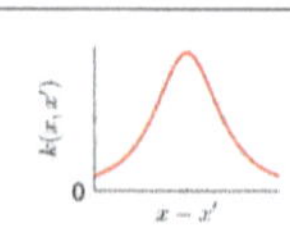	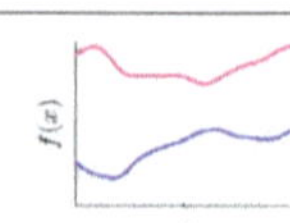
NN	Rapid or large variations with non-stationary spatial correlation, suitable for the irregular surfaces with random features, such as the complex terrain	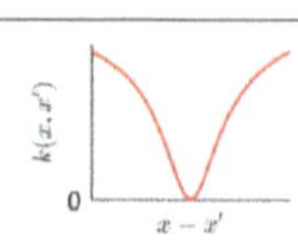	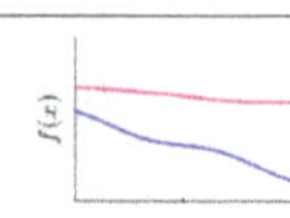
PP	Finite continuously differentiable, suitable for large continuous or fast-changing shape	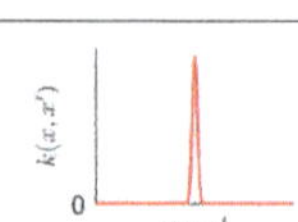	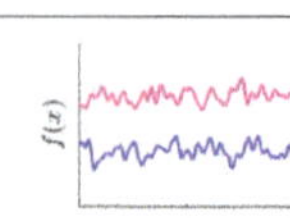

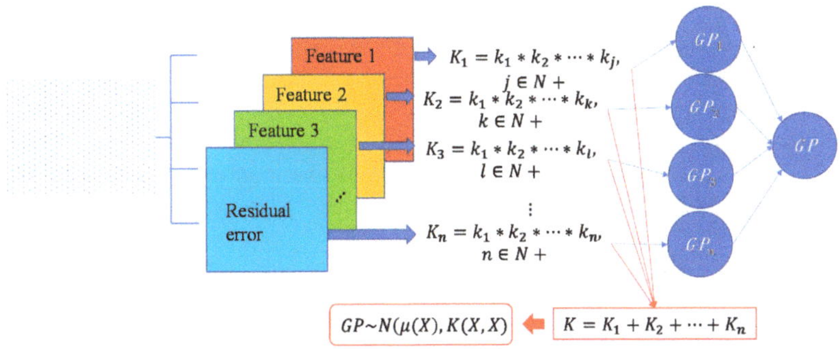

Figure 3. Design of composite covariance kernel functions.

3.3. Sampling Strategy Adaptation via Multi-Dataset Regression

One of the important technical merits of multi-sensor measurement is capability in improving the quality of the measurement results via data fusion process. Considering that all the measured datasets are from a common surface, they should be inherently correlated. Hence, besides the spatial correlation within each dataset, the correlation across the datasets should also be taken into account. In GP modelling, this can be formulated by modelling the auto-covariances and cross-covariances among the measured datasets [34]. For arbitrary two given dataset, the mean and the covariance can be given by Equations (6) and (7).

$$m^* = [K(\mathbf{x}^*, \mathbf{X}_1), K(\mathbf{x}^*, \mathbf{X}_2)]K(\mathbf{X}_{12}, \mathbf{X}_{12})[\mathbf{Z}_1, \mathbf{Z}_2]^T \tag{6}$$

$$\text{cov}(f^*) = K(\mathbf{x}^*, \mathbf{x}^*) - [K(\mathbf{x}^*, \mathbf{X}_1), K(\mathbf{x}^*, \mathbf{X}_2)]K(\mathbf{X}_{12}, \mathbf{X}_{12})^{-1}[K(\mathbf{X}_1, \mathbf{x}^*), K(\mathbf{X}_2, \mathbf{x}^*)] \tag{7}$$

where

$$K(\mathbf{X}_{12}, \mathbf{X}_{12}) = \begin{bmatrix} K(\mathbf{X}_1, \mathbf{X}_1) + \sigma_{\varepsilon 1}^2 \mathbf{I} & K(\mathbf{X}_1, \mathbf{X}_2) \\ K(\mathbf{X}_1, \mathbf{X}_2) & K(\mathbf{X}_2, \mathbf{X}_2) + \sigma_{\varepsilon 2}^2 \mathbf{I} \end{bmatrix} \tag{8}$$

X_1 and X_2 are the locations of the measured datasets Z_1 and Z_2 from two different sensors, $K(X_1, X_1)$ and $K(X_2, X_2)$ are the auto-covariances of X_1 and X_2 respectively, $K(X_1, X_2)$ is the cross-covariance between the X_1 and X_2, $\sigma_{\varepsilon 1}$ and $\sigma_{\varepsilon 2}$ are the hyperparameters representing the variance of the random part of the X_1 and X_2 respectively.

One of immediate advantage of above multi-dataset regression is that it preserves the full Bayesian inference during the fusion process so that credibility can still be assigned to the fused model using the covariance of the established GP model as given in Equation (7). Hence, the measurement process can generally be considered as a time-series learning process that the credibility of the fused model can be served as critical sampling criterion to perform on-line sampling adaptation. As shown in Figure 2, it starts from extracting a set of data from the surface using the initial sampling strategy obtained in multi-feature classification module, and multi-dataset regression is then performed to establish a fused GP from the measured datasets and evaluate the covariance of the model. If the covariance exceeds a pre-specified threshold, it is considered that the sampling is insufficient to lean the surface geometry accurately and more data are sampled at the regions where the maximum covariance is located. By such a consecutive learning and adaptive sampling process, a measured surface model with acceptable reliability can be obtained efficiently.

4. Experimental Study

4.1. Computer Simulation on Surface Modelling

The proposed GP-BIS has been implemented in MATLAB 2018a (from The MathWorks, Natick, USA) and tested on several surfaces with variety of complexities. Firstly, case studies have been conducted on multiple features enriched complex surfaces to study the capability of the GP modelling. A set of 121 × 121 grid points are extracted from the designed surface in each case study, and are added Gaussian noise with 2 μm standard deviation to simulate the measurement error. Table 2 shows complex surfaces which are made of a parabola and different scales of sinusoidal waves. It starts from a simple parabola given by Equation (10). Since the surface is generally a simple smooth surface, GP-BIS selects the SE as the covariance kernel functions. It is clearly seen from the residual map that the surface has been accurately modelled and the contained random errors has been successfully rejected. This can be explained from the principle of the GP which is basically a process to find a correlation that best explains the spatial distribution of the sampled points. Hence, the uncorrelated part, i.e., the added random errors will be considered as measurement errors, which is incorporated in the modelling process by $\sigma_\varepsilon \boldsymbol{I}$ in Equation (3). In experiment, σ_ε has been evaluated to be 1.99 μm which is very close to the true value.

$$z = -0.3(0.0625x^2 + 0.0625y^2)x, y \in [-10, 10] \text{ mm} \tag{9}$$

Table 2. GP modelling of various complex surfaces.

Designed Complex Surfaces	Covariance Kernel Functions	Residual Maps
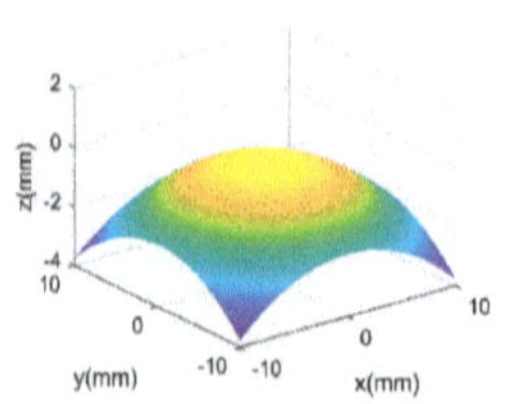	SE	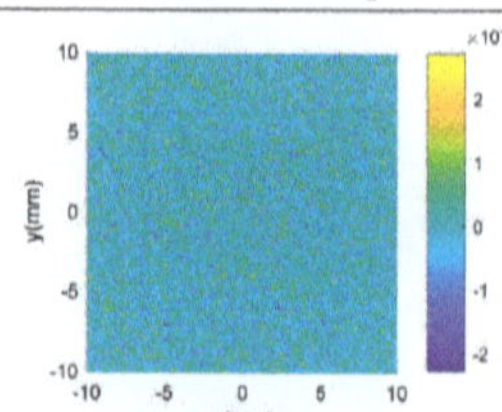
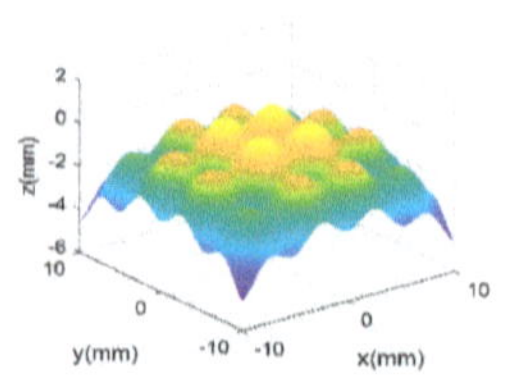	SE + PER	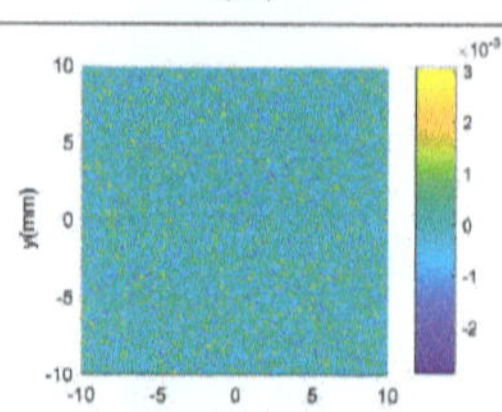
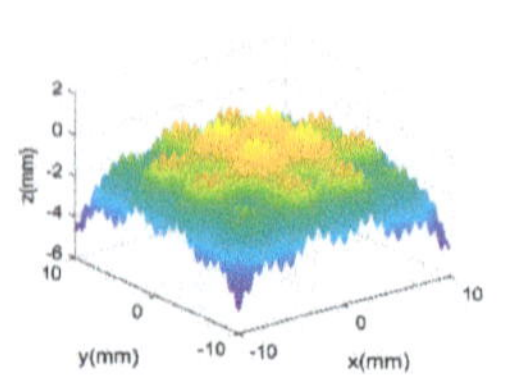	SE + PER + MC	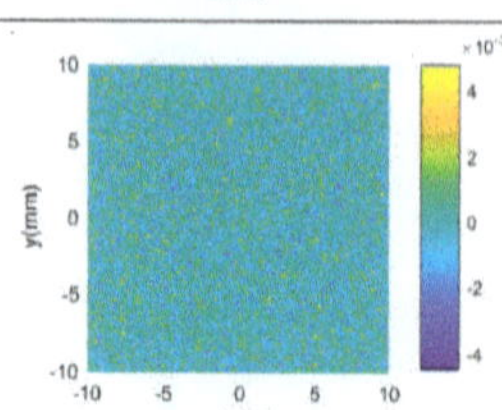
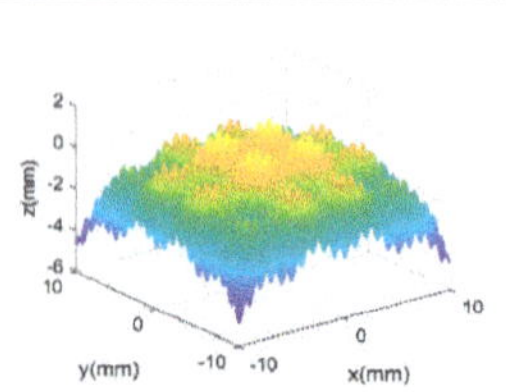	SE + PER + PER	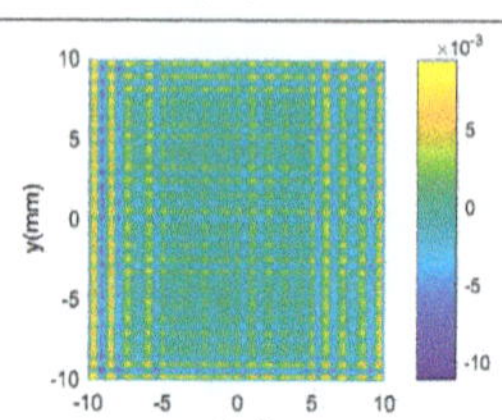

Then, a sinusoidal wave with amplitude 0.4 mm and period 5 mm in both two X and Y directions is superimposed to the parabola to make the surface more complex. The designed surface now has two radically different topographies in two different scales. Hence, in the multi-classification process, the GP-BIS identifies two different types of topographies and a new composite kernel functions has been designed using SE and PER (see Table 1). The measurement component σ_ε has been identified to be 2.18 µm and hence the surface has also been accurately modelled. To further improve the complexity of the surface, another sinusoidal wave with amplitude 0.2 mm and period 1 mm in both two X and Y directions is further superimposed to the surface designed in second case study. GP-BIS identifies the third different topographies, and a three level of composite kernel SE + PER + MC has been designed. The measurement error part is identified to be 2.23 µm, which further verifies the accuracy of the established GP model. Intuitively, considering the prior knowledge of three different features, SE + PER + PER seems also a good choice for the surface. The results, i.e., the last row of Table 2 show that SE + PER + PER also achieved reasonably good result while it is still not compatible with SE + PER + MC. This is due to the reason that, after the use of SE + PER, the scale of the remaining feature may be affected by the modelling residual. Hence more flexible MC would be better choice in modelling the remaining part which is a combination of the third scales of feature and the modelling residual of the two larger scales of features.

4.2. Actual Application on Multi-Sensor Instrument

The proposed GP-BIS has applied to an instrument equipped with a touch trigger probe and a laser scanner. Trigger probe is one of typical sensors which can achieve high accurate measurement with generally very low efficiency. The laser scanner, on the other hand, is capable of realizing high efficient measurement while with relatively low accuracy. For the measurement of complex surfaces, more intelligent multi-sensor measurement strategy is required to achieve both high accurate and high efficient measurement. The probing error and the linearity of the laser scanner are identified to be 0.9 µm (1σ, normal) and 15.2 µm (uniform).

Measurement has been conducted on a machined workpiece which is designed by a peak function defined by Equation (10).

$$z = 6\left(1-\tfrac{x}{16}\right)^2 \cdot \exp\left(-\left(\tfrac{x}{16}\right)^2-\left(\tfrac{y}{16}+1\right)^2\right) - 20\left(\tfrac{x}{5\times 16}-\left(\tfrac{x}{16}\right)^3-\left(\tfrac{y}{16}\right)^5\right)\cdot \exp\left(-\left(\tfrac{x}{16}\right)^2-\left(\tfrac{y}{16}\right)^2\right) \\ -\tfrac{2}{3}\exp\left(-\left(\tfrac{x}{16}+1\right)^2-\left(\tfrac{y}{16}\right)^2\right),\ x,y\in[-40,40] \quad (10)$$

Figure 4a shows the machined workpiece. The measurement has been conducted in four steps. In the first step, touch trigger probe is used to extract a set of dense points, i.e., 6456 points with uniform spacing over the entire surface. The results are considered sufficiently accurate and is served as benchmarking to verify other more efficient measurement. Secondly, the laser scanner is used to measure the surface with high density. The measurement is very fast while the accuracy would be low. Thirdly, touch trigger probe is used to perform adaptive sampling based on multi-dataset regression module. Since there is only a single data source, the multi-dataset regression becomes a single GP modelling. It has been noted in published literature [18,32] that GP based adaptive sampling has superior performance than other sampling methods in sampling complex surfaces. Hence, the results are also used as benchmarking to examine the efficiency of the proposed GP-BIS. Fourthly, by taking the measurement result of the laser scanner as input, adaptive sampling is conducted using touch trigger probe. Hence, in the multi-dataset regression, the datasets obtained by the laser scanner and the trigger probe should be fused to establish a GP model in each adaptation. Hence, the GP-BIS becomes producing multi-sensor measurement strategy in the forth experiment.

Figure 4b,c shows the evaluated results of benchmarking and multi-sensor fused dataset by using best fitting method [35] which is one of the typical methods in form characterization of complex surfaces. It performs the form characterization of the measured dataset by best fitting the dataset to the corresponding designed surface via rigid body transformation, and the error of the dataset can be evaluated by projecting it onto the designed surface. The measured number of points and the root-mean-square (RMS) of the height error are used to characterize the efficiency and accuracy of different measurement methods. A summary of the results is given in Table 3.

Although laser scanner has extremely good measurement efficiency, the accuracy is very low comparing with touch trigger probe. With the aid of the GP-BIS, the trigger probe can perform adaptive sampling, and the efficiency has been dramatically improved and the accuracy has been preserved at the same time. With the aid of the laser scanner, multi-dataset regression is used to further improve the measurement efficiency around 40% comparing with pure triggering measurement. The results demonstrated the capability of the proposed GP-BIS in enhancing the performance of the multi-sensor instrument in measuring complex surfaces.

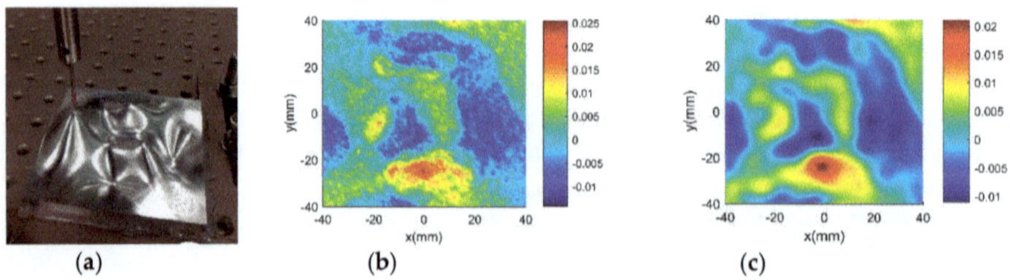

Figure 4. Measurement of a designed complex surfaces on multi-sensor instrument. (**a**) machined workpiece; (**b**) Benchmarking form error; (**c**) multi-sensor fused form error.

Table 3. A summary of the actual measurement result.

Measurement Strategy	Number of Points	RMS (μm)
Trigger probe dense measurement	6456	5.9
Laser scanner dense measurement	more than 40,000	11.2
Trigger probe adaptive measurement	493	5.7
Multi-sensor adaptive measurement	281	5.5

5. Conclusions

This paper presented a GP-BIS for realizing automatic and intelligent surface measurement on multi-sensor instruments. The system considers the surface measurement as a consecutive learning process and makes use of Gaussian process to model the process based on two key modules including multi-feature classification and multi-dataset regression. The system has been implemented and applied to produce multi-sensor measurement strategy for an instrument which is equipped with a trigger probe and a laser scanner. The results shows that the performance of the multi-sensor instrument has been dramatically improved around 40% in terms of measurement efficiency while maintains the measurement accuracy at the save level. The proposed study should provide new insight for the realization of intelligent measurement of complex surfaces. One of the problem of GP modelling is the high computation complexity which would lead huge time for the computation when large number of data is involved in the inference process. Future work considers the implementation of approximated GP inference with the aid of graphics processing unit (GPU) for the practical use of the proposed method especially in manipulating vision based measurement dataset.

Author Contributions: Conceptualization, M.J.R.; Methodology, C.F.C. and G.B.X.; Validation, M.J.R.; Data Curation, C.F.C. and G.B.X.; Writing-Original Draft Preparation, M.J.R.; Writing-Review & Editing, C.F.C. and G.B.X.; Funding Acquisition, M.J.R. and C.F.C.

Funding: The work was substantially supported by the National Natural Science Foundation of China (Grant No. 51505404, and No. 51675456), and the financial support from the State key laboratory of precision measuring technology and instruments (Tianjin University). The work described in this paper was also supported by a grant from the Research Grants Council of the Government of the Hong Kong Special Administrative Region, China (Project No. 15202717).

Conflicts of Interest: The funders had no role in the design of the study; in the collection, analyses, or interpretation of data; in the writing of the manuscript, and in the decision to publish the results.

References

1. Brinksmeier, E.; Schönemann, L. Generation of discontinuous microstructures by Diamond Micro Chiseling. *CIRP Ann. Manuf. Technol.* **2014**, *63*, 49–52. [CrossRef]
2. Jiang, X.J.; Whitehouse, D.J. Technological shifts in surface metrology. *CIRP Ann. Manuf. Technol.* **2012**, *61*, 815–836. [CrossRef]
3. Whitehouse, D.J. Surface geometry, miniaturization and metrology. *Philos. Trans. R. Soc. A Math. Phys. Eng. Sci.* **2012**, *370*, 4042–4065. [CrossRef] [PubMed]
4. Fang, F.Z.; Zhang, X.D.; Weckenmann, A. Manufacturing and measurement of freeform optics. *CIRP Ann. Manuf. Technol.* **2013**, *62*, 823–846. [CrossRef]
5. Ramasamy, S.K. Multi-Scale Data Fusion for Surface Metrology. ProQuest Dissertations, The University of North Carolina at Charlotte, Charlotte, NC, USA, 2011.
6. Leach, R.K.; Boyd, R.D.; Burke, T.; Danzebrink, H.; Dirscherl, K.; Dziomba, T.; Gee, M.G.; Koenders, L.; Morazzani, V.; Pidduck, A.; Unger, W.E.S.; Yacoot, A. The European nanometrology landscape. *Nanotechnology* **2011**, *2011. 22*, 062001. [CrossRef]
7. Weckenmann, A.; Peggs, G.; Hoffmann, J. Probing systems for dimensional micro-and nano-metrology. *Meas. Sci. Technol.* **2006**, *17*, 504–509. [CrossRef]
8. Weckenmann, A. Application of modern high resolution tactile sensors for micro-objects. *Int. J. Precis. Technol.* **2011**, *2*, 266–288. [CrossRef]
9. ISARA 400 Ultra Precision Coordinate Measuring Machine. Available online: http://www.ibspe.com/ibs_precision_engineering_uk/ibs_isara_400.html (access on 7 August 2018).
10. Schwenke, H.; Neuschaeferrube, U.; Pfeifer, T.; Kunzmann, H. Optical Methods for Dimensional Metrology in Production Engineering. *CIRP Ann. Manuf. Technol.* **2002**, *51*, 685–699. [CrossRef]
11. Yang, W.; Liu, X.; Lu, W.; Hu, C.; Guo, X. Towards a traceable probe calibration method for white light interference based AFM. *Precis. Eng.* **2017**, *51*, 40–47. [CrossRef]
12. Goodhew, P.J.; Humphreys, J.; Beanland, R. *Electron Microscopy and Analysis*, 3rd ed.; Taylor & Francis: New York, NY, USA, 2000; ISBN 9780748409686.
13. Hu, C.; Liu, X.; Yang, W.; Lu, W.; Yu, N.; Chang, S. Improved zero-order fringe positioning algorithms in white light interference based atomic force microscopy. *Optics Lasers Eng.* **2018**, *100*, 71–76. [CrossRef]
14. Yang, W.; Liu, X.; Lu, W.; Guo, X. Influence of probe dynamic characteristics on the scanning speed for white light interference based AFM. *Precis. Eng.* **2017**, *52*, 348–352. [CrossRef]
15. Yang, W.; Yang, X.; Lu, W.; Yu, N.; Chen, L.; Zhou, L.; Chang, S. A Novel White Light Interference Based AFM Head. *J. Lightwave Technol.* **2017**, *35*, 3604–3610. [CrossRef]
16. GmbH, W.M. Multisensor Coordinate Metrology: Dimensional Measurement Using Optics, Probes, and X-ray Tomography. Available online: http://www.werthmesstechnik.de (accessed on 11 March 2018).
17. GmbH, W.M. Modular Product Line. Available online: http://www.witec.de/products/ (accessed on 11 March 2018).
18. Colosimo, B.M.; Pacella, M.; Senin, N. Multisensor data fusion via Gaussian process models for dimensional and geometric verification. *Precis. Eng.* **2015**, *40*, 199–213. [CrossRef]
19. Whitehouse, D.J.; Phillips, M.J. Sampling in a two-dimensional plane. *J. Phys. A Gen. Phys.* **1985**, *18*, 2465. [CrossRef]
20. Colosimo, B.M.; Moroni, G.; Petrò, S. A tolerance interval based criterion for optimizing discrete point sampling strategies. *Precis. Eng.* **2010**, *34*, 745–754. [CrossRef]
21. Wang, J.; Jiang, X.; Blunt, L.; Leach, R.K.; Scott, P.J. Intelligent sampling for the measurement of structured surfaces. *Meas. Sci. Technol.* **2012**, *23*, 755–766. [CrossRef]
22. Ferrari, S.; Frosio, I.; Piuri, V.; Borghese, N.A. Automatic multiscale meshing through HRBF networks. *IEEE Trans. Instrum. Meas.* **2005**, *54*, 1463–1470. [CrossRef]
23. Fan, K.C.; Ming, C.L. Intelligent planning of CAD-directed inspection for coordinate measuring machines. *Comput. Integr. Manuf. Syst.* **1998**, *11*, 43–51. [CrossRef]

24. Stojadinovic, S.M.; Majstorovic, V.D.; Durakbasa, N.M.; Sibalija, T.V. Towards an intelligent approach for CMM inspection planning of prismatic parts. *Measurement* **2016**, *92*, 326–339. [CrossRef]
25. Roberts, S.; Aigrain, S. Gaussian processes for time-series modelling. *Philos. Trans.* **2012**, *371*, 20110550. [CrossRef] [PubMed]
26. Galetto, M.; Mastrogiacomo, L.; Maisano, D.A.; Franceschini, F. Cooperative fusion of distributed multi-sensor LVM (Large Volume Metrology) systems. *CIRP Ann. Manuf. Technol.* **2015**, *64*, 483–486. [CrossRef]
27. Besl, P.J.; Mckay, N.D. A Method for Registration of 3-D Shapes. *IEEE Comput. Soc.* **1992**, *14*, 239–256. [CrossRef]
28. Wang, J.; Leach, R.K.; Jiang, X. Review of the mathematical foundations of data fusion techniques in surface metrology. *Surf. Topograp. Metrol. Prop.* **2015**, *3*, 023001. [CrossRef]
29. Qian, P.Z.G.; Wu, C.F.J. Bayesian Hierarchical Modeling for Integrating Low-Accuracy and High-Accuracy Experiments. *Technometrics* **2008**, *50*, 192–204. [CrossRef]
30. Brentzen, J.A.; Gravesen, J.; Anton, F.; Aanæs, H. *Guide to Computational Geometry Processing. Foundations Algorithms, and Methods*; Springer: London, UK, 2012; pp. 287–308.
31. Rasmussen, C.E.; Williams, C.K.I. *Gaussian Processes for Machine Learning. Adaptive Computation and Machine Learning*; The MIT Press: Cambridge, MA, USA, 2005; pp. 69–106. ISBN 0-262-18253-X.
32. Dumas, A.; Echard, B.; Gayton, N.; Rochat, O.; Dantan, J.Y.; Van Der Veen, S. AK-ILS: An Active learning method based on Kriging for the Inspection of Large Surfaces. *Precis. Eng.* **2013**, *37*, 1–9. [CrossRef]
33. Caldara, S.; Nuccio, S.; Spataro, C. Measurement uncertainty estimation of a virtual instrument. In Proceesings of the 17th IEEE Instrumentation and Measurement Technology Conference, Baltimore, MD, USA, 1–4 May 2000.
34. Vasudevan, S. Data fusion with gaussian processes. *Rob. Autonom. Syst.* **2012**, *60*, 1528–1544. [CrossRef]
35. Ren, M.J.; Cheung, C.F.; Kong, L.B.; Jiang, X.Q. Invariant Feature Pattern Based Form Characterization for the Measurement of Ultra-precision Freeform Surfaces. *IEEE Trans. Instrum. Meas.* **2012**, *4*, 963–973. [CrossRef]

sensors

MDPI

Article

Multitarget Tracking Algorithm Using Multiple GMPHD Filter Data Fusion for Sonar Networks

Xueli Sheng [1,2,3], Yang Chen [1,2,3], Longxiang Guo [1,2,3,*], Jingwei Yin [1,2,3] and Xiao Han [1,2,3]

1 Acoustic Science and Technology Laboratory, Harbin Engineering University, Harbin 150001, China; shengxueli@hrbeu.edu.cn (X.S.); cy5311@hrbeu.edu.cn (Y.C.); yinjingwei@hrbeu.edu.cn (J.Y.); hanxiao1322@hrbeu.edu.cn (X.H.)

2 Key Laboratory of Marine Information Acquisition and Security (Harbin Engineering University), Ministry of Industry and Information Technology, Harbin 150001, China

3 College of Underwater Acoustic Engineering, Harbin Engineering University, Harbin 150001, China

* Correspondence: heu503@hrbeu.edu.cn; Tel.: +86-137-9667-1095

Received: 8 August 2018; Accepted: 19 September 2018; Published: 21 September 2018

Abstract: Multitarget tracking algorithms based on sonar usually run into detection uncertainty, complex channel and more clutters, which cause lower detection probability, single sonar sensors failing to measure when the target is in an acoustic shadow zone, and computational bottlenecks. This paper proposes a novel tracking algorithm based on multisensor data fusion to solve the above problems. Firstly, under more clutters and lower detection probability condition, a Gaussian Mixture Probability Hypothesis Density (GMPHD) filter with computational advantages was used to get local estimations. Secondly, this paper provided a maximum-detection capability multitarget track fusion algorithm to deal with the problems caused by low detection probability and the target being in acoustic shadow zones. Lastly, a novel feedback algorithm was proposed to improve the GMPHD filter tracking performance, which fed the global estimations as a random finite set (RFS). In the end, the statistical characteristics of OSPA were used as evaluation criteria in Monte Carlo simulations, which showed this algorithm's performance against those sonar tracking problems. When the detection probability is 0.7, compared with the GMPHD filter, the OSPA mean of two sensor and three sensor fusion was decrease almost by 40% and 55%, respectively. Moreover, this algorithm successfully tracks targets in acoustic shadow zones.

Keywords: multisensor data fusion; multitarget tracking; GMPHD; sonar network; RFS

1. Introduction

The issue of multiple target tracking (MTT) has emerged as an area of interest in radar, sonar, etc. Traditionally, there are many classical MTT algorithms based on explicit data association information, such as probability data association (PDA) [1,2], joint probability data association (JPDA) [3–5], multiple hypothesis tracking (MHT) [6] and derivative algorithms [7,8]. As the key of these MTT algorithms is data association, the data association algorithm usually causes computational bottlenecks when the number of targets is too large. Therefore, these algorithms usually perform poorly when the number of targets is large.

In response, the random finite set (RFS) [9,10] has attracted the attention of scholars engaged in MTT algorithm research. As no explicit data association is required, MTT algorithms based on RFS have a computational advantage [11,12]. In the last 15 years, the probability hypothesis density (PHD) [10], cardinalized PHD (CPHD) filter [13], sequential Monte Carlo PHD (SMCPHD) [14], Gaussian Mixture PHD (GMPHD) [15] and multi-Bernoulli filters [16] have been proposed for MTT. In 2013, the notion of labeled RFS [17] was introduced to address target trajectories and their uniqueness. Thus, by utilizing the labeled RFS theory, the labeled multi-Bernoulli (LMB) filter [18,19] and generalized labeled multi-Bernoulli (GLMB) [20] filter have advantages in target track estimation and low signal to noise ratio (SNR). Vo proposed an efficient implementation of the GLMB filter based on Gibbs sampling, which has linear complexity in the number of measurements, but at least quadratic in the number of targets [21].

In recent years, people have increased the research and development of ocean resources, so the collection of ocean information has consequently become important. In addition, considering the unpredictable dangers of underwater and harsh working conditions, a growing number of buoy sonar and underwater unmanned vehicles (UUVs) are responsible for underwater information collection. Since these sonar devices are powered by batteries and transmit the preprocessing results of collected information periodically to communication buoys, efficient information processing is particularly important. Moreover, more clutters, poor detection accuracy and complex channels all make sonar detection difficult, weakening the accuracy of MTT.

Unluckily, the PHD filter is designed for high SNR, while the sonar work environments are lower SNR. Distributed multisensor data fusion not only compensates for the lack of information caused by low SNR, but also improves the tracking accuracy [22–24]. Distributed fusion architectures composed of tracker, data association and fusion are characterized by low communication bandwidth demands, high system reliability and strong survivability. On the other hand, distributed sensor networks also have another advantage in detection coverage (e.g., acoustic shadow zones).

The purpose of this paper is to propose an efficient MTT algorithm for sonar detection systems. The structure of this paper is as follows: in the Section 2, we analyze the problems of sonar detection systems. Section 3 presents the classical GMPHD filter algorithm. The maximum-detection capability multitarget track fusion (MDC-MTF) algorithm is proposed in Section 4, and Monte Carlo simulations are provided in Section 5. In the Section 6, the conclusions are presented.

2. Problem Analysis and Solutions

In order to make sure this algorithm could successfully solve above multitarget problems based on sonar sensors, this algorithm framework and three analyses are provided in this section.

2.1. Computational Bottle-Neck

Many papers [11,12,18–20] have analyzed in depth the computational complexity of PHD filter and others MTT algorithms. The explicit data association-based algorithms (e.g., MHT, JPDA) suffer from prohibitive computational complexity with increasing number of targets and measurements. For example, the amount of computation will increase exponentially with the increase of the number of targets. However, without explicit data association algorithms, the PHD filter has a linear computational complexity $O(mn)$, where m is the number of detections and n is number of targets. Hence, PHD filter can solve the computational bottleneck problem better.

2.2. Lower Probability Detection and Acoustic Shadow Zone

It is well known that sonar is always working with noise. A simple active sonar detection schematic is shown in Figure 1. A detecting signal is emitted by a sonar sensor array with sound level SL. After the transmission loss of TL_1, the signal reaches the target. When the target's scattering strength is TS, the sound level of the scattering signal is $SL - TL_1 + TL$. After the transmission loss of TL_2 the signal is received by the receiver sonar sensor array. Let the receiver noise level be denoted as NL, and DI denotes the receiver directivity index. When the received signal of sound level is not less than the detection threshold DT, the target can be detected:

$$SL - TL_1 - TL_2 + TS - (NL - DI) \geq DT \tag{1}$$

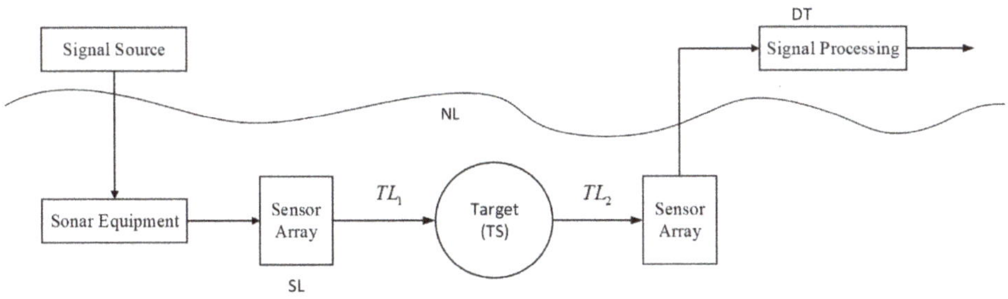

Figure 1. A simple schematic of active sonar detection.

In the Figure 2, we can see that there are many factors influencing target detection. However, these influences can be reflected by detection probability, number of clutters and measurement errors in the MTT algorithm.

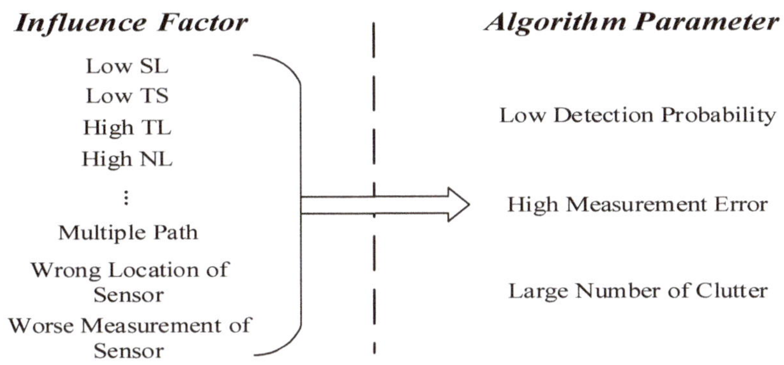

Figure 2. The relationship between influence factors and algorithm parameters.

For example, as shown in Figure 3, when a target is located in the acoustic shadow zone (red circle) of the sensor it can be deemed that the transmission loss of the target is very large. Thus, the detection probability of the target is very small, which results in little effective measurement data for the target. This is also the reason for studying the MTT algorithm for low SNR situations.

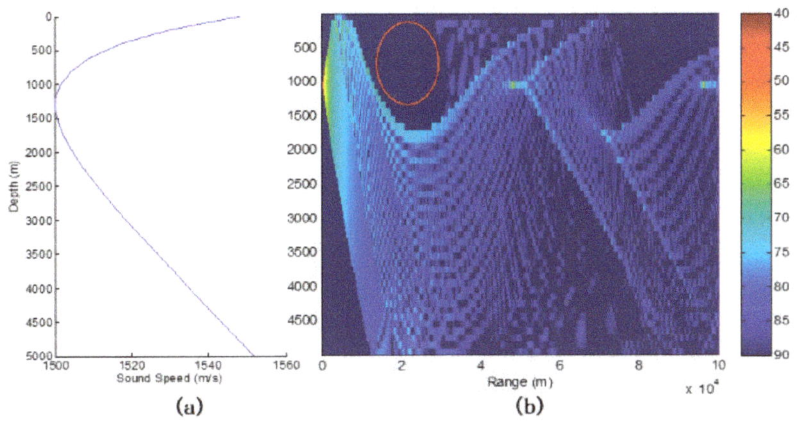

Figure 3. Example of sound acoustic zone: (**a**) the speed of sound; (**b**) the transmission loss, where the red circle is an acoustic shadow zone.

Luckily, distributed multiple sensor networks have a huge advantage in MTT by fusing multisensor data. For example, although a target is located in the acoustic shadow zone of sensor 1, the MTT algorithm can still track the target when it can be detected by sensor 2. In addition, when a target can be detected by multiple sensors at the same time, the MTT algorithm can achieve statistical accuracy improvement.

2.3. Framework of Maximum-Detection Capability Multitarget Track Fusion Algorithm

Distributed fusion structure is a common fusion method in which each local sensor has a tracker, and the local track calculated from the tracker is sent to the fusion center. In the fusion center, all the tracks will be associated and fused to estimate global tracks. The distributed fusion structure has the following advantages: low communication burden, high reliability, easy implementation and computational balance. Moreover, MTT algorithms based on distributed fusion structure have the capability of local tracking and global monitoring. In sonar detection networks, MTT algorithms based on a distributed fusion structure could also track targets in acoustic shadow zones.

Therefore, this paper proposes a MDC-MTF algorithm. As Figure 4 shows, the GMPHD filter is firstly used to get a local estimation from local sonar sensor measurements. Secondly, association and fusion algorithms are used to estimate the global tracking result. Thirdly, a novel feedback algorithm is used to improve the local sensor tracking performance.

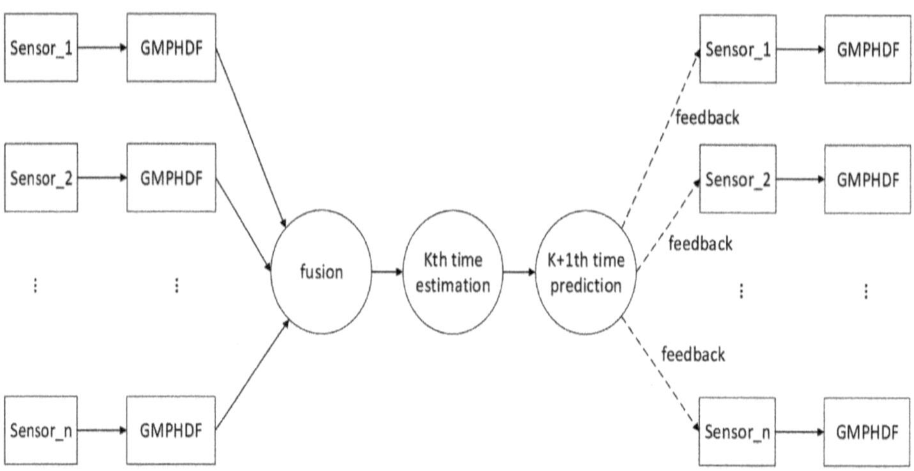

Figure 4. Framework of the maximum-detection capability multitarget track algorithm.

3. GMPHD Filter Theory

In a multitarget tracking environment, target states finite sets X_k and measurements finite sets Z_k are determined as follows:

$$X_k = \left\{x_{k,1}, x_{k,2}, \cdots, x_{k,N(k)}\right\} \tag{2}$$

$$Z_k = \left\{z_{k,1}, z_{k,2}, \cdots, z_{k,M(k)}\right\} \tag{3}$$

where $M(k)$ and $N(k)$ are respective number of targets state $x_{k,1}, \ldots, x_{k,M(k)} \in X_k$ and measurements $z_{k,1}, \ldots, z_{k,N(k)} \in Z_k$ at time k.

For a given multitarget state X_{k-1} at time $k-1$, each $x_{k-1} \in X_{k-1}$ either continues to exist at time k with probability $P_{S,k}(x_{k-1})$, or dies with probability $1 - P_{S,k}(x_{k-1})$. Hence this behavior could be modelled as a RFS $S_{k|k-1}(x_{k-1})$. At time k, a new target can arise by spontaneous birth or by spawning from an exist target at time $k-1$. Also, they could be modelled spontaneous births sets Γ_k

and spawned target sets $B_{k|k-1}(x_{k-1})$ as a RFS at time k. Therefore, at time k, a given multitarget state X_k consists of three sets of $S_{k|k-1}(St)$, $B_{k|k-1}(St)$ and Γ_k:

$$X_k = [\bigcup_{St\in X_{k-1}} S_{k|k-1}(St)]\bigcup[\bigcup_{St\in X_{k-1}} B_{k|k-1}(St)]\bigcup\Gamma_k \tag{4}$$

Moreover, at time k, each target could be detected by sensor with probability $P_{D,k}(x_k)$. Each target state $x_k \in X_k$ could generate a RFS $\Theta_k(x_k)$ at time k. In addition, the sensor also could receive some false measurements or clutter at time k. They can be modelled as a RFS K_k. Consequently, the RFS measurement set Z_k can be described as follows:

$$Z_k = K_k\bigcup[\bigcup_{x\in X_k} \Theta_k(x)] \tag{5}$$

Let $p_k(\bullet|Z_{1:k})$ denote the multitarget posterior density, $f_{k|k-1}(\bullet|\bullet)$ denote the multitarget transition density, and $g_k(\bullet|\bullet)$ denote the multitarget likelihood. Then, based on optimal multitarget Bayes filter theory, the multitarget posterior can be propagated by the recursion:

$$p_{k|k-1}(X_k|Z_{1:k-1}) = \int f_{k|k-1}(X_k|X)p_{k-1}(X|Z_{1:k-1})\mu_s(dX) \tag{6}$$

$$p_k(X_k|Z_{1:k}) = \frac{g_k(Z_k|X_k)p_{k|k-1}(X_k|Z_{1:k-1})}{\int g_k(Z_k|X)p_{k|k-1}(X|Z_{1:k-1})\mu_s(dX)} \tag{7}$$

where μ_s is an appropriate reference measure on $F(\chi)$ [14].

We assume that each target evolves and generates observations independently of one another, the clutter is independent of target-originated measurements, and the clutter and predicted multitarget RFS follow a Poisson distribution. Then, let $v_k(\bullet)$ denote the multitarget posterior density intensity, the posterior intensity can be propagated by the PHD recursion:

$$v_{k|k-1}(x) = \int P_{S,k}(St)f_{k|k-1}(x|St)v_{k-1}(St)dSt + \int \beta_{k|k-1}(x|St)v_{k-1}(St)dSt + \gamma_k(x) \tag{8}$$

$$v_k(x) = [1 - P_{D,k}(x)]v_{k|k-1}(x) + \sum_{z\in Z_k} \frac{P_{D,k}(x)g_k(z|x)v_{k|k-1}(x)}{\kappa_k(z) + \int P_{D,k}(\xi)g_k(z|\xi)v_{k|k-1(\xi)d\xi}} \tag{9}$$

According to Gaussian mixture model (GMM) theory and GMPHD algorithm [15], the Equations (10) and (11) could be substituted by Equations (8) and (9):

$$v_{k-1}(x) = \sum_{i=1}^{J_{k-1}} \omega_{k-1}^{(i)} N(x; m_{k-1}^{(i)}, P_{k-1}^{(i)}) \tag{10}$$

$$v_{k|k-1}(x) = \sum_{i=1}^{J_{k|k-1}} \omega_{k|k-1}^{(i)} N(x; m_{k|k-1}^{(i)}, P_{k|k-1}^{(i)}) \tag{11}$$

where, ω is the weight of Gaussian distribution, $N(\bullet; m, P)$ denotes a Gaussian density with mean m and covariance P, J is the number of components of the intensity. Therefore, the prediction updating and estimation of the target can be implemented. For the implementation process, please refer to the paper [15].

4. The MDC-MTF Algorithm

In practical applications, sonar equipment usually only acquires the target location information, MTT algorithms need to start tracking according to the target location information, while the target speed information is essential for most MTT Bayes trackers. However, these track initiation algorithms are sensitive to SNR. The performance of tracker may be severely degraded when the SNR decreases. Thus, a MDC-MTF algorithm and a novel feedback algorithm were proposed in this paper to improve the GMPHD filter performance.

4.1. Maximum Detection Capability Fusion Strategy

In the distributed fusion structure, most of the clutter has been filtered out by GMPHD filter. Then, all local estimations will be associated and fused at a fusion center. In order to ensure the maximum detection capability, we divide the local estimates into two categories. One is the correlated local estimates, and the other is the uncorrelated local estimates. For the correlated local estimates, we associate and fuse those local estimates. For the uncorrelated local estimates, we treat them as a global estimation. Usually, there are three cases in associating two local estimates:

Case 1: the local estimation from sensor i can match with sensor j;
Case 2: the local estimation from sensor i mismatch with sensor j;
Case 3: the target state does not exist in sensor i and sensor j;

Therefore, this paper proposes a fusion strategy:

(a) For case 1, the local estimation from two sensors association, and the most possible data fusion to estimate target state;
(b) For case 2, the local estimation is retained as a global estimation;
(c) For case 3, treated them as missing detection.

4.2. Data Association Algorithm

Unlike the JPDA algorithm, the target density has been significantly reduced after GMPHD filtering, so the association algorithm does not result in a heavier computational burden in MDC-MTF. Today, there are many classical data association algorithms, such as nearest neighbor (NN) [25], weighted track association (WTA) [26], modified weighted track association (m-WTA) [27], k-nearest neighbor (k-NN), modified k-nearest neighbor (MK-NN) [28], independent and dependent sequential track correlation criteria (STCC), independent and dependent binary track correlation (BTC) algorithms, and fuzzy synthetic track correlation criterion (FSTCC). A detailed analysis of association performance was presented in the paper [29]. The performance comparison of track correlation algorithms is shown in Table 1.

Table 1. Performance comparison of track correlation algorithm for distributed multisensor systems.

Name	Computing Time (Second)	Communication Burden	Correct Correlation Probability (Medium Target Density)	Correct Correlation Probability (High Target Density)
NN	48	low	0.6449	0.4284
k-NN	307	low	0.8922	0.7526
MK-NN	291	low	0.8956	0.7694
WTA	47	medium	0.7315	0.4755
m-WTA	138	high	0.7384	0.4901
independent-STCC	470	medium	0.9065	0.7735
dependent-STCC	1406	high	0.8294	0.7009
independent-BTC	284	medium	0.9319	0.8067
dependent-BTC	818	high	0.9143	0.7958
FSTCC	352	medium	0.9218	0.7786

As shown in Table 1, the weighted track association algorithm is an optimal choice considering the computational cost, correct correlation probability (medium target density) and communication burden. The weighted track association algorithm is described as follows:

At time k, $t_{gh}(k)$ is defined as the difference in the value of two sensors:

$$t_{gh}(k) = \hat{X}_g^i(k) - \hat{X}_h^j(k) \tag{12}$$

where $\hat{X}_g^i$ and $\hat{X}_h^j$ were respectively the g-th local estimation of sensor i and j the h-th local estimation of sensor. Also, P_g^i and P_h^j are defined as covariance. When the errors of sensors are uncorrelated, the Mahalanobis distance α_{gh} could be calculated by Equations (13) and (14):

$$\alpha_{gh}(k) = t'_{gh}(k)C_{gh}(k)^{-1}t_{gh}(k) \tag{13}$$

$$C_{gh}(k) = P_g^i(k) + P_h^j(k) \tag{14}$$

According to [30], the $\hat{X}_g^i$ and $\hat{X}_h^j$ have been associated when $\alpha_{gh} \leq T_a$. T_a is an association threshold. Even when there are multiple $\hat{X}_h^j$ satisfying $\alpha_{gh} \leq T_a$, a smaller α_{gh} means a higher correlation. Thus, we fuse the data when α_{gh} is the minimum.

4.3. Multisensor Data Fusion

The convex combination fusion [30] is an optimal fusion algorithm when there is no process noise and, the local estimation of two tracks is not correlated. While the local estimation is correlated [31], the Bar-Shalom-Campo fusion algorithm [32] is better. However, due to the slow motion of targets, the effect of process noise is usually smaller. Thus, we assumed the local estimation is not uncorrelated in this paper. The two sensor estimations and covariance matrix are respectively X^m and P^m, $m = i, j$. According to the convex combination fusion theory, the global estimation is obtained via Equation (15):

$$\begin{cases} \hat{X} = [(P^i)^{-1} + (P^j)^{-1}]^{-1}(P^i)^{-1}\hat{X}^i + [(P^i)^{-1} + (P^j)^{-1}]^{-1}(P^j)^{-1}\hat{X}^j \\ \hat{P}^{-1} = (P^i)^{-1} + (P^i)^{-1} \end{cases} \tag{15}$$

Extending to multisensor ($N > 2$) conditions, the multisensor global estimation could be derived from Equation (16):

$$\begin{cases} \hat{X} = [\sum_{u=1}^{N} (P^u)^{-1}]^{-1} \sum_{u=1}^{N} ((P^u)^{-1}\hat{X}^u) \\ \hat{P}^{-1} = \sum_{u=1}^{N} (P^u)^{-1} \end{cases} \tag{16}$$

4.4. Feedback Algorithm Based on RFS Theory

Since GMPHD is statistically unbiased, it is possible to bias GMPHD by feeding other target information to GMPHD. We thought that the method of independent implement of feedback algorithm and GMPHD was a good way to avoid bias problem. As shown in Figure 5, at time $k-1$, global estimations were calculated by fusion algorithm, and then modeled as a RFS. Based on the feedback target information, the state of target can be predicted for time k. After prediction, the prediction information will be fed back to local sensors. After that, at time k, we referred to the RFS theory to make a feedback estimate. Finally, the estimation of the local sensor is obtained by fusing the estimation of feedback with estimation of GMPHD filter.

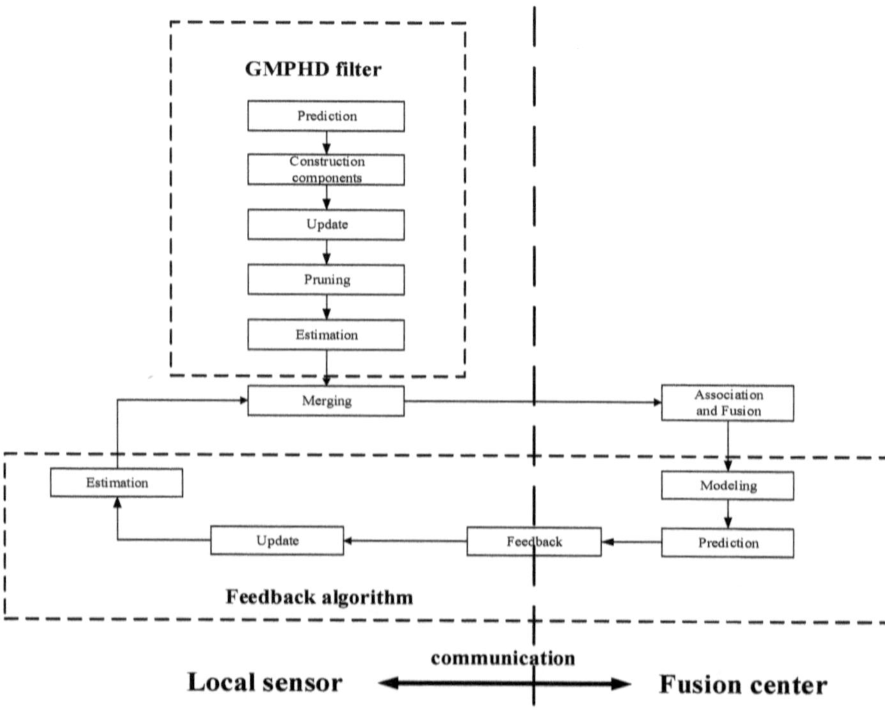

Figure 5. Structure of feedback algorithm.

This paper proposes a feedback algorithm that offered two advantages. The feedback algorithm could improve the detection ability of local sensor without biasing GMPHD filter. On the other hand, we have noted that feedback algorithm might expand the error estimation effect of GMPHD filter.

The details of the feedback algorithm are described as follows:

(1) Modeling. According to RFS theory, the feedback is modeled as a RFS:

$$\Psi_{F,k-1|k-1} = \left\{ \hat{x}_{k-1|k-1,1}, \hat{x}_{k-1|k-1,2}, \cdots, \hat{x}_{k-1|k-1,J_{F,k-1}} \right\} \quad (17)$$

where $J_{F,k-1}$ is the number of global estimations, $\hat{x}$ is the state of global estimation at time $k-1$.

(2) Prediction. Based on the transfer matrix $F_{k|k-1}$ and process noise Q, the target state and covariance are predicted via Equations (18) and (19):

$$\tilde{x}_{k|k-1,j} = F_{k|k-1}\hat{x}_{k-1|k-1,j} \quad (18)$$

$$P_{F,k|k-1,j} = Q + F_{k|k-1}P_{F,k-1|k-1,j}F_{k|k-1}^{T} \quad (19)$$

(3) Feedback. The prediction state and covariance are fed back to the local sensors at time $k-1$.

(4) Update. By Equations (20)–(22), we could update the target state and covariance at time k:

$$K_{F,k,j} = P_{F,k|k-1,j}H_k^T\left(H_kP_{F,k|k-1,j}H_k^T + R_k\right)^{-1} \quad (20)$$

$$\hat{x}_{F,k,j} = \tilde{x}_{F,k|k-1,j} + K_{F,k,j}(Z_{k,i} - H_k\tilde{x}_{F,k|k-1,j}) \quad (21)$$

$$P_{F,k,j} = (I - K_{F,k,j}H_k)P_{F,k|k-1,j} \quad (22)$$

(5) Estimation. At time $k-1$, assuming the intensity of the feedback target is a Gaussian mixture form:

$$v_{F,k-1}(x) = \sum_{i=1}^{J_{F,k-1}} \omega_{F,k-1}^{(i)} N(x; m_{F,k-1}^{(i)}, P_{F,k-1}^{(i)}) \tag{23}$$

where, $N(\bullet; m_{F,k-1}, P_{F,k-1})$ denotes a Gaussian density with mean $m_{F,k-1}$ and covariance $P_{F,k-1}$. The $\omega_{F,k-1}$ is the weight of the Gaussian density. Thus, the predicted intensity is a Gaussian mixture of form:

$$\widetilde{v}_{F,k|k-1}(x) = \sum_{i=1}^{J_{F,k-1}} \omega_{F,k|k-1}^{(i)} N(x; \widetilde{m}_{F,k|k-1}^{(i)}, \widetilde{P}_{F,k|k-1}^{(i)}) \tag{24}$$

Then, at time k, the intensity of measured target is a Gaussian mixture of form:

$$\overline{v}_{F,k}(z) = \sum_{i=1}^{J_{F,k-1}} \omega_{F,k}^{(i)} N(z; \widetilde{m}_{F,k|k-1}^{(i)}, \widetilde{P}_{F,k|k-1}^{(i)}) \tag{25}$$

When the ratio of intensity of measured target to target birth intensity exceeds the feedback threshold T_F, the target state will be extracted. In general, we recommend that the correlation threshold be the same as the merge threshold. All extracted targets are the feedback estimation of feedback algorithm.

$$\hat{X}_{F,k} = \begin{cases} [\hat{X}_{F,k}, z], \frac{\overline{v}_{F,k-1}(z)}{\gamma_k(z)} > T_F \\ \hat{X}_{F,k}, else \end{cases} \tag{26}$$

(6) Merging. At time k, the GMPHD filter estimations $\hat{X}_k$ and feedback estimations $\hat{X}_{F,k}$ are merged as the local estimations via Equations (9)–(13).

5. Simulation

In this section, there are four examples. All sensors could detect the targets except Example C. Considering a two-dimensional scenario, the number of measurements (contains targets and clutters) is time-varying and unknown over the surveillance region $[-1000, 1000] \times [-1000, 1000]$ (in m). At time k, each measurement contains location $(p_{x,k}, p_{y,k})$ and velocity $(v_{x,k}, v_{y,k})$, and is represented by $x_k = [p_{x,k}, p_{y,k}, v_{x,k}, v_{y,k}]^T x_k = [p_{x,k}, p_{y,k}, v_{x,k}, v_{y,k}]^T$. Each target has survival probability $P_{S,k} = 0.99$ and follows the linear Gaussian Model. The transfer model F_k and process noise Q_k are represented as follows:

$$F_k = \begin{bmatrix} I_2 & \Delta I_2 \\ 0_2 & I_2 \end{bmatrix}, Q_k = \sigma_v^2 \begin{bmatrix} \frac{\Delta^4}{4} I_2 & \frac{\Delta^2}{2} I_2 \\ \frac{\Delta^3}{2} I_2 & \Delta^2 I_2 \end{bmatrix}$$

where I_n and 0_n are $n \times n$ identity and zero matrices, $\Delta = 1s$ is the sampling period, and $\sigma_v = 5(\mathrm{m/s^2})$. $H = [I_2, 0_2]$ is the observation model and observation noise is $R_k = \sigma_\varepsilon^2 I_2, \sigma_\varepsilon = 10(m)$.

There are three targets and clutters (less than 50) over the surveillance region. Target 1 and target 2 are born at time $k = 0$, the target 3 is spawned by target 2 at time $k = 66$. All the targets are straight uniform motion as shown in Figure 6. The number of clutter varies randomly with time.

The intensity of birth and spawn target are represented by $\gamma_k(x) = 0.1 \sum_{i=1}^{J_{\gamma,k}} N(x; x, P_\gamma)$ and $\beta_{k|k-1}(x|St) = 0.05N(x; St, Q_\beta)$, where the ς is the previous state, $P_\gamma = diag([100, 100, 25, 25]^T)$, $Q_\beta = diag([100, 100, 400, 400]^T)$. The intensity of clutter follows uniform distribution.

In addition, the GMPHD filter parameters with detecting threshold $T_\omega = 0.5$, merging threshold $U = 4$, maximum allowable number of Gaussian terms $J_{max} = 100$. The feedback threshold is $T_F = 0.5$, the association threshold is $T_a = 4$.

The OSPA [33] is a good index for evaluating the performance of MTT algorithms. Therefore, the statistical characteristics of OSPA was used to evaluate the performance of the algorithm in this paper. The OSPA order $p = 1$ and truncation distance $c = 200$. We carried out 100 Monte Carlo simulations, each with 100 steps.

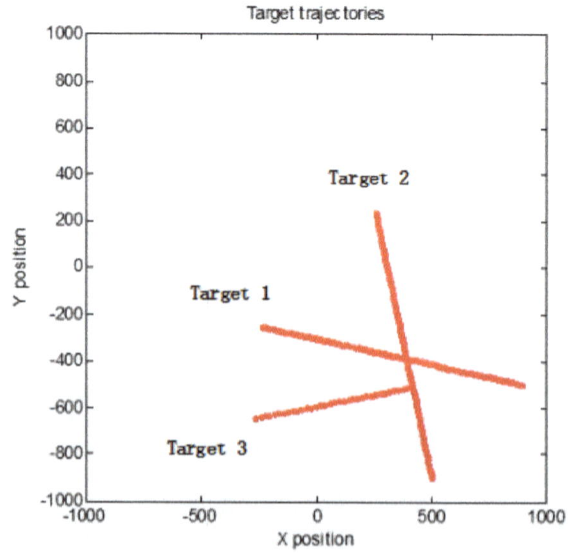

Figure 6. Targets trajectories.

5.1. Example A. The Tracking Performance at Different Detection Probabilities

The purpose of this example was to evaluate the performance between this tracking framework and GMPHD filter at different detection probabilities (0.9, 0.8 and 0.7). Figure 7a–c are GMPHD filter tracking results at different detection probabilities; Figure 7d–f are the global estimation based on two sensor fusion; Figure 7g–i are the global estimation based on three sensor fusion. Their corresponding OSPA results are shown in the Figure 8. The red line is the real track of the target; the blue 'o' is the algorithm estimation; the black 'x' is clutters. In order to test the robustness of this algorithm, this paper performed a Monte Carlo simulation, and calculated the OSPA statistical natures as shown in the Figure 9.

From Figures 7–9, it is obvious that this algorithm has a better performance than GMPHD filter when the detection probability is low. When detection probability is 0.7, the mean of OSPA could decrease almost by 40% for two sensors fusion and 55% for three sensors fusion.

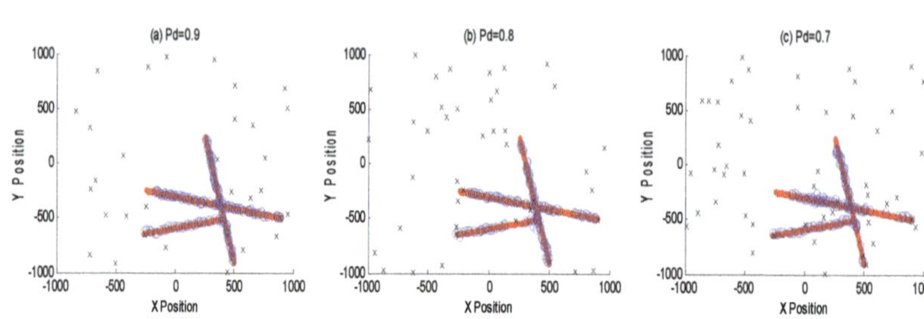

Figure 7. *Cont.*

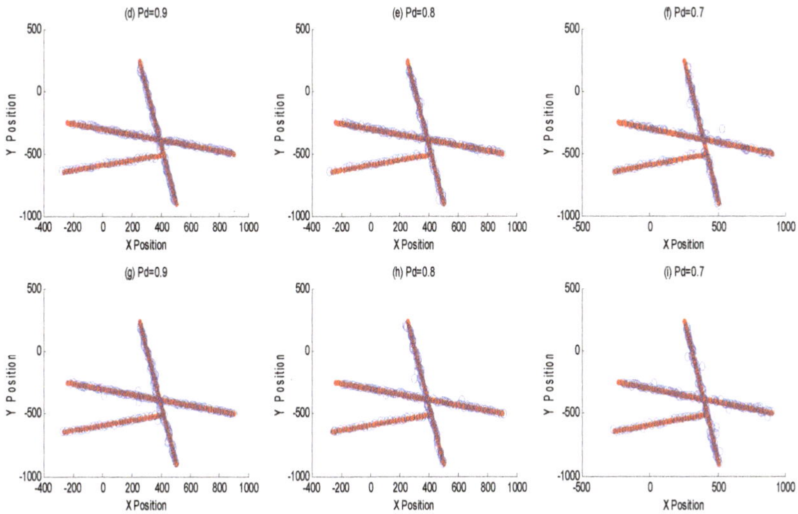

Figure 7. Multisensor tracking results at different detection probabilities. (**a**–**c**) are classical GMPHD filter tracking results at different detection probabilities; (**d**–**f**) are two sensors global estimations by MDC-MTF algorithm; (**g**–**i**) are three sensors global estimations by MDC-MTF algorithm.

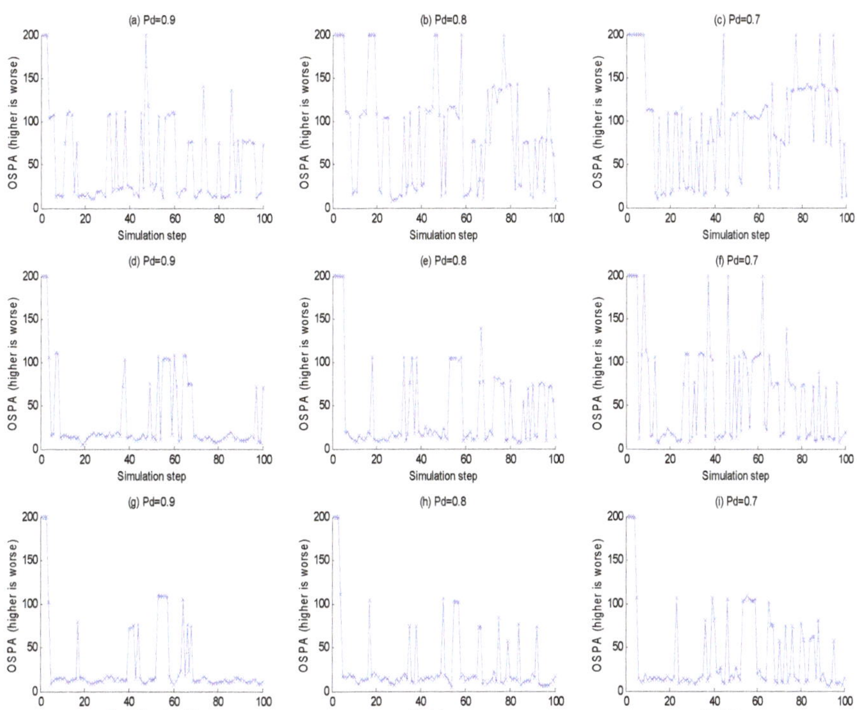

Figure 8. The OSPA of multisensor data fusion. (**a**–**c**) are the OSPA of GMPHD filter tracking results at different detection probabilities; (**d**–**f**) are the OSPA of two sensors global estimations by the MDC-MTF algorithm; (**g**–**i**) are the OSPA of three sensors global estimations by MDC-MTF algorithm.

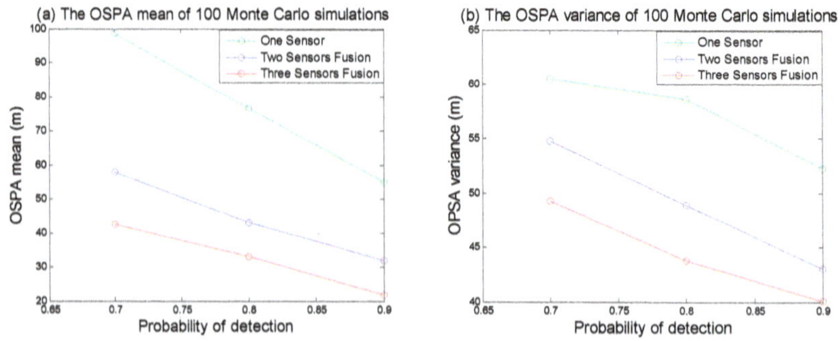

Figure 9. The OSPA statistical nature of Monte Carlo simulations. (**a**) is the OSPA mean of 100 times Monte Carlo simulations; (**b**) is the OSPA variance of 100 times Monte Carlo simulations.

5.2. Example B. Feedback/No Feedback Effect

In this example, the performance of feedback algorithm was evaluated. As the Figure 10 shown, the local sensor tracking results with feedback and no feedback. Figure 10a–c are no feedback results; Figure 10d–f are the local sensor tracking results in the two sensor fusion structure; Figure 10g–i are the local sensor tracking results in the three sensor fusion structure. The OSPA statistical nature of the Monte Carlo simulation is shown in the Figure 11. We could see the multitarget tracking performance of the local GMPHD filter with feedback algorithm is better than without feedback algorithm. That is because feedback algorithm could help GMPHD filter track those targets with no predicted information. Meanwhile, in Figure 11a, when the detection probability is low, the red line is higher than the blue line, this indicates the feedback algorithm may expand the impact of estimation error.

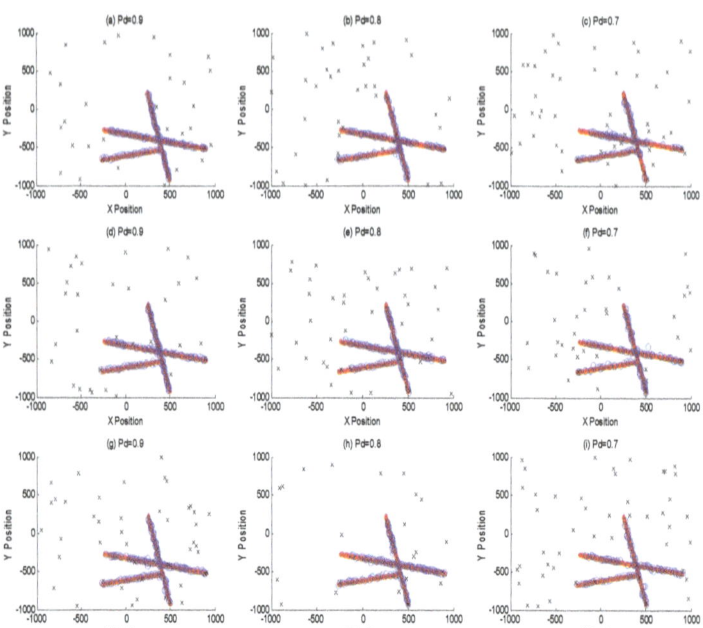

Figure 10. Feedback/ no feedback tracking results. (**a–c**) are no feedback results; (**d–f**) are local sensors tracking results after feeding back two sensors fusion results; (**g–i**) are local sensors tracking results after feeding back three sensors fusion results.

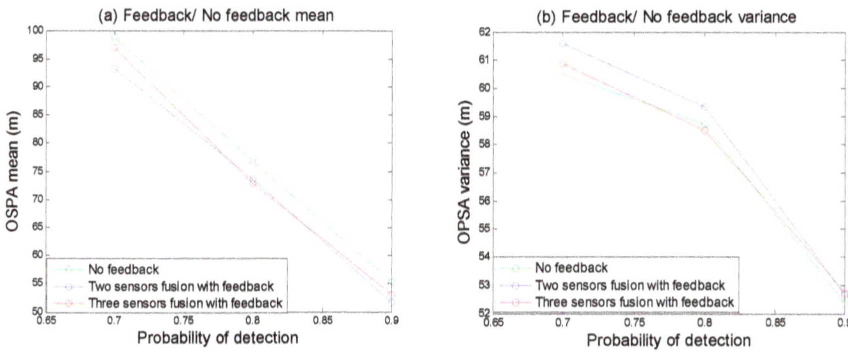

Figure 11. The OSPA of feedback/ no feedback. (**a**) is the OSPA mean of feedback or no feedback; (**b**) is the OSPA variance of feedback or no feedback.

5.3. *Example C. Simulation of a Target is in Acoustic Shadow Zone*

Considering that a wrong sensor location (e.g., an acoustic shadow zone) may make the sensor unable to detect a target, we performed a simulation based on three sensor data fusion, where the detection probability was 0.9. In this simulation, sensor 1 failed to detect the target 1 from step = 20 to step = 80. Thus, from step = 20 to step = 80, the measurement data of the target did not exist in sensor 1. As shown in Figure 12, Figure 12a–c are the local sensor tracking results, Figure 12d is the global estimation, respectively. In Figure 12, we could see the MDC-MTF algorithm can track the target in an acoustic shadow zone.

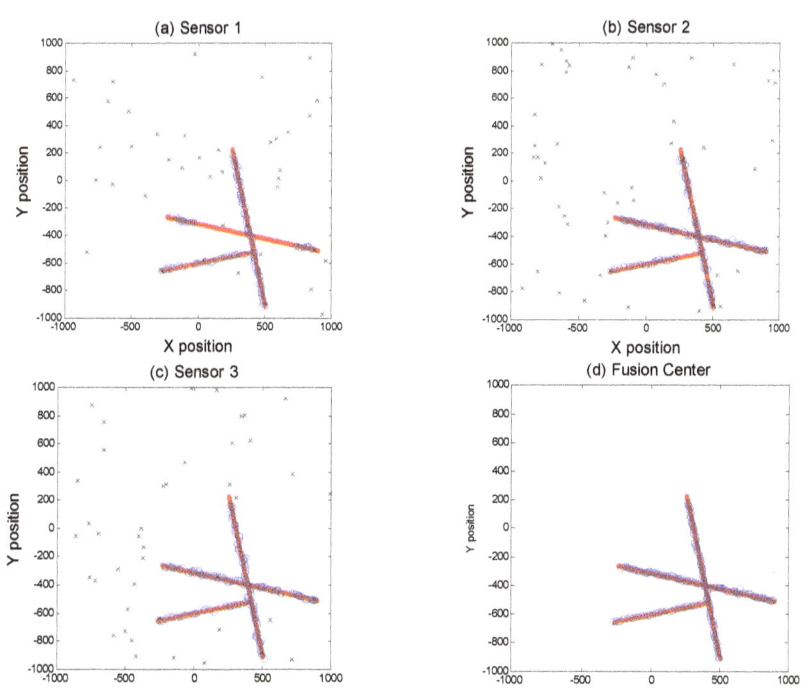

Figure 12. Tracking simulation of the target in acoustic shadow zone.

5.4. Example D. Analysis of the Influence of Some Important Parameters on Performance

As some thresholds are important to the performance of the tracking algorithm, we will analyze the influence of the feedback threshold, merging and correlation threshold on the performance of the algorithm when the probability of detection is 0.8. As shown in Figure 13a,b are the statistics of the OSPA with different feedback thresholds, (c) and (d) are the statistics of the OSPA with different association thresholds. Figure 13 illustrated two issues: (1) A small feedback threshold means greater tolerance for measured error. However, a small feedback threshold also means the risk of clutter or error estimation increases, though feedback threshold could improve MTT tracking performance. (2) The essence of association threshold based on Mahalanobis distance is the correlation of data sets. In this paper, within a certain range, a lager threshold can improve MTT performance.

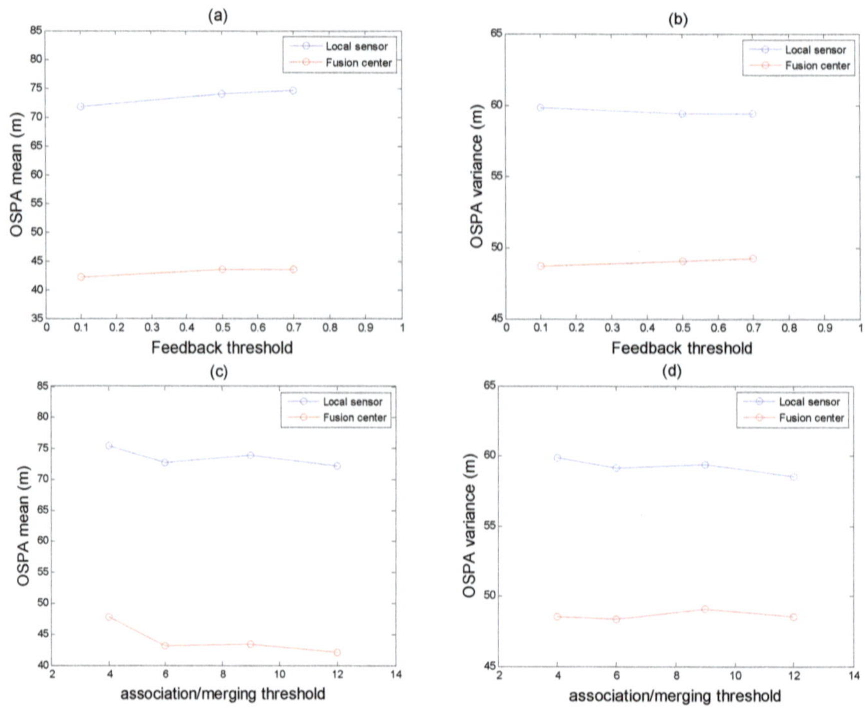

Figure 13. Simulation analysis of the effect of threshold on algorithm performance.

6. Conclusions

In this paper, a maximum-detection capability multitarget track fusion (MDC-MTF) algorithm was proposed, which contains a maximum detection capability fusion strategy, data association, multisensor data fusion and a novel feedback algorithm based on RFS theory. In the distributed sensor network, considering the complexity of computation and the sonar working environment, the GMPHD filter was selected to track local sensors. To deal with the problem that GMPHD was designed for high SNR, we associated and fused multisensor data. Moreover, this algorithm also successfully solved the problem of target tracking in acoustic shadow zones. Monte Carlo simulations have proved this algorithm's performance. Firstly, when detection probability is 0.7, the OSPA mean of two sensors fusion could be decreased almost by 40% with GMPHD, and three sensor fusion could be decreased by almost 55%. Secondly, the feedback algorithm could improve the detection ability of local sensors without biasing the GMPHD filter. On the other hand, we have noted that feedback algorithm might expand the error estimation effect of the GMPHD filter. Thirdly, by fusing multisensor data, the MDC-MTF algorithm could track targets which were in the acoustic shadow zones of a sonar sensor.

Author Contributions: Conceptualization, X.S. and Y.C.; Methodology, L.G.; Software, Y.C.; Validation, X.S., L.G. and J.Y.; Formal Analysis, Y.C.; Investigation, Y.C.; Resources, L.G.; Data Curation, X.H.; Writing-Original Draft Preparation, Y.C.; Writing-Review & Editing, X.S.; Visualization, X.H.; Supervision, J.Y.; Project Administration, J.Y.; Funding Acquisition, L.G.

Funding: This research was funded by the National Key R&D Program of China under Grant No. 2018YFC1405900, by the National Natural Science Foundation of China under Grant No. 61631008, No. 50509059, No. 51779061, and Ministry of Land and Resources Industry research special funds for public welfare projects under grant No. 201511037-02.

Acknowledgments: We thank the College of Underwater Acoustic Engineering of Harbin Engineering University for assistance and the anonymous reviewers for their instructive comments.

Conflicts of Interest: The authors declare no conflict of interest.

References

1. Barshalom, Y.; Kirubarajan, T.; Lin, X. Probabilistic data association techniques for target tracking with applications to sonar, radar and EO sensors. *Aerosp. Electron. Syst. Mag. IEEE* **2005**, *20*, 37–56. [CrossRef]
2. Yong, J.; Teng, T.; Gao, L.; Yu, J. Research on distributed multistatic sonar target tracking technique. In Proceedings of the 2016 IEEE/OES China Ocean Acoustics (COA), Harbin, China, 9–11 January 2016; Institute of Electrical and Electronics Engineers Inc.: Interlaken, Switzerland, 2016.
3. Musicki, D.; Evans, R. Joint integrated probabilistic data association: JIPDA. *IEEE Trans. Aerosp. Electron. Syst.* **2004**, *40*, 1093–1099. [CrossRef]
4. Fortmann, T.E.; Barshalom, Y.; Scheffe, M. Sonar tracking of multiple targets using joint probabilistic data association. *IEEE J. Ocean. Eng.* **2003**, *8*, 173–184. [CrossRef]
5. He, S.; Shin, H.S.; Tsourdos, A. Joint probabilistic data association filter with unknown detection probability and clutter rate. *Sensors* **2018**, *18*, 269.
6. Blackman, S.S. Multiple hypothesis tracking for multiple target tracking. *IEEE Aerosp. Electron. Syst. Mag.* **2004**, *19*, 5–18. [CrossRef]
7. Han, Y.; Zheng, C.; Sun, D. Underwater Node Localization Using Modified Hough Transform and Least Square Method. In Proceedings of the 10th International Conference on Underwater Networks & Systems, Washington, DC, USA, 22–24 October 2015.
8. Yao, L.; Liu, Y.; He, Y. A Novel Ship-Tracking Method for GF-4 Satellite Sequential Images. *Sensors* **2018**, *18*, 2007. [CrossRef] [PubMed]
9. Mahler, R.P.S. *Random Sets in Information Fusion. An Overview*; Springer: New York, NY, USA, 1997; pp. 129–164, ISBN 978-1-4612-7350-9.
10. Mahler, R.P.S. Multitarget Bayes filtering via first-order multitarget moments. *IEEE Trans. Aerosp. Electron. Syst.* **2004**, *39*, 1152–1178. [CrossRef]
11. Mahler, R.P.S. "Statistics 101" for multisensor, multitarget data fusion. *IEEE Aerosp. Electron. Syst. Mag.* **2004**, *19*, 53–64. [CrossRef]
12. Mahler, R.P.S. *Statistical Multisource-Multitarget Information Fusion*; Artech House, Inc.: Norwood, MA, USA, 2007; ISBN 1596930926-9781596930926.
13. Mahler, R. PHD filters of higher order in target number. *IEEE Trans. Aerosp. Electron. Syst.* **2007**, *43*, 1523–1543. [CrossRef]
14. Vo, B.N.; Singh, S.; Doucet, A. Sequential Monte Carlo methods for multitarget filtering with random finite sets. *IEEE Trans. Aerosp. Electron. Syst.* **2005**, *41*, 1224–1245.
15. Vo, B.N.; Ma, W.K. The Gaussian Mixture Probability Hypothesis Density Filter. *IEEE Trans. Signal Process.* **2006**, *54*, 4091–4104. [CrossRef]
16. Vo, B.-T.; Vo, B.-N.; Cantoni, A. The cardinality balanced multi-target multi-Bernoulli filter and its implementations. *IEEE Trans. Signal Process.* **2009**, *57*, 409–423.
17. Vo, B.-T.; Vo, B.-N. Labeled random finite sets and multi-object conjugate priors. *IEEE Trans. Signal Process.* **2013**, *61*, 3460–3475. [CrossRef]

18. Reuter, S.; Vo, B.T.; Vo, B.N.; Dietmayer, K. The Labeled Multi-Bernoulli Filter. *IEEE Trans. Signal Process.* **2014**, *62*, 3246–3260.
19. Vo, B.N.; Vo, B.T.; Phung, D. Labeled Random Finite Sets and the Bayes Multi-Target Tracking Filter. *IEEE Trans. Signal Process.* **2014**, *62*, 6554–6567. [CrossRef]
20. Papi, F.; Vo, B.N.; Vo, B.T.; Fantacci, C.; Beard, M. Generalized Labeled Multi-Bernoulli Approximation of Multi-Object Densities. *IEEE Trans. Signal Process.* **2015**, *63*, 5487–5497. [CrossRef]
21. Vo, B.-N.; Vo, B.-T.; Hoang, H. An Efficient Implementation of the Generalized Labeled Multi-Bernoulli Filter. *IEEE Trans. Signal Process.* **2017**, *65*, 1975–1987. [CrossRef]
22. Zhang, L.; Wang, T.; Zhang, F.; Xu, D. Cooperative Localization for Multi-AUVs Based on GM-PHD Filters and Information Entropy Theory. *Sensors* **2017**, *17*, 2286. [CrossRef] [PubMed]
23. Subedi, S.; Zhang, Y.D.; Amin, M.G.; Himed, B. Group sparsity based multi-target tracking in passive multi-static radar systems using Doppler-only measurements. *IEEE Trans. Signal Process.* **2016**, *64*, 3619–3634. [CrossRef]
24. Zhang, F.; Buckl, C.; Knoll, A. Multiple Vehicle Cooperative Localization with Spatial Registration Based on a Probability Hypothesis Density Filter. *Sensors* **2014**, *14*, 995–1009. [CrossRef] [PubMed]
25. Kosaka, M.; Miyamoto, S.; Ihara, H. A track correlation algorithm for multi-sensor integration. *J. Guid. Control Dyn.* **1987**, *10*, 166–171. [CrossRef]
26. Singer, R.A.; Kanyuck, A.J. *Computer Control of Multiple Site Track Correlation*; Pergamon Press, Inc.: Tarrytown, NY, USA, 1971; pp. 455–463.
27. Gul, E. On the track similarity test in track splitting algorithm. *IEEE Trans. Aerosp. Electron. Syst.* **1994**, *30*, 604–606. [CrossRef]
28. Jian, G.; You, H.; Ying-Ning, P. Distributed CFAR detector based on local test statistic. *Signal Process.* **2000**, *80*, 373–379. [CrossRef]
29. He, Y.; Wang, G.H. Review of Track Correlation Algorithm in Distributed Multi-sensor System. *Syst. Eng. Electron.* **1999**, *10*, 15–18.
30. You, H.; Guohong, W.; Xin, G. *Information Fusion Theory with Applications*; Publishing House of Electronics Industry: Beijing, China, 2010; ISBN 978-7-121-10323-0.
31. Bar-Shalom, Y.; Campo, L. The effect of the common process noise on the two-sensor fused-track covariance. *IEEE Trans. Aerosp. Electron. Syst.* **1986**, *6*, 803–805. [CrossRef]
32. Bar-Shalom, Y. On the track-to-track correlation problem. *IEEE Trans. Autom. Control* **1981**, *26*, 571–572. [CrossRef]
33. Schuhmacher, D.; Vo, B.T.; Vo, B.N. A Consistent Metric for Performance Evaluation of Multi-Object Filters. *IEEE Trans. Signal Process.* **2008**, *56*, 3447–3457. [CrossRef]

sensors

MDPI

Article

Globally Optimal Distributed Kalman Filtering for Multisensor Systems with Unknown Inputs

Yali Ruan, Yingting Luo * and Yunmin Zhu

College of Mathematics, Sichuan University, Chengdu 610064, Sichuan, China; ruanyali2018@163.com (Y.R.); ymzhu@scu.edu.cn (Y.Z.)

* Correspondence: ytluo@scu.edu.cn; Tel.: +86-138-8227-9920

Received: 10 July 2018; Accepted: 3 September 2018; Published: 6 September 2018

Abstract: In this paper, the state estimation for dynamic system with unknown inputs modeled as an autoregressive AR (1) process is considered. We propose an optimal algorithm in mean square error sense by using difference method to eliminate the unknown inputs. Moreover, we consider the state estimation for multisensor dynamic systems with unknown inputs. It is proved that the distributed fused state estimate is equivalent to the centralized Kalman filtering using all sensor measurement; therefore, it achieves the best performance. The computation complexity of the traditional augmented state algorithm increases with the augmented state dimension. While, the new algorithm shows good performance with much less computations compared to that of the traditional augmented state algorithms. Moreover, numerical examples show that the performances of the traditional algorithms greatly depend on the initial value of the unknown inputs, if the estimation of initial value of the unknown input is largely biased, the performances of the traditional algorithms become quite worse. However, the new algorithm still works well because it is independent of the initial value of the unknown input.

Keywords: optimal estimate; unknown inputs; distributed fusion; augmented state Kalman filtering (ASKF)

1. Introduction

The classic Kalman filtering (KF) [1] requires the model of the dynamic system is accurate. However, in many realistic situations, the model may contain unknown inputs in process or measurement equations. The issue concerning estimating the state of a linear time-varying discrete time system with unknown inputs is widely studied by researchers. One widely adopted approach is to consider the unknown inputs as part of the system state and then estimate both of them. This leads to an augmented state Kalman filtering (ASKF). Its computational cost increases due to the augmented state dimension. It is proposed by Friedland [2] in 1969 a two-stage Kalman filtering (TSKF) to reduce the computation complexity of the ASKF, which is optimal for the situation of a constant unknown input. On the basis of the work in [2], it is proposed by Hsieh et al. an optimal two-stage algorithm (OTSKF) for the dynamic system with random bias and a robust two-stage algorithm for the dynamic system with unknown inputs in 1999 [3] and 2000 [4] respectively. It is assumed in [3–5] that the unknown inputs were an autoregressive AR (1) process, with the two-stage algorithms being optimal in the mean square error (MSE) sense. However, the optimality of the ASKF and OTSKF depends on the premise that the initial value of the unknown measurement can be estimated correctly. Under the condition of incorrect initial value of the unknown measurement, the ASKF and OTSKF will have poor performance (see Examples 1 and 2 in Section 5), especially, when the unknown measurement is not stationary as regarded in [4,5]. Due to the difficulty of knowing the exact initial value of the unknown measurement, improvements should be made on these approaches. Many other researchers have focused on the problem of unknown inputs [6–8] in recent years.

A large number of sensors are now used in practical applications in numerous advanced systems. With the processing center receiving all measurements from the local sensors in time, centralized Kalman filtering (CKF) can be accomplished, and the resulting state estimates are optimal in the MSE. Nevertheless, because of limited communication bandwidth, and relatively low survivability of the system in unfavorable conditions, like martial circumstances, Kalman filtering is required to be carried on every local sensor upon its own observation first for local requirement, and then the processed data-local state estimate is transmitted to a fusion center. Therefore, the fusion center now needs to fuse all the local estimates received to produce a globally optimal or suboptimal state estimate. In the existing research literatures, a large number of researches on distributed Kalman filtering (DKF) have been done. Under certain common conditions, particularly, the supposition of cross-independent sensor noises, an optimal DKF fusion was proposed in [9–11] respectively, which was proved to be the same as the CKF adopting all sensor measurements, illustrating that it is universally optimal. Besides, a Kalman filtering fusion with feedback was also suggested there. Then, it was presented in [12] a rigorous performance analysis for Kalman filtering fusion with feedback. The results mentioned above are effective only for conditions with uncoupled observation noises across sensors. It is demonstrated by Song et al. [13] that when the sensor noises are cross-correlated, the fused state estimate was also equal to the CKF under a mild condition. Similarly with [13], Luo et al. [14] posed a distributed Kalman filtering fusion with random state transition and measurement matrices, i.e., random parameter matrices Kalman filtering in 2008. Moreover, they proved that under a mild condition the distributed fusion state estimate is equivalent to the centralized random parameter matrices Kalman filtering using all sensor measurement, which under the assumption that the expectation of all sensor measurement matrices are of full row rank. As far as we know, few studies have been done for multisensor system with unknown inputs by the above mentioned augmented methods. The main reason is that the augmented methods greatly increase the state dimension and computation complexity for the multisensor system.

In this paper, an optimal estimation for dynamic system with unknown inputs in the MSE sense is proposed. Different from the work in [2–5], the approach of eliminating instead of estimating the unknown inputs is used. The unknown inputs are assumed to be an autoregressive AR (1) process and are eliminated by measurement difference method. Then the original dynamic system is converted to a remodeled system with correlated process and measurement noises. The new measurement noise of the remodeled system in this paper is not only one-step correlated in time but also correlated with the process noise. We propose a globally optimal recursive state estimate algorithm for this remodeled system. Compared with the ASKF and OTSKF, the new algorithm is still optimal in the MSE sense but with less computation stress. Additionally, it is showed that the performance of the new algorithm does not rely on the initial value of the unknown input. For the multisensor system with unknown inputs, we show that the centralized filtering can still be expressed by a linear combination of the local estimates. Therefore, the performance of the distributed filtering fusion is the same as that of the centralized fusion. The new algorithm is optimal in the MSE sense with low computation complexity. Numerical examples are given to support our analysis.

The remainder of this paper is organized as follows: the problem formulation is discussed in Section 2, followed by an optimal estimation algorithm for dynamic system with unknown inputs being put forward in Section 3. In Section 4, a distributed algorithm for multisensor system with unknown inputs will be given, demonstrating that the fused state estimate is equal to the centralized Kalman filtering with all sensor measurements. Several simulation examples are given in Section 5. Section 6 is the summary of our analysis and possible future work.

2. Problem Formulation

Consider a discrete time dynamic system:

$$x_{k+1} = F_k x_k + \nu_k, \tag{1}$$

$$y_k = H_k x_k + A_k d_k + \omega_k, \tag{2}$$

where $x_k \in \mathbf{R}^m$ is the system state, $\nu_k \in \mathbf{R}^m$ is the measurement vector, the process noise and measurement noise $\omega_k \in \mathbf{R}^n$ are zero-mean white noise sequences with the following covariances:

$$E(\nu_k \nu_j^T) = R_{\nu_k} \delta_{k-j}, \tag{3}$$

$$E(\omega_k \omega_j^T) = R_{\omega_k} \delta_{k-j}, \tag{4}$$

$$E(\nu_k \omega_j^T) = 0, \qquad \forall k, j. \tag{5}$$

where:

$$\delta_{k-j} = \begin{cases} 1, & k = j \\ 0, & k \neq j \end{cases},$$

$d_k \in \mathbf{R}^p$ is the unknown input. Matrices F_k, H_k and A_k are of appropriate dimensions by assuming that $A_k \in \mathbf{R}^{n \times p}$ is of full column rank, i.e., $rank(A_k) = p$. Therefore, we have $(A_k)^{\dagger} A_k = I$, where the superscript "†" denotes Moore-Penrose pseudo inverse. It is assumed d_k follows an autoregressive AR (1), i.e.,:

$$d_{k+1} = B_k d_k + \omega_{d_k}, \tag{6}$$

where B_k is nonsingular and ω_{d_k} is a zero-mean white noise sequences with covariance:

$$E(\omega_{d_k} \omega_{d_j}^T) = R_{d_k} \delta_{k-j}. \tag{7}$$

This model is widely considered in [2–5]. For example, in radar systems, the measurement often contains a fixed unknown deviation or an unknown deviation that gradually increases as the distance becomes longer. Such deviations can be described by Equation (6).

ASKF and OTSKF are two classic algorithms to handle this problem. The ASKF regards x_k and d_k as an augmented state and estimates them together, while the OTSKF estimates x_k and d_k respectively at first and then fusions them to achieve the optimal estimation. As a matter of fact, the unknown inputs can be eliminated easily by difference method. Denote:

$$z_k = B_k^{-1} A_{k+1}^{\dagger} y_{k+1} - A_k^{\dagger} y_k. \tag{8}$$

Equations (1) and (2) can be represented as:

$$x_{k+1} = F_k x_k + \nu_k, \tag{9}$$

$$z_k = M_k x_k + u_k, \tag{10}$$

where:

$$M_k = B_k^{-1} A_{k+1}^{\dagger} H_{k+1} F_k - A_k^{\dagger} H_k, \tag{11}$$

$$u_k = B_k^{-1} A_{k+1}^{\dagger} H_{k+1} \nu_k + B_k^{-1} \omega_{d_k} + B_k^{-1} A_{k+1}^{\dagger} \omega_{k+1} - A_k^{\dagger} \omega_k. \tag{12}$$

From Equation (12), it is not difficult to find out that the new measurement noise u_k is one-step correlated and correlates with the process noise, i.e.,:

$$E(u_k \nu_j^T) = B_k^{-1} A_{k+1}^{\dagger} H_{k+1} R_{\nu_k} \delta_{k-j}, \tag{13}$$

$$\begin{aligned} E(u_k u_k^T) &= B_k^{-1} A_{k+1}^{\dagger} H_{k+1} R_{v_k} H_{k+1}^T (A_{k+1}^{\dagger})^T (B_k^{-1})^T + B_k^{-1} R_{d_k} (B_k^{-1})^T \\ &+ B_k^{-1} A_{k+1}^{\dagger} R_{\omega_{k+1}} (A_{k+1}^{\dagger})^T (B_k^{-1})^T + A_k^{\dagger} R_{\omega_k} (A_k^{\dagger})^T, \end{aligned} \tag{14}$$

$$E(u_k u_j^T) = -A_k^{\dagger} R_{\omega_k} (A_k^{\dagger})^T (B_{k-1}^{-1})^T \delta_{k-j+1}, k \neq j. \tag{15}$$

3. Optimal Estimation for the Remodeled System

It is assumed in the classic Kalman filtering that the process noises and measurement noises are uncorrelated temporally; in the meantime, both noises are mutually uncorrelated except at the same time instant. The noises in Equations (13)–(15) apparently violate these assumptions. Using the latest research achievements about Kalman filtering with correlated noises in [15–20], we can give an optimal estimation for the remodeled system (9) and (10) in the MSE sense. The recursive state estimate of the new system is presented in the following theorem.

Theorem 1. *The globally optimal estimate for the remodeled system (9) and (10) is given by:*

$$x_{k|k} = x_{k|k-1} + J_k L_k^{\dagger} \Delta z_k,$$

$$P_{k|k} = P_{k|k-1} - J_k L_k^{\dagger} J_k^T,$$

where:

$$x_{k|k-1} = F_{k-1} x_{k-1|k-1} + R_{v_{k-1}} H_k^T (A_k^{\dagger})^T (B_{k-1}^{-1})^T L_{k-1}^{\dagger} \Delta z_{k-1}, \tag{16}$$

$$\begin{aligned} P_{k|k-1} &= E(x_k - x_{k|k-1})(x_k - x_{k|k-1})^T \\ &= F_{k-1} P_{k-1|k-1} F_{k-1}^T - F_{k-1} J_{k-1} L_{k-1}^{\dagger} B_{k-1}^{-1} A_k^{\dagger} H_k R_{v_{k-1}} \\ &\quad - R_{v_{k-1}} H_k^T (A_k^{\dagger})^T (B_{k-1}^{-1})^T L_{k-1}^{\dagger} J_{k-1}^T F_{k-1}^T + R_{v_{k-1}} \\ &\quad - R_{v_{k-1}} H_k^T (A_k^{\dagger})^T (B_{k-1}^{-1})^T L_{k-1}^{\dagger} B_{k-1}^{-1} A_k^{\dagger} H_k R_{v_{k-1}} \end{aligned} \tag{17}$$

$$\begin{aligned} \Delta z_k &= z_k - z_{k|k-1}, \\ &= z_k - M_k x_{k|k-1} + A_k^{\dagger} R_{\omega_k} (A_k^{\dagger})^T (B_{k-1}^{-1})^T L_{k-1}^{\dagger} \Delta z_{k-1}, \end{aligned} \tag{18}$$

$$\begin{aligned} J_k &= E(x_k - x_{k|k-1})(z_k - z_{k|k-1})^T, \\ &= P_{k|k-1} M_k^T + (F_{k-1} J_{k-1} L_{k-1}^{\dagger} + R_{v_{k-1}} H_k^T (A_k^{\dagger})^T (B_{k-1}^{-1})^T L_{k-1}^{\dagger}) \\ &\quad \cdot B_{k-1}^{-1} A_k^{\dagger} R_{\omega_k} (A_k^{\dagger})^T, \end{aligned} \tag{19}$$

$$\begin{aligned} L_k &= E(z_k - z_{k|k-1})(z_k - z_{k|k-1})^T, \\ &= M_k J_k + A_k^{\dagger} R_{\omega_k} (A_k^{\dagger})^T (B_{k-1}^{-1})^T L_{k-1}^{\dagger} (J_{k-1}^T F_{k-1}^T + B_{k-1}^{-1} A_k^{\dagger} H_k R_{v_{k-1}}) \\ &\quad \cdot M_k^T + R_{u_k} - A_k^{\dagger} R_{\omega_k} (A_k^{\dagger})^T (B_{k-1}^{-1})^T L_{k-1}^{\dagger} B_{k-1}^{-1} A_k^{\dagger} R_{\omega_k} (A_k^{\dagger})^T. \end{aligned} \tag{20}$$

Remark 1. *From Theorem 1, the new algorithm presented in this section is optimal in the MSE sense. In theory, the ASKF and OTSKF are also optimal in the MSE sense (see [2,3]). Nevertheless, the optimality of the ASKF and OTSKF depends on the assumption that the initial condition of the unknown measurement $d_{0|0} = E(d_0)$, which is difficult to meet in real situations. It will be demonstrated by numerical examples in the Section 5 that if the initial value of the unknown input is wrong, their performances will be greatly influenced. By contrast, the new algorithm will continue its good performance as it does not rely on the initial value of the unknown input.*

Remark 2. *A flop is defined as one addition, subtraction and multiplication. To estimate the complexity of an algorithm, the total number of flops is counted, expressing it as a polynomial of the dimensions of the matrices and vectors involved, and simplifying the expression by ignoring all terms except the leading terms. Then the complexities of the ASKF, OTSKF and the new algorithm are equivalent to $O(m^3 + n^3 + p^3 + m^2 n + mn^2 + m^2 p + mp^2 + n^2 p + np^2 + mnp)$. The evaluation complexities of the three algorithms are the same order polynomials. We will compare their complexities more precisely by numerical examples in Section 5.*

4. Multisensor Fusion

The l-sensor dynamic system is given by:

$$\begin{aligned} x_{k+1} &= F_k x_k + v_k, \qquad k = 0, 1, \ldots \\ y_k^i &= H_k^i x_k + A_k^i d_k^i + \omega_k^i, \\ d_{k+1}^i &= B_k^i d_k^i + \omega_{d_k}^i, \qquad i = 1, \ldots, l \end{aligned} \tag{21}$$

where $x_k \in \mathbf{R}^m$ is the system state, $y_k^i \in \mathbf{R}^{n_i}$ is the measurement vector in the i-th sensor, $v_k \in \mathbf{R}^m$ is the process noise and $\omega_k^i \in \mathbf{R}^{n_i}$ is measurement noise, $d_k^i \in \mathbf{R}^{p_i}$ is the unknown input in i-th sensor. Matrices F_k, H_k^i and A_k^i are of appropriate dimensions.

We assume the system has the following statistical properties:

(1) Every single sensor satisfies the assumption in Section 2.

(2) $A_k^i \in \mathbf{R}^{n_i \times p_i}$ is of full column rank, then $(A_k^i)^{\dagger} A_k^i = I$.

(3) $\{v_k, \omega_k^j, k = 0, 1, 2, \ldots\}, i, j = 1, \ldots, l$ is a sequence of independent variables.

Similarly to Equations (9) and (10), Equation (21) could be converted to:

$$x_{k+1} = F_k x_k + v_k, \qquad k = 0, 1, \ldots \tag{22}$$

$$z_k^i = M_k^i x_k + u_k^i, \qquad i = 1, \ldots, l \tag{23}$$

where:

$$M_k^i = B_k^{i^{-1}} A_{k+1}^{i^{\dagger}} H_{k+1}^i F_k - A_k^{i^{\dagger}} H_k^i,$$

$$u_k^i = B_k^{i^{-1}} A_{k+1}^{i^{\dagger}} H_{k+1}^i v_k + B_k^{i^{-1}} \omega_{d_k}^i + B_k^{i^{-1}} A_{k+1}^{i^{\dagger}} \omega_{k+1}^i - A_k^{i^{\dagger}} \omega_k^i.$$

The stacked measurement equation is written as:

$$z_k = M_k x_k + u_k$$

where:

$$z_k = (z_k^{1^T}, \ldots, z_k^{l^T})^T, M_k = (M_k^{1^T}, \ldots, M_k^{l^T})^T, u_k = (u_k^{1^T}, \ldots, u_k^{l^T})^T.$$

According to Theorem 1, the local Kalman filtering at the i-th sensor is:

$$x_{k|k}^i = x_{k|k-1}^i + J_k^i L_k^{i^{\dagger}} \Delta z_k^i, \tag{24}$$

$$x_{k|k-1}^i = F_{k-1} x_{k-1|k-1}^i + R_{v_{k-1}} H_k^{i^T} (A_k^{i^{\dagger}})^T (B_{k-1}^{i^{-1}})^T L_{k-1}^{i^{\dagger}} \Delta z_{k-1}^i, \tag{25}$$

with covariances of filtering error given by:

$$P_{k|k}^i = P_{k|k-1}^i - J_k^i L_k^{i^{\dagger}} J_k^{i^T},$$

where:

$$\Delta z_k^i = z_k^i - M_k^i x_{k|k-1}^i + A_k^{i^{\dagger}} R_{\omega_k}^i (A_k^{i^{\dagger}})^T (B_{k-1}^{i^{-1}})^T L_{k-1}^{i^{\dagger}} \Delta z_{k-1}^i, \tag{26}$$

$$J_k^i = E(x_k^i - x_{k|k-1}^i)(z_k^i - z_{k|k-1}^i)^T,$$

$$L_k^i = E(z_k^i - z_{k|k-1}^i)(z_k^i - z_{k|k-1}^i)^T,$$

$$P_{k|k-1}^i = E(x_k^i - x_{k|k-1}^i)(x_k^i - x_{k|k-1}^i)^T.$$

According to Theorem 1, the centralized Kalman filtering with all sensor data is given by:

$$x_{k|k} = x_{k|k-1} + J_k L_k^{\dagger} \Delta z_k, \tag{27}$$

$$x_{k|k-1} = F_{k-1} x_{k-1|k-1} + R_{v_{k-1}} H_k^T (A_k^{\dagger})^T (B_{k-1}^{-1})^T L_{k-1}^{\dagger} \Delta z_{k-1}, \tag{28}$$

The covariance of filtering error given by:

$$P_{k|k} = P_{k|k-1} - J_k L_k^{\dagger} J_k^T,$$

where:

$$\begin{aligned} A_k &= diag(A_k^1, \ldots, A_k^l), B_k = diag(B_k^1, \ldots, B_k^l), \\ \Delta z_k &= z_k - M_k x_{k|k-1} + A_k^{\dagger} R_{\omega_k} (A_k^{\dagger})^T (B_{k-1}^{-1})^T L_{k-1}^{\dagger} \Delta z_{k-1}, \\ J_k &= E(x_k - x_{k|k-1})(z_k - z_{k|k-1})^T, \\ L_k &= E(z_k - z_{k|k-1})(z_k - z_{k|k-1})^T, \\ P_{k|k-1} &= E(x_k - x_{k|k-1})(x_k - x_{k|k-1})^T. \end{aligned} \tag{29}$$

$diag$ is the diagonalization of a matrix.

Remark 3. *There are two key points to express the centralized filtering Equations (27) and (28) in terms of the local filtering:*

(1) Taking into consideration the measurement noise of single sensor in new system Equations (22) and (23), it can be observed that the sensor noises of the converted system are cross-correlated even if the original sensor noises are mutually independent.

(2) Δz_k in Equation (27) is not stacked by local Δz_k^i in Equation (26) directly and includes Δz_{k-1} in its expression, which makes our problem more complicated than the previous distributed problem in [9–14,21].

Next, we will solve these two problems to express the centralized filtering Equation (28) in terms of the local filtering. We assume that H_k^T is of full column rank. Thus, we have $(H_k^T)^{\dagger} H_k^T = I$. Using (28), we can get:

$$\Delta z_{k-1} = L_{k-1} [(B_{k-1}^{-1})^T]^{-1} [(A_k^{\dagger})^T]^{\dagger} (H_k^T)^{\dagger} R_{v_{k-1}}^{-1} (x_{k|k-1} - F_{k-1} x_{k-1|k-1}). \tag{30}$$

Substituting (29) and (30) into (27), we have:

$$\begin{aligned} x_{k|k} &= x_{k|k-1} + J_k L_k^{\dagger} \Delta z_k \\ &= x_{k|k-1} + J_k L_k^{\dagger} (z_k - M_k x_{k|k-1} + A_k^{\dagger} R_{\omega_k} (A_k^{\dagger})^T (B_{k-1}^{-1})^T L_{k-1}^{\dagger} \Delta z_{k-1}) \\ &= x_{k|k-1} + J_k L_k^{\dagger} (z_k - M_k x_{k|k-1} + A_k^{\dagger} R_{\omega_k} (A_k^{\dagger})^T (B_{k-1}^{-1})^T L_{k-1}^{\dagger} L_{k-1} ((B_{k-1}^{-1})^T))^{-1} ((A_k^{\dagger})^T)^{\dagger} \\ &\quad \cdot (H_k^T)^{\dagger} R_{v_{k-1}}^{-1} (x_{k|k-1} - F_{k-1} x_{k-1|k-1})) \\ &= x_{k|k-1} + J_k L_k^{\dagger} (z_k - M_k x_{k|k-1} + A_k^{\dagger} R_{\omega_k} (H_k^T)^{\dagger} R_{v_{k-1}}^{-1} (x_{k|k-1} - F_{k-1} x_{k-1|k-1})) \\ &= x_{k|k-1} - J_k L_k^{\dagger} M_k x_{k|k-1} + J_k L_k^{\dagger} A_k^{\dagger} R_{\omega_k} (H_k^T)^{\dagger} R_{v_{k-1}}^{-1} (x_{k|k-1} - F_{k-1} x_{k-1|k-1}) + J_k L_k^{\dagger} z_k. \end{aligned} \tag{31}$$

Using (26), we have:

$$z_k^i = \Delta z_k^i + M_k^i x_{k|k-1}^i - A_k^{i\dagger} R_{\omega_k}^i (A_k^{i\dagger})^T (B_{k-1}^{i-1})^T L_{k-1}^{i\dagger} \Delta z_{k-1}^i. \tag{32}$$

We assume that $J_k^i \in \mathbf{R}^{m \times n_i}$ is of full column rank, i.e., $rankJ_k^i = n_i$. Thus, we have $(J_k^i)^{\dagger} J_k^i = I$. Then, using (24), we can get:

$$\Delta z_k^i = L_k^i J_k^{i^{\dagger}} (x_{k|k}^i - x_{k|k-1}^i). \tag{33}$$

To express the centralized filtering $x_{k|k}$ in terms of the local filtering, by (25), (32) and (33), we have:

$$\begin{aligned} J_k L_k^{\dagger} z_k &= J_k \sum_{i=1}^{l} L_k^{\dagger}(*i) z_k^i \\ &= J_k \sum_{i=1}^{l} L_k^{\dagger}(*i)(\Delta z_k^i + M_k^i x_{k|k-1}^i - A_k^{i^{\dagger}} R_{\omega_k}^i (A_k^{i^{\dagger}})^T (B_{k-1}^{i-1})^T L_{k-1}^{i^{\dagger}} \Delta z_{k-1}^i) \\ &= J_k \sum_{i=1}^{l} L_k^{\dagger}(*i)(L_k^i J_k^{i^{\dagger}} (x_{k|k}^i - x_{k|k-1}^i) + M_k^i x_{k|k-1}^i - A_k^{i^{\dagger}} R_{\omega_k}^i (A_k^{i^{\dagger}})^T (B_{k-1}^{i-1})^T L_{k-1}^{i^{\dagger}} \\ &\quad \cdot L_{k-1}^i ((B_{k-1}^{i-1})^T)^{-1} ((A_k^{i^{\dagger}})^T)^{\dagger} (H_k^{i^T})^{\dagger} R_{v_{k-1}}^{-1} (x_{k|k-1}^i - F_{k-1} x_{k-1|k-1}^i)) \\ &= J_k \sum_{i=1}^{l} L_k^{\dagger}(*i)(L_k^i J_k^{i^{\dagger}} (x_{k|k}^i - x_{k|k-1}^i) + M_k^i x_{k|k-1}^i - A_k^{i^{\dagger}} R_{\omega_k}^i (H_k^{i^T})^{\dagger} R_{v_{k-1}}^{-1} \\ &\quad \cdot (x_{k|k-1}^i - F_{k-1} x_{k-1|k-1}^i)), \end{aligned} \tag{34}$$

where $L_k^{\dagger}(*i)$ is the i-th column block of $L_k^{\dagger}$.

Thus, substituting (34) into (31) yields:

$$\begin{aligned} x_{k|k} &= x_{k|k-1} - J_k L_k^{\dagger} M_k x_{k|k-1} + J_k L_k^{\dagger} A_k^{\dagger} R_{\omega_k} (H_k^{\dagger})^T R_{v_{k-1}}^{-1} (x_{k|k-1} - F_{k-1} x_{k-1|k-1}) \\ &\quad + J_k \sum_{i=1}^{l} L_k^{\dagger}(*i)(L_k^i J_k^{i^{\dagger}} (x_{k|k}^i - x_{k|k-1}^i) + M_k^i x_{k|k-1}^i - A_k^{i^{\dagger}} R_{\omega_k}^i (H_k^{i^T})^{\dagger} R_{v_{k-1}}^{-1} \\ &\quad \cdot (x_{k|k-1}^i - F_{k-1} x_{k-1|k-1}^i)) \\ &= (I - J_k L_k^{\dagger} M_k + J_k L_k^{\dagger} A_k^{\dagger} R_{\omega_k} (H_k^{\dagger})^T R_{v_{k-1}}^{-1}) x_{k|k-1} - J_k L_k^{\dagger} A_k^{\dagger} R_{\omega_k} (H_k^{\dagger})^T R_{v_{k-1}}^{-1} F_{k-1} x_{k-1|k-1} \\ &\quad + J_k \sum_{i=1}^{l} L_k^{\dagger}(*i)(L_k^i J_k^{i^{\dagger}} x_{k|k}^i + (M^i - L_k^i J_k^{i^{\dagger}} - A_k^{i^{\dagger}} R_{\omega_k}^i (H_k^{i^T})^{\dagger} R_{v_{k-1}}^{-1}) x_{k|k-1}^i \\ &\quad + A_k^{i^{\dagger}} R_{\omega_k}^i (H_k^{i^T})^{\dagger} R_{v_{k-1}}^{-1} F_{k-1} x_{k-1|k-1}^i). \end{aligned} \tag{35}$$

Similarly to Equation (35), using Equations (24), (26), (29) and (32), we have:

$$\begin{aligned} x_{k|k-1} &= F_{k-1} x_{k-1|k-1} - R_{v_{k-1}} H_k^T (A_k^{\dagger})^T (B_{k-1}^{-1})^T L_{k-1}^{\dagger} (M_{k-1} x_{k-1|k-2} - A_{k-1}^{\dagger} R_{\omega_{k-1}} \\ &\quad \cdot (H_{k-1}^T)^{\dagger} R_{v_{k-2}}^{-1} (x_{k-1|k-2} - F_{k-2} x_{k-2|k-2})) + R_{v_{k-1}} H_k^T (A_k^{\dagger})^T (B_{k-1}^{-1})^T \\ &\quad \cdot \sum_{i=1}^{l} L_{k-1}^{\dagger}(*i)(L_{k-1}^i ((B_{k-1}^{i-1})^T)^{-1} ((A_k^{i^{\dagger}})^T)^{\dagger} (H_k^{i^T})^{\dagger} R_{v_{k-1}}^{-1} (x_{k|k-1}^i - F_{k-1} x_{k-1|k-1}^i) \\ &\quad + M_{k-1}^i x_{k-1|k-2}^i - A_{k-1}^{i^{\dagger}} R_{\omega_{k-1}}^i (H_{k-1}^{i^T})^{\dagger} R_{v_{k-2}}^{-1} (x_{k-1|k-2}^i - F_{k-2} x_{k-2|k-2}^i)) \\ &= F_{k-1} x_{k-1|k-1} - R_{v_{k-1}} H_k^T (A_k^{\dagger})^T (B_{k-1}^{-1})^T L_{k-1}^{\dagger} (M_{k-1} - A_{k-1}^{\dagger} R_{\omega_{k-1}} (H_{k-1}^T)^{\dagger} R_{v_{k-2}}^{-1}) \\ &\quad \cdot x_{k-1|k-2} - R_{v_{k-1}} H_k^T (A_k^{\dagger})^T (B_{k-1}^{-1})^T L_{k-1}^{\dagger} A_{k-1}^{\dagger} R_{\omega_{k-1}} (H_{k-1}^T)^{\dagger} R_{v_{k-2}}^{-1} F_{k-2} x_{k-2|k-2} \\ &\quad + R_{v_{k-1}} H_k^T (A_k^{\dagger})^T (B_{k-1}^{-1})^T \sum_{i=1}^{l} L_{k-1}^{\dagger}(*i)(L_{k-1}^i ((B_{k-1}^{i-1})^T)^{-1} ((A_k^{i^{\dagger}})^T)^{\dagger} (H_k^{i^T})^{\dagger} \\ &\quad \cdot R_{v_{k-1}}^{-1} x_{k|k-1}^i - L_{k-1}^i ((B_{k-1}^{i-1})^T)^{-1} ((A_k^{i^{\dagger}})^T)^{\dagger} (H_k^{i^T})^{\dagger} R_{v_{k-1}}^{-1} F_{k-1} x_{k-1|k-1}^i \\ &\quad + (M_{k-1}^i - A_{k-1}^{i^{\dagger}} R_{\omega_{k-1}}^i (H_{k-1}^{i^T})^{\dagger} R_{v_{k-2}}^{-1}) x_{k-1|k-2}^i + A_{k-1}^{i^{\dagger}} R_{\omega_{k-1}}^i \\ &\quad \cdot (H_{k-1}^{i^T})^{\dagger} R_{v_{k-2}}^{-1} F_{k-2} x_{k-2|k-2}^i). \end{aligned} \tag{36}$$

That means the centralized filtering is expressed in terms of the local filtering. Therefore, the distributed fused state estimate is equal to the centralized Kalman filtering adopting all sensor measurements, which means the distributed fused state estimate achieves the best performance.

Remark 4. *From this new algorithm, it is easy to see that local sensors should transmit* $x^i_{k|k}$, $x^i_{k|k-1}$, $P^i_{k|k}$ *and* $P^i_{k|k-1}$ *to the fusion center to get global fusion result. The augmented methods greatly increase the state dimension and computation complexity for the multisensor system. Since the difference method does not increase the state dimension, the computation complexity is lower than that of the augmented method for the multisensory system.*

5. Numerical Examples

In this section, several simulations will be carried out for dynamic system with unknown inputs. It is assumed that the unknown input $d_{k+1} = B_k d_k + \omega_k$ in this paper. Actually, the unknown measurement d_k is a stationary time series if the eigenvalue of B_k is less than 1, or else the unknown measurement d_k is a non-stationary time series. The performances of the new algorithm (denoted as Difference KF) in these two cases are discussed in Example 1 and 2, respectively:

Example 1. *A two dimension target tracking problem is considered. The target dynamic models are given as Equations (1)–(7). The state transition matrices:*

$$F_k = \begin{pmatrix} 1 & 1 & 0 & 0 \\ 0 & 1 & 0 & 0 \\ 0 & 0 & 1 & 1 \\ 0 & 0 & 0 & 1 \end{pmatrix}$$

and the measurement matrix is given by:

$$H_k = \begin{pmatrix} 1 & 0 & 0 & 0 \\ 0 & 0 & 1 & 0 \end{pmatrix}$$

Suppose A_k *is an identity matrix with appropriate dimensions,* $B_k = 0.9I$. *In this case,* d_k *is a stationary time series. The targets start at* $x_0 = (50, 1, 50, 1)^T$ *and the initial value* $d_0 = (5, 5)^T$. *The covariance matrices of the noises are given by:*

$$R_{v_k} = \begin{pmatrix} 1 & 0 & 0 & 0 \\ 0 & 0.1 & 0 & 0 \\ 0 & 0 & 1 & 0 \\ 0 & 0 & 0 & 0.1 \end{pmatrix},$$

$$R_{\omega_k} = \begin{pmatrix} 1 & 0 \\ 0 & 1 \end{pmatrix}, \; R_{d_k} = \begin{pmatrix} 1 & 0 \\ 0 & 1 \end{pmatrix}.$$

In the following, the computer time and performances of the ASKF, OTSKF and Difference KF will be compared respectively.

- Computer time

The complexities of the three algorithms are analyzed in Remark 2, which shows the complexities of the three algorithms are the equivalent order polynomials. Now let us compare their computer time by this example. Table 1 illustrates the computer time of the three algorithms with 1000 Monte-Carlo runs respectively, through which we can find out that the new algorithm is the fastest algorithm in this example.

Table 1. The computer time of the three algorithms.

Algorithm	Computer Time (seconds)
ASKF	16.163026
OTSKF	11.684104
Difference KF	9.274128

- Estimation Performances

In [3], Hsieh et al. has proved that the OTSKF is equivalent to the ASKF, so the tracking results of the two algorithms are the same. In order to make the figure clearer, we will only compare the performances of the following six algorithms:

Algorithm 1: KF without considering unknown input.

Algorithm 2: ASKF with accurate initial value of unknown input ($d_0 = (5,5)^T$).

Algorithm 3: OTSKF with accurate initial value of unknown input ($d_0 = (5,5)^T$).

Algorithm 4: ASKF with wrong inaccurate initial value of unknown input ($d_0 = (0,0)^T$).

Algorithm 5: ASKF with inaccurate initial value of unknown input ($d_0 = (20,20)^T$).

Algorithm 6: Difference KF without any information about initial value of unknown input.

The initial states of the six algorithms are set at $x_{0|0} = x_0$, the initial $P_{x_{0|0}} = R_{v_0}$, $P_{d_{0|0}} = R_{d_0}$. Using 100 Monte-Carlo runs, we can evaluate estimation performance of an algorithm by estimating the second moment of the tracking error:

$$E_k^2 = \frac{1}{100}\sum_{j=1}^{100} ||x_{k|k}^{(j)} - x_k||^2, \quad k = 1,2,\cdots,100.$$

It must be noticed that Difference KF uses $(y_1, y_2, \cdots, y_k, y_{k+1})$ to estimate x_k at step k. However, the KF, ASKF and OTSKF only use $(y_1, y_2, \cdots, y_k)$ to estimate x_k at step k. To make an equal comparison, $x_{k|k-1}$ in Difference KF with $x_{k|k}$ in the other five algorithms is compared. As $d_{k+1} = 0.9d_k + \omega_k$, d_k is almost equal to a random white noise with small covariance after several steps and the influence of the initial value d_0 will be gradually weakened. The tracking errors of the six methods are compared in Figure 1 and Table 2. It can be noticed that no matter whether the initial values of the unknown input in ASKF and OTSKF are accurate or wrong, the tracking results of the six algorithms are almost the same after about 25 steps. However, it should be noticed that the Difference KF performs better than the ASKF with inaccurate initial value of unknown measurement in the first stage, which is important for some practical conditions, for instance, in multi-target tracking problems, due to data association errors and heavy clutters, tracking has to restart very often. Therefore, in order to derive an entirely good tracking, initial work status at each tracking restarting should be as good as possible.

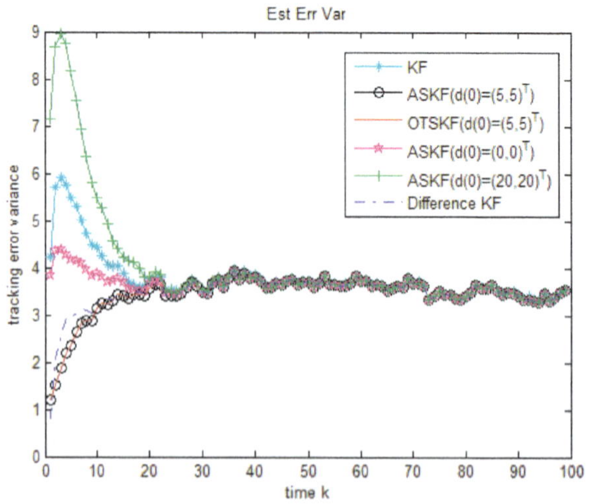

Figure 1. Comparison of the six algorithms when d_k is a stationary time series.

Table 2. The average tracking errors of the six methods.

Algorithm	KF	ASKF $d_0 = (5,5)^T$	OTSKF $d_0 = (5,5)^T$	ASKF $d_0 = (0,0)^T$	ASKF $d_0 = (20,20)^T$	Difference KF
Average Tracking Error	3.6843	3.3261	3.3261	3.5335	3.9229	3.3540

Example 2. *The dynamic equations are the same as Example 1. Assume $B_k = I$. This model has been considered in [4,5]. d_k is a non-stationary time series here. The non-stationary unknown measurement is common in practice. For instance, for an air target, the unknown radar bias is frequently increasing with distance changing between the target and radar.*

The targets start at $\mathbf{x}_0 = (50, 1, 50, 1)^T$ and the initial value $d_0 = (5,5)^T$.The performances of the following six algorithms are compared:

Algorithm 1: KF without considering unknown input.

Algorithm 2: ASKF with accurate initial value of unknown input ($d_0 = (5,5)^T$).

Algorithm 3: OTSKF with accurate initial value of unknown input ($d_0 = (5,5)^T$).

Algorithm 4: ASKF with wrong inaccurate initial value of unknown input ($d_0 = (0,0)^T$).

Algorithm 5: ASKF with inaccurate initial value of unknown input ($d_0 = (20,20)^T$).

Algorithm 6: Difference KF without any information about initial value of unknown input.

Figure 2 and Table 3 compare the tracking errors of the six methods. As the new algorithm, ASKF and OTSKF with accurate initial value of unknown input are optimal in the MSE sense. Their performances are almost of no difference. The KF without considering unknown input is worse because it does not use any information of the unknown input. Numerical examples also demonstrate that once the initial value of the unknown input is inaccurate, the performance of the ASKF becomes poorer. We can also see that if the initial value of the unknown input is largely biased, the performance of ASKF is even poorer than KF ignoring unknown input. This is because $d_{k+1} = d_k + \omega_k$ in this example, the influence of the incorrect initial value d_0 will always exist. Nevertheless, the new algorithm is independent of the initial value of the unknown input and yet performs well.

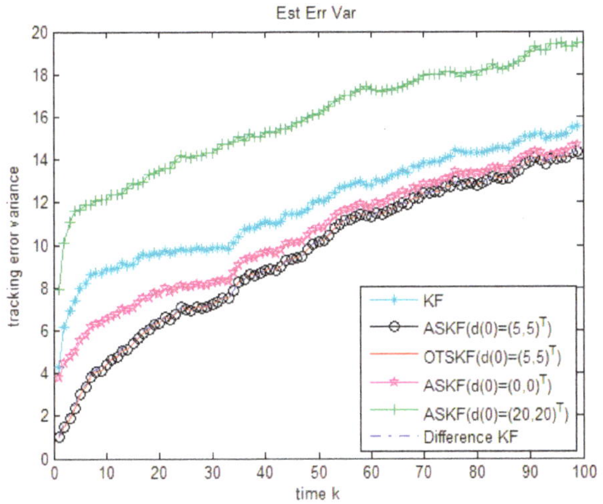

Figure 2. Comparison of the six algorithms when d_k is a non-stationary time series.

Table 3. The average tracking errors of the six methods.

Algorithm	KF	ASKF $d_0 = (5,5)^T$	OTSKF $d_0 = (5,5)^T$	ASKF $d_0 = (0,0)^T$	ASKF $d_0 = (20,20)^T$	Difference KF
Average Tracking Error	11.8091	9.5470	9.5470	10.4493	15.8539	9.5787

From Examples 1 and 2, we can see that the performance of the difference KF is almost the same to that of the ASKF and OTSKF with accurate initial value of unknown input. If the initial value of the unknown measurement is largely biased, performances of the ASKF and OTSKF will be badly influenced. Due to the fact that it is not easy to get the exact initial value of the unknown measurement, Difference KF is a better option than the ASKF and OTSKF in practice.

Example 3. *A two-sensor Kalman filtering fusion problem with unknown inputs is considered. The object dynamics and measurement equation are modeled as follows:*

$$x_{k+1} = F_k x_k + v_k, \qquad k = 0, 1, \ldots, 100$$

$$y_k^i = H_k^i x_k + A_k^i d_k^i + \omega_k^i,$$

$$d_{k+1}^i = B_k^i d_k^i + \omega_{d_k}^i, \qquad i = 1, 2.$$

The state transition matrix F_k and the measurement matrices H_k^i are the same as Example 1, A_k^i and B_k^i are identity matrix with appropriate dimensions. The targets start at $x_0 = (50, 1, 50, 1)^T$ and the initial value $d_0^i = (5, 5)^T$.

The covariance matrices of the process noises is given by:

$$R_{v_k} = \begin{pmatrix} 5 & 0 & 0 & 0 \\ 0 & 0.1 & 0 & 0 \\ 0 & 0 & 5 & 0 \\ 0 & 0 & 0 & 0.1 \end{pmatrix}$$

The covariance matrices of the measurement noises and the unknown inputs are diagonal given by $R_{\omega_k}^i = 1, R_{d_k}^i = 1, i = 1, 2.$

The performances of the following three algorithms are compared as follows:
Algorithm 1: Centralized KF without considering unknown input.
Algorithm 2: The centralized fusion of the Difference KF.
Algorithm 3: The distributed fusion of the Difference KF.

The initial states of the three algorithms are set at $x_{0|0} = x_0$, the initial $P^i_{x_{0|0}} = I$. Using 100 Monte-Carlo runs, we can evaluate estimation performance of an algorithm by estimating the second moment of the tracking error.

It is illustrated in Figure 3 and Table 4 that the simulation outcome of distributed fusion and centralized fusion of the new algorithm are exactly the same. Additionally, the new algorithm fusion gives better performance than the KF. Thus, the distributed algorithm has not only the global optimality, but also the good survivability in a poor situation.

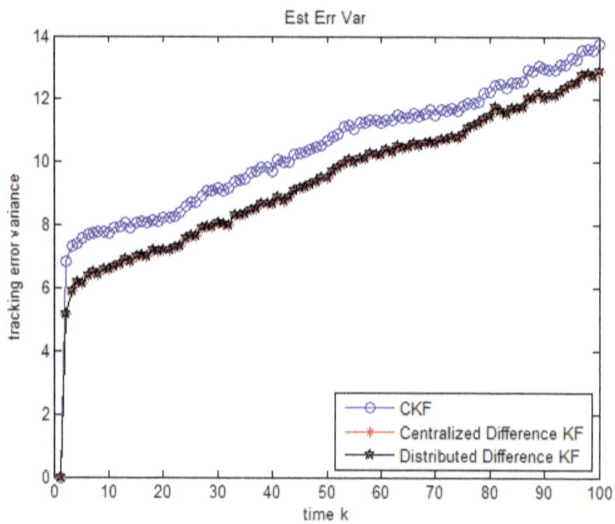

Figure 3. Comparison of the three algorithms in multisensor case.

Table 4. The average tracking errors of the three methods.

Algorithm	CKF	Centralized Difference KF	Distributed Difference KF
Average Tracking Error	10.3720	9.3582	9.3582

6. Conclusions

In this paper, the state estimation for dynamic system with unknown inputs modeled as an autoregressive AR (1) process is considered. The main contributions are: (1) A novel optimal algorithm for dynamic system with unknown inputs in the MSE sense is proposed by differential method. The computational burden of the new algorithm is lower than that of ASKF. The performance of the new algorithm is independent of the initial value of the unknown input. (2) A distributed fusion algorithm is proposed for the multisensor dynamic system with unknown inputs, the result of which is equal to the centralized difference Kalman filtering adopting all sensor measurements.

However, it should be noticed that the new algorithm uses y_k and y_{k+1} to estimate x_k, which leads the new algorithm to be one-step delayed. The new algorithm can only cope with the unknown inputs in measurement equation, while the ASKF can handle the unknown inputs in both state and measurement equation. Besides, it is assumed throughout the paper that the model of the unknown inputs d_k follows an autoregressive AR (1) process. As for the future research, one interesting direction is to extend the difference method to dynamic system with more general unknown inputs.

Author Contributions: Conceptualization, Y.L. and Y.Z.; Methodology, Y.L.; Software, Y.R.; Validation, Y.L. and Y.Z.; Formal Analysis, Y.R.; Investigation, Y.R.; Resources, Y.L.; Data Curation, Y.R.; Writing-Original Draft Preparation, Y.R.; Writing-Review & Editing, Y.L.; Visualization, Y.R.; Supervision, Y.L.; Project Administration, Y.Z.; Funding Acquisition, Y.L. and Y.Z.

Funding: This research was funded by the NNSF of China (61201065 and 61273074).

Conflicts of Interest: The authors declare that there are no conflicts of interest regarding the publication of this paper.

References

1. Kalman, R.E. A new approach to linear filtering and prediction problems. *J. Basic Eng.* **1960**, *82*, 35–55. [CrossRef]
2. Friedland, B. Treatment of bias in recursive filtering. *IEEE Trans. Autom. Control* **1969**, *14*, 359–367. [CrossRef]
3. Hsieh, C.-S.; Chen, F.-C. Optimal solution of the two-stage Kalman estimator. *IEEE Trans. Autom. Control* **1999**, *11*, 194–199. [CrossRef]
4. Hsieh, C.-S. Robust two-stage Kalman filters for systems with unknown inputs. *IEEE Trans. Autom. Control* **2000**, *45*, 2374–2378. [CrossRef]
5. Alouani, A.T.; Xia, P.; Rice, T.R.; Blair, W.D. A Two-Stage Kalman Estimator For State Estimation In The Presence of Random Bias And For Tracking Maneuvering Targets. In Proceedings of the 30th IEEE Conference on Decision and Control, Brighton, England, UK, 11–13 December 1991.
6. Hsieh, C.S. Extension of unbiased minimum-variance input and state estimation for systems with unknown inputs. *Automatica* **2009**, *45*, 2149–2153. [CrossRef]
7. Hsieh, C.S. Optimal filtering for systems with unknown inputs via the descriptor Kalman filtering method. *Automatica* **2011**, *47*, 2313–2318. [CrossRef]
8. Hsieh, C.S. State estimation for descriptor systems via the unknown input filtering method. *Automatica* **2013**, *49*, 1281–1286. [CrossRef]
9. Chong, C.Y.; Chang, K.C.; Mori, S. Distributed Tracking in Distributed Sensor Networks. In Proceedings of the 1986 American Control Conference, Seattle, WA, USA, 18–20 June 1986.
10. Chong, C.Y.; Mori, S.; Chang, K.C. Distributed Multitarget Multisensor Tracking. In *Multitarget-Multisensor Tracking: Advanced Applications*; Artech House: Norwood, MA, USA, 1990.
11. Hashmipour, H.R.; Roy, S.; Laub, A.J. Decentralized Structures for Parallel Kalman Filtering. *IEEE Trans. Autom. Control* **1988**, *33*, 88–93. [CrossRef]
12. Zhu, Y.M.; You, Z.S.; Zhao, J.; Zhang, K.S.; Li, X.R. The Optimality for the Distributed Kalman Filter with Feedback. *Automatica* **2001**, *37*, 1489–1493. [CrossRef]
13. Song, E.; Zhu, Y.; Zhou, J.; You, Z. Optimality Kalman Filtering fusion with cross-correlated sensor noises. *Automatica* **2007**, *43*, 1450–1456. [CrossRef]
14. Luo, Y.; Zhu, Y.; Luo, D.; Zhou, J.; Song, E.; Wang, D. Globally Optimal Multisensor Distributed Random Parameter Matrices Kalman Filtering Fusion with Applications. *Sensors* **2008**, *8*, 8086–8103. [CrossRef] [PubMed]
15. Feng, J.; Wang, Z.; Zeng, M. Distributed weighted robust Kalman filter fusion for uncertain systems with autocorrelated and cross-correlated noises. *Inf. Fusion* **2013**, *14*, 78–86. [CrossRef]
16. Jiang, P.; Zhou, J.; Zhu, Y. Globally optimal Kalman filtering with finite-time correlated noises. In Proceedings of the 49th IEEE Conference on Decision and Control, Atlanta, GA, USA, 15–17 December 2010.
17. Li, F.; Zhou, J.; Wu, D. Optimal filtering for systems with finite-step autocorrelated noises and multiple packet dropouts. *Aerosp. Sci. Technol.* **2013**, *24*, 255–263. [CrossRef]
18. Ren, L.; Luo, Y. Optimal Kalman filtering for systems with unknown inputs. In Proceedings of the 25th Chinese Control and Decision Conference, Guiyang, China, 25–27 May 2013.
19. Sun, S.; Tian, T.; Lin, H. State estimators for systems with random parameter matrices, stochastic nonlinearities, fading measurements and correlated noises. *Inf. Sci.* **2017**, *397–398*, 118–136. [CrossRef]

20. Tian, T.; Sun, S.; Li, N. Multi-sensor information fusion estimators for stochastic uncertain systems with correlated noises. *Inf. Fusion* **2016**, *27*, 126–137. [CrossRef]
21. Zhu, Y.; Zhou, J.; Shen, X.; Song, E.; Luo, Y. *Networked Multisensor Decision and Estimation Fusion Based on Advanced Mathematical Methods*; CRC Press: Boca Raton, FL, USA, 2012.

sensors

MDPI

Article

Centralized Fusion Approach to the Estimation Problem with Multi-Packet Processing under Uncertainty in Outputs and Transmissions

Raquel Caballero-Águila [1,*,†], Aurora Hermoso-Carazo [2,†] and Josefa Linares-Pérez [2,†]

1 Departamento de Estadística, Universidad de Jaén, Paraje Las Lagunillas, 23071 Jaén, Spain
2 Departamento de Estadística, Universidad de Granada, Avda. Fuentenueva, 18071 Granada, Spain; ahermoso@ugr.es (A.H.-C.); jlinares@ugr.es (J.L.-P.)
* Correspondence: raguila@ujaen.es; Tel.: +34-95-321-2926
† These authors contributed equally to this work.

Received: 8 July 2018; Accepted: 13 August 2018; Published: 16 August 2018

check for updates

Abstract: This paper is concerned with the least-squares linear centralized estimation problem in multi-sensor network systems from measured outputs with uncertainties modeled by random parameter matrices. These measurements are transmitted to a central processor over different communication channels, and owing to the unreliability of the network, random one-step delays and packet dropouts are assumed to occur during the transmissions. In order to avoid network congestion, at each sampling time, each sensor's data packet is transmitted just once, but due to the uncertainty of the transmissions, the processing center may receive either one packet, two packets, or nothing. Different white sequences of Bernoulli random variables are introduced to describe the observations used to update the estimators at each sampling time. To address the centralized estimation problem, augmented observation vectors are defined by accumulating the raw measurements from the different sensors, and when the current measurement of a sensor does not arrive on time, the corresponding component of the augmented measured output predictor is used as compensation in the estimator design. Through an innovation approach, centralized fusion estimators, including predictors, filters, and smoothers are obtained by recursive algorithms without requiring the signal evolution model. A numerical example is presented to show how uncertain systems with state-dependent multiplicative noise can be covered by the proposed model and how the estimation accuracy is influenced by both sensor uncertainties and transmission failures.

Keywords: least-squares filtering; least-squares smoothing; networked systems; random parameter matrices; random delays; packet dropouts

1. Introduction

1.1. Background and Motivation

With the active development of computer and communication technologies, the estimation problem in multi-sensor network stochastic systems has become an important research topic in the last few years. The significant advantages of multi-sensor systems in practical situations (low cost, remote operation, simple installation, and maintenance) are obvious, and have triggered wide use of these systems in many areas, such as target tracking, communications, the manufacturing industry, etc. Moreover, they usually provide more information than traditional communication systems with a single sensor alone. In spite of these advantages, a sensor network is not generally a reliable communication medium, and together with the communication capacity limitations (network bandwidths or service capabilities, among others), may yield different uncertainties during data transmission, such as missing measurements, random delays, and packet dropouts.

The development of sensor networks motivates the necessity to desig fusion estimation algorithms which integrate the information from the different sensors and take these network-induced uncertainties into account to achieve a satisfactory performance. Depending on the way the information fusion is performed, there are two fundamental fusion techniques: the centralized fusion approach, in which the measurements from all sensors are sent to a central processor where the fusion is performed, and the distributed fusion approach, in which the measurements from each sensor are processed independently to obtain local estimators before being sent to the fusion center. The survey papers [1–3] can be examined for a wide view of these and other multi-sensor data fusion techniques.

As already indicated, centralized fusion architecture is based on a fusion centre that is able to receive, fuse, and process the data coming from every sensor; hence, centralized fusion estimation algorithms provide optimal signal estimators based on the measured outputs from all sensors and, consequently, when all of the sensors work correctly and the connections are perfect, they have the optimal estimation accuracy. In light of these concerns, it is not surprising that the study of the centralized and distributed fusion estimation problems in multi-sensor systems with network-induced uncertainties (in both the sensor measured outputs and the data transmission) has become an active research area in recent years. The estimation problem in systems with uncertainties in the sensor outputs (such as missing measurements, stochastic sensor gain degradation and fading measurements) is addressed in refs. [4–6], among others. In refs. [7–10], systems with failures during transmission (such as uncertain observations, random delays, and packet dropouts) are considered. Also, recent advances in the estimation, filtering, and fusion of networked systems with network-induced phenomena can be reviewed in refs. [11,12], where detailed overviews on this field are presented.

Since our aim in this paper is the design of centralized fusion estimators in multi-sensor network systems with measurements perturbed by random parameter matrices subject to random transmission failures (delays and packet dropouts), and multi-packet processing is considered, we discuss the research status of the estimation problem in networked systems with some of these characteristics.

1.2. Multi-Sensor Measured Outputs with Random Parameter Matrices

It is well known that in sensor-network environments, the measured outputs can be subject not only to additive noises, but also to multiplicative noise uncertainties due to several reasons, such as the presence of an intermittent sensor or hardware failure, natural or human-made interference, etc. For example, measurement equations that model the above-mentioned situations involving degradation of the sensor gain, or missing or fading measurements must include multiplicative noises described by random variables with values of $[0,1]$. So, random measurement parameter matrices provide a unified framework to address different simultaneous network-induced phenomena, and networked systems with random parameter matrices are used in different areas of science (see, e.g., refs. [13,14]). Also, systems with random sensor delays and/or multiple packet dropouts are transformed into equivalent observation models with random measurement matrices (see, e.g., ref. [15]). Hence, the estimation problem for systems with random parameter matrices has experienced increasing interest due to its diverse applications, and many estimation algorithms for such systems have been proposed over the last few years (see, e.g., refs. [16–24], and references therein).

1.3. Transmission Random Delays and Losses: Observation Predictor Compensation

Random delays and packet dropouts in the measurement transmissions are usually unavoidable and clearly deteriorate the performance of the estimators. For this reason, much effort has been made towards the study of the estimation problem to incorporate the effects of these transmission uncertainties, and several modifications of the standard estimation algorithms have been proposed (see, e.g., refs. [25–27], and references therein). In the estimation problem from measurements subject to transmission losses, when a packet drops out, the processor does not recieve a valid measurement and the most common compensation procedure is the hold-input mechanism which consists of processing the last measurement that was successfully transmitted. Unlike the approach to deal with losses, in ref. [28] the estimator of the lost measurement based on the information received previously is proposed as the compensator; this methodology significantly improves the estimations, since in cases of loss, all the previously received measurements are considered, instead of using only the last one. In view of this clear improvement of the estimators, the compensation strategy developed in ref. [28] has been adopted in some other recent investigations (see, e.g., refs. [29,30], and references therein).

1.4. Multi-Packet Processing

Another concern at the forefront of research in networked systems subject to random delays and packet dropouts is the number of packets that are processed to update the estimator at each moment, and different observation models have been proposed to deal with this issue. For example, to avoid losses as much as possible, in ref. [16] it is assumed that each packet is transmitted several times. In contrast, to avoid the network congestion that may be caused by multiple transmissions, ref. [31] the packets are sent just once. These papers also assume that each packet is either received on time, delayed for, at most, one sampling time, or lost, and only one packet or no packets are processed to update the estimator at each moment. However, in refs. [32–34] two packets were able to arrive at each sampling time, in which case, both were used to update the estimators, thus improving their performance. In these papers, different packet dropout compensation procedures have been employed. The last available measurement was used as compensation in refs. [32,34], while the observation predictor was considered in ref. [34].

1.5. Addressed Problem and Paper Contributions

Based on the considerations made above, we were motivated to address the study of the centralized fusion estimation problem for multi-sensor networked systems perturbed by random parameter matrices. This problem is discussed under the following assumptions: (*a*) Each sensor transmits their measured outputs to a central processor over different communication channels and random delays, and packet dropouts are assumed to occur during the transmission; (*b*) in order to avoid the network congestion, at each time instant, the different sensors send their packets only once, but due to the transmission random failures, the processing center can receive more than one packet; specifically, either one packet, two packets, or nothing; and (*c*) the measurement output predictor is used as a loss compensation strategy.

The main contributions and advantages of the current work are summarized as follows: (*1*) A unified framework to deal with different network-induced phenomena in the measured outputs, such as missing measurements or sensor gain degradation, is provided by the use of random measurement matrices. (*2*) Besides the uncertainties in the measured outputs, random one-step delays and packet dropouts are assumed to occur during the transmission at different rates at each sensor. Unlike previous authors' papers concerning random measurement matrices and random transmission delays and losses where only one packet is processed to update the estimator at each moment, in this paper, the estimation algorithms are obtained under the assumption that either one packet, two packets, or nothing may arrive at each sampling time. (*3*) Concerning the compensation strategy, the use of the measurement predictor as the loss compensator combined with the simultaneous processing of delayed packets provides better estimators in comparison to other approaches where the last measurement successfully received is used to compensate the data packets and only one packet is processed to update the estimator at each moment. (*4*) The centralized fusion estimation problem is addressed using covariance information, without requiring full knowledge of the state-space model generating the signal process, thus providing a general approach to deal with different kinds of signal processes. (*5*) The innovation approach is used to obtain recursive prediction, filtering, and fixed-point smoothing algorithms which are recursive and computationally simple, and thus aresuitable for online implementation. In contrast to the approaches where the state augmentation technique is used, the proposed algorithms are deduced without making use of augmented systems; therefore, they have lower computational costs than those based on the augmentation method.

1.6. Paper Structure and Notation

The remaining sections of the paper are organized as follows. Section 2 presents the assumptions for the signal process, the mathematical models of the multi-sensor measured outputs with random parameter matrices, and the measurements received by the central processor with random delays and packet losses. Section 3 provides the main results of the research, namely, the covariance-based centralized least-squares linear prediction and filtering algorithm (Theorem 1) and fixed-point smoothing algorithm (Theorem 2). A numerical example is presented in Section 4 to show the performance of the proposed centralized estimators, and some concluding remarks are drawn in Section 5. The proofs of Theorems 1 and 2 are presented in the Appendix A and Appendix B, respectively.

The notations used throughout the paper are standard. $\mathbb{R}^n$ and $\mathbb{R}^{m\times n}$ denote the n-dimensional Euclidean space and the set of all $m \times n$ real matrices, respectively. A^T and A^{-1} denote the transpose and inverse of a matrix (A), respectively. I_n and 0_n denote the $n \times n$ identity matrix and zero matrix, respectively. $\mathbf{1}_n$ denotes the all-ones vector. Finally, $\otimes$ and $\circ$ are the Kronecker and Hadamard products, respectively, and $\delta_{k,s}$ is the Kronecker delta function.

2. Observation Model and Preliminaries

The aim of this section is to design a mathematical model to allow the observations to be processed in the least-squares (LS) linear estimation problem of discrete-time signal processes from multi-sensor noisy measurements transmitted through imperfect communication channels where random one-step delay and packet dropouts may arise in the transmission process. More specifically, in order to avoid the network congestion, at every sampling time, it is assumed that the measured outputs from each sensor, which are perturbed by random parameter matrices, are transmitted just once to a central processor, and due to random delays and losses, the processing center (PC) may receive, from each sensor, either one packet, two packets, or nothing at each time instant.

In this context, our goal is to find recursive algorithms for the LS linear prediction, filtering, and fixed-point smoothing problems using the centralized fusion method. We assume that only information about the mean and covariance functions of the signal process is available, and this information is specified in the following hypothesis:

(H1) The n_x-dimensional signal process, $\{x_k\}_{k\geq 1}$, has a zero-mean, and its autocovariance function is expressed in a separable form, $E[x_k x_s^T] = A_k B_s^T$, $s \leq k$, where $A_k, B_s \in \mathbb{R}^{n_x \times M}$ are known matrices.

2.1. Multi-Sensor Measured Outputs with Random Parameter Matrices

We assume that there are m sensors which provide measured outputs of the signal process that are affected by random parameter matrices according to the following model:

$$z_k^{(i)} = H_k^{(i)} x_k + v_k^{(i)}, \quad k \geq 1, \quad i = 1, \dots, m, \tag{1}$$

where $z_k^{(i)} \in \mathbb{R}^{n_z}$ is the signal measurement in the i-th sensor at time k, $H_k^{(i)}$ are random parameter matrices, and $v_k^{(i)}$ are the measurement noises. We assume the following hypotheses for these proceses:

(H2) $\{H_k^{(i)}\}_{k\geq 1}$, for $i = 1, \dots, m$, are independent sequences of independent random parameter matrices. For $p = 1, \dots, n_z$ and $q = 1, \dots, n_x$, we denote $h_{pq}^{(i)}(k)$ as the (p,q)-th entry of $H_k^{(i)}$, which has known first and second order moments, and $\overline{H}_k^{(i)} = E[H_k^{(i)}]$.

(H3) The measurement noises $\{v_k^{(i)}\}_{k\geq 1}$, $i = 1, \ldots, m$, are zero-mean second-order white processes with $E[v_k^{(i)} v_s^{(j)T}] = R_k^{(ij)} \delta_{k,s}$.

2.2. Observation Model. Properties

To address the estimation problem with the centralized fusion method, the observations coming from the different sensors are gathered and jointly processed at each sampling time to yield the optimal signal estimator. So, the problem is addressed by considering, at each time ($k \geq 1$), the vector constituted by the measurements received from all sensors and for this purpose, Equations (1) were combined to yield the following stacked measured output equation:

$$z_k = H_k x_k + v_k, \quad k \geq 1, \tag{2}$$

where $z_k = \big(z_k^{(1)T}, \ldots, z_k^{(m)T}\big)^T$, $H_k = \big(H_k^{(1)T}, \ldots, H_k^{(m)T}\big)^T$, $v_k = \big(v_k^{(1)T}, \ldots, v_k^{(m)T}\big)^T$.

As already indicated, random one-step delays and packet dropouts occur during the transmissions to the PC. To model these failures, we introduced the following sequences of random variables:

- $\{\gamma_k^{(i)}\}_{k\geq 1}$, $i = 1, \ldots, m$, are sequences of Bernoulli random variables. Each $i = 1, \ldots, m$, $\gamma_k^{(i)} = 0$ means that the output at the current sampling time, $z_k^{(i)}$, arrives on time to be processed for the estimation, while $\gamma_k^{(i)} = 1$ means that this output is either delayed or dropped out; and
- $\{\psi_k^{(i)}\}_{k\geq 2}$, $i = 1, \ldots, m$, are sequences of Bernoulli random variables. For each $i = 1, \ldots, m$, $\psi_k^{(i)} = 1$ means that $z_{k-1}^{(i)}$ is processed at sampling time k (because it was one-step delayed) and $\psi_k^{(i)} = 0$ means that $z_{k-1}^{(i)}$ is not processed at sampling time k (because it was either received at time $k-1$ or dropped out). Since $\gamma_{k-1}^{(i)} = 0$ implies $\psi_k^{(i)} = 0$, it is clear that the value of $\psi_k^{(i)}$ is conditioned by that of $\gamma_{k-1}^{(i)}$.

For the previous sequences of Bernoulli variables, we assume the following hypothesis:

(H4) $\Big\{\big(\gamma_k^{(i)}, \psi_{k+1}^{(i)}\big)^T\Big\}_{k\geq 1}$, $i = 1, \ldots, m$, are independent sequences of independent random vectors, such that

- $\{\gamma_k^{(i)}\}_{k\geq 1}$, $i = 1, \ldots, m$, are sequences of Bernoulli random variables with known probabilities, $P(\gamma_k^{(i)} = 1) = \overline{\gamma}_k^{(i)}$, $k \geq 1$; and
- $\{\psi_k^{(i)}\}_{k\geq 2}$, $i = 1, \ldots, m$, are sequences of Bernoulli random variables such that the conditional probabilities $(P(\psi_k^{(i)} = 1/\gamma_{k-1}^{(i)} = 1))$ are known. Thus,

$$\overline{\psi}_k^{(i)} \equiv P(\psi_k^{(i)} = 1) = P(\psi_k^{(i)} = 1/\gamma_{k-1}^{(i)} = 1)\overline{\gamma}_{k-1}^{(i)}, \quad k \geq 2.$$

Moreover, the mutual independence hypothesis of the involved processes is also necessary:

(H5) For $i = 1, \ldots, m$, the signal process $\{x_k\}_{k\geq 1}$, the random matrices $\{H_k^{(i)}\}_{k\geq 1}$, and the noises $\{v_k^{(i)}\}_{k\geq 1}$ and $\Big\{\big(\gamma_k^{(i)}, \psi_{k+1}^{(i)}\big)^T\Big\}_{k\geq 1}$ are mutually independent.

Remark 1. *From hypothesis (H4), for $i, j = 1, \ldots, m$, the following correlations are clear:*

$$E[\gamma_k^{(i)}\gamma_k^{(j)}] = \begin{cases} \overline{\gamma}_k^{(i)}, & i = j, \\ \overline{\gamma}_k^{(i)}\overline{\gamma}_k^{(j)}, & i \neq j. \end{cases} \qquad E[\psi_k^{(i)}\psi_k^{(j)}] = \begin{cases} \overline{\psi}_k^{(i)}, & i = j, \\ \overline{\psi}_k^{(i)}\overline{\psi}_s^{(j)}, & i \neq j. \end{cases}$$
$$E[\psi_{k+1}^{(i)}(1 - \gamma_k^{(j)})] = \begin{cases} 0, & i = j, \\ \overline{\psi}_{k+1}^{(i)}(1 - \overline{\gamma}_k^{(j)}), & i \neq j. \end{cases} \tag{3}$$

In order to write jointly the sensor measurements to be processed at each sampling time, we efined the matrices $\Gamma_k \equiv Diag(\gamma_k^{(1)}, \ldots, \gamma_k^{(m)}) \otimes I_{n_z}$. and $\Psi_k \equiv Diag(\psi_k^{(1)}, \ldots, \psi_k^{(m)}) \otimes I_{n_z}$, $k \geq 1$. From the definition of variables $\gamma_k^{(i)}$, $i = 1, \ldots, m$,, it is clear that the non-zero components of vector $(I_{mn_z} - \Gamma_k)z_k$ are those of z_k that arrive on time at the PC and, consequently, those processed at time k. The other components of z_k are delayed or lost, and as compensation, the corresponding components of the predictor $\widehat{z}_{k/k-1}$, specified in $\Gamma_k \widehat{z}_{k/k-1}$, are processed. Similarly, the non-zero components of $\Psi_k z_{k-1}$ are those of z_{k-1} that are affected by one-step delay, and consequently, they are also processed at time k. Hence, the processed observations at each time are expressed by the following model:

$$y_k = \begin{pmatrix} (I_{mn_z} - \Gamma_k)z_k + \Gamma_k \widehat{z}_{k/k-1} \\ \Psi_k z_{k-1} \end{pmatrix}, \ k \geq 2; \qquad y_1 = \begin{pmatrix} (I_{mn_z} - \Gamma_1)z_1 \\ 0 \end{pmatrix}, \tag{4}$$

or equivalently,

$$\begin{aligned} y_k &= \mathbb{C}_0(I_{mn_z} - \Gamma_k)z_k + \mathbb{C}_0\Gamma_k\widehat{z}_{k/k-1} + \mathbb{C}_1\Psi_k z_{k-1}, \ k \geq 2; \\ y_1 &= \mathbb{C}_0(I_{mn_z} - \Gamma_1)z_1, \end{aligned} \tag{5}$$

where $\mathbb{C}_0 = (\ I_{mn_z}\ ,\ 0_{mn_z}\ )^T$ and $\mathbb{C}_1 = (\ 0_{mn_z}\ ,\ I_{mn_z}\ )^T$.

Remark 2. *For a better understanding of Model (4) for the measurements processed after the possible transmission one-step delays and losses, a single sensor is considered in the following comments. On the one hand, note that $\gamma_k = 0$ means that the output at the current sampling time (z_k) arrives on time to be processed. Then, if $\psi_k = 1$, the measurement processed at time k is $y_k = (\ z_k^T \ \ z_{k-1}^T\ )^T$, while if $\psi_k = 0$, then $y_k = (\ z_k^T \ \ 0\ )^T$. On the other hand, if $\gamma_k = 1$, the output z_k is either delayed or dropped out, and its predictor $\widehat{z}_{k/k-1}$ is processed at time k. Then, if $\psi_k = 1$, the measurement processed at time k is $y_k = (\ \widehat{z}_{k/k-1}^T \ \ z_{k-1}^T\ )^T$, while if $\psi_k = 0$, then $y_k = (\ \widehat{z}_{k/k-1}^T \ \ 0\ )^T$. Table 1 displays ten iterations of a specific simulation of packet transmission.*

From Table 1, it can be observed that z_1, z_3, z_6, z_7, and z_9 arrive on time to be processed; z_2, z_4 and z_8 are one-step delayed; and z_5 and z_{10} are lost. So, Model (4) describes possible one-step random transmission delays and packet dropouts in networked systems, where one or two packets can be processed for the estimation. Finally, note that the predictors $\widehat{z}_{k/k-1}$, $k = 2, 4, 5, 8, 10$ are used to compensate for the measurements that do not arrive on time.

Table 1. Measurements processed to update the estimators.

Time k	1	2	3	4	5	6	7	8	9	10
γ_k	0	1	0	1	1	0	0	1	0	1
ψ_k		0	1	0	1	0	0	0	1	0
y_k	$\begin{pmatrix} z_1 \\ 0 \end{pmatrix}$	$\begin{pmatrix} \widehat{z}_{1/0} \\ 0 \end{pmatrix}$	$\begin{pmatrix} z_3 \\ z_2 \end{pmatrix}$	$\begin{pmatrix} \widehat{z}_{4/3} \\ 0 \end{pmatrix}$	$\begin{pmatrix} \widehat{z}_{5/4} \\ z_4 \end{pmatrix}$	$\begin{pmatrix} z_6 \\ 0 \end{pmatrix}$	$\begin{pmatrix} z_7 \\ 0 \end{pmatrix}$	$\begin{pmatrix} \widehat{z}_{8/7} \\ 0 \end{pmatrix}$	$\begin{pmatrix} z_9 \\ z_8 \end{pmatrix}$	$\begin{pmatrix} \widehat{z}_{10/9} \\ 0 \end{pmatrix}$

The problem is then formulated as that of obtaining the LS linear estimator of the signal, x_k based on the observations $\{y_1, \ldots, y_L\}$ given in (5). Next, some statistical properties of the processes involved in observation models (2) and (5), which are necessary to address the LS linear estimation problem, are specified:

(P1) $\{H_k\}_{k \geq 1}$ is a sequence of independent random matrices with known means: $\overline{H}_k \equiv E[H_k] = (\overline{H}_k^{(1)T}, \ldots, \overline{H}_k^{(m)T})^T$, $k \geq 1$.

(P2) The sequence $\{v_k\}_{k \geq 1}$ is a zero-mean second-order process with $E[v_k v_s^T] = R_k \delta_{k,s}$, where $R_k = (R_k^{(ij)})_{i,j=1,\ldots,m}$.

(P3) The random matrices $\{(\Gamma_k, \Psi_{k+1})\}_{k\geq 1}$ are independent, and their means are given by

$$\overline{\Gamma}_k \equiv E[\Gamma_k] = Diag(\overline{\gamma}_k^{(1)}, \dots, \overline{\gamma}_k^{(m)}) \otimes I_{n_z}, \quad \overline{\Psi}_k \equiv E[\Psi_k] = Diag(\overline{\psi}_k^{(1)}, \dots, \overline{\psi}_k^{(m)}) \otimes I_{n_z}.$$

(P4) The signal process, $\{x_k\}_{k\geq 1}$ and the processes $\{H_k\}_{k\geq 1}$ $\{v_k\}_{k\geq 1}$ and $\{(\Gamma_k, \Psi_{k+1})\}_{k\geq 1}$ are mutually independent.

(P5) $\{z_k\}_{k\geq 1}$ is a zero-mean process with covariance matrices $\Sigma^z_{k,s} \equiv E[z_k z_s^T]$, for $s \leq k$ which, from (P4), are given by

$$\Sigma^z_{k,s} = E\left[H_k A_k B_s^T H_s^T\right] + R_k \delta_{k,s}, \; s \leq k,$$

with $E[H_k A_k B_s^T H_s^T] = \overline{H}_k A_k B_s^T \overline{H}_s^T$, for $s < k$, and

$$E[H_k A_k B_k^T H_k^T] = \left(E[H_k^{(i)} A_k B_k^T H_k^{(j)T}]\right)_{i,j=1,\dots,m},$$

where the (p,q)-th entries of the matrices $E[H_k^{(i)} A_k B_k^T H_k^{(j)T}]$ are given by

$$\left(E[H_k^{(i)} A_k B_k^T H_k^{(j)T}]\right)_{pq} = \sum_{a=1}^{n_x}\sum_{b=1}^{n_x} E[h_{pa}^{(i)}(k) h_{qb}^{(j)}(k)]\left(A_k B_k^T\right)_{ab}, \; p,q = 1,\dots,n_z.$$

Remark 3. *By denoting* $\gamma_k = (\gamma_k^{(1)}, \dots, \gamma_k^{(m)})^T \otimes \mathbf{1}_{n_z}$ *and* $\psi_k = (\psi_k^{(1)}, \dots, \psi_k^{(m)})^T \otimes \mathbf{1}_{n_z}$, *it is clear that* $K_k^{\mathbf{1}-\gamma} \equiv E\left[(\mathbf{1}_{mn_z} - \gamma_k)(\mathbf{1}_{mn_z} - \gamma_k)^T\right]$ *and* $K_k^{\psi} \equiv E\left[\psi_k \psi_k^T\right]$ *are known matrices whose entries are given in (3). Now, by defining*

$$\xi_k = \mathbb{C}_0 \left(I_{mn_z} - \Gamma_k\right) z_k + \mathbb{C}_1 \Psi_k z_{k-1}, \; k \geq 2; \quad \xi_1 = \mathbb{C}_0 \left(I_{mn_z} - \Gamma_1\right) z_1, \tag{6}$$

and taking the Hadamard product properties into account, it is easy to check that the covariance matrices ($\Sigma_k^{\xi} \equiv E\left[\xi_k \xi_k^T\right]$) *are given by*

$$\begin{aligned}\Sigma_k^{\xi} &= \mathbb{C}_0(K_k^{\mathbf{1}-\gamma} \circ \Sigma_k^z)\mathbb{C}_0^T + \mathbb{C}_1(K_k^{\psi} \circ \Sigma_{k-1}^z)\mathbb{C}_1^T \\ &\quad + \mathbb{C}_0(I_{mn_z} - \overline{\Gamma}_k)\Sigma^z_{k,k-1}\overline{\Psi}_k\mathbb{C}_1^T + \mathbb{C}_1\overline{\Psi}_k\Sigma^{zT}_{k,k-1}(I_{mn_z} - \overline{\Gamma}_k)\mathbb{C}_0^T, \; k \geq 2; \\ \Sigma_1^{\xi} &= \mathbb{C}_0(K_1^{\mathbf{1}-\gamma} \circ \Sigma_k^1)\mathbb{C}_0^T.\end{aligned} \tag{7}$$

3. Centralized Fusion Estimators

This section is concerned with the problem of obtaining recursive algorithms for the LS linear centralized fusion prediction, filtering, and fixed-point smoothing estimators. For this purpose, we used an innovation approach. Also the estimation error covariance matrices, which are used to measure the accuracy of the proposed estimators when the LS optimality criterion is used, were derived.

The centralized fusion structure for the considered networked systems with random uncertainties in the measured outputs and transmission is illustrated in Figure 1.

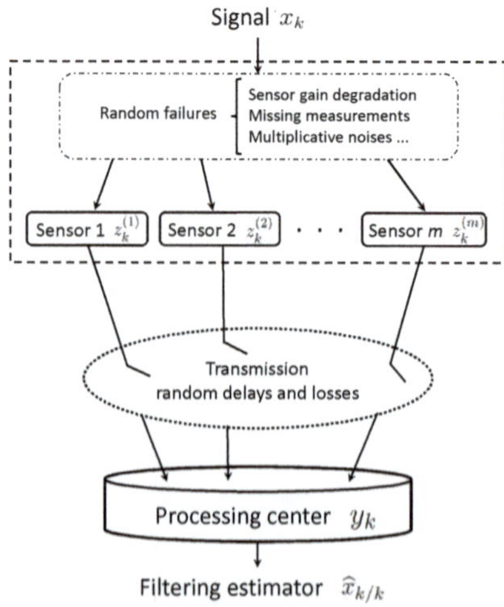

Figure 1. Centralized fusion filtering estimation with random uncertainties in measured outputs and transmission.

3.1. Innovation Technique

As indicated above, our aim was to obtain the optimal LS linear estimators, $\widehat{x}_{k/L}$, of the signal x_k based on the measurements $\{y_1,\ldots,y_L\}$, given in (5), by recursive algorithms. Since the estimator $\widehat{x}_{k/L}$ is the orthogonal projection of the signal x_k onto the linear space spanned by the nonorthogonal vectors $\{y_1,\ldots,y_L\}$, we used an innovation approach in which the observation process $\{y_k\}_{k\geq 1}$ was transformed into an equivalent one (*innovation process*) of orthogonal vectors $\{\mu_k\}_{k\geq 1}$; the equivalence means that each set $\{\mu_1,\ldots,\mu_L\}$ spans the same linear subspace as $\{y_1,\ldots,y_L\}$.

The innovation at time k is defined as $\mu_k = y_k - \widehat{y}_{k/k-1}$, where $\widehat{y}_{1/0} = E[y_1] = 0$ and, for $k \geq 2$, $\widehat{y}_{k/k-1}$, the one-stage linear predictor of y_k is the projection of y_k onto the linear space generated by $\{\mu_1,\ldots,\mu_{k-1}\}$. Due to the orthogonality property of the innovations and since the innovation process is uniquely determined by the observations, by replacing the observation process by the innovation one, the following general expression for the LS linear estimators of any vector w_k based on the observations $\{y_1,\ldots,y_L\}$ was obtained

$$\widehat{w}_{k/L} = \sum_{h=1}^{L} E\left[w_k\mu_h^T\right]\left(E\left[\mu_h\mu_h^T\right]\right)^{-1}\mu_h. \tag{8}$$

This expression is derived from the uncorrelation property of the estimation error with all of the innovations, which is guaranteed by the Orthogonal Projection Lemma (OPL). As shown by (8), the first step to obtain the signal estimators is to find an explicit formula for the innovation or, equivalently, for the one-stage linear predictor of the observation.

One-Stage Observation Predictor

To obtain $\widehat{y}_{k/k-1}$, the projection of y_k onto the linear space generated by $\{\mu_1,\ldots,\mu_{k-1}\}$, we used (5) and we note that Ψ_k and H_{k-1} are correlated with the innovation μ_{k-1}. So, to simplify the derivation of $\widehat{y}_{k/k-1}$, the observations (5) were rewritten as follows:

$$\begin{aligned} y_k &= \mathbb{C}_0\left(I_{mn_z} - \Gamma_k\right)z_k + \mathbb{C}_1\overline{\Psi}_k\overline{H}_{k-1}x_{k-1} + \mathbb{C}_0\Gamma_k\widehat{z}_{k/k-1} + V_{k-1}, \quad k \geq 2, \\ V_k &= \mathbb{C}_1\Psi_{k+1}z_k - \mathbb{C}_1\overline{\Psi}_{k+1}\overline{H}_k x_k, \quad k \geq 1. \end{aligned} \tag{9}$$

Taking into account that Ψ_{k+1} and H_k are independent of μ_h, for $h \leq k-1$, it is easy to see that $E\left[V_k\mu_h^T\right] = 0$ for $h \leq k-1$. So, from the general expression (8), we obtained $\widehat{V}_{k/k} = \mathcal{V}_k\Pi_k^{-1}\mu_k,\ k \geq 1$, where $\mathcal{V}_k \equiv E\left[V_k\mu_k^T\right] = E\left[V_k y_k^T\right]$. Hence, according to the projection theory, $\widehat{y}_{k/k-1}$ satisfies

$$\widehat{y}_{k/k-1} = \mathbb{C}_0\overline{H}_k\widehat{x}_{k/k-1} + \mathbb{C}_1\overline{\Psi}_k\overline{H}_{k-1}\widehat{x}_{k-1/k-1} + \mathcal{V}_{k-1}\Pi_{k-1}^{-1}\mu_{k-1},\ k \geq 2. \tag{10}$$

This expression for the one-stage observation predictor along with (8) for the LS linear estimators are the starting points to get the recursive prediction, filtering, and fixed-point smoothing algorithms.

3.2. Centralized Fusion Prediction, Filtering, and Smoothing Algorithms

The following theorem presents a recursive algorithm for the LS linear centralized fusion prediction and filtering estimators $\widehat{x}_{k/L}$, $L \leq k$, of the signal x_k based on the observations $\{y_1, \ldots, y_L\}$ given in (5) or equivalently, in (9).

Theorem 1. *Under hypotheses (H1)–(H5), the LS linear centralized predictors and filter $\widehat{x}_{k/L},\ L \leq k$ and the corresponding error covariance matrices $\widehat{\Sigma}_{k/L} \equiv E[(x_k - \widehat{x}_{k/L})(x_k - \widehat{x}_{k/L}^T)]$ are obtained by*

$$\widehat{x}_{k/L} = A_k e_L, \qquad \widehat{\Sigma}_{k/L} = A_k\left(B_k - A_k\Sigma_L^e\right)^T,\ L \leq k, \tag{11}$$

where the vectors e_L and the matrices $\Sigma_L^e \equiv E\left[e_L e_L^T\right]$ are recursively obtained from

$$e_L = e_{L-1} + \mathcal{E}_L\Pi_L^{-1}\mu_L,\ L \geq 1;\ \ e_0 = 0, \tag{12}$$

$$\Sigma_L^e = \Sigma_{L-1}^e + \mathcal{E}_L\Pi_L^{-1}\mathcal{E}_L^T,\ L \geq 1;\ \Sigma_0^e = 0, \tag{13}$$

and the matrices $\mathcal{E}_L \equiv E\left[e_L\mu_L^T\right]$ satisfy

$$\mathcal{E}_L = \overline{\mathcal{H}}_{B_L}^T - \Sigma_{L-1}^e\overline{\mathcal{H}}_{A_L}^T - \mathcal{E}_{L-1}\Pi_{L-1}^{-1}\mathcal{V}_{L-1}^T,\ L \geq 2;\ \ \mathcal{E}_1 = \overline{\mathcal{H}}_{B_1}^T, \tag{14}$$

where $\overline{\mathcal{H}}_{D_L}$, for $D = A,\ B$, is defined by

$$\overline{\mathcal{H}}_{D_L} = \mathbb{C}_0\left(I_{mn_z} - \overline{\Gamma}_L\right)\overline{H}_L D_L + \mathbb{C}_1\overline{\Psi}_L\overline{H}_{L-1}D_{L-1},\ L \geq 2;\ \overline{\mathcal{H}}_{D_1} = \mathbb{C}_0(I_{mn_z} - \overline{\Gamma}_1)\overline{H}_1 D_1. \tag{15}$$

The innovations $\mu_L = y_L - \widehat{y}_{L/L-1}$ are given by

$$\mu_L = y_L - \left(\overline{\mathcal{H}}_{A_L} + \mathbb{C}_0\overline{\Gamma}_L\overline{H}_L A_L\right)e_{L-1} - \mathcal{V}_{L-1}\Pi_{L-1}^{-1}\mu_{L-1},\ L \geq 2;\ \ \mu_1 = y_1, \tag{16}$$

and the coefficients $\mathcal{V}_L = E\left[V_L\mu_L^T\right]$, are obtained by

$$\begin{aligned}\mathcal{V}_L = \mathbb{C}_1\Big(&K_{L+1,L}^{\psi(\mathbf{1}-\gamma)} \circ \left(\Sigma_L^z - \overline{H}_L A_L\Sigma_{L-1}^e A_L^T\overline{H}_L^T\right)\\ &-\overline{\Psi}_{L+1}\overline{H}_L A_L\left(B_L - A_L\Sigma_{L-1}^e\right)^T\overline{H}_L^T(I - \overline{\Gamma}_L)\Big)\mathbb{C}_0^T,\ L \geq 1,\end{aligned} \tag{17}$$

where $K_{L+1,L}^{\psi(\mathbf{1}-\gamma)} \equiv E\left[\psi_{L+1}(\mathbf{1}_{mn_z} - \gamma_L)^T\right]$, whose entries are given in (3).

The innovation covariance matrices $\Pi_L \equiv E\left[\mu_L\mu_L^T\right]$ satisfy

$$\begin{aligned}\Pi_L = &\Sigma_L^\xi - \mathbb{C}_0\left[K_L^\gamma \circ \left(\overline{H}_L A_L\Sigma_{L-1}^e A_L^T\overline{H}_L^T\right)\right]\mathbb{C}_0^T + \mathcal{O}_{L,L-1}A_L^T\overline{H}_L^T\overline{\Gamma}_L\mathbb{C}_0^T\\ &-\overline{\mathcal{H}}_{A_L}\mathcal{O}_{L,L-1}^T - \mathcal{V}_{L-1}\Pi_{L-1}^{-1}\mathcal{Y}_{L,L-1}^T,\ L \geq 2;\ \ \Pi_1 = \Sigma_1^\xi,\end{aligned} \tag{18}$$

where the matrices Σ_L^{ξ} *are given in (7),* $K_L^{\gamma} \equiv E\left[\gamma_L \gamma_L^T\right]$, *whose entries are obtained by (3), and the matrices* $\mathcal{O}_{L,L-1} \equiv E[y_L e_{L-1}^T]$ *and* $\mathcal{Y}_{L,L-1} \equiv E[y_L \mu_{L-1}^T]$ *are given by*

$$\begin{aligned} \mathcal{O}_{L,L-1} &= (\overline{\mathcal{H}}_{A_L} + \mathbb{C}_0 \overline{\Gamma}_L \overline{H}_L A_L)\Sigma_{L-1}^e + \mathcal{V}_{L-1}\Pi_{L-1}^{-1}\mathcal{E}_{L-1}^T, \quad L \geq 2. \\ \mathcal{Y}_{L,L-1} &= (\overline{\mathcal{H}}_{A_L} + \mathbb{C}_0 \overline{\Gamma}_L \overline{H}_L A_L)\mathcal{E}_{L-1} + \mathcal{V}_{L-1}, \quad L \geq 2. \end{aligned} \tag{19}$$

Proof. See Appendix A. □

Next, a recursive algorithm for the LS linear centralized fusion smoothers $\widehat{x}_{k/k+h}$ at the fixed point k for any $h \geq 1$ is established in the following theorem.

Theorem 2. *Under hypotheses (H1)–(H5), the LS linear centralized fixed-point smoothers* $\widehat{x}_{k/k+h}$ *are calculated by*

$$\widehat{x}_{k/k+h} = \widehat{x}_{k/k+h-1} + \mathcal{X}_{k,k+h}\Pi_{k+h}^{-1}\mu_{k+h}, \quad k \geq 1, \quad h \geq 1, \tag{20}$$

whose initial condition is given by the centralized filter $\widehat{x}_{k/k}$, *and the matrices* $\mathcal{X}_{k,k+h} = E\left[x_k \mu_{k+h}^T\right]$ *are obtained by*

$$\begin{aligned} \mathcal{X}_{k,k+h} &= (B_k - E_{k,k+h-1})\overline{\mathcal{H}}_{A_{k+h}}^T - \mathcal{X}_{k,k+h-1}\Pi_{k+h-1}^{-1}\mathcal{V}_{k+h-1}^T, \quad h \geq 1; \\ \mathcal{X}_{k,k} &= A_k \mathcal{E}_k. \end{aligned} \tag{21}$$

The matrices $E_{k,k+h} = E[x_k e_{k+h}^T]$ *satisfy the following recursive formula:*

$$E_{k,k+h} = E_{k,k+h-1} + \mathcal{X}_{k,k+h}\Pi_{k+h}^{-1}E_{k+h}^T, \quad h \geq 1; \quad E_{k,k} = A_k \Sigma_k^e. \tag{22}$$

The fixed-point smoothing error covariance matrices, $\widehat{\Sigma}_{k/k+h} \equiv E\left[(x_k - \widehat{x}_{k/k+h})(x_k - \widehat{x}_{k/k+h})^T\right]$, *are calculated by*

$$\widehat{\Sigma}_{k/k+h} = \widehat{\Sigma}_{k/k+h-1} - \mathcal{X}_{k,k+h}\Pi_{k+h}^{-1}\mathcal{X}_{k,k+h}^T, \quad k \geq 1, \quad h \geq 1,$$

with the initial condition given by the filtering error covariance matrix $\widehat{\Sigma}_{k/k}$.

The filter $\widehat{x}_{k/k}$, *the innovations* μ_{k+h}, *their covariance matrices* $\widehat{\Sigma}_{k/k}$ *and* Π_{k+h}, *and the matrices* $\mathcal{E}_{k+h}$ *and* Σ_k^e *were obtained from Theorem 1.*

Proof. See Appendix B. □

4. Numerical Simulation Example

The performance of the proposed centralized filtering and fixed-point smoothing algorithms was analyzed in a numerical simulation example which also shows how some of the sensor uncertainties covered by the current measurement model (1) with random parameter matrices influence the accuracy of the estimators. Also, the effect of the random transmission delays and packet losses on the performance of the estimators was analyzed.

4.1. Signal Process

Consider a discrete-time scalar signal process generated by the following model with the state-dependent multiplicative noise

$$x_{k+1} = (0.9 + 0.01\varepsilon_k)x_k + w_k, \quad k \geq 0,$$

where x_0 is a standard Gaussian variable, and $\{w_k\}_{k\geq 0}$, $\{\varepsilon_k\}_{k\geq 0}$ are zero-mean Gaussian white processes with unit variance. Assuming that x_0, $\{w_k\}_{k\geq 0}$ and $\{\varepsilon_k\}_{k\geq 0}$ are mutually independent, the signal covariance is given by $E[x_k x_s] = 0.9^{k-s}D_s$, $s \leq k$, where $D_s = E[x_s^2]$ is obtained by

$$D_s = (0.9^2 + 0.01^2)D_{s-1} + 1, \ s \geq 1; \ D_0 = 1.$$

Hence, the hypothesis (H1) is satisfied with, for example, $A_k = 0.9^k$ y $B_s = 0.9^{-s} D_s$.

This signal process has been considered in the current authors' previous papers and shows that hypothesis (H1) regarding the signal autocovariance function is satisfied for uncertain systems with state-dependent multiplicative noise. Also, situations where the system matrix in the state-space model is singular are covered by hypothesis (H1) (see, e.g., ref. [9]). Hence, this hypothesis provides a unified general context to deal with different situations, thus avoiding the derivation of specific algorithms for each of them.

4.2. Sensor Measured Outputs

As in ref. [20], let us consider four sensors that provide scalar measurements with different random failures, which are described using random parameters according to the theoretical model (1). Namely, sensor 1 has continuous gain degradation, sensor 2 has discrete gain degradation, sensor 3 has missing measurements and sensor 4 has both missing measurements and multiplicative noise. Specifically, the scalar measured outputs are described according to the model

$$z_k^{(i)} = H_k^{(i)} x_k + v_k^{(i)}, \quad k \geq 1, \quad i = 1,2,3,4,$$

where the additive noises are defined as $v_k^{(i)} = c_i \eta_k$, with $c_1 = 1$, $c_2 = 0.5$, $c_3 = c_4 = 0.75$, and $\{\eta_k\}_{k\geq 1}$ is a Gaussian white sequence with a mean of 0 and variance of 0.5. The additive noises are correlated with $R_k^{(ij)} = 0.5c_i c_j$, $k \geq 1$; $i,j = 1,2,3,4$. The random measurement matrices are defined by $H_k^{(i)} = \theta_k^{(i)} C_k^{(i)}$, for $i = 1,2,3$, where $C_k^{(1)} = 0.82$, $C_k^{(2)} = 0.75$, $C_k^{(3)} = 0.74$, and $H_k^{(4)} = \theta_k^{(4)}(0.75 + 0.95\varphi_k)$, where the sequence $\{\varphi_k\}_{k\geq 1}$ is a standard Gaussian white process, and $\{\theta_k^{(i)}\}_{k\geq 1}$, $i = 1,2,3,4$, are also white processes with time-invariant probability distributions that are given as follows:

- $\{\theta_k^{(1)}\}_{k\geq 1}$ are uniformly distributed over $[0.2, 0.7]$.
- $P(\theta_k^{(2)} = 0) = 0.3$, $P(\theta_k^{(2)} = 0.5) = 0.3$, $P(\theta_k^{(2)} = 1) = 0.4$, $k \geq 1$.
- For $i = 3,4$, $\{\theta_k^{(i)}\}_{k\geq 1}$ are Bernoulli random variables with $P(\theta_k^{(i)} = 1) = \overline{\theta}^{(i)}$, $k \geq 1$.

4.3. Model for the Measurements Processed

Now, according to the theoretical study, we assume that the sensor measurements, y_k, that are processed to update the estimators are modeled by

$$y_k = \begin{pmatrix} (I_4 - \Gamma_k) z_k + \Gamma_k \widehat{z}_{k/k-1} \\ \Psi_k z_{k-1} \end{pmatrix}, \quad k \geq 2; \quad y_1 = \begin{pmatrix} (I_4 - \Gamma_1) z_1 \\ 0 \end{pmatrix},$$

where $\Gamma_k = Diag(\gamma_k^{(1)}, \gamma_k^{(2)}, \gamma_k^{(3)}, \gamma_k^{(4)})$ and $\Psi_k = Diag(\psi_k^{(1)}, \psi_k^{(2)}, \psi_k^{(3)}, \psi_k^{(4)})$, and for $i = 1,2,3,4$, $\{\gamma_k^{(i)}\}_{k\geq 1}$ and $\{\psi_k^{(i)}\}_{k\geq 2}$ are sequences of independent Bernoulli random variables whose distributions are determined by the following probabilities:

- $\overline{\gamma}^{(i)} \equiv P(\gamma_k^{(i)} = 1)$, $k \geq 1$: probability that the measurement $z_k^{(i)}$ is not received at time k because it is delayed or lost.
- $\overline{\psi}_\gamma^{(i)} \equiv P(\psi_k^{(i)} = 1/\gamma_{k-1}^{(i)} = 1)$, $k \geq 1$: probability that the measurement $z_{k-1}^{(i)}$ is received at the current time (k), conditioned to the fact that it is not received on time.
- $\overline{\psi}^{(i)} \equiv P(\psi_k^{(i)} = 1) = \overline{\psi}_\gamma^{(i)} \overline{\gamma}^{(i)}$, $k \geq 1$: probability that the measurement $z_{k-1}^{(i)}$ is received and processed at the current time k.

Finally, in order to apply the proposed algorithms, it was assumed that all the processes involved in the observation equations satisfy the independence hypotheses imposed on the theoretical model.

To illustrate the feasibility and effectiveness of the proposed algorithms, they were implemented in MATLAB, and fifty iterations of the filtering and fixed-point smoothing algorithms were performed. The estimation accuracy was examined by analyzing the error variances for different probabilities of the Bernoulli variables modeling the random failures in sensors 3 and 4 ($\overline{\theta}^{(i)}$, $i = 3,4$). Also, different values of the probabilities $\overline{\gamma}^{(i)}$, corresponding to the transmission uncertainties, and different conditional probabilities $\overline{\psi}_{\gamma}^{(i)}$, $i = 1,2,3,4$, were considered in order to analyze the effect of these failures on the estimation accuracy.

In the study of the performance of the centralized estimators, they were compared with local ones, which were computed using only the measurements received from each single sensor. In that case, the measurements processed at each local processor can be described by

$$y_k^{(i)} = \begin{pmatrix} (1-\gamma_k^{(i)})z_k + \gamma_k^{(i)}\widehat{z}_{k/k-1}^{(i)} \\ \psi_k^{(i)} z_{k-1} \end{pmatrix}, \ k \geq 2; \ \ y_1^{(i)} = \begin{pmatrix} (1-\gamma_1^{(i)})z_1^{(i)} \\ 0 \end{pmatrix}, \quad i = 1,2,3,4,$$

where $\widehat{z}_{k/k-1}^{(i)}$ is the one-stage predictor of $z_k^{(i)}$ based on $y_1^{(i)}, \ldots, y_{k-1}^{(i)}$, and the corresponding local estimators are obtained via recursive algorithms similar to those in Theorems 1 and 2.

4.4. *Performance of the Centralized Fusion Filtering and Smoothing Estimators*

For $i = 1,2,3,4$, we assumed that $\overline{\gamma}^{(i)} = \overline{\psi}_{\gamma}^{(i)} = 0.1i$, and that the missing probabilities $1 - \overline{\theta}^{(i)}$ had the same value in sensors $i = 3,4$, namely, $\overline{\theta}^{(i)} = 0.5$, $i = 3,4$. The error variances of the local filtering estimators and both the centralized filtering and smoothing error variances are displayed in Figure 2. This figure shows, on the one hand, that the error variances of the centralized fusion filtering estimators are significantly smaller than those of every local estimator. Consequently, agreeing with what is theoretically expected, the centralized fusion filter has better accuracy than the local ones, as it is the optimal one based on the information from all the contributing sensors. On the other hand, Figure 2 also shows that as more observations are considered in the estimation, the error variances are lower and consequently, the performance of the centralized estimators becomes better. In other words, the smoothing estimators are more accuracy than the filtering ones, and the accuracy of the smoothers at each fixed-point k is better as the number of available observations $k + h$ increases, although this improvement is practically imperceptible for $h > 3$. Similar results were obtained for other values of the probabilities $\overline{\theta}^{(i)}$, $\overline{\gamma}^{(i)}$ and $\overline{\psi}_{\gamma}^{(i)}$.

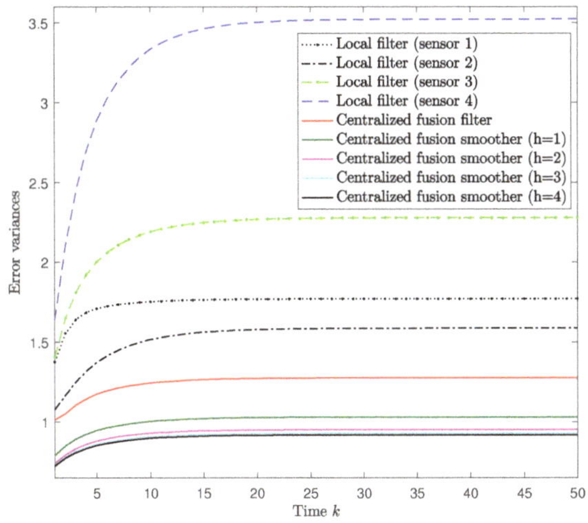

Figure 2. Error variance comparison of the local filters and centralized fusion filter and smoothers.

4.5. *Influence of the Missing Measurement Phenomenon in Sensors 3 and 4*

Considering $\overline{\gamma}^{(i)} = \overline{\psi}_{\gamma}^{(i)} = 0.1i$, $i = 1,2,3,4$, again, in order to show the effect of the missing probabilities in sensors $i = 3,4$, the centralized filtering error variances are displayed in Figure 3 for different values of these probabilities $1 - \overline{\theta}^{(i)}$. Specifically, in Figure 3a, it is assumed that $\overline{\theta}^{(3)} = \overline{\theta}^{(4)}$ with a range of values from 0.5 to 0.9, and in Figure 3b, $\overline{\theta}^{(3)} = 0.5$ and $\overline{\theta}^{(4)}$ varies from 0.6 to 0.9. From these figures, it is clear that the performance of the centralized fusion filter is indeed influenced by the probabilities $\overline{\theta}^{(i)}$, $i = 3,4$. Specifically, the performance of the centralized filter is poorer as $\overline{\theta}^{(i)}$ decreases, which means that, as expected, the lower the probability of missing measurements is, the better performance the filter has. Analogous results were obtained for the centralized smoothers and for other values of the probabilities.

Considering that the behavior of the error variances was analogous in all of the iterations, only the results at a specific iteration ($k = 50$) are displayed in the following figures.

4.6. *Influence of the Probabilities $\overline{\gamma}^{(i)}$ and $\overline{\psi}_{\gamma}^{(i)}$*

Considering $\overline{\theta}^{(i)} = 0.5$, $i = 3,4$, as in Figure 2, we analyze the influences of the random delays and packet dropouts on the performance of the centralized filtering estimators. We assume that the four sensors have the same probability of measurements not arriving on time ($\overline{\gamma}^{(i)} = \overline{\gamma}$, $i = 1,2,3,4$) and also the same conditional probability ($\overline{\psi}_{\gamma}^{(i)} = \overline{\psi}_{\gamma}$, $i = 1,2,3,4$). Figure 4 displays the centralized filtering error variances at $k = 50$ versus $\overline{\psi}_{\gamma}$ for $\overline{\gamma}$ varying from 0.1 to 0.9. This figure shows that for each value of $\overline{\gamma}$, the error variances decrease when the conditional probability increases. This result was expected since, for a fixed arbitrary value of $\overline{\gamma}$, the increase in $\overline{\psi}_{\gamma}$ entails that of $\overline{\psi}$, which is the probability of processing the delayed measurement at the previous time at the current time. Also, we observed that a decrease in the error variances was more evident for higher values of $\overline{\gamma}$, which was also expected since $\overline{\psi} = \overline{\psi}_{\gamma}\overline{\gamma}$ and hence, $\overline{\gamma}$ specifies the increasing rate of $\overline{\psi}$ with respect to $\overline{\psi}_{\gamma}$.

Similar results to the previous ones and consequently, analogous conclusions, were deduced for the smoothing estimators and for different values of the probabilities $\overline{\gamma}^{(i)}$ and $\overline{\psi}_{\gamma}^{(i)}$ at each sensor.

By way of example, the smoothing error variances $\widehat{\Sigma}_{50/51}$ are displayed in Figure 5 for some of the situations considered above.

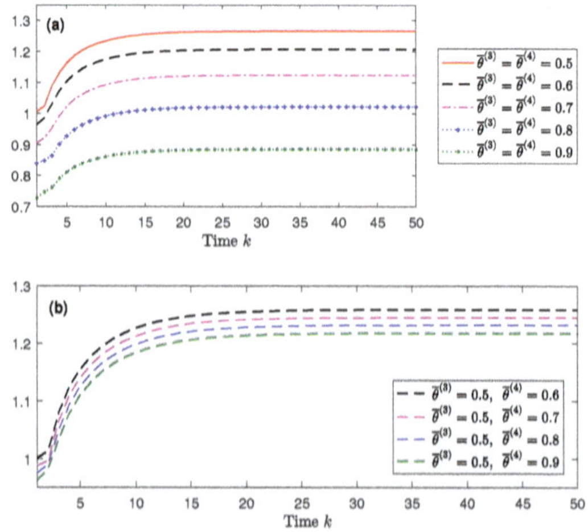

Figure 3. Centralized fusion filtering error variances for different values of $\overline{\theta}^{(3)}$ and $\overline{\theta}^{(4)}$: (**a**) $\overline{\theta}^{(3)} = \overline{\theta}^{(4)}$ from 0.5 to 0.9; (**b**) $\overline{\theta}^{(3)} = 0.5$ and $\overline{\theta}^{(4)}$ from 0.6 to 0.9.

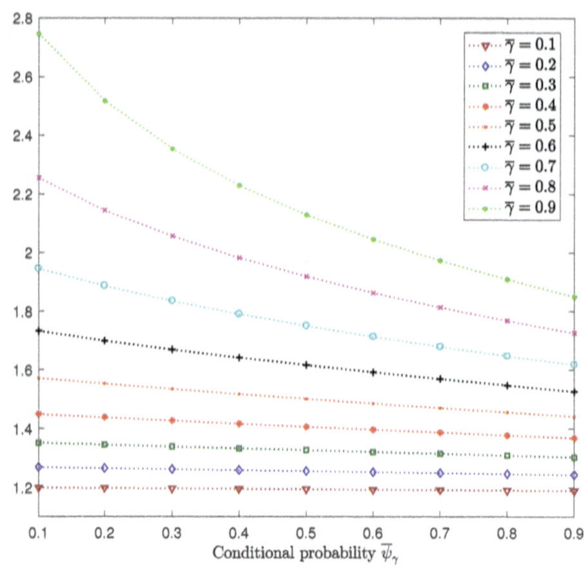

Figure 4. Centralized filtering error variances at $k = 50$ versus $\overline{\psi}_\gamma$, for $\overline{\gamma}$, varying from 0.1 to 0.9 when $\overline{\theta}^{(i)} = 0.5$, $i = 3, 4$.

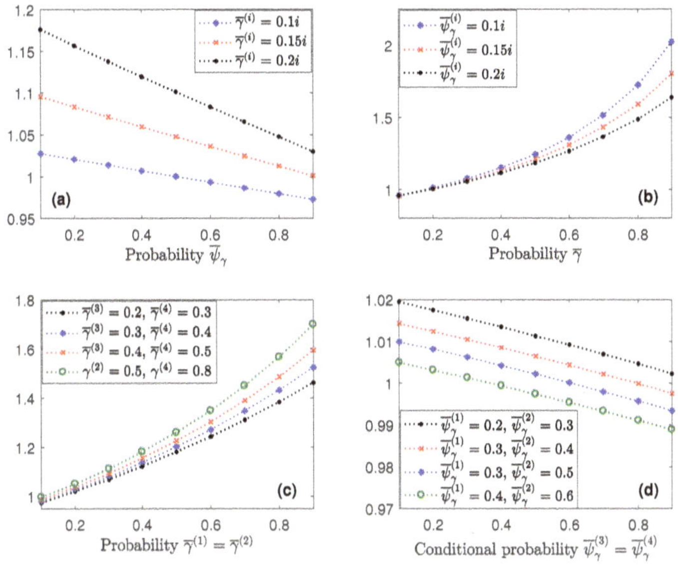

Figure 5. Centralized smoothing error variances ($\widehat{\Sigma}_{50/51}$) when $\overline{\theta}^{(i)} = 0.5$ $i = 3, 4$, for different values of the probabilities $\overline{\gamma}^{(i)}$ and $\overline{\psi}_{\gamma}^{(i)}$, $i = 1, 2, 3, 4$: (**a**) versus $\overline{\psi}_{\gamma}$, for $\overline{\gamma}^{(i)} = 0.1i,\ 0.15i,\ 0.2i$; (**b**) versus $\overline{\gamma}$, for $\overline{\psi}_{\gamma}^{(i)} = 0.1i,\ 0.15i,\ 0.2i$; (**c**) versus $\overline{\gamma}^{(1)} = \overline{\gamma}^{(2)}$ for $\overline{\psi}_{\gamma}^{(i)} = 0.1i$ and different values of $\overline{\gamma}^{(3)}$ and $\overline{\gamma}^{(4)}$; and (**d**) versus $\overline{\psi}_{\gamma}^{(3)} = \overline{\psi}_{\gamma}^{(4)}$, for $\overline{\gamma}^{(i)} = 0.1i$ and different values of $\overline{\psi}_{\gamma}^{(1)}$ and $\overline{\psi}_{\gamma}^{(2)}$.

5. Concluding Remarks

In this paper, recursive algorithms were designed for the LS linear centralized fusion prediction, filtering, and smoothing problems in networked systems with random parameter matrices in the measured outputs. At each sampling time, every sensor sends its measured output to the fusion centre where the data packets coming from all the sensors are gathered. Every data packet is assumed to be transmitted just once, but random delays and packet dropouts can occur during this transmission, so the estimator may receive either one packet, two packets, or nothing. When the current measurement of a sensor does not arrive punctually, the corresponding component of the stacked measured output predictor is used as the compensator in the design of the estimators.

Some of the main advantages of the current approach are the following ones:

- The consideration of random measurement matrices provides a general framework to address different uncertainties, such as missing measurements, multiplicative noise, or sensor gain degradation, as has been illustrated by a simulation example.
- The covariance-based approach used to design the estimation algorithms does not require the knowledge of the state-space model, even though it is also applicable to the classical formulation using this model.
- In contrast to most estimation algorithms dealing with random delays and packet dropouts in the literature, the proposed ones do not require any state vector augmentation technique, and thus are computationally more simple.
- The current estimation algorithms were designed using the LS optimality criterion by a innovation approach and no particular structure of the estimators is required.

Author Contributions: All authors have contributed equally to this work. R.C.-Á., A.H.-C. and J.L.-P. provided original ideas for the proposed model and they all collaborated in the derivation of the estimation algorithms; they participated equally in the design and analysis of the simulation results; and the paper was also written and reviewed cooperatively.

Funding: This research is supported by *Ministerio de Economía, Industria y Competitividad, Agencia Estatal de Investigación* and *Fondo Europeo de Desarrollo Regional FEDER* (grant no. MTM2017-84199-P).

Conflicts of Interest: The authors declare no conflict of interest.

Appendix A

Proof of Theorem 1. Based on the general expression (8), to obtain the LS linear estimators $\widehat{x}_{k/L}$, $L \le k$, it is necessary to calculate the coefficients

$$\mathcal{X}_{k,h} \equiv E\left[x_k \mu_h^T\right] = E\left[x_k y_h^T\right] - E\left[x_k \widehat{y}_{h/h-1}^T\right], \quad h \le k.$$

Using (5) for y_h, the independence hypotheses and the factorization of the signal covariance (H1) lead to $E[x_k y_h^T] = A_k \overline{\mathcal{H}}_{B_h}^T + E[x_k \widehat{x}_{h/h-1}^T] \overline{H}_h^T \overline{\Gamma}_h \mathbb{C}_0^T$, $2 \le h \le k$, and $E[x_k y_1^T] = A_k \overline{\mathcal{H}}_{B_1}^T$, with $\overline{\mathcal{H}}_{B_h}$ given in (15). Now, using expression (10) for $\widehat{y}_{h/h-1}$, together with (8) for $\widehat{x}_{h/h-1}$ and $\widehat{x}_{h-1/h-1}$, the coefficients $\mathcal{X}_{k,h}$, $1 \le h \le k$, are expressed as follows:

$$\mathcal{X}_{k,h} = A_k \overline{\mathcal{H}}_{B_h}^T - \sum_{j=1}^{h-1} \mathcal{X}_{k,j} \Pi_j^{-1} \left(\mathcal{X}_{h,j}^T \overline{H}_h^T (I_{mn_z} - \overline{\Gamma}_h) \mathbb{C}_0^T + \mathcal{X}_{h-1,j}^T \overline{H}_{h-1}^T \overline{\Psi}_h \mathbb{C}_1^T \right) - \mathcal{X}_{k,h-1} \Pi_{h-1}^{-1} \mathcal{V}_{h-1}^T, \quad 2 \le h \le k;$$
$$\mathcal{X}_{k,1} = A_k \overline{\mathcal{H}}_{B_1}^T,$$

which guarantees that $\mathcal{X}_{k,h} = A_k \mathcal{E}_h$, $1 \le h \le k$, with $\mathcal{E}_h$ given by

$$\mathcal{E}_h = \overline{\mathcal{H}}_{B_h}^T - \sum_{j=1}^{h-1} \mathcal{E}_j \Pi_j^{-1} \mathcal{E}_j \mathcal{H}_{A_h}^T - \mathcal{E}_{h-1} \Pi_{h-1}^{-1} \mathcal{V}_{h-1}^T, \quad h \ge 2; \quad \mathcal{E}_1 = \overline{\mathcal{H}}_{B_1}^T.$$

Then, by defining $e_L \equiv \sum_{h=1}^{L} \mathcal{E}_h \Pi_h^{-1} \mu_h$ and $\Sigma_L^e \equiv E\left[e_L e_L^T\right] = \sum_{h=1}^{L} \mathcal{E}_h \Pi_h^{-1} \mathcal{E}_h$, for $L \ge 1$, and taking into account that $E\left[e_L \mu_h^T\right] = \mathcal{E}_h$, for $h \le L$, it is easy to obtain expressions (11)–(16).

Next, the expression (17) for $\mathcal{V}_L = E\left[V_L \mu_L^T\right] = E\left[V_L y_L^T\right]$ is derived. Using (9) for V_L, we write $\mathcal{V}_L = \mathbb{C}_1 \left(E[\Psi_{L+1} z_L y_L^T] - \overline{\Psi}_{L+1} \overline{H}_L E[x_L y_L^T] \right)$, and we calculate each of these expectations:

- From (5), we write $y_L = \mathbb{C}_0 (I_{mn_z} - \Gamma_L)(z_L - \widehat{z}_{L/L-1}) + \mathbb{C}_1 \Psi_L z_{L-1} + \mathbb{C}_0 \widehat{z}_{L/L-1}$, and from the independence properties, it is clear that

$$\begin{aligned} E[\Psi_{L+1} z_L y_L^T] = {} & E[\Psi_{k+1} z_L (z_L - \widehat{z}_{L/L-1})^T (I_{mn_z} - \Gamma_L)] \mathbb{C}_0^T \\ & + \overline{\Psi}_{L+1} E[z_L z_{L-1}^T] \overline{\Psi}_L \mathbb{C}_1^T + \overline{\Psi}_{L+1} E[z_L \widehat{z}_{L/L-1}^T] \mathbb{C}_0^T. \end{aligned}$$

Now, from the Hadamard product properties, we obtain $E[\Psi_{L+1} z_L (z_L - \widehat{z}_{L/L-1})^T (I_{mn_z} - \Gamma_L)] = (K_{L+1,L}^{\psi(\mathbf{1}-\gamma)} \circ (\Sigma_L^z - E[z_L \widehat{z}_{L/L-1}]^T)$; from property (P5), $E[z_L z_{L-1}^T] = \overline{H}_L A_L B_{L-1}^T \overline{H}_{L-1}^T$, and using the OPL and (2), $E[z_L \widehat{z}_{L/L-1}^T] = E[\widehat{z}_{L/L-1} \widehat{z}_{L/L-1}^T] = \overline{H}_L E[\widehat{x}_{L/L-1} \widehat{x}_{L/L-1}^T] \overline{H}_L^T$. Then, using (11) and the definition of Σ_L^e, the following expression is obtained:

$$\begin{aligned} E[\Psi_{L+1} z_L y_L^T] = {} & (K_{L+1,L}^{\psi(\mathbf{1}-\gamma)} \circ (\Sigma_L^z - \overline{H}_L A_L \Sigma_{L-1}^e A_L^T \overline{H}_L^T)) \mathbb{C}_0^T \\ & + \overline{\Psi}_{L+1} \overline{H}_L A_L \left(B_{L-1}^T \overline{H}_{L-1}^T \overline{\Psi}_L \mathbb{C}_1^T + \Sigma_{L-1}^e A_L^T \overline{H}_L^T \mathbb{C}_0^T \right). \end{aligned}$$

- Using (2), (5) again and the OPL, together with Hypothesis (H1) and (15) for $\overline{\mathcal{H}}_{B_L}$, we have

$$\overline{\Psi}_{L+1}\overline{H}_L E[x_L y_L^T] = \overline{\Psi}_{L+1}\overline{H}_L A_L \left(\overline{\mathcal{H}}_{B_L}^T + \Sigma_{L-1}^e A_L^T \overline{H}_L^T \overline{\Gamma}_L \mathbb{C}_0^T\right).$$

From the above items and using (15), $B_{L-1}^T \overline{H}_{L-1}^T \overline{\Psi}_L \mathbb{C}_1^T - \overline{\mathcal{H}}_{B_L}^T = -B_L^T \overline{H}_L^T (I_{mn_z} - \overline{\Gamma}_L)\mathbb{C}_0^T$, expression (17) is deduced with no difficulty.

To obtain expression (18) for $\Pi_L = E[\mu_L \mu_L^T]$, we apply the OPL to write $\Pi_L = E[y_L y_L^T] - E[\widehat{y}_{L/L-1}\widehat{y}_{L/L-1}^T]$.

On the one hand, using the OPL again, we express $E[\widehat{y}_{L/L-1}\widehat{y}_{L/L-1}^T] = E[\widehat{y}_{L/L-1} y_L^T]$ which, takes (16) into account for $\widehat{y}_{L/L-1}$, and the definitions of $\mathcal{O}_{L,L-1}$ and $\mathcal{Y}_{L,L-1}$, clearly satisfy

$$E[\widehat{y}_{L/L-1} y_L^T] = (\overline{\mathcal{H}}_{A_L} + \mathbb{C}_0 \overline{\Gamma}_L \overline{H}_L A_L)\mathcal{O}_{L,L-1}^T + \mathcal{V}_{L-1}\Pi_{L-1}^{-1}\mathcal{Y}_{L,L-1}^T, \quad L \geq 2.$$

On the other hand, to obtain $E[y_L y_L^T]$, we use (9) and (6) to write $y_L = \xi_L + \mathbb{C}_0 \Gamma_L \widehat{z}_{L/L-1}$, and since $\widehat{z}_{L/L-1} = \overline{H}_L A_L e_{L-1}$, the following expression is obtained from the definition of Σ_L^e after some manipulations:

$$E[y_L y_L^T] = \Sigma_L^{\xi} - \mathbb{C}_0[K_L^{\gamma} \circ (\overline{H}_L A_L \Sigma_{L-1}^e A_L^T \overline{H}_L^T)]\mathbb{C}_0^T + \mathbb{C}_0 \overline{\Gamma}_L \overline{H}_L A_L \mathcal{O}_{L,L-1}^T + \mathcal{O}_{L,L-1} A_L^T \overline{H}_L^T \overline{\Gamma}_L \mathbb{C}_0^T, \quad L \geq 2.$$

From the above expectations, again, after some manipulations, expression (18) for Π_L is obtained.

To complete the proof, expression (19) for $\mathcal{O}_{L,L-1} = E[y_L e_{L-1}^T]$ and $\mathcal{Y}_{L,L-1} = E[y_L \mu_{L-1}^T]$ is derived. Using the OPL, we have $E[y_L e_{L-1}^T] = E[\widehat{y}_{L/L-1} e_{L-1}^T]$ and $E[y_L \mu_{L-1}^T] = E[\widehat{y}_{L/L-1}\mu_{L-1}^T]$, and from (16) for $\widehat{y}_{L/L-1}$, expression (19) is straightforward. Then, the proof of Theorem 1 is complete. □

Appendix B

Proof of Theorem 2. Using (8), the signal estimators are written as $\widehat{x}_{k/k+h} = \sum_{l=1}^{k+h} \mathcal{X}_{k,l}\Pi_l^{-1}\mu_l, \ h \geq 1$, from which it is immediately deduced that the smoothers are recursively obtained by (20) from the filter $\widehat{x}_{k/k}$.

Taking into account that $\mathcal{X}_{k,k+h} = E[x_k y_{k+h}^T] - E[x_k \widehat{y}_{k+h/k+h-1}^T]$, $h \geq 1$, the recursive relation (21) is derived by just calculating each of these expectations as follows:

- Hypothesis (H1) together with (15), leads to

$$E[x_k y_{k+h}^T] = B_k \overline{\mathcal{H}}_{A_{k+h}}^T + E[x_k e_{k+h-1}^T] A_{k+h}^T \overline{H}_{k+h}^T \overline{\Gamma}_{k+h} \mathbb{C}_0^T, \quad h \geq 1.$$

- From (16) for $\widehat{y}_{k+h/k+h-1}$, it is clear that

$$E[x_k \widehat{y}_{k+h/k+h-1}^T] = E[x_k e_{k+h-1}^T](\overline{\mathcal{H}}_{A_{k+h}} + \mathbb{C}_0 \overline{\Gamma}_{k+h}\overline{H}_{k+h}A_{k+h})^T + \mathcal{X}_{k,k+h-1}\Pi_{k+h-1}^{-1}\mathcal{V}_{k+h-1}^T, \quad h \geq 1.$$

From the above items, (21) is proven simply by denoting $E_{k,k+h} = E[x_k e_{k+h}^T]$, whose recursive expression (22) is also obvious by using (12) for e_{k+h}.

Finally, using (20) for the smoothers $\widehat{x}_{k/k+h}$, the recursive formula for the fixed-point smoothing error covariance matrices $\widehat{\Sigma}_{k/k+h}$ is immediately deduced. □

References

1. Castanedo, F. A review of data fusion techniques. *Sci. World J.* **2013**, *2013*, 704504. [CrossRef] [PubMed]
2. Khaleghi, B.; Khamis, A.; Karray, F.O.; Razavi, S.N. Multisensor data fusion: A review of the state-of-the-art. *Inform. Fusion* **2013**, *14*, 28–44. [CrossRef]

3. Bark, M.; Lee, S. Distributed multisensor data fusion under unknown correlation and data inconsistency. *Sensors* **2017**, *17*, 2472.
4. Lin, H.; Sun, S. State estimation for a class of non-uniform sampling systems with missing measurements. *Sensors* **2016**, *16*, 1155. [CrossRef] [PubMed]
5. Liu, Y.; Wang, Z.; He, X.; Zhou, D.H. Minimum-variance recursive filtering over sensor networks with stochastic sensor gain degradation: Algorithms and performance analysis. *IEEE Trans. Control Netw. Syst.* **2016**, *3*, 265–274. [CrossRef]
6. Li, W.; Jia, Y.; Du, J. Distributed filtering for discrete-time linear systems with fading measurements and time-correlated noise. *Digit. Signal Process.* **2017**, *60*, 211–219. [CrossRef]
7. Ma, J.; Sun, S. Centralized fusion estimators for multisensor systems with random sensor delays, multiple packet dropouts and uncertain observations. *IEEE Sens. J.* **2013**, *13*, 1228–1235. [CrossRef]
8. Chen, B.; Zhang, W.; Yu, L. Distributed fusion estimation with missing measurements, random transmission delays and packet dropouts. *IEEE Trans. Autom. Control* **2014**, *59*, 1961–1967. [CrossRef]
9. Caballero-Águila, R.; Hermoso-Carazo, A.; Linares-Pérez, J. Fusion estimation using measured outputs with random parameter matrices subject to random delays and packet dropouts. *Signal Process.* **2016**, *127*, 12–23. [CrossRef]
10. Caballero-Águila, R.; Hermoso-Carazo, A.; Linares-Pérez, J. Least-squares estimation in sensor networks with noise correlation and multiple random failures in transmission. *Math. Probl. Eng.* **2017**, *2017*, 1570719. [CrossRef]
11. Hu, J.; Wang, Z.; Chen, D.; Alsaadi, F.E. Estimation, filtering and fusion for networked systems with network-induced phenomena: New progress and prospects. *Inform. Fusion* **2016**, *31*, 65–75. [CrossRef]
12. Sun, S.; Lin, H.; Ma, J.; Li, X. Multi-sensor distributed fusion estimation with applications in networked systems: A review paper. *Inform. Fusion* **2017**, *38*, 122–134. [CrossRef]
13. Luo, Y.; Zhu, Y.; Luo, D.; Zhou, J.; Song, E.; Wang, D. Globally optimal multisensor distributed random parameter matrices Kalman filtering fusion with applications. *Sensors* **2008**, *8*, 8086–8103. [CrossRef] [PubMed]
14. Shen, X.J.; Luo, Y.T.; Zhu, Y.M.; Song, E.B. Globally optimal distributed Kalman filtering fusion. *Sci. China Inf. Sci.* **2012**, *55*, 512–529. [CrossRef]
15. Wang, S.; Fang, H.; Tian, X. Minimum variance estimation for linear uncertain systems with one-step correlated noises and incomplete measurements. *Digit. Signal Process.* **2016**, *49*, 126–136. [CrossRef]
16. Hu, J.; Wang, Z.; Gao, H. Recursive filtering with random parameter matrices, multiple fading measurements and correlated noises. *Automatica* **2013**, *49*, 3440–3448. [CrossRef]
17. Linares-Pérez, J.; Caballero-Águila, R.; García-Garrido, I. Optimal linear filter design for systems with correlation in the measurement matrices and noises: Recursive algorithm and applications. *Int. J. Syst. Sci.* **2014**, *45*, 1548–1562. [CrossRef]
18. Yang, Y.; Liang, Y.; Pan, Q.; Qin, Y.; Yang, F. Distributed fusion estimation with square-root array implementation for Markovian jump linear systems with random parameter matrices and cross-correlated noises. *Inf. Sci.* **2016**, *370–371*, 446–462. [CrossRef]
19. Caballero-Águila, R.; Hermoso-Carazo, A.; Linares-Pérez, J. Networked fusion filtering from outputs with stochastic uncertainties and correlated random transmission delays. *Sensors* **2016**, *16*, 847. [CrossRef] [PubMed]
20. Caballero-Águila, R.; Hermoso-Carazo, A.; Linares-Pérez, J. Optimal fusion estimation with multi-step random delays and losses in transmission. *Sensors* **2017**, *17*, 1151. [CrossRef] [PubMed]
21. Sun, S.; Tian, T.; Honglei, L. State estimators for systems with random parameter matrices, stochastic nonlinearities, fading measurements and correlated noises. *Inf. Sci.* **2017**, *397–398*, 118–136. [CrossRef]
22. Wang, W.; Zhou, J. Optimal linear filtering design for discrete time systems with cross-correlated stochastic parameter matrices and noises. *IET Control Theory Appl.* **2017**, *11*, 3353–3362. [CrossRef]
23. Han, F.; Dong, H.; Wang, Z.; Li, G.; Alsaadi, F.E. Improved Tobit Kalman filtering for systems with random parameters via conditional expectation. *Signal Process.* **2018**, *147*, 35–45. [CrossRef]
24. Caballero-Águila, R.; Hermoso-Carazo, A.; Linares-Pérez, J.; Wang, Z. A new approach to distributed fusion filtering for networked systems with random parameter matrices and correlated noises. *Inform. Fusion* **2019**, *45*, 324–332. [CrossRef]

25. Guo, Y. Switched filtering for networked systems with multiple packet dropouts. *J. Franklin Inst.* **2017**, *354*, 3134–3151. [CrossRef]
26. Yang, C.; Deng, Z. Robust time-varying Kalman estimators for systems with packet dropouts and uncertain-variance multiplicative and linearly correlated additive white noises. *Int. J. Adapt. Control Signal Process.* **2018**, *32*, 147–169.
27. Xing, Z.; Xia, Y.; Yan, L.; Lu, K.; Gong, Q. Multisensor distributed weighted Kalman filter fusion with network delays, stochastic uncertainties, autocorrelated, and cross-correlated noises. *IEEE Trans. Syst. Man Cybern. Syst.* **2018**, *48*, 716–726. [CrossRef]
28. Silva, E.I.; Solis, M.A. An alternative look at the constant-gain Kalman filter for state estimation over erasure channels. *IEEE Trans. Autom. Control* **2013**, *58*, 3259–3265. [CrossRef]
29. Caballero-Águila, R.; Hermoso-Carazo, A.; Linares-Pérez, J. New distributed fusion filtering algorithm based on covariances over sensor networks with random packet dropouts. *Int. J. Syst. Sci.* **2017**, *48*, 1805–1817. [CrossRef]
30. Ding, J.; Sun, S.; Ma, J.; Li, N. Fusion estimation for multi-sensor networked systems with packet loss compensation. *Inform. Fusion* **2019**, *45*, 138–149. [CrossRef]
31. Caballero-Águila, R.; Hermoso-Carazo, A.; Linares-Pérez, J. Covariance-based fusion filtering for networked systems with random transmission delays and non-consecutive losses. *Int. J. Gen. Syst.* **2017**, *46*, 752–771. [CrossRef]
32. Zhu, C.; Xia, Y.; Xie, L.; Yan, L. Optimal linear estimation for systems with transmission delays and packet dropouts. *IET Signal Process.* **2013**, *7*, 814–823. [CrossRef]
33. Ma, J.; Sun, S. Linear estimators for networked systems with one-step random delay and multiple packet dropouts based on prediction compensation. *IET Signal Process.* **2017**, *1*, 197–204. [CrossRef]
34. Ma, J.; Sun, S. Distributed fusion filter for networked stochastic uncertain systems with transmission delays and packet dropouts. *Signal Process.* **2017**, *130*, 268–278. [CrossRef]

sensors

Article

Square-Root Unscented Information Filter and Its Application in SINS/DVL Integrated Navigation

Yan Guo *, Meiping Wu *, Kanghua Tang and Lu Zhang

National University of Defense Technology, Deya Road No. 109, Kaifu District, Changsha 410073, China; tt_kanghua@hotmail.com (K.T.); zhanglu_nudter@163.com (L.Z.)

* Correspondence: guoyan15@nudt.edu.cn (Y.G.); meipingwu@263.net (M.W.); Tel.: +86-0731-84575601 (Y.G.)

Received: 21 May 2018; Accepted: 25 June 2018; Published: 28 June 2018

Abstract: To address the problem of low accuracy for the regular filter algorithm in SINS/DVL integrated navigation, a square-root unscented information filter (SR-UIF) is presented in this paper. The proposed method: (1) adopts the state probability approximation instead of the Taylor model linearization in EKF algorithm to improve the accuracy of filtering estimation; (2) selects the most suitable parameter form at each filtering stage to simply the calculation complexity; (3) transforms the square root to ensure the symmetry and positive definiteness of the covariance matrix or information matrix, and then to enhance the stability of the filter. The simulation results indicate that the estimation accuracy of SR-UIF is higher than that of EKF, and similar to UKF; meanwhile the computational complexity of SR-UIF is lower than that of UKF.

Keywords: SINS/DVL integrated navigation; unscented information filter; square root; state probability approximation; most suitable parameter form

1. Introduction

Most underwater or surface navigation applications employ a Strapdown Inertial Navigation System (SINS) as their main navigation sensor, since SINS is a standalone system that can provide all of the required navigation data: position, velocity and orientation [1–3]. However, even with high precision SINS, the navigation solutions drift in time due to measurement errors of its inertial sensors. The Doppler velocity log (DVL) is a good acoustic-based device in marine applications, which can provide three-dimensional velocities to mitigate the errors of marine SINS [4–6]. Therefore, the integrated SINS and DVL navigation system is a common navigation method for underwater or surface navigation during long voyages [7–9]. The SINS system error model is nonlinear, and a nonlinear filtering algorithm is generally used for state estimation. The earliest nonlinear filtering method used in engineering was the Extend Kalman Filter (EKF) algorithm [10,11]. Its core idea is to approximate a linear expansion of the current nonlinear state equation (namely a Taylor series expansion, truncating high-order terms, retaining first-order terms) to apply the rules of EKF. However, the EKF algorithm is only applicable to weakly nonlinear systems. The stronger the nonlinearity of the estimated object is, the larger the estimation error will be, and it will even cause filter divergence.

Some scholars have proposed a probabilistic approximation of the nonlinear filtering construction idea [12], that is, using a deterministic sampling method to replace the Taylor series expansion linearization of the system model in EKF algorithm, and to approach the mean and variance of the Gauss state distribution by utilizing deterministic sampling points through a nonlinear system equation transformation propagation. The Unscented Kalman Filter (UKF) algorithm proposed by Uhlman is the first nonlinear algorithm to practice this idea [13–15]. It adopts an Unscented Transformation (UT) to obtain 2n + 1 Sigma sampling points with different weights, and uses the abovementioned Sigma sampling points to generate new points after transforming the nonlinear system equation for estimating the mean and the variance of the system states at the next moment. The theoretical deduction proves that the estimation accuracy of UKF algorithm can reach the third-order terms of Taylor series expansion for the nonlinear system [16–18].

The information filtering algorithm realizes a state estimation by transforming information parameters (including information matrix and information vector) [19,20]. It is equivalent to a series of Kalman filter algorithms which pass the moment parameters (covariance matrix and state vector) [21,22]. The whole filtering process can be divided into two processes: time updating and measurement updating. The time updating process involves the calculation of marginal probability, and the retrieval of the moment parameter form is relatively simple. The measurement updating process involves conditional probability, and the information parameter form is more effective. Therefore, in order to optimize the performance of the algorithm, the Square-Root Unscented Information Filter (SR-UIF) algorithm is proposed in this paper, which is applied to a nonlinear integrated SINS/DVL navigation system. This algorithm adopts the form of moment parameters in the process of time updating according the characteristic of the parameter form, respectively, the form of information parameters in the process of measurement updating; and ensures the symmetry and positive definite of the information matrix or covariance matrix by propagating their square root, and alleviates the problems such as divergence and data precision degradation caused by rounding errors in the filtering process.

The remainder of this paper is organized as follows: Section 2 studies the duality of Gaussian distribution; Section 3 focuses on the theoretical derivation of the SR-UIF algorithm; Section 4 briefly introduces the SINS/DVL model used in this paper; Section 5 verifies our findings through simulations and the conclusions are given in Section 6.

2. Duality of a Gaussian Distribution

We set $\boldsymbol{X}$ as the state vector of the integrated navigation system, which obeys a Gaussian distribution of multi-dimensional random variables, namely $p(\boldsymbol{X}) \sim N(\hat{\boldsymbol{X}}, \boldsymbol{P})$, $\hat{\boldsymbol{X}}$ is its mean value and $\boldsymbol{P}$ is its variance. Through the expansion of the Gauss distribution index term, it can be obtained that:

$$\begin{aligned} p(\boldsymbol{X}) &= N(\hat{\boldsymbol{X}}, \boldsymbol{P}) \\ &= \frac{1}{\sqrt{|2\pi \boldsymbol{P}|}} \exp\left\{-\frac{1}{2}(\boldsymbol{X}-\hat{\boldsymbol{X}})^T \boldsymbol{P}^{-1}(\boldsymbol{X}-\hat{\boldsymbol{X}})\right\} \\ &= \frac{e^{-\frac{1}{2}\boldsymbol{y}^T \boldsymbol{Y}^{-1}\boldsymbol{y}}}{\sqrt{|2\pi \boldsymbol{Y}^{-1}|}} \exp\left\{-\frac{1}{2}\boldsymbol{X}^T \boldsymbol{Y}\boldsymbol{X} + \boldsymbol{y}^T \boldsymbol{X}\right\} \\ &= N^{-1}(\boldsymbol{y}, \boldsymbol{Y}) \end{aligned} \tag{1}$$

where $\boldsymbol{Y}$ is the information matrix, and $\boldsymbol{y}$ is the information vector. It is usually called $\hat{\boldsymbol{X}}$ and $\boldsymbol{P}$ as the moment parameter form, $\boldsymbol{y}$ and $\boldsymbol{Y}$ as the information parameter form, and their transformation relations are:

$$\begin{cases} \boldsymbol{Y} = \boldsymbol{P}^{-1} \\ \boldsymbol{y} = \boldsymbol{Y}\hat{\boldsymbol{X}} = \boldsymbol{P}^{-1}\hat{\boldsymbol{X}} \end{cases} \tag{2}$$

From Equations (1) and (2), the information parameter form is also an expression of the Gaussian distribution, which is equivalent to the representation of the moment parameter form, and the two forms can be converted to each other. These are collectively referred to as the duality of the Gaussian distribution. The whole filtering process can be divided into time updating and measurement updating:

(1) the time updating process involves the calculation of marginal probability:

$$\rho(\boldsymbol{X}_k/\boldsymbol{U}^{k-1}, \boldsymbol{Z}^{k-1}) = \int \rho(\boldsymbol{X}_k, \boldsymbol{X}_{k-1}/\boldsymbol{U}^{k-1}, \boldsymbol{Z}^{k-1}) d\boldsymbol{X}_{k-1}$$

(2) The measurement updating process involves the calculation of conditional probability:

$$\rho(\mathbf{X}_k/\mathbf{U}^{k-1},\mathbf{Z}^k)=\frac{\rho(\mathbf{Z}_k/\mathbf{X}_k)\rho(\mathbf{X}_k/\mathbf{U}^{k-1},\mathbf{Z}^{k-1})}{\int\rho(\mathbf{Z}_k/\mathbf{X}_k)\rho(\mathbf{X}_k/\mathbf{U}^{k-1},\mathbf{Z}^{k-1})d\mathbf{X}_k}$$

Suppose that $\boldsymbol{X}$ and $\boldsymbol{Z}$ satisfy the Gauss distribution as follows:

$$\rho(\mathbf{X},\mathbf{Z})=N\left(\begin{bmatrix}\hat{x}_X\\ \hat{x}_Z\end{bmatrix};\begin{bmatrix}P_{X,X} & P_{X,Z}\\ P_{Z,X} & P_{Z,Z}\end{bmatrix}\right)=N^{-1}\left(\begin{bmatrix}\hat{y}_X\\ \hat{y}_Z\end{bmatrix};\begin{bmatrix}Y_{X,X} & Y_{X,Z}\\ Y_{Z,X} & Y_{Z,Z}\end{bmatrix}\right)$$

The condition probability and marginal probability of the moment parameter form and information parameter are as shown in Table 1 below. It can be seen when using the moment parameters and information parameters to calculate the marginal probability and conditional probability, the two equivalent expressions have completely different calculation characteristics. It is relatively simple to calculate the marginal probability in the moment parameter form. On the contrary, when calculating conditional probability, it is relatively effective to calculate the conditional probability in the information parameter form.

Table 1. Expressions for the condition probability and marginal probability of the different form.

Parameter Form	Marginal Probability	Conditional Probability
Moment parameter	$\hat{x}=\hat{x}_X$ $P=P_{X,X}$	$\hat{x}=\hat{x}_X+P_{X,Z}P_{Z,Z}^{-1}(\mathbf{Z}-\hat{x}_Z)$ $P=P_{X,X}-P_{X,Z}P_{Z,Z}^{-1}P_{Z,X}$
Information parameter	$\hat{y}=\hat{y}_X-Y_{X,Z}Y_{Z,Z}^{-1}\hat{y}_Z$ $Y=Y_{X,X}-Y_{X,Z}Y_{Z,Z}^{-1}Y_{Z,X}$	$\hat{y}=\hat{y}_X-Y_{X,Z}\mathbf{Z}$ $P=Y_{X,X}$

3. Square-Root Unscented Information Filter (SR-UIF) Algorithm

Square-Root Unscented Information Filter (SR-UIF) algorithm is a nonlinear filtering algorithm based on probabilistic approximation, which has the same structure as the well known UKF algorithm. It adopts an Unscented Transformation (UT) to obtain 2n + 1 Sigma sampling points with different weights, and uses the abovementioned Sigma sampling points to generate new points after transforming the nonlinear system equation for estimating the mean and the variance of the system states at the next moment. However, compared with the UKF algorithm, the SR-UIF algorithm selects the most suitable expression form at each filtering stage, that is, using the moment parameter form in the time updating, using the information parameter form in the measurement updating, for simplifying the computational complexity. Meanwhile, SR-UIF algorithm ensures the symmetry and positive definiteness of the information matrix or covariance matrix by propagating their square root, for improving the stability of the algorithm.

The discretized integrated navigation system model is:

$$\begin{cases}\mathbf{X}_k=f(\mathbf{X}_{k-1})+\mathbf{G}_{k-1}\mathbf{W}_{k-1}\\ \mathbf{Z}_k=h(\mathbf{X}_k)+\mathbf{V}_k\end{cases}\tag{3}$$

The following is analyzed for the implementation steps of SR-UIF algorithm. First, it is assumed that the initial filter estimation state is as follows:

$$\begin{cases}\hat{\mathbf{X}}_0=E[\mathbf{X}_0]\\ \mathbf{P}_0=E[(\mathbf{X}_0-\hat{\mathbf{X}}_0)(\mathbf{X}_0-\hat{\mathbf{X}}_0)^T]\end{cases}\tag{4}$$

Combined with the duality of Gaussian distribution, the initial information parameters can be obtained:

$$\begin{cases} y_0 = Y_0 \hat{X}_0 \\ Y_0^{-1} = P_0 = \left\{ E[(X_0 - \hat{X}_0)(X_0 - \hat{X}_0)^T] \right\}^{-1} \end{cases} \tag{5}$$

3.1. Time Updating

The calculating of the edge probabilities involved in the time updating is simpler by using the moment parameter form. Therefore, the mean and variance are applied as the iterative factors in the time updating process of SR-UIF algorithm. The specific process is as follow:

(1) Decompose the covariance matrix $\hat{P}_{k-1}$ at $k-1$ time by using the Cholesky algorithm:

$$\hat{P}_{k-1} = S_{k-1} S_{k-1}^T \tag{6}$$

where S_{k-1} is the Cholesky factor of the covariance matrix $\hat{P}_{k-1}$, and is the lower triangular matrix.

(2) Calculate the Sigma point sets and their weights according to the Unscented Transformation (UT):

$$\begin{cases} \xi_i = 0,\ W_i = W_i^m = W_i^c = \frac{\kappa}{n+\kappa} & ,\ i = 0 \\ \xi_i = \sqrt{\frac{n+\kappa}{2}}[1]_i,\ W_i = W_i^m = W_i^c = \frac{1}{2(n+\kappa)} & ,\ i = 1, 2, \cdots, 2n \end{cases} \tag{7}$$

where n is the system dimension; κ is a free parameter; W_i^m is the weighted value corresponding to each Sigma point, W_i^c is the weighted value corresponding to variance matrix, and satisfies $\sum W_i^c = \sum W_i^m = 1$; $[1]_i$ indicates the i column or line of identity matrix [1]. Further, the state sample points at k time can be obtained as follows:

$$X_{i,k-1} = S_{k-1}\xi_i + \hat{x}_{k-1} \tag{8}$$

(3) One-step state prediction augmented sample points $X^*_{i,k|k-1}$ can be obtained through a nonlinear transformation:

$$X^*_{i,k|k-1} = F(X_{i,k-1}) \tag{9}$$

Then their mean and covariance matrix at k time are

$$\hat{x}_{k|k-1} = \sum_{i=0}^{m} W_i^m X^*_{i,k|k-1} \tag{10}$$

$$P_{k/k-1} = \sum_{i=0}^{m} W_i^c X^*_{i,k|k-1} X^{*T}_{i,k|k-1} - \hat{x}_{k/k-1}\hat{x}^T_{k/k-1} + Q_{k-1} \tag{11}$$

where Q_{k-1} is the variance matrix of system noise.

3.2. Measurement Updating

As it is relatively simple to calculate the conditional probabilities by utilizing the information parameter form, the information vector and information matrix are applied as the iterative factors in the measurement updating process of SR-UIF algorithm, for achieving the optimal design of the algorithm. According to the theoretically deduced knowledge in the Appendix, it can be seen that the update equation of the information matrix and the information vector at k time:

$$\begin{cases} \hat{Y}_k = Y_{k/k-1} + I_k \\ \hat{y}_k = \hat{y}_{k/k-1} + i_k \end{cases} \tag{12}$$

where $Y_{k/k-1}$ is the one-step prediction information matrix, and $Y_{k/k-1} = P_{k/k-1}^{-1}$; $\hat{y}_{k/k-1}$ is the one-step prediction information vector, and $\hat{y}_{k/k-1} = Y_{k/k-1}\hat{x}_{k/k-1}$; I_k and i_k are the contributions of the information matrix and the information parameters, respectively, and:

$$\begin{cases} I_k = (Y_{k/k-1}P_{xz,k/k-1})R_k^{-1}(Y_{k/k-1}P_{xz,k/k-1})^T \\ i_k = (Y_{k/k-1}P_{xz,k/k-1})R_k^{-1}\left(z_k - h(\hat{x}_{k/k-1}) + P_{xz,k/k-1}^T\hat{y}_{k/k-1}\right) \end{cases} \tag{13}$$

R_k is the variance matrix of measurement noise, $P_{xz,k/k-1}$ is the cross-covariance matrix between x and z.

It can be seen from Equation (13) that the measurement update states require the known $P_{xz,k/k-1}$. Now the specific solution process of $P_{xz,k/k-1}$ is given as follows:

(1) Decompose the covariance matrix $P_{k/k-1}$ by using the Cholesky algorithm again,

$$P_{k/k-1} = S_{k/k-1}S_{k/k-1}^T \tag{14}$$

where $S_{k/k-1}$ is the Cholesky factor of the covariance matrix $P_{k/k-1}$, and is the lower triangular matrix.

(2) Calculates the one-step prediction state sample points:

$$X_{i,k|k-1} = S_{k|k-1}\xi_i + \hat{x}_{k|k-1} \tag{15}$$

(3) One-step measurement prediction augmented sample points can be obtained through nonlinear transformation:

$$Z_{i,k|k-1} = h\left(X_{i,k|k-1}\right) \tag{16}$$

then their mean at k time is:

$$\hat{z}_{k|k-1} = \sum_{i=0}^{m} W_i^m Z_{i,k|k-1} \tag{17}$$

and the cross-covariance matrix is:

$$P_{xz,k/k-1} = \sum_{i=0}^{m} W_i^c X_{i,k|k-1}^* Z_{i,k|k-1}^{*T} - \hat{x}_{k/k-1}\hat{z}_{k/k-1}^T \tag{18}$$

In order to facilitate the experiment recording and observation, the filtering finally needs to transform the information parameter form into the moment parameter form. However, the state quantities involved in the inversion of large information matrix are undoubtedly huge, which makes the algorithm difficult to process. For this, this paper utilizes the Cholesky decomposition to deal with it, that is, firstly:

$$\hat{Y}_k = L_k L_k^T \tag{19}$$

where L_k is the Cholesky factor of the information matrix $\hat{Y}_k$.

Since the Cholesky factor L_k is a lower triangular matrix and its upper half elements are all zero, the computational complexity of the inverse operation of L_k is much less than that of $\hat{Y}_k$ when both L_k and $\hat{Y}_k$ are the same dimension. It can realize the optimization design of the algorithm for using Cholesky factor L_k to solve the state vector estimate at k time:

$$\hat{x}_k = \left(L_k^T\right)^{-1} f \tag{20}$$

where f is the forward vector of Cholesky, and $f = L_k^{-1}\hat{y}_k$.

Equation (6) shows that the covariance matrix $\hat{P}_k$ needs to be decomposed during the next time updating process, and Equation (19) has been decomposed to the information matrix $\hat{Y}_k$ at the next moment, which is a duplicate calculation. In view of this, the paper applies Equation (19) instead of Equation (6) to achieve further reduce the amount of the filtering calculation according to the transformation relation between $\hat{P}_k$ and $\hat{Y}_k$. The specific operation is as follows:

(1) According to the transformation relation between $\hat{P}_k$ and $\hat{Y}_k$, the equation relationship between S_k and L_k is:

$$\begin{cases} L_k = \left(S_k^T\right)^{-1} \\ L_k^T = S_k^{-1} \end{cases} \quad (21)$$

(2) Then L_k is used to instead of S_k in Equation (7) for iteration:

$$X_{i,k} = \left(L_k^T\right)^{-1} \xi_i + \hat{x}_k \quad (22)$$

3.3. Summary

Based on the above analysis, the frame diagram of SR-UIF algorithm is shown in the following Figure 1.

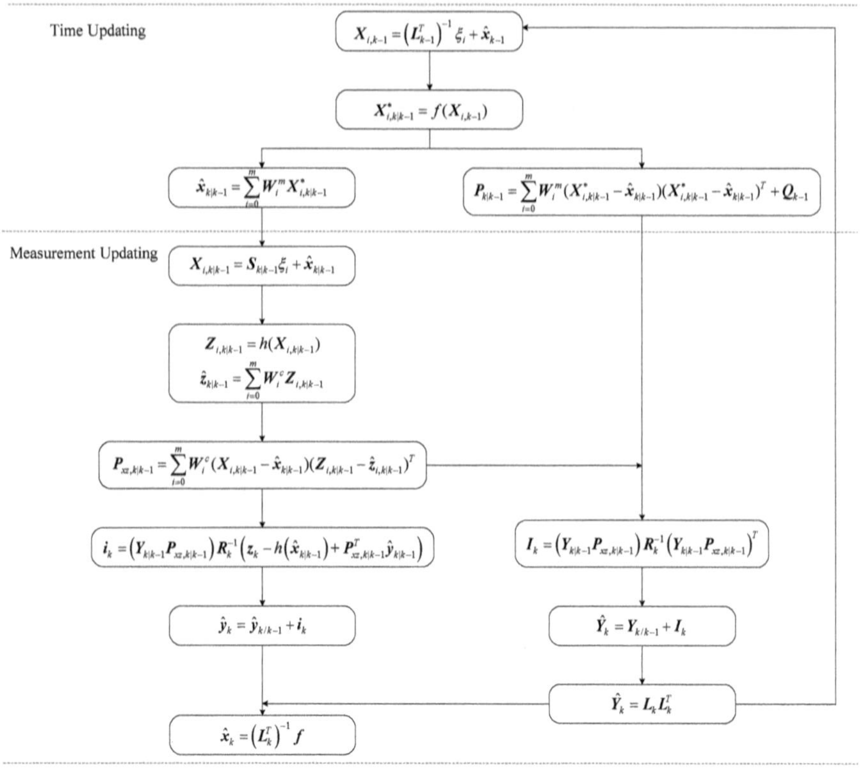

Figure 1. Frame diagram of SR-UIF algorithm.

4. Nonlinear SINS/DVL Integrated Navigation Model

Set the local geographic coordinate frame t as the navigation frame of SINS, the error propagation equation about misalignment angles $\boldsymbol{\varphi}^t$ are constructed as:

$$\dot{\boldsymbol{\varphi}}^t = \delta\boldsymbol{\omega}_{ie}^t + \delta\boldsymbol{\omega}_{et}^t - \left(\boldsymbol{\omega}_{ie}^t + \boldsymbol{\omega}_{et}^t\right) \times \boldsymbol{\varphi}^t + \boldsymbol{C}_b^t\boldsymbol{\varepsilon}^b \tag{23}$$

where e, b, and i denote the Earth, body, and inertial frames, respectively; $\boldsymbol{C}_b^t$ is the attitude transformation matrix from the body coordinate frame b to frame t; $\boldsymbol{\varepsilon}^b$ is constant gyro drift in body frame; $\boldsymbol{\omega}_{ie}^t$ is the earth rotational angular rate in frame t, and $\delta\boldsymbol{\omega}_{ie}^t$ is its calculation error; $\boldsymbol{\omega}_{et}^t$ is the rotational angular rate from frame t to frame e in frame t, and $\delta\boldsymbol{\omega}_{et}^t$ is it calculation error.

Meanwhile, the vector expression of velocity equation is:

$$\delta\dot{\boldsymbol{v}}^t = \boldsymbol{f}^t \times \boldsymbol{\varphi}^t - \left(2\boldsymbol{\omega}_{ie}^t + \boldsymbol{\omega}_{et}^t\right) \times \delta\boldsymbol{v}^t - \left(2\delta\boldsymbol{\omega}_{ie}^t + \delta\boldsymbol{\omega}_{et}^t\right) \times \boldsymbol{v}^t + \boldsymbol{C}_b^t\nabla^b \tag{24}$$

where $\delta\boldsymbol{v}^t$ is the velocity error vector; $\boldsymbol{f}^t$ is the specific force measured by the accelerometer in frame t; ∇^b is the errors of accelerometers in body frame.

Ignoring the height channel, the equations describing position errors are two equations as follow:

$$\begin{bmatrix} \delta\dot{L} \\ \delta\dot{\lambda} \end{bmatrix} = \begin{bmatrix} \frac{\delta v_y}{R+h} \\ \frac{\delta v_x}{R+h}\sec L + \frac{\delta v_x}{R+h}\sec L \tan L\delta L \end{bmatrix} \tag{25}$$

where δL and $\delta\lambda$ are the latitude and longitude error, respectively.

In this paper, a four-beam phased array DVL is used. The measurement error includes the velocity offset error δV_d, drift angle error $\delta\Delta$, and calibration coefficient error δC. Assuming that δC is a random constant, δV_d and $\delta\Delta$ are represented by the first-order Markov process, the error equations of DVL are expressed as:

$$\begin{cases} \delta\dot{V}_d = -\beta_d\delta V_d + \omega_d \\ \delta\dot{\Delta} = -\beta_\Delta\delta\Delta + \omega_\Delta \\ \delta\dot{C} = 0 \end{cases} \tag{26}$$

where β_d^{-1}, β_Δ^{-1} are expressed as the correlation time of DVL velocity offset error and bias angle error, respectively; ω_d and ω_Δ are their Gauss white noise.

The system vector is defined as $\boldsymbol{X} = \begin{bmatrix} \delta L & \delta\lambda & \delta v_e & \delta v_n & \varphi_e & \varphi_n & \varphi_u & \delta V_d & \delta\Delta & \delta C^{\,T} \end{bmatrix}$, and the model of the system state equation is applied in this paper:

$$\dot{\boldsymbol{X}} = f(\boldsymbol{X}) + \boldsymbol{GW} \tag{27}$$

where $\boldsymbol{W} = \begin{bmatrix} 0 & 0 & \omega_{ae} & \omega_{an} & 0 & 0 & 0 & \omega_d & \omega_\Delta & 0 \end{bmatrix}^T$ is the process noise sequence; the specific expressions of and refer to Equations (23)–(26).

Assuming that the DVL measurement is the ground velocity V_d', then the components of V_d' in the east and north directions are:

$$\begin{cases} V_{de}' = (1+\delta C)(V_d + \delta V_d)\sin(K_d + \phi_u + \delta\Delta) \\ V_{dn}' = (1+\delta C)(V_d + \delta V_d)\cos(K_d + \phi_u + \delta\Delta) \end{cases} \tag{28}$$

where ϕ_u is the azimuth misalignment angle; K_d is the heading angle added to the drift angle. The Taylor series expansion is performed on $x = K_d$ for Equation (28):

$$\begin{cases} V_{de}' \approx V_e + V_n(\phi_u + \delta\Delta) + \delta V_d \sin K_d + \delta C V_e \\ V_{dn}' \approx V_n - V_e(\phi_u + \delta\Delta) + \delta V_d \cos K_d + \delta C V_n \end{cases} \tag{29}$$

Meanwhile, the computing velocity of SINS can be expressed as:

$$\begin{cases} V_{se} = V_e + \delta V_e \\ V_{sn} = V_n + \delta V_n \end{cases} \tag{30}$$

The difference between the SINS computing velocity and the DVL measurement velocity is taken as the measurement vectors, namely

$$\begin{cases} \delta V_e = \delta V_e - V_n(\phi_u + \delta\Delta) - \delta V_d \sin K_d - \delta C V_e \\ \delta V_n = \delta V_n + V_e(\phi_u + \delta\Delta) - \delta V_d \cos K_d - \delta C V_n \end{cases} \tag{31}$$

Then the measurement equation is set up as follows:

$$\boldsymbol{Z} = \boldsymbol{H}\boldsymbol{X} + \boldsymbol{V} \tag{32}$$

where the measurement noise is taken as $\boldsymbol{V} = \begin{bmatrix} v_e & v_n \end{bmatrix}^T$; $\boldsymbol{H}$ is the measurement matrix, and

$$\boldsymbol{H} = \begin{bmatrix} 0 & 0 & 1 & 0 & 0 & 0 & -V_n & -\sin K_d & -V_n & -V_e \\ 0 & 0 & 0 & 1 & 0 & 0 & V_e & -\cos K_d & V_e & -V_n \end{bmatrix}$$

5. Results

In the simulation experiments, two classical motion models are designed: uniform linear motion and uniform circular motion. First of all, the accuracy of position and velocity for purely inertial navigation, filtering with the SR-UIF algorithm, UKF algorithm, and EKF algorithm are compared and analyzed in these two modes of motion. Then, the computational complexity of SR-UIF algorithm and UKF algorithm are also compared and analyzed, which is determined by the average time consumed by a once filtering operation and the total elapsed time of each simulation experiment. Finally, the filtering performance of 1 h off-line data is used to further analyze the performance advantages of SR-UIF algorithm in terms of estimation accuracy and computational complexity. Finally, the filtering performance of 1 h off-line data is used to further analyze the performance advantages of SR-UIF algorithm in terms of estimation accuracy and computational complexity.

5.1. Simulation Analysis

Set the simulation conditions: initial latitude $\varphi_0 = 45.7796°$, and initial longitude $\lambda_0 = 126.6705°$; initial position error $\delta L = \delta\lambda = 100/R$ rad, initial velocity error 0.01 m/s; initial misalignment angles $\phi_{e0} = \phi_{n0} = \phi_{u0} = 1°$; the gyro constant drifts along three axes of body frame are 0.01°/h with white noise 0.005°/h; the accelerometer biases along three axes of body frame are 1×10^{-4} g with white noise 0.5×10^{-4} g; for DVL, the velocity offset error $\delta V_d = 0.01$ m/s, the drift angle error $\delta\Delta = 1'$, the calibration coefficient error $\delta C = 0.001$, the correlation time of DVL velocity offset error and bias angle error $\beta_d^{-1} = 5$ min, $\beta_\Delta^{-1} = 15$ min. The measurement data are obtained from IMU at a rate of 100 Hz and from DVL at a rate of 1 Hz. The filtering period is 1 s and the simulation time is 6284 s.

The initial parameters are set as:

$$\hat{\boldsymbol{X}}_0 = \begin{bmatrix} 0 & 0 & 0 & 0 & 0 & 0 & 0 & 0 & 0 & 0 \end{bmatrix}^T$$

$$\boldsymbol{P}_0 = diag\left\{ (100/R)^2 \quad (100/R)^2 \quad (0.1\,\text{m/s})^2 \quad (0.1\,\text{m/s})^2 \quad (1°)^2 \quad (1°)^2 \quad (1°)^2 \quad (0.005\,\text{m/s})^2 \quad (1')^2 \quad (0.001)^2 \right\}$$

$$\boldsymbol{Q}_0 = diag\left\{ 0 \quad 0 \quad (50\,\text{ug})^2 \quad (50\,\text{ug})^2 \quad (0.005°/\text{h})^2 \quad (0.005°/\text{h})^2 \quad (0.005°/\text{h})^2 \quad q\delta V_d \quad q\delta\Delta \quad 0 \right\}$$

$$\boldsymbol{R}_0 = diag\left\{ (0.01\,\text{m/s})^2 \quad (0.01\,\text{m/s})^2 \right\}$$

where $q\delta V_d = (0.005\,\mathrm{m/s})^2(1 - e^{-2\beta_d T})$, $q\delta\Delta = (1')^2(1 - e^{-2\beta_\Delta T})$.

5.1.1. Uniform Linear Motion

The system is in uniform linear motion at 10 m/s with an initial heading angle of 45°, and the simulation time is 24 h. Figures 2–5 show the position and velocity estimation error curves for purely inertial navigation, filtering with EKF, UKF, SR-UIF. Table 2 gives the estimation error values.

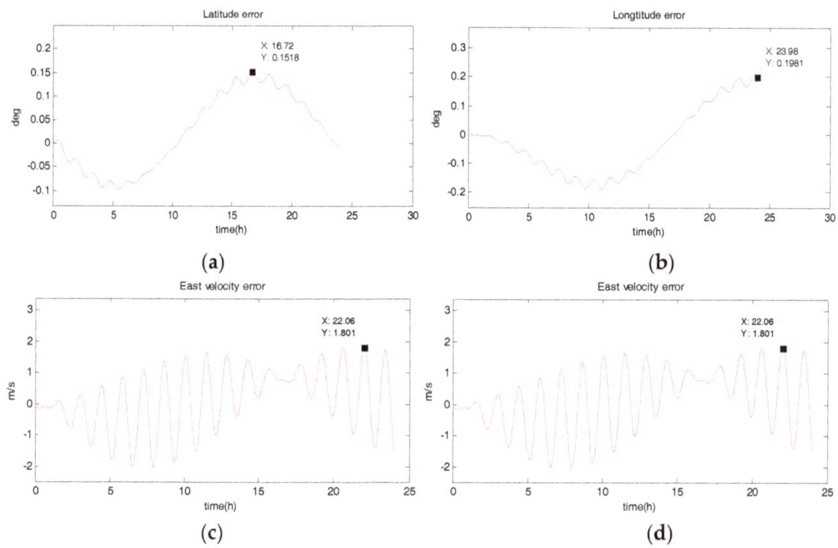

Figure 2. Estimation error curves for purely SINS, where X axis represents simulation time, and Y axis represents: (**a**) latitude error; (**b**) longitude error; (**c**) east velocity error; (**d**) north velocity error.

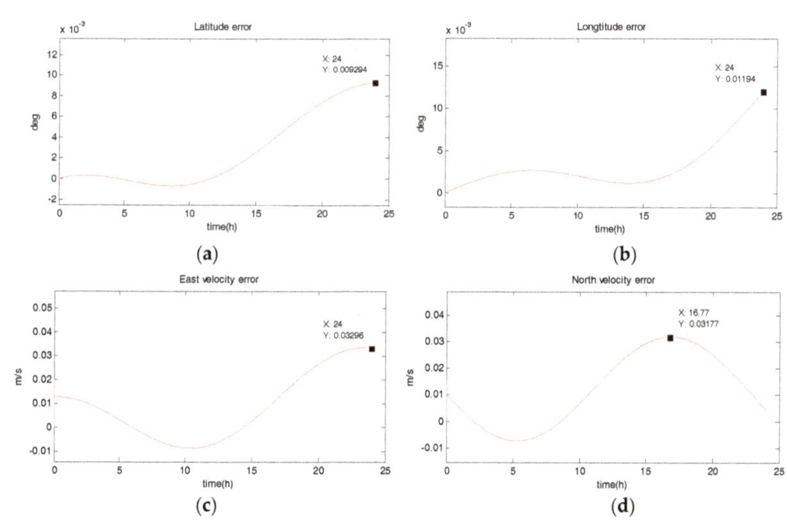

Figure 3. Estimation error curves after EKF, where X axis represents simulation time, and Y axis represents: (**a**) latitude error; (**b**) longitude error; (**c**) east velocity error; (**d**) north velocity error.

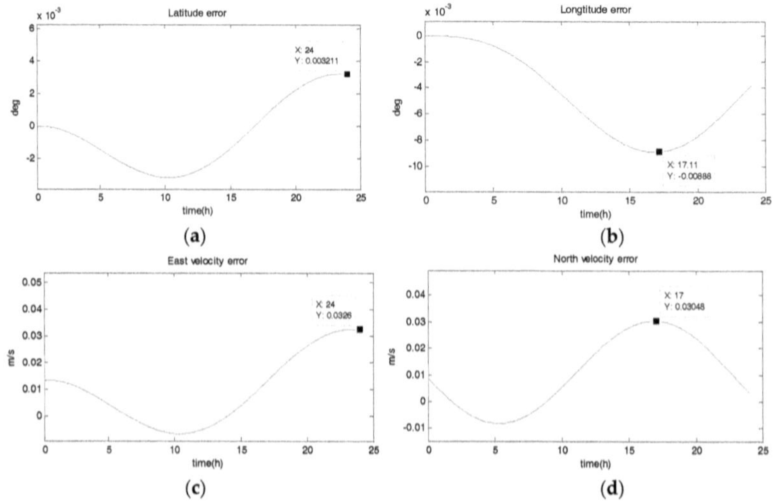

Figure 4. Estimation error curves after UKF, where X axis represents simulation time, and Y axis represents: (**a**) latitude error; (**b**) longitude error; (**c**) east velocity error; (**d**) north velocity error.

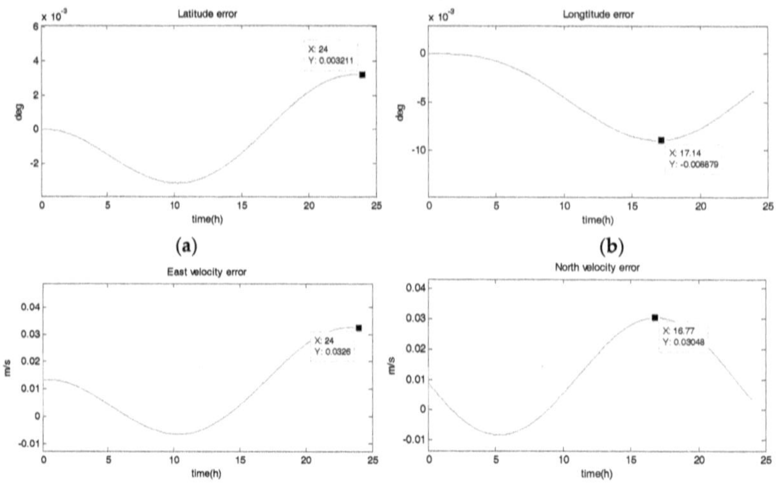

Figure 5. Estimation error curves after SR-UIF, where X axis represents simulation time, and Y axis represents: (**a**) latitude error; (**b**) longitude error; (**c**) east velocity error; (**d**) north velocity error.

Table 2. Comparison of position and velocity errors for purely inertial navigation, filtering with EKF, UKF, SR-UIF in uniform linear motion.

Filtering Method	Maximum Latitude Error (deg)	Maximum Longitude Error (deg)	Maximum East Velocity Error (m/s)	Maximum North Velocity Error (m/s)
purely SINS	0.1518	0.1981	1.801	−2.517
EKF	9.294×10^{-3}	1.194×10^{-2}	0.03296	0.03177
UKF	3.211×10^{-3}	8.888×10^{-3}	0.0326	0.03048
SR-UIF	3.211×10^{-3}	8.879×10^{-3}	0.0326	0.03048

From the above simulation results in the uniform linear motion, it can be seen that:

(1) The position and velocity errors for purely inertial navigation output diverge with time. The maximum latitude error and longitude error after 24 h are 0.1518° and 0.1981°, meaning 16,888 m and 23,000 m. Meanwhile, the maximum east velocity error and north velocity error reach 1.801 m/s and −2.517 m/s, respectively.

(2) The position estimation errors in SINS/DVL integrated navigation are suppressed after filtering. The maximum latitude error and longitude error after EKF filtering are 9.294×10^{-30} and 1.194×10^{-20}, meaning 720 m and 1322 m. The position estimation accuracy of UKF and SR-UIF algorithm is approximately the same, both higher than that of EKF algorithm. Specifically, the maximum latitude error and longitude error after SR-UIF filtering are 3.211×10^{-30} and 8.879×10^{-20}, meaning 249 m and 986 m.

(3) The velocity estimation errors in SINS/DVL integrated navigation are also suppressed after filtering. The maximum velocity error can be maintained in 10^{-2} order of magnitude, and the velocity estimation accuracy of UKF or SR-UIF algorithm is slightly higher than that of EKF.

5.1.2. Uniform Circular Motion

The system has a radius of 10 km, a velocity of 10 m/s, an angular rate of $\omega = 10^{-3}$ rad/s, and performs a uniform circular motion counterclockwise. The simulation results are shown in the following figure. Figures 6–9 show the position and velocity estimation error curves for purely inertial navigation, filtering with EKF, UKF, SR-UIF. Table 3 gives the estimation error values.

Table 3. Comparison of position and velocity errors for purely inertial navigation, filtering with EKF, UKF, SR-UIF in uniform circular motion.

Filtering Method	Maximum Latitude Error (deg)	Maximum Longitude Error (deg)	Maximum East Velocity Error (m/s)	Maximum North Velocity Error (m/s)
purely SINS	0.2792	−0.2703	−7.544	12.88
EKF	-3.594×10^{-4}	-5.332×10^{-4}	0.0178	0.2058
UKF	9.561×10^{-5}	1.3254×10^{-5}	−0.003045	0.003076
SR-UIF	9.561×10^{-5}	1.3235×10^{-5}	−0.003043	0.003071

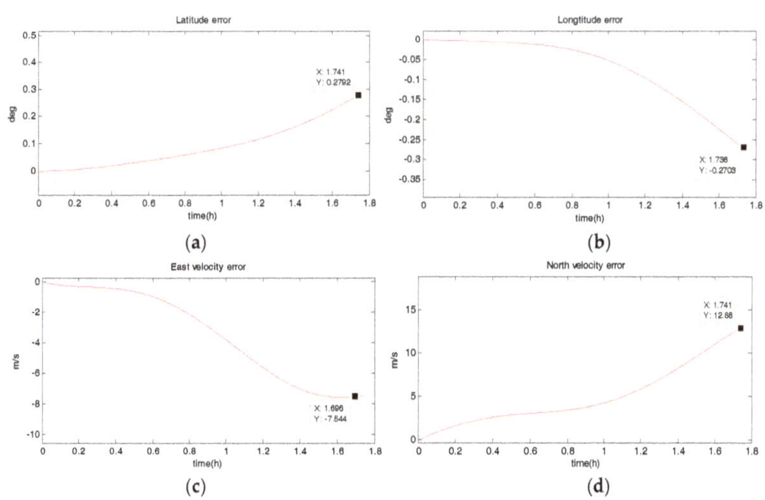

Figure 6. Estimation error curves for purely SINS, where X axis represents simulation time, and Y axis represents: (**a**) latitude error; (**b**) longitude error; (**c**) east velocity error; (**d**) north velocity error.

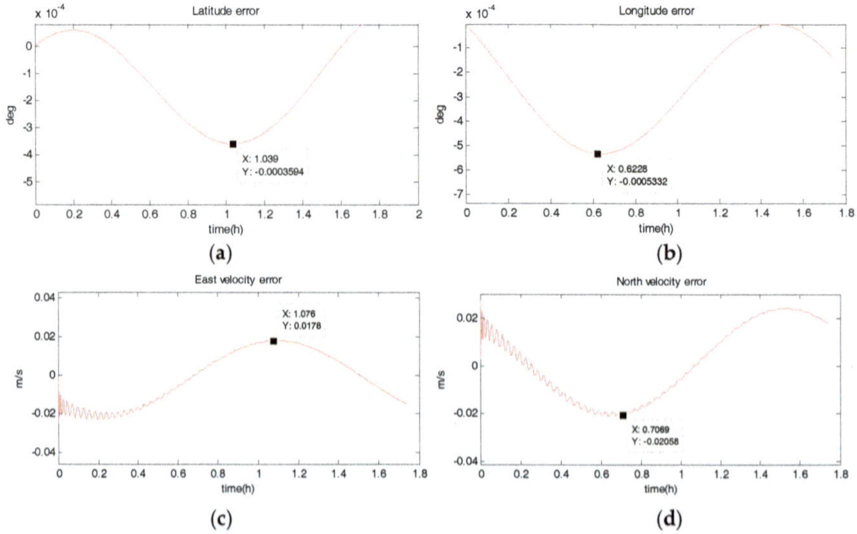

Figure 7. Estimation error curves after EKF, where X axis represents simulation time, and Y axis represents: (**a**) latitude error; (**b**) longitude error; (**c**) east velocity error; (**d**) north velocity error.

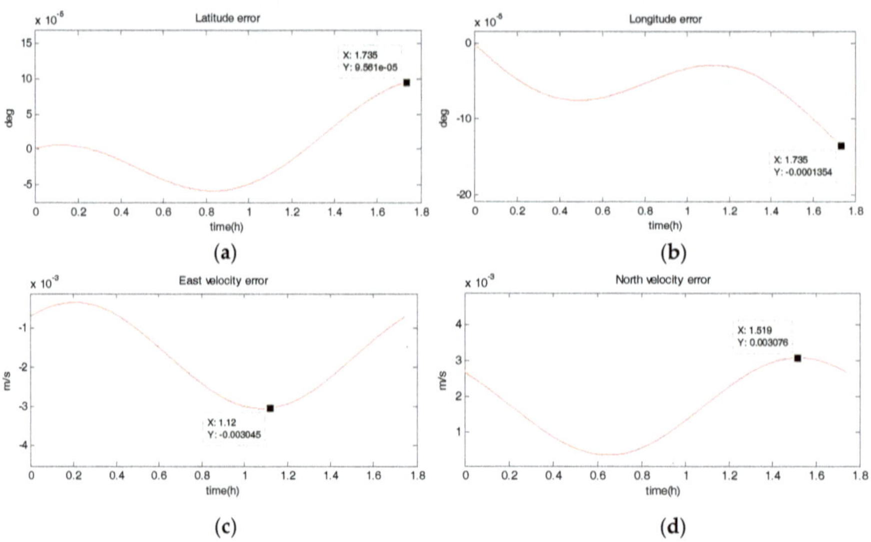

Figure 8. Estimation error curves for purely SINS, where X axis represents simulation time, and Y axis represents: (**a**) latitude error; (**b**) longitude error; (**c**) east velocity error; (**d**) north velocity error.

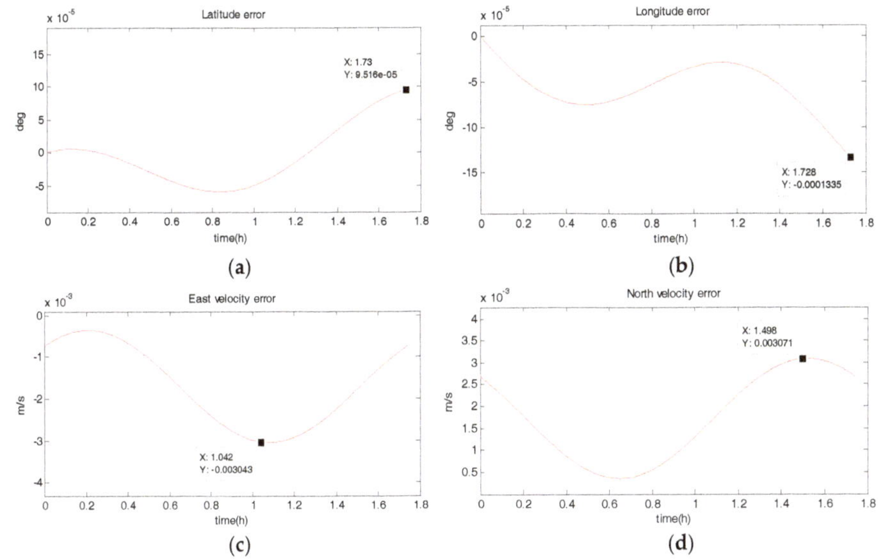

Figure 9. Estimation error curves after EKF, where X axis represents simulation time, and Y axis represents: (**a**) latitude error; (**b**) longitude error; (**c**) east velocity error; (**d**) north velocity error.

From the above simulation results, it can be seen that:

(1) Compared to the uniform linear motion, the uniform circular motion is more complex. Then the accuracy of pure inertial navigation is lower in the uniform circular motion for a short time, of which the maximum latitude error and longitude error are 0.2792° and −0.2684°, meaning 31,056 m and 30,302 m; and the maximum east velocity error and north velocity error reach −7.477 m/s and 12.88 m/s, respectively.

(2) The position accuracy is obviously improved after filtering, the maximum latitude error and longitude error after EKF filtering are -3.594×10^{-40} and -5.332×10^{-40}, meaning 28 m and 59 m. The position estimation accuracy of UKF and SR-UIF algorithm is approximately the same, which can be controlled within the same order of magnitude and both higher than that of EKF algorithm.

(3) The velocity error can converge to a small range. The maximum velocity error after EKF filtering can be maintained in 10^{-2} order of magnitude, and the initial filtering stage has obvious oscillatory process. The velocity estimation accuracy of UKF and SR-UIF algorithm advance by an order of magnitude compared to that of EKF, which can be maintained in 10^{-3} order of magnitude and the entire filtering process is smooth.

5.1.3. Performance Analysis

In order to analyze the superiority of SR-UIF algorithm performance, the simulation effect diagram of the time consumed by once filtering operation with SR-UIF and UKF in uniform linear motion or uniform circular motion is given in Figure 10.

We simulate multiple uniform linear motion and uniform circular motion experiments to calculate the average time consumed by once filtering operation with SR-UIF and UKF, and the total elapsed time of each simulation experiment. The results are shown in the following Table 4.

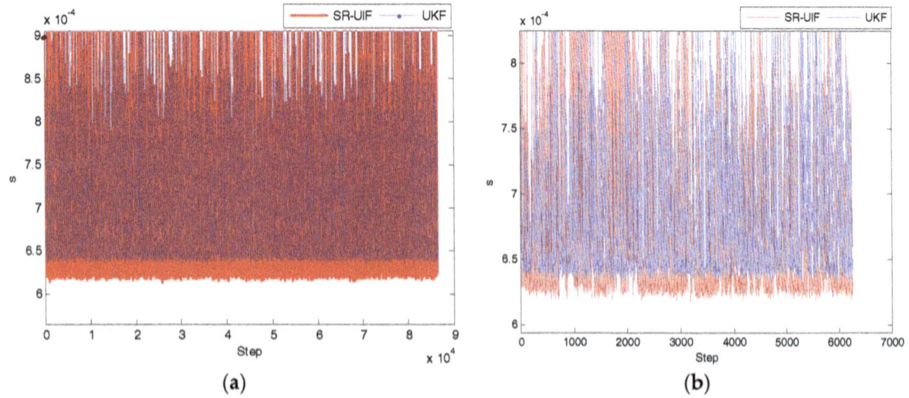

Figure 10. Time comparison consumed by once filtering operation with SR-UIF and UKF: (**a**) in uniform linear motion; (**b**) in uniform circular motion, where X axis represents simulation iterative step, and Y axis represents the time consumed by once filtering operation.

Table 4. Comparison of the average time consumed by once filtering operation, and the total elapsed time of each simulation experiment with UKF, SR-UIF in uniform linear motion or uniform circular motion.

Filtering Method	Uniform Linear Motion (24 h)		Uniform Circular Motion (1.67 h)	
	Average Time Consumed by Once Filtering Operation (s)	Total Elapsed Time of Each Simulation Experiment (s)	Average Time Consumed by Once Filtering Operation (s)	Total Elapsed Time of Each Simulation Experiment (s)
UKF	6.413×10^{-4}	55.40832	6.396×10^{-4}	4.0083732
SR-UIF	6.212×10^{-4}	53.67168	6.195×10^{-4}	3.8824065

From the above simulation results, it can be seen that whether the system is in a uniform linear motion or a uniform circular motion, the average time consumed by once filtering operation and the total elapsed time of each simulation experiment with SR-UIF algorithm are slightly smaller than with the UKF algorithm, which indicates that in view of the face that SR-UIF algorithm adopts the most suitable parameter form in each filtering stage, the computation amount of filtering is less than that of UKF algorithm.

5.2. Measured Data Analysis

Based on a ship-borne experiment using laboratory fiber optic gyro (FOG) strapdown inertial navigation equipment and a RDI Workhorse-type DVL instrument, the accuracy of the SR-UIF algorithm is analyzed. Figure 11 shows the velocity and position error curves of SINS/DVL integrated navigation system using SR-UIF algorithm and EKF algorithm. It can be seen that in the one-hour data analysis process, the latitude error and longitude error only reach -1.5×10^{-40} and -1×10^{-40} with SR-UIF algorithm, while these errors reach -2.8×10^{-40} and -2.3×10^{-40} with EKF algorithm. Meanwhile, the velocity error with SR-UIF algorithm has converged to within 0.02 m/sand remains stable, while within 0.1 m/s with EKF algorithm. The above results show that the positioning accuracy and velocity accuracy of SR-UIF algorithm are higher than that of EKF algorithm.

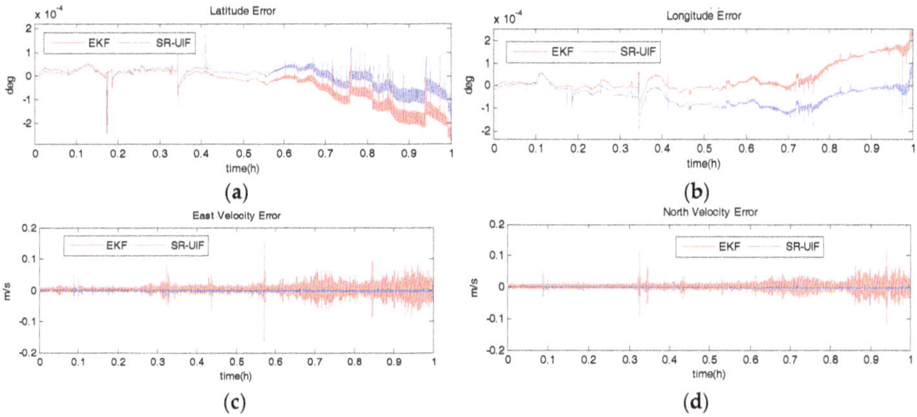

Figure 11. Error curves of SINS/DVL integrated navigation system with SR-UIF and EKF, where X axis represents simulation time, and Y axis represents: (**a**) latitude error; (**b**) longitude error; (**c**) east velocity error; (**d**) north velocity error.

The estimation accuracy and the time consumed for data analysis of SR-UIF algorithm and UKF algorithm are given below in different data collection time, as shown in Tables 5–7.

Table 5. Performance comparison of SR-UIF and UKF in one hour collection data.

Filtering Method	Maximum Latitude Error (deg)	Maximum Longitude Error (deg)	Maximum East Velocity Error (m/s)	Maximum North Velocity Error (m/s)	Total Elapsed Time of Data Analysis (s)
UKF	-2.707×10^{-5}	1.081×10^{-4}	−0.02059	0.05312	1.841076
SR-UIF	-2.693×10^{-4}	9.264×10^{-5}	−0.02052	0.05304	1.834968

Table 6. Performance comparison of SR-UIF algorithm and UKF algorithm in one three collection data.

Filtering Method	Maximum Latitude Error (deg)	Maximum Longitude Error (deg)	Maximum East Velocity Error (m/s)	Maximum North Velocity Error (m/s)	Total Elapsed Time of Data Analysis (s)
UKF	-1.055×10^{-3}	1.103×10^{-3}	0.02574	0.05312	3.294395
SR-UIF	-9.494×10^{-4}	1.0843×10^{-3}	0.02569	0.05304	3.267851

Table 7. Performance comparison of SR-UIF algorithm and UKF algorithm in one five collection data.

Filtering Method	Maximum Latitude Error (deg)	Maximum Longitude Error (deg)	Maximum East Velocity Error (m/s)	Maximum North Velocity Error (m/s)	Total Elapsed Time of Data Analysis (s)
UKF	-1.888×10^{-3}	1.501×10^{-3}	−0.03409	0.05312	4.7329875
SR-UIF	-1.795×10^{-3}	1.446×10^{-3}	−0.03272	0.05304	4.7236778

As can be seen from the data in Tables 5–7, the positioning accuracy and velocity accuracy of the SR-UIF algorithm are nearly identical as with the UKF algorithm, but for the total elapsed time in the same collection time, the SR-UIF algorithm is significantly lower than the UKF algorithm.

5.3. Discussion

The simulation results and the measured data analysis results show that:

(1) The position and velocity estimation accuracy of the nonlinear filtering based on the probabilistic approximation (such as UKF algorithm and SR-UIF algorithm) is higher than that of EKF algorithm based on the model Taylor series expansion.
(2) SR-UIF algorithm and UKF algorithm are different expressions of the same filtering algorithm based on the probabilistic approximation, and they are equivalent in the filtering estimation value due to the duality of Gaussian distribution. The difference is that SR-UIF algorithm adopts the most suitable parameter form in each filtering stage, making its computational complexity lower than that of UKF algorithm. The simulation results are reflected that the average time consumed by once filtering operation and the total elapsed time of each experiment with SR-UIF algorithm are slightly smaller than UKF algorithm.

6. Conclusions and Future Works

In this paper, a square-root unscented information filter (SR-UIF) algorithm is applied to a SINS/DVL integrated navigation system. It is shown that: (1) the algorithm based on probability approximation has a relatively high estimation accuracy, and (2) the performance advantage of the algorithm is optimized by utilizing the most suitable parameter form in each filtering stage, and 3) using the square root of the variance matrix as the iterative factor to ensure the symmetry and positive definite of the information matrix or covariance matrix and thereby enhance the stability of the filtering. Finally, the simulation experiments show that: (1) the positioning and velocity measurement accuracy of SR-UIF algorithm are obviously higher than that of EKF algorithm, equaling to that of UKF algorithm; (2) and the computational complexity of SR-UIF algorithm is slightly lower than that of UKF algorithm. The simulation results provide a good theoretical basis and solution for popularizing the application of SINS/DVL integrated navigation filtering algorithm.

What we touched on in this paper is just the beginning, and there are many places where SR-UIF algorithm can wait for mining applications. The SR-UIF algorithm adopts the form of moment parameters in the process of time updating and the form of information parameters in the process of measurement updating, which makes it convenient to realize distributed or decentralized design of measurement data structure. The potential benefits of this data structure are: (1) improving the fault tolerance of the algorithm by detecting the accuracy of data measured from different sensors; (2) fusing the time unsynchronized measurement information by processing the distributed data. At the same time, the application scene of SR-UIF algorithm can also be transferred from underwater navigation of SINS/DVL to the cooperative navigation system on the ground or in the air, which will also become a direction of the future SR-UIF extension application.

Author Contributions: Y.G. and M.W. conceived and designed the experiments; K.T. and L.Z. preformed the experiments; Y.G. and L.Z. analyzed the data; M.W. and K.T. contributed some materials; and Y.G. wrote the paper.

Funding: This research received the funding of "Hunan Provincial Science and Technology Plan Project", innovation platform and talent plan (leading talent), contract No. 2017RS3045; and is supported by the National Key R&D Program of China, contract No. 2016YFC0303002.

Acknowledgments: This work was co-funded by the "Remote Sensing Observation Technology for High frequency Subsequent Unmanned Aircraft Regional Networking" in the National Key Research and Development Program of Chinaon Earth observation and navigation item. The authors would like to thank all the editors and anonymous reviewers for improving this article.

Conflicts of Interest: The authors declare no conflict of interest.

Appendix A

A. Verify that the update variance of the information matrix at $k+1$ time is:

$$\hat{Y}_k = Y_{k/k-1} + I_k \tag{A1}$$

where $I_k = (Y_{k/k-1} P_{xz,k/k-1}) R_{k-1}^{-1} (Y_{k/k-1} P_{xz,k/k-1})^T$.

Proof. The approximate estimate form of the covariance matrix $\boldsymbol{P}_k$ based on the linear minimum variance is:

$$\boldsymbol{P}_k = \boldsymbol{P}_{k/k-1} - \boldsymbol{P}_{xz,k/k-1}\boldsymbol{P}_{zz,k/k-1}^{-1}\boldsymbol{P}_{xz,k/k-1}^{T} \tag{A2}$$

Perform inverse calculations on both sides of Equation (A2),

$$\boldsymbol{P}_k^{-1} = \left(\boldsymbol{P}_{k/k-1} - \boldsymbol{P}_{xz,k/k-1}\boldsymbol{P}_{zz,k/k-1}^{-1}\boldsymbol{P}_{xz,k/k-1}^{T}\right)^{-1}$$

According to the matrix inversion formula, it can be obtained:

$$\begin{aligned}\boldsymbol{P}_k^{-1} &= \left(\boldsymbol{P}_{k/k-1} - \boldsymbol{P}_{xz,k/k-1}\boldsymbol{P}_{zz,k/k-1}^{-1}\boldsymbol{P}_{xz,k/k-1}^{T}\right)^{-1} \\ &= \boldsymbol{P}_{k/k-1}^{-1} + \boldsymbol{P}_{k/k-1}^{-1}\boldsymbol{P}_{xz,k/k-1}\left(\boldsymbol{P}_{zz,k/k-1} - \boldsymbol{P}_{xz,k/k-1}^{T}\boldsymbol{P}_{k/k-1}^{-1}\boldsymbol{P}_{xz,k/k-1}\right)\boldsymbol{P}_{xz,k/k-1}^{T}\boldsymbol{P}_{k/k-1}^{-1}\end{aligned} \tag{A3}$$

Simultaneously it exits:

$$\begin{aligned}\boldsymbol{P}_{xz,k/k-1} &= \sum_{i=0}^{m} W_i^c \boldsymbol{X}_{i,k|k-1}^{*}\boldsymbol{Z}_{i,k|k-1}^{*T} - \hat{\boldsymbol{x}}_{k/k-1}\hat{\boldsymbol{z}}_{k/k-1}^{T} \\ \boldsymbol{P}_{k/k-1} &= \sum_{i=0}^{m} W_i^c \boldsymbol{X}_{i,k|k-1}^{*}\boldsymbol{X}_{i,k|k-1}^{*T} - \hat{\boldsymbol{x}}_{k/k-1}\hat{\boldsymbol{x}}_{k/k-1}^{T} + \boldsymbol{Q}_{k-1}\end{aligned}$$

Since $\boldsymbol{X}$, $\boldsymbol{Z}$ and W are not related to each other, then

$$\begin{aligned}&\boldsymbol{P}_{xz,k/k-1}^{T}\boldsymbol{P}_{k/k-1}^{-1}\boldsymbol{P}_{xz,k/k-1} \\ &= \left(\sum_{i=1}^{m} W_i^c \boldsymbol{Z}_{i,k/k-1}^{*}\boldsymbol{X}_{i,k/k-1}^{*T} - \hat{\boldsymbol{z}}_{k/k-1}\hat{\boldsymbol{x}}_{k/k-1}^{T}\right)\cdot\left(\sum_{i=1}^{m} W_i^c X_{i,k/k-1}^{*}\boldsymbol{X}_{i,k/k-1}^{*T} - \hat{\boldsymbol{x}}_{k/k-1}\hat{\boldsymbol{x}}_{k/k-1}^{T} + \boldsymbol{Q}_{k-1}\right)^{-1}\cdot\left(\sum_{i=1}^{m} W_i^c \boldsymbol{X}_{i,k/k-1}^{*}\boldsymbol{Z}_{i,k/k-1}^{*T} - \hat{\boldsymbol{x}}_{k/k-1}\hat{\boldsymbol{z}}_{k/k-1}^{T}\right) \\ &= \left[W_1^c \boldsymbol{Z}_{1,k/k-1}^{*}\boldsymbol{X}_{i,k/k-1}^{*T}\left(\boldsymbol{X}_{1,k/k-1}^{*T}\right)^{-1}\left(\boldsymbol{X}_{1,k/k-1}^{*}\right)^{-1}\boldsymbol{X}_{1,k/k-1}^{*}\boldsymbol{Z}_{1,k/k-1}^{*T}\right]\cdots\left[W_m^c Z_{m,k/k-1}^{*}\boldsymbol{X}_{m,k/k-1}^{*T}\left(\boldsymbol{X}_{m,k/k-1}^{*T}\right)^{-1}\left(\boldsymbol{X}_{m,k/k-1}^{*}\right)^{-1}\boldsymbol{X}_{m,k/k-1}^{*}\boldsymbol{Z}_{m,k/k-1}^{*T}\right] \\ &\quad -\hat{\boldsymbol{z}}_{k/k-1}\hat{\boldsymbol{x}}_{k/k-1}^{T}\left(\hat{\boldsymbol{x}}_{k/k-1}^{T}\right)^{-1}\left(\hat{\boldsymbol{x}}_{k/k-1}\right)^{-1}\hat{\boldsymbol{x}}_{k/k-1}\hat{\boldsymbol{z}}_{k/k-1}^{T} \\ &= \sum_{i=1}^{m} W_1^c \boldsymbol{Z}_{i,k/k-1}^{*}\boldsymbol{Z}_{i,k/k-1}^{*T} - \hat{\boldsymbol{z}}_{k/k-1}\hat{\boldsymbol{z}}_{k/k-1}^{T}\end{aligned}$$

While

$$\boldsymbol{P}_{zz,k/k-1} = \sum_{i=1}^{m} W_1^c \boldsymbol{Z}_{i,k/k-1}^{*}\boldsymbol{Z}_{i,k/k-1}^{*T} - \hat{\boldsymbol{z}}_{k/k-1}\hat{\boldsymbol{z}}_{k/k-1}^{T} + \boldsymbol{R}_k$$

then,

$$\boldsymbol{P}_{zz,k/k-1} - \boldsymbol{P}_{xz,k/k-1}^{T}\boldsymbol{P}_{k/k-1}^{-1}\boldsymbol{P}_{xz,k/k-1} = \boldsymbol{R}_k \tag{A4}$$

Substitute Equation (A4) into Equation (A3), and due to $\boldsymbol{P}_k^{-1} = \boldsymbol{Y}_k$ and $\boldsymbol{P}_{k/k-1}^{-1} = \boldsymbol{Y}_{k/k-1}$, it can be get:

$$\boldsymbol{Y}_k = \boldsymbol{Y}_{k/k-1} + (\boldsymbol{Y}_{k/k-1}\boldsymbol{P}_{xz,k/k-1})\boldsymbol{R}_k^{-1}(\boldsymbol{Y}_{k/k-1}\boldsymbol{P}_{xz,k/k-1})^{T}$$

□

Appendix B

B. Verify that the update variance of the information vector at $k+1$ time is:

$$\hat{\boldsymbol{y}}_k = \boldsymbol{y}_{k/k-1} + \boldsymbol{i}_k \tag{A5}$$

where $\boldsymbol{i}_k = (\boldsymbol{Y}_{k/k-1}\boldsymbol{P}_{xz,k/k-1})\boldsymbol{R}_{k-1}^{-1}\left(\boldsymbol{v}_k + \boldsymbol{P}_{xz,k/k-1}^{T}\hat{\boldsymbol{y}}_{k/k-1}\right)$.

Proof. The approximate estimate form of the gain matrix $\boldsymbol{K}_k$ based on the linear minimum variance is:

$$\boldsymbol{K}_k = \boldsymbol{P}_{xz,k/k-1}\boldsymbol{P}_{zz,k/k-1}^{-1} = \boldsymbol{Y}_{k/k-1}^{-1}\boldsymbol{P}_{yz,k/k-1}\boldsymbol{P}_{zz,k/k-1}^{-1} \tag{A6}$$

Meanwhile the state vector estimate is:

$$\begin{aligned}\hat{\boldsymbol{x}}_k &= \hat{\boldsymbol{x}}_{k/k-1} + \boldsymbol{K}_k(\boldsymbol{Z}_k - \hat{\boldsymbol{z}}_{k/k-1}) \\ &= \hat{\boldsymbol{x}}_{k/k-1} + \boldsymbol{Y}_{k/k-1}^{-1}\boldsymbol{P}_{yz,k/k-1}\boldsymbol{P}_{zz,k/k-1}^{-1}(\boldsymbol{Z}_k - \hat{\boldsymbol{z}}_{k/k-1})\end{aligned} \tag{A7}$$

Multiply $\boldsymbol{Y}_{k/k-1}$ on both sides of Equation (A2), namely

$$\boldsymbol{Y}_{k/k-1}\hat{\boldsymbol{x}}_k = \boldsymbol{Y}_{k/k-1}\hat{\boldsymbol{x}}_{k/k-1} + \boldsymbol{P}_{yz,k/k-1}\boldsymbol{P}_{zz,k/k-1}^{-1}(\boldsymbol{Z}_k - \hat{\boldsymbol{z}}_{k/k-1}) \tag{A8}$$

Substitute Equation (A1) into Equation (A8), it can be obtained:

$$(\boldsymbol{Y}_k - \boldsymbol{I}_k)\hat{\boldsymbol{x}}_k = \boldsymbol{Y}_{k/k-1}\hat{\boldsymbol{x}}_{k/k-1} + \boldsymbol{P}_{yz,k/k-1}\boldsymbol{P}_{zz,k/k-1}^{-1}(\boldsymbol{Z}_k - \hat{\boldsymbol{z}}_{k/k-1})$$

Due to $\boldsymbol{Y}_{k/k-1}\hat{\boldsymbol{x}}_{k/k-1} = \hat{\boldsymbol{y}}_{k/k-1}$ and $\boldsymbol{Y}_k\hat{\boldsymbol{x}}_k = \hat{\boldsymbol{y}}_k$, then

$$\begin{aligned}\hat{\boldsymbol{y}}_k &= \hat{\boldsymbol{y}}_{k/k-1} + \boldsymbol{I}_k\hat{\boldsymbol{x}}_k + \boldsymbol{P}_{yz,k/k-1}\boldsymbol{P}_{zz,k/k-1}^{-1}(\boldsymbol{Z}_k - \hat{\boldsymbol{z}}_{k/k-1}) \\ &= \hat{\boldsymbol{y}}_{k/k-1} + (\boldsymbol{Y}_{k/k-1}\boldsymbol{P}_{xz,k/k-1})\boldsymbol{R}_k^{-1}(\boldsymbol{Y}_{k/k-1}\boldsymbol{P}_{xz,k/k-1})^T\hat{\boldsymbol{x}}_k + \boldsymbol{P}_{yz,k/k-1}\boldsymbol{P}_{zz,k/k-1}^{-1}(\boldsymbol{Z}_k - \hat{\boldsymbol{z}}_{k/k-1}) \\ &= \hat{\boldsymbol{y}}_{k/k-1} + (\boldsymbol{Y}_{k/k-1}\boldsymbol{P}_{xz,k/k-1})\boldsymbol{R}_k^{-1}(\boldsymbol{Y}_{k/k-1}\boldsymbol{P}_{xz,k/k-1})^T\hat{\boldsymbol{x}}_k + (\boldsymbol{Y}_{k/k-1}\boldsymbol{P}_{xz,k/k-1})\boldsymbol{P}_{zz,k/k-1}^{-1}(\boldsymbol{Z}_k - \hat{\boldsymbol{z}}_{k/k-1}) \\ &= \hat{\boldsymbol{y}}_{k/k-1} + (\boldsymbol{Y}_{k/k-1}\boldsymbol{P}_{xz,k/k-1})\boldsymbol{R}_k^{-1}\left[(\boldsymbol{Y}_{k/k-1}\boldsymbol{P}_{xz,k/k-1})^T\hat{\boldsymbol{x}}_k + \boldsymbol{R}_k\boldsymbol{P}_{zz,k/k-1}^{-1}(\boldsymbol{Z}_k - \hat{\boldsymbol{z}}_{k/k-1})\right] \\ &= \hat{\boldsymbol{y}}_{k/k-1} + (\boldsymbol{Y}_{k/k-1}\boldsymbol{P}_{xz,k/k-1})\boldsymbol{R}_k^{-1}\left[(\boldsymbol{Y}_{k/k-1}\boldsymbol{P}_{xz,k/k-1})^T\left(\hat{\boldsymbol{x}}_{k/k-1} + \boldsymbol{Y}_{k/k-1}^{-1}\boldsymbol{P}_{yz,k/k-1}\boldsymbol{P}_{zz,k/k-1}^{-1}(\boldsymbol{Z}_k - \hat{\boldsymbol{z}}_{k/k-1})\right) + \boldsymbol{R}_k\boldsymbol{P}_{zz,k/k-1}^{-1}(\boldsymbol{Z}_k - \hat{\boldsymbol{z}}_{k/k-1})\right] \\ &= \hat{\boldsymbol{y}}_{k/k-1} + (\boldsymbol{Y}_{k/k-1}\boldsymbol{P}_{xz,k/k-1})\boldsymbol{R}_k^{-1}\left\{(\boldsymbol{Y}_{k/k-1}\boldsymbol{P}_{xz,k/k-1})^T\hat{\boldsymbol{x}}_{k/k-1} + \left[(\boldsymbol{Y}_{k/k-1}\boldsymbol{P}_{xz,k/k-1})^T\boldsymbol{Y}_{k/k-1}^{-1}\boldsymbol{Y}_{k/k-1}\boldsymbol{P}_{xz,k/k-1} + \boldsymbol{R}_k\right]\boldsymbol{P}_{zz,k/k-1}^{-1}(\boldsymbol{Z}_k - \hat{\boldsymbol{z}}_{k/k-1})\right\} \\ &= \hat{\boldsymbol{y}}_{k/k-1} + (\boldsymbol{Y}_{k/k-1}\boldsymbol{P}_{xz,k/k-1})\boldsymbol{R}_k^{-1}\left[\boldsymbol{P}_{xz,k/k-1}^T\hat{\boldsymbol{y}}_{k/k-1} + \left[\boldsymbol{P}_{xz,k/k-1}^T\boldsymbol{P}_{k/k-1}^{-1}\boldsymbol{P}_{xz,k/k-1} + \boldsymbol{R}_k\right]\boldsymbol{P}_{zz,k/k-1}^{-1}(\boldsymbol{Z}_k - \hat{\boldsymbol{z}}_{k/k-1})\right]\end{aligned}$$

According to Equation (A8), further simplification can be obtained:

$$\hat{\boldsymbol{y}}_k = \hat{\boldsymbol{y}}_{k/k-1} + (\boldsymbol{Y}_{k/k-1}\boldsymbol{P}_{xz,k/k-1})\boldsymbol{R}_k^{-1}\left[\boldsymbol{P}_{xz,k/k-1}^T\hat{\boldsymbol{y}}_{k/k-1} + \boldsymbol{v}_k\right]$$

□

References

1. Chen, G.; Li, K.; Wang, W.; Li, P. A novel redundant INS based on triple rotary inertial measurement units. *Meas. Sci. Technol.* **2016**, *27*, 105102. [CrossRef]
2. Nourmohammadi, H.; Keighobadi, J. Fuzzy adaptive integration scheme for low-cost SINS/GPS navigation system. *Mech. Syst. Signal Process.* **2018**, *99*, 434–449. [CrossRef]
3. Liu, Y.; Fan, X.; Lv, C.; Wu, J.; Li, L. An innovative information fusion method with adaptive Kalman filter for integrated INS/GPS navigation of autonomous vehicles. *Mech. Syst. Signal Process.* **2017**, *100*, 605–616. [CrossRef]
4. Yao, Y.; Xu, X.; Xu, X. An IMM-Aided ZUPT Methodology for an INS/DVL Integrated Navigation System. *Sensors* **2017**, *17*, 2030. [CrossRef]
5. Alahyari, A.; Rozbahani, S.G.; Habibzadeh, A.; Alahyari, R.; Dousti, M. INS/DVL positioning system using Kalman filter. *Aust. J. Basic Appl. Sci.* **2011**, *5*, 1123–1129.
6. Zhang, Y.; Guo, Y.; Yang, T.; Li, C.; Wang, Z. A novel separation and calibration method for DVL and compass error in dead reckoning navigation systems. *Meas. Sci. Technol.* **2016**, *27*, 065003. [CrossRef]
7. Kang, Y.; Zhao, L.; Cheng, J.; Wu, M.; Fan, X. A Novel Grid SINS/DVL Integrated Navigation Algorithm for Marine Application. *Sensors* **2018**, *18*, 364. [CrossRef] [PubMed]
8. Liu, P.; Wang, B.; Deng, Z.; Fu, M. A Correction Method for DVL Measurement Errors by Attitude Dynamics. *IEEE Sens. J.* **2017**, *17*, 4628–4638. [CrossRef]

9. Tal, A.; Klein, I.; Katz, R. Inertial Navigation System/Doppler Velocity Log (INS/DVL) Fusion with Partial DVL Measurements. *Sensors* **2017**, *17*, 415. [CrossRef] [PubMed]
10. Luo, C.; Mcclean, S.I.; Parr, G.; Teacy, L.; Nardi, R.D. UAV Position Estimation and Collision Avoidance Using the Extended Kalman Filter. *IEEE Trans. Veh. Technol.* **2013**, *62*, 2749–2762. [CrossRef]
11. Zhao, L.Y.; Liu, X.J.; Wang, L.; Zhu, Y.H.; Liu, X.X. A Pretreatment Method for the Velocity of DVL Based on the Motion Constraint for the Integrated SINS/DVL. *Appl. Sci.* **2016**, *6*, 79. [CrossRef]
12. Uhlmann, J.K. *Simultaneous Map Building and Localization for Real Time Applications*; Technical Report; University of Oxford: Oxford, UK, 1994.
13. Julier, S.J.; Uhlmann, J.K. Unscented filtering and nonlinear estimation. *Proc. IEEE* **2004**, *92*, 401–422. [CrossRef]
14. Julier, S.J.; Uhlmann, J.K. Reduced sigma point filters for propagation of means and covariances through nonlinear transformations. In Proceedings of the 2002 American Control Conference (IEEE Cat. No.CH37301), Anchorage, AK, USA, 8–10 May 2002; pp. 887–892.
15. Julier, S.; Uhlmann, J.; Durrantwhyte, H.F. A new method for the nonlinear transformation of means and covariances in filters and estimators. *IEEE Trans. Automat. Control* **2001**, *45*, 477–482. [CrossRef]
16. Li, W.; Wang, J.; Lu, L.; Wu, W. A Novel Scheme for DVL-Aided SINS In-Motion Alignment Using UKF Techniques. *Sensors* **2013**, *13*, 1046–1063. [CrossRef] [PubMed]
17. Tang, K.; Wang, J.; Li, W.; Wu, W. A novel INS and Doppler sensors calibration method for long range underwater vehicle navigation. *Sensors* **2013**, *13*, 14583–14600. [CrossRef] [PubMed]
18. Karimi, M.; Bozorg, M.; Khayatian, A.R. A comparison of DVL/INS fusion by UKF and EKF to localize an autonomous underwater vehicle. In Proceedings of the 2013 First RSI/ISM International Conference on Robotics and Mechatronics (ICRoM), Tehran, Iran, 13–15 February 2013; pp. 62–67.
19. Assimakis, N.; Adam, M.; Douladiris, A. Information Filter and Kalman Filter Comparison: Selection of the Faster Filter. *Int. J. Inf. Eng.* **2012**, *2*, 1–5.
20. Webster, S.E.; Walls, J.M.; Whitcomb, L.L.; Eustice, R.M. Decentralized Extended Information Filter for Single-Beacon Cooperative Acoustic Navigation: Theory and Experiments. *IEEE Trans. Robot.* **2013**, *29*, 957–974. [CrossRef]
21. Li, S.S.; Xu, Y.H.; Xu, X.L.; Zhu, N.H. Simulation Research on Multi-Robot SLAM of Information Filter. *Appl. Mech. Mater.* **2013**, *278–280*, 660–663. [CrossRef]
22. Zhang, H.; He, B.; Luan, N. Sparse Extended Information Filter for AUV SLAM: Insights into the Optimal Sparse Time. *Appl. Mech. Mater.* **2013**, *427–429*, 1670–1673. [CrossRef]

sensors

MDPI

Article

Kalman Filtering for Attitude Estimation with Quaternions and Concepts from Manifold Theory

Pablo Bernal-Polo * and Humberto Martínez-Barberá

Department of Information and Communication Engineering, University of Murcia, 30100 Murcia, Spain; humberto@um.es

* Correspondence: pablo.bernal.polo@gmail.com

Received: 5 December 2018; Accepted: 27 December 2018; Published: 3 January 2019

Abstract: The problem of attitude estimation is broadly addressed using the Kalman filter formalism and unit quaternions to represent attitudes. This paper is also included in this framework, but introduces a new viewpoint from which the notions of "multiplicative update" and "covariance correction step" are conceived in a natural way. Concepts from manifold theory are used to define the moments of a distribution in a manifold. In particular, the mean and the covariance matrix of a distribution of unit quaternions are defined. Non-linear versions of the Kalman filter are developed applying these definitions. A simulation is designed to test the accuracy of the developed algorithms. The results of the simulation are analyzed and the best attitude estimator is selected according to the adopted performance metric.

Keywords: attitude; orientation; estimation; Kalman filter; quaternion; manifold

---

1. Introduction

Mechanical state estimation of a vehicle is a field of interest. A vehicle is considered a rigid body, and its state of motion is represented by 4 mathematical objects: two of them represent its position and velocity, and the other two represent its orientation, and angular velocity. This paper is focused on the estimation of the angular state, composed of orientation, and angular velocity.

Although there are other mathematical tools used for estimation [1], the Kalman Filter [2] has become the algorithm par excellence in this area. Because of its simplicity, the rigor and elegance in its mathematical derivation, and its recursive nature it is very attractive for many practical applications. Its non-linear versions have been widely used in orientation estimation: the Extended Kalman Filter (EKF), and the Unscented Kalman Filter (UKF) [3]. However, there are problems arising from the used parametrization to represent the orientation.

The orientation of a system is represented by the rotation transformation that relates two reference frames: a reference frame anchored to that system, and an external reference frame. A thorough survey of attitude representations is provided in Reference [4]. The parametrization used to represent the rotation transformation could be singular, or present discontinuities among others. Table 1 summarizes the main characteristics of the most used parametrizations.

Having in mind that the *special orthogonal group* SO(3) has dimension three, ideally we would seek for a continuous and non-singular representation expressed by 3 parameters. However, since 1964 we know that "...it is topologically impossible to have a global 3-dimensional parametrization without singular points for the rotation group" [5]. Knowing this, we would not be wrong to say that unit quaternions are the most convenient representation we have, and that we will have for orientations. In Reference [6] the literature on attitude estimation is reviewed until 1982, when other parametrizations like Euler angles were common, and founds the basis of modern quaternion-based attitude estimation, in which this paper is supported. After that work, many others have explored this viewpoint, and have demonstrated its superiority [7–12].

Table 1. Main characteristics of the most used parametrizations to represent an orientation.

Representation	Parameters	Continuous	Non-Singular	Linear Evolution Equation
Euler angles	3	✗	✗	✗
Axis-angle	3–4	✗	✗	✗
Rotation matrix	9	✓	✓	✓
Unit quaternion	4	✓	✓	✓

Quaternions are 4-dimensional entities, but only those having unit norm represent a rotation transformation. This fact implies a problem in applying the ordinary Kalman Filter, so different approaches have emerged. Since a quaternion is of dimension 4, one tends to think at first on a 4×4 covariance matrix, and in the direct application of the Kalman Filter [13]. Given that all predictions are contained in the surface defined by the unit constraint, the covariance matrix shrinks in the orthogonal direction to this surface, which leads to a singular covariance matrix after several updates. A second perspective was firstly approached in Reference [6] and was after named as "Multiplicative Extended Kalman Filter" [8,11,12]. In this second approach we define an "error-quaternion" that is transformed to a 3-vector. We use this vector to build the covariance matrix, and we talk about a "3×3 representation of the quaternion covariance matrix". However, there are still details in this adaptation that are currently being developed. Namely, the "covariance correction step" [14].

This paper presents a new viewpoint on the problem of attitude estimation using Kalman filters when the orientation is represented by unit quaternions. Noticing that unit quaternions live in a manifold (the unit sphere in $\mathbb{R}^4$), we use basic concepts from manifold theory to define the mean and covariance matrix of a distribution of unit quaternions. With these definitions we develop two estimators based on the Kalman filter (one EKF-based and another UKF-based) arriving at the concepts of "multiplicative update" and "covariance correction step" in a natural and satisfying way. The inartificial emergence of these ideas establishes a solid foundation for the development of general navigation algorithms. Lastly, we also analyze the accuracy in the estimations of these two estimators using simulations.

The organization of this paper is as follows. In Section 2 we review quaternion basics. We also expose the new viewpoint on the definition of the quaternion mean and covariance matrix. In Section 3 we present the developed estimation algorithms. In Section 4 we define the performance metric, describe the simulation scheme, and present the results of the simulations. We also discuss the results. Finally, Section 5 concludes the paper.

2. Quaternions Describing Orientations

2.1. Quaternions

Quaternions are hypercomplex numbers composed of a real part and an imaginary part. The imaginary part is expressed using three different imaginary units $\{i, j, k\}$ satisfying the Hamilton axiom:

$$i^2 = j^2 = k^2 = i * j * k = -1. \tag{1}$$

A quaternion q can be represented with 4 real numbers, and using several notations:

$$\begin{aligned} q &= q_0 + q_1\, i + q_2\, j + q_3\, k \equiv & \text{(2a)} \\ &\equiv (q_0, q_1, q_2, q_3)^T \equiv & \text{(2b)} \\ &\equiv (q_0, \mathbf{q})^T. & \text{(2c)} \end{aligned}$$

We will denote quaternions with bold italic symbols ($\boldsymbol{q}$), while vectors will be denoted with bold upright symbols ($\mathbf{q}$). Vectors will be written in matrix form, and the transposed of a matrix $\mathbf{M}$ will be denoted as $\mathbf{M}^T$.

Quaternion product is defined by Equation (1) which produces the multiplication rule

$$\boldsymbol{p} * \boldsymbol{q} = \begin{pmatrix} p_0\, q_0 - \mathbf{p} \cdot \mathbf{q} \\ p_0\, \mathbf{q} + q_0\, \mathbf{p} + \mathbf{p} \times \mathbf{q} \end{pmatrix}, \tag{3}$$

where $(\cdot)$ represents the usual *dot product*, and $(\times)$ represents the 3-vector *cross product*. Note that the quaternion product $(*)$ is different from the product denoted by $(\otimes)$ in reference [4]. Given this multiplication rule, the inverse of a quaternion $\boldsymbol{q}$ (the one for which $\boldsymbol{q} * \boldsymbol{q}^{-1} = \boldsymbol{q}^{-1} * \boldsymbol{q} = 1$) is given by

$$\boldsymbol{q}^{-1} = \frac{1}{\|\boldsymbol{q}\|^2}\, \boldsymbol{q}^* = \frac{1}{\|\boldsymbol{q}\|^2}\, (q_0, -\mathbf{q})^T, \tag{4}$$

where $\boldsymbol{q}^*$ represents the *complex conjugate quaternion*. Note that if $\boldsymbol{q}$ is a unit quaternion (a quaternion with $\|\boldsymbol{q}\| = 1$), then $\boldsymbol{q}^{-1} = \boldsymbol{q}^*$.

2.2. Quaternions Representing Rotations

Each rotation transformation is mapped with a rotation matrix $\mathbf{R}$, and with two unit quaternions $\boldsymbol{q}$ and $-\boldsymbol{q}$ all of them related through

$$\mathbf{R}(\boldsymbol{q}) = \begin{pmatrix} 1 - 2q_2^2 - 2q_3^2 & 2\,(q_1\, q_2 - q_3\, q_0) & 2\,(q_1\, q_3 + q_2\, q_0) \\ 2\,(q_1\, q_2 + q_3\, q_0) & 1 - 2q_1^2 - 2q_3^2 & 2\,(q_2\, q_3 - q_1\, q_0) \\ 2\,(q_1\, q_3 - q_2\, q_0) & 2\,(q_2\, q_3 + q_1\, q_0) & 1 - 2q_1^2 - 2q_2^2 \end{pmatrix}. \tag{5}$$

Note that $\mathbf{R}(\boldsymbol{q}) = \mathbf{R}(-\boldsymbol{q})$.

Quaternions representing rotations have the form

$$\boldsymbol{q} = \left(\cos(\theta/2)\,,\ \hat{\mathbf{q}}\, \sin(\theta/2) \right)^T, \tag{6}$$

where $\hat{\mathbf{q}}$ denotes the unit vector that defines the rotation axis, and θ the angle of rotation. Having this form, they satisfy the restriction

$$\|\boldsymbol{q}\|^2 = q_0^2 + q_1^2 + q_2^2 + q_3^2 = 1. \tag{7}$$

This means that quaternions describing rotations live in the unit sphere of $\mathbb{R}^4$, S^3. This space is a manifold, and some concepts regarding these mathematical objects are useful in our context. In particular, the concept of chart is of special interest.

2.3. Distributions of Unit Quaternions

When dealing with the Kalman filter, the distribution of a random variable $\mathbf{x}$ is encoded by its mean $\overline{\mathbf{x}} = \mathrm{E}[\,\mathbf{x}\,]$ and its covariance matrix $\mathbf{P}$ defined as

$$\mathbf{P} = \mathrm{E}\Big[\, (\mathbf{x} - \overline{\mathbf{x}})\, (\mathbf{x} - \overline{\mathbf{x}})^T \,\Big]. \tag{8}$$

This definition makes sense when our random variables are defined in the Euclidean space. But how do we define the covariance matrix of a random variable living in a manifold like ours? How can we define the covariance for unit quaternions if $\boldsymbol{q} - \overline{\boldsymbol{q}}$ does not represent a rotation? (Unit quaternions form a group under multiplication, but not under addition. This means that the addition of two unit quaternions may not result in another unit quaternion. Therefore, the addition of two unit quaternions may not represent a rotation.) What would be the covariance matrix if each quaternion was equiprobable in the unit sphere? We cannot redefine the covariance matrix, because the Kalman filter uses this precise form in its derivations, but we can take advantage of the properties of a manifold. Let us retrieve some important definitions:

Definition 1 (Homeomorphism). *A* homeomorphism *is a function $f : X \to Y$ between two topological spaces X and Y satisfying the next properties:*

- *f is a bijection,*
- *f is continuous,*
- *its inverse function f^{-1} is continuous.*

If such a function exists, we say that X and Y are homeomorphic.

Definition 2 (Manifold). *A n-*manifold *M^n is a topological space in which each point is locally homeomorphic to the Euclidean space $\mathbb{R}^n$. This is, each point $x \in M^n$ has a neighborhood $N \subset M^n$ for which we can define a homeomorphism $f : N \to B_n$, with B_n the unit ball of $\mathbb{R}^n$.*

Definition 3 (Chart). *A* chart *for a manifold M^n, is a homeomorphism φ from an open subset $U \subset M^n$, to an open subset of the Euclidean space $V \subset \mathbb{R}^n$. This is, a chart is a function*

$$\varphi : U \subset M^n \to V \subset \mathbb{R}^n, \tag{9}$$

with φ a homeomorphism. Traditionally, a chart is expressed as the pair (U, φ).

Given these definitions we can continue our reasoning.

In Reference [8] it talks about four "attitude error representations". Namely, the one we will call *Orthographic* (O), the *Rodrigues Parameters* (RP), the *Modified Rodrigues Parameters* (MRP), and the *Rotation Vector* (RV). The first three are what we know as *stereographic projections* (and are called *Orthographic*, *Gnomonic*, and *Stereographic* respectively). The last one is a projection called *Equidistant*. But all four are charts defining a homeomorphism from the manifold S^3 to the Euclidean space $\mathbb{R}^3$. This is, they map a point $\boldsymbol{q}$ in the manifold with a point $\mathbf{e}$ in $\mathbb{R}^3$. Table 2 arranges these chart definitions, together with their domain and image. We must ensure the charts to be bijections so that they properly define a homeomorphism, and that they do not map $\boldsymbol{q}$ and $-\boldsymbol{q}$ with different points of $\mathbb{R}^3$ since they represent the same rotation. We achieve this by the given definitions of the domain and image for each chart.

Figure 1 shows how points in the sphere S^2 (subspace of the sphere S^3, where quaternions live) are mapped to points in $\mathbb{R}^2$ (subspace of $\mathbb{R}^3$, where the images of the charts are contained) through each one of the named charts. Since our charts are homeomorphisms, it is possible to invert the functions. Figure 2 shows how points from $\mathbb{R}^2$ are mapped to points in the manifold through the inverted charts. As pointed in Reference [8], all four charts provide the same second-order approximation for a point $\mathbf{e} \in \mathbb{R}^3$ near the origin, to a quaternion $\boldsymbol{q} \in S^3$:

$$\varphi^{-1}(\mathbf{e}) \approx \left(1 - \frac{\|\mathbf{e}\|^2}{8}, \frac{\mathbf{e}}{2}\right)^T. \tag{10}$$

Table 2. Main characteristics of the charts studied.

Chart	Domain	Image	$\mathbf{e} = \varphi(q)$	$q = \varphi^{-1}(\mathbf{e})$
O	$\{q \in S^3 : q_0 \geq 0\}$	$\{\mathbf{e} \in \mathbb{R}^3 : \lVert\mathbf{e}\rVert \leq 2\}$	$2\,\mathbf{q}$	$\begin{pmatrix} \sqrt{1 - \frac{\lVert\mathbf{e}\rVert^2}{4}} \\ \mathbf{e}/2 \end{pmatrix}$
RP	$\{q \in S^3 : q_0 > 0\}$	$\mathbb{R}^3$	$2\,\frac{\mathbf{q}}{q_0}$	$\frac{1}{\sqrt{4 + \lVert\mathbf{e}\rVert^2}} \begin{pmatrix} 2 \\ \mathbf{e} \end{pmatrix}$
MRP	$\{q \in S^3 : q_0 \geq 0\}$	$\{\mathbf{e} \in \mathbb{R}^3 : \lVert\mathbf{e}\rVert \leq 4\}$	$4\,\frac{\mathbf{q}}{1 + q_0}$	$\frac{1}{16 + \lVert\mathbf{e}\rVert^2} \begin{pmatrix} 16 - \lVert\mathbf{e}\rVert^2 \\ 8\,\mathbf{e} \end{pmatrix}$
RV	$\{q \in S^3 : q_0 \geq 0\}$	$\{\mathbf{e} \in \mathbb{R}^3 : \lVert\mathbf{e}\rVert \leq \pi\}$	$2\,\hat{\mathbf{q}}\,\arcsin(\lVert\mathbf{q}\rVert)$	$\begin{pmatrix} \cos\left(\frac{\lVert\mathbf{e}\rVert}{2}\right) \\ \hat{\mathbf{e}}\,\sin\left(\frac{\lVert\mathbf{e}\rVert}{2}\right) \end{pmatrix}$

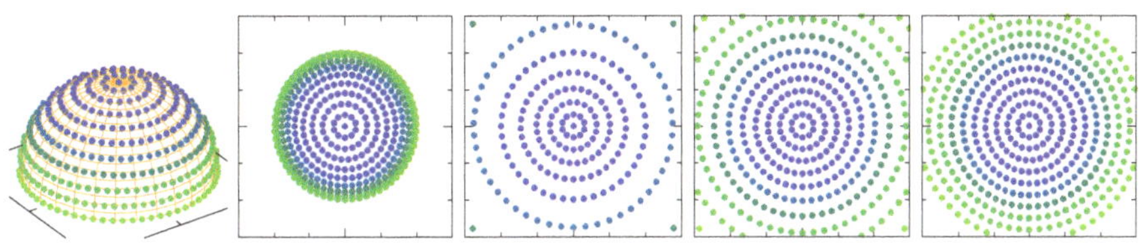

Figure 1. Points in the manifold with $q_3 = 0$ are mapped with points in the Euclidean space through each chart φ. (**a**) S^2; (**b**) O; (**c**) RP; (**d**) MRP; (**e**) RV.

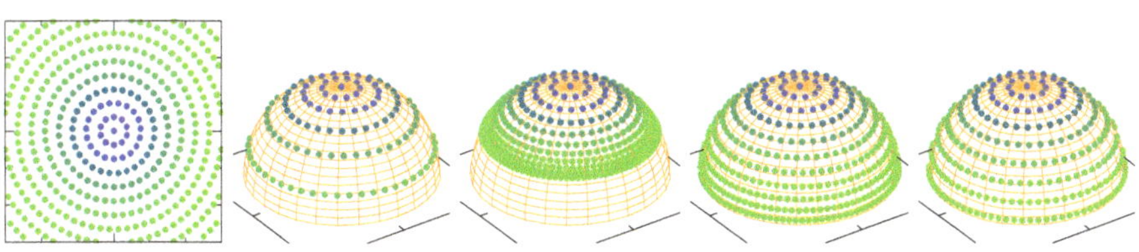

Figure 2. Points in the Euclidean space with $e_3 = 0$ are mapped with points in the manifold through each chart inverse φ^{-1}. (**a**) $\mathbb{R}^2$; (**b**) O; (**c**) RP; (**d**) MRP; (**e**) RV.

We should notice that having $\mathbb{R}^3$ and S^3 different metrics, a chart φ will inevitably produce a deformation of the space. However, for quaternions in the neighborhood of the identity quaternion (top of the sphere), our charts behave like the identity transformation between the imaginary part of these quaternions, and the points near the origin in $\mathbb{R}^3$, as suggested by (10). This is a desirable property, as this means that the space around the identity quaternion closely resembles the Euclidean space, which is the space for which the Kalman filter is designed. But this just happens in the neighborhood of the identity quaternion. However, we can extend this property for any quaternion $\overline{q} \in S^3$ noting that any quaternion $q \in S^3$ can be expressed as a "deviation" from the first one through the quaternion product:

$$q = \overline{q} * \delta^{\overline{q}}, \tag{11}$$

where $\delta^{\overline{q}}$ represents such a deviation. (This definition is arbitrary: we could have chosen to relate the quaternions through $q = \delta^{\overline{q}} * \overline{q}$, but it is important to establish one of these definitions, and then be consequent with it. However, (11) entails a computational advantage for the computation of (37).) Then, we define a chart $\varphi_{\overline{q}}$ for each quaternion $\overline{q} \in S^3$ as

$$\mathbf{e}^{\overline{q}} = \varphi_{\overline{q}}(q) = \varphi(\delta^{\overline{q}}), \tag{12}$$

where $\delta^{\overline{q}} = \overline{q}^* * q$, and where we have denoted the point of the Euclidean space mapped with the quaternion $q \in S^3$ through the chart $\varphi_{\overline{q}}$ as $\mathbf{e}^{\overline{q}}$. Then, we will have a set of charts $\{\varphi_{\overline{q}}\}_{\overline{q}}$, each one resembling the Euclidean space around a quaternion $\overline{q} \in S^3$, and mapping this last quaternion to the origin of $\mathbb{R}^3$. We will refer to the Euclidean space associated with the chart $\varphi_{\overline{q}}$ as the $\overline{q}$-centered chart. Thus, the homeomorphism $\varphi_{\overline{q}}^{-1}$ takes a point $\mathbf{e}^{\overline{q}}$ in the $\overline{q}$-centered chart and maps it to a point q in the manifold through

$$q = \varphi_{\overline{q}}^{-1}(\mathbf{e}^{\overline{q}}) = \overline{q} * \varphi^{-1}(\mathbf{e}^{\overline{q}}). \quad (13)$$

After reviewing these concepts, we can define the covariance matrix of a distribution of unit quaternions.

Given a unit quaternion $\overline{q}$ and a chart φ, we will define the expected value of a distribution of unit quaternions in the $\overline{q}$-centered chart as

$$\overline{\mathbf{e}}^{\overline{q}} = \mathrm{E}[\mathbf{e}^{\overline{q}}], \quad (14)$$

and its covariance matrix as

$$\mathbf{P}^{\overline{q}} = \mathrm{E}\Big[(\mathbf{e}^{\overline{q}} - \overline{\mathbf{e}}^{\overline{q}})(\mathbf{e}^{\overline{q}} - \overline{\mathbf{e}}^{\overline{q}})^T\Big], \quad (15)$$

and the probability density of each unit quaternion q would be defined through the homeomorphism $q = \varphi_{\overline{q}}^{-1}(\mathbf{e}^{\overline{q}})$. Then, a distribution of unit quaternions needs of four mathematical objects to be encoded: $(\varphi, \overline{q}, \overline{\mathbf{e}}^{\overline{q}}, \mathbf{P}^{\overline{q}})$. Although a distribution of unit quaternions is unique, given this definition, its expected value $\overline{\mathbf{e}}^{\overline{q}}$ and its covariance matrix $\mathbf{P}^{\overline{q}}$ may take different values depending on the chosen quaternion $\overline{q}$ and chart φ. However, knowing that the Kalman filter is designed for the Euclidean space, it will be convenient to choose a unit quaternion $\overline{q}$ central in the distribution, in order that the manifold space around it closely resembles the most significant region for the covariance matrix in the $\overline{q}$-centered chart. It is particularly convenient to choose a quaternion $\overline{q}$ such that $\overline{\mathbf{e}}^{\overline{q}} = \mathbf{0}$, so that the covariance matrix is centered in the origin of the $\overline{q}$-centered chart.

2.4. Transition Maps

At some step of the Kalman filter, we will have a distribution of unit quaternions defined in a $\overline{q}$-centered chart, and we will be interested in expressing our distribution in another $\overline{p}$-centered chart. The concept of transition map is relevant for this purpose.

Definition 4 (Transition map). *Given two charts $(U_\alpha, \varphi_\alpha)$ and (U_β, φ_β) for a manifold M, with $U_{\alpha\beta} = U_\alpha \cap U_\beta \neq \emptyset$, we can define a function $\varphi_{\alpha\beta} : \varphi_\alpha(U_{\alpha\beta}) \to \varphi_\beta(U_{\alpha\beta})$ as*

$$\varphi_{\alpha\beta}(x) = \varphi_\beta\left(\varphi_\alpha^{-1}(x)\right), \quad (16)$$

with $x \in \varphi_\alpha(U_{\alpha\beta})$. The function $\varphi_{\alpha\beta}$ is called a transition map. *Being that φ_α and φ_β are homeomorphisms, so is $\varphi_{\alpha\beta}$.*

For the present case, let us consider two unit quaternions $\overline{p}$ and $\overline{q}$ both related through

$$\overline{p} = \overline{q} * \overline{\delta}. \quad (17)$$

These two quaternions define the charts $\varphi_{\overline{p}}$ and $\varphi_{\overline{q}}$. We build the transition map that relates a point $\mathbf{e}^{\overline{q}}$ expressed in the $\overline{q}$-centered chart with a point $\mathbf{e}^{\overline{p}}$ expressed in the $\overline{p}$-centered chart doing

$$\mathbf{e}^{\overline{p}} = \varphi_{\overline{p}}\left(\varphi_{\overline{q}}^{-1}(\mathbf{e}^{\overline{q}})\right) = \tag{18a}$$

$$= \varphi\left(\overline{p}^* * \overline{q} * \varphi^{-1}(\mathbf{e}^{\overline{q}})\right) = \tag{18b}$$

$$= \varphi\left(\overline{\delta}^* * \varphi^{-1}(\mathbf{e}^{\overline{q}})\right). \tag{18c}$$

That is to say, first we take the point $\mathbf{e}^{\overline{q}}$ in the $\overline{q}$-centered chart, and we obtain its associated quaternion q in the manifold using $\varphi_{\overline{q}}^{-1}$. Then, we transform this quaternion q to a point $\mathbf{e}^{\overline{p}}$ in the $\overline{p}$-centered chart. Nevertheless, knowing the quaternion $\overline{\delta}$ we do not need to explicitly compute q. In fact, being able to express the same quaternion q as two different deviations,

$$\left.\begin{aligned} q &= \overline{q} * \delta^{\overline{q}} \\ q &= \overline{p} * \delta^{\overline{p}} \end{aligned}\right\} \implies \delta^{\overline{p}} = \underbrace{\overline{p}^* * \overline{q}}_{\overline{\delta}^*} * \delta^{\overline{q}}. \tag{19}$$

Note the equivalence of expressions (18c) and (19).

Table 3 displays the transition maps for the charts studied. The detailed derivations of these transition maps can be found in Appendix A. Figure 3 attempts to provide some insight into how points are transformed through the transition map of each chart.

Table 3. Transition maps for the charts studied.

Chart	Transition Map $\mathbf{e}^{\overline{p}}(\mathbf{e}^{\overline{q}})$
O	$\overline{\delta}_0\,\mathbf{e}^{\overline{q}} - \sqrt{4 - \lVert\mathbf{e}^{\overline{q}}\rVert^2}\,\overline{\boldsymbol{\delta}} - \overline{\boldsymbol{\delta}} \times \mathbf{e}^{\overline{q}}$
RP	$2\,\dfrac{\overline{\delta}_0\,\mathbf{e}^{\overline{q}} - 2\overline{\boldsymbol{\delta}} - \overline{\boldsymbol{\delta}} \times \mathbf{e}^{\overline{q}}}{2\overline{\delta}_0 + \overline{\boldsymbol{\delta}} \cdot \mathbf{e}^{\overline{q}}}$
MRP	$4\,\dfrac{8\overline{\delta}_0\,\mathbf{e}^{\overline{q}} - (16 - \lVert\mathbf{e}^{\overline{q}}\rVert^2)\,\overline{\boldsymbol{\delta}} - 8\,\overline{\boldsymbol{\delta}} \times \mathbf{e}^{\overline{q}}}{16 + \lVert\mathbf{e}^{\overline{q}}\rVert^2 + \overline{\delta}_0\,(16 - \lVert\mathbf{e}^{\overline{q}}\rVert^2) + 8\,\overline{\boldsymbol{\delta}} \cdot \mathbf{e}^{\overline{q}}}$
RV	$2\,\dfrac{\boldsymbol{\delta}^{\overline{p}}}{\lVert\boldsymbol{\delta}^{\overline{p}}\rVert} \arcsin\lVert\boldsymbol{\delta}^{\overline{p}}\rVert$, with $\boldsymbol{\delta}^{\overline{p}} = \overline{\delta}_0\,\hat{\mathbf{e}}^{\overline{q}} \sin\left(\frac{\lVert\mathbf{e}^{\overline{q}}\rVert}{2}\right) - \cos\left(\frac{\lVert\mathbf{e}^{\overline{q}}\rVert}{2}\right)\overline{\boldsymbol{\delta}} - \overline{\boldsymbol{\delta}} \times \hat{\mathbf{e}}^{\overline{q}} \sin\left(\frac{\lVert\mathbf{e}^{\overline{q}}\rVert}{2}\right)$

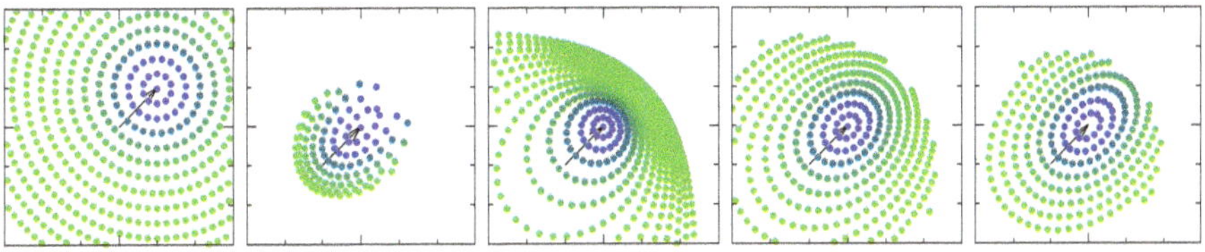

(a) (b) (c) (d) (e)

Figure 3. Points in a $\overline{q}$-centered chart are transformed using the transition map defined by each chart, and travel to the chart centered in the quaternion mapped with $\mathbf{e}^{\overline{q}} = (1,1,0)^T$ in the previous chart. **(a)** $\mathbb{R}^2$; **(b)** O; **(c)** RP; **(d)** MRP; **(e)** RV.

3. Manifold Kalman Filters

In this section we present the models adopted for the Manifold Kalman Filters (MKF), and we display the resulting algorithms.

The state of the system at a time t is defined by an orientation, encoded with a unit quaternion q_t, and by an angular velocity ω'_t. We will consider them to be random variables, and we will try to estimate their value using a Kalman filter.

Our unit quaternions $\{q_t \in \mathbb{H} : \|q_t\| = 1\}$ will define the rotation transformation that relates a vector $\mathbf{v}'_t$ expressed in a reference frame $\mathcal{S}'$ attached to the solid whose state we want to describe, with the same vector $\mathbf{v}_t$ expressed in an external reference frame $\mathcal{S}$:

$$\mathbf{v}_t = \mathbf{R}(q_t)\,\mathbf{v}'_t \quad \equiv \quad v_t = q_t * v'_t * q_t^*. \tag{20}$$

For example, if we measure an acceleration $\mathbf{a}'_t$ in reference frame $\mathcal{S}'$, the acceleration in the inertial reference frame $\mathcal{S}$ would be given by $\mathbf{a}_t = \mathbf{R}(q_t)\,\mathbf{a}'_t$. This acceleration would be the one that we would have to integrate to obtain the position estimated by an accelerometer.

The vector ω'_t will define the angular velocity of the solid measured in $\mathcal{S}'$. Note that we do not include the bias of the sensors in the state of our system. We will assume that our sensors are calibrated, so the biases are zero.

We can predict the value of the random variables that describe the state of our system through the following motion equations:

$$\frac{d\,\omega'(t)}{dt} = \mathbf{q}^{\omega}(t), \tag{21}$$

$$\frac{d\,q(t)}{dt} = \frac{1}{2}\,q(t) * \omega'(t) = \frac{1}{2}\,q(t) * \begin{pmatrix} 0 \\ \omega'(t) \end{pmatrix}, \tag{22}$$

where $\mathbf{q}^{\omega}(t)$ is a random variable that represents the process noise, and is associated with the torque acting on the system, and with its inertia tensor. Its expected value at a given time t will be denoted as $\overline{\mathbf{q}}_t^{\omega}$, and its covariance matrix will be denoted as $\mathbf{Q}_t^{\omega}$.

We will assume that we have sensors giving measurements of angular velocity ω_t^m (which provide information about the relative change in orientation), and of a vector $\mathbf{v}_t^m$ whose value $\mathbf{v}_t$, expressed in the external reference frame $\mathcal{S}$, is known (this provides information about absolute orientation). Examples of such sensors could be a gyroscope giving angular velocity measurements, an accelerometer measuring the gravity vector near the Earth surface ($\mathbf{v}_t := -\mathbf{g}$), or a magnetometer measuring the Earth magnetic field ($\mathbf{v}_t := \mathbf{B}$). The measurement model relates these measurements with the variables that describe the state of the system:

$$\mathbf{v}_t^m = \mathbf{R}^T(q_t)\,(\mathbf{q}_t^{\mathbf{v}} + \mathbf{v}_t) + \mathbf{r}_t^{\mathbf{v}}, \tag{23}$$

$$\omega_t^m = \omega'_t + \mathbf{r}_t^{\omega}, \tag{24}$$

where $\mathbf{r}_t^{\omega}$ and $\mathbf{r}_t^{\mathbf{v}}$ are random variables with zero mean and covariance matrices $\mathbf{R}_t^{\omega}$ and $\mathbf{R}_t^{\mathbf{v}}$ respectively that represent the measurement noises, and $\mathbf{q}_t^{\mathbf{v}}$ is another random variable with mean $\overline{\mathbf{q}}_t^{\mathbf{v}}$ and covariance matrix $\mathbf{Q}_t^{\mathbf{v}}$ representing external disturbances in the measurement of the vector $\mathbf{v}_t$. For example, it could represent accelerations others than gravity for an accelerometer, or magnetic disturbances produced by moving irons for a magnetometer.

We will assume that the measurements arrive at discrete times $\{t_n\}_n$. The format $x_{t|t_n}$ will be used to denote a variable x at a time t, having included measurements up to a time t_n with $t > t_n$. For the n-th time stamp, in which a measurement arrives, we will write $x_{t|n}$ for the sake of simplicity. Then, our knowledge about the state at a time t, having included measurements up to a time t_n with $t > t_n$, is described by a distribution encoded in the collection of mathematical objects $\left(\varphi, \overline{p}, \overline{\mathbf{x}}^{\overline{p}}_{t|n}, \mathbf{P}^{\overline{p}}_{t|n}\right)$ as described in Section 2.3. For the present case, $\overline{\mathbf{x}}^{\overline{p}}_{t|n} = \left(\overline{\mathbf{e}}^{\overline{p}}_{t|n}, \overline{\omega}'_{t|n}\right)^T$ is the expected value of the distribution, and $\mathbf{P}^{\overline{p}}_{t|n}$ is its 6×6 covariance matrix, both expressing the quaternion distribution in the $\overline{p}$-centered chart. Preferably, $\overline{p}$ will be a unit quaternion central in the distribution, so that the mapping of points from the $\overline{p}$-centered chart to the manifold causes minimal deformation in such distribution. The unit quaternion $\overline{q}_{t|n} = \varphi^{-1}_{\overline{p}}\left(\overline{\mathbf{e}}^{\overline{p}}_{t|n}\right)$ will be our best estimation of the real quaternion q_t that defines the orientation of the system with respect to the external reference frame $\mathcal{S}$ at time t.

The following subsections present the developed Kalman filters: one version based on the EKF and another version based on the UKF. The EKF is based on the linearization of the non-linear models to calculate the predicted covariance matrices. That is, the EKF approximates non-linear functions using their Jacobian matrices. To apply the EKF, our functions must be differentiable. On the other hand, the UKF is based on a deterministic sampling to approximate the distribution of our random variables. We select a minimal set of samples whose mean and covariance matrix are those of the state distribution. Then, they are transformed by the non-linear models, and the resulting set of points is used to compute the means and covariance matrices necessary to perform the Kalman update. This second approach does not need the functions to be differentiable.

3.1. Manifold Extended Kalman Filter

In this section we present the EKF-based estimator: the Manifold Extended Kalman Filter (MEKF). We offer here the main results of the more detailed derivation given in Appendix B.

A measurement

$$\mathbf{z}_n = \begin{pmatrix} \mathbf{v}^m_n \\ \omega^m_n \end{pmatrix} \tag{25}$$

arrives at time t_n. Our knowledge about the orientation at a previous time $t_{n\text{-}1}$ is described by a distribution expressed in the $\overline{q}_{n\text{-}1|n\text{-}1}$-centered chart. We assume that this distribution has mean

$$\overline{\mathbf{x}}^{\overline{q}_{n\text{-}1|n\text{-}1}}_{n\text{-}1|n\text{-}1} = \begin{pmatrix} \overline{\mathbf{e}}^{\overline{q}_{n\text{-}1|n\text{-}1}}_{n\text{-}1|n\text{-}1} = \mathbf{0} \\ \overline{\omega}'_{n\text{-}1|n\text{-}1} \end{pmatrix}, \tag{26}$$

and covariance matrix $\mathbf{P}^{\overline{q}_{n\text{-}1|n\text{-}1}}_{n\text{-}1|n\text{-}1}$. This is, we have an initial four

$$\left(\varphi, \overline{q}_{n\text{-}1|n\text{-}1}, \overline{\omega}'_{n\text{-}1|n\text{-}1}, \mathbf{P}^{\overline{q}_{n\text{-}1|n\text{-}1}}_{n\text{-}1|n\text{-}1}\right). \tag{27}$$

The state prediction at time t_n given all the information up to $t_{n\text{-}1}$ is computed through

$$\overline{\omega}'_{n|n\text{-}1} = \overline{\omega}'_{n\text{-}1|n\text{-}1}, \tag{28}$$

$$\delta^{\omega}_n = \begin{pmatrix} \cos\left(\frac{\|\overline{\omega}'_{n|n\text{-}1}\|\,\Delta t_n}{2}\right) \\ \frac{\overline{\omega}'_{n|n\text{-}1}}{\|\overline{\omega}'_{n|n\text{-}1}\|}\sin\left(\frac{\|\overline{\omega}'_{n|n\text{-}1}\|\,\Delta t_n}{2}\right) \end{pmatrix}, \tag{29}$$

$$\overline{q}_{n|n\text{-}1} = \overline{q}_{n\text{-}1|n\text{-}1} * \delta^{\omega}_n, \tag{30}$$

$$\mathbf{F}_n = \begin{pmatrix} \mathbf{R}^T(\delta^{\omega}_n) & \mathbf{I}\,\Delta t_n \\ \mathbf{0} & \mathbf{I} \end{pmatrix}, \tag{31}$$

$$\mathbf{P}^{\overline{q}_{n|n\text{-}1}}_{n|n\text{-}1} = \mathbf{F}_n \left[\mathbf{P}^{\overline{q}_{n\text{-}1|n\text{-}1}}_{n\text{-}1|n\text{-}1} + \mathbf{Q}_n\right] \mathbf{F}^T_n, \tag{32}$$

with

$$\mathbf{Q}_n = \begin{pmatrix} \mathbf{Q}^{\omega}_n \frac{(\Delta t_n)^3}{3} & -\mathbf{Q}^{\omega}_n \frac{(\Delta t_n)^2}{2} \\ -\mathbf{Q}^{\omega}_n \frac{(\Delta t_n)^2}{2} & \mathbf{Q}^{\omega}_n \Delta t_n \end{pmatrix}. \quad (33)$$

The measurement prediction at the same time is given by

$$\overline{\mathbf{v}}^m_{n|n\text{-}1} = \mathbf{R}^T(\overline{q}_{n|n\text{-}1})\,(\overline{\mathbf{q}}^{\mathbf{v}}_n + \mathbf{v}_n), \quad (34)$$

$$\overline{\omega}^m_{n|n\text{-}1} = \overline{\omega}'_{n|n\text{-}1}, \quad (35)$$

$$\overline{\mathbf{z}}_{n|n\text{-}1} = \begin{pmatrix} \overline{\mathbf{v}}^m_{n|n\text{-}1} \\ \overline{\omega}^m_{n|n\text{-}1} \end{pmatrix}, \quad (36)$$

$$\mathbf{H}_n = \begin{pmatrix} [\overline{\mathbf{v}}^m_{n|n\text{-}1}]_\times & \mathbf{0} \\ \mathbf{0} & \mathbf{I} \end{pmatrix}, \quad (37)$$

$$\mathbf{S}_{n|n\text{-}1} = \mathbf{H}_n\,\mathbf{P}^{\overline{q}_{n|n\text{-}1}}_{n|n\text{-}1}\,\mathbf{H}^T_n + \begin{pmatrix} \mathbf{R}^T(\overline{q}_{n|n\text{-}1})\,\mathbf{Q}^{\mathbf{v}}_n\,\mathbf{R}(\overline{q}_{n|n\text{-}1}) + \mathbf{R}^{\mathbf{v}}_n & \mathbf{0} \\ \mathbf{0} & \mathbf{R}^{\omega}_n \end{pmatrix}, \quad (38)$$

where $[\mathbf{v}]_\times$ stands for

$$[\mathbf{v}]_\times = \begin{pmatrix} 0 & -v_3 & v_2 \\ v_3 & 0 & -v_1 \\ -v_2 & v_1 & 0 \end{pmatrix}. \quad (39)$$

At this point, we compute the Kalman gain $\mathbf{K}_n$ and use it to obtain the optimal estimation of the state:

$$\mathbf{K}_n = \mathbf{P}^{\overline{q}_{n|n\text{-}1}}_{n|n\text{-}1}\,\mathbf{H}^T_n\,\mathbf{S}^{-1}_{n|n\text{-}1}, \quad (40)$$

$$\overline{\mathbf{x}}^{\overline{q}_{n|n\text{-}1}}_{n|n} = \overline{\mathbf{x}}^{\overline{q}_{n|n\text{-}1}}_{n|n\text{-}1} + \mathbf{K}_n\left(\mathbf{z}_n - \overline{\mathbf{z}}_{n|n\text{-}1}\right), \quad (41)$$

$$\mathbf{P}^{\overline{q}_{n|n\text{-}1}}_{n|n} = (\mathbf{I} - \mathbf{K}_n\,\mathbf{H}_n)\,\mathbf{P}^{\overline{q}_{n|n\text{-}1}}_{n|n\text{-}1}, \quad (42)$$

where $\overline{\mathbf{x}}^{\overline{q}_{n|n\text{-}1}}_{n|n\text{-}1} = (\overline{\mathbf{e}}^{\overline{q}_{n|n\text{-}1}}_{n|n\text{-}1} = \mathbf{0}, \overline{\omega}'_{n|n\text{-}1})^T$. Finally, we need to obtain the updated unit quaternion, $\overline{q}_{n|n}$, and compute the mean and the covariance matrix in the $\overline{q}_{n|n}$-centered chart, so that the distribution is expressed in the same conditions as at the beginning of the iteration. The point $\overline{\mathbf{e}}^{\overline{q}_{n|n\text{-}1}}_{n|n}$ that results from (41), and that is defined in the $\overline{q}_{n|n\text{-}1}$-centered chart, correspond to a unit quaternion in the manifold. This is the updated unit quaternion $\overline{q}_{n|n}$ which we are looking for:

$$\overline{q}_{n|n} = \varphi^{-1}_{\overline{q}_{n|n\text{-}1}}\left(\overline{\mathbf{e}}^{\overline{q}_{n|n\text{-}1}}_{n|n}\right) = \quad (43a)$$

$$= \overline{q}_{n|n\text{-}1} * \varphi^{-1}\left(\overline{\mathbf{e}}^{\overline{q}_{n|n\text{-}1}}_{n|n}\right) = \quad (43b)$$

$$= \overline{q}_{n|n\text{-}1} * \overline{\delta}_n. \quad (43c)$$

Knowing that the Kalman update (41) could produce any point in the $\overline{q}_{n|n\text{-}1}$-centered chart we will need to "saturate" to the closest point contained in the image of each chart. The point $\overline{\mathbf{e}}^{\overline{q}_{n|n\text{-}1}}_{n|n}$ in the $\overline{q}_{n|n\text{-}1}$-centered chart is the origin in the $\overline{q}_{n|n}$-centered chart. Then, the expected value of the state in this new chart will be given by $\overline{\mathbf{x}}^{\overline{q}_{n|n}}_{n|n} = (\overline{\mathbf{e}}^{\overline{q}_{n|n}}_{n|n} = \mathbf{0}, \overline{\omega}'_{n|n})^T$, as at the beginning of the iteration.

To update the covariance matrix we need to consider its definition (15). We want to compute $\mathbf{P}^{\overline{q}_{n|n}}$ having $\mathbf{P}^{\overline{q}_{n|n\text{-}1}}$ and knowing the relation $\mathbf{e}^{\overline{p}}(\mathbf{e}^{\overline{q}})$ provided by the transition maps in Table 3. Continuing with the EKF philosophy, the update for the covariance matrix will be found by linearizing

$\mathbf{e}^{\overline{p}}(\mathbf{e}^{\overline{q}})$ around the point where the majority of information is comprised (in our case, the point $\overline{\mathbf{e}}^{\overline{q}} = \overline{\mathbf{e}}_{n|n}^{\overline{q}_{n|n\text{-}1}}$):

$$e_i^{\overline{p}}(\mathbf{e}^{\overline{q}}) = e_i^{\overline{p}}(\overline{\mathbf{e}}^{\overline{q}}) + \sum_j \left.\frac{\partial e_i^{\overline{p}}(\mathbf{e}^{\overline{q}})}{\partial e_j^{\overline{q}}}\right|_{\mathbf{e}^{\overline{q}}=\overline{\mathbf{e}}^{\overline{q}}} (e_j^{\overline{q}} - \overline{e}_j^{\overline{q}}) + \mathcal{O}\left(\|\mathbf{e}^{\overline{q}} - \overline{\mathbf{e}}^{\overline{q}}\|^2\right), \tag{44}$$

where we have used the *big O notation* to describe the limiting behavior of the error term of the approximation as $\mathbf{e}^{\overline{q}} \to \overline{\mathbf{e}}^{\overline{q}}$. In particular, if we define

$$(\mathbf{T})_{ij} = \left.\frac{\partial e_i^{\overline{p}}(\mathbf{e}^{\overline{q}})}{\partial e_j^{\overline{q}}}\right|_{\mathbf{e}^{\overline{q}}=\overline{\mathbf{e}}^{\overline{q}}}, \tag{45}$$

then,

$$\mathbf{e}^{\overline{p}} - \overline{\mathbf{e}}^{\overline{p}} \approx \mathbf{e}^{\overline{p}}(\mathbf{e}^{\overline{q}}) - \mathbf{e}^{\overline{p}}(\overline{\mathbf{e}}^{\overline{q}}) \approx \mathbf{T}\left(\mathbf{e}^{\overline{q}} - \overline{\mathbf{e}}^{\overline{q}}\right), \tag{46}$$

and the final update for the covariance matrix will be computed through

$$\mathbf{P}_{n|n}^{\overline{q}_{n|n}} = \mathrm{E}\left[(\mathbf{x}_{n|n}^{\overline{q}_{n|n}} - \overline{\mathbf{x}}_{n|n}^{\overline{q}_{n|n}})(\mathbf{x}_{n|n}^{\overline{q}_{n|n}} - \overline{\mathbf{x}}_{n|n}^{\overline{q}_{n|n}})^T\right] \approx \tag{47a}$$

$$\approx \begin{pmatrix} \mathbf{T}(\overline{\delta}_n) & \mathbf{0} \\ \mathbf{0} & \mathbf{I} \end{pmatrix} \mathbf{P}_{n|n}^{\overline{q}_{n|n\text{-}1}} \begin{pmatrix} \mathbf{T}(\overline{\delta}_n) & \mathbf{0} \\ \mathbf{0} & \mathbf{I} \end{pmatrix}^T. \tag{47b}$$

Table 4 summarizes the resulting **T**-matrix for each chart, along with their application domain. A detailed derivation of these **T**-matrices can be found in Appendix C.

Table 4. **T**-matrices for the transition maps of the charts studied.

Chart	$\mathbf{T}(\overline{\delta})$ Matrix	Domain
O	$\overline{\delta}_0\mathbf{I} - [\overline{\boldsymbol{\delta}}]_\times + \frac{\overline{\boldsymbol{\delta}}\,\overline{\boldsymbol{\delta}}^T}{\overline{\delta}_0}$	$\{\overline{\delta} \in S^3 : \overline{\delta}_0 > 0\}$
RP	$\overline{\delta}_0\left(\overline{\delta}_0\mathbf{I} - [\overline{\boldsymbol{\delta}}]_\times\right)$	$\{\overline{\delta} \in S^3 : \overline{\delta}_0 \neq 0\}$
MRP	$\frac{1}{2}\left[(1+\overline{\delta}_0)\left(\overline{\delta}_0\mathbf{I} - [\overline{\boldsymbol{\delta}}]_\times\right) + \overline{\boldsymbol{\delta}}\,\overline{\boldsymbol{\delta}}^T\right]$	$\{\overline{\delta} \in S^3 : \overline{\delta}_0 \geq 0\}$
RV	$\left[\overline{\delta}_0\left(\mathbf{I} - \hat{\overline{\boldsymbol{\delta}}}\hat{\overline{\boldsymbol{\delta}}}^T\right) - [\overline{\boldsymbol{\delta}}]_\times\right]\frac{\|\overline{\boldsymbol{\delta}}\|}{\arcsin\|\overline{\boldsymbol{\delta}}\|} + \hat{\overline{\boldsymbol{\delta}}}\hat{\overline{\boldsymbol{\delta}}}^T$	$\{\overline{\delta} \in S^3 : \overline{\delta}_0 \geq 0, \|\overline{\boldsymbol{\delta}}\| \neq 0\}$

After the final computation we obtain the four

$$(\varphi, \overline{q}_{n|n}, \overline{\omega}'_{n|n}, \mathbf{P}_{n|n}^{\overline{q}_{n|n}}), \tag{48}$$

that is a condition equivalent to (27) in which we started the iteration.

3.2. Manifold Unscented Kalman Filter

In this section we present the UKF-based estimator: the Manifold Unscented Kalman Filter (MUKF).

A measurement $\mathbf{z}_n$ arrives at time t_n. Our knowledge about the orientation at a previous time $t_{n\text{-}1}$ is described by a distribution expressed in the $\overline{q}_{n\text{-}1|n\text{-}2}$-centered chart. This distribution is encoded in the four

$$\left(\varphi \, , \, \overline{q}_{n\text{-}1|n\text{-}2} \, , \, \overline{\mathbf{x}}^{\overline{q}_{n\text{-}1|n\text{-}2}}_{n\text{-}1|n\text{-}1} \, , \, \mathbf{P}^{\overline{q}_{n\text{-}1|n\text{-}2}}_{n\text{-}1|n\text{-}1} \right) . \tag{49}$$

The first step in the UKF is to create the augmented $N\times 1$ mean $\widetilde{\mathbf{x}}_n$ and $N\times N$ covariance matrix $\widetilde{\mathbf{P}}_n$. Since the measurement equations are linear for the random variables $\mathbf{r}^{\omega}_t$ and $\mathbf{r}^{\mathbf{v}}_t$, we can leave their covariance matrices out of the augmented one and add them later:

$$\widetilde{\mathbf{x}}_n = \begin{pmatrix} \overline{\mathbf{x}}^{\overline{q}_{n\text{-}1|n\text{-}2}}_{n\text{-}1|n\text{-}1} \\ \overline{\mathbf{q}}^{\omega}_n \\ \overline{\mathbf{q}}^{\mathbf{v}}_n \end{pmatrix}, \tag{50}$$

$$\widetilde{\mathbf{P}}_n = \begin{pmatrix} \mathbf{P}^{\overline{q}_{n\text{-}1|n\text{-}2}}_{n\text{-}1|n\text{-}1} & \mathbf{0} & \mathbf{0} \\ \mathbf{0} & \mathbf{Q}^{\omega}_n & \mathbf{0} \\ \mathbf{0} & \mathbf{0} & \mathbf{Q}^{\mathbf{v}}_n \end{pmatrix}. \tag{51}$$

Then, we obtain the matrix $\mathbf{L}_n$ which satisfies $\mathbf{L}_n \, \mathbf{L}_n^T = \widetilde{\mathbf{P}}_n$ and we use it to generate the $2N+1$ sigma points $\{\boldsymbol{\mathcal{X}}_j\}_{j=0}^{2N}$ as described in Ref. [15]:

$$\mathcal{X}_{i,0} = (\widetilde{\mathbf{x}}_n)_i, \tag{52a}$$

$$\mathcal{X}_{i,j} = (\widetilde{\mathbf{x}}_n)_i + \frac{(\mathbf{L}_n)_{ij}}{\sqrt{2W_j}} \quad \text{for} \quad j = 1, \ldots, N, \tag{52b}$$

$$\mathcal{X}_{i,j+N} = (\widetilde{\mathbf{x}}_n)_i - \frac{(\mathbf{L}_n)_{ij}}{\sqrt{2W_j}} \quad \text{for} \quad j = 1, \ldots, N, \tag{52c}$$

being $W_j = (1 - W_0)/(2N)$ for $j \neq 0$ where W_0 regulates the importance given to the sigma point $\boldsymbol{\mathcal{X}}_0$ in the computation of the mean. These sigma points $\{\boldsymbol{\mathcal{X}}_j\}_j$ are expressed in the $\overline{q}_{n\text{-}1|n\text{-}2}$-centered chart. We need to express them in the manifold before applying the evolution equations and the measurement equations:

$$\boldsymbol{\mathcal{X}}^{q}_j = \varphi^{-1}_{\overline{q}_{n\text{-}1|n\text{-}2}}(\boldsymbol{\mathcal{X}}^{\mathbf{e}}_j) = \overline{q}_{n\text{-}1|n\text{-}2} * \varphi^{-1}(\boldsymbol{\mathcal{X}}^{\mathbf{e}}_j), \tag{53}$$

$$\boldsymbol{\mathcal{Y}}^{\omega}_j = \boldsymbol{\mathcal{X}}^{\omega}_j + \boldsymbol{\mathcal{X}}^{\mathbf{q}^{\omega}}_j \Delta t_n, \tag{54}$$

$$\boldsymbol{\mathcal{Y}}^{q}_j = \boldsymbol{\mathcal{X}}^{q}_j * \begin{pmatrix} \cos\left(\frac{\|\boldsymbol{\mathcal{Y}}^{\omega}_j\| \, \Delta t_n}{2}\right) \\ \widehat{\boldsymbol{\mathcal{Y}}}^{\omega}_j \sin\left(\frac{\|\boldsymbol{\mathcal{Y}}^{\omega}_j\| \, \Delta t_n}{2}\right) \end{pmatrix}, \tag{55}$$

$$\boldsymbol{\mathcal{Z}}^{\mathbf{v}}_j = \mathbf{R}^T\left(\boldsymbol{\mathcal{X}}^{q}_j\right)\left(\boldsymbol{\mathcal{X}}^{\mathbf{v}}_j + \mathbf{v}_t\right), \tag{56}$$

$$\boldsymbol{\mathcal{Z}}^{\omega}_j = \boldsymbol{\mathcal{Y}}^{\omega}_j, \tag{57}$$

where for the j-th sigma point, $\boldsymbol{\mathcal{X}}^{\mathbf{e}}_j$ is its chart point part and $\boldsymbol{\mathcal{X}}^{q}_j$ is the quaternion with which it is mapped, $\boldsymbol{\mathcal{X}}^{\omega}_j$ is its angular velocity part, $\boldsymbol{\mathcal{X}}^{\mathbf{q}^{\omega}}_j$ is its angular velocity noise part, $\boldsymbol{\mathcal{Y}}^{\omega}_j$ is its angular velocity prediction, $\boldsymbol{\mathcal{Y}}^{q}_j$ is the quaternion part of its prediction (we have assumed that the angular velocity $\boldsymbol{\mathcal{Y}}^{\omega}_j$ is constant in the time interval $[t_{n\text{-}1}, t_n)$ so that we can use (A20)), $\boldsymbol{\mathcal{X}}^{\mathbf{v}}_j$ is the vector process noise part, $\boldsymbol{\mathcal{Z}}^{\mathbf{v}}_j$ is its vector measurement prediction, $\boldsymbol{\mathcal{Z}}^{\omega}_j$ is its angular velocity measurement prediction, and $\Delta t_n = t_n - t_{n\text{-}1}$. Note that when applying the inverse chart φ^{-1} we will need to "saturate" $\boldsymbol{\mathcal{X}}^{\mathbf{e}}_j$ to the closest point in the image of φ. Having these new sigma points, we can obtain the means and covariance matrices of the distributions present in the UKF. First, defining $\mathcal{Z}_j := \left(\mathcal{Z}_j^{\mathrm{v}}, \mathcal{Z}_j^{\omega} \right)^T$, the means are computed through

$$
\begin{aligned}
\overline{q}_{n|n\text{-}1} &= \frac{\sum_j W_j \, \mathcal{Y}_j^q}{\| \sum_j W_j \, \mathcal{Y}_j^q \|}, && (58) \\
\overline{\omega}'_{n|n\text{-}1} &= \sum_j W_j \, \mathcal{Y}_j^{\omega}, && (59) \\
\overline{\mathbf{x}}_{n|n\text{-}1}^{\overline{q}_{n|n\text{-}1}} &= \begin{pmatrix} \varphi_{\overline{q}_{n|n\text{-}1}}(\overline{q}_{n|n\text{-}1}) = \mathbf{0} \\ \overline{\omega}'_{n|n\text{-}1} \end{pmatrix}, && (60) \\
\overline{\mathbf{z}}_{n|n\text{-}1} &= \sum_j W_j \, \mathcal{Z}_j. && (61)
\end{aligned}
$$

where we have used a variation of the result provided in Ref. [16]. Namely,

$$
\overline{q} \approx \frac{\sum_j q_j}{\| \sum_j q_j \|}, \tag{62}
$$

with $q_j \cdot q_k > 0$ for $j, k = 0, \ldots, 2N$. This result is shown to minimize the fourth order approximation of the distance defined as the sum of squared angles between the rotation transformation represented by each quaternion q_j, and the one represented by $\overline{q}$. This approach to compute the mean quaternion is extremely efficient, and its derivation is elegant and simple. In order to ensure that $q_j \cdot q_k > 0$, it is useful to remember the property that both q and $-q$ represent the same rotation. This property is also useful for introducing the quaternions in the domain of φ to execute the next step of the filter.

After this, we use the obtained mean quaternion $\overline{q}_{n|n\text{-}1}$ to express each sigma point in the $\overline{q}_{n|n\text{-}1}$-centered chart, and compute the covariance matrices:

$$
\begin{aligned}
\mathcal{Y}_j^{\mathbf{e}} &= \varphi_{\overline{q}_{n|n\text{-}1}}(\mathcal{Y}_j^q) = \varphi(\overline{q}_{n|n\text{-}1}^* * \mathcal{Y}_j^q), && (63) \\
\mathbf{P}_{n|n\text{-}1}^{\overline{q}_{n|n\text{-}1}} &= \sum_j W_j \, \mathcal{Y}_j \, \mathcal{Y}_j^T, && (64) \\
\mathbf{P}_{n|n\text{-}1}^{yz} &= \sum_j W_j \, \mathcal{Y}_j \left(\mathcal{Z}_j - \overline{\mathbf{z}}_{n|n\text{-}1} \right)^T, && (65) \\
\mathbf{S}_{n|n\text{-}1} &= \sum_j W_j \left(\mathcal{Z}_j - \overline{\mathbf{z}}_{n|n\text{-}1} \right) \left(\mathcal{Z}_j - \overline{\mathbf{z}}_{n|n\text{-}1} \right)^T + \begin{pmatrix} \mathbf{R}_n^{\mathrm{v}} & \mathbf{0} \\ \mathbf{0} & \mathbf{R}_n^{\omega} \end{pmatrix}, && (66)
\end{aligned}
$$

where we have denoted $\mathcal{Y}_j := (\mathcal{Y}_j^{\mathbf{e}}, \mathcal{Y}_j^{\omega} - \overline{\omega}'_{n|n\text{-}1})^T$. Finally, we compute the UKF version of the Kalman gain $\mathbf{K}_n$, and we use it to obtain the optimal estimation of the state:

$$
\begin{aligned}
\mathbf{K}_n &= \mathbf{P}_{n|n\text{-}1}^{yz} \, \mathbf{S}_{n|n\text{-}1}^{-1}, && (67) \\
\overline{\mathbf{x}}_{n|n}^{\overline{q}_{n|n\text{-}1}} &= \overline{\mathbf{x}}_{n|n\text{-}1}^{\overline{q}_{n|n\text{-}1}} + \mathbf{K}_n \left(\mathbf{z}_n - \overline{\mathbf{z}}_{n|n\text{-}1} \right), && (68) \\
\mathbf{P}_{n|n}^{\overline{q}_{n|n\text{-}1}} &= \mathbf{P}_{n|n\text{-}1}^{\overline{q}_{n|n\text{-}1}} - \mathbf{K}_n \, \mathbf{S}_{n|n\text{-}1} \, \mathbf{K}_n^T, && (69)
\end{aligned}
$$

arriving at the same conditions in which we began the iteration, with a distribution expressed in the $\overline{q}_{n|n\text{-}1}$-centered chart, and encoded by the four

$$
(\varphi, \overline{q}_{n|n\text{-}1}, \overline{\mathbf{x}}_{n|n}^{\overline{q}_{n|n\text{-}1}}, \mathbf{P}_{n|n}^{\overline{q}_{n|n\text{-}1}}). \tag{70}
$$

Our best estimation for the orientation at this time is

$$\overline{q}_{n|n} = \varphi^{-1}_{\overline{q}_{n|n\text{-}1}}\left(\overline{\mathbf{e}}^{\overline{q}_{n|n\text{-}1}}_{n|n}\right) = \overline{q}_{n|n\text{-}1} * \varphi^{-1}\left(\overline{\mathbf{e}}^{\overline{q}_{n|n\text{-}1}}_{n|n}\right), \tag{71}$$

being $\overline{\mathbf{e}}^{\overline{q}_{n|n\text{-}1}}_{n|n}$ the part of the mean $\overline{\mathbf{x}}^{\overline{q}_{n|n\text{-}1}}_{n|n}$ that represent the quaternion in the $\overline{q}_{n|n\text{-}1}$-centered chart.

Note that setting $\overline{q}_{n\text{-}1|n\text{-}2} := \overline{q}_{n\text{-}1|n\text{-}1}$ and $\overline{\mathbf{e}}^{\overline{q}_{n\text{-}1|n\text{-}2}}_{n\text{-}1|n\text{-}1} := \mathbf{0}$ at the beginning of each iteration yields the traditional version of the algorithm, where a "reset operation" is performed instead of the covariance matrix update.

4. Simulation Results

This section presents the results of the simulations used to measure the accuracy of each estimator. Simulations are chosen instead of real experiments because a real system entails an uncertainty in the measurement of the true attitude: the attitude that is used to compare with that estimated by the algorithms. There are sources of error ranging from a miscalibration of the measurement system to a possible bias in the "true attitude" produced by another attitude estimator, which makes it problematic to define an adequate metric to measure the accuracy of the algorithms. For this reason, the authors consider that using a simulation is more reliable to avoid possible biases in the results due to said sources of error. Others have performed similar types of tests [7,17]. However, the results do not seem to be statistically conclusive: only the estimations of some orientation trajectories are shown.

We perform our comparison through a simulation in which we do have an absolute knowledge of the attitude of the system: a true oracle exists in a simulation. Therefore, we can compare the real orientation with the attitude estimated by the algorithms having fed them only with simulated measurements that we obtain from such known orientations. We will extract our performance metrics from a wide set of orientation trajectories in order to obtain statistically conclusive results.

We try to answer three questions with the simulation test. The first question is, is there a chart for which we get a greater accuracy in attitude estimation? The second one is, what algorithm produces the most accurate attitude estimation, the MEKF or the MUKF? The last question stems from the fact that previous algorithms on attitude estimation, such as the Multiplicative Extended Kalman Filter, did not contemplate updating the distribution from one chart to another as done at (47b) in the MEKF. However, their estimators performed well [6,7,12]. Then the third question is, does this "chart update" imply an improvement in the accuracy of the attitude estimation?

Although a simulation has been used to compare our algorithms, these have also been tested with a real IMU. In the Supplementary Materials one can find a demonstration video, the source code used in the video, the source code used to generate the simulations, and the source code used to obtain the computational cost of the algorithms in each platform.

4.1. Performance Metric

We have already described a quaternion q as a deviation from another quaternion $\overline{q}$ as $q = \overline{q} * \delta$. Now we define the instantaneous error between an estimated attitude, represented by a unit quaternion $\overline{q}$, and the real attitude, represented by the unit quaternion $\overset{\star}{q}$, as the angle we have to rotate one of them to transform it into the other. This is, the angle of the rotation transformation defined by the quaternion δ_e such that $\overset{\star}{q} = \overline{q} * \delta_e$. Recalling (6), this angle can be computed as:

$$\theta_e = 2 \arccos\left[\left(\overline{q}^* * \overset{\star}{q}\right)_0\right] = \tag{72a}$$

$$= 2 \arccos\left(\overline{q} \cdot \overset{\star}{q}\right), \tag{72b}$$

having previously ensured that $\overline{q} \cdot \overset{\star}{q} \geq 0$ using the fact that both q and $-q$ represent the same rotation transformation.

Angle θ_e will vary along an orientation trajectory. Then, we will define the mean error in orientation estimation for a given trajectory starting at time $t = 0$ and ending at time $t = T$ as

$$e_\theta \;=\; \frac{1}{T}\int_0^T \theta_e(t)\,dt. \tag{73}$$

Finally, e_θ will depend on the followed trajectory, and on the set of taken measurements. We will need to generate several orientation trajectories to obtain the mean value $\bar{e}_\theta$ and the variance $\sigma^2_{\bar{e}_\theta}$ that characterize the distribution of the error in orientation estimation e_θ for each algorithm. We will define the confidence interval for the computed $\bar{e}_\theta$ as

$$\left[\,\bar{e}_\theta - 3\,\sigma_{\bar{e}_\theta}/\sqrt{N_s}\,,\; \bar{e}_\theta + 3\,\sigma_{\bar{e}_\theta}/\sqrt{N_s}\,\right], \tag{74}$$

where N_s is the number of samples taken for the $\bar{e}_\theta$ computation, so that $\sigma^2_{\bar{e}_\theta}/N_s$ is the variance of the sample mean distribution.

Being that the lower the better, the value of $\bar{e}_\theta$ gives us a measure of how well an algorithm estimates the orientation. We will consider that the performance of an algorithm A is better than the performance of other algorithm B if $\bar{e}_\theta(\mathrm{A}) < \bar{e}_\theta(\mathrm{B})$ and their confidence intervals do not overlap.

4.2. Simulation Scheme

To compute the performance metrics we will need to generate a large number of simulations. Each independent simulation will consist of three steps: initialization, convergence, and estimation.

In the initialization step we set up the initial conditions accordingly to the chosen simulation parameters. This includes generating the initial unit quaternion $\overset{\star}{q}_0$ from a uniform distribution in S^3, setting the initial angular velocity $\overset{\star}{\omega}{}'_0$ to zero, setting the update frequency f_{update}, generating the variances of the process noises σ^2_ω and $\sigma^2_{\mathbf{v}}$ from a uniform distribution in the intervals $(0, Q^\omega_{\text{max}}]$ and $(0, Q^{\mathbf{v}}_{\text{max}}]$ respectively, and initializing the estimation algorithm. The initialization of the MEKF includes setting $\bar{q}_{0|0} = 1$, $\bar{\omega}'_{0|0} = \mathbf{0}\,\text{rad/s}$, and $\mathbf{P}^{\bar{q}_{0|0}}_{0|0} = 10^2\,\mathbf{I}$. On the other hand, the initialization of the MUKF includes setting $\bar{q}_{0|-1} = 1$, $\mathbf{e}^{\bar{q}_{0|-1}}_{0|0} = \mathbf{0}$, $\bar{\omega}'_{0|0} = (1,1,1)^T\,\text{rad/s}$, and $\mathbf{P}^{\bar{q}_{0|-1}}_{0|0} = 10^2\,\mathbf{I}$. The angular velocity is not initialized to $\mathbf{0}$ in the MUKF because it has been observed that it is sometimes necessary to "break the symmetry" for the algorithm to converge; especially when we do not apply the chart update (when we perform the "reset operation") for the RV chart. The covariance matrices that appear in both algorithms are initialized as $\mathbf{Q}^\omega_n = \mathbf{I}\,\text{rads}^2/\text{s}^4$, $\mathbf{Q}^{\mathbf{v}}_n = 10^{-2}\,\mathbf{I}\,\text{p.d.u.}$ ("p.d.u." stands for "Procedure Defined Unit". In the present case it depends on the definition of the vector $\mathbf{v}$), $\mathbf{R}^\omega_n = R^\omega\,\mathbf{I}\,\text{rads}^2/\text{s}^2$, $\mathbf{R}^{\mathbf{v}}_n = R^{\mathbf{v}}\,\mathbf{I}\,\text{p.d.u.}$, where R^ω and $R^{\mathbf{v}}$ are the variances of the measurement noise that will be used in the simulation. We give this information about the measurement noise to the algorithms because it can be obtained offline, while the information about the process noise cannot. Given that a priori we cannot know how the system will behave, the values of $\mathbf{Q}^\omega_n$ and $\mathbf{Q}^{\mathbf{v}}_n$ have been chosen according to what we understand could be normal. Choosing these values we are assuming that after a second it is normal for the angular velocity to have changed by $1\,\text{rad/s}$, and also that it is normal to find external noises added to the vector $\mathbf{v}_t$ of magnitude $10^{-1}\,\text{p.d.u.}$. For the mean values we set $\bar{\mathbf{q}}^\omega_n = \mathbf{0}\,\text{rads/s}^2$, and $\bar{\mathbf{q}}^{\mathbf{v}} = \mathbf{0}\,\text{p.d.u.}$.

In the convergence step we keep the system in the initial orientation $\overset{\star}{q}_0$. Simulated measurements are generated using (23) and (24). For each measurement, a different $\mathbf{v}_t$ is sampled from a uniform distribution in the unit sphere of $\mathbb{R}^3$. The values for each component of $\mathbf{q}^{\mathbf{v}}_t$, $\mathbf{r}^{\mathbf{v}}_t$, and $\mathbf{r}^\omega_t$ are obtained from normal distributions with zero mean and variances $\sigma^2_{\mathbf{v}}$, $R^{\mathbf{v}}$, and R^ω respectively. The term $\mathbf{R}^T(\,q_t\,)$ in (23) is obtained from the true attitude $\overset{\star}{q}_t$, which in the convergence step takes the value of $\overset{\star}{q}_t = \overset{\star}{q}_0$. The term ω'_t in (24) is the true angular velocity, which in the convergence step takes the value $\overset{\star}{\omega}{}'_t = \mathbf{0}$. The tested algorithm updates its state estimation until the inequality $\theta_e(t) < \theta^0_e$ is satisfied, where $\theta_e(t)$ is the value of the error (72), and θ^0_e is a parameter in the simulation. The convergence step could have been replaced by an initialization of the attitude estimated by the algorithm $\overline{q}_t$ to the real value $\overset{\star}{q}_t$, but then it would have also been necessary to fix a certain covariance matrix. Since the metric of the space generated by each chart is different, it is difficult to set a covariance matrix that provides the same information for each chart. It seemed more natural to the authors to allow the algorithm to find the true attitude by its own means, and for the covariance matrix to converge to a value in each case.

Finally, in the estimation step we generate a random but continuous orientation sequence using a Wiener process for the angular velocity:

$$\overset{\star}{\omega}'_t = \overset{\star}{\omega}'_{t\text{-}\delta t} + \mathbf{n}_t \sqrt{\delta t}, \tag{75}$$

$$\overset{\star}{q}_t = \overset{\star}{q}_{t\text{-}\delta t} * \begin{pmatrix} \cos\left(\frac{\|\overset{\star}{\omega}'_t\|\,\delta t}{2}\right) \\ \frac{\overset{\star}{\omega}'_t}{\|\overset{\star}{\omega}'_t\|} \sin\left(\frac{\|\overset{\star}{\omega}'_t\|\,\delta t}{2}\right) \end{pmatrix}, \tag{76}$$

where $\mathbf{n}_t$ is a random vector whose components are sampled from a normal distribution with zero mean and variance σ_{ω}^2, and δt is the simulation time step that is related to the algorithm time step Δt trough $\texttt{dtdtsim}\,\delta t = \Delta t$, being `dtdtsim` an integer parameter that determines the simulation updates per algorithm update. Note that we multiply $\mathbf{n}_t$ by $\sqrt{\delta t}$ and not by δt. We do it this way so that the covariance matrix after k steps does not depend on the simulation time step δt. In fact, after a time $T = k\,\delta t$ the covariance matrix of the angular velocity will have grown by $\Delta \mathbf{P}^{\omega} = k\,\mathbf{I}\,\sigma_{\omega}^2\,\delta t = \mathbf{I}\,\sigma_{\omega}^2\,T$, and not by $(\Delta \mathbf{P}^{\omega})' = k\,\mathbf{I}\,\sigma_{\omega}^2\,(\delta t)^2 = \mathbf{I}\,\sigma_{\omega}^2\,T\,\delta t$. After each `dtdtsim` simulation updates, a simulated measurement is generated in the same way it was done in the convergence step, and the algorithm is updated with it. The simulation will run for a time $\texttt{Tsim} = k'\,\Delta t$, where k' is an integer number. This way we will perform the last algorithm update at the end of the simulation. The error (72) will be evaluated after each algorithm update, and it will be added up through the simulation to obtain the averaged error (73). After each simulation, we will obtain a sample for the computation of $\overline{e}_\theta$ and $\sigma_{\overline{e}_\theta}^2$. We will perform N_s of these simulations to obtain the confidence interval (74).

4.3. Results

In this section we present the results of the simulations. The algorithms are tested for update frequencies $f_{\text{update}} = 1/\Delta t$ in the interval $[2, 1000]$ Hz. This range has been chosen thinking about the possible limitations of a real system. For example, the maximum data rate of a low cost IMU is around 1000 Hz. On the other hand, the update frequency may be limited by processing. The computational cost of each estimator has been evaluated in two platforms: an Arduino MEGA 2560, and a Raspberry Pi 3 Model B. The code has been written in c++. The resulting maximum update frequencies are presented in Figure 4, which indicates that the MEKF can be executed approximately 3 times faster than the MUKF.

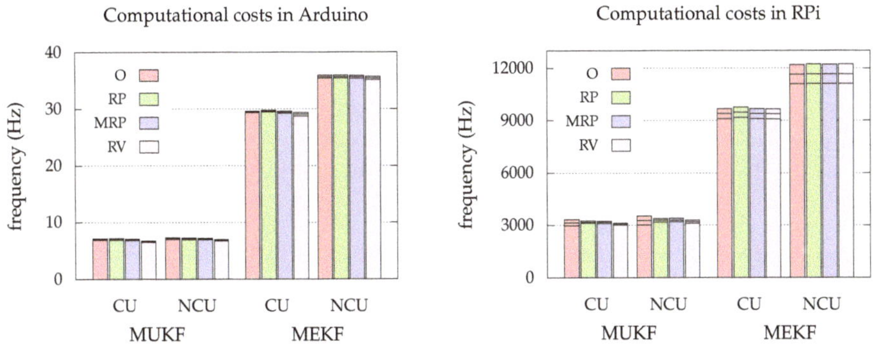

Figure 4. Maximum update frequency for each approach. The lines at the top represent the mean and the deviation (3σ) of the distribution of maximum update frequencies. "CU" stands for Chart Update, while "NCU" stands for No Chart Update.

Although the algorithms have been developed allowing a different Δt_n for each update, the simulations are performed using a constant Δt, and the simulation parameters depicted in Table 5.

Table 5. Parameters used in the simulations.

Parameter	Value
θ_e^0	$1°$
`Tsim`	10 s
`dtdtsim`	100
N_s	1000
Q_{max}^{ω}	10^2 rads^2/s^3
$Q_{\text{max}}^{\mathbf{v}}$	1 p.d.u.
R	$\{10^{-2}, 10^{-4}, 10^{-6}\}$
R^{ω}	R rads^2/s^2
$R^{\mathbf{v}}$	R p.d.u.
W_0	1/25

The parameters θ_e^0, `Tsim`, `dtdtsim`, and N_s have been chosen trying to reach a compromise between the precision of the results, and the execution time of the simulation. The values for Q_{max}^{ω} and $Q_{\text{max}}^{\mathbf{v}}$ have been chosen in such a way that the estimation algorithms face both normal situations ($\mathbf{Q}_n^{\omega} \approx \sigma_\omega^2 \mathbf{I}$ and $\mathbf{Q}_n^{\mathbf{v}} \approx \sigma_{\mathbf{v}}^2 \mathbf{I}$) and situations that were not foreseen ($\mathbf{Q}_n^{\omega} \neq \sigma_\omega^2 \mathbf{I}$ or $\mathbf{Q}_n^{\mathbf{v}} \neq \sigma_{\mathbf{v}}^2 \mathbf{I}$). A typical low cost IMU has $R^{\omega} \approx 10^{-4}\,\text{rad}^2/\text{s}^2$ and $R^{\mathbf{v}} \approx 10^{-4} g^2$. The values chosen for R represent an imprecise sensor (10^{-2}), a normal sensor (10^{-4}), and a precise sensor (10^{-6}). The value of W_0 has been chosen so that all sigma points have the same importance, but very similar results, if not identical, have been obtained for other selections of W_0.

4.3.1. Chart Choice

The results of the simulation are presented in Figure 5. The average of the performance metric is shown along with its confidence interval for each of the selected update frequencies. The results of the MEKF and the MUKF are shown in different graphs, but drawn in the same one are the results for each chart and for a given MKF. In this way we are able to distinguish if a chart has an advantage over the others.

We observe that there is no chart that is especially advantageous. All things being equal, we would opt for the RP chart. For this chart it is not necessary to worry about the domain since it maps q

and $-q$ with the same point of $\mathbb{R}^3$ and with the same $\mathbf{T}$-matrix; or of the image since it is all $\mathbb{R}^3$. In addition, the expressions of φ^{-1} and the $\mathbf{T}$-matrix for the MEKF are simpler for the RP chart. These computational advantages make us prefer the RP chart over the others.

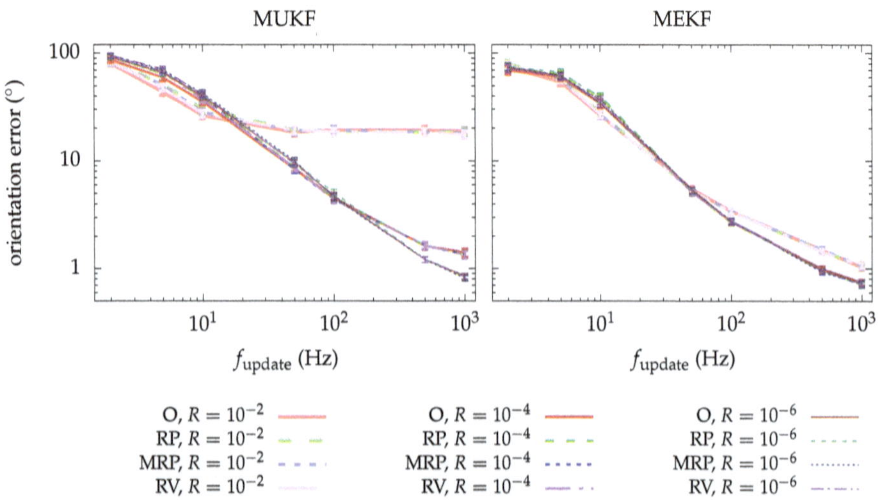

Figure 5. Mean of the performance metric for each approach. Results from different charts are plotted in the same graph. Results from different MKF are plotted in different graphs. Bars represent the confidence interval (3σ) for the mean computation.

4.3.2. MEKF vs. MUKF

Figure 6 also presents the results of the simulations. This time, we display on the same graph the resulting performance metrics for the MUKF and the MEKF when the RP chart is used. In this way, we can distinguish if one MKF has an advantage over the other.

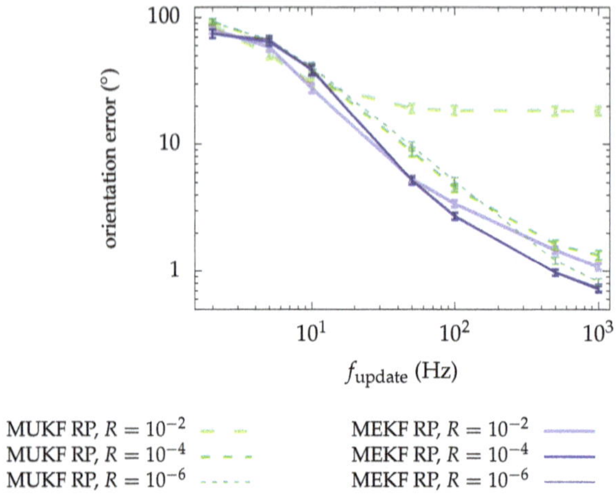

Figure 6. Mean of the performance metric for each MKF. Only results for the RP chart are plotted. Bars represent the confidence interval (3σ) in the mean computation.

We note that the MEKF performs the same or better than the MUKF. This differs from the usual experience, in which the UKF outperforms the EKF in traditional non-linear estimation applications. The fact that the charts resemble the Euclidean space near the origin (see Section 2.3) might be favoring the MEKF, since the Jacobian matrices, used to approximate the non-linear functions, are defined at that point. However, the sigma points generated for the MUKF are sampled far from the origin of the chart, where the non-linearities become notorious. We are facing a very particular scenario in which the model is approximately linear for the MEKF, while for the MUKF it is not. In addition, due to the difference in computational cost (see Figure 4), the MUKF update frequencies will generally be lower than those of the MEKF, which will imply worse accuracy in its estimations. Then, the MEKF with the RP chart seems to be our best option.

4.3.3. Chart Update vs. No Chart Update

Figure 7 presents the results of each MKF with each chart in a different graph, but displayed in the same one are the results using the "chart update" and the results without using it.

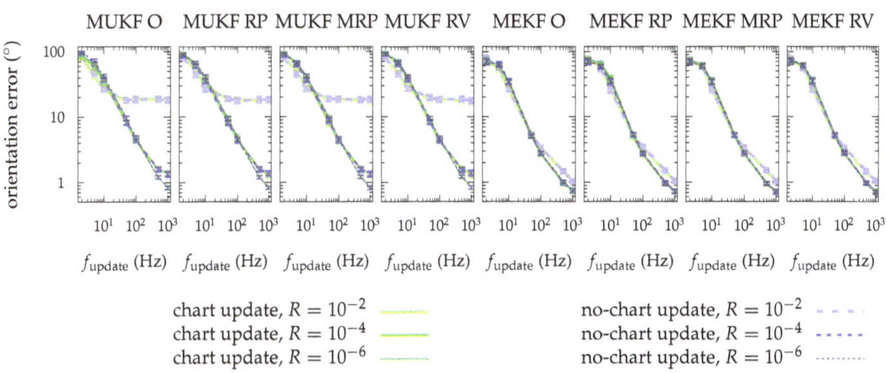

Figure 7. Mean of the performance metric for each approach. Results from the approach in which we apply the chart update, and those of which we do not apply it are plotted together. Bars represent the confidence interval (3σ) in the mean computation.

We can observe that there is almost no difference between using the "chart update" and not using it. The concepts used in this paper have helped us to understand the mechanisms of the MKF, and ultimately to arrive to the concepts of "multiplicative update", and of "covariance correction step" with the **T**-matrix definition. However, it is not necessary to apply the latest update (47b) in practice: we will obtain essentially the same accuracy in our estimations.

5. Conclusions

We have used concepts from manifold theory to define the expected value and the covariance matrix of a distribution in a manifold. In particular, we have defined the expected value and covariance matrix of a distribution of unit quaternions in S^3, the unit sphere in $\mathbb{R}^4$, using the concept of chart. These definitions have helped us to develop Kalman filters for orientation estimation, where the attitude has been represented by a unit quaternion. They have also helped us solve the problem of the "covariance correction step". Two estimators have been developed: one based on the EKF (the MEKF), and another based on the UKF (the MUKF). The MEKF and the MUKF have been tested in simulations, and some results have been obtained. The conclusions of the simulations are:

- There is no chart that presents a clear advantage over the others, but the RP chart has some characteristics that motivate us to prefer it.

- The MEKF is preferable to the MUKF due to its lower computational cost and its greater accuracy in orientation estimation.
- The "chart update" is not necessary for the MKF in practice.

Then, the MEKF with the RP chart and without applying the "chart update" is our best attitude estimator according to the adopted performance metric. This algorithm resembles the conventional "Multiplicative Extended Kalman Filter", but we have obtained the MEKF without having to redefine any aspect of the classic Kalman filter.

Supplementary Materials: The following are available online at http://www.mdpi.com/1424-8220/19/1/149/s1: SupplementaryMaterials.zip.

Author Contributions: conceptualization, P.B.-P.; methodology, P.B.-P.; software, P.B.-P.; validation, P.B.-P. and H.M.-B.; formal analysis, P.B.-P.; investigation, P.B.-P.; resources, H.M.-B.; data curation, P.B.-P.; writing–original draft preparation, P.B.-P.; writing–review and editing, P.B.-P. and H.M.-B.; visualization, P.B.-P.; supervision, H.M.-B.; project administration, P.B.-P. and H.M.-B.

Funding: This research received no external funding.

Conflicts of Interest: The authors declare no conflict of interest.

Abbreviations

The following abbreviations are used in this manuscript:

EKF	Extended Kalman Filter
UKF	Unscented Kalman Filter
MKF	Manifold Kalman Filter
MEKF	Manifold Extended Kalman Filter
MUKF	Manifold Unscented Kalman Filter
O	Orthographic
RP	Rodrigues Parameters
MRP	Modified Rodrigues Parameters
RV	Rotation Vector

Appendix A. Derivation of Transition Maps

This appendix contains the derivation of the transition map for each chart.

Appendix A.1. Orthographic

Using the inverse of the transformation that defines the chart, φ^{-1},

$$\delta^{\overline{q}} = \begin{pmatrix} \sqrt{1 - \frac{\|\mathbf{e}^{\overline{q}}\|^2}{4}} \\ \mathbf{e}^{\overline{q}}/2 \end{pmatrix}. \tag{A1}$$

Introducing (A1) into (19),

$$\delta^{\overline{p}} = \overline{\delta}^* * \delta^{\overline{q}} = \begin{pmatrix} \overline{\delta}_0 \sqrt{1 - \frac{\|\mathbf{e}^{\overline{q}}\|^2}{4}} + \overline{\boldsymbol{\delta}} \cdot \frac{\mathbf{e}^{\overline{q}}}{2} \\ \overline{\delta}_0 \frac{\mathbf{e}^{\overline{q}}}{2} - \sqrt{1 - \frac{\|\mathbf{e}^{\overline{q}}\|^2}{4}}\, \overline{\boldsymbol{\delta}} - \overline{\boldsymbol{\delta}} \times \frac{\mathbf{e}^{\overline{q}}}{2} \end{pmatrix}. \tag{A2}$$

Finally, applying the chart definition,

$$\mathbf{e}^{\overline{p}} = 2\,\boldsymbol{\delta}^{\overline{p}} = \overline{\delta}_0\, \mathbf{e}^{\overline{q}} - \sqrt{4 - \|\mathbf{e}^{\overline{q}}\|^2}\, \overline{\boldsymbol{\delta}} - \overline{\boldsymbol{\delta}} \times \mathbf{e}^{\overline{q}}. \tag{A3}$$

Appendix A.2. Rodrigues Parameters

Using the inverse of the transformation that defines the chart, φ^{-1},

$$\delta^{\overline{q}} = \frac{1}{\sqrt{4 + \|\mathbf{e}^{\overline{q}}\|^2}} \begin{pmatrix} 2 \\ \mathbf{e}^{\overline{q}} \end{pmatrix}. \tag{A4}$$

Introducing (A4) into (19),

$$\delta^{\overline{p}} = \overline{\delta}^* * \delta^{\overline{q}} = \frac{1}{\sqrt{4 + \|\mathbf{e}^{\overline{q}}\|^2}} \begin{pmatrix} 2\,\overline{\delta}_0 + \overline{\boldsymbol{\delta}} \cdot \mathbf{e}^{\overline{q}} \\ \overline{\delta}_0\, \mathbf{e}^{\overline{q}} - 2\,\overline{\boldsymbol{\delta}} - \overline{\boldsymbol{\delta}} \times \mathbf{e}^{\overline{q}} \end{pmatrix}. \tag{A5}$$

Finally, applying the chart definition,

$$\mathbf{e}^{\overline{p}} = 2\,\frac{\boldsymbol{\delta}^{\overline{p}}}{\delta_0^{\overline{p}}} = 2\,\frac{\overline{\delta}_0\, \mathbf{e}^{\overline{q}} - 2\,\overline{\boldsymbol{\delta}} - \overline{\boldsymbol{\delta}} \times \mathbf{e}^{\overline{q}}}{2\,\overline{\delta}_0 + \overline{\boldsymbol{\delta}} \cdot \mathbf{e}^{\overline{q}}}. \tag{A6}$$

Appendix A.3. Modified Rodrigues Parameters

Using the inverse of the transformation that defines the chart, φ^{-1},

$$\delta^{\overline{q}} = \frac{1}{16 + \|\mathbf{e}^{\overline{q}}\|^2} \begin{pmatrix} 16 - \|\mathbf{e}^{\overline{q}}\|^2 \\ 8\,\mathbf{e}^{\overline{q}} \end{pmatrix}. \tag{A7}$$

Introducing (A7) into (19),

$$\delta^{\overline{p}} = \overline{\delta}^* * \delta^{\overline{q}} = \frac{1}{16 + \|\mathbf{e}^{\overline{q}}\|^2} \begin{pmatrix} \overline{\delta}_0\,(16 - \|\mathbf{e}^{\overline{q}}\|^2) + 8\,\overline{\boldsymbol{\delta}} \cdot \mathbf{e}^{\overline{q}} \\ 8\,\overline{\delta}_0\, \mathbf{e}^{\overline{q}} - (16 - \|\mathbf{e}^{\overline{q}}\|^2)\,\overline{\boldsymbol{\delta}} - 8\,\overline{\boldsymbol{\delta}} \times \mathbf{e}^{\overline{q}} \end{pmatrix}. \tag{A8}$$

Finally, applying the chart definition,

$$\mathbf{e}^{\overline{p}} = 4\,\frac{\boldsymbol{\delta}^{\overline{p}}}{1 + \delta_0^{\overline{p}}} = 4\,\frac{8\,\overline{\delta}_0\, \mathbf{e}^{\overline{q}} - (16 - \|\mathbf{e}^{\overline{q}}\|^2)\,\overline{\boldsymbol{\delta}} - 8\,\overline{\boldsymbol{\delta}} \times \mathbf{e}^{\overline{q}}}{16 + \|\mathbf{e}^{\overline{q}}\|^2 + \overline{\delta}_0\,(16 - \|\mathbf{e}^{\overline{q}}\|^2) + 8\,\overline{\boldsymbol{\delta}} \cdot \mathbf{e}^{\overline{q}}}. \tag{A9}$$

Appendix A.4. Rotation Vector

Using the inverse of the transformation that defines the chart, φ^{-1},

$$\delta^{\overline{q}} = \begin{pmatrix} \cos\left(\frac{\|\mathbf{e}^{\overline{q}}\|}{2}\right) \\ \hat{\mathbf{e}}^{\overline{q}} \sin\left(\frac{\|\mathbf{e}^{\overline{q}}\|}{2}\right) \end{pmatrix}. \tag{A10}$$

Introducing (A10) into (19),

$$\delta^{\overline{p}} = \overline{\delta}^* * \delta^{\overline{q}} = \begin{pmatrix} \overline{\delta}_0 \cos\left(\frac{\|\mathbf{e}^{\overline{q}}\|}{2}\right) + \overline{\boldsymbol{\delta}} \cdot \hat{\mathbf{e}}^{\overline{q}} \sin\left(\frac{\|\mathbf{e}^{\overline{q}}\|}{2}\right) \\ \overline{\delta}_0\, \hat{\mathbf{e}}^{\overline{q}} \sin\left(\frac{\|\mathbf{e}^{\overline{q}}\|}{2}\right) - \cos\left(\frac{\|\mathbf{e}^{\overline{q}}\|}{2}\right) \overline{\boldsymbol{\delta}} - \overline{\boldsymbol{\delta}} \times \hat{\mathbf{e}}^{\overline{q}} \sin\left(\frac{\|\mathbf{e}^{\overline{q}}\|}{2}\right) \end{pmatrix}. \tag{A11}$$

Finally, applying the chart definition,

$$\mathbf{e}^{\overline{p}} = 2\,\frac{\boldsymbol{\delta}^{\overline{p}}}{\|\boldsymbol{\delta}^{\overline{p}}\|} \arcsin \|\boldsymbol{\delta}^{\overline{p}}\|, \tag{A12}$$

with

$$\delta^{\overline{p}} = \overline{\delta}_0\, \widehat{\mathbf{e}}^{\overline{q}} \sin\left(\frac{\|\mathbf{e}^{\overline{q}}\|}{2}\right) - \cos\left(\frac{\|\mathbf{e}^{\overline{q}}\|}{2}\right) \overline{\delta} - \overline{\delta} \times \widehat{\mathbf{e}}^{\overline{q}} \sin\left(\frac{\|\mathbf{e}^{\overline{q}}\|}{2}\right). \tag{A13}$$

Note that all transition maps are expressed using the $\overline{\delta}$ quaternion. Given that $\overline{\mathbf{e}} = \varphi(\overline{\delta})$ we could also have expressed them using $\overline{\mathbf{e}}$ which is what we get after applying the Kalman update (41). However, our choice makes transition maps to take a simpler form. In addition, having to compute the quaternion $\overline{\delta}$ to perform (43c), this choice does not imply a computational overhead.

Appendix B. Details in the Derivation of the MEKF

This appendix contains the details in the derivation of the Manifold Extended Kalman Filter used in this study.

Appendix B.1. State Prediction

This subsection contains the derivation of the equations for the state prediction.

Appendix B.1.1. Evolution of the Expected Value of the State

Taking expected values in Equation (21) we obtain

$$\frac{d\overline{\omega}'}{dt} = \overline{\mathbf{q}}^{\omega} \quad \Longrightarrow \quad \overline{\omega}'(t) = \overline{\omega}'_0 + \overline{\mathbf{q}}^{\omega}_t\, t, \tag{A14}$$

with $\overline{\mathbf{q}}^{\omega}_t$ the expected value of the random variable $\mathbf{q}^{\omega}$ at time t. Doing $\overline{\omega}'_0 = \overline{\omega}'_{n\text{-}1|n\text{-}1}$ we arrive at

$$\overline{\omega}'_{t|n\text{-}1} = \overline{\omega}'_{n\text{-}1|n\text{-}1} + \overline{\mathbf{q}}^{\omega}_t\, t. \tag{A15}$$

On the other hand, approximating (22) with its Taylor series up to first order around the current state $(\overline{q}, \overline{\omega}')$, and taking its expected value we obtain

$$\mathrm{E}\left[\frac{d\,q(t)}{dt}\right] \approx \frac{1}{2}\,\overline{q}(t) * \overline{\omega}'(t). \tag{A16}$$

This differential equation has no general closed solution. But if we assume that the expected value of the process noise $\overline{\mathbf{q}}^{\omega}(t)$ is zero when $t \in (t_{n\text{-}1}, t_n)$, so that $\overline{\omega}'(t)$ is constant in that interval, then we will have the matrix differential equation

$$\frac{d\,\overline{q}(t)}{dt} = \check{\Omega}_n\, \overline{q}(t), \tag{A17}$$

with

$$\check{\Omega}_n := \frac{1}{2}\begin{pmatrix} 0 & -\overline{\omega}'_1 & -\overline{\omega}'_2 & -\overline{\omega}'_3 \\ \overline{\omega}'_1 & 0 & \overline{\omega}'_3 & -\overline{\omega}'_2 \\ \overline{\omega}'_2 & -\overline{\omega}'_3 & 0 & \overline{\omega}'_1 \\ \overline{\omega}'_3 & \overline{\omega}'_2 & -\overline{\omega}'_1 & 0 \end{pmatrix}_{n|n\text{-}1}. \tag{A18}$$

This differential equation has the solution

$$\overline{q}(t) = e^{\check{\Omega}\, t}\, \overline{q}_0, \tag{A19}$$

where $\overline{q}_0$ represents the initial conditions. After taking $\overline{q}_0 = \overline{q}_{n\text{-}1|n\text{-}1}$, we obtain the prediction $\overline{q}_{t|n\text{-}1}$, that can be expressed using the quaternion product as

$$\overline{q}_{t|n\text{-}1} = \overline{q}_{n\text{-}1|n\text{-}1} * \delta_t^{\omega} = \overline{q}_{n\text{-}1|n\text{-}1} * \begin{pmatrix} \cos\left(\frac{\|\overline{\omega}'_{t|n\text{-}1}\|\,\Delta t}{2}\right) \\ \frac{\overline{\omega}'_{t|n\text{-}1}}{\|\overline{\omega}'_{t|n\text{-}1}\|}\sin\left(\frac{\|\overline{\omega}'_{t|n\text{-}1}\|\,\Delta t}{2}\right) \end{pmatrix}, \tag{A20}$$

with $\Delta t = t - t_{n\text{-}1}$.

Appendix B.1.2. Evolution of the State Covariance Matrix

For a continuous nonlinear system of the form

$$\frac{d\,\mathbf{x}}{dt} = \mathbf{f}(\mathbf{x}, t) + \mathbf{g}(\mathbf{q}), \tag{A21}$$

we know [18] that the covariance matrix satisfies the following differential equation:

$$\frac{d\,\mathbf{P}}{dt} = \mathbf{F}\,\mathbf{P} + \mathbf{P}\,\mathbf{F}^T + \mathbf{G}\,\mathbf{Q}\,\mathbf{G}^T, \tag{A22}$$

where $\mathbf{F} = \frac{\partial \mathbf{f}}{\partial \mathbf{x}}$, and $\mathbf{G} = \frac{\partial \mathbf{g}}{\partial \mathbf{q}}$. This is so because the evolution equation for $\Delta\mathbf{x} = \mathbf{x} - \overline{\mathbf{x}}$ is approximately given by:

$$\frac{d\,\mathbf{x}}{dt} \approx \frac{d\,\overline{\mathbf{x}}}{dt} + \left.\frac{\partial\,\mathbf{f}}{\partial\mathbf{x}}\right|_{\mathbf{x}=\overline{\mathbf{x}}} (\mathbf{x} - \overline{\mathbf{x}}) + \left.\frac{\partial\,\mathbf{g}}{\partial\mathbf{q}}\right|_{\mathbf{q}=\overline{\mathbf{q}}} (\mathbf{q} - \overline{\mathbf{q}}), \tag{A23}$$

and $\mathbf{P}$ is defined as $\mathbf{P} = \mathrm{E}\left[\Delta\mathbf{x}\,(\Delta\mathbf{x})^T\right]$. However, we have a different definition for $\mathbf{P}$:

$$\mathbf{P} = \mathrm{E}\left[\begin{pmatrix} \mathbf{e}^{\overline{q}} - \overline{\mathbf{e}}^{\overline{q}} \\ \omega' - \overline{\omega}' \end{pmatrix} \begin{pmatrix} \mathbf{e}^{\overline{q}} - \overline{\mathbf{e}}^{\overline{q}} \\ \omega' - \overline{\omega}' \end{pmatrix}^T\right]. \tag{A24}$$

Then we need to find the evolution equation for $\mathbf{e}^{\overline{q}}$. Recall that we are assuming $\overline{\mathbf{e}}^{\overline{q}} = \mathbf{0}$ at the beginning of the iteration. Knowing that any quaternion in the unit sphere can be expressed as a deviation from a central quaternion $\overline{q}$ as $q = \overline{q} * \delta$, and using the differential Equations (22) and (A16), we can find a differential equation for the quaternion δ:

$$q = \overline{q} * \delta \implies \tag{A25a}$$

$$\implies \dot{q} = \dot{\overline{q}} * \delta + \overline{q} * \dot{\delta} \implies \tag{A25b}$$

$$\implies \frac{1}{2} q * \omega' \approx \frac{1}{2}\overline{q} * \overline{\omega}' * \delta + \overline{q} * \dot{\delta}, \tag{A25c}$$

where a dot over a symbol represents time derivative, and we have obviated the time dependence. Isolating the time derivative $\dot{\delta}$,

$$\dot{\delta} \approx \frac{1}{2}\overbrace{\overline{q}^* * q}^{\delta} * \omega' - \frac{1}{2}\overbrace{\overline{q}^* * \overline{q}}^{1} * \overline{\omega}' * \delta = \tag{A26a}$$

$$= \frac{1}{2}\left[\delta * \omega' - \overline{\omega}' * \delta\right] = \tag{A26b}$$

$$= \frac{1}{2}\left[\begin{pmatrix}\delta_0\\ \boldsymbol{\delta}\end{pmatrix} * \begin{pmatrix}0\\ \boldsymbol{\omega}'\end{pmatrix} - \begin{pmatrix}0\\ \overline{\boldsymbol{\omega}}'\end{pmatrix} * \begin{pmatrix}\delta_0\\ \boldsymbol{\delta}\end{pmatrix}\right] = \tag{A26c}$$

$$= \frac{1}{2}\begin{pmatrix} -(\boldsymbol{\omega}' - \overline{\boldsymbol{\omega}}')\cdot\boldsymbol{\delta} \\ \delta_0\,(\boldsymbol{\omega}' - \overline{\boldsymbol{\omega}}') - (\boldsymbol{\omega}' + \overline{\boldsymbol{\omega}}')\times\boldsymbol{\delta}\end{pmatrix} = \tag{A26d}$$

$$= \frac{1}{2}\begin{pmatrix} -\Delta\boldsymbol{\omega}'\cdot\boldsymbol{\delta} \\ \delta_0\,\Delta\boldsymbol{\omega}' - (2\overline{\boldsymbol{\omega}}' + \Delta\boldsymbol{\omega}')\times\boldsymbol{\delta}\end{pmatrix}. \tag{A26e}$$

Knowing that, for each of our charts, the δ quaternion can be approximated by (10) as $\mathbf{e} \to \mathbf{0}$, then we can obtain an approximate differential equation for a point $\mathbf{e}$ expressed in the $\overline{q}$-centered chart. Note that we have not explicitly denoted $\mathbf{e}^{\overline{q}}$ or $\delta^{\overline{q}}$. This will be assumed implicitly, since these quantities will always be expressed in the $\overline{q}$-centered chart in this appendix. Using the chain rule for a time derivative and expression (A26e),

$$\dot{e}_i = \sum_j \overbrace{\frac{\partial e_i}{\partial \delta_j}}^{\approx 2\,\delta_{ij}} \overbrace{\frac{\partial \delta_j}{\partial t}}^{\dot{\delta}_j} \equiv \tag{A27a}$$

$$\equiv \quad \dot{\mathbf{e}} \approx \delta_0\,\Delta\boldsymbol{\omega}' - (2\overline{\boldsymbol{\omega}}' + \Delta\boldsymbol{\omega}')\times\boldsymbol{\delta} \approx \tag{A27b}$$

$$\approx \left(1 - \frac{\|\mathbf{e}\|^2}{8}\right)\Delta\boldsymbol{\omega}' - (2\overline{\boldsymbol{\omega}}' + \Delta\boldsymbol{\omega}')\times\frac{\mathbf{e}}{2}. \tag{A27c}$$

Then, the first order approximation to differential Equation (A27c) would be

$$\dot{\mathbf{e}} \approx \Delta\boldsymbol{\omega}' - \overline{\boldsymbol{\omega}}'\times\mathbf{e}. \tag{A28}$$

On the other hand, combining Equations (21) and (A14) we obtain

$$\frac{d\,\Delta\boldsymbol{\omega}'}{dt} = \frac{d\,(\boldsymbol{\omega}' - \overline{\boldsymbol{\omega}}')}{dt} = \mathbf{q}^{\omega} - \overline{\mathbf{q}}^{\omega} = \Delta\mathbf{q}^{\omega}. \tag{A29}$$

Summarizing,

$$\frac{d}{dt}\begin{pmatrix}\mathbf{e}\\ \Delta\boldsymbol{\omega}'\end{pmatrix} \approx \begin{pmatrix} -[\overline{\boldsymbol{\omega}}']_\times & \mathbf{I} \\ \mathbf{0} & \mathbf{0}\end{pmatrix}\begin{pmatrix}\mathbf{e}\\ \Delta\boldsymbol{\omega}'\end{pmatrix} + \begin{pmatrix}\mathbf{0}\\ \Delta\mathbf{q}^{\omega}\end{pmatrix}, \tag{A30}$$

therefore matrices $\mathbf{F}$, $\mathbf{G}$, and $\mathbf{Q}$ in (A22) are in our case

$$\mathbf{F} = \begin{pmatrix} -[\overline{\omega}']_\times & \mathbf{I} \\ \mathbf{0} & \mathbf{0} \end{pmatrix}, \tag{A31}$$

$$\mathbf{G} = \mathbf{I}, \tag{A32}$$

$$\mathbf{Q} = \begin{pmatrix} \mathbf{0} & \mathbf{0} \\ \mathbf{0} & \mathrm{E}[\,\Delta\mathbf{q}^{\omega}\,(\Delta\mathbf{q}^{\omega})^T\,] \end{pmatrix}. \tag{A33}$$

We are now in a position to solve the differential Equation (A22). Let us consider its homogeneous version first:

$$\frac{d\,\mathbf{P}_H}{dt} = \mathbf{F}\,\mathbf{P}_H + \mathbf{P}_H\,\mathbf{F}^T, \tag{A34}$$

which has as solution

$$\mathbf{P}_H = e^{\mathbf{F}\,t}\,\mathbf{C}_0\,e^{\mathbf{F}^T\,t}. \tag{A35}$$

Taking into account the definition of matrix exponential, and after computing the powers of $\mathbf{F}$ we obtain

$$e^{\mathbf{F}\,t} = \begin{pmatrix} \sum\limits_{n=0}^{\infty} \frac{(-\boldsymbol{\Omega})^n\,t^n}{n!} & \sum\limits_{n=1}^{\infty} \frac{(-\boldsymbol{\Omega})^{n-1}\,t^n}{n!} \\ \mathbf{0} & \mathbf{I} \end{pmatrix} \approx \begin{pmatrix} \mathbf{R}^T(\,\delta^{\omega}\,) & \mathbf{I}\,t \\ \mathbf{0} & \mathbf{I} \end{pmatrix}, \tag{A36}$$

where we have denoted $\boldsymbol{\Omega} = [\overline{\omega}']_\times$, and $\delta^{\omega} = \left(\cos\frac{\|\overline{\omega}'\|\,t}{2},\ \frac{\overline{\omega}'}{\|\overline{\omega}'\|}\sin\frac{\|\overline{\omega}'\|\,t}{2}\right)$. We also have assumed that t takes small values so we can approximate the infinite sums truncating in the first term. To find the solution of the non-homogeneous differential equation we use the *variation of constants method*:

$$\mathbf{P} = e^{\mathbf{F}\,t}\,\mathbf{C}(t)\,e^{\mathbf{F}^T\,t} \quad \Longrightarrow \tag{A37}$$

$$\Longrightarrow \quad \frac{d\,\mathbf{P}}{dt} = \mathbf{F}\,e^{\mathbf{F}\,t}\,\mathbf{C}(t)\,e^{\mathbf{F}^T\,t} + e^{\mathbf{F}\,t}\,\mathbf{C}(t)\,e^{\mathbf{F}^T\,t}\,\mathbf{F}^T + e^{\mathbf{F}\,t}\,\frac{d\,\mathbf{C}(t)}{dt}\,e^{\mathbf{F}^T\,t} = \tag{A38}$$

$$= \mathbf{F}\,\mathbf{P} + \mathbf{P}\,\mathbf{F}^T + e^{\mathbf{F}\,t}\,\frac{d\,\mathbf{C}(t)}{dt}\,e^{\mathbf{F}^T\,t}. \tag{A39}$$

Identifying terms with (A22) we obtain that

$$e^{\mathbf{F}\,t}\,\frac{d\,\mathbf{C}(t)}{dt}\,e^{\mathbf{F}^T\,t} = \mathbf{G}\,\mathbf{Q}\,\mathbf{G}^T \quad \Longrightarrow \tag{A40}$$

$$\Longrightarrow \quad \frac{d\,\mathbf{C}(t)}{dt} = e^{-\mathbf{F}\,t}\,\mathbf{Q}\,e^{-\mathbf{F}^T\,t} = \sum_{n=0}^{\infty}\sum_{m=0}^{\infty} \frac{(-\mathbf{F})^n\,t^n}{n!}\,\mathbf{Q}\,\frac{(-\mathbf{F}^T)^m\,t^m}{m!} \quad \Longrightarrow \tag{A41}$$

$$\Longrightarrow \quad \mathbf{C}(t) = \mathbf{C}_0 + \sum_{n=0}^{\infty}\sum_{m=0}^{\infty} \frac{(-\mathbf{F})^n}{n!}\,\mathbf{Q}\,\frac{(-\mathbf{F}^T)^m}{m!}\,\frac{t^{n+m+1}}{n+m+1}. \tag{A42}$$

Finally, truncating the summation in (A42) at the first non-zero elements, and inserting the result into (A37), we obtain (32) where we have identified $\mathbf{C}_0 = \mathbf{P}(0)$ through the initial conditions.

Appendix B.2. Measurement Prediction

This subsection contains the derivation of the equations for the measurement prediction.

Appendix B.2.1. Expected Value of the Measurement Prediction

Taking expected values on (24), and assuming $\overline{\mathbf{r}}_t^{\omega} = \mathbf{0}$ we arrive at (35). On the other hand, approximating (23) with its Taylor series up to first order around the current estimation of the state $(\overline{q}, \overline{\omega}')$, taking its expected value, and assuming $\overline{\mathbf{r}}_t^{\mathbf{v}} = \mathbf{0}$ we obtain (34).

Appendix B.2.2. Covariance Matrix of the Measurement Prediction

In order to find the covariance matrix of the measurement prediction we need the linear approximation of the vector measurement around the point $\mathbf{x}_0 := (\mathbf{e} = \mathbf{0}, \mathbf{q}^{\mathbf{v}} = \overline{\mathbf{q}}^{\mathbf{v}}, \mathbf{r}^{\mathbf{v}} = \overline{\mathbf{r}}^{\mathbf{v}})$:

$$\mathbf{v}^m \approx \overline{\mathbf{v}}^m + \left.\frac{\partial \mathbf{v}^m}{\partial \mathbf{e}}\right|_{\mathbf{x}_0} \mathbf{e} + \left.\frac{\partial \mathbf{v}^m}{\partial \mathbf{q}^{\mathbf{v}}}\right|_{\mathbf{x}_0} (\mathbf{q}^{\mathbf{v}} - \overline{\mathbf{q}}^{\mathbf{v}}) + \left.\frac{\partial \mathbf{v}^m}{\partial \mathbf{r}^{\mathbf{v}}}\right|_{\mathbf{x}_0} (\mathbf{r}^{\mathbf{v}} - \overline{\mathbf{r}}^{\mathbf{v}}). \tag{A43}$$

It is direct to identify

$$\left.\frac{\partial \mathbf{v}^m}{\partial \mathbf{q}^{\mathbf{v}}}\right|_{\mathbf{x}_0} = \mathbf{R}^T(\overline{q}), \tag{A44}$$

$$\left.\frac{\partial \mathbf{v}^m}{\partial \mathbf{r}^{\mathbf{v}}}\right|_{\mathbf{x}_0} = \mathbf{I}. \tag{A45}$$

On the other hand, rewriting (23) as

$$v^m = \delta^* * \overline{q}^* * (q^{\mathbf{v}} + v) * \overline{q} * \delta + r^{\mathbf{v}} \equiv \tag{A46}$$

$$\equiv \mathbf{v}^m = \mathbf{R}^T(\delta)\, \mathbf{R}^T(\overline{q})\, (\mathbf{q}^{\mathbf{v}} + \mathbf{v}) + \mathbf{r}^{\mathbf{v}}, \tag{A47}$$

and noting that setting $\mathbf{e} = \mathbf{0}$ is equivalent to do $\delta = 1$,

$$\left.\frac{\partial v_i^m}{\partial e_j}\right|_{\mathbf{x}_0} = \sum_k \left.\frac{\partial v_i^m}{\partial \delta_k}\right|_{\mathbf{x}_0} \left.\frac{\partial \delta_k}{\partial e_j}\right|_{\mathbf{e}=\mathbf{0}} = \sum_{kl} \left.\frac{\partial R_{il}^T(\delta)}{\partial \delta_k}\right|_{\delta=1} \left[\mathbf{R}^T(\overline{q})\, (\overline{\mathbf{q}}^{\mathbf{v}} + \mathbf{v})\right]_l \left.\frac{\partial \delta_k}{\partial e_j}\right|_{\mathbf{e}=\mathbf{0}}. \tag{A48}$$

Now, recalling (5) we have

$$\left.\frac{\partial \mathbf{R}^T(\delta)}{\partial \delta_k}\right|_{\delta=1} = 2 \begin{pmatrix} 0 & \delta_{3k} & -\delta_{2k} \\ -\delta_{3k} & 0 & \delta_{1k} \\ \delta_{2k} & -\delta_{1k} & 0 \end{pmatrix} \equiv -2 \sum_n \varepsilon_{inl}\, \delta_{nk}, \tag{A49}$$

with ε_{inl} the Levi-Civita symbol, and δ_{nk} the Kronecker delta. Recalling (10) we also have

$$\frac{\partial \delta}{\partial e_j} = \left.\begin{pmatrix} -e_j/4 \\ \delta_{1j}/2 \\ \delta_{2j}/2 \\ \delta_{3j}/2 \end{pmatrix}\right|_{\mathbf{e}=\mathbf{0}} \equiv (1 - \delta_{0k})\, \delta_{kj}/2. \tag{A50}$$

Then, introducing in (A48) Equations (A49), (34), and (A50),

$$\left.\frac{\partial v_i^m}{\partial e_j}\right|_{\mathbf{x}_0} \approx -\sum_{kln} \varepsilon_{inl}\, \delta_{nk}\, \overline{v}_l^m\, (1 - \delta_{0k})\, \delta_{kj} = \sum_l \varepsilon_{ilj}\, \overline{v}_l^m \equiv [\overline{\mathbf{v}}^m]_\times, \tag{A51}$$

where we have used $\varepsilon_{ijl} = -\varepsilon_{ilj}$. Finally, assuming the independence of the random variables $\mathbf{x}_t = (\mathbf{e}_t, \omega_t')^T$, $\mathbf{q}_t^{\mathbf{v}}$, $\mathbf{r}_t^{\mathbf{v}}$, and $\mathbf{r}_t^{\omega}$, and computing the covariance matrix $\mathbf{S}_t = \mathrm{E}[(\mathbf{z}_t - \overline{\mathbf{z}}_t)(\mathbf{z}_t - \overline{\mathbf{z}}_t)^T]$ with $\mathbf{z}_t = (\mathbf{v}_t^m, \omega_t^m)^T$ and (A43), we arrive at (37) and (38).

Appendix C. Derivation of the T-matrices

This appendix contains the derivation of the **T**-matrix for each chart.

Appendix C.1. Orthographic

Our transition map (A3) can be written as

$$e_i^{\overline{p}}(\mathbf{e}^{\overline{q}}) \quad = \quad \overline{\delta}_0\, e_i^{\overline{q}} \,-\, \sqrt{4 - \textstyle\sum_k (e_k^{\overline{q}})^2}\; \overline{\delta}_i \,-\, \sum_{lm} \varepsilon_{ilm}\, \overline{\delta}_l\, e_m^{\overline{q}}, \tag{A52}$$

being ε_{ilm} the Levi-Civita symbol. Finding (45) for (A52) we obtain

$$(\mathbf{T})_{ij} \quad = \quad \left[\overline{\delta}_0\, \delta_{ij} \,-\, \frac{-\sum_k e_k^{\overline{q}}\, \delta_{kj}}{\sqrt{4 - \sum_k (e_k^{\overline{q}})^2}}\, \overline{\delta}_i \,-\, \sum_{lm} \varepsilon_{ilm}\, \overline{\delta}_l\, \delta_{mj} \right]_{\mathbf{e}^{\overline{q}} = \overline{\mathbf{e}}^{\overline{q}}} \quad = \tag{A53a}$$

$$= \quad \overline{\delta}_0\, \delta_{ij} \,+\, \frac{\overline{e}_j^{\overline{q}}}{\sqrt{4 - \sum_k (\overline{e}_k^{\overline{q}})^2}}\, \overline{\delta}_i \,-\, \sum_l \varepsilon_{ilj}\, \overline{\delta}_l, \tag{A53b}$$

This expression can be rewritten in matrix form as

$$\mathbf{T} \quad = \quad \overline{\delta}_0\, \mathbf{I} \,+\, \frac{\overline{\boldsymbol{\delta}}\, (\overline{\mathbf{e}}^{\overline{q}})^T}{\sqrt{4 - \|\overline{\mathbf{e}}^{\overline{q}}\|^2}} \,-\, [\,\overline{\boldsymbol{\delta}}\,]_\times. \tag{A54}$$

Finally, recalling that for this chart $\overline{\delta}_0 = \sqrt{1 - \|\overline{\mathbf{e}}^{\overline{q}}\|^2/4}$ and $\overline{\boldsymbol{\delta}} = \overline{\mathbf{e}}^{\overline{q}}/2$, we arrive at the final expression

$$\mathbf{T} \quad = \quad \overline{\delta}_0\, \mathbf{I} \,+\, \frac{\overline{\boldsymbol{\delta}}\, \overline{\boldsymbol{\delta}}^T}{\overline{\delta}_0} \,-\, [\,\overline{\boldsymbol{\delta}}\,]_\times. \tag{A55}$$

Appendix C.2. Rodrigues Parameters

First, let us denote the numerator of (A6) as $\mathbf{N}(\mathbf{e}^{\overline{q}})$, and its denominator as $D(\mathbf{e}^{\overline{q}})$:

$$\mathbf{N}(\,\mathbf{e}^{\overline{q}}\,) \quad := \quad \overline{\delta}_0\, \mathbf{e}^{\overline{q}} \,-\, 2\,\overline{\boldsymbol{\delta}} \,-\, \overline{\boldsymbol{\delta}} \times \mathbf{e}^{\overline{q}}, \tag{A56}$$

$$D(\,\mathbf{e}^{\overline{q}}\,) \quad := \quad 2\,\overline{\delta}_0 \,+\, \overline{\boldsymbol{\delta}} \cdot \mathbf{e}^{\overline{q}}. \tag{A57}$$

Now let us evaluate (A56) at $\overline{\mathbf{e}}^{\overline{q}}$:

$$\mathbf{N}(\,\overline{\mathbf{e}}^{\overline{q}}\,) \quad = \quad \overbrace{\overline{\delta}_0\, \underbrace{\overline{\mathbf{e}}^{\overline{q}}}_{2\overline{\boldsymbol{\delta}}/\overline{\delta}_0} - 2\,\overline{\boldsymbol{\delta}}}^{0} \,-\, \overbrace{\overline{\boldsymbol{\delta}} \times \overline{\mathbf{e}}^{\overline{q}}}^{=\,\mathbf{0}\ (\overline{\boldsymbol{\delta}} \,\|\, \overline{\mathbf{e}}^{\overline{q}}).} \quad = \quad \mathbf{0}. \tag{A58}$$

Then, the approximation of $\mathbf{N}(\,\mathbf{e}^{\overline{q}}\,)$ does not have terms of order $\mathcal{O}(1)$. This means that we will only need to approximate $D\,(\mathbf{e}^{\overline{q}})$ to the zeroth order. Any further approximation would produce, after multiplying by the linear approximation of $\mathbf{N}(\,\mathbf{e}^{\overline{q}}\,)$, a higher order term. Let us then calculate each approximation.

We can rewrite (A56) as

$$N_i(\mathbf{e}^{\overline{q}}) \quad = \quad \overline{\delta}_0\, e_i^{\overline{q}} \,-\, 2\,\overline{\delta}_i \,-\, \sum_{kl} \varepsilon_{ikl}\, \overline{\delta}_k\, e_l^{\overline{q}}, \tag{A59}$$

with ε_{ikl} the Levi-Civita symbol. Applying (44) to (A59),

$$N_i(\mathbf{e}^{\overline{q}}) \approx \sum_j \left[\overline{\delta}_0 \delta_{ij} - \sum_{kl} \varepsilon_{ikl} \overline{\delta}_k \delta_{lj} \right]_{\mathbf{e}^{\overline{q}}=\overline{\mathbf{e}}^{\overline{q}}} \left(e_j^{\overline{q}} - \overline{e}_j^{\overline{q}} \right) = \tag{A60a}$$

$$= \sum_j \left[\overline{\delta}_0 \delta_{ij} - \sum_k \varepsilon_{ikj} \overline{\delta}_k \right] \left(e_j^{\overline{q}} - \overline{e}_j^{\overline{q}} \right), \tag{A60b}$$

being δ_{ij} the Kronecker delta. Returning to matrix notation, the linear approximation of $\mathbf{N}\left(\mathbf{e}^{\overline{q}}\right)$ is

$$\mathbf{N}(\mathbf{e}^{\overline{q}}) = \left[\overline{\delta}_0 \mathbf{I} - [\overline{\boldsymbol{\delta}}]_\times \right] \left(\mathbf{e}^{\overline{q}} - \overline{\mathbf{e}}^{\overline{q}} \right) + \mathcal{O}\left(\|\mathbf{e}^{\overline{q}} - \overline{\mathbf{e}}^{\overline{q}}\|^2 \right). \tag{A61}$$

On the other hand, evaluating (A57) at $\overline{\mathbf{e}}^{\overline{q}}$ we obtain the zeroth order approximation:

$$D(\overline{\mathbf{e}}^{\overline{q}}) = 2\overline{\delta}_0 + \overline{\boldsymbol{\delta}} \cdot 2\frac{\overline{\boldsymbol{\delta}}}{\overline{\delta}_0} + \mathcal{O}\left(\|\mathbf{e}^{\overline{q}} - \overline{\mathbf{e}}^{\overline{q}}\| \right) = \tag{A62a}$$

$$= \frac{2}{\overline{\delta}_0} \underbrace{\left(\overline{\delta}_0^2 + \|\overline{\boldsymbol{\delta}}\|^2 \right)}_{1} + \mathcal{O}\left(\|\mathbf{e}^{\overline{q}} - \overline{\mathbf{e}}^{\overline{q}}\| \right) = \tag{A62b}$$

$$= \frac{2}{\overline{\delta}_0} + \mathcal{O}\left(\|\mathbf{e}^{\overline{q}} - \overline{\mathbf{e}}^{\overline{q}}\| \right). \tag{A62c}$$

Finally, combining (A61) and (A62c) we can compute the linear approximation of (A6):

$$\mathbf{e}^{\overline{p}}(\mathbf{e}^{\overline{q}}) = 2\left\{ \left[\overline{\delta}_0 \mathbf{I} - [\overline{\boldsymbol{\delta}}]_\times \right] (\mathbf{e}^{\overline{q}} - \overline{\mathbf{e}}^{\overline{q}}) + \mathcal{O}\left(\|\mathbf{e}^{\overline{q}} - \overline{\mathbf{e}}^{\overline{q}}\|^2 \right) \right\} \left[\frac{\overline{\delta}_0}{2} + \mathcal{O}\left(\|\mathbf{e}^{\overline{q}} - \overline{\mathbf{e}}^{\overline{q}}\| \right) \right] = \tag{A63a}$$

$$= \overline{\delta}_0 \left[\overline{\delta}_0 \mathbf{I} - [\overline{\boldsymbol{\delta}}]_\times \right] (\mathbf{e}^{\overline{q}} - \overline{\mathbf{e}}^{\overline{q}}) + \mathcal{O}\left(\|\mathbf{e}^{\overline{q}} - \overline{\mathbf{e}}^{\overline{q}}\|^2 \right). \tag{A63b}$$

Appendix C.3. Modified Rodrigues Parameters

First, let us denote the numerator of (A9) as $\mathbf{N}(\mathbf{e}^{\overline{q}})$, and its denominator as $D(\mathbf{e}^{\overline{q}})$:

$$\mathbf{N}(\mathbf{e}^{\overline{q}}) = 8\overline{\delta}_0 \mathbf{e}^{\overline{q}} - \left(16 - \|\mathbf{e}^{\overline{q}}\|^2 \right) \overline{\boldsymbol{\delta}} - 8\overline{\boldsymbol{\delta}} \times \mathbf{e}^{\overline{q}}, \tag{A64}$$

$$D(\mathbf{e}^{\overline{q}}) = 16 + \|\mathbf{e}^{\overline{q}}\|^2 + \overline{\delta}_0 \left(16 - \|\mathbf{e}^{\overline{q}}\|^2 \right) + 8\overline{\boldsymbol{\delta}} \cdot \mathbf{e}^{\overline{q}}. \tag{A65}$$

Now let us evaluate (A64) at $\overline{\mathbf{e}}^{\overline{q}}$:

$$\mathbf{N}(\overline{\mathbf{e}}^{\overline{q}}) = 8\overline{\delta}_0 \overbrace{\overline{\mathbf{e}}^{\overline{q}}}^{4\overline{\boldsymbol{\delta}}/(1+\overline{\delta}_0)} - (16 - \overbrace{\|\overline{\mathbf{e}}^{\overline{q}}\|^2}^{16\|\overline{\boldsymbol{\delta}}\|^2/(1+\overline{\delta}_0)^2})\,\overline{\boldsymbol{\delta}} - 8\overbrace{\overline{\boldsymbol{\delta}} \times \overline{\mathbf{e}}^{\overline{q}}}^{=\mathbf{0}\,(\overline{\mathbf{e}}^{\overline{q}} \| \overline{\boldsymbol{\delta}})} = \tag{A66a}$$

$$= \frac{16\overline{\boldsymbol{\delta}}}{1+\overline{\delta}_0} \left[2\overline{\delta}_0 - \left(1 + \overline{\delta}_0 - \frac{\overbrace{\|\overline{\boldsymbol{\delta}}\|^2}^{1-\overline{\delta}_0^2}}{1+\overline{\delta}_0} \right) \right] = \mathbf{0}. \tag{A66b}$$

Then, as with the RP chart, the approximation of $\mathbf{N}\left(\mathbf{e}^{\overline{q}}\right)$ does not have terms of order $\mathcal{O}(1)$, and we will only need to approximate $D\left(\mathbf{e}^{\overline{q}}\right)$ to the zeroth order.

We can write (A64) as

$$N_i(\mathbf{e}^{\overline{q}}) = 8\overline{\delta}_0 e_i^{\overline{q}} - \left(16 - \sum_k (e_k^{\overline{q}})^2 \right) \overline{\delta}_i - 8\sum_{lm} \varepsilon_{ilm} \overline{\delta}_l e_m^{\overline{q}}, \tag{A67}$$

with ε_{ilm} the Levi-Civita symbol. Applying (44) to (A67),

$$N_i(\mathbf{e}^{q}) \approx \sum_j \left[8\,\bar{\delta}_0\,\delta_{ij} + \sum_k 2\,e^{q}_k\,\delta_{kj}\,\bar{\delta}_i - 8\sum_{lm}\varepsilon_{ilm}\,\bar{\delta}_l\,\delta_{mj} \right]_{\mathbf{e}^{q}=\bar{\mathbf{e}}^{q}} \left(e^{q}_j - \bar{e}^{q}_j \right) = \tag{A68a}$$

$$= \sum_j \left[8\,\bar{\delta}_0\,\delta_{ij} + 2\,\bar{e}^{q}_j\,\bar{\delta}_i - 8\sum_l \varepsilon_{ilj}\,\bar{\delta}_l \right] \left(e^{q}_j - \bar{e}^{q}_j \right), \tag{A68b}$$

being δ_{ij} the Kronecker delta. Returning to matrix notation, the linear approximation of $\mathbf{N}(\mathbf{e}^{q})$ is

$$\mathbf{N}(\mathbf{e}^{q}) = \left[8\,\bar{\delta}_0\,\mathbf{I} + 2\,\bar{\boldsymbol{\delta}}\,(\bar{\mathbf{e}}^{q})^T - 8\,[\bar{\boldsymbol{\delta}}]_\times \right] \left(\mathbf{e}^{q} - \bar{\mathbf{e}}^{q} \right) + \mathcal{O}\left(\|\mathbf{e}^{q} - \bar{\mathbf{e}}^{q}\|^2 \right) = \tag{A69a}$$

$$= 8\left[\bar{\delta}_0\,\mathbf{I} + \frac{\bar{\boldsymbol{\delta}}\,\bar{\boldsymbol{\delta}}^T}{1+\bar{\delta}_0} - [\bar{\boldsymbol{\delta}}]_\times \right] \left(\mathbf{e}^{q} - \bar{\mathbf{e}}^{q} \right) + \mathcal{O}\left(\|\mathbf{e}^{q} - \bar{\mathbf{e}}^{q}\|^2 \right). \tag{A69b}$$

On the other hand, evaluating (A65) at $\bar{\mathbf{e}}^{q}$ we obtain the zeroth order approximation:

$$D(\bar{\mathbf{e}}^{q}) \approx 16 + 16\,\frac{\|\bar{\boldsymbol{\delta}}\|^2}{(1+\bar{\delta}_0)^2} + \bar{\delta}_0 \left(16 - 16\,\frac{\|\bar{\boldsymbol{\delta}}\|^2}{(1+\bar{\delta}_0)^2} \right) + 8\,\bar{\boldsymbol{\delta}}\cdot 4\,\frac{\bar{\boldsymbol{\delta}}}{1+\bar{\delta}_0} = \tag{A70a}$$

$$= \frac{16}{1+\bar{\delta}_0} \left[(1+\bar{\delta}_0) + \frac{\|\bar{\boldsymbol{\delta}}\|^2}{1+\bar{\delta}_0} + \bar{\delta}_0 \left((1+\bar{\delta}_0) - \frac{\|\bar{\boldsymbol{\delta}}\|^2}{1+\bar{\delta}_0} \right) + 2\,\|\bar{\boldsymbol{\delta}}\|^2 \right] = \tag{A70b}$$

$$= \frac{16}{1+\bar{\delta}_0} \left[2 + \bar{\delta}_0\,(2\,\bar{\delta}_0) + 2\left(1 - \bar{\delta}_0^2\right) \right] = \frac{64}{1+\bar{\delta}_0}, \tag{A70c}$$

where we have used the equality $\|\bar{\boldsymbol{\delta}}\|^2 = 1 - \bar{\delta}_0^2$ for unit quaternions. Finally, combining (A69b) and (A70c) we can compute the linear approximation of (A9):

$$\mathbf{e}^{\bar{p}}(\mathbf{e}^{q}) = 4\left\{ 8\left[\bar{\delta}_0\,\mathbf{I} + \frac{\bar{\boldsymbol{\delta}}\,\bar{\boldsymbol{\delta}}^T}{1+\bar{\delta}_0} - [\bar{\boldsymbol{\delta}}]_\times \right] (\mathbf{e}^{q} - \bar{\mathbf{e}}^{q}) + \mathcal{O}\left(\|\mathbf{e}^{q} - \bar{\mathbf{e}}^{q}\|^2 \right) \right\} \left[\frac{1+\bar{\delta}_0}{64} + \mathcal{O}\left(\|\mathbf{e}^{q} - \bar{\mathbf{e}}^{q}\| \right) \right] = \tag{A71a}$$

$$= \frac{1+\bar{\delta}_0}{2} \left[\bar{\delta}_0\,\mathbf{I} + \frac{\bar{\boldsymbol{\delta}}\,\bar{\boldsymbol{\delta}}^T}{1+\bar{\delta}_0} - [\bar{\boldsymbol{\delta}}]_\times \right] (\mathbf{e}^{q} - \bar{\mathbf{e}}^{q}) + \mathcal{O}\left(\|\mathbf{e}^{q} - \bar{\mathbf{e}}^{q}\|^2 \right) = \tag{A71b}$$

$$= \frac{1}{2}\left[(1+\bar{\delta}_0)\left(\bar{\delta}_0\,\mathbf{I} - [\bar{\boldsymbol{\delta}}]_\times \right) + \bar{\boldsymbol{\delta}}\,\bar{\boldsymbol{\delta}}^T \right] (\mathbf{e}^{q} - \bar{\mathbf{e}}^{q}) + \mathcal{O}\left(\|\mathbf{e}^{q} - \bar{\mathbf{e}}^{q}\|^2 \right). \tag{A71c}$$

Appendix C.4. Rotation Vector

Let us start evaluating the vector $\boldsymbol{\delta}^{\bar{p}}$ in (A12) and (A13) at the point $\bar{\mathbf{e}}^{q}$:

$$\boldsymbol{\delta}^{\bar{p}}(\bar{\mathbf{e}}^{q}) = \overbrace{\underbrace{\bar{\delta}_0\,\hat{\bar{\mathbf{e}}}^{q}\sin\left(\frac{\|\bar{\mathbf{e}}^{q}\|}{2}\right)}_{\bar{\boldsymbol{\delta}}} - \underbrace{\cos\left(\frac{\|\bar{\mathbf{e}}^{q}\|}{2}\right)}_{\bar{\delta}_0}\bar{\boldsymbol{\delta}}}^{=0} - \overbrace{\bar{\boldsymbol{\delta}}\times\underbrace{\hat{\bar{\mathbf{e}}}^{q}\sin\left(\frac{\|\bar{\mathbf{e}}^{q}\|}{2}\right)}_{\bar{\boldsymbol{\delta}}}}^{=0} = \mathbf{0}. \tag{A72}$$

Then, the first order approximation of $\boldsymbol{\delta}^{\bar{p}}$ around $\bar{\mathbf{e}}^{q}$ will have the form

$$\boldsymbol{\delta}^{\bar{p}} = \tilde{\mathbf{T}}\left(\mathbf{e}^{q} - \bar{\mathbf{e}}^{q} \right) + \mathcal{O}\left(\|\mathbf{e}^{q} - \bar{\mathbf{e}}^{q}\|^2 \right), \tag{A73}$$

and $\|\boldsymbol{\delta}^{\bar{p}}\| \to 0$ as $\mathbf{e}^{q} \to \bar{\mathbf{e}}^{q}$. Taking the Taylor series of the $\arcsin x$,

$$\frac{\arcsin\|\boldsymbol{\delta}^{\bar{p}}\|}{\|\boldsymbol{\delta}^{\bar{p}}\|} = \frac{\|\boldsymbol{\delta}^{\bar{p}}\| + \mathcal{O}\left(\|\boldsymbol{\delta}^{\bar{p}}\|^3 \right)}{\|\boldsymbol{\delta}^{\bar{p}}\|} = 1 + \mathcal{O}\left(\|\boldsymbol{\delta}^{\bar{p}}\|^2 \right) = 1 + \mathcal{O}\left(\|\mathbf{e}^{q} - \bar{\mathbf{e}}^{q}\|^2 \right), \tag{A74}$$

so that (A12) is linearized as

$$2\,\frac{\delta^{\overline{p}}}{\|\delta^{\overline{p}}\|}\arcsin\|\delta^{\overline{p}}\| \quad = \quad 2\,\widetilde{\mathbf{T}}\left(\mathbf{e}^{\overline{q}}-\overline{\mathbf{e}}^{\overline{q}}\right)+\mathcal{O}\left(\|\mathbf{e}^{\overline{q}}-\overline{\mathbf{e}}^{\overline{q}}\|^2\right). \tag{A75}$$

We only lack the $\widetilde{\mathbf{T}}$ matrix. We will need the linear approximations of $\cos\left(\|\mathbf{e}^{\overline{q}}\|/2\right)$ and $\widehat{\mathbf{e}}^{\overline{q}}\sin\left(\|\mathbf{e}^{\overline{q}}\|/2\right)$ around $\overline{\mathbf{e}}^{\overline{q}}$. To this end we will first obtain the linear approximation of $\|\mathbf{x}\|$:

$$\|\mathbf{x}\| \quad = \quad \sqrt{\textstyle\sum_k x_k^2} \quad = \tag{A76a}$$

$$= \quad \|\overline{\mathbf{x}}\| + \sum_j\left[\frac{\sum_k x_k\,\delta_{kj}}{\sqrt{\sum_k x_k^2}}\right]_{\mathbf{x}=\overline{\mathbf{x}}}(x_j-\overline{x}_j)+\mathcal{O}\left(\|\mathbf{x}-\overline{\mathbf{x}}\|^2\right) \quad = \tag{A76b}$$

$$= \quad \|\overline{\mathbf{x}}\| + \sum_j\left[\frac{\overline{x}_j}{\sqrt{\sum_k \overline{x}_k^2}}\right](x_j-\overline{x}_j)+\mathcal{O}\left(\|\mathbf{x}-\overline{\mathbf{x}}\|^2\right) \quad = \tag{A76c}$$

$$= \quad \|\overline{\mathbf{x}}\| + \widehat{\overline{\mathbf{x}}}^T(\mathbf{x}-\overline{\mathbf{x}})+\mathcal{O}\left(\|\mathbf{x}-\overline{\mathbf{x}}\|^2\right). \tag{A76d}$$

Noticing that

$$\frac{\partial\,\|\mathbf{x}\|}{\partial\mathbf{x}} \quad = \quad \widehat{\overline{\mathbf{x}}}^T+\mathcal{O}\left(\|\mathbf{x}-\overline{\mathbf{x}}\|\right), \tag{A77}$$

our computations are straightforward:

$$\cos\left(\frac{\|\mathbf{x}\|}{2}\right) = \cos\left(\frac{\|\overline{\mathbf{x}}\|}{2}\right)-\left[\sin\left(\frac{\|\mathbf{x}\|}{2}\right)\frac{1}{2}\left[\widehat{\overline{\mathbf{x}}}^T+\mathcal{O}\left(\|\mathbf{x}-\overline{\mathbf{x}}\|\right)\right]\right]_{\mathbf{x}=\overline{\mathbf{x}}}(\mathbf{x}-\overline{\mathbf{x}})+\mathcal{O}\left(\|\mathbf{x}-\overline{\mathbf{x}}\|^2\right) = \tag{A78a}$$

$$= \cos\left(\frac{\|\overline{\mathbf{x}}\|}{2}\right)-\frac{1}{2}\sin\left(\frac{\|\overline{\mathbf{x}}\|}{2}\right)\widehat{\overline{\mathbf{x}}}^T(\mathbf{x}-\overline{\mathbf{x}})+\mathcal{O}\left(\|\mathbf{x}-\overline{\mathbf{x}}\|^2\right). \tag{A78b}$$

For our particular case,

$$\cos\left(\frac{\|\mathbf{e}^{\overline{q}}\|}{2}\right) \quad = \quad \overline{\delta}_0-\frac{1}{2}\,\overline{\boldsymbol{\delta}}^T\left(\mathbf{e}^{\overline{q}}-\overline{\mathbf{e}}^{\overline{q}}\right)+\mathcal{O}\left(\|\mathbf{e}^{\overline{q}}-\overline{\mathbf{e}}^{\overline{q}}\|^2\right). \tag{A79}$$

On the other hand,

$$\frac{\sin\left(\frac{\|\mathbf{x}\|}{2}\right)}{\|\mathbf{x}\|} = \tag{A80a}$$

$$= \frac{\sin\left(\frac{\|\overline{\mathbf{x}}\|}{2}\right)}{\|\overline{\mathbf{x}}\|}+\left[\left(\frac{\cos\left(\frac{\|\mathbf{x}\|}{2}\right)}{\|\mathbf{x}\|}\frac{1}{2}-\frac{\sin\left(\frac{\|\mathbf{x}\|}{2}\right)}{\|\mathbf{x}\|^2}\right)\left[\widehat{\overline{\mathbf{x}}}^T+\mathcal{O}\left(\|\mathbf{x}-\overline{\mathbf{x}}\|\right)\right]\right]_{\mathbf{x}=\overline{\mathbf{x}}}(\mathbf{x}-\overline{\mathbf{x}})+\mathcal{O}\left(\|\mathbf{x}-\overline{\mathbf{x}}\|^2\right) = \tag{A80b}$$

$$= \frac{\sin\left(\frac{\|\overline{\mathbf{x}}\|}{2}\right)}{\|\overline{\mathbf{x}}\|}+\left[\left(\frac{\cos\left(\frac{\|\overline{\mathbf{x}}\|}{2}\right)}{\|\overline{\mathbf{x}}\|}\frac{1}{2}-\frac{\sin\left(\frac{\|\overline{\mathbf{x}}\|}{2}\right)}{\|\overline{\mathbf{x}}\|^2}\right)\widehat{\overline{\mathbf{x}}}^T\right](\mathbf{x}-\overline{\mathbf{x}})+\mathcal{O}\left(\|\mathbf{x}-\overline{\mathbf{x}}\|^2\right). \tag{A80c}$$

Now, taking $\mathbf{x}=\overline{\mathbf{x}}+(\mathbf{x}-\overline{\mathbf{x}})$ we arrive at

$$\frac{\mathbf{x}}{\|\mathbf{x}\|}\sin\left(\frac{\|\mathbf{x}\|}{2}\right) = [\overline{\mathbf{x}}+(\mathbf{x}-\overline{\mathbf{x}})]\frac{\sin\left(\frac{\|\mathbf{x}\|}{2}\right)}{\|\mathbf{x}\|} = \tag{A81a}$$

$$= \widehat{\overline{\mathbf{x}}}\sin\left(\frac{\|\overline{\mathbf{x}}\|}{2}\right)+\left[\frac{\sin\left(\frac{\|\overline{\mathbf{x}}\|}{2}\right)}{\|\overline{\mathbf{x}}\|}+\frac{1}{2}\cos\left(\frac{\|\overline{\mathbf{x}}\|}{2}\right)\widehat{\overline{\mathbf{x}}}\widehat{\overline{\mathbf{x}}}^T-\frac{\sin\left(\frac{\|\overline{\mathbf{x}}\|}{2}\right)}{\|\overline{\mathbf{x}}\|}\widehat{\overline{\mathbf{x}}}\widehat{\overline{\mathbf{x}}}^T\right](\mathbf{x}-\overline{\mathbf{x}})+\mathcal{O}\left(\|\mathbf{x}-\overline{\mathbf{x}}\|^2\right). \tag{A81b}$$

For our particular case,

$$\frac{\mathbf{e}^{\overline{q}}}{\|\mathbf{e}^{\overline{q}}\|} \sin\left(\frac{\|\mathbf{e}^{\overline{q}}\|}{2}\right) = \overline{\delta} + \left[\frac{\|\overline{\delta}\|}{2\arcsin\|\overline{\delta}\|} + \frac{\overline{\delta}_0}{2}\hat{\overline{\delta}}\hat{\overline{\delta}}^T - \frac{\|\overline{\delta}\|}{2\arcsin\|\overline{\delta}\|}\hat{\overline{\delta}}\hat{\overline{\delta}}^T\right](\mathbf{e}^{\overline{q}} - \overline{\mathbf{e}}^{\overline{q}}) + \mathcal{O}\left(\|\mathbf{e}^{\overline{q}} - \overline{\mathbf{e}}^{\overline{q}}\|^2\right) = \tag{A82a}$$

$$= \overline{\delta} + \frac{1}{2}\left[\left(\mathbf{I} - \hat{\overline{\delta}}\hat{\overline{\delta}}^T\right)\frac{\|\overline{\delta}\|}{\arcsin\|\overline{\delta}\|} + \overline{\delta}_0\hat{\overline{\delta}}\hat{\overline{\delta}}^T\right](\mathbf{e}^{\overline{q}} - \overline{\mathbf{e}}^{\overline{q}}) + \mathcal{O}\left(\|\mathbf{e}^{\overline{q}} - \overline{\mathbf{e}}^{\overline{q}}\|^2\right). \tag{A82b}$$

Finally, we just have to replace (A79) and (A82b) in (A13) to obtain the required linear approximation. Returning to the original notation we have

$$2\frac{\delta^{\overline{p}}}{\|\delta^{\overline{p}}\|}\arcsin\|\delta^{\overline{p}}\| = 2\delta^{\overline{p}} + \mathcal{O}\left(\|\mathbf{e}^{\overline{q}} - \overline{\mathbf{e}}^{\overline{q}}\|^2\right) = \tag{A83a}$$

$$= 2\overline{\delta}_0\hat{\mathbf{e}}^{\overline{q}}\sin\left(\frac{\|\mathbf{e}^{\overline{q}}\|}{2}\right) - 2\cos\left(\frac{\|\mathbf{e}^{\overline{q}}\|}{2}\right)\overline{\delta} - 2\overline{\delta}\times\hat{\mathbf{e}}^{\overline{q}}\sin\left(\frac{\|\mathbf{e}^{\overline{q}}\|}{2}\right) + \mathcal{O}\left(\|\mathbf{e}^{\overline{q}} - \overline{\mathbf{e}}^{\overline{q}}\|^2\right) = \tag{A83b}$$

$$\begin{aligned} = & \; 2\overline{\delta}_0\overline{\delta} + \overline{\delta}_0\left[\left(\mathbf{I} - \hat{\overline{\delta}}\hat{\overline{\delta}}^T\right)\frac{\|\overline{\delta}\|}{\arcsin\|\overline{\delta}\|} + \overline{\delta}_0\hat{\overline{\delta}}\hat{\overline{\delta}}^T\right]\left(\mathbf{e}^{\overline{q}} - \overline{\mathbf{e}}^{\overline{q}}\right) - \left(2\overline{\delta}_0 - \overline{\delta}^T\left(\mathbf{e}^{\overline{q}} - \overline{\mathbf{e}}^{\overline{q}}\right)\right)\overline{\delta} + \\ & - \overline{\delta}\times\left\{2\overline{\delta} + \left[\left(\mathbf{I} - \hat{\overline{\delta}}\hat{\overline{\delta}}^T\right)\frac{\|\overline{\delta}\|}{\arcsin\|\overline{\delta}\|} + \overline{\delta}_0\hat{\overline{\delta}}\hat{\overline{\delta}}^T\right]\left(\mathbf{e}^{\overline{q}} - \overline{\mathbf{e}}^{\overline{q}}\right)\right\} + \mathcal{O}\left(\|\mathbf{e}^{\overline{q}} - \overline{\mathbf{e}}^{\overline{q}}\|^2\right) = \end{aligned} \tag{A83c}$$

$$= \left[\left(\overline{\delta}_0\left(\mathbf{I} - \hat{\overline{\delta}}\hat{\overline{\delta}}^T\right) - [\overline{\delta}]_\times\right)\frac{\|\overline{\delta}\|}{\arcsin\|\overline{\delta}\|} + \overline{\delta}_0^2\hat{\overline{\delta}}\hat{\overline{\delta}}^T + \overline{\delta}\overline{\delta}^T\right]\left(\mathbf{e}^{\overline{q}} - \overline{\mathbf{e}}^{\overline{q}}\right) + \mathcal{O}\left(\|\mathbf{e}^{\overline{q}} - \overline{\mathbf{e}}^{\overline{q}}\|^2\right) = \tag{A83d}$$

$$= \left[\left(\overline{\delta}_0\left(\mathbf{I} - \hat{\overline{\delta}}\hat{\overline{\delta}}^T\right) - [\overline{\delta}]_\times\right)\frac{\|\overline{\delta}\|}{\arcsin\|\overline{\delta}\|} + \hat{\overline{\delta}}\hat{\overline{\delta}}^T\right]\left(\mathbf{e}^{\overline{q}} - \overline{\mathbf{e}}^{\overline{q}}\right) + \mathcal{O}\left(\|\mathbf{e}^{\overline{q}} - \overline{\mathbf{e}}^{\overline{q}}\|^2\right). \tag{A83e}$$

Note that the linear approximations of our transition maps are valid for $\mathbf{e}^{\overline{q}}$ near of $\overline{\mathbf{e}}^{\overline{q}}$. However, we have not made any assumption about the $\overline{\delta}$ quaternion. This means that our linear approximations are exact for any $\overline{\delta} = \varphi^{-1}(\overline{\mathbf{e}}^{\overline{q}})$ in the domain of each **T**-matrix, provided that $\mathbf{e}^{\overline{q}}$ is close enough to $\overline{\mathbf{e}}^{\overline{q}}$.

References

1. Crassidis, J.L.; Markley, F.L.; Cheng, Y. Survey of nonlinear attitude estimation methods. *J. Guid. Control Dyn.* **2007**, *30*, 12–28. [CrossRef]
2. Kalman, R.E. A new approach to linear filtering and prediction problems. *J. Basic Eng.* **1960**, *82*, 35–45. [CrossRef]
3. Julier, S.J.; Uhlmann, J.K. *New Extension of the Kalman Filter to Nonlinear Systems*; AeroSense'97; International Society for Optics and Photonics: Bellingham, WA, USA 1997, pp. 182–193.
4. Shuster, M.D. A survey of attitude representations. *Navigation* **1993**, *8*, 439–517.
5. Stuelpnagel, J. On the parametrization of the three-dimensional rotation group. *SIAM Rev.* **1964**, *6*, 422–430. [CrossRef]
6. Lefferts, E.J.; Markley, F.L.; Shuster, M.D. Kalman filtering for spacecraft attitude estimation. *J. Guid. Control Dyn.* **1982**, *5*, 417–429. [CrossRef]
7. Crassidis, J.L.; Markley, F.L. Unscented filtering for spacecraft attitude estimation. *J. Guid. Control Dyn.* **2003**, *26*, 536–542. [CrossRef]
8. Markley, F.L. Attitude error representations for Kalman filtering. *J. Guid. Control Dyn.* **2003**, *26*, 311–317. [CrossRef]
9. Hall, J.K.; Knoebel, N.B.; McLain, T.W. Quaternion attitude estimation for miniature air vehicles using a multiplicative extended Kalman filter. In Proceedings of the 2008 IEEE/ION Position, Location and Navigation Symposium, Monterey, CA, USA, 5–8 May 2008; IEEE: Piscataway, NJ, USA; pp. 1230–1237.
10. VanDyke, M.C.; Schwartz, J.L.; Hall, C.D. Unscented Kalman filtering for spacecraft attitude state and parameter estimation. *Adv. Astronaut. Sci.* **2004**, *118*, 217–228.

11. Markley, F.L. Multiplicative vs. additive filtering for spacecraft attitude determination. *Dyn. Control Syst. Struct. Space* **2004**, *6*, 311–317.
12. Crassidis, J.L.; Markley, F.L. Attitude Estimation Using Modified Rodrigues Parameters. Avaliable online: https://ntrs.nasa.gov/archive/nasa/casi.ntrs.nasa.gov/19960035754.pdf (accessed on 1 January 2019).
13. Bar-Itzhack, I.; Oshman, Y. Attitude determination from vector observations: Quaternion estimation. *IEEE Trans. Aerosp. Electr. Syst.* **1985**, *AES-21*, 128–136. [CrossRef]
14. Mueller, M.W.; Hehn, M.; D'Andrea, R. Covariance correction step for kalman filtering with an attitude. *J. Guid. Control Dyn.* **2016**, *40*, 2301–2306. [CrossRef]
15. Julier, S.J.; Uhlmann, J.K. Unscented filtering and nonlinear estimation. *Proc. IEEE* **2004**, *92*, 401–422. [CrossRef]
16. Gramkow, C. On averaging rotations. *J. Math. Imag. Vision* **2001**, *15*, 7–16. [CrossRef]
17. LaViola, J.J. A comparison of unscented and extended Kalman filtering for estimating quaternion motion. In Proceedings of the American Control Conference, Denver, CO, USA, 4–6 June 2003; IEEE: Piscataway, NJ, USA, 2003, Volume 3; pp. 2435–2440.
18. Xie, L.; Popa, D.; Lewis, F.L. *Optimal and Robust Estimation: With an Introduction to Stochastic Control Theory*; CRC Press: Boca Raton, FL, USA, 2007.

sensors

MDPI

Article

A New Image Registration Algorithm Based on Evidential Reasoning

Zhe Zhang [1], Deqiang Han [1,*], Jean Dezert [2] and Yi Yang [3]

1 MOE KLINNS Lab, Institute of Integrated Automation, School of Electronic and Information Engineering, Xi'an Jiaotong University, Xi'an 710049, China; zhangzsmg@gmail.com
2 ONERA, The French Aerospace Lab, Chemin de la Hunière, F-91761 Palaiseau, France; jean.dezert@onera.fr
3 SKLSVMS, School of Aerospace, Xi'an Jiaotong University, Xi'an 710049, China; jiafeiyy@mail.xjtu.edu.cn
* Correspondence: deqhan@gmail.com; Tel.: +86-131-1911-5975

Received: 30 November 2018; Accepted: 26 February 2019; Published: 4 March 2019

Abstract: Image registration is a crucial and fundamental problem in image processing and computer vision, which aims to align two or more images of the same scene acquired from different views or at different times. In image registration, since different keypoints (e.g., corners) or similarity measures might lead to different registration results, the selection of keypoint detection algorithms or similarity measures would bring uncertainty. These different keypoint detectors or similarity measures have their own pros and cons and can be jointly used to expect a better registration result. In this paper, the uncertainty caused by the selection of keypoint detector or similarity measure is addressed using the theory of belief functions, and image information at different levels are jointly used to achieve a more accurate image registration. Experimental results and related analyses show that our proposed algorithm can achieve more precise image registration results compared to several prevailing algorithms.

Keywords: image registration; evidential reasoning; belief functions; uncertainty

---

1. Introduction

Image registration is a fundamental problem encountered in image processing, e.g., image fusion [1] and image change detection [2]. It refers to the alignment of two or more images of the same scene taken at different time, from different sensors, or from different viewpoints. Image registration plays an increasingly important role in applications of surveillance [3], remote-sensing [4] and medical imaging [5].

For a collection of images to be registered, one is chosen as the reference image and the others are selected as sensed images. Image registration align each sensed image to the reference image by finding the correspondence between all pixels in the image pair and estimating the spatial transformation from the sensed image to the reference image. In this paper, we just consider the image registration between two images, i.e., there is only one sensed image together with a given reference image.

Current image registration techniques that based on image domain can be generally divided into two categories [6]: the sparse methods and dense methods. There are also some methods based on transform domain, like Fourier-Mellin transformation method [7]. The transform domain based methods are often used for image registration with similarity transformation model. In this paper, we focus on the image domain based methods.

The sparse methods [8] extracts and matches salient features from the reference image and sensed image and then estimates the spatial transformation between the two images based on these matched features. Line features (e.g., edges) and point features (corners, line intersections and gravities of regions) all can be used for image registration. Corner features are the mostly used features and can be manually selected or automatically detected by Harris [9], FAST (Features from Accelerated Segment Test) [10], SIFT (Scale-Invariant Feature Transform) [11], SURF (Speeded-Up Robust Features) [12], DAISY [13], ORB (Oriented FAST and Rotated BRIEF) [14], KAZE [15], etc.

In contrast to the sparse methods, the dense methods [16] do not detect features from the image pair but search the optimal spatial transformation directly that can best match all the pixels in the image pair. Similarity (resp. dissimilarity) measures are defined to quantify the independency (resp. dependency) between the pair of images. Various similarity and dissimilarity measures have been proposed [17] such as RMSE (Root-Mean-Squared Error), PSNR (Peak Signal to Noise Ratio), Spearman's Rho [18], NCC (Normalized Cross-correlation Coefficient) and MI (Mutual Information). It should be noted that dense methods based on RMSE or PSNR cannot handle the cases with illumination variation since these two similarity/dissmilarity measures are very sensitive to illumination changes.

Both the sparse methods and dense methods involve uncertainty problems. For the sparse methods, keypoints obtained from different keypoint detectors describe different corner features of the image. Therefore, image registrations based on different keypoint detectors would obtain different spatial transformations. For the dense methods, different similarity (dissimilarity) measures quantify the difference between the pair of images from different aspects so that image registrations based on different similarity (dissimilarity) measures would obtain different spatial transformations. These different spatial transformations obtained have their own pros and cons, and the selection of the spatial transformation (the selection of the feature detector or similarity measure indeed) would bring uncertainty.

To deal with the uncertainty caused by the particular selection of feature detector or similarity (dissimilarity) measure, one feasible way is to combine these registration transformations obtained from different feature detection methods or similarity measures to obtain a better registration result. The belief functions introduced in Demspter–Shafer Theory (DST) [19] of evidence offer a powerful theoretical tool for uncertainty modeling and reasoning; therefore, we propose a fusion based image registration method using belief functions. In this paper, the spatial transformations obtained from different feature detection algorithms or similarity measures compose the frame of discernment (FOD) and their uncertainties are modeled using belief functions. In uncertainty modeling, image information at different levels, i.e., image's intensities, edges and phase angles, are jointly used to evaluate the beliefs about image transformations. Then, these uncertainties are further handled through the evidence combination of the above multiple information. The final registration result is obtained according to the combined evidence.

This paper is an extension of our previous work in [20] where the basic idea is briefly presented. The main added values with respect to [20] are as follows. First, the transformation model between the reference image and sensed image is more comprehensive. We use similarity transformation model in [20] but use projective transformation model in this paper, which is more general since all similarity transformations are examples of projective transformations. Second, the keypoints used in the sparse approach in [20] are manually selected. To reduce the subjective influence to the registration result, in this paper, the keypoints are generated from detection algorithms. Accordingly, feature matching and mismatching removal are added after the keypoint detection. Third, when modeling uncertainties, one more information source, i.e., image's phase angle information, is considered in this work. Fourth, more experiments and analyses are provided for performance evaluation and analysis.

The rest of this paper is organized as follows. The basics of image registration are introduced in Section 2. The basics of evidence theory are introduced in Section 4. The proposed image registration method is introduced in Section 4.1 with emphasis of uncertainty modeling and handling. Evaluation method is introduced in Section 5. Experiment results of the proposed method and other registration methods are presented and compared in Section 6.1. Concluding remarks are given in Section 7.

2. Basics of Image Registration

For two (or more) images of the same scene taken at different time, from different sensors, or from different viewpoints, one is chosen as the reference image (R) and the other one is chosen as the sensed image (S). In this paper, we focus on the projective transformation model between the reference image and sensed image, which is a commonly used model in image registration [16].

Denote pixel coordinates in the reference image R as (v, w) and their mapping counterparts in the sensed image S as (g, h). The projective transformation from R to S can be expressed based on the homogeneous coordinates (Homogeneous coordinates can easily express the translation transformation as matrix multiplications while Cartesian coordinates cannot) as

$$[g\ h\ 1] = [v\ w\ 1]\,T = [v\ w\ 1]\begin{bmatrix} t_{11} & t_{12} & t_{13} \\ t_{21} & t_{22} & t_{23} \\ t_{31} & t_{32} & t_{33} \end{bmatrix} \tag{1}$$

The similarity transformation and affine transformation are important specializations of the projective transformation, as illustrated in Table 1.

Table 1. Projective transformation and its two specializations.

Similarity	Affine	Projective
$\begin{bmatrix} s\cos\theta & s\sin\theta & 0 \\ -s\sin\theta & s\cos\theta & 0 \\ t_{31} & t_{32} & 1 \end{bmatrix}$	$\begin{bmatrix} t_{11} & t_{12} & 0 \\ t_{21} & t_{22} & 0 \\ t_{31} & t_{32} & 1 \end{bmatrix}$	$\begin{bmatrix} t_{11} & t_{12} & t_{13} \\ t_{21} & t_{22} & t_{23} \\ t_{31} & t_{32} & t_{33} \end{bmatrix}$

The purpose of image registration is to estimate the transformation T to align the sensed image S with the reference image R by

$$[v'\ w'\ 1] = [g\ h\ 1]\,T^{-1}, \tag{2}$$

where (v', w') and (g, h) denote pixel coordinates in registered sensed image S' and sensed image S, respectively. Current image registration techniques can be divided into two categories [6] in general, including the sparse method and dense method. Basics of these two methods are introduced below.

2.1. Sparse Image Registration and Its Uncertainty

The feature detection and feature matching are two critical steps in the sparse methods. The flow chart of the sparse approach is illustrated in Figure 1, where each functional block is detailed in the sequel.

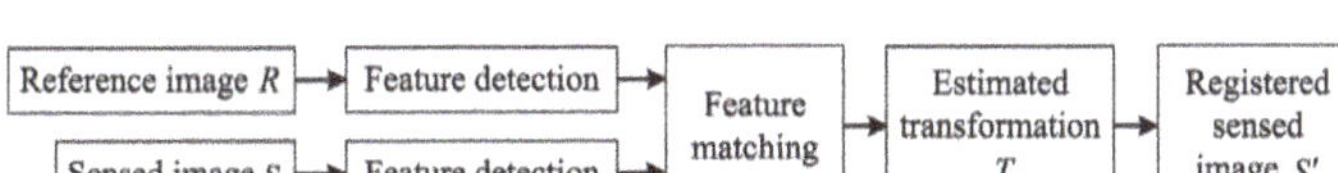

Figure 1. Flow chart of sparse approach.

2.1.1. Feature Detection

Corner features are the mostly used features in image registration due to their invariance to imaging geometry [6]. Some early keypoint detectors, like Harris and FAST, are very sensitive to image scale changes so that have poor performance when the sensed images have different scales with the reference image. The most well-known SIFT detector shows good robustness to illumination, orientation and scale changes. Most scale invariant detectors, like SIFT, SURF, ORB and BRISK, detect and describe features at different scale levels by building or approximating the Gaussian scale space of the image. In a different way, KAZE detects features in a nonlinear scale space built using efficient additive operator splitting techniques and variable conductance diffusion.

2.1.2. Feature Matching

To align the sensed image and the reference image, the detected keypoints in the two images are matched first by comparing their local feature characterized by descriptors. Generally, if the two keypoints' descriptors are similar, the two keypoints are likely to be a matched pair. Given a keypoint t in the reference image, there might be a set of candidates in the sensed image having similar descriptor with t. Among these candidates, t's real counterpart should have the closest distance with t, and at the same time its distance should be much closer than other candidates' distances.

The accuracy of the keypoints' matching affects the accuracy of the transformation's estimation. The mismatched keypoint pairs should be further removed before estimating the transformation. RANSAC (RANdom SAmple Consensus) [21] and MSAC (M-estimator SAmple and Consensus) [22] are often used to deal with this problem. A recent RANRESAC (RANdom RESAmple Consensus) [23] algorithm has been proposed to remove mismatched keypoint pairs for noisy image registration. Besides the accuracy of the keypoints' matching, the distribution of matched pairs over the image space is another key factor to obtain a high-quality estimation of transformation.

2.1.3. Transformation Estimation

With all the matched keypoint pairs, the transformation matrix T can be estimated using Equation (1). Since T has eight degrees of freedom, four point correspondences (with no three collinear) are needed to obtain the unique solution of T according to Cramer's rule.

Normally, the amount of the matched keypoint pairs is more than four and T can be estimated using the least squares (LS) fitting technique [6] by searching the minimum sum of the Euclidean distances between all the matched keypoints:

$$\hat{T} = \arg\min \sum_i d\left(cor_i^R, cor_i^{S'}\right), \tag{3}$$

where $cor_i^R = (v_i, w_i)$ represents the coordinate of the ith matched keypoint in the reference image and $cor_i^{S'} = (v_i', w_i')$ represents the coordinate of the ith matched keypoint in the registered sensed image transformed from the sensed image using Equation (2).

2.1.4. Uncertainty Encountered in Sparse Approach

Since different keypoint detection algorithms detect different kinds of corner features, the detected keypoints are usually different, as shown in Figure 2.

Image registrations based on different matched keypoint pairs would in general yield different spatial transformations to align two images. Different transformations obtained have their own pros and cons. Therefore, the selection of keypoint detection algorithms would bring uncertainty problem to the registration results.

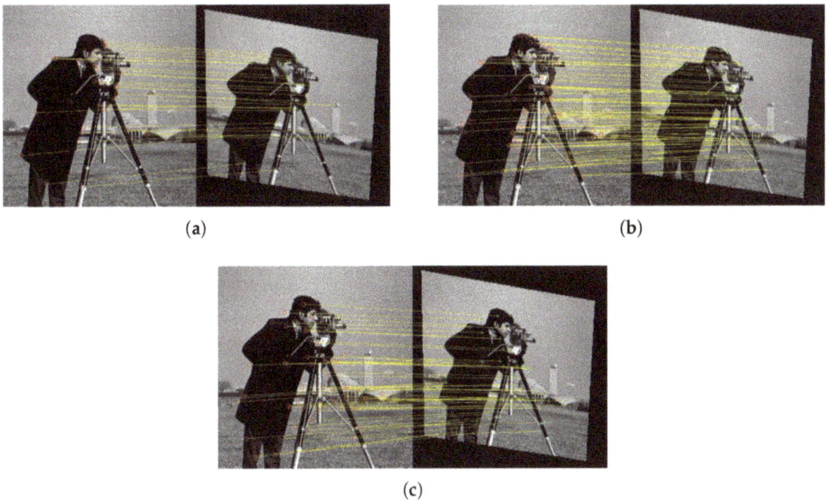

Figure 2. Different keypoint pairs detected by different keypoint detectors. (**a**) BRISK. (**b**) KAZE. (**c**) SURF.

2.2. Dense Image Registration and Its Uncertainty

The dense image registration estimates the optimal transformation T by searching the largest similarity (or the smallest dissimilarity) between the reference image R and the registered sensed image $S' = T(S)$:

$$\hat{T} = \arg\max Sim\,(R, T\,(S)) \tag{4}$$

where Sim is a chosen similarity measure. The flow chart of the dense approach is illustrated in Figure 3, where each functional block is detailed in the sequel.

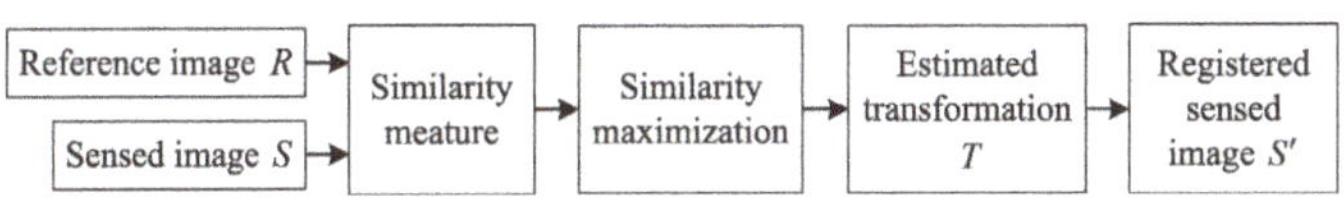

Figure 3. Flow chart of dense approach.

2.2.1. Similarity Measure

Various similarity (or dissimilarity) measures have been proposed. Here we briefly introduce the commonly used MI, NCC and PSNR measures.

(1) MI

MI measure between images A and B is

$$\mathrm{MI}(A, B) = \sum_{a=0}^{255} \sum_{b=0}^{255} p_{AB}(a, b) \log \frac{P_{AB}(a, b)}{P_A(a) P_B(b)}, \tag{5}$$

where p_{AB} is the joint probability distribution function (PDF) of images A and B, and p_A and p_B are the marginal PDFs of A and B, respectively. $\mathrm{MI}(A, B)$ is larger when A and B are more similar.

(2) NCC

For given images A and B with size of $M \times N$, NCC measure between them is

$$\mathrm{NCC}(A,B) = \sum_{x=1}^{M}\sum_{y=1}^{N} \frac{(A(x,y)-\mu_A)(B(x,y)-\mu_B)}{\sigma_A\sigma_B}, \quad (6)$$

where $A(x,y)$ and $B(x,y)$ are the pixels' intensities in images A and B at (x,y), respectively; μ_A and μ_B are the mean intensities of A and B, respectively; σ_A and σ_B are the standard deviation intensities of A and B, respectively. $\mathrm{NCC}(A,B)$ is larger when A and B are more similar.

(3) PSNR

PSNR measure between images A and B is

$$\mathrm{PSNR}(A,B) = 10 \times \log_{10}\left(\frac{255^2}{\mathrm{MSE}(A,B)}\right), \quad (7)$$

where $\mathrm{MSE}(A,B) = \frac{1}{M \times N}\sum_{x=1}^{M}\sum_{y=1}^{N}[A(x,y)-B(x,y)]^2$. $\mathrm{PSNR}(A,B)$ is larger when A and B are more similar. Since PSNR measure is very sensitive to illumination changes, it cannot be used for image registration when there are illumination variations between image pairs.

2.2.2. Transformation Estimation

The estimation for transformation T, i.e., Equation (4), is always a non-convex problem and is not so easy to obtain the global maximum [24]. Therefore, advanced optimization methods [25], or intelligent optimization approaches (like genetic, or particle swarm algorithms, etc.) are often used to estimate the optimal transformation T.

2.2.3. Uncertainty Encountered in Dense Approach

Since different similarity (dissimilarity) measures compare two images from different aspects, their calculated similarities (dissimilarities) between the reference image and registered sensed image are different. Image registration based on different measures would obtain different spatial transformations to align two images and they have their own pros and cons. Therefore, the selection of similarity (dissimilarity) measure would bring uncertainty problem to the registration results.

To deal with the uncertainty caused by the selection of feature detection algorithms or similarity measures, one feasible way is to combine the registration transformations ($T_1, T_2, \ldots, T_Q$) obtained from different feature detection algorithms (or different similarity measures) to expect a better registration result. We propose an evidential reasoning [19] based image registration algorithm to generate a combined transformation from $T_1, T_2, \ldots, T_Q$ thanks to the ability of belief functions for uncertainty modeling and reasoning. Basics of the theory of belief functions are recalled first below.

3. Basics of Evidence Theory

Dempster–Shafer evidence theory (DST) [19] is a theoretical framework for uncertainty modeling and reasoning. In DST, elements in the frame of discernment (FOD) $\Theta = \{\theta_1, \theta_2, \ldots, \theta_Q\}$ are mutually exclusive and exhaustive. The power set of Θ, i.e., 2^Θ, is the set of all subsets of Θ. For example, if $\Theta = \{\theta_1, \theta_2, \theta_3\}$, then $2^\Theta = \{\{\varnothing\}, \{\theta_1\}, \{\theta_2\}, \{\theta_3\}, \{\theta_1, \theta_2\}, \{\theta_1, \theta_3\}, \{\theta_2, \theta_3\}, \{\theta_1, \theta_2, \theta_3\}\}$. The basic belief assignment (BBA, also called mass function) is defined by a function m: $2^\Theta \mapsto [0,1]$, satisfying

$$\textstyle\sum_{A \subseteq \Theta} m(A) = 1\ ,\ m(\varnothing) = 0, \quad (8)$$

where $m(A)$ depicts the evidence support to the proposition A. A is called a focal element when $m(A) > 0$. If there is only one element in A, like $\{\theta_1\}$ and $\{\theta_2\}$, A is called the singleton element;

if there are more than one element in A, e.g., $\{\theta_1, \theta_2\}$ and $\{\theta_1, \theta_2, \theta_3\}$, A is called the compound element. The belief assigned to a compound element represents the degree of ambiguity for the multiple elements.

The plausibility function (Pl) and belief function (Bel) are defined as follows:

$$Pl(A) = \sum\nolimits_{A \cap B \neq \emptyset} m(B), \tag{9}$$

$$Bel(A) = \sum\nolimits_{B \subseteq A} m(B). \tag{10}$$

Dempster's combination rule [19] for combining two distinct pieces of evidence is defined as

$$(m_1 \oplus m_2)(A) = \begin{cases} 0, \ A = \emptyset \\ \frac{1}{1-K} \sum\limits_{B \cap C = A} m_1(B)m_2(C), \ A \neq \emptyset \end{cases} \tag{11}$$

Here, $K = \sum_{B \cap C = \emptyset} m_1(B)m_2(C)$ denotes the total conflict or contradictory mass assignments.

An alternative fusion rule PCR6 [26] for the combination of two sources is defined as

$$m_{12}^{PCR6}(A) = m_{12}^{Conj}(A) + \sum\limits_{A \cap Y = \emptyset} [\frac{m_1(A)^2 m_2(Y)}{m_1(A) + m_2(Y)} + \frac{m_2(A)^2 m_1(Y)}{m_2(A) + m_1(Y)}] \tag{12}$$

where $m_{12}^{Conj}(A)$ is the conjunctive rule defined as

$$m_{12}^{Conj}(A) = \sum\limits_{B \cap C = A} m_1(B)m_2(C) \tag{13}$$

General PCR6 formula for the combination of more than two sources is given in [26].

For a probabilistic decision-making, Smets defined the pignistic probability transformation [27] to obtain the probability measure $BetP$ from a BBA

$$BetP(\theta_i) \triangleq \sum\nolimits_{\theta_i \in A} \frac{m(A)}{|A|} \quad \forall \theta_i \in \Theta, \tag{14}$$

where $|A|$ is the cardinality of A. The decision can be made by choosing the element in FOD whose $BetP$ value is the highest one and higher than a preset threshold. Other types of probability transformation methods can be found in [26,28].

4. Image Registration Based on Evidential Reasoning

To deal with the uncertainty caused by the choice of keypoint detectors in the sparse approach or the choice of similarity measure in the dense approach, we propose an image registration method based on evidential reasoning. Suppose that the spatial transformation between the reference image and sensed image is projective. Our purpose is to estimate the transformation matrix to align two images. Unlike the prevailing methods estimating the transformation matrix from single method of keypoint detection or similarity (dissimilarity) measure, we estimate the transformation matrix by jointly utilizing different keypoint detection methods or similarity measures.

To use belief functions for image registration, one should define the frame of discernment (FOD) first. The FOD $\Theta = \{T_1, T_2, \ldots, T_Q\}$, where Q is the amount of transformations obtained from different single feature detection algorithms or different single similarity measures. We first model the beliefs for every proposition $A \subseteq \Theta$ using BBAs. A can be single transformation in FOD or a set of transformations in FOD. One BBA depicts the support to each proposition A from one evidence source. The BBA allocations from different evidence sources describes the uncertainty of the transformations in FOD. Next, the BBAs are combined to generate the combined BBA m_c depicting the fused support to each proposition A. Then, the combined transformation T_c is generated from the combined BBA m_c. Finally, the registered sensed image S'_c is transformed from the sensed image using Equation (2). During this process, the resampling [29] is needed to determine the intensity of each pixel in S'_c. Figure 4 illustrates the flow chart of this new proposed method. It should be noted that the classical interpretation of BFT assumes that the final estimation should be in the FOD. In this work, we relax this assumption and the final transformation is a combination result of those in the FOD.

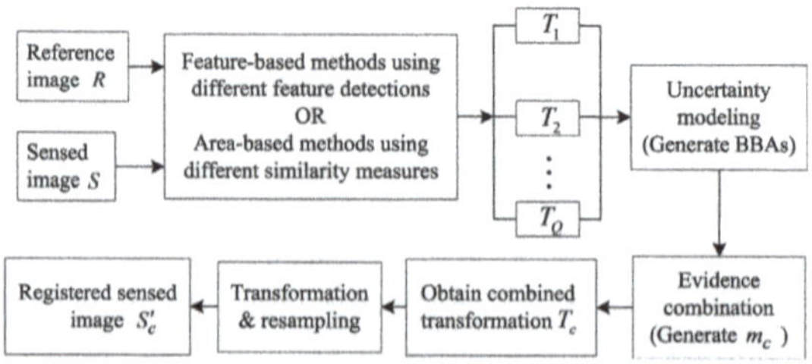

Figure 4. Flow chart of the proposed image registration.

4.1. Uncertainty Modeling

If the similarity between the reference image R and registered sensed image $S_i{}'$ is large, the corresponding transformation T_i is quite accurate and should be allocated large support ($S_i{}'$ is transformed from sensed image S by T_i^{-1}). Here we use NCC (Other similarity or dissimilarity measures, e.g., MI, are also appropriate to quantify the similarity here) to measure the similarity between R and $S_i{}'$:

$$\mathrm{NCC}_i = \sum_{x=1}^{M}\sum_{y=1}^{N} \frac{(R(x,y) - \mu_R)(S'_i(x,y) - \mu_{S'_i})}{\sigma_R \sigma_{S'_i}} \tag{15}$$

where μ_R and $\mu_{S'_i}$ are the mean intensities of R and S'_i, respectively; σ_R and $\sigma_{S'_i}$ are the standard deviation intensities of R and S'_i, respectively.

Since multi-source information can help to reduce the uncertainty through evidence combination, we use different levels of image information to quantify the similarity between R and S'_i. The similarity can be calculated from the gray images, edge feature images or reconstructed images using phase angle as shown in Figure 5. Their corresponding NCC_i are denoted as $\mathrm{NCC}_i(G)$, $\mathrm{NCC}_i(E)$ and $\mathrm{NCC}_i(P)$, respectively. The edge detection method used in Figure 5b is the Canny detector [30]. More details of the image reconstruction from phase angle information can be found in [29].

(a) (b) (c)

Figure 5. Image information at different levels. (**a**) Gray image. (**b**) Edge feature image. (**c**) Reconstructed image using phase angle.

The value range of $\mathrm{NCC}_i(\cdot)$ is $[-1,1]$. According to our experiments, most values of $\mathrm{NCC}_i(\cdot)$ are larger than 0. Before allocating BBAs, we first enlarge the differences of $\mathrm{NCC}_i(\cdot)$ within $[0,1]$ using function $y = e^{x-1}$, as illustrated in Figure 6.

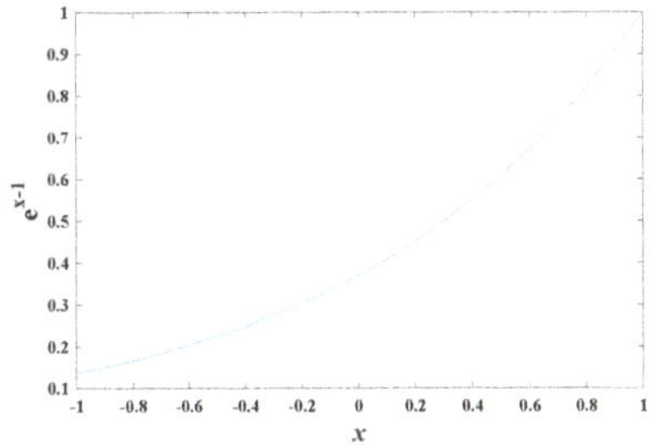

Figure 6. The curve of function e^{x-1}.

Each level of image information (gray images (G), edge feature images (E) and reconstructed images using phase angle (P)) can be viewed as one evidence source and their corresponding $e^{NCC_i(\cdot)-1}$ can be used to assign beliefs for transformation T_i:

$$\begin{cases} m_G(T_i) = \frac{e^{NCC_i(G)-1}}{\sum_{j=1}^{Q} e^{NCC_j(G)-1}} \\ m_E(T_i) = \frac{e^{NCC_i(E)-1}}{\sum_{j=1}^{Q} e^{NCC_j(E)-1}} \\ m_P(T_i) = \frac{e^{NCC_i(P)-1}}{\sum_{j=1}^{Q} e^{NCC_j(P)-1}} \end{cases} \quad (16)$$

4.2. Fusion-Based Registration

After obtaining BBAs m_G, m_E and m_P, we generate the combined BBA m_c using a combination rule denoted symbolically with $\oplus$:

$$m_c(\cdot) = [m_G \oplus m_E \oplus m_P](\cdot) \quad (17)$$

$m_c(T_i)$ describes the combined evidence support to T_i (a 3×3 matrix with 6 unknown parameters). The combined transformation T_c is computed by

$$T_c^{-1} = \sum_{i=1}^{Q} m_c(T_i) T_i^{-1}. \quad (18)$$

Finally, the registered sensed image $S_c{}'$ can be obtained using Equation (2) following the resampling.

5. Evaluation of Image Registration

Since the purpose of image registration is to align the reference image R and sensed image S to a single coordinate frame, one popular evaluation method for the registration result is to quantify the difference (usually quantified by Root-Mean-Squared Error (RMSE)) between R and the registered sensed image $S_c{}'$ [31,32]. However, since $S_c{}'$ is transformed from the sensed image S, which may have less information than R (S may be part of R or have lower resolution than R since R and S can be taken from different views or taken by different cameras), the difference between R and $S_c{}'$ could be large even when the estimated transformation T_c equals to the true transformation T_{true} from the reference image R to the sensed image S, as shown in Figure 7. Therefore, this kind of evaluation method is not accurate enough.

Figure 7. Relationship among R, S, $R_c{}'$ and $S_c{}'$.

Another popular evaluation method is to quantify the difference between the reference image R and image $R_c{}'$, which is transformed from R by the transformation matrix $T_{\text{true}}T_c^{-1}$ [16,33], as shown in Figure 7. The mapping relationship between pixel at (v, w) in image R and pixel at (v', w') in image $R_c{}'$ satisfies

$$\begin{bmatrix} v' & w' & 1 \end{bmatrix} = \begin{bmatrix} v & w & 1 \end{bmatrix} T_{\text{true}}T_c^{-1}, \tag{19}$$

when the registration is absolutely accurate, $T_c = T_{\text{true}}$ and $R_c{}' = R$.

In this paper, we evaluate the registration performance by quantifying the difference between R and $R_c{}'$ using AAID (average absolute intensity difference) [16]:

$$\text{AAID}(R, R_c{}') = \frac{1}{MN} \sum_{x=1}^{M} \sum_{y=1}^{N} \left| R(x,y) - R_c{}'(x,y) \right|. \tag{20}$$

$\text{AAID}(R, R_c{}')$ is smaller when the registration result is better.

6. Experiments

To verify the performance of our new proposed image registration method, we provide experiments on noise-free images and noisy images, respectively. Image registration under the noisy condition is difficult since the noise pixels bring difficulties for keypoints' detection and matching and reduce the accuracy for similarity measure. For the sparse method, experiment results based on BRISK [34], KAZE [15] and SURF [12] feature detection algorithms are provided for comparison. For the dense method, experiment results based on MI, PSNR and NCC similarity measures are provided for comparison. For the noisy image registration, the experiment result of RANRESAC (a recently proposed method for noisy image registration) [23] is also provided for comparison.

6.1. Sparse Image Registration Results

We first do experiments on actual data to illustrate the effectiveness of the proposed method. The reference image and sensed image are taken from different cameras with different views, as shown in Figure 8. BRISK, KAZE and SURF feature detections are used for generating transformations T_1, T_2 and T_3, respectively. When deriving combined BBAs in Equation (17), an alternative fusion rule PCR6, which is more robust than Dempster's rule [26], is also used for comparison.

(**a**) (**b**)

Figure 8. Fence image pair. (**a**) Reference image. (**b**) Sensed image.

The registered results of the proposed method are illustrated in Figure 9. From Figure 9, the proposed method can successfully align the sensed image to the reference image illustrating that the proposed method is effective for actual data.

(**a**) (**b**)

Figure 9. Registered results of the proposed methods for Fence image. (**a**) Dempster's rule. (**b**) PCR6.

To quantify the accuracy of the registration results, the actual transformation between the reference image and sensed image is needed and we do experiments on simulated images. We first do experiments on Boats image (The reference image can be found at https://imagej.nih.gov/ij/images/boats.gif.) and Foosball image (sample image from the MATLAB), as shown in Figure 10.

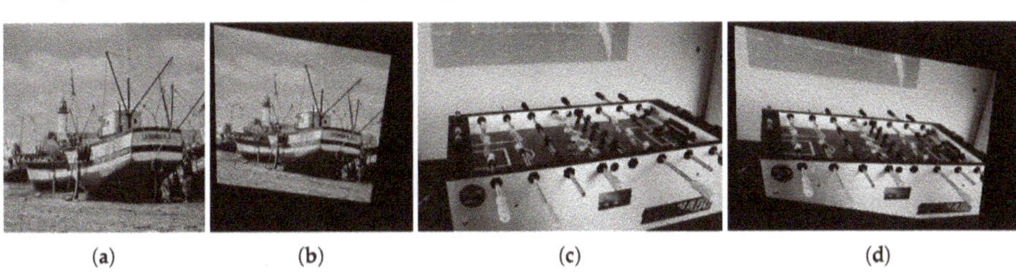

(**a**) (**b**) (**c**) (**d**)

Figure 10. Boats image pair and Foosball image pair. (**a**) Boats reference image. (**b**) Boats sensed image. (**c**) Foosball reference image. (**d**) Foosball sensed image.

The AAID evaluations of these registration results for Boats image and Foosball image are compared in Figure 11, where Demp represents the Dempster's combination rule. According to

Figure 11, the proposed fusion-based method achieves much better registration result (smaller AAID) than algorithms based on BRISK, KAZE or SURF feature detections, respectively.

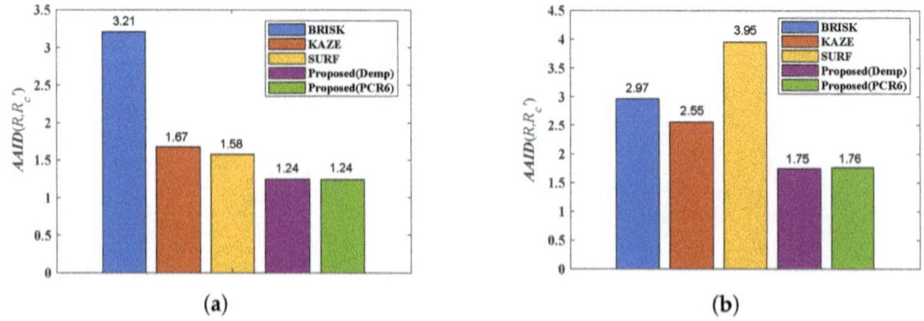

Figure 11. AAID evaluations of registration results for Boats image pair and Foosball image pair. (**a**) Boats. (**b**) Foosball.

Furthermore, we also analyzed the spatial partition of the AAID evaluation for each result by evenly dividing the reference image into 5 × 5 parts (as shown in Figure 12a) and calculating the AAID between the reference image and the registration result in each part. The AAID spatial partition results for Boats image and Foosball image are illustrated in Figures 12 and 13, respectively. For Boats image, the AAID of BRISK and KAZE results varies significantly for different parts while the SURF result is relatively uniform; the proposed methods have low and similar AAID in most parts while the rightmost parts (parts 5, 10, 15, 20 and 25) have significant larger AAID. For Foosball image, the AAID spatial partition of all these results are uneven.

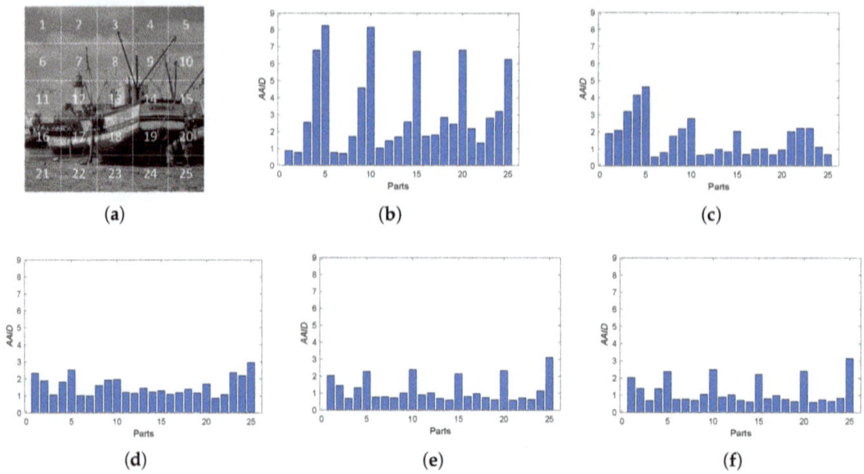

Figure 12. Spatial partition of the AAID evaluation for Boats image. (**a**) Partition method. (**b**) BRISK. (**c**) KAZE. (**d**) SURF. (**e**) Proposed (Demp). (**f**) Proposed (PCR6).

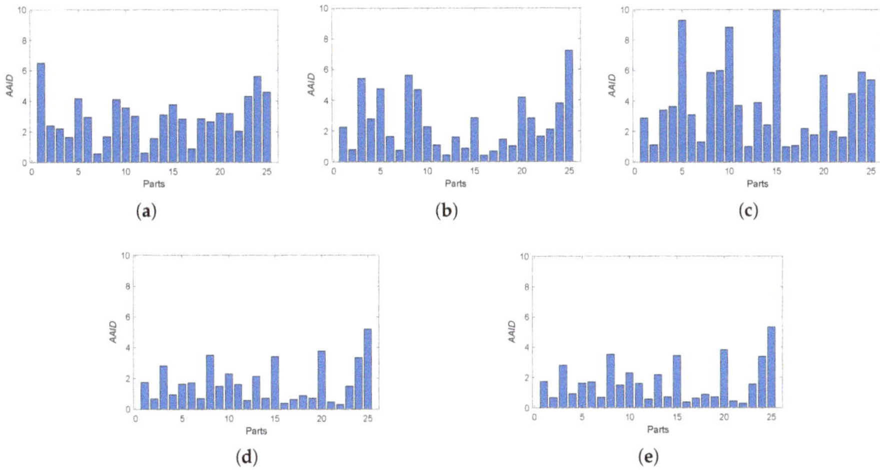

Figure 13. Spatial partition of the AAID evaluation for Foosball image. (**a**) BRISK. (**b**) KAZE. (**c**) SURF. (**d**) Proposed (Demp). (**e**) Proposed (PCR6).

Then, we consider the noisy image registration and do experiments on West Concord image pair (sample image from the MATLAB) with zero-mean Gaussian noise (variance is 0.01), as shown in Figure 14. The AAID evaluations for these registration results are compared in Figure 15, where the proposed fusion-based methods achieve better performance (smaller AAID) than RANTESAC and methods based on BRISK, KAZE and SURF feature detections, respectively. The spatial partition of the AAID evaluation for each result is illustrated in Figure 16, where the KAZE result is the most uneven one.

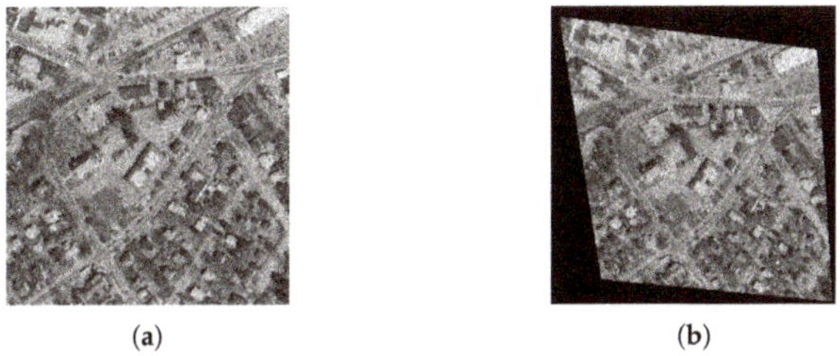

Figure 14. West Concord image pair. (**a**) Reference image. (**b**) Sensed image.

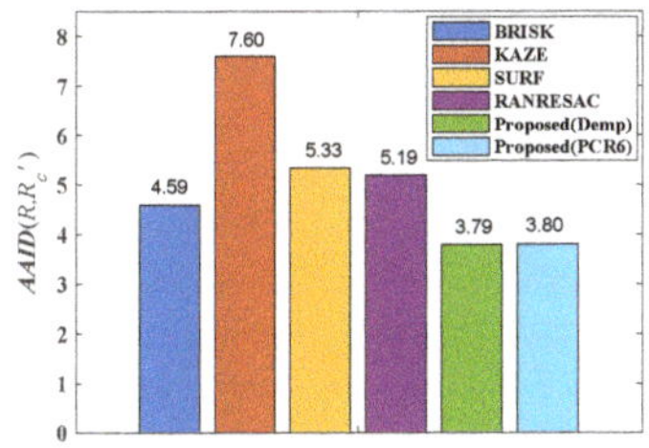

Figure 15. AAID evaluations of registration results for West Concord image pair.

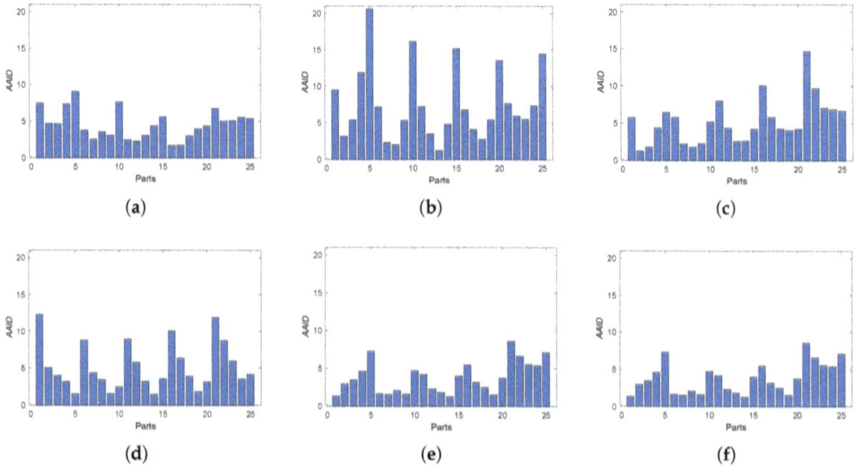

Figure 16. Spatial partition of the AAID evaluation for West Concord image. (**a**) BRISK. (**b**) KAZE. (**c**) SURF. (**d**) RANRESAC. (**e**) Proposed (Demp). (**f**) Proposed(PCR6).

6.2. Dense Image Registration Results

Since the optimization of dense registration is intractable when the solution space has high dimensions, we simplify the transformation model to rigid transformation here. The solution space for rigid model only has three dimensions: one for rotation and two for translations in horizontal and vertical directions, respectively. We first provide experiments on Concord image and Hestain image (sample images from the MATLAB) as shown in Figure 17, where the sensed image is transformed from the reference image through the rotation ($\theta = 10^\circ$ in anticlockwise) and translation ($(t_v, t_h) = (-10, 5)$) successively.

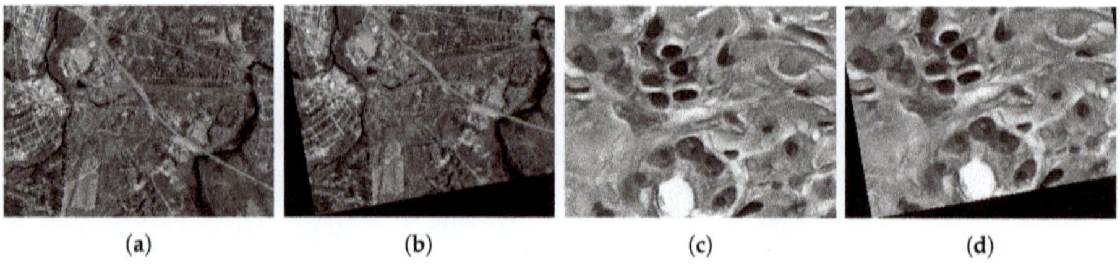

Figure 17. Concord image pair and Hestain image pair. (**a**) Concord reference image. (**b**) Concord sensed image. (**c**) Hestain reference image. (**d**) Hestain sensed image.

In the proposed dense approach, MI, PSNR and NCC similarity measures are used for generating transformations T_1, T_2 and T_3, respectively. The AAID evaluations of these registration results for the Concord image and Hestain image are compared in Figure 18, where the proposed fusion-based methods achieve much better registration results (smaller AAID) than algorithms based on MI, PSNR or NCC similarities, respectively. The AAID spatial partition results for Concord image and Hestain image are illustrated in Figures 19 and 20, respectively. For these two images, the AAID results of the proposed methods are smaller in the downside parts compared with those in upside parts.

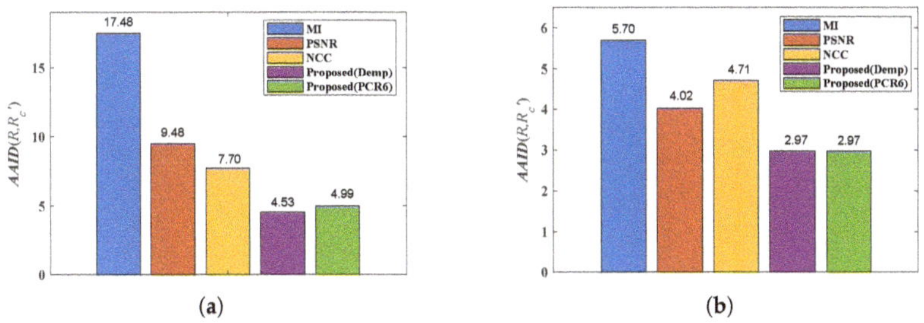

Figure 18. AAID evaluations of registration results for Concord image pair and Hestain image pair. (**a**) Concord. (**b**) Hestain.

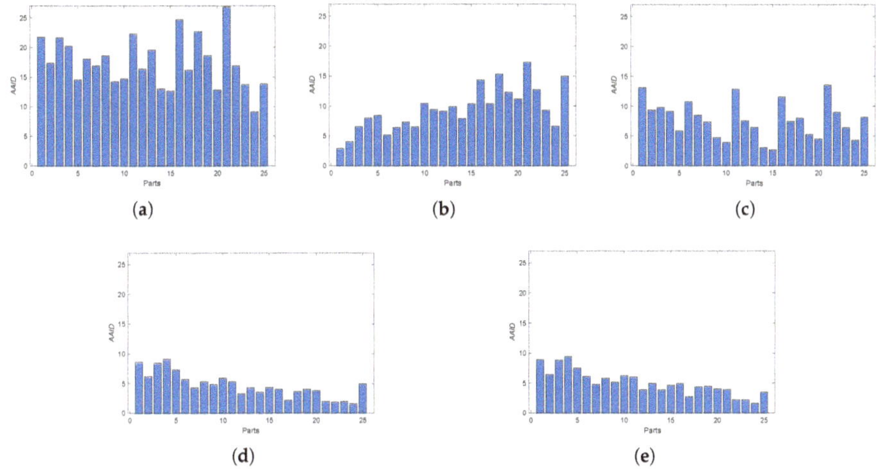

Figure 19. Spatial partition of the AAID evaluation for Concord image. (**a**) MI. (**b**) PSNR. (**c**) NCC. (**d**) Proposed (Demp). (**e**) Proposed(PCR6).

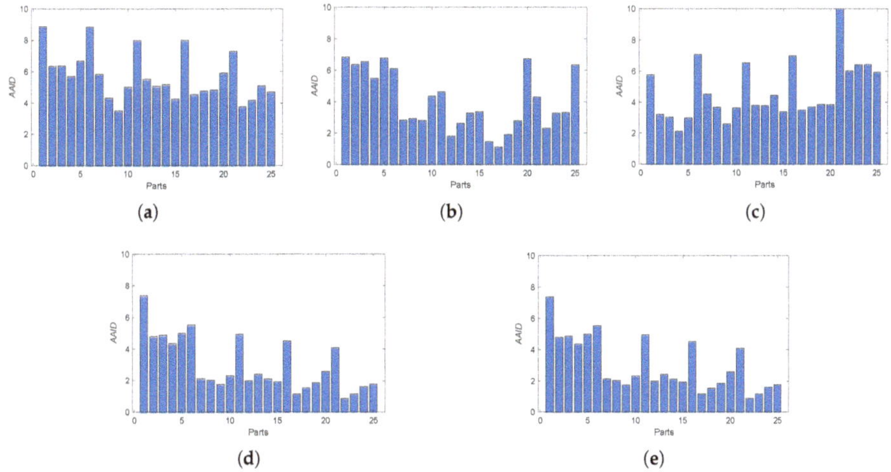

Figure 20. Spatial partition of the AAID evaluation for Hestain image. (**a**) MI. (**b**) PSNR. (**c**) NCC. (**d**) Proposed (Demp). (**e**) Proposed (PCR6).

Then, we consider the noisy image condition and implement experiments on Lifting Body image pair (sample image from the MATLAB) with zero-mean Gaussian noise (variance is 0.01), as shown in Figure 21. The sensed image is transformed from the reference image through the rotation ($\theta = -10°$) and translation ($(t_v, t_h) = (-10, 5)$), successively.

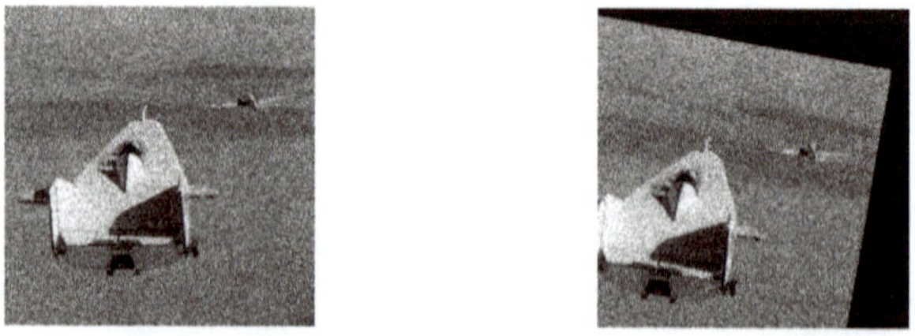

(a) (b)

Figure 21. Lifting Body image pair. (**a**) Reference image. (**b**) Sensed image.

The AAID evaluations for these registration results are compared in Figure 22 and the spatial partition of the AAID evaluation for each result is illustrated in Figure 23, From these two figures, the proposed fusion-based methods achieve better performance and the rightmost parts (parts 5, 10, 15, 20 and 25) have larger AAID than other parts.

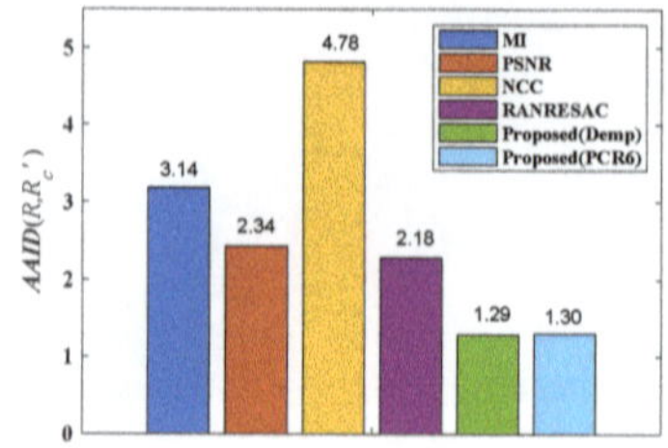

Figure 22. AAID evaluations of registration results for Lifting Body image pair.

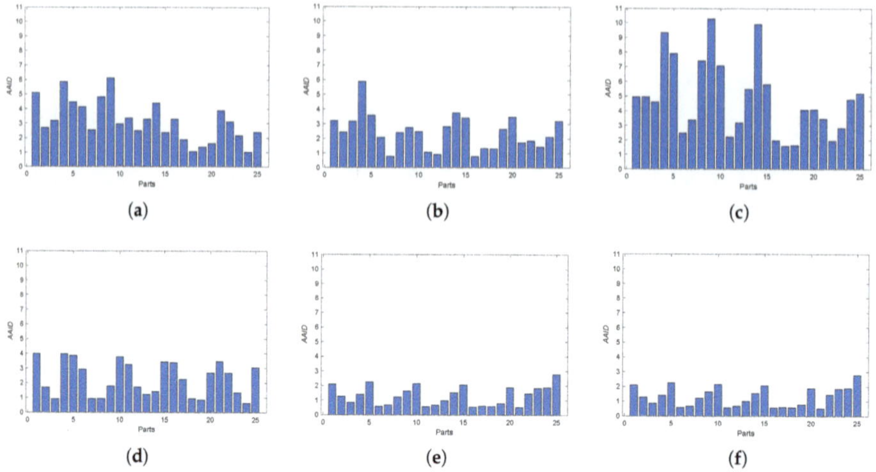

Figure 23. Spatial partition of the AAID evaluation for Lifting Body image. (**a**) MI. (**b**) PSNR. (**c**) NCC. (**d**) RANRESAC. (**e**) Proposed (Demp). (**f**) Proposed (PCR6).

According to all the experiments, the proposed fusion-based methods achieve better registration results than those prevailing ones (BRISK, KAZE, SURF, MI, PSNR and NCC). For noisy image registration, the proposed methods also obtain better performance than RANRESAC. This indicates that the theory of belief function can well deal with the uncertainty brought by the selection of keypoint detection algorithms or similarity measures, and the jointly use of the different keypoint detections or similarity measures is effective. Furthermore, from the above provided experiments one sees that the choice of combination rule does not affect the registration performance that much.

6.3. Computational Cost

The computational cost is an important criterion to evaluate an algorithm. We counted the computational costs of the above sparse algorithms and dense algorithms for Cameraman image (Figure 5a) on a Windows 10 Enterprise system equipped with Intel Core i7-7700HQ CPU at 2.80 GHz and 16.00 GB RAM. The platform is MATLAB R2018a. The average execution time comparisons for the sparse algorithms and dense algorithms are provided in Tables 2 and 3, respectively. Each average execution time is calculated from 100 runs of experiments.

Table 2. Average execution time comparison for sparse algorithms (unit second).

Method	Noise-Free Images	Noisy Images
BRISK	0.2847	0.2832
KAZE	0.1348	0.1304
SURF	0.0431	0.0437
RANRESAC	–	6.2648
Proposed (Demp)	0.3934	0.3933
Proposed (PCR6)	0.3938	0.3989

Table 3. Average execution time comparison for dense algorithms (unit second).

Method	Noise-Free Images	Noisy Images
MI	16.7622	16.4789
PSNR	12.8583	13.4214
NCC	14.7187	15.1050
Proposed (Demp)	17.0734	16.7945.
Proposed (PCR6)	17.0812	16.8729

From Tables 2 and 3, the dense algorithms need more execution time than the sparse algorithms. Furthermore, since the proposed fusion-based method combines the registration transformations generated from the three sparse methods (BRISK, KAZE and SURF) or the three dense methods (PSNR, MI and NCC) and these three methods can be parallely executed, the execution time of the proposed fusion-based method is longer than the most time-consuming one among the three methods.

6.4. Discussion of BBA Generation

The BBA generated in Equation (16) is Bayesian BBA, where all its focal elements are singletons. People in the community of belief function theory may prefer to use the compound focal elements, which usually seems better than only using singletons in Bayesian BBAs. We have also designed experiments of generating non-Bayesian BBAs for image registration using FCOWA-ER (Fuzzy-Cautious Ordered Weighted Averaging with Evidential Reasoning) [35] method. In detail, when multiple image information (image's intensities, edges and phase angle) are simultaneously considered, image registration can be viewed as a multi-criteria decision making problem. FCOWA-ER (Fuzzy-Cautious Ordered Weighted Averaging with Evidential Reasoning) [35] is a decision making approach under multi-criteria with uncertainty and it generates non-Bayesian BBAs using α-cut method (The α-cut method used in FCOWA-ER boils down to the Dubois and Prade allocation [36] in this case) when modeling uncertainties. According to the experimental results, non-Bayesian BBAs obtain similar registration results with Bayesian BBAs. Since Bayesian BBAs are easier to generate than non-Bayesian BBAs, we recommend Bayesian BBAs for image registration and do not provide the non-Bayesian BBA based method in this work.

7. Conclusions

In this paper, we proposed a new image registration algorithm based on evidential reasoning. The uncertainty encountered in image registration is taken into account and modeled by belief functions. Image information at different levels are jointly used to achieve a more effective registration. Experimental results show that the proposed algorithm can improve the precision of image registration.

The generation of BBA is crucial in evidential reasoning and most methods are proposed based on applications. In this paper, we generate BBAs from three different image information, i.e., intensity, edge and phase angle. In future work, other image information, such as texture feature and gradient feature, will also be considered and jointly used in image registration. Furthermore, we will attempt to apply the proposed method to color image registration. Different color channels of the color image provide different image information and can be jointly used in image registration. We will also focus on the comparison with the state-of-the-art approaches based on convolutional neural networks (CNN).

Author Contributions: Z.Z., D.H., J.D. and Y.Y. conceived and designed the experiments; Z.Z. performed the experiments; Z.Z., D.H., J.D. and Y.Y. analyzed the data; D.H. and Y.Y. contributed analysis tools; Z.Z., D.H. and J.D. wrote the paper.

Funding: This research was funded by National Natural Science Foundation (Nos. 61573275, 61671370), Postdoctoral Science Foundation of China (No. 2016M592790), Postdoctoral Science Foundation of Shaanxi Province (No. 2016BSHEDZZ46), and Fundamental Research Funds for the Central Universities (No. xjj2016066).

Conflicts of Interest: The authors declare no conflict of interest.

References

1. Meher, B.; Agrawal, S.; Panda, R.; Abraham, A. A survey on region based image fusion methods. *Inf. Fusion* **2019**, *48*, 119–132. [CrossRef]
2. Krylov, V.A.; Moser, G.; Serpico, S.B.; Zerubia, J. False discovery rate approach to unsupervised image change detection. *IEEE Trans. Image Process.* **2016**, *25*, 4704–4718. [CrossRef] [PubMed]
3. Torabi, A.; Massé, G.; Bilodeau, G. An iterative integrated framework for thermal-visible image registration, sensor fusion, and people tracking for video surveillance applications. *Comput. Vis. Image Underst.* **2012**, *116*, 210–221. [CrossRef]
4. Zhang, G.; Wu, Q.; Wang, T.; Zhao, R.; Deng, M.; Jiang, B.; Li, X.; Wang, H.; Zhu, Y.; Li, F. Block Adjustment without GCPs for Chinese Spaceborne SAR GF-3 Imagery. *Sensors* **2018**, *18*, 4023. [CrossRef] [PubMed]
5. Saygili, G.; Staring, M.; Hendriks, E.A. Confidence estimation for medical image registration based on stereo confidences. *IEEE Trans. Med. Imaging* **2016**, *35*, 539–549. [CrossRef] [PubMed]
6. Zitová, B.; Flusser, J. Image registration methods: A survey. *Image Vis. Comput.* **2003**, *21*, 977–1000.
7. Guo, X.; Xu, Z.; Lu, Y.; Pang, Y. An Application of Fourier-Mellin Transform in Image Registration. In Proceedings of the International Conference on Computer and Information Technology, Shanghai, China, 21–23 September 2005.
8. Ask, E.; Enqvist, O.; Svärm, L.; Kahl, F.; Lippolis, G. Tractable and reliable registration of 2D point sets. In Proceedings of the 13th European Conference on Computer Vision, Zurich, Switzerland, 6–12 September 2014; pp. 393–406.
9. Harris, C.; Stephens, M. A combined corner and edge detector. In Proceedings of the 4th Alvey Vision Conference, Manchester, UK, 31 August–2 September 1988; pp. 147–151.
10. Rosten, E.; Drummond, T. Machine learning for high-speed corner detection. In Proceedings of the 9th European Conference on Computer Vision, Graz, Austria, 7–13 May 2006; pp. 430–443.
11. Lowe, D.G. Object recognition from local scale-invariant features. In Proceedings of the International Conference on Computer Vision, Corfu, Greece, 20–25 September 1999; pp. 1150–1157.

12. Bay, H.; Tuytelaars, T.; Gool, L.V. Surf: Speeded up robust features. *Comput. Vis. Image Underst.* **2008**, *110*, 346–359. [CrossRef]
13. Tola, E.; Lepetit, V.; Fua, P. Daisy: An efficient dense descriptor applied to wide baseline stereo. *IEEE Trans. Pattern Anal. Mach. Intell.* **2010**, *32*, 815–830. [CrossRef] [PubMed]
14. Rublee, E.; Rabaud, V.; Konolige, K.; Bradski, G. ORB: An efficient alternative to SIFT or SURF. In Proceedings of the IEEE International Conference on Computer Vision, Barcelona, Spain, 6–13 November 2011.
15. Alcantarilla, P.F.; Bartoli, A.; Davison, A.J. KAZE Features. In Proceedings of the European Conference on Computer Vision, Florence, Italy, 7–13 October 2012; pp. 214–227.
16. Goshtasby, A.A. *Image Registration: Principles, Tools and Methods*; Springer: Berlin, Germany, 2012.
17. Santini, S.; Jain, R. Similarity measures. *IEEE Trans. Pattern Anal. Mach. Intell.* **1999**, *21*, 871–883. [CrossRef]
18. Spearman, C. The proof and measurement of association between two things. *Int. J. Epidemiol.* **2010**, *39*, 1137–1150. [CrossRef] [PubMed]
19. Shafer, G. *A Mathematical Theory of Evidence*; Princeton University Press: Princeton, NJ, USA, 1976.
20. Han, D.; Dezert, J.; Li, S.; Han, C.; Yang, Y. Image registration based on evidential reasoning. In Proceedings of the 16th International Conference on Information Fusion, Istanbul, Turkey, 9–12 July 2013; pp. 1143–1150.
21. Fischler, M.A.; Bolles, R.C. *Readings in Computer Vision: Issues, Problem, Principles, and Paradigms*; Morgan Kaufmann: Burlington, VT, USA, 1987.
22. Torr, P.H.S.; Murray, D.W. The development and comparison of robust methods for estimating the fundamental matrix. *Int. J. Comput. Vis.* **1997**, *24*, 271–300. [CrossRef]
23. Nakazawa, A. Noise stable image registration using random resample consensus. In Proceedings of the 23rd International Conference on Pattern Recognition, Cancun, Mexico, 4–8 December 2016; pp. 853–858.
24. Jain, P.; Kar, P. Nonconvex Optimization for Machine Learning. *Found. Trends Mach. Learn.* **2017**, *10*, 142–336. [CrossRef]
25. Pham, D.T. *Intelligent Optimisation Techniques*; Springer: Berlin, Germany, 2000.
26. Smarandache, F.; Dezert, J. *Advances and Applications of DSmT for Information Fusion: Collected Works IV*; American Research Press: Rehoboth, DE, USA, 2015.
27. Smets, P. The transferable belief model. *Artif. Intell.* **1994**, *66*, 191–234. [CrossRef]
28. Han, D.; Dezert, J.; Duan, Z. Evaluation of probability transformations of belief functions for decision making. *IEEE Trans. Syst. Man Cybern. Syst.* **2016**, *46*, 93–108. [CrossRef]
29. Gonzalez, R.C.; Woods, R.E. *Digital Image Processing*, 3rd ed.; Pearson Prentice Hall: Upper Saddle River, NJ, USA, 2008.
30. Canny, J. A computational approach to edge detection. *IEEE Trans. Pattern Anal. Mach. Intell.* **1986**, *8*, 679–698. [CrossRef] [PubMed]
31. Ferreira, D.P.L.; Ribeiro, E.; Barcelos, C.A.Z. A variational approach to non-rigid image registration with bregman divergences and multiple features. *Pattern Recognit.* **2018**, *77*, 237–247. [CrossRef]
32. Xia, X.; Dang, G.; Yao, Y.; Liang, J. Image registration model and algorithm for multi-focus images. *Pattern Recognit. Lett.* **2017**, *86*, 26–30. [CrossRef]
33. Pluim, J.P.W.; Likar, B.; Gerritsen, F.A. *Biomedical Image Registration*; Springer: Berlin, Germany, 2006.
34. Leutenegger, S.; Chli, M.; Siegwart, R.Y. BRISK: Binary Robust invariant scalable keypoints. In Proceedings of the IEEE International Conference on Computer Vision, Barcelona, Spain, 6–13 November 2011.
35. Han, D.; Dezert, J.; Tacnet, J.M.; Han, C. A fuzzy-cautious OWA approach with evidential reasoning. In Proceedings of the International Conference on Information Fusion, Singapore, 9–12 July 2012; pp. 278–285.
36. Dubois, D.; Prade, H. Representation and Combination of Uncertainty with Belief Functions and Possibility Measures. *Comput. Intell.* **1988**, *1*, 244–264. [CrossRef]

sensors

MDPI

Article

Research about DoS Attack against ICPS

Jianlei Gao, Senchun Chai *, Baihai Zhang and Yuanqing Xia

School of Automation, Beijing Institute of Technology, Beijing 100081, China; jianleixinye@163.com (J.G.); smczhang@bit.edu.cn (B.Z.); xia_yuanqing@163.net (Y.X.)

* Correspondence: chaisc97@bit.edu.cn; Tel.: +86-1391-145-7765

Received: 13 February 2019; Accepted: 24 March 2019; Published: 29 March 2019

Abstract: This paper studies denial-of-services (DoS) attacks against industrial cyber-physical systems (ICPSs) for which we built a proper ICPS model and attack model. According to the impact of different attack rates on systems, instead of directly studying the time delay caused by the attacks some security zones are identified, which display how a DoS attack destroys the stable status of the ICPS. Research on security zone division is consistent with the fact that ICPSs' communication devices actually have some capacity for large network traffic. The research on DoS attacks' impacts on ICPSs by studying their operation conditions in different security zones is simplified further. Then, a detection method and a mimicry security switch strategy are proposed to defend against malicious DoS attacks and bring the ICPS under attack back to normal. Lastly, practical implementation experiments have been carried out to illustrate the effectiveness and efficiency of the method we propose.

Keywords: DoS attack; industrial cyber-physical system (ICPS); security zones; mimicry security switch strategy

1. Introduction

A cyber-physical system (CPS) is a physical system that combines physical plants with network systems for data transmission and control functions, which has attracted worldwide attention after it was put forward in 2006 by the U.S. National Natural Science Foundation [1]. The CPS usually integrates some physical processes, data communication capabilities, sensors, data calculation and process control. It utilizes computers and networks to monitor physical process and control production parameters. It realizes combined calculations with physical processes depending on the real-time data interaction. CPSs are ubiquitous in modern life, ranging from current sweeping robots to global energy power system networks, which include smart cities, medical systems, military command systems, etc.

There are a lot of different types of CPSs, whose most typical application is the industrial control system (ICS), also called industrial cyber-physical system (ICPS), such as a supervisory control and data acquisition (SCADA) system, or programmable logic controller (PLC) system. They are widely used in a variety of industries, especially those related to critical national infrastructures, such as smart grids, energy production and transmission, smart cities, municipal engineering, the petrochemical industry and so on [2].

In recent decades, the corresponding technology has been developed dramatically. In order to enhance the facilities, reduce the complexity and cut down costs, more and more ICPSs are being upgraded with the latest communication and control technology, such as network communication, wireless sensors networks, multi-agent systems and so on. Generally speaking, ICPSs with integration modern cyber-technologies, which include Internet technology, cloud technology, Internet of Things and so on, have been using those technologies to communicate with each part, monitor plants and control physical processes. However, network attacks and vulnerabilities that have produced great risks and a large number of information incidents have been triggered due to the open networks protocols, which have already resulted in serious damage.

In recent years, many ICPS incidents have happened around the world. In 2010, the Iranian nuclear incident where the country's nuclear energy program was attacked by the "Stunex Virus"and "Duqu Trojan" that was detected in many countries in 2011, was considered the first premeditated destruction aimed at critical ICPSs. In 2012 some security experts found that the "Flame virus" not only attacked Iran, but also affected the entire Middle East region. A German steelworks suffered from a cyber-attack, which resulted in the control systems and production systems being forced to stop in 2014, but the most striking example was the collapse of the Ukrainian electricity grid in December, 2016.

The report [3] published by ICS-CERT of China provides statistics of ICS information security incidents from the Industrial Control Systems Cyber Emergency Response Team (ICS-CERT), which shows that more and more ICPSs are being attacked by various malicious cyber actions as depicted in Figure 1. As is seen from the chart, obviously the incident occurrence is on the rise.

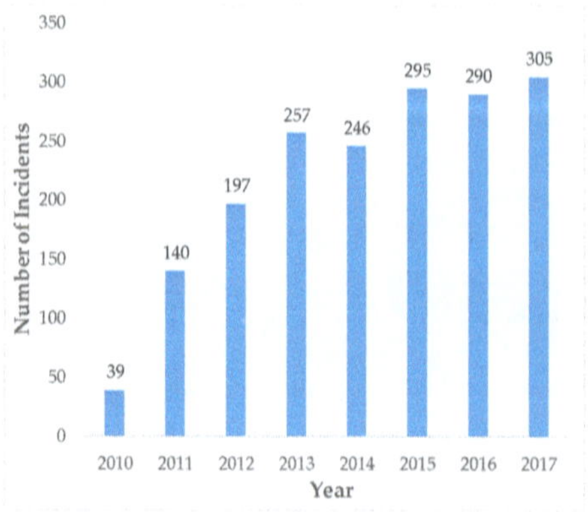

Figure 1. ICPS security incidents by year [3].

The ICPSs, especially the control systems of ICPSs, are connected to the public Internet, which raises greater security challenges than pure information systems when they are under cyberattack. As ICPSs are media to bridge physical process and virtual world of information, so the availability of data, including control data and communication data, is more important because of its effect on real-time performance. That is to say, once an ICPS is attacked by malicious attacks, it will cause more serious consequences than attacks on pure information systems.

The rest of paper is organized as follows: in Section 2, some related works are presented. Section 3 discusses some basic knowledge and mathematic models, which include an ICPS model, DoS attack model, DoS attack effect on ICPS model and so on. The detection model and defense strategy are designed in Section 4. We present some experimental results and analyses in the following Section 5. The last part presents the conclusions.

2. Related Work

As is known to all, unlike traditional information security where more attention is paid to the protection of data, cyber-attacks on CPSs' control networks usually wreck physical processes because of the existence of feedback networks, so the research and analysis on CPS must take both the cyber area and physical parts into consideration. Currently, there are various modes of attack against CPS, such as denial-of-service (DoS) attacks [4,5], bias injection attacks [6–8], zero dynamics attacks [9], convert attacks [10], zero response attacks [9], eavesdropping attacks [10] and so on. According to reference [10], the authors created a three-dimensional space, illustrated in Figure 2, to quantify them.

There have been a great deal of algorithms designed to analyze and solve these malicious attacks in CPSs [7,10–14]. They typically provide an explanation, system model and analysis, and control system experiments against different attack ways [10]. In [15], the authors do a lot of work about cyber-physical systems, supply a mathematical framework of the systems' attacks and monitors, present some fundamental monitoring limitations from a system-theoretic and graph-theoretic perspective, and design a distributed attack detector and identification monitors. Reference [12] studies a general convex optimization method of estimation which demonstrates generic sufficient and necessary conditions instead of specific estimators. The current detection methods against cyber-attacks are based on statistical learning algorithms which could cause misleading alarms. Reference [13] adopts a mulit-order Markov chain framework based on supervised statistical learning to solve the above shortcoming. Besides, it designs an optimal attack strategy to destroy wireless sensor network control systems and worsens the cost function to maximum value and find a coping strategy in this way [14]. Among these efforts aimed at studying specific malicious attacks, the DoS attacks (including DDoS) has been widely studied because of their easiest implementation, most serious consequences and least system knowledge that is needed to destroy the communication channel between a system's parts.

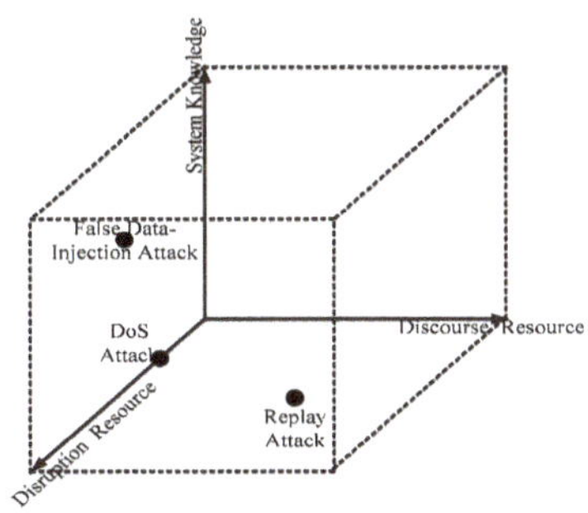

Figure 2. Three-dimensional attack space.

Many people have been devoted to studying DoS attacks against CPSs (including ICPSs), whose focus can mainly be divided into two parts: information security and control science. From the viewpoint of computer information security, people usually design an intrusion detection system (IDS) to protect targets from DoS attacks. The most typical methods are based on machine learning algorithms. For example: [16] proposes a new method based on support vector machine (SVM) which is motivated by the fact that the cloud environment is changeable/dynamic to detect DoS attacks. However, current IDS can't detect two specific hardware Trojans (HT)-assisted DoS attacks (sinkhole and blackhole attacks) which is explained by quantifying the effects of attacks as packet loss rates [17]. In order to deal with this problem in embedded systems designed with Multiprocessor-System-on-Chip (MPSoC) architectures, the utilization of pipelined MPSoC (PMPSoCs) is selected and improved to detect DoS attack-based hardware Trojan attacks [18]. Although all the studies provide some reasonable and useful methods to prevent, detect, defend and eliminate DoS attacks, they all have their limitations and deficiencies, especially in dealing with the carefully designed network DoS packets. What is worse is that they all don't consider the impact of attack on the physical part when they only fix their attention on the cyber layer. However, the physical part of ICPS is especially in need of strong real-time control data.

Many people study DoS attacks on ICPSs from the consideration of control theory. Some attack models and scenarios are given by reference [4] whose analyses are shown in Figure 2. Reference [5] uses the Tennessee Eastman challenge process to study the DoS attack through modeling the problem of DoS attacks as optimal stopping problems, which cause a change of the timing parameter in a physical process. The authors [14] analyze the problem of DoS attacks from the viewpoint of an attacker to study the optimal DoS attack strategy which can maximize the cost function of the linear quadratic regulator (LQR) controller. What' more, Yuan et al. [19] use a unified game theory to improve the robustness by designing a resilient control network system. Obviously, these algorithms can validly address the influences of attacks against physical layer such as control systems, but most of them lack any study on the cyber layer, which cannot eliminate DoS attacks.

In this paper, we try our best to combine the two aspects of ICPS security research to detect and eliminate DoS attacks against networks, and effectively solve the impacts of the attacks in the physical area to maintain ICPSs' normal operation.

The main contribution of this work are: (1) we try to combine and information security method with a control theory method to study the DoS attacks against industrial cyber-physical systems (ICPSs), and propose a mathematical model of DoS attacks with a detailed explanation; (2) according to the influence of different attack rates against ICPSs, we study the time delay caused by attacks dividing the ICPS into security zones instead of studying it directly, which displays how a DoS attack destroys the stable status of the ICPS; (3) a detection method and a mimicry security switch strategy are proposed to defend against this malicious DoS attack and bring the abnormal operation of ICPS back to a normal status; (4) a practical implementation has been carried out to illustrate the effectiveness and efficiency of the proposed method, which gives us an inspiration to protect our critical ICPSs with multiple sets of redundant sub-control systems.

3. Preliminary Knowledge

3.1. ICPS Structure

With the improvement of information technology, more and more ICPSs adopt Ethernet technology based on the TCP/IP protocol, which makes the control system more integrated, improves information transfer rate and the compatibility between different systems and enhances the range of application. A typical ICPS structure is shown in Figure 3.

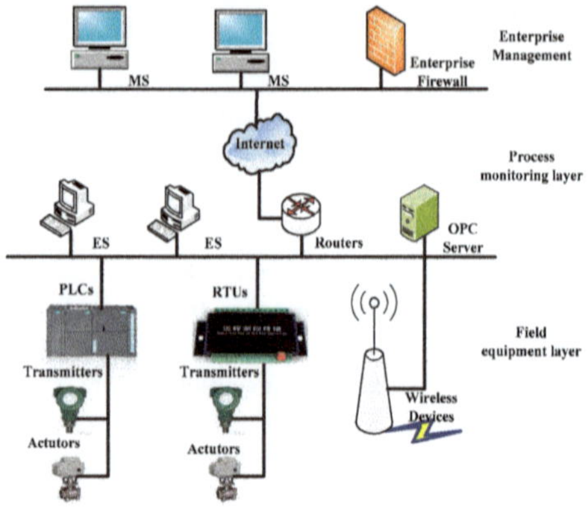

Figure 3. The structure of an ICPS.

3.2. ICPS Model

Consider the following ICPS with multi-subsystems:

$$\begin{cases} g = g(o_{i,}) \\ o_i = o_i(p_i, f_i) \end{cases} \quad (i = 1, 2, \cdots, N) \tag{1}$$

where i is the index of sub-control system, o_i is the i-th-sub-system, N is the sum of sub-control systems, p_i is the transfer function, f_i is the network characteristic function for which a detailed explanation will be given below. Consider the following physical system which is assumed a continuous linear dynamic system:

$$p_i(x) = \begin{cases} \dot{x}(t) = A_s x(t) + B_s u(t) \\ y(t) = C_s x(t) + D_s u(t) \end{cases} \tag{2}$$

where $x(t) \in \mathbb{R}^n$ and $y(t) \in \mathbb{R}^m$ show the system states and system output, respectively, at time $t \in \mathbb{N}$. Besides, the matrix of A_s, B_s, C_s and D_s are constant matrixes with related ranks.

Assumption 1. *Only one sub-system o_i is running at one moment, and other sub-systems $o_j (j \neq i)$ are listening and in standby mode at the same time.*

3.3. DoS Attack Model

In this subsection, in order to build a DoS attack model, we need to provide some assumptions and a definition firstly:

Assumption 2. *The time delay caused by the network's background is not considerable.*

Assumption 3. *The ICPS has more than one sub-system $f = \{f_i | i = 1, 2 \cdots, m\}$ (m is the sum of sub-system), and each of them has different network parameters which include an IP address (l_1) and communication port (l_2).*

Definition 1. *There exists an attack function $a = \{a_i | i = 1, 2, \cdots, m\}$ (m is the sum of sub-system), an attack operator $\otimes$ and attack set $I_H = \{0, 1\}$.*

A Denial-of-Service (DoS) attack is defined as a means to send lots of network data packets to targets, which will shut down users' computers and make the paralyze the communication network. DoS attacks accomplish this by flooding the target with traffic, or sending malicious information that triggers a crash. However, explicit DoS attack models are not given in [20], which lacks real network information, so we will define an attack model to explain how a DoS attacks a system.

According to Assumption 3, the attack function $a = \{a_i | i = 1, 2, \cdots, m\}$ and the attack object (the ICPS with one running subject) have two parameters, we can get $a_i = a_i(l_1, l_2)$ and $f_i = f_i(l_1, l_2)$.

Applying Definition 1 and Equation (1), it can be obtained that:

$$I_H = a_i \otimes f_j = a_i \otimes o(f_i, *) \tag{3}$$

Theorem 1. *Consider a DoS attack against ICPS, the attack a_i is independent of each other and the attack object f_i is independent of each other. Therefore, the attack set $I_H = \{0, 1\}$ has:*

$$\begin{cases} I_H = a_i \otimes o(f_j, *) = a_i \otimes f_j = 1, i = j \\ I_H = a_i \otimes o(f_j, *) = a_i \otimes f_j = 0, i \neq j \end{cases} \tag{4}$$

3.4. DoS Attack Effect on ICPS

As we all know, a DoS attack will affect a system's normal operation. As for how it affects the system, modelling system service performance from an information security perspective is relatively plausible in traditional information systems but not reasonable in an ICPS without consideration of stability of its physical parts. Studies in control science suggest the DoS attacks can increase the delay of control processes, and thus this will degrade the performance of the control system which is lacking details. Therefore, we try our best to explain how a DoS attack affects an ICPS's performance with Definition 2.

Definition 2. *We define a packets rate function fr:*

$$fr = \frac{sum\,(packet)}{sum\,(time)} \tag{5}$$

Actually, the DoS attack destroys the system's performance by increasing the time delay and this undoubtedly reduces the real-time control performance when the DoS attack lasts for a certain period. According to the network's features and working principle, for a more intuitive explanation, we define a time delay τ, dangerous zone Ω and their relation with *fr* to explain the details here.

Definition 3: *The ICPS has running zones $\Omega = \{\Omega_i | i = 1,2,3,4\}$, and Ω_{ICPS} represents the current running zone, which is shown in Figure 4.*

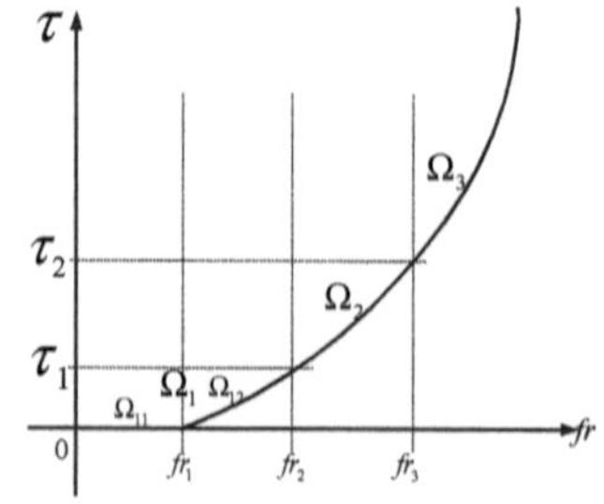

Figure 4. Relation between attack rate and time delay.

$$S = \begin{cases} \Omega_{11}, & 0 \le fr < fr_1 \\ \Omega_{12}, & fr_1 \le fr < fr_2 \\ \Omega_2, & fr_2 \le fr < fr_3 \\ \Omega_3, & fr_3 \le fr \end{cases} \tag{6}$$

where Ω_{11} is an absolutely secure zone which indicates that no attack can affect the system's normal operation. That is to say no attack is launched. Ω_{12} is a related security zone which indicates the system can still operate normally under attack. Ω_2 is a hazardous zone which indicates the system runs abnormally under attack but does not crash. Ω_3 is an absolutely hazardous zone which indicates the system has collapsed under attack. τ_1 is the resilience time delay which means this time delay can be accommodated by the network under attack and does not have any negative effects. τ_2 is the maximum time delay that the system can sustain. τ_{num} is the sum of the delays in the current communication network.

Merging Ω_{11} and Ω_{12}, we can get:

$$\Omega = \begin{cases} \Omega_1, & 0 \le fr < fr_2 \\ \Omega_2, & fr_2 \le fr < fr_3 \\ \Omega_3, & fr_3 \le fr \end{cases} \tag{7}$$

From the above, it is known that Ω_1 is a security area; Ω_2 is a transient-normal area; Ω_3 is an abnormal operation area.

Explanation. *Nowadays, network devices are built with an inherent time delay, which does not affect the normal operation of the ICPS. Meanwhile, they also have an elasticity feature that has some capacity to bear a bit of large network traffic to maintain the normal operation of the ICPS. This is in conformity with the actual situation. Therefore, our assumption is reasonable and the experimental data provided later will prove it too.*

Remark 1. *When the ICPS was attacked, every zone has following property:*

$$\begin{cases} \bigcup_{i=1}^{3} \Omega_i = 1 \\ \bigcap_{i=1}^{3} \Omega_i = 0 \end{cases} \quad (8)$$

Remark 2. *When the ICPS was attacked, every zone had the following migration process as is shown in Figure 5:*

Figure 5. Attack zone migration process.

Theorem 2. *The necessary and sufficient conditions for the system to run normally are:*

(1) $\Omega_{ICPS} \notin \Omega_3$ *and* $\Omega_{ICPS} \notin \Omega_2$
(2) $\Omega_{ICPS} \in \Omega_2$ *but* $T_{ICPS} < \tau_2 - \tau_1$

Proof of Theorem 2.

Necessary condition:

About condition (1), if $\Omega_{ICPS} \notin \Omega_2$ and $\Omega_{ICPS} \notin \Omega_3 \rightarrow \Omega_{ICPS} \in \Omega_1$.

About condition (2), if $\Omega_{ICPS} \in \Omega_2$ but $T_{ICPS} \leq \tau_2 - \tau_1 \rightarrow \tau_{num} = 0 + \tau_1 + T_{ICPS} < \tau_2 \rightarrow$ runs normally.

Sufficient condition:

About the proof of sufficient condition, we can use opposite to prove it. Firstly we assume the system is abnormal, so the time delay $\tau_{num} \geq \tau_2$.

Obviously, there are only two conditions which can satisfy it: $\Omega_{ICPS} \in \Omega_3$ or $\Omega_{ICPS} \in \Omega_2$ and $T_{ICPS} + \tau_1 \geq \tau_2$

From the analysis of an attacker's perspective, it is intended to implement the DoS attack plan that transforms the running zone Ω_1 into Ω_2, or even Ω_3.

Problem 3.1 (Attackers' Purpose)

$$\begin{cases} \max \quad T_{ICPS} \\ s.t. \quad I_H = 1 \end{cases} \quad (9)$$

The Problem 3.1 means that this malicious DoS attack is launched ($I_H = 1$) *to increase the maximum communication time delay of data packets* (T_{ICPS})*. According to the abovementioned Theorem 2, the increasing* T_{ICPS} *to deteriorate ICPS's normal running is equivalent to making the ICPS run in* Ω_2 *and* Ω_3*, so the Problem 3.1 can be equal to:*

$$\begin{cases} \Omega_{ICPS} \in (\Omega_2 \cup \Omega_3) \\ s.t. \quad I_H = 1 \end{cases} \quad (10)$$

3.5. Mimicry Security Policy

Assumption 4. *The time interval of switching between two different sub-systems is 0.*

In Nature a large number of creatures, the most typical example of which is the octopus, simulate other creatures through morphology, behavior and color, thus deceiving possible attackers and protecting themselves. This phenomenon is called mimicry, and it gives many living creatures a way to survive. Inspired by this ability, a large number of scholars began to study this mimetic defense strategy in the field of information security, and they have achieved excellent results. What's more important is that this strategy is effective against many methods of network attack. Besides, the majority of important infrastructures have several sets of stand-by sub-systems, which provides good conditions for the application of this strategy, so we try to use a mimicry security policy to defend DoS attack against ICPS. First, we define a mimicry defense strategy $\sigma(\bullet)$, following modal transfer as is shown in Figure 6:

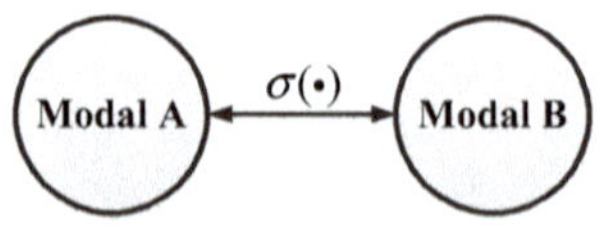

Figure 6. Mimicry policy.

That means the ICPS with multi-sub-systems will change its running-sub-system which is the modality of the current moment using a mimicry defense strategy $\sigma(\bullet)$ when it is attacked by a DoS attack.

Replace $\sigma(\bullet)$ into ICPS system's function:

$$\begin{cases} g = g(o_i) \\ \sigma(\bullet) \end{cases} = g_{\sigma(\bullet)}(o_i)$$

Applying Equation (1):

$$g_{\sigma(\bullet)}(o_i) = \begin{cases} f_{\sigma(\bullet)}(t) \\ p_{\sigma(\bullet)}(t) \end{cases} = \begin{cases} f_{\sigma(\bullet)}(t) \\ \begin{cases} \dot{x}(t) = A_{\sigma(\bullet)}x(t) + B_{\sigma(\bullet)}u(t) \\ z(t) = C_{\sigma(\bullet)}x(t) \end{cases} \end{cases}$$

Adding the constraint condition attack set I_H, we can get the following Equation (11):

$$\begin{cases} g_{\sigma(\bullet)}(o_i) \\ s.t \quad I_H = 1 \end{cases} = \begin{cases} \begin{cases} f_{\sigma(\bullet)}(t) \\ p_{\sigma(\bullet)}(t) \end{cases} \\ s.t \quad I_H \end{cases} = \begin{cases} \begin{cases} f_{\sigma(\bullet)}(t) \\ \begin{cases} \dot{x}(t) = A_{\sigma(\bullet)}x(t) + B_{\sigma(\bullet)}u(t) \\ y(t) = C_{\sigma(\bullet)}x(t) \end{cases} \end{cases} \\ s.t \quad I_H \end{cases} \tag{11}$$

Equation (11) shows that the function of mimicry switch strategy is to change the sub-system of the ICPS under DoS (I_H) to another sub-system to protect the ICPS. From the above analysis and assumption, we know that the cyber layer with a new network configuration has a natural immunity ability against DoS attacks after changing its sub-system using the mimicry defense strategy. These malicious attack packets cannot reach the ICPSS because of the new network configuration with a different IP address and port parameters. What's more, the physical layer with a new sub-system which has a new control system has the capability to keep the system stable and reduce the time delay to ensure the real-time character after the ICPS changes its old model.

In short, the role of these security tactics is that ICPS's running zone is transferred into a security zone from a transient-normal area as shown in Figure 7.

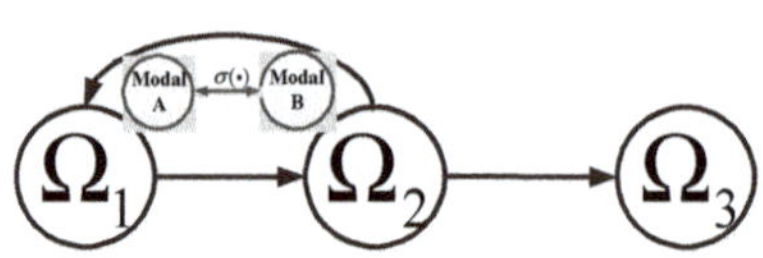

Figure 7. Zone transfer using mimicry strategy when under attack.

4. The Defense Strategy for DoS Attacks

In this section, some methods are developed to defend from DoS attacks, which contain a detection method and a mimicry security strategy to avoid the adverse effects of the attack.

4.1. The Detection of a DoS Attack

According to the physical system of ICPS shown in formula 2, we can get the model discretized by shift operator:

$$\begin{cases} x_{k+1} = Ax_k + Bu_k + w_k \\ y_k = C_k + v_k \end{cases} \tag{12}$$

where $x_k \in R^n$ is the n-dimensional vector of state variable at time k, $u(k) \in R^m$ is the m-dimensional system input vector at time k, y_k^T is an m-dimensional observation vector at time k, $w_k \in R^n$ and $v_k \in R^m$ are measurable white noises whose means are 0 at time k, respectively.

It is assumed that w_k and v_k are independent:

$$\begin{cases} w_k \sim \mathrm{N}(0, Q) \\ v_k \sim \mathrm{N}(0, R) \end{cases} \tag{13}$$

As we all know, the purpose of attacks against ICPS is to downgrade a stable operation state to a target state [21] and to evade detection, which could cause some indicators to deviate from the normal range. Different studies choose different standard indexes to identify anomalies, such as 2-normal, Chi-square, cost function, etc.

Although they are different methods, they are essentially all based on related errors. References [8,22,23] use a Chi-square detector to detect a CPS abnormity if the error between an estimated value and the real value exceeds a threshold. Although this method is unable to detect false data-injection attacks, it is applicable for other types of attack, for example: DoS attacks. The error covariance between the state value and estimated value, which we call minimum mean-square error (MMSE), is used in [24] to detect abnormal actions caused by network attacks because DoS attacks will break the system balance, which will increase system regulation cost. References [2,21] adopt a cost function to judge whether a system is under attack or not.

In many studies, a Kalman filter was utilized to perform state estimation, distinguish deviations and detect mistakes from observations under malicious attacks. In our research, we formulate a cost function that penalizes deviations from normal to abnormal states, and detects whether a DoS attack has happened. In this section, we model the physical part of an ICPS as a time-varying linear control system, which is equipped with a Kalman filter, LQR controller and failure detector:

$$\begin{cases} \widetilde{x}_{k+1|k} = A\widetilde{x}_k + Bu_k \\ P_{k+1|k} = AP_kA^T + Q \\ K_k = P_{k|k-1}C^T[CP_{k|k-1}C^T + R]^{-1} \\ P_k = P_{k|k-1} - K_kCP_{k|k-1} \\ \widetilde{x}_{k+1} = \widetilde{x}_{k+1|k} + K_k[y_k - C\widetilde{x}_{k+1|k}] \end{cases} \tag{14}$$

where $\widetilde{x}_k$ is the a posteriori state estimation value at time k, the error $(x_k - \widetilde{x}_{k+1|k})$ is between the estimation and real value, $P_k = \mathrm{cov}(x_k - \widetilde{x}_k)$ is the error covariance that shows the accuracy of the a priori estimation, A^T is the transposed matrix of A, and $X_{k=0} = X_0$, $P_{k=0} = P_0$.

According to [23,25], although the gain of Kalman filter K_k is time-varying, it always converges in a few steps to guarantee the system is detectable, so it can be defined as follows:

$$K \triangleq K_k = P_{k|k-1}C^T[CP_{k|k-1}C^T + R]^{-1} \tag{15}$$

At the same time, in order to simplify the analysis, it is usually assumed that the initial state of an ICPS with a linear state feedback controller is stable. Based on the LQR controller used in control systems, we assumed it is used in the ICPS to minimize the cost function, and the usefulness of the controller is to minimize the cost function J' as much as possible as follows:

$$J \triangleq \min \lim_{T\to\infty} E\frac{1}{T}\left[\sum_{k=0}^{T-1}(e_k^T W e_k + u_k^T U u_k)\right] \tag{16}$$

where $e_k = x_k - \widetilde{x}_k$ and the matrices of W and U are assumed as positive semi-definite matrices.

When an ICPS is attacked by a DoS attack, the attacker's intention is to transfer the running zone Ω_1 to other zones which will certainly increase the time delay. The increase of time delay means increasing the cost function, which also adds to the system's operation cost. Obviously, the communication time delay located in the network layer caused by a DoS attack will increase the cost function of the control system located in the physical layer. That is to say, the purpose of a malicious attacker is to degrade the stable running state of the control system of the physical layer by attacking the network layer, so Problem 3.1 can be rewritten as:

Problem 4.1 (Attacker's Purpose):

$$\begin{cases} \max \quad J \\ s.t. \quad I_H = 1 \end{cases} \tag{17}$$

It uses a Kalman filter to provide a system optimal state estimate of $\widetilde{x}_k$, so it can be obtained that:

$$\widetilde{x}_k = \widetilde{x}_{k|k-1} + K[y_k - C\widetilde{x}_{k|k-1}] \tag{18}$$

and we can get the optimal control law of LQR with fixed gain:

$$u_k = -(B^T S B + U)^{-1} B^T S A \widetilde{x}_k \tag{19}$$

where matrix S satisfies the Riccati equation:

$$S = A^T S A + W + A^T S B (B^T S B + U)^{-1} B^T S A \tag{20}$$

If we want to keep the ICPS running stably, we must make sure both J and the error are not unbounded. That is to say, it can be determined whether there is a DoS attack from whether the cost function J is bounded.

Simultaneously, it also defines a threshold function J_{th} J_{th}:

$$J_{th} \triangleq \max \lim_{T\to\infty} E\frac{1}{T}\left[\sum_{k=0}^{T-1}(e_k^T W e_k + u_k^T U u_k)\right] \tag{21}$$

so the detector works successfully with the following condition:

$$\begin{cases} J > J_{th} \quad , \quad alarm \\ J \leq J_{th} \quad , \quad no \quad alarm \end{cases} \tag{22}$$

This can trigger an attack alarm under DoS attack when the cost function exceeds the threshold.

4.2. Mimicry Security of Defense Policy

According to the DoS attack model, when an ICPS with one sub-system running is under attack, it could increase the time delay or even the rise of control cost as depicted in Problem 3.1 or Problem 4.1. In this sub-section, a mimicry security method is presented to solve this problem, which includes a state management and a mimicry switch strategy. It requires every sub-system to be waiting for running in real time. The state management is listening to all sub-systems' running states, inputs, outputs, detection of DoS attacks and other running state variables. The mimicry switch strategy is responsible for switching the running sub-system equipped with different network configurations and same control algorithm on the basis of switch rules from the attack detection of state management, which is depicted in the following Figure 8.

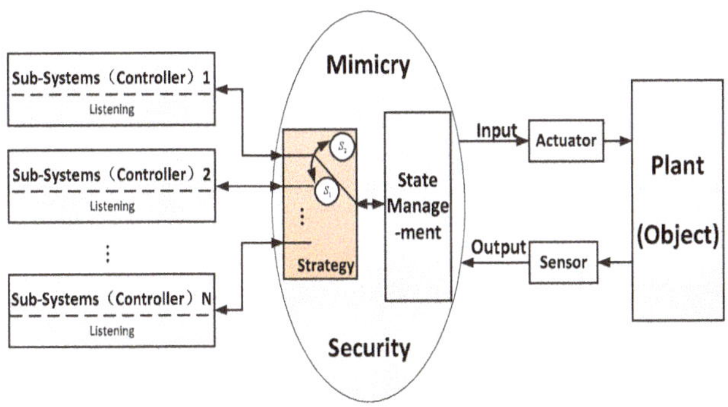

Figure 8. Mimicry security defense strategy.

From Subsection 3.3, it is known that the ICPS is secure when it is running in Ω_1. That means we can keep the ICPS running normally, whether the defense strategy is used or not. Only when the ICPS is running in the Ω_2 state under attack, we must adopt the mimicry defense strategy in time, keep the physical part stable, and ensure it is not operating in Ω_3, so the function of the defense strategy is to solve Problem 3.1 (or Problem 4.1) which can be rewritten as the following Problem 4.2 under the condition of the mimicry transformation time $T_{\sigma(\cdot)}$:

Problem 4.2

$$\begin{cases} g_{\sigma(\bullet)}(o_i) \\ s.t. \quad I_H = 1 \\ s.t. \quad T_{\sigma(\bullet)} \end{cases} \tag{23}$$

According to Equation (10), when $I_H = 1$*, which means* $\Omega_{ICPS} \in (\Omega_2 \cup \Omega_3)$ *Combined with Remark 2, we know* $\Omega_{ICPS} \in \Omega_2$. $\because P(\Omega_{ICPS} \in \Omega_2 | I_H = 1) = P(\Omega_{ICPS} \in \Omega_2)$*, so Equation (23) becomes:*

$$\begin{cases} g_{\sigma(\bullet)}(o_i) \\ s.t. \quad (\Omega_{ICPS} \in \Omega_2) \\ s.t. \quad T_{\sigma(\bullet)} \end{cases} \tag{24}$$

$\because \sup \quad T_{ICPS} = \tau_1 \therefore \inf \quad (\tau_2 - T_{ICPS}) = \tau_2 - \tau_1 = \tau$. *Combined with equations (1) and (2), Equation (24) becomes:*

$$\begin{cases} f_{\sigma(t)}(t) \\ \begin{cases} \dot{x}(t) = A_{\sigma(t)}x(t) + B_{\sigma(t)}u(t) \\ y(t) = C_{\sigma(t)}x(t) \end{cases} \\ s.t. \quad (\Omega_{ICPS} \in \Omega_2) \\ s.t. \quad T_{\sigma(\cdot)} \le \tau \end{cases} \tag{25}$$

From the above analysis, we know a new sub-system with new network configuration is waiting to run. Based on Equation (4), once the ICPS adopted a mimicry security to defend a DoS attack, the cyber part can work normally immediately, which will make the attack useless by transferring the ICPS to the Ω_1 from the Ω_2.

As is known to us, the sub-systems of ICPSs are changed after eliminating the impact of the cyber layer. However, if we want to protect the whole ICPS, we must keep the balance of the physical part. Hence, the problem that the mimicry security strategy needs to solve is changed from Problem 4.2 to Problem 4.3:

Problem 4.3

$$\begin{cases} \dot{x}(t) = A_{\sigma(t)}x(t) + B_{\sigma(t)}u(t) \\ y(t) = C_{\sigma(t)}x(t) \\ s.t. \quad T_{\sigma(\bullet)} \leq \tau \\ \sigma(\bullet) : J \geq J_{th} \end{cases} \tag{26}$$

If we want to make the ICPS free from paralysis, we should not only eliminate adverse effects from the network part, but also ensure the physical plant keeps running normally. When we adopt mimicry security strategy to switch sub-systems with different cyber parameters, it makes the system's cyber part be free from malicious actions instantly, so we need to design a switching controller to guarantee the every sub-system is running after switching.

Combined with equations (1), (2), (11) and (22), the switching signal $\sigma(\bullet) : J \geq J_{th}$ can be converted into $\sigma(\bullet) : \lim\limits_{T\to\infty} E\frac{1}{T}\left[\sum\limits_{k=0}^{T-1}(e_k^T W e_k + u_k^T U u_k)\right] \geq J_{th}$.

Define a switching sequence: $\{x_k; i_1, i_2, \cdots, i_k, \cdots | i_k \in N, k = 0, 1 \ldots.\}$. Therefore, the physical plant with feedback gain is:

$$\begin{cases} \dot{x}(t) = A_{\sigma(t)}x(t) + B_{\sigma(t)}u(t) \\ y(t) = C_{\sigma(t)}x(t) \\ s.t. \quad T_{\sigma(\cdot)} \leq \tau \\ \sigma(\bullet) : J \geq J_{th} \end{cases} \tag{27}$$

To keep the ICPS running normally, it is also needed to make $J \leq J_{th}$ after using the mimicry switch strategy, so the solution to Problem 4.3 becomes how to design a switching feedback gain $K = \{K_i \mid i = 0,1....,N\}$. If we want to keep the new sub-system stable after a mimicry switch, we need a positive definite matrix P [26,27]. At the same time, according to Theorem 1 in paper [28], the system must satisfy a bound to achieve a guaranteed cost function:

$$\begin{cases} A_i^T P A_i - P + Q + K_i^T R K_i < 0 \\ J \leq X_0^T P X_0 \end{cases} \tag{28}$$

Proof of Equation 28. Considering a known definite matrix P, and a Lyapunov function $V(x(k))$,

Make $V(x(k)) = x(k)^T P x(k) = \sum\limits_{i=1}^{N} V_i(x(k))$

Obviously: $V(x(k)) = 0$ only $x(k) = 0$; and $V(x(k)) > 0$ when $x(k) \neq 0$. Then:

$$
\begin{aligned}
\Delta V(x(k)) = \; & x(k+1)^T P x(k+1) - x(k)^T P x(k) \\
& = \sum_{i=1}^{N} \Delta V_i(x(k)) = \sum_{i=1}^{N} (V_i(x(k+1)) - V_i(x(k))) \\
& = \sum_{i=1}^{N} [((A_i + B_i K_i)x(k))^T P((A_i + B_i K_i)x(k)) - x(k)^T P x(k)] \\
& = \sum_{i=1}^{N} [(\overline{A}_i x(k))^T P(\overline{A}_i x(k)) - x(k)^T P x(k)] \\
& = \sum_{i=1}^{N} [x(k)^T \overline{A}_i^T P \overline{A}_i x(k) - x(k)^T P x(k)] \\
& = \sum_{i=1}^{N} [x(k)^T (\overline{A}_i^T P \overline{A}_i - P) x(k)]
\end{aligned}
$$

From (28): $\Delta V(x(k)) < 0$.

5. Experiments

In this section, we do some experiments to test our algorithms on a platform as depicted in Figure 8 which used a real industrial control system equipped with some industrial computers, sensors, electric motors, programmable logic controllers, a network server, cloud server and so on. Besides, this typical ICPS communicated by a network as shown in Figure 9.

On this experimental platform, we use two Siemens' programmable logic controllers (PLCs), which were set to two different IP addresses and two communication ports. The workflow of this platform is that the pump will pump some water into Tank 2 and keep a certain liquid level when the valve F_1 between Tank 1 and Tank 2 and the valve F_2 are opened. In our paper, we use the state of liquid level H that is the system response as an indicator to show whether the ICPS is being attacked by a DoS attack. The steady state of this liquid level H and sampling time are set as 400 mm and 0.2 s, respectively.

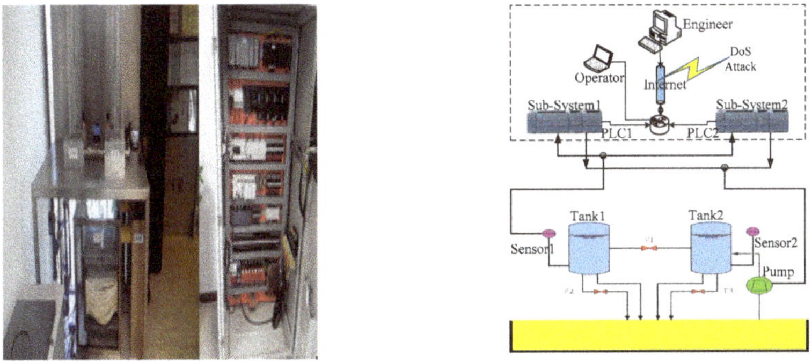

(a) (b)

Figure 9. This is the platform used in our experiments. (**a**) Test Platform Entity; (**b**) Test Platform Framework.

5.1. Related Network Feature

In this subsection, we analyze the effect on the ICPS's cyber part caused by DoS attacks with different attack rates. We repeat the DoS attacks against the PLC controller of ICPS 100 by the Monte Carlo method to get details of the network features used in the communication network. Firstly, we designed a probe to test the communication time delay between the PLC controller and the upper computer located in the Alibaba cloud server. Secondly, we used the hping3 network tool to implement DoS attacks with different attack rates for at least 1 minute each time. Then, we made use of the probe to randomly test the time delay for 60 to 600 s. What's more, this experiment was repeated 100 times. Finally, it the statistical data was achieved and the relation between attack rate and time delay (TD) were obtained, as displayed in Table 1 and Figures 10–14.

Table 1. The time delay caused by different DoS attack rates.

Attack Rate	0	0.1	1	10	100	1000	10000	100000
Min TD (ms)	29.170	30.876	30.539	30.812	30.892	179.614	4850.917	5000
Max TD (ms)	157.017	156.437	130.180	327.192	232.517	369.547		
Average TD (ms)	42.224	43.352	42.251	42.139	42.430	196.638	∞	
Average Packet/s	23.647	22.552	23.081	23.162	22.962	20.807	0.691	0

Actually, the configuration software (for example: Intouch) used in industrial control systems always has a default time delay (5 s, 10 s, or 15 s and so on), which means once the time delay of data packets from the sender exceeds a default value, the system will trigger an alarm. In our paper, we set 5 s as a default value, which is the time delay threshold.

On this experimental platform, the PLC controllers send data to the upper computer and receive data from the upper computer. A 5 s socket timeout was designed, which means that if new data was not sent or new data was not accepted for more than five seconds, the communication connection was considered broken. In Table 1, ∞ indicates that the PLC controller's network has crashed due to a high attack DoS attack rate.

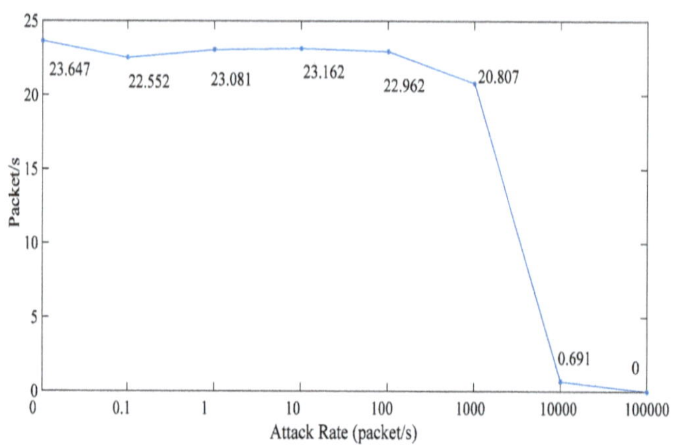

Figure 10. Test packet number per-second under different DoS attack rates.

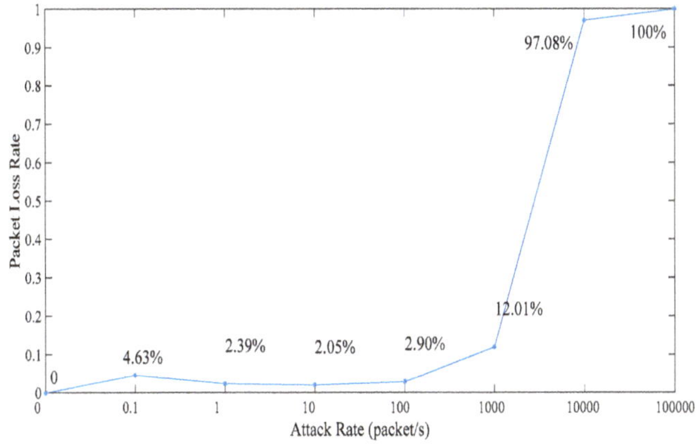

Figure 11. Test packet loss rate under different DoS attack rates.

The number of test packets from the probe is relatively stable when the attack rate is less than 1000. However, the change is sharply reduced when the attack rate is more than 1000 as seen in Figure 10. It can be seen that the packet loss rate is opposite to the above test packets numbers from the probe in Figure 11. When the attack rate is more than 1000, the packet loss rate will sharply increase until no packet data exists.

We can get that the maximum, minimum and average time delay of ICPS's network from Figures 12–14, respectively. No matter which the time delay it, its data trend is basically the same. They all reflect that the communication delay will increase with the rise of DoS attack rates until the network services has crashed undoubtedly, but every kind of time delay has only a little change when that attack rate is less than 1000.

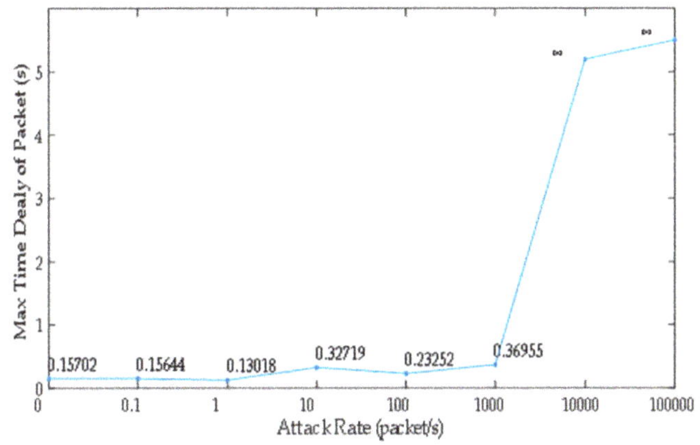

Figure 12. Max transmission time delay under different DoS attack rates.

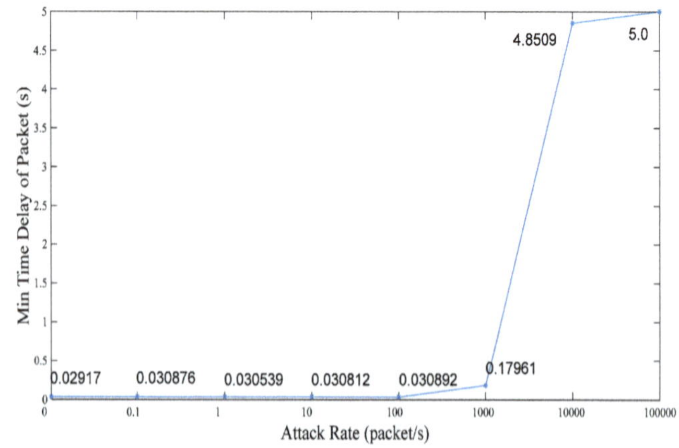

Figure 13. Min transmission time delay under different DoS attack rates.

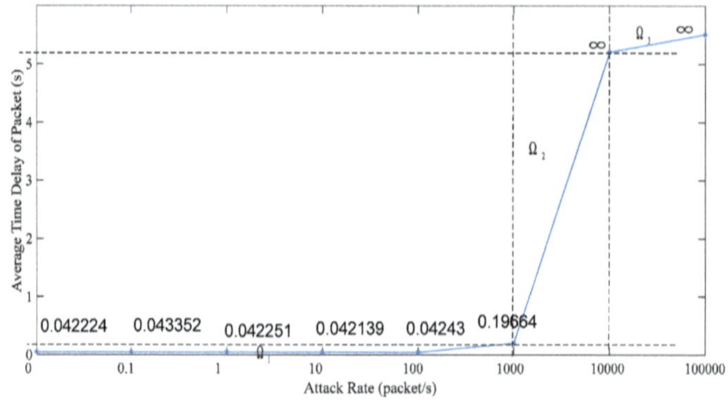

Figure 14. Average transmission time delay under different DoS attack rates.

The above table and charts show that the normal running of ICPS will not be affected by malicious attacks when the attack rate is less than a certain value. Usually, if the ICPS is not attacked by a DoS attack or the ICPS is under the DoS attack with an attack rate lower than fr_2, the time delay must be in the range Min TD to Max TD. Due to the performance limitations of network devices, the time delay can't be less than Min TD. Once the time delay exceeds Max TD, the performance of the ICPS will be destroyed. The randomness of time delay makes us select Average TD as an indicator to show the network performance. That is to say, we don't have to consider DoS attacks when the ICPS is running in the Ω_1 zone, which also demonstrates the correctness of Section 4.2.

5.2. Mimicry Security Strategy

In this subsection, we analyze the effect on an ICPS's physical part against DoS attack with different attack rates. The liquid level H (system response) and cost function of ICPS are illustrated in detail here, when it is in stable status.

Figure 15 depicts the system response without a DoS attack. We can see that no matter which sub-system was used, the ICPS whose sub-systems had different IP addresses and ports could stabilize the liquid level of platform at the same height H = 400 mm without a DoS attack.

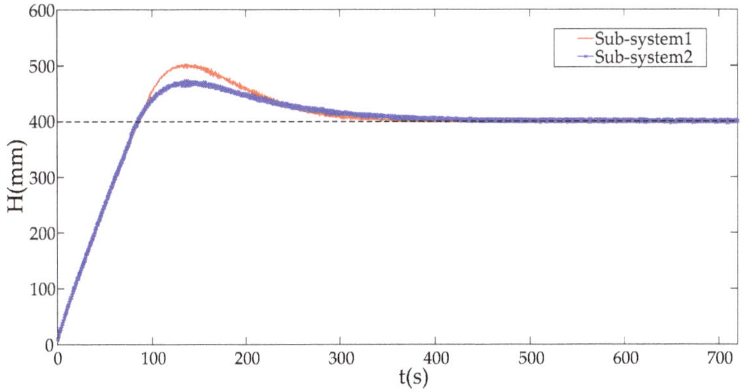

Figure 15. The system responses with different sub-system without DoS attack.

Figure 16 shows that system can keep running normally in the Ω_1 zone (including Ω_{11} and Ω_{12}) and Ω_2, but not in the Ω_3 zone. That is to say, the physical plant will not function well once that DoS attack rate exceeds a certain value. This has proved the validity of security zones.

Combining Figures 10–16, we can conclude that this ICPS is not affected by DoS attacks within a certain range of attack rates. However, once the DoS attack rate is more than a threshold, this malicious action will seriously damage the natural communication function of the cyber part, and will also affect the normal operation of the physical plant seriously in turn.

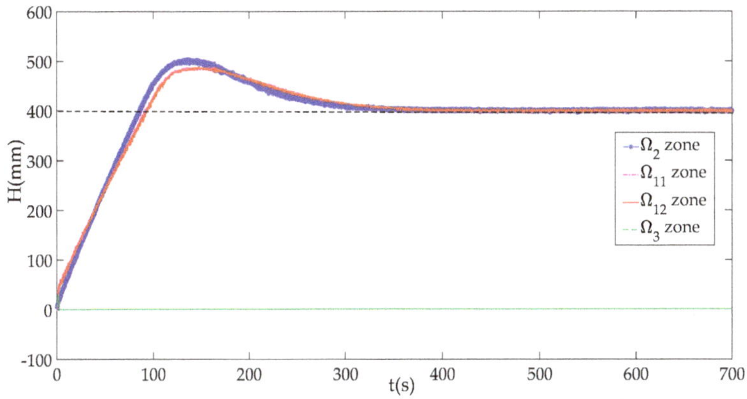

Figure 16. System responses in different zones.

We can see that the cost function value is very small and relatively stable in the stable running state of the control system in Figure 17, but, it will increase sharply under DoS attack with an attack rate of 1000 as shown in Figure 18.

Comparing Figure 17 with Figure 18, it can be seen that the cost function value J of ICPS will be enlarged more times. Obviously, once the DoS attack is launched by malicious attackers, we must have $J > J_{th}$, which will trigger an alarm.

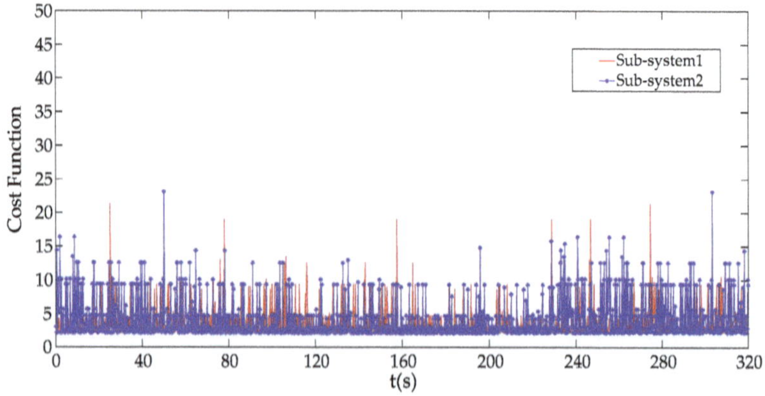

Figure 17. The cost function without DoS attack.

When an alarm is triggered, the mimicry security strategy will be used to protect the ICPS against the DoS attack. In our paper, we take sub-system 1 and sub-system 2 as an example. When the ICPS is under a DoS attack with an attack rate greater than 1000, the liquid tank level H controlled by the physical plant with sub-system1 begins to become unstable, and the same situation happens to sub-system2; However, when we use the mimicry security switch strategy to switch sub-system 2 with IP2 and port 2, the original ICPS equipped with sub-system 1 with IP1 and port1 become stable again, which is the same as switching ICPS's sub-system 2 to sub-system 1 as displayed in Figure 19.

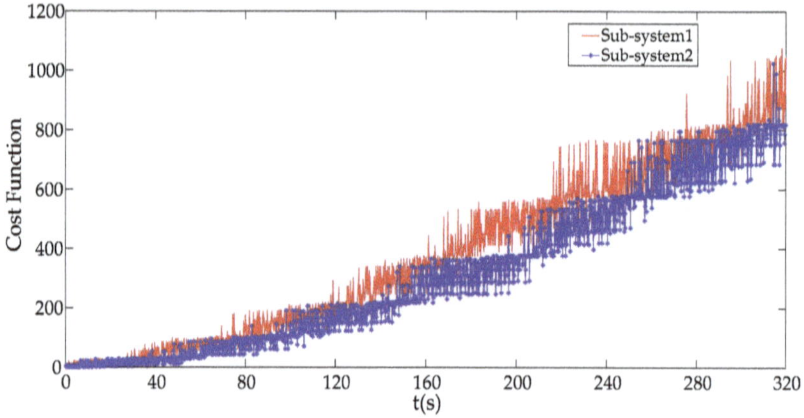

Figure 18. The cost function under DoS attack with an attack rate of 1000.

There is no doubt that the mimicry security strategy can solve the DoS attack against ICPS, which proves that this defense strategy is effective.

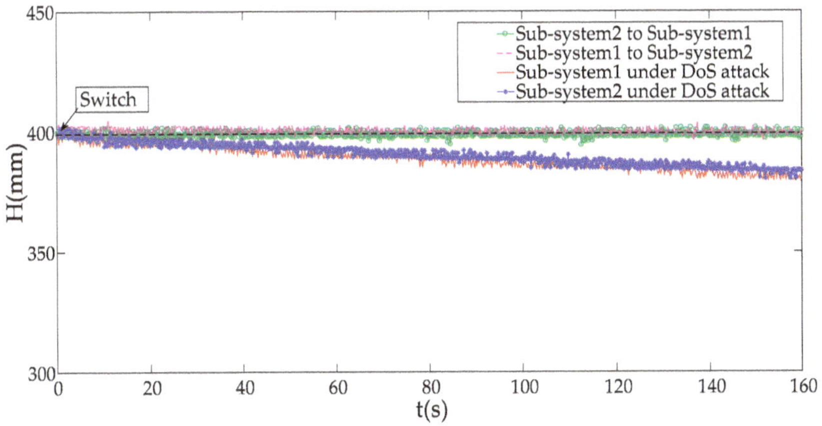

Figure 19. Different system responses before and after using the switch strategy.

5.3. Results Comparison

In this subsection, we compare with some experimental results using different methods to show details about the effectiveness of method proposed by us. In our paper, a predicted model-based algorithm [29,30] is selected as a contrast to show the usefulness of our proposed method. Figure 20 shows the comparisons of experimental results under different DoS attack rates. Obviously, both our method and the predicted method can eliminate the impacts on the physical plant caused by DoS attacks under the attack a rate of 1000 as is shown in Figure 20a. However, when the ICPS is under DoS attack with a rate of 10,000, our method can still work to keep the operation stable, while the method based on model prediction is invalid.

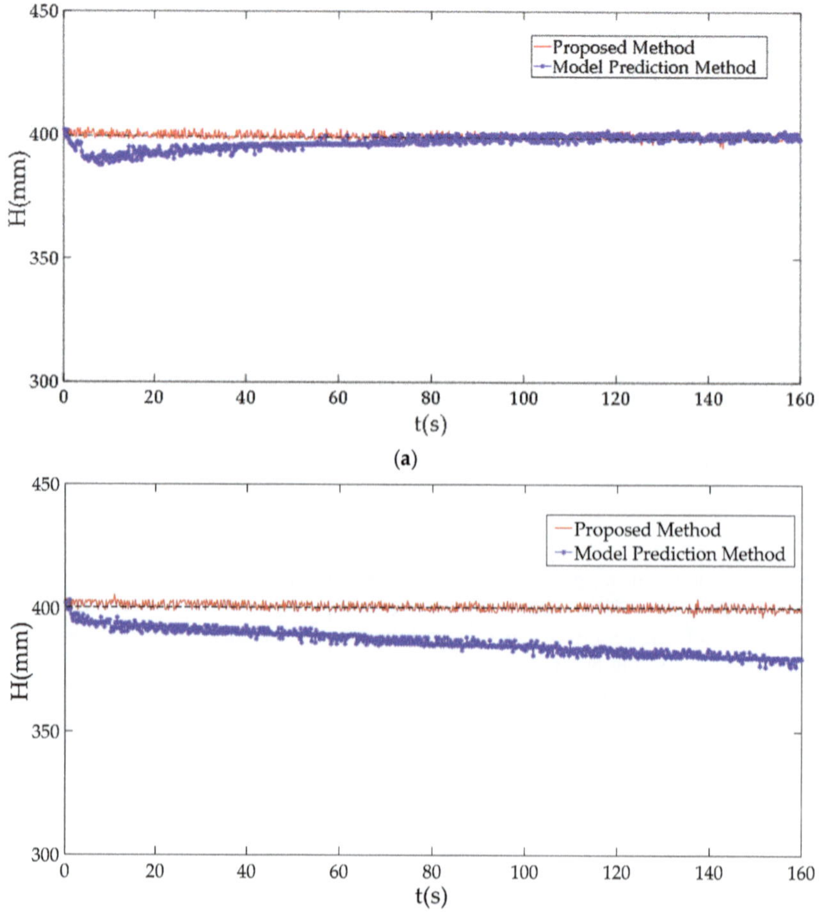

(b)

Figure 20. The comparison of results using different methods. (**a**) Experimental results under a DoS attack rate of 1000; (**b**) Experimental results under a DoS attack rate of 10,000.

Figure 21 shows that the comparisons of different packet loss of experimental results when we use different methods. It is obvious that this malicious attack can't affect the packet loss when the ICPS was under a DoS attack with an attack rate of less than 1000, which is equivalent to saying that that the ICPS is running in Ω_1. However, when the attack rate is more than 1000 ($\Omega_2 \cup \Omega_3$), and our method can solve this serious network problem through switching to a new sub-system equipped with a new IP address and communication port. The method based on model prediction does not have this function.

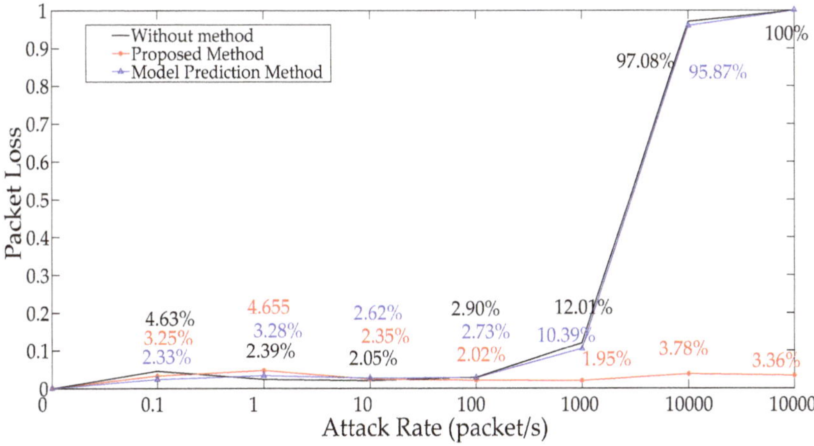

Figure 21. Different packet losses using different methods.

Figure 22 depicts the comparison of different average time delays of experimental results when we use the different methods. It can be obtained that our method can deal with the huge time delays caused by the DoS attack when the attack rate is more than 1000 and less than 10,000 (Ω_2), especially the controller crash problem when the attack rate is more than 10,000 (Ω_3), but, the predicted method based on the model cannot clear up it.

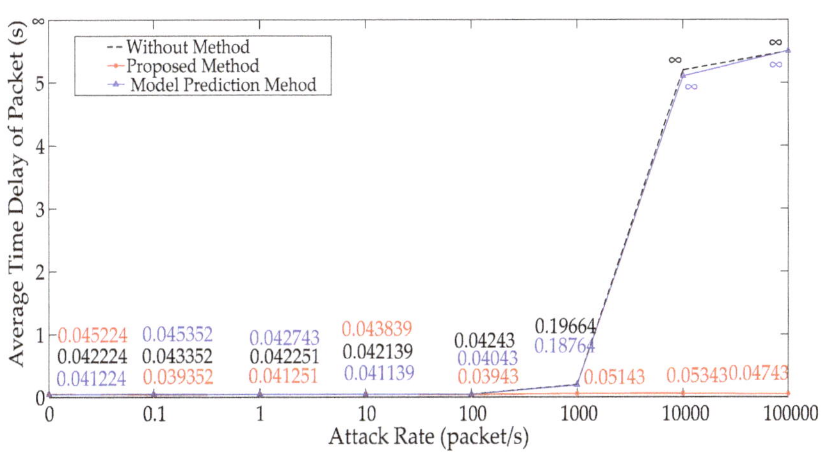

Figure 22. Different time delays using different methods.

From Figures 20–22, we know that the method proposed by us not only solves the network problems caused by DoS attacks, but also can maintain the normal operation of the ICPS. The algorithm based on model prediction is still unable to handle the cyber problem caused by this DoS attack with great attack rates. Even the physical process can't remain stable when the attack rate is too large ($\Omega_2 \cup \Omega_3$).

Obviously due to the resilient control ability of network devices, some DoS attacks with low attack rates cannot affect the normal operation of an ICPS. Previous works will produce some false alarms and reduce the detection accuracy, because of the existing resilient ability of network devices. Most models built by previous works don't consider specific DoS attacks, which is equivalent to studying the situation of ICPS running in zones Ω_1, Ω_2 and Ω_3. However, our paper takes this robustness of network devices into consideration and it is simplified further by studying their operation conditions in different security zones. That is to say, we don't need the DoS attacks of Ω_1, so all we have to deal with is the DoS attacks of Ω_2 and Ω_3, which need to be detected no matter which method is used. When the DoS attacks which belong to zones of Ω_2 and Ω_3 are launched, our method not only detects this malicious action, but also maintains the physical process stable and eliminates the serious impact on the cyber layer, while the method based on model prediction cannot deal with the two aspects of the problem at the same time.

6. Conclusions

In this paper, we study DoS attack problems and build related mathematic models to explain how DoS attacks affect the stable operation of ICPSs with different attack rates, which are based on studying the impacts of attacks on the cyber part and physical plant, respectively.

According to different attack rates, we divide them into different running zones firstly, which is consistent with facts. Then, we build a DoS attack model and explain the effect on an ICPS against attack actions using the above zones instead of analyzing the time delay from ICPSs' control data directly, which also shows clearly that the ICPS has a defense ability against malicious DoS attacks. The time delays ought to be negligible and the impact is fatal once the DoS attack rate exceeds a threshold. What's more, we chose the cost function value as a norm to detect anomalous actions and propose a mimicry security switch strategy to defend against such malicious attacks. Finally, we modeled a lot of DoS attacks and used a mimicry switch strategy repeatedly. From the above table and charts we can obviously see the impacts on the ICPS's cyber part and physical plant caused by this malicious action. The comparisons with different experimental results also verify our model's correctness and our method's effectiveness.

Author Contributions: J.G. established experiment platform, provided experimental data after conducting numerous experiments and then composed the first draft of this paper; S.C. provided research methods for this paper and also improved it; B.Z. helped revise and finalize the paper; Y.X. revised the paper, and provided the research methods.

Funding: This work has been supported by National Natural Science Foundation of China (No.61573061).

Conflicts of Interest: The authors declare no conflict of interest.

References

1. Lee, E.A. Cyber Physical Systems: Design Challenges. In Proceedings of the 11th IEEE International Symposium on Object and Component-Oriented Real-Time Distributed Computing (ISORC), Orlando, FL, USA, 5–7 May 2008.
2. Shang, W.; Zeng, P.; Wan, M.; Li, L.; An, P. Intrusion detection algorithm based on OCSVM in industrial control system. *Secur. Commun. Netw.* **2016**, *9*, 1040–1049. [CrossRef]
3. ICS-CERT Analysis on the Current Situation of Industrial Information Security in China in 2018. Available online: https://www.ics-cert.org.cn/portal/page/132/72be03492e944039a7a750bdb17bd42e.html (accessed on 28 July 2018).
4. Teixeira, A.; Perez, D.; Sandberg, H.; Johansson, K.H. Attack models and scenarios for networked control systems. In Proceedings of the 1st ACM International Conference on High Confidence Networked Systems, HiCoNS'12, Beijing, China, 17–19 April 2012; Association for Computing Machinery: Beijing, China, 2012; pp. 55–64.
5. Krotofil, M.; Manning, B.; Larsen, J. CPS: Driving cyber-physical systems to unsafe operating conditions by timing DoS attacks on sensor signals. In Proceedings of the Computer Security Applications Conference, New Orleans, LA, USA, 8–12 December 2014.
6. Bayou, L.; Espes, D.; Cuppens-Boulahia, N.; Cuppens, F. A Prediction-Based Method for False Data Injection Attacks Detection in Industrial Control Systems. In Proceedings of the International Conference on Risks and Security of Internet and Systems, Arcachon, France, 16–18 October 2018.
7. Manandhar, K.; Cao, X.; Hu, F.; Liu, Y. Detection of Faults and Attacks Including False Data Injection Attack in Smart Grid Using Kalman Filter. *IEEE Trans. Control Netw. Syst.* **2014**, *1*, 370–379. [CrossRef]

8. Yang, Q.; Yang, J.; Yu, W.; An, D.; Zhang, N.; Zhao, W. On False Data-Injection Attacks against Power System State Estimation: Modeling and Countermeasures. *IEEE Trans. Parallel Distrib. Syst.* **2014**, *25*, 717–729. [CrossRef]
9. Yuan, C.; Kar, S.; Moura, J.M.F. Dynamic Attack Detection in Cyber-Physical Systems with Side Initial State Information. *IEEE Trans. Autom. Control* **2017**, *6*, 4618–4624.
10. Teixeira, A.; Shames, I.; Sandberg, H.; Johansson, K.H. A Secure Control Framework for Resource-Limited Adversaries. *Automatica* **2015**, *51*, 135–148. [CrossRef]
11. Pasqualetti, F.; Dorfler, F.; Bullo, F. Attack Detection and Identification in Cyber-Physical Systems. *IEEE Trans. Autom. Control* **2013**, *58*, 2715–2729. [CrossRef]
12. Han, D.; Mo, Y.; Xie, L. Convex Optimization Based State Estimation against Sparse Integrity Attacks. *arXiv*, 2015; arXiv:1511.07218.
13. Sha, W.; Zhu, Y.; Chen, M.; Huang, T. Statistical Learning for Anomaly Detection in Cloud Server Systems: A Multi-Order Markov Chain Framework. *IEEE Trans. Cloud Comput.* **2015**, *6*, 401–413. [CrossRef]
14. Zhang, H.; Cheng, P.; Shi, L.; Chen, J. Optimal DoS Attack Scheduling in Wireless Networked Control System. *IEEE Trans. Control Syst. Technol.* **2016**, *24*, 843–852. [CrossRef]
15. Pasqualetti, F.; Dorfler, F.; Bullo, F. Attack Detection and Identification in Cyber-Physical Systems—Part II: Centralized and Distributed Monitor Design. *arXiv*, 2012; arXiv:1202.6049.
16. Abusitta, A.; Bellaiche, M.; Dagenais, M. An SVM-based framework for detecting DoS attacks in virtualized clouds under changing environment. *J. Cloud Comput.* **2018**, *7*, 9. [CrossRef]
17. Zhang, L.; Wang, X.; Jiang, Y.; Yang, M.; Mak, T.; Singh, A. Effectiveness of HT-assisted Sinkhole and Blackhole Denial of Service Attacks Targeting Mesh Networks-on-chip. *J. Syst. Arch.* **2018**, *89*, 84–94. [CrossRef]
18. Malekpour, A.; Ragel, R.; Ignjatovic, A.; Parameswaran, S. DoSGuard: Protecting pipelined MPSoCs against hardware Trojan based DoS attacks. In Proceedings of the IEEE International Conference on Application-specific Systems, Seattle, WA, USA, 10–12 July 2017.
19. Yuan, Y.; Yuan, H.; Lei, G.; Yang, H.; Sun, S. Resilient Control of Networked Control System under DoS Attacks: A Unified Game Approach. *IEEE Trans. Ind. Inform.* **2016**, *12*, 1786–1794. [CrossRef]
20. Zhou, W.; Jia, W.; Wen, S.; Xiang, Y.; Zhou, W. Detection and defense of application-layer DDoS attacks in backbone web traffic. *Future Gener. Comput. Syst.* **2014**, *38*, 36–46. [CrossRef]
21. Chen, Y.; Kar, S.; Moura, J.M.F. Optimal Attack Strategies Subject to Detection Constraints Against Cyber-Physical Systems. *IEEE Trans. Control. Netw. Syst.* **2017**, *5*, 1157–1168. [CrossRef]
22. Fawzi, H.; Tabuada, P.; Diggavi, S.N. Secure Estimation and Control for Cyber-Physical Systems Under Adversarial Attacks. *IEEE Trans. Autom. Control.* **2014**, *59*, 1454–1467. [CrossRef]
23. Mo, Y.; Garone, E.; Casavola, A.; Sinopoli, B. In False data injection attacks against state estimation in wireless sensor networks. In Proceedings of the Conference on Decision and Control, Atlanta, GA, USA, 15–17 December 2010; pp. 5967–5972.
24. Pang, Z.; Liu, G.; Zhou, D.; Hou, F.; Sun, D. Two-Channel False Data Injection Attacks Against Output Tracking Control of Networked Systems. *IEEE Trans. Ind. Electron.* **2016**, *63*, 3242–3251. [CrossRef]
25. Mo, Y.; Chabukswar, R.; Sinopoli, B. Detecting Integrity Attacks on SCADA Systems. *IEEE Trans. Control. Syst. Technol.* **2014**, *22*, 1396–1407.
26. Malmborg, J.; Bernhardsson, B.; Astrom, K.J. A Stabilizing Switching Scheme for Multi Controller Systems. *IFAC Proc. Vol.* **1996**, *29*, 2627–2632. [CrossRef]
27. Gao, H.; Liu, X.; Lam, J. Stability Analysis and Stabilization for Discrete-Time Fuzzy Systems with Time-Varying Delay. *Syst. Man Cybern.* **2009**, *39*, 306–317.
28. Wang, R.; Liu, G.; Wang, W.; Rees, D.; Zhao, Y.B. Guaranteed Cost Control for Networked Control Systems Based on an Improved Predictive Control Method. *IEEE Trans. Control. Syst. Technol.* **2010**, *18*, 1226–1232. [CrossRef]
29. Pang, Z.H.; Liu, G.P.; Dong, Z. Secure Networked Control Systems under Denial of Service Attacks. *IFAC Proc. Vol.* **2011**, *44*, 8908–8913. [CrossRef]
30. Pang, Z.H.; Liu, G.P. Design and Implementation of Secure Networked Predictive Control Systems under Deception Attacks. *IEEE Trans. Control. Syst. Technol.* **2012**, *20*, 1334–1342. [CrossRef]

sensors

Article

An EKF-Based Fixed-Point Iterative Filter for Nonlinear Systems

Xiaoliang Feng [1,*], Yuxin Feng [1] and Chenglin Wen [1,2]

1 College of Electrical Engineering, Henan University of Technology, Zhengzhou 450001, China; fyxxq0126@163.com (Y.F.); wencl@hdu.edu.cn (C.W.)
2 School of Automatic, Hangzhou Dianzi University, Hangzhou 310018, China
* Correspondence: fxl@haut.edu.cn

Received: 13 March 2019; Accepted: 15 April 2019; Published: 21 April 2019

Abstract: In this paper, a fixed-point iterative filter developed from the classical extended Kalman filter (EKF) was proposed for general nonlinear systems. As a nonlinear filter developed from EKF, the state estimate was obtained by applying the Kalman filter to the linearized system by discarding the higher-order Taylor series items of the original nonlinear system. In order to reduce the influence of the discarded higher-order Taylor series items and improve the filtering accuracy of the obtained state estimate of the steady-state EKF, a fixed-point function was solved though a nested iterative method, which resulted in a fixed-point iterative filter. The convergence of the fixed-point function is also discussed, which provided the existing conditions of the fixed-point iterative filter. Then, Steffensen's iterative method is presented to accelerate the solution of the fixed-point function. The final simulation is provided to illustrate the feasibility and the effectiveness of the proposed nonlinear filtering method.

Keywords: fixed-point filter; extended Kalman filter; nested iterative method; Steffensen's iterative method; convergence condition

1. Introduction

Interesting but unavailable signal variables can be estimated using a proper filter. In recent decades, filter design for various theoretical and applied systems has been a popular research topic in the fields of automatic control, target tracking, fault diagnosis, etc. [1–5].

The Kalman filter is designed for linear systems with noise-satisfying Gaussian distribution, and is an optimal filter due to the minimum mean square error (MMSE) [6]. The Kalman filter provides an estimate and prediction of the system state in real time. However, the classical Kalman filter must be applied to linear systems. For the filtering problem of nonlinear systems, a large number of effective filters have been developed on the basis of the classical Kalman filter.

The filters for nonlinear systems are mainly designed according to two principles. The first kind of nonlinear filter is designed by linearizing the system function, such as the celebrated extended Kalman filter (EKF) [7]. The second kind of nonlinear filter is designed on the basis of approximating the state statistics, such as the Unscented Kalman filter (UKF) [8–10], and the Cubature Kalman filter (CKF) [11–13]. Various nonlinear filters are designed in different ways to predict and update nonlinear system state estimates. In the state estimate prediction stage, according to EKF, the process function is linearized by discarding the second or higher order Taylor series items at the state estimate of the previous time instant. Then, the Kalman filter is utilized to predict the state estimate in terms of the linearized process function. In the state estimate updating stage, the measurement function is linearized by discarding the higher order items of its Taylor series at the predicted state estimate of the current time instant. Then, the Kalman filter is utilized to update the state estimate in terms of the linearized measurement function [14,15]. Although the EKF can estimate the state of a nonlinear system, the estimation accuracy of the EKF is always limited, due to the discarded second and higher order Taylor series terms of the process function and the measurement function in their linearization processes. In this context, a few nonlinear filters have been developed from the EKF. Based on the orthogonal principle, a fading factor is introduced to adjust the prediction error variance and to improve the robustness of the system modeling error and the filtering accuracy. The so-called strong tracking filter is proposed in [16,17], and developed in [18,19]. The multiple model method introduced in [20] simultaneously considers multiple possible state estimates, then improves the final state estimation accuracy through distributed extended Kalman filtering fusion. However, the discarded higher-order terms of the Taylor series in the linearization processes still inevitably affect the estimation accuracy of the nonlinear filters developed from the EKF.

The fixed-point theory is an effective numerical solution method for nonlinear equations. Therefore, the fixed-point theory is also utilized to design linear filters. In [21], the fixed-point theory is used for the identification of Wiener systems, which includes an infinite impulse response (IIR) system and a nonlinear static function. In [22,23], the maximum correntropy Kalman filter is designed by solving a fixed-point equation. The fixed-point iteration Gaussian sum filtering estimator with unknown time-varying non-Gaussian measurement noise is proposed in [24]. A characteristic function filter is developed in [25] for a class of non-Gaussian nonlinear dynamical systems with the linear measurement model.

Motivated by the discussion above, in this paper, a novel fixed-point nonlinear filtering method is studied to update the state estimate obtained by the EKF. Firstly, taking advantage of the fixed-point theory, a fixed-point function is constructed to solve the filtering problem. Since the state estimate and the filter gain in the fixed-point function are both unknown, a nested iterative method is presented to solve the fixed-point function and obtain the fixed-point filter. The convergence and the existing of the fixed-point filter are also discussed. Steffensen's iterative method is presented to accelerate the solution of the fixed-point iterative function.

The main contributions of this paper are threefold. (1) A fixed-point function was utilized to reduce the influence of discarded second- and higher-order Taylor series terms of nonlinear systems and to improve the filtering accuracy of the steady-state EKF. (2) Two kinds of fixed-point function solution methods were used and resulted in two fixed-point iterative filters. (3) The convergence of the fixed-point function and the existing conditions of the fixed-point iterative filter are also provided.

The remainder of this article is organized as follows. In Section 2, the general discrete time nonlinear system is formulated and the motivation of this paper is analyzed. In Section 3, a fixed-point function is provided to update the state estimate obtained by the EKF. The fixed-point filter is designed using two iterative solution methods, which iteratively update the state estimates and the filter gains. The convergence and existing conditions of the fixed-point filter are also discussed. The final illustrated simulation is provided in Section 4. Section 5 concludes the paper.

2. Problem Formulation

Considering the following nonlinear discrete-time system:

$$x(k+1) = f(x(k),k) + w(k+1,k) \tag{1}$$

$$y(k+1) = h(x(k+1),k) + v(k+1) \tag{2}$$

where $x(k)$ is the system state at the discrete time instant k, $f(*)$ is the nonlinear state evolution function and $h(*)$ represents the nonlinear measurement function. The process noise $w(k+1,k)$ and the measurement noise $v(k+1)$ satisfy $w(k+1,k) \sim N(0,Q(k+1,k))$ and $v(k+1) \sim N(0,R(k))$.

Remark 1. *The existing nonlinear filtering methods designed for the system above mainly include two types: (1) EKF-based nonlinear filtering methods, in which the nonlinear system model is always approximated by an* *appropriate linear function; and (2) sample point-based nonlinear filtering methods, in which the statistics of the state estimate are usually approximated by the statistical result of the sample points, for instance. However, it is unavoidable that certain errors exist in the above two kinds of approximations and result in estimation errors of the above two kinds of nonlinear filters. In order to improve the nonlinear filtering accuracy of the EKF-based nonlinear filtering methods, taking advantage of fixed-point theory, an EKF-based fixed-point iterative filter is proposed in this paper.*

3. EKF-Based Fixed-Point Filter

In this section, for the filtering problem of the general nonlinear system described in Equations (1) and (2), the classical EKF will firstly be introduced. Then, a fixed-point function will be provided to update the obtained state estimate by solving the fixed-point function. In Section 3.3, a nested iterative method is derived to solve the fixed-point filter. The convergence and existing condition of the fixed-point filter will be discussed in Section 3.4. Then, Steffensen's iterative method will be presented to accelerate the solution of the fixed-point function in Section 3.5.

3.1. Classic Extended Kalman Filter

In the EKF-based nonlinear filtering methods for the system described in Equations (1) and (2), the nonlinear system model is always represented by the Taylor series at a working point and linearized by discarding the second and higher order terms. Then, the well-known Kalman filter is used to estimate the state of the linearized system. The classical EKF can be summarized as follows.

Assumption 1. *The estimate and the estimation error variance of the state at the time instant k have been obtained and are denoted as $\hat{x}(k|k), P(k|k)$, then the state at the time instant $k+1$* will be estimated as follows.

(1) *Time Updating*

Taking the state estimate as the working point, the Taylor series expression of the nonlinear process function is

$$\begin{aligned} x(k+1) &= f(x(k),k) + w(k+1,k) \\ &= f(\hat{x}(k|k),k) + \left.\frac{\partial f(x(k),k)}{\partial x(k)}\right|_{x(k)=\hat{x}(k|k)} (x(k)-\hat{x}(k|k)) + O\left((x(k)-\hat{x}(k|k))^2\right) + w(k+1,k) \\ &\approx f(\hat{x}(k|k),k) + \left.\frac{\partial f(x(k),k)}{\partial x(k)}\right|_{x(k)=\hat{x}(k|k)} (x(k)-\hat{x}(k|k)) + w(k+1,k) \end{aligned} \tag{3}$$

Applying the Kalman filter to the linearized system shown as the second approximate formula above, the state prediction at the time instant $k+1$ and the prediction error variance can be obtained by

$$\begin{cases} \hat{x}(k+1|k) = f(\hat{x}(k|k),k), \\ P(k+1|k) = F(k+1,k)P(k|k)F^T(k+1,k) + Q(k+1,k), \end{cases} \tag{4}$$

where $F(k+1,k) = \left.\frac{\partial f(x(k),k)}{\partial x(k)}\right|_{x(k)=\hat{x}(k|k)}$ is the Jacobian of $f(x(k),k)$ at the working point $\hat{x}(k|k)$.

(2) *Measurement Updating:*

Taking the state prediction obtained above as the working point, the measurement function can be rewritten as

$$\begin{aligned} y(k+1) &= h(x(k+1),k+1) + v(k+1) \\ &= h(\hat{x}(k+1|k),k+1) + \left.\frac{\partial h(x(k+1),k+1)}{\partial x(k+1)}\right|_{x(k+1)=\hat{x}(k+1|k)} (x(k+1)-\hat{x}(k+1|k)) \\ &+ O\left((x(k+1)-\hat{x}(k+1|k))^2\right) + v(k+1) \\ &\approx h(\hat{x}(k+1|k),k+1) + \left.\frac{\partial h(x(k+1),k+1)}{\partial x(k+1)}\right|_{x(k+1)=\hat{x}(k+1|k)} (x(k+1)-\hat{x}(k+1|k)) + v(k+1) \end{aligned} \tag{5}$$

The state estimate and the estimate error variance at time instant $k+1$ can be derived as

$$\begin{cases} \hat{x}(k+1|k+1) = \hat{x}(k+1|k) + K(k+1)(y(k+1) - h(\hat{x}(k+1|k))), \\ P(k+1|k+1) = P(k+1|k) - K(k+1)H(k+1)P(k+1|k), \end{cases} \tag{6}$$

where $H(k+1) = \left.\frac{\partial h(x(k+1),k+1)}{\partial x(k+1)}\right|_{x(k+1)=\hat{x}(k+1|k)}$ *is the Jacobian of* $h(x(k+1),k+1)$ *at the working point* $\hat{x}(k+1|k)$ *and* $K(k+1)$ *is the filter gain, which is represented as*

$$K(k+1) = P(k+1|k)H^T(k+1)\Big(H(k+1)P(k+1|k)H^T(k+1) + R(k+1)\Big)^{-1} \tag{7}$$

3.2. Numerical Updating Method Based on the Fixed-Point Theory

As mentioned above, in the EKF, $\hat{x}(k+1|k+1)$ is obtained to estimate the system state $x(k+1)$ on the basis of the measurements $\{y(1), y(2), \cdots, y(k+1)\}$. Due to the discarded second and higher order Taylor series items of the process function and the measurement function, not all of the information contained in the measurement could be utilized to obtain $\hat{x}(k+1|k+1)$. In order to update this state estimate and make more full use of the unused information in the measurement, taking advantage of fixed-point theory, the following fixed-point function is constructed for a steady-state filter

$$\begin{cases} \hat{\hat{x}}(k+1|k+1) = \hat{\hat{x}}(k+1|k+1) + K^*(k+1)[y(k+1) - \hat{y}^*(k+1)] \\ \hat{y}^*(k+1) = h(\hat{\hat{x}}(k+1|k+1), k+1) + E\{v(k+1)\} \end{cases} \tag{8}$$

where $\hat{\hat{x}}(k+1|k+1)$ is the state of the above fixed-point filter, $K^*(k+1)$ is the filter gain to be determined, and $E\{v(k+1)\}$ is the mean of the measurement noise, which was found to be zero in this paper.

Remark 2. *In (8), the* $\hat{\hat{x}}(k+1|k+1)$ *will be estimated on the basis of the measurements* $\{y(1), y(2), \cdots, y(k+1)\}$*, such that the solution* $\hat{\hat{x}}(k+1|k+1)$ *is close to the actual state value at time instant* $k+1$*. This aims to solve* $\hat{\hat{x}}(k+1|k+1)$ *such that* $y(k+1) = h(\hat{\hat{x}}(k+1|k+1), k+1)$*. In order to solve* $\hat{\hat{x}}(k+1|k+1)$*, such that* $y(k+1) = h(\hat{\hat{x}}(k+1|k+1), k+1) + v(k+1)$*, the fixed-point function method is used, and (8) is the structured fixed-point function. However, as the state estimate and the filter gain are both unknown, it is difficult to solve the fixed-point function (8) by the traditional fixed-point iterative solution method.*

3.3. A Nested Iterative Solution for the Fixed-Point Function

In the fixed-point function (8), two unknown items must be determined simultaneously—The unknown state of the above fixed-point filter, $\hat{\hat{x}}(k+1|k+1)$ and the filter gain to be determined $K^*(k+1)$. In this section, a fixed-point filter is presented by solving the above fixed-point function (8), according to the following nested iterative computation process.

In the first iteration, take the state estimate $\hat{x}(k+1|k+1)$ obtained by the EKF as the initial value on the right side of the fixed-point function. Denote the initial state of the first iteration as $\hat{\hat{x}}_0^0(k+1|k+1) = \hat{x}(k+1|k+1)$, where the subscript 0 means the first iteration, while the superscript 0 is the initial value of this iteration, which increases in this iteration. Then we have the following iterative solution process of the fixed-point function with the initial filter gain $K_0^*(k+1) = K(k+1)$.

$$\begin{cases} \hat{\hat{x}}_0^1(k+1|k+1) = \hat{\hat{x}}_0^0(k+1|k+1) + K_0^*(k+1)\Big(y(k+1) - h(\hat{\hat{x}}_0^0(k+1|k+1), k+1)\Big) \\ \hat{\hat{x}}_0^2(k+1|k+1) = \hat{\hat{x}}_0^1(k+1|k+1) + K_0^*(k+1)\Big(y(k+1) - h(\hat{\hat{x}}_0^1(k+1|k+1), k+1)\Big) \\ \vdots \\ \hat{\hat{x}}_0^{i_0}(k+1|k+1) = \hat{\hat{x}}_0^{i_0-1}(k+1|k+1) + K_0^*(k+1)\Big(y(k+1) - h(\hat{\hat{x}}_0^{i_0-1}(k+1|k+1), k+1)\Big) \end{cases} \tag{9}$$

If $\left|\hat{x}_0^{i_0}(k+1|k+1) - \hat{x}_0^{i_0-1}(k+1|k+1)\right| < \varepsilon$, take $\hat{x}_0^{i_0}(k+1|k+1)$ as the state estimate in the first iterative process.

Due to

$$\hat{x}(k+1|k+1) = \hat{x}(k+1|k+1) + K^*(k+1)\big(y(k+1) - h(\hat{x}(k+1|k+1), k+1)\big) \tag{10}$$

the filter gain can be updated in the first iteration by

$$K_1^*(k+1) = \frac{\hat{x}_0^{i_0}(k+1|k+1) - \hat{x}_0^{i_0-1}(k+1|k+1)}{y(k+1) - h(\hat{x}_0^{i_0-1}(k+1|k+1), k+1)} \tag{11}$$

In the second iteration, the initial value in the fixed-point numerical solving process is $\hat{x}_1^0(k+1|k+1) = \hat{x}_0^{i_0}(k+1|k+1)$, then we have

$$\begin{cases} \hat{x}_1^1(k+1|k+1) = \hat{x}_1^0(k+1|k+1) + K_1^*(k+1)(y(k+1) - h(\hat{x}_1^0(k+1|k+1), k+1)) \\ \hat{x}_1^2(k+1|k+1) = \hat{x}_1^1(k+1|k+1) + K_1^*(k+1)(y(k+1) - h(\hat{x}_1^1(k+1|k+1), k+1)) \\ \vdots \\ \hat{x}_1^{i_1}(k+1|k+1) = \hat{x}_1^{i_1-1}(k+1|k+1) + K_1^*(k+1)(y(k+1) - h(\hat{x}_1^{i_1-1}(k+1|k+1), k+1)) \end{cases} \tag{12}$$

Similarly, if $\left|\hat{x}_1^{i_1}(k+1|k+1) - \hat{x}_1^{i_1-1}(k+1|k+1)\right| < \varepsilon$, $\hat{x}_1^{i_1}(k+1|k+1)$ is taken as the state estimate in the second iterative process, then the filter gain in the second iteration can be updated by

$$K_2^*(k+1) = \frac{\hat{x}_1^{i_1}(k+1|k+1) - \hat{x}_1^{i_1-1}(k+1|k+1)}{y(k+1) - h(\hat{x}_1^{i_1}(k+1|k+1), k+1)} \tag{13}$$

For the lth iteration, $\hat{x}_l^0(k+1|k+1) = \hat{x}_{l-1}^{i_{l-1}}(k+1|k+1)$ is taken as the initial value to solve the fixed-point numerical and is iteratively updated by the following

$$\begin{cases} \hat{x}_l^1(k+1|k+1) = \hat{x}_l^0(k+1|k+1) + K_l^*(k+1)(y(k+1) - h(\hat{x}_l^0(k+1|k+1), k+1)) \\ \hat{x}_l^2(k+1|k+1) = \hat{x}_l^1(k+1|k+1) + K_l^*(k+1)(y(k+1) - h(\hat{x}_l^1(k+1|k+1), k+1)) \\ \vdots \\ \hat{x}_l^{i_l}(k+1|k+1) = \hat{x}_l^{i_l-1}(k+1|k+1) + K_l^*(k+1)(y(k+1) - h(\hat{x}_l^{i_l-1}(k+1|k+1), k+1)) \end{cases} \tag{14}$$

Similarly, if $\left|\hat{x}_l^{i_l}(k+1|k+1) - \hat{x}_l^{i_l-1}(k+1|k+1)\right| < \varepsilon$, $\hat{x}_l^{i_l}(k+1|k+1)$ is taken as the state estimate in the lth iterative process, then the filter gain in the lth iteration can be updated by

$$K_{l+1}^*(k+1) = \frac{\hat{x}_l^{i_l}(k+1|k+1) - \hat{x}_l^{i_l-1}(k+1|k+1)}{y(k+1) - h(\hat{x}_l^{i_l}(k+1|k+1), k+1)} \tag{15}$$

For the $\alpha(k+1)$th iteration, if $\left|K_{\alpha(k+1)+1}^*(k+1) - K_{\alpha(k+1)}^*(k+1)\right| < \varepsilon_K$ (ε_K is a given scalar), the final estimate output of the state at the time instant $k+1$ is given by

$$\hat{x}(k+1|k+1) = \hat{x}_{\alpha(k+1)}^{i_{l\alpha(k+1)}}(k+1|k+1) \tag{16}$$

The flowchart of the iterative computational process of the estimates is illustrated in Figure 1.

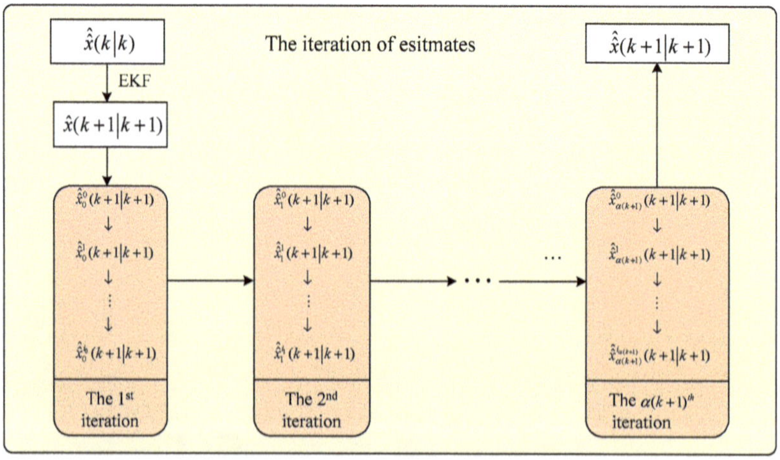

Figure 1. Flowchart of the iterative computational process of the estimates.

3.4. Convergence of the Fixed-Point Filter

In order to ensure that the constructed fixed-point filter is convergent, it is necessary to conditionally constrain the fixed-point equation. In this subsection, the convergence of the fixed-point function (8) will be discussed.

Firstly, a necessary lemma is given as follows [26,27].

Lemma 1. *An iterative function $\varphi(x)$ has a unique fixed point x^* in $[a,b]$, if*

(1) $x \in [a,b]$*, then* $a \le \varphi(x) \le b$*;*
(2) *There exists a positive scalar L, such that* $\left|\frac{\partial\varphi(x)}{\partial x}\right| \le L$*, for* $x \in [a,b]$*.*

Based on this lemma, the convergence conditions of the fixed-point function (8) are given as follows.

Theorem 1. *For a given filter $K^*(k+1)$, the fixed-point function (8) has a fixed point in [a,b], if the measurement function satisfies the following two conditions*

(1) $a + \underline{\hat{y}}_{[a,b]} \le K^*(k+1)y(k+1) \le b + \overline{\hat{y}}_{[a,b]}$, *where*

$$\underline{\hat{y}}_{[a,b]} = \min\left\{K^*(k+1)h(\hat{\hat{x}}(k+1|k+1),k+1) - \hat{\hat{x}}(k+1|k+1), where\ \hat{\hat{x}}(k+1|k+1) \in [a,b]\right\} \quad (17)$$

$$\overline{\hat{y}}_{[a,b]} = \max\left\{K^*(k+1)h(\hat{\hat{x}}(k+1|k+1),k+1) - \hat{\hat{x}}(k+1|k+1), where\ \hat{\hat{x}}(k+1|k+1) \in [a,b]\right\} \quad (18)$$

(2) $\left|1 - K^*(k+1)\frac{\partial h(\hat{\hat{x}}(k+1|k+1),k+1)}{\partial\hat{\hat{x}}(k+1|k+1)}\right| \le L, \quad \hat{\hat{x}}(k+1|k+1) \in [a,b].$

Proof. Denote

$$\varphi(\hat{\hat{x}}(k+1|k+1)) = \hat{\hat{x}}(k+1|k+1) + K^*(k+1)[y(k+1) - h(\hat{\hat{x}}(k+1|k+1),k+1)] \quad (19)$$

If $\hat{\hat{x}}(k+1|k+1) \in [a,b]$, according to Lemma 1, one has $a \le \varphi(\hat{\hat{x}}(k+1|k+1)) \le b$. It means that

$$a \le \hat{\hat{x}}(k+1|k+1) + K^*(k+1)[y(k+1) - h(\hat{\hat{x}}(k+1|k+1),k+1)] \le b \quad (20)$$

Then, the measurement function satisfies

$$a \leq K^*(k+1)y(k+1) - \hat{y}_{[a,b]} \leq b \tag{21}$$

where

$$\hat{y}_{[a,b]} = K^*(k+1)h(\hat{x}(k+1|k+1), k+1) - \hat{x}(k+1|k+1), \quad \hat{x}(k+1|k+1) \in [a,b] \tag{22}$$

Using the upper bound in (17) and the lower bound in (18), one gets

$$a + \underline{\hat{y}}_{[a,b]} \leq K^*(k+1)y(k+1) \leq b + \overline{\hat{y}}_{[a,b]} \tag{23}$$

The Jacobian value of $\varphi(\hat{x}(k+1|k+1))$ is obtained by

$$\frac{\partial\varphi(\hat{x}(k+1|k+1))}{\partial\hat{x}(k+1|k+1)} = 1 - K^*(k+1)\frac{\partial h(\hat{x}(k+1|k+1), k+1)}{\partial\hat{x}(k+1|k+1)} \tag{24}$$

Then, according to Lemma 1, the measurement function should satisfy

$$\left| 1 - K^*(k+1)\frac{\partial h(\hat{x}(k+1|k+1), k+1)}{\partial\hat{x}(k+1|k+1)} \right| \leq L \tag{25}$$

Therefore, the fixed-point function (8) has a unique fixed point in [a,b], if the measurement function satisfies (23) and (25). The proof is completed. □

Remark 3. *It is commonly known that not all fixed-point equations are convergent. The convergence conditions of the fixed-point function (8) are presented in Theorem 1, which is also the existing condition of the fixed-point iterative filter given in Section 3.3. If the convergence conditions in Theorem 1 are satisfied, the state estimate will be further updated by the fixed-point iterative filter shown in Section 3.3. Otherwise, the state estimate obtained by the EKF need not be updated and denoted as the final estimate at a certain time instant. The interval [a,b] can be set according to the actual demand, or by using the state prediction value with the 3-Delta rule. The norm could be 1-noum for the vector parameters.*

3.5. Steffensen's Iterative Solution of the Fixed-Point Function

In Section 3.3, a nested iterative solution for the fixed-point filter was presented. In this section, Steffensen's iterative method is used to accelerate the solution of the fixed-point function (8). Steffensen's iterative method in the lth iteration is given as follows

$$\hat{x}_l^{j+1}(k+1|k+1) = \hat{x}_l^j(k+1|k+1) - \frac{(\phi(\hat{x}_l^j(k+1|k+1)) - \hat{x}_l^j(k+1|k+1))^2}{\phi(\phi(\hat{x}_l^j(k+1|k+1))) - 2\phi(\hat{x}_l^j(k+1|k+1)) + \hat{x}_l^j(k+1|k+1)}, j = 1, 2, \cdots \tag{26}$$

Let $K_0^*(k+1) = K(k+1)$, then Steffensen's iterative method can be given by

$$\begin{cases} \hat{x}_0^1(k+1|k+1) = \hat{x}_0^0(k+1|k+1) - \dfrac{(\phi(\hat{x}_0^0(k+1|k+1)) - \hat{x}_0^0(k+1|k+1))^2}{\phi((\phi(\hat{x}_0^0(k+1|k+1))) - 2\phi(\hat{x}_0^0(k+1|k+1)) + \hat{x}_0^0(k+1|k+1)} \\ \vdots \\ \hat{x}_0^{n_0-1}(k+1|k+1) = \hat{x}_0^{n_0-2}(k+1|k+1) - \dfrac{(\phi(\hat{x}_0^{n_0-2}(k+1|k+1)) - \hat{x}_0^{n_0-2}(k+1|k+1))^2}{\phi((\phi(\hat{x}_0^{n_0-2}(k+1|k+1))) - 2\phi(\hat{x}_0^{n_0-2}(k+1|k+1)) + \hat{x}_0^{n_0-2}(k+1|k+1)} \\ \hat{x}_0^{n_0}(k+1|k+1) = \hat{x}_0^{n_0-1}(k+1|k+1) - \dfrac{(\phi(\hat{x}_0^{n_0-1}(k+1|k+1)) - \hat{x}_0^{n_0-1}(k+1|k+1))^2}{\phi((\phi(\hat{x}_0^{n_0-1}(k+1|k+1))) - 2\phi(\hat{x}_0^{n_0-1}(k+1|k+1)) + \hat{x}_0^{n_0-1}(k+1|k+1)} \end{cases} \tag{27}$$

where

$$\phi(\hat{x}_0^i(k+1|k+1)) = \hat{x}_0^i(k+1|k+1) + K_0^*(k+1)(y(k+1) - h(\hat{x}_0^i(k+1|k+1))), i = 0, 1, \cdots, n_0 \quad (28)$$

$$\phi((\phi(\hat{x}_0^i(k+1|k+1))) = \phi(\hat{x}_0^i(k+1|k+1) + K_0^*(k+1)(y(k+1) - h((\phi(\hat{x}_0^i(k+1|k+1)))), i = 0, 1, \cdots, n_0 \quad (29)$$

If $\left|\hat{x}_0^{n_0}(k+1|k+1) - \hat{x}_0^{n_0-1}(k+1|k+1)\right| < \varepsilon$, take $\hat{x}_0^{n_0}(k+1|k+1)$ as the state estimate of the fixed-point filter with the filter gain $K_0^*(k+1)$.

Similar to (11), the filter gain can be updated by

$$K^*{}_1(k+1) = \frac{\hat{x}_0^{n_0}(k+1|k+1) - \hat{x}_0^{n_0-1}(k+1|k+1)}{y(k+1) - h(\hat{x}_0^{n_0}(k+1|k+1), k+1)} \quad (30)$$

Substituting the new filter gain above, the fixed-point function is represented as

$$\hat{x}(k+1|k+1) = \hat{x}(k+1|k+1) + K^*{}_1(k+1)[y(k+1) - h(\hat{x}(k+1|k+1), k+1) - E\{v(k+1)\}] \quad (31)$$

According to Steffensen's iterative method, one has

$$\begin{cases} \hat{x}_1^1(k+1|k+1) = \hat{x}_1^0(k+1|k+1) - \frac{(\phi(\hat{x}_1^0(k+1|k+1)) - \hat{x}_1^0(k+1|k+1))^2}{\phi((\phi(\hat{x}_1^0(k+1|k+1))) - 2\phi(\hat{x}_1^0(k+1|k+1)) + \hat{x}_1^0(k+1|k+1)} \\ \vdots \\ \hat{x}_1^{n_1-1}(k+1|k+1) = \hat{x}_1^{n_1-2}(k+1|k+1) - \frac{(\phi(\hat{x}_1^{n_1-2}(k+1|k+1)) - \hat{x}_1^{n_1-2}(k+1|k+1))^2}{\phi((\phi(\hat{x}_1^{n_1-2}(k+1|k+1))) - 2\phi(\hat{x}_1^{n_1-2}(k+1|k+1)) + \hat{x}_1^{n_1-2}(k+1|k+1)} \\ \hat{x}_1^{n_1}(k+1|k+1) = \hat{x}_1^{n_1-1}(k+1|k+1) - \frac{(\phi(\hat{x}_1^{n_1-1}(k+1|k+1)) - \hat{x}_1^{n_1-1}(k+1|k+1))^2}{\phi((\phi(\hat{x}_1^{n_1-1}(k+1|k+1))) - 2\phi(\hat{x}_1^{n_1-1}(k+1|k+1)) + \hat{x}_1^{n_1-1}(k+1|k+1)} \end{cases} \quad (32)$$

where

$$\phi(\hat{x}_1^i(k+1|k+1)) = \hat{x}_1^i(k+1|k+1) + K_1^*(k+1)(y(k+1) - h(\hat{x}_1^i(k+1|k+1))), i = 0, 1, \cdots, n_1 \quad (33)$$

$$\phi((\phi(\hat{x}_1^i(k+1|k+1))) = \phi(\hat{x}_1^i(k+1|k+1) + K_1^*(k+1)(y(k+1) - h((\phi(\hat{x}_1^i(k+1|k+1)))), i = 0, 1, \cdots n_1 \quad (34)$$

If $\left|\hat{x}_1^{n_1}(k+1|k+1) - \hat{x}_1^{n_1-1}(k+1|k+1)\right| < \varepsilon$, take $\hat{x}_1^{n_1}(k+1|k+1)$ as the state estimate of the fixed-point filter with the filter gain $K^*{}_1(k+1)$.

Similarly, the filter gain can be updated by

$$K_j^*(k+1) = \frac{\hat{x}_{j-1}^{n_{j-1}}(k+1|k+1) - \hat{x}_{j-1}^{n_{j-1}-1}(k+1|k+1)}{y(k+1) - h(\hat{x}_{j-1}^{n_{j-1}}(k+1|k+1), k+1)} \quad (35)$$

The fixed-point filter is constructed as

$$\hat{x}(k+1|k+1) = \hat{x}(k+1|k+1) + K_j^*(k+1)[y(k+1) - h(\hat{x}(k+1|k+1), k+1) - E\{v(k+1)\}] \quad (36)$$

Steffensen's iterative solution of the fixed-point filter above is obtained by

$$\begin{cases} \hat{x}_j^1(k+1|k+1) = \hat{x}_j^0(k+1|k+1) - \frac{(\phi(\hat{x}_j^0(k+1|k+1)) - \hat{x}_j^0(k+1|k+1))^2}{\phi((\phi(\hat{x}_j^0(k+1|k+1))) - 2\phi(\hat{x}_j^0(k+1|k+1)) + \hat{x}_j^0(k+1|k+1)} \\ \vdots \\ \hat{x}_j^{n_j-1}(k+1|k+1) = \hat{x}_j^{n_j-2}(k+1|k+1) - \frac{(\phi(\hat{x}_j^{n_j-2}(k+1|k+1)) - \hat{x}_j^{n_j-2}(k+1|k+1))^2}{\phi((\phi(\hat{x}_j^{n_j-2}(k+1|k+1))) - 2\phi(\hat{x}_j^{n_j-2}(k+1|k+1)) + \hat{x}_j^{n_j-2}(k+1|k+1)} \\ \hat{x}_j^{n_j}(k+1|k+1) = \hat{x}_j^{n_j-1}(k+1|k+1) - \frac{(\phi(\hat{x}_j^{n_j-1}(k+1|k+1)) - \hat{x}_j^{n_j-1}(k+1|k+1))^2}{\phi((\phi(\hat{x}_j^{n_j-1}(k+1|k+1))) - 2\phi(\hat{x}_j^{n_j-1}(k+1|k+1)) + \hat{x}_j^{n_j-1}(k+1|k+1)} \end{cases} \quad (37)$$

where

$$\phi(\hat{x}_j^i(k+1|k+1)) = \hat{x}_j^i(k+1|k+1) + K_j^*(k+1)(y(k+1) - h(\hat{x}_j^i(k+1|k+1))), i = 0, 1, \cdots, n_j \quad (38)$$

$$\phi((\phi(\hat{x}_j^i(k+1|k+1))) = \phi(\hat{x}_j^i(k+1|k+1) + K_j^*(k+1)(y(k+1) - h((\phi(\hat{x}_j^i(k+1|k+1)))), i = 0, 1, \cdots, n_j \quad (39)$$

If $\left|\hat{x}_j^{n_j}(k+1|k+1) - \hat{x}_j^{n_j-1}(k+1|k+1)\right| < \varepsilon$, take $\hat{x}_j^{n_j}(k+1|k+1)$ as the solution of the fixed-point filter with the filter gain $K_j^*(k+1)$.

If $\left|K_t^*(k+1) - K_{t-1}^*(k+1)\right| < \varepsilon_K$, the state estimate $\hat{x}_t^{n_t}(k+1|k+1)$ obtained by the fixed-point iteration process with the filter gain $K_t^*(k+1)$ is the final solution of the fixed-point filter.

Remark 4. *Compared with the nested iterative solution method in Section 3.3, Steffensen's iterative method can accelerate the solution process of the fixed-point function (8). Because of this, the filtering result of the two fixed-point filters may be slightly different.*

Remark 5. *It should be noted that the conditions for judging whether the iterative process ends in the solution methods of the fixed-point function are to compare the differences between the results of two adjacent iterations. However, the final purposes of the updating methods are reducing the differences between the estimates and the active values of the system states. It is a pity that the active values of the system states are unavailable. Therefore, the difference between the active value and the prediction values of the measurement by different methods are compared in the simulation to determine whether the updating is necessary.*

4. Simulation

In this section, two comparison simulation examples are provided to illustrate the feasibility and the effectiveness of the proposed EKF-based fixed-point iterative filters (the one provided in Section 3.3 is denoted as FEKF, while the one provided in Section 3.5 is denoted SFEKF), which are compared with the classical EKF and UKF methods.

Simulation I. Considering the following non-linear system

$$\begin{cases} x(k+1) = 0.5x(k) + 2.5x(k)/(1+x^2(k)) + w(k+1,k) \\ y(k+1) = x^2(k+1) + v(k+1) + 0.2\cos(\mathrm{k}/\pi) \end{cases} \quad (40)$$

where $w(k+1,k)$, $v(k+1)$ are respectively the process noise and the measurement noise, which satisfy the zero-mean Gaussian distributions with the variances $Q = 0.1$, $R = 0.001$. The initial state estimation and its error variance are respectively $x(0) = 2$, $p(0) = 0.01$.

In order to reduce the effects of random noise on the comparison results, a Monte Carlo simulation was repeated 50 times. The simulation results were as follows.

As shown in Figure 2, compared with the absolute error curves of the classical EKF and UKF, the error curves of the proposed iterative fixed-point filters were lower at almost all 80 simulation time instants. The main cause of the higher absolute error curve of the EKF was the discarded higher order terms of the Taylor series in the linearization processes. Since the proposed fixed-point filters iteratively updated the estimates of the EKF, the absolute error curve was lower than that of the EKF. The UKF approximated the statistics of the state estimates by using unscented transformation, and the estimate accuracy of UKF was related to the filter parameters. The estimate accuracy of the UKF was the lowest, although several possible filter parameters were tested. The mean weightings W_i^m and variance weightings W_i^c of the UKF were respectively set as $W_i^m = \begin{cases} \lambda/(n+\lambda),\ i=1 \\ 1/(2n+2\lambda),\ i=2,3 \end{cases}$, $W_i^c = \begin{cases} \lambda/(n+\lambda)+(1-\alpha^2+\beta),\ i=1 \\ 1/(2+2\lambda),\ i=2,3 \end{cases}$, where $\lambda = \alpha^2(n+\kappa)-n$, $\alpha = 1$, $\beta = 2$, $\kappa = 3-n$, $n = 1$.

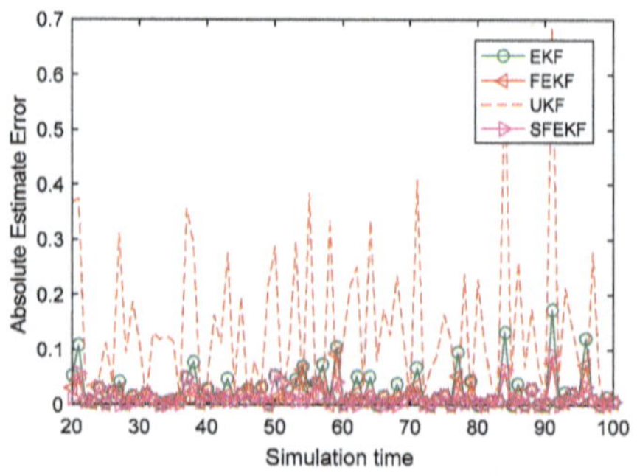

Figure 2. The absolute error curves of the four methods.

The fixed-point iterative filter based on the extended Kalman filter can compensate the influence of discarded higher order terms of the Taylor series. As a result, the filtering results of the methods proposed in this paper were better than others. The mean values of the absolute error of the four comparison methods in Table 1 illustrated the feasibility and effectiveness of the proposed methods. In the same simulation conditions, the mean absolute error of the FEKF was 0.0178 and the mean absolute error of the SFEKF was 0.0141, while the others were 0.0278 for the EKF, and 0.1426 for the UKF, respectively. As shown in Table 2, the computation complexity of the EKF was the smallest, and that of the UKF was the largest, due to the computation of the sigma points. The FEKF and SFEKF were designed to update the estimation results of the EKF, and accordingly, their computation complexities were larger than that of the EKF. Steffensen's iterative method was used to accelerate the solution of the fixed-point function (8), therefore, the CPU time cost of repeating Monte Carlo simulations 50 times was smaller with the SFEKF than with the FEKF.

Table 1. The mean absolute errors (MAEs) of four comparison methods.

Methods	EKF	UKF	FEKF	SFEKF
MAEs	0.0278	0.1426	0.0178	0.0141

Table 2. The CPU time cost of 50 Monte Carlo simulations for the four comparison methods.

Methods	EKF	UKF	FEKF	SFEKF
CPU time	0.1406	2.4063	2.2656	0.3594

It is noted that the estimates obtained by the proposed methods were the closest to the actual values, as shown in Figure 3. Due to the influence of the discarded higher order items of the Taylor series system functions, the estimate accuracy of the EKF method was relatively poor. The influence was reduced using a fixed-point iterative filter. The measurement prediction results in Figure 4 also illustrated the feasibility and the effectiveness of the proposed fixed-point iterative filters.

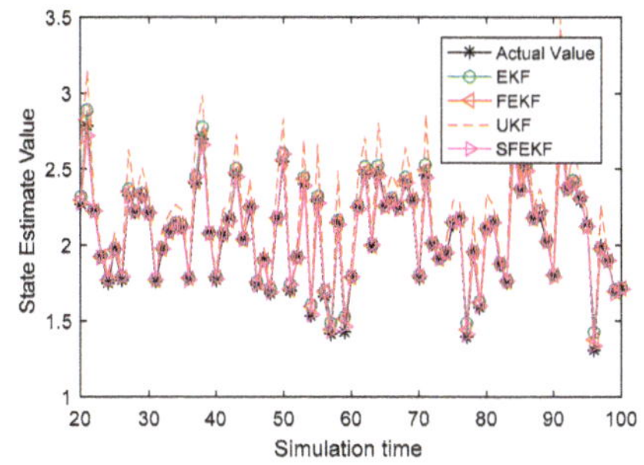

Figure 3. The estimate curves of the four methods and the actual value curve.

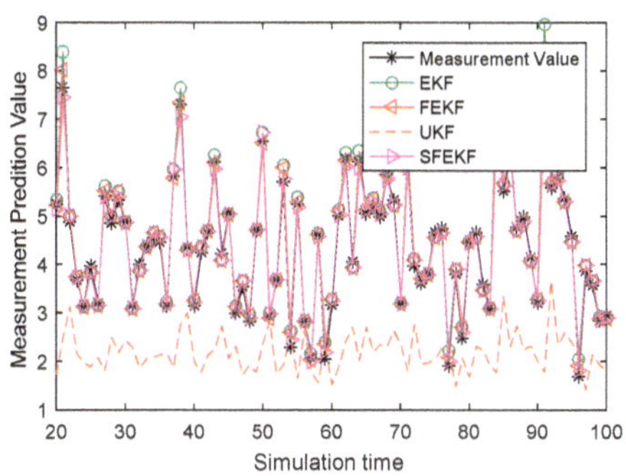

Figure 4. The measurement prediction curves of the four methods and the actual observation curve.

Simulation II. Consider the following target tracking system, in which the target is in nearly constant velocity:

$$\begin{bmatrix} x(k+1) \\ \dot{x}(k+1) \\ y(k+1) \\ \dot{y}(k+1) \end{bmatrix} = \begin{bmatrix} 1 & T & 0 & 0 \\ 0 & 1 & 0 & 0 \\ 0 & 0 & 1 & T \\ 0 & 0 & 0 & 1 \end{bmatrix} \begin{bmatrix} x(k) \\ \dot{x}(k) \\ y(k) \\ \dot{y}(k) \end{bmatrix} + \begin{bmatrix} T^2/2 & 0 \\ T & 0 \\ 0 & T^2/2 \\ 0 & T \end{bmatrix} \begin{bmatrix} w_x(k+1,k) \\ w_y(k+1,k) \end{bmatrix} \tag{41}$$

The target was initially located at (0, 1400), with an initial velocity of (1.8, −9.5). A radar was located at the origin of the polar coordinate. The measurement equation was given by

$$\begin{cases} \theta(k) = \arctan\left(\frac{y(k)}{x(k)}\right) + v_\theta(k) \\ r(k) = \sqrt{(x(k))^2 + (y(k))^2} + v_r(k) \end{cases} \tag{42}$$

The variances of $\begin{bmatrix} w_x(k+1,k) \\ w_y(k+1,k) \end{bmatrix}$ and $\begin{bmatrix} v_\theta(k) \\ v_r(k) \end{bmatrix}$ were $\begin{bmatrix} 10^{-4} & 0 \\ 0 & 10^{-4} \end{bmatrix}$ and $\begin{bmatrix} 10^{-2} & 0 \\ 0 & 10 \end{bmatrix}$. The sampling period was 1. The simulation results are shown in Figures 5 and 6.

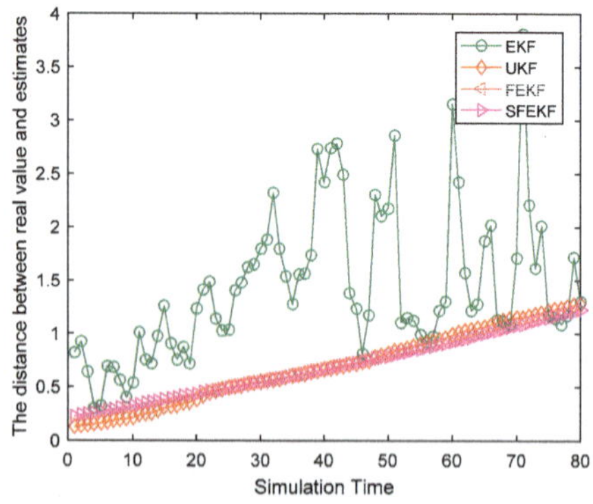

Figure 5. The distance between the real value and estimates for the four methods.

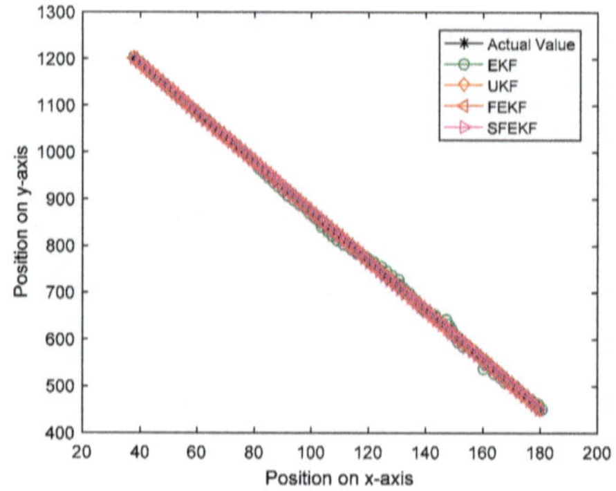

Figure 6. The real position and estimates for the four methods.

As shown in Figure 6, the proposed FEKF and SFEKF methods effectively tracked the target with the radar measurements. The two proposed methods obtained more accurate estimates of the target than the EKF for the updating method with the fixed-point theory, as shown in Figure 5 and Table 3.

Table 3. The mean distances between the real value and estimates for the four comparison methods.

Methods	EKF	UKF	FEKF	SFEKF
Mean Distance	1.1913	0.5573	0.5530	0.5530

5. Conclusions

As a development of the classic extended Kalman filter, the fixed-point iterative updating method was studied in this paper, drawing on fixed-point theory. On the basis of the extended Kalman filter, a fixed-point function was provided to update the state estimate obtained by the EKF and reduce the influence of the discarded higher order items of the Taylor series. The fixed-point function was solved by the nested iterative method and Steffensen's iterative method and resulted in two fixed-point iterative filters. The convergence conditions of the fixed-point function were also studied, as were the existing conditions of fixed-point iterative filters.

Author Contributions: Formal analysis, Y.F.; Methodology, X.F. and C.W.; Writing–original draft, X.F. and Y.F.; Writing–review & editing, C.W.

Funding: This work was supported by the National Natural Science Foundation of China under grant nos. U1804163, 61673160 and 61603108.

Conflicts of Interest: The authors declare no conflict of interest.

References

1. Wen, C.; Cai, Y.; Liu, Y.; Wen, C. A Reduced-order Approach to Filtering for Systems with Linear Equality Constraints. *Neurocomputing* **2016**, *193*, 219–226. [CrossRef]
2. Qiu, A.; Al-Dabbagh, A.W.; Chen, T. A Tradeoff Approach for Optimal Event-Triggered Fault Detection. *IEEE Trans. Ind. Electron.* **2018**, *66*, 2111–2121. [CrossRef]
3. Shi, Y.; Tang, X.; Feng, X.; Bian, D.; Zhou, X. Hybrid Adaptive Cubature Kalman Filter with Unknown Variance of Measurement Noise. *Sensors* **2018**, *18*, 4335. [CrossRef]
4. Xu, Y.; Lu, R.; Shi, P.; Li, H.; Xie, S. Finite-Time Distributed State Estimation Over Sensor Networks with Round-Robin Protocol and Fading Channels. *IEEE Trans. Cybern.* **2018**, *48*, 336–345. [CrossRef]
5. Feng, X.; Wen, C.; Park, J.H. Sequential fusion H∞ filtering for multi-rate multi-sensor time-varying systems–a Krein space approach. *IET Control Theory Appl.* **2017**, *11*, 369–381. [CrossRef]
6. Kalman, R.E. A new approach to linear filtering and prediction problems. *Trans. Asme–J. Basic Eng.* **1960**, *82*, 35–45. [CrossRef]
7. Brown, R.; Hwang, P. *Introduction to Random Signals and Applied Kalman Filtering*, 3rd ed.; John Wiley & Sons: New York, NY, USA, 1997.
8. Julier, S.J.; Uhlmann, J.K.; Durrant-Whyte, H.F. A new method for nonlinear transformation of means and covariances in filters and estimators. *IEEE Trans. Autom. Control* **2000**, *45*, 477–482. [CrossRef]
9. Julier, S.J.; Uhlmann, J.K. Unscented filtering and nonlinear estimation. *Proc. IEEE* **2004**, *92*, 401–422. [CrossRef]
10. Kulikov, G.Y.; Kulikova, M.V. Accurate continuous–discrete unscented Kalman filtering for estimation of nonlinear continuous–time stochastic models in radar tracking. *Signal Process.* **2017**, *139*, 25–35. [CrossRef]
11. Arasaratnam, I.; Haykin, S. Cubature Kalman filters. *IEEE Trans. Autom. Control* **2009**, *54*, 1254–1269. [CrossRef]
12. Ge, Q.; Xu, D.; Wen, C. Cubature information filters with correlated noises and their applications in decentralized fusion. *Signal Process.* **2014**, *94*, 434–444. [CrossRef]
13. Ge, Q.; Shao, T.; Yang, Q.; Shen, X.; Wen, C. Multisensor Nonlinear Fusion Methods Based on Adaptive Ensemble Fifth-Degree Iterated Cubature Information Filter for Biomechatronics. *IEEE Trans. Syst. Man Cybern. Syst.* **2016**, *46*, 912–925. [CrossRef]
14. Reif, K.; Gunther, S.; Yaz, E. Stochastic stability of the discrete-time extended Kalman filter. *IEEE Trans. Autom. Control* **1999**, *44*, 714–728. [CrossRef]
15. Xiao, M.; Zhang, Y.; Wang, Z.; Fu, H. An adaptive three-stage extended Kalman filter for nonlinear discretetime system in presence of unknown inputs. *ISA Trans.* **2018**, *75*, 101–117. [CrossRef]
16. Zhou, D.; Xi, Y.; Zhang, Z. A suboptimal multiple fading extended Kalman filter. *Acta Autom. Sin.* **1991**, *17*, 689–695.
17. Zhou, D.; Xi, Y.; Zhang, Z. Suboptimal fading extended Kalman filtering for nonlinear systems. *Control Decis.* **1990**, *12*, 1–6.
18. Ge, Q.; Shao, T.; Chen, S.; Wen, C. Carrier tracking estimation analysis by using the extended strong tracking filtering. *IEEE Trans. Ind. Electron.* **2017**, *64*, 1415–1424. [CrossRef]
19. Yin, Z.; Li, G.; Zhang, Y.; Liu, J. Symmetric-strong-tracking-extended-Kalman-filter-based sensorless control of induction motor drives for modeling error reduction. *IEEE Trans. Ind. Inform.* **2018**, *15*, 650–662. [CrossRef]
20. Duan, Z.; Li, X. A new nonlinear state estimator using the fusion of multiple extended Kalman filters. In Proceedings of the International Conference on Information Fusion, Washington, DC, USA, 6–9 July 2015; pp. 90–97.
21. Li, G.; Wen, C. Identification of Wiener systems based on fixed point theory. In Proceedings of the 2010 11th International Conference on Control Automation Robotics & Vision, Singapore, 7–10 December 2010; pp. 491–496.
22. Chen, B.; Liu, X.; Zhao, H.; Principe, J. Maximum correntropy Kalman filter. *Automatica* **2017**, *76*, 70–77. [CrossRef]
23. Wang, G.; Xue, R.; Wang, J. A distributed maximum correntropy Kalman filter. *Signal Process.* **2019**, *160*, 247–251. [CrossRef]

24. Xu, H.; Xie, W.; Yuan, H.; Duan, K.; Liu, W.; Wang, Y. Fixed-point iteration Gaussian sum filtering estimator with unknown time-varying non-Gaussian measurement noise. *Signal Process.* **2018**, *153*, 132–142. [CrossRef]
25. Chen, W.; Wen, C.; Ren, Y. Multi-dimensional Observation Characteristic Function Filtering Based on Fixed Point Equation. In Proceedings of the 2018 International Conference on Control, Automation and Information Sciences, Hangzhou, China, 24–27 October 2018; pp. 118–123.
26. Guan, Z.; Lu, J. *Foundation of Numerical Analysis*; Higher Education Press: Beijing, China, 1998.
27. Li, Q.; Guan, Z.; Bai, F. *Principles of Numerical Calculation*; Tsinghua University Press: Beijing, China, 2000.

sensors

Article

A Unified Multiple-Target Positioning Framework for Intelligent Connected Vehicles

Zhongyang Xiao [1], Diange Yang [1,*], Fuxi Wen [1,2] and Kun Jiang [1]

[1] State Key Laboratory of Automotive Safety and Energy, School of Vehicle and Mobility, Tsinghua University, Beijing 100084, China; xiaozy15@mails.tsinghua.edu.cn (Z.X.); fuxi@chalmers.se (F.W.); jiangkun@mail.tsinghua.edu.cn (K.J.)

[2] Department of Electrical Engineering, Chalmers University of Technology, SE-412 96 Gothenburg, Sweden

* Correspondence: ydg@mail.tsinghua.edu.cn

Received: 11 April 2019; Accepted: 24 April 2019; Published: 26 April 2019

Abstract: Future intelligent transport systems depend on the accurate positioning of multiple targets in the road scene, including vehicles and all other moving or static elements. The existing self-positioning capability of individual vehicles remains insufficient. Also, bottlenecks in developing on-board perception systems stymie further improvements in the precision and integrity of positioning targets. Vehicle-to-everything (V2X) communication, which is fast becoming a standard component of intelligent and connected vehicles, renders new sources of information such as dynamically updated high-definition (HD) maps accessible. In this paper, we propose a unified theoretical framework for multiple-target positioning by fusing multi-source heterogeneous information from the on-board sensors and V2X technology of vehicles. Numerical and theoretical studies are conducted to evaluate the performance of the framework proposed. With a low-cost global navigation satellite system (GNSS) coupled with an initial navigation system (INS), on-board sensors, and a normally equipped HD map, the precision of multiple-target positioning attained can meet the requirements of high-level automated vehicles. Meanwhile, the integrity of target sensing is significantly improved by the sharing of sensor information and exploitation of map data. Furthermore, our framework is more adaptable to traffic scenarios when compared with state-of-the-art techniques.

Keywords: vehicular localization; target positioning; high-definition map; vehicle-to-everything; intelligent and connected vehicles; intelligent transport system

1. Introduction

The intelligent transportation system (ITS) is one of the most indispensable components of the smart city concept that integrates sensing, control, information, and communication technologies into transportation [1]. In recent years, with the emergence of cutting-edge applications of ITS, the positioning of multiple targets, including vehicles and other elements has been playing an increasingly important role in improving safety, mobility, and efficiency [2–4]. For example, future intelligent connected vehicles (ICVs) require the positioning of their own real-time location with centimeter-level precision [5], and the awareness of all objects such as surrounding vehicles and vulnerable road users with significant integrity and confidence. In ITS, the positioning of vehicles and other targets are usually referred to as vehicular self-positioning and target localization, respectively. Although attention has been paid in these topics [6–9], there still exist many limitations that need to be eliminated.

1.1. Self-Positioning

Multiple self-positioning technologies are already present in the market, but none are effective under all road conditions and scenarios [10]. GNSS systems are widely employed in ITS devices, but they can support only low-precision navigation. Researchers have tried to integrate information from the base station and on-board sensors for error compensation. However, in dense urban environments where the signal is disturbed by surrounding buildings [5], even the most accurate GNSS with real-time kinematic correction and INS fusion schemes [11] cannot provide localization with adequate accuracy and stability.

Introducing new sources of information is an effective way of improving vehicular self-positioning. V2X communication, which has drawn increasing interest in recent years, renders information easily accessible to the vehicles connected [12–14]. The V2X-based (or cooperative) method aids in improving vehicular localization capability by employing the position information of other vehicles and relative measurements from their on-board sensors [15,16]. The integration of on-board sensors and V2X communication is shown to be more cost-effective than approaches based on high-quality sensors [17]. A general framework for multi-vehicular localization using pose graph optimization is proposed in [18], using vehicle-vehicle (V-V) measurements to improve the precision of vehicular localization. More recently, an implicit cooperative positioning algorithm that exploits the joint sensing of passive features is proposed in [19,20], and precludes the use of explicit V-V measurements. In addition to the use of ranging sensors, angle measurement-based cooperative localization is proposed in [21].

Maps are additional sources of information, and the locations of static elements can be used as references to improve the vehicular positioning capability [22]. In contrast to simultaneous localization and mapping (SLAM), in which a map is generated real-time [23], the map-based method assumes that maps are available in advance and aligns landmarks in the maps with on-board sensors to achieve independent positioning or as an aid to GNSS with INS system (GNSS/INS). As shown in Table 1, over the past few years, with the development of V2X, HD maps, which are characterized by high-accuracy and real-time updates, have grown to become standard and indispensable components of intelligent vehicles [4]. This also enables centimeter-level precision to be achieved in map-based localization [24].

Table 1. Maps for different levels of Intelligent Connected Vehicles.

Grade	**Title**	**Map**	**Accuracy**	**Typical Condition**
	Driver Scenario			
1 (DA)	Driver Assistance	ADAS	Submeter	Optional
2 (PA)	Partial Autopilot	ADAS	Submeter	Optional
Automatic Driving Sys. Scenario				
3 (CA)	Conditional Autopilot	ADAS + HD	Submeter Centimeter	Optional
4 (HA)	High-Level Automated Driving	ADAS + HD	Submeter Centimeter	Essential
5 (FA)	Completely Automated Driving	HD	Centimeter	Essential (auto updated)

The HD map-based method benefits from the high precision of the map used. For example, the digital map used in [25–27] is created from light detection and ranging (LiDAR) data and has a precision of up to 10 cm. A high-accuracy localization technique using urban environment maps for vehicles in motion is proposed in [28], and these maps are generated by integrating GNSS, LiDAR data, and on-board sensors. In addition, more features are used in the process of map alignment, which contributes to higher self-positioning precision. Traffic lights and visual lane markers are used as landmarks in [5,29], respectively. In a more recent study [26], a unified framework using more references in addition to the abovementioned features (lamp poles, traffic signs, etc.) is proposed, and self-localization with an accuracy of within 30 cm is achieved with merely a low-cost camera. While the abovementioned studies conduct map-based localization independently, there is much research that integrates it with other on-board sensors. A lane determination system that fuses on-board sensors, GNSS, and commercially available road network maps is proposed in [30]. A proof-of-concept study using INS and maps for vehicular localization in GNSS-denied environments is conducted in [31].

1.2. Target Localization

Apart from vehicular self-positioning, the relative localization of targets in the surrounding environment is another fundamental technology underpinning ICVs. This task is mainly undertaken by vehicle perception modules, and the positioning result is obtained in the vehicular coordinate instead of in the world coordinate. Although the recent decades have witnessed the rapid development of on-board sensors, the current on-board sensing technology still faces the following problems [4]. First, there is a trade-off between localization accuracy and cost. For example, low-cost cameras and radars can achieve accuracies of only several centimeters, while LiDAR systems with centimeter-level ranging accuracy are expensive [32]. Second, all sensors have limited sensing ranges, and the occlusion of sensors by other vehicles and objects is a frequent occurrence [33]. Irrespective of the number of sensors equipped in the vehicle, the perception of the environment remains incomplete. Attempts to improve the perception of bicycles have already encountered bottlenecks to a certain extent.

Studies using the perceptual information of other vehicles to improve the integrity of target localization have proven to be effective. This is because other vehicles in the network may have seen a target that cannot be seen by the ego-vehicle because of occlusion or limited field of view. A vehicle-to-vehicle (V2V) communication and map merging-based cooperative perception system to extend the perception range beyond line of sight and field of view is proposed in [34]. In [35], the results of the awareness of other vehicles are integrated into the ego-vehicle's perception system as virtual sensors to achieve perception enhancement. In [36], a multi-vehicle perception framework combining image and semantic features is proposed, and experiments have proved that the problem of front-vehicle occlusion can be solved. In these studies, the problems of self-positioning and localization of other targets were considered separately, which rendered the effect of the fusion very sensitive to their relative positioning. In addition, these articles do not provide quantitative analyses of the integrity of the results perceived.

Maps also contribute to the relative positioning of targets. For example, the geometry of intersection can be directly extracted from a HD map for motion planning and control [37]. This reduces the pressure on the vehicle-mounted sensing system, but relies on vehicular self-positioning. Incorrect self-positioning greatly affects decision-making. Other works integrate the semantic and geometric prior knowledge in HD maps with the on-board sensing system to improve positioning confidence. In [38,39], a prior probability map is generated in a bird's-eye view or image plane to aid understanding of the scene. Recently, a neural network incorporating prior knowledge with on-board sensors was presented in [40]. However, these studies treat the map only as an auxiliary tool for perception to improve the recognition result without improving the integrity of perception. Moreover, these studies also rely on the accuracy of vehicular positioning.

1.3. Contributions

In general, recent research exploits additional sources of information to improve vehicular self-positioning and localization of other targets. However, to the best of our knowledge, V2X and HD maps are considered separately in the literature, while the positioning of the vehicles themselves and those of other targets are usually treated as different modules. In this paper, we propose a unified theoretical positioning framework for multiple targets for ICVs. The bottlenecks of vehicular self-positioning and target localization can be eliminated. Our main contributions are summarized as follows.

- A unified theoretical framework for vehicular self-positioning and relative localization of targets based on V2X is proposed, and it can integrate data from the on-board sensors in the vehicular network and HD maps with GNSS/INS measurements into a unified system.
- By cooperative positioning, accuracy of under 0.2 m can be achieved in terms of self-positioning and relative localization of targets in urban areas using low-cost GNSS/INS, on-board sensors, and widely equipped HD maps. Simultaneously, the target sensing range is extended beyond the line of sight and field of view, and this greatly improves the integrity of perception.
- Furthermore, compared with state-of-the-art techniques, the proposed framework places fewer demands on vehicular network nodes' density and the amount of vehicle-to-target measurements.

The remainder of the paper is organized as follows. In Section 2, the system model is provided. The development of the proposed joint multiple-target positioning for ICVs is detailed in Section 3. Detailed implementation aspects are introduced in Section 4. Theoretical studies are explained in Section 5. Numerical results are given in Section 6. Finally, we conclude the paper in Section 7.

2. Problem Formulation

Firstly, we describe the targets in a traffic scene in this section.

- Targets: All objects related to vehicle driving, including the connected vehicles themselves and the elements that constitute the environment.
- Connected vehicles: Vehicles in the vehicular network that can obtain information from other vehicles and HD maps.
- Features: Static targets that can be associated with HD maps, e.g., lamps, trees, traffic lights, and traffic signs.
- Objects: Targets, both static and moving, that do not exist in HD maps. These can be pedestrians, bicycles, and disconnected vehicles, all of which are unlabeled on the map.

Consider a vehicular network scenario with a set of N_v interconnected vehicles $\mathcal{V} = \{1, 2, \cdots, N_v\}$, as shown in Figure 1. At time t, let $\mathbf{x}_{i,t}^{(V)}$ be the position and orientation of connected vehicle i in the global coordinate (see Equation (1)).

$$\mathbf{x}_{i,t}^{(V)} = [\mathbf{p}_{i,t}^{(V)}, \theta_{i,t}^{(V)}] = [x_{i,t}^{(V)}, y_{i,t}^{(V)}, \theta_{i,t}^{(V)}]^T, i \in \mathcal{V} \tag{1}$$

A set of N_f static features $\mathcal{F} = \{1, 2, \cdots, N_f\}$ also exists in the scene. Their positions are stored on a HD map with noise and, although not necessary, can be captured by the on-board sensors. We use Equation (2) to describe the two-dimensional position of the jth feature.

$$\mathbf{x}_{j,t}^{(F)} = \mathbf{p}_{j,t}^{(F)} = [x_{j,t}^{(F)}, y_{j,t}^{(F)}]^T, j \in \mathcal{F} \tag{2}$$

In addition, we also consider a set of N_o objects, moving or static, $\mathcal{O} = \{1, 2, \cdots, N_o\}$, which is described in Equation (3).

$$\mathbf{x}_{k,t}^{(O)} = \mathbf{p}_{k,t}^{(O)} = [x_{k,t}^{(O)}, y_{k,t}^{(O)}]^T, k \in \mathcal{O} \tag{3}$$

It must be noted that we do not estimate the orientations of the features and objects because the planning module usually does not require this information.

Our task is to estimate the localization and orientation of all the connected vehicles,

$$\mathbf{X}_t^{(V)} = \left[\mathbf{x}_{1,t}^{(V)} \cdots \mathbf{x}_{N_v,t}^{(V)}\right], \tag{4}$$

We also attempt to localize the features and objects (see Equation (5))

$$\mathbf{X}_t^{(F)} = \left[\mathbf{x}_{1,t}^{(F)} \cdots \mathbf{x}_{N_f,t}^{(F)}\right]$$
$$\mathbf{X}_t^{(O)} = \left[\mathbf{x}_{1,t}^{(O)} \cdots \mathbf{x}_{N_o,t}^{(O)}\right] \tag{5}$$

Based on Equations (4) and (5), we can obtain the relative localization of other targets by transforming their location into the vehicles' coordinate system.

In terms of the measurements, a target may be captured by a vehicle if it is within the vehicle's sensing range and without any occlusions. As shown in Figure 1, the target can be a connected vehicle, a feature, or an object, which are indicated with red, blue, and brown arrows, respectively. Its measurement model is described as Equation (6).

$$\mathbf{z}_{i,j,t}^{(\Xi)} = h^{(S)}(\mathbf{p}_{j,t}^{(\Xi)}, \mathbf{x}_{i,t}^{(V)}) + \mathbf{v}_{i,j,t}^{(S)} \tag{6}$$

where $\Xi \in \{V, F, O\}$ and $\mathbf{v}_{k,j,t}^{(S)} \sim \mathcal{N}(\mathbf{0}, \mathbf{R}_{k,j,t}^{(S)})$ is additive white Gaussian measurement noise with covariance $\mathbf{R}_{k,j,t}^{(S)}$, $h^{(S)}(\mathbf{p}_{j,t}^{(\Xi)}, \mathbf{x}_{k,t}^{(V)})$ is a function which denotes the measurement of target at position $\mathbf{p}_{j,t}^{(\Xi)}$ from vehicle $\mathbf{x}_{k,t}^{(V)}$.

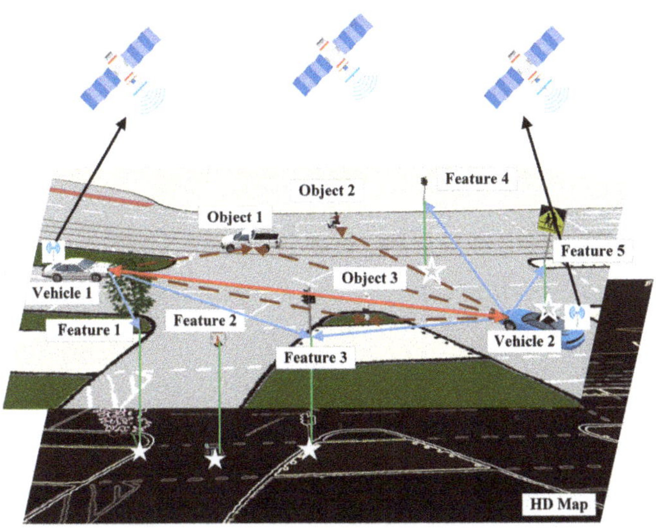

Figure 1. A demonstration of the multiple targets positioning scenario.

The connected vehicles are also equipped with GNSS/INS, which can provide measurements of their localization and orientation. The corresponding measurements are indicated with black arrows in Figure 1. Similarly, we treat the map as a virtual sensor with measurements pertaining to an associated feature as shown with green arrows. The measurement of GNSS/INS on vehicle i and that of HD map on feature j is indicated by Equations (7) and (8).

$$\mathbf{z}_{i,t}^{(G)} = \mathbf{h}^{(G)}(\mathbf{p}_{i,t}^{(V)}) + \mathbf{v}_{i,t}^{(G)} \tag{7}$$

and

$$\mathbf{z}_{j,t}^{(M)} = \mathbf{h}^{(M)}(\mathbf{p}_{j,t}^{(F)}) + \mathbf{v}_{j,t}^{(M)} \tag{8}$$

where $\mathbf{v}_{i,t}^{(G)} \sim \mathcal{N}(\mathbf{0}, \mathbf{R}_{i,t}^{(G)})$ is the measurement noise of GNSS/INS, and $\mathbf{v}_{j,t}^{(M)} \sim \mathcal{N}(\mathbf{0}, \mathbf{R}_{j,t}^{(M)})$ denotes the measurement noise from the map.

If we consider this problem as analogous to a distributed sensor network, the features and connected vehicles are static and mobile anchors, respectively, and their locations are constrained by the HD map and GNSS/INS. The objects are static or mobile nodes, and the on-board sensors generate constraints between the vehicle and nodes. For the vehicles, additional constraints arise from the V-V measurements.

3. The Unified Multiple-Target Positioning Framework

The objective of multiple-target positioning in this paper is to estimate the states $\mathbf{X}_t = \left[\mathbf{X}_t^{(\mathbf{V})}, \mathbf{X}_t^{(F)}, \mathbf{X}_t^{(O)}\right]$, from measurements,

$$\mathbf{Z}_t = \left[\mathbf{z}_{k,j,t}^{(S)}, \mathbf{z}_{k,t}^{(G)}, \mathbf{z}_{l,t}^{(M)}\right], \tag{9}$$

where $k \in \mathcal{V}, j \in \{\mathcal{V}, \mathcal{F}, \mathcal{O}\}, l \in \mathcal{F}$. From a probabilistic perspective, the maximum likelihood estimation of $\mathbf{X}_t$ is given by Equation (10).

$$\mathbf{X}_t^* = \arg\max P\left(\mathbf{Z}_t | \mathbf{X}_t\right) \tag{10}$$

where $P\left(\mathbf{Z}_t | \mathbf{X}_t\right)$ is the likelihood of the measurements $\mathbf{Z}_t$ given the states $\mathbf{X}_t$. The conditional distribution of the on-board sensor measurements in Equation (6) is given by Equation (11), where $P\left(\mathbf{z}_{k,j,t}^{(S)} | \mathbf{p}_{j,t}^{(\Xi)}, \mathbf{x}_{k,t}^{(V)}\right)$ denotes the probability distribution of measurement $\mathbf{z}_{k,j,t}^{(S)}$ given the states $\mathbf{p}_{j,t}^{(\Xi)}$ and $\mathbf{x}_{k,t}^{(V)}$. $\mathcal{N}\left(h^{(S)}\left(\mathbf{p}_{j,t}^{(\Xi)}, \mathbf{x}_{k,t}^{(V)}\right), \mathbf{R}_{k,j,t}^{(S)}\right)$ denotes a normal distribution with expectation $h^{(S)}$ and variance $\mathbf{R}_{k,j,t}^{(S)}$.

$$P\left(\mathbf{z}_{k,j,t}^{(S)} | \mathbf{p}_{j,t}^{(\Xi)}, \mathbf{x}_{(k,t)}^{(V)}\right) = \mathcal{N}\left(h^{(S)}\left(\mathbf{p}_{j,t}^{(\Xi)}, \mathbf{x}_{k,t}^{(V)}\right), \mathbf{R}_{k,j,t}^{(S)}\right) \tag{11}$$

Similarly, we get the conditional distributions of the measurement from GNSS/INS $P\left(\mathbf{z}_{i,t}^{(G)} | \mathbf{p}_{i,t}^{(V)}\right)$ and map $P\left(\mathbf{z}_{j,t}^{(M)} | \mathbf{p}_{j,t}^{(F)}\right)$ given states of vehicles $\mathbf{p}_{i,t}^{(V)}$ or feature states $\mathbf{p}_{j,t}^{(F)}$ (see Equations (12) and (13)). Their distributions are also normal with perception model h as expectation.

$$P\left(\mathbf{z}_{i,t}^{(G)} | \mathbf{p}_{i,t}^{(V)}\right) = \mathcal{N}\left(h^{(G)}\left(\mathbf{p}_{i,t}^{(V)}\right), \mathbf{R}_{i,t}^{(G)}\right) \tag{12}$$

and

$$P\left(\mathbf{z}_{j,t}^{(M)} | \mathbf{p}_{j,t}^{(F)}\right) = \mathcal{N}\left(h^{(M)}\left(\mathbf{p}_{j,t}^{(F)}\right), \mathbf{R}_{j,t}^{(M)}\right) \tag{13}$$

Remark 1. *Given a set of independent and identically distributed (i.i.d.) data $\mathcal{D} = \{\mathbf{x}_n, n = 1, 2, \cdots, N\}$, where observation $\mathbf{x}_n \in \mathbb{R}^{D\times 1}$ is drawn from a multivariate Gaussian distribution $\mathcal{N}(\mathbf{x}_n; \boldsymbol{\mu}_n, \mathbf{R}_n)$. The log-likelihood of the data set can be written as Equation (14).*

$$\mathcal{L}(\mathcal{D}) = -\frac{1}{2}\sum_{n=1}^{N}\left(\ln\left((2\pi)^D \det(\mathbf{R}_n)\right) + \boldsymbol{e}_n{}^T \mathbf{R}_n^{-1} \boldsymbol{e}_n\right) \tag{14}$$

where $\boldsymbol{e}_n = \boldsymbol{x}_n - \boldsymbol{\mu}_n$.

According to Remark 1, the maximization of $\mathcal{L}(\mathcal{D})$ is equivalent to the minimization of $\mathcal{J}(\mathcal{D})$ (see Equation (15)). The problem is solved with optimization as described in Section 4.

$$\mathcal{J}(\mathcal{D}) = \sum_{n=1}^{N} \boldsymbol{e}_n^T \mathbf{R}_n{}^{-1} \boldsymbol{e}_n \tag{15}$$

Considering (11)–(13) and assuming that the three types of measurements are independent, the joint probability density can be factorized as given in Equation (16).

$$\begin{aligned} P\left(\mathbf{Z}_t|\mathbf{X}_t\right) =& P\left(\mathbf{Z}_\mathbf{t}^{(\mathbf{V})}, \mathbf{Z}_t^{(F)}, \mathbf{Z}_t^{(O)}|\mathbf{X}_t\right) \\ =& \prod_{k,j} P\left(\mathbf{z}_{k,j,t}^{(S)}|\mathbf{p}_{j,t}^{(\Xi)}, \mathbf{x}_{(k,t)}^{(V)}\right) \\ & \prod_i P\left(\mathbf{z}_{i,t}^{(G)}|\mathbf{p}_{i,t}^{(V)}\right) \prod_j P\left(\mathbf{z}_{j,t}^{(M)}|\mathbf{p}_{j,t}^{(F)}\right) \end{aligned} \tag{16}$$

The maximization of $P(\mathbf{Z}_t|\mathbf{X}_t)$ can be reformulated as the following nonlinear least squares problem (see Equation (17)).

$$\begin{aligned} \mathbf{X}_t^* = \arg\min & \sum_k \sum_j \left(\mathbf{e}_{k,j,t}^{(S)}\right)^T \left(\mathbf{R}_{k,j,t}^{(S)}\right)^{-1} \mathbf{e}_{k,j,t}^{(S)} \\ & + \sum_i \left(\mathbf{e}_{i,t}^{(G)}\right)^T \left(\mathbf{R}_{i,t}^{(G)}\right)^{-1} \mathbf{e}_{i,t}^{(G)} \\ & + \sum_j \left(\mathbf{e}_{j,t}^{(M)}\right)^T \left(\mathbf{R}_{j,t}^{(M)}\right)^{-1} \mathbf{e}_{j,t}^{(M)} \end{aligned} \tag{17}$$

To enable insightful visualization, the nonlinear least–squares problem is interpreted in terms of inference over a factor graph [41]. This graph consists of 2 types of nodes: variable nodes, which represent the state $\mathbf{X}_t$, and factor nodes, which represent the constraints to on the variables. The factor nodes can be further divided into bi-directed nodes, which denote the constraints for 2 states (from the on-board sensor measurements), and directed prior nodes, which denote the constraints from the map and GNSS/INS.

As shown in Figure 2, for each measurement, we have the following factors.

- Factor between the variables $\mathcal{V}$ and $\Xi = \{\mathcal{V}, \mathcal{F}, \mathcal{O}\}$, on behalf of the constraints of V-V, vehicle-feature (V-F), vehicle-object (V-O), as expressed in Equation (18).

$$\phi_{k,j,t} = P\left(\mathbf{z}_{k,j,t}^{(S)}|\mathbf{p}_{j,t}^{(\Xi)}, \mathbf{x}_{(k,t)}^{(V)}\right) \tag{18}$$

- Factor between the variables $\mathcal{V}$ and GNSS/INS, on behalf of the constraints from GNSS/INS, as expressed in Equation (19).

$$\phi_{i,t} = P\left(\mathbf{z}_{i,t}^{(G)}|\mathbf{p}_{i,t}^{(V)}\right) \tag{19}$$

- Factor between the variables $\mathcal{F}$ and the map, on behalf of the constraints from the HD map, as expressed in Equation (20).

$$\phi_{j,t} = P\left(\mathbf{z}_{j,t}^{(M)}|\mathbf{p}_{j,t}^{(F)}\right) \tag{20}$$

The joint probability in Equation (16) can then be rewritten as the product of all the factors.

$$P\left(\mathbf{Z}_t|\mathbf{X}_t\right) = \prod_{k,j,t} \phi_{k,j,t} \prod_{i,t} \phi_{i,t} \prod_{j,t} \phi_{j,t} \tag{21}$$

We can clearly see the constraints applied on each node in the factor graph.

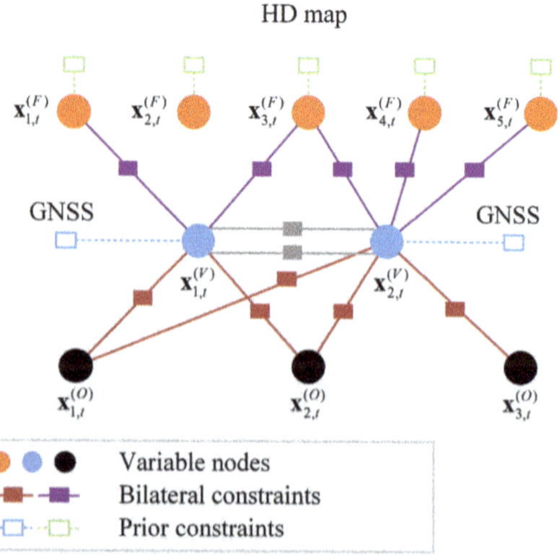

Figure 2. The proposed framework interpreted as inference on factor graphs.

4. Implementation Aspects

In this section, we introduce the implementation aspects related to the hypothesis on vehicle perception, the measurement model, optimization, and data association.

4.1. Perception Demands and Sensing Capability of Vehicles

We assume that due to occlusion and the limitations of perception range, the vehicle cannot completely locate the desired target. In this section, we explain the hypothesis of this work. It should be noted that our hypothesis is based on typical perceptual systems, but can be easily adapted to other forms.

As shown in Figure 3, we identify the scope of targets that need to be localized by a vehicle as "demanding space" and assume that it is a rectangle that can be quantitatively described by l_f and l_r, i.e., the distances that the vehicle requires to sense ahead of and behind itself, respectively, and W_d, the range that should be perceived laterally. We assume that the vehicle sensing range is a forward-facing cone with a radius of R_s, and the field of view is θ_{FOV}.

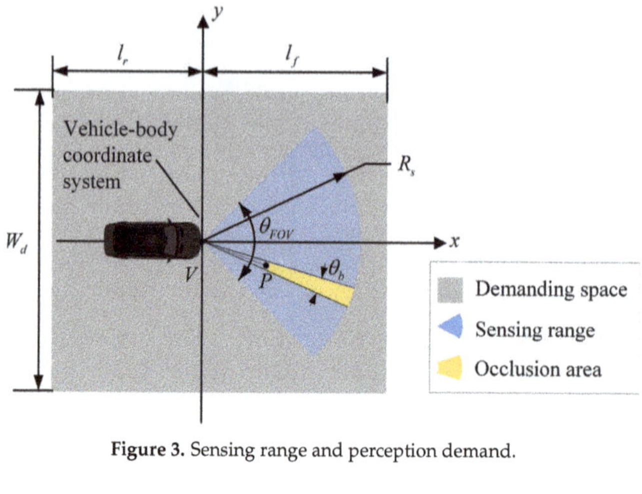

Figure 3. Sensing range and perception demand.

To consider situations of occlusion, we assume that there is an object P in the sensing range, and that the area outside P in the sector with line VP as the axis of symmetry is regarded as the occlusion area. Thus, only the blue area in Figure 3 can be perceived. The limitations in sensing range and occlusion constitute the blind spots of environment perception.

4.2. Measurement Model

In this work, we assume that data from the on-board sensors are in a 2D vehicular coordinate fashion where $h^{(S)} \sim \mathbb{R}^{2\times 1}$, correspond to the measurements from low-cost cameras. This can be easily adapted to other measurement types, such as polar coordinates in LiDAR measurements. The measurement model is expressed as,

$$h^{(S)}(\mathbf{p}_{j,t}^{(\Xi)}, \mathbf{x}_{i,t}^{(V)}) = \mathbf{R}^{-1}(\theta_{i,t}^{(V)})\left[\mathbf{p}_{j,t}^{(\Xi)} - \mathbf{p}_{i,t}^{(V)}\right], \tag{22}$$

where the rotation matrix is expressed as,

$$\mathbf{R}(\theta_{i,t}^{(V)}) = \begin{bmatrix} \cos(\theta_{i,t}^{(V)}) & -\sin(\theta_{i,t}^{(V)}) \\ \sin(\theta_{i,t}^{(V)}) & \cos(\theta_{i,t}^{(V)}) \end{bmatrix}, \tag{23}$$

and the covariance of the on-board sensors' measurements is given in Equation (24).

$$\mathbf{R}_{k,j,t}^{(S)} = \begin{bmatrix} \delta_{sensor}^2 & 0 \\ 0 & \delta_{sensor}^2 \end{bmatrix} \tag{24}$$

We assume that the GNSS/INS can provide the measurements of the coordinates and angles of the vehicles. There are many studies on modeling the measurements noise of GNSS/INS [42,43]. In this paper, we simplify the error of GNSS to Gaussian distribution, and the measurement model and uncertainty of the GNSS/INS on vehicle i are indicated as Equations (25) and (26). Our framework is also suitable for extending to other error assumptions.

$$\mathbf{h}^{(G)}(\mathbf{p}_{i,t}^{(V)}) = \mathbf{p}_{i,t}^{(V)} \tag{25}$$

and

$$\mathbf{R}_{i,t}^{(G)} = \begin{bmatrix} \delta_{GNSS,l}^2 & 0 & 0 \\ 0 & \delta_{GNSS,l}^2 & 0 \\ 0 & 0 & \delta_{GNSS,\theta}^2 \end{bmatrix} \tag{26}$$

In commercial HD maps, the coordinates of the features are provided along with noise, so we formulate the measurement mode and covariance matrix as expressed by Equations (27) and (28).

$$\mathbf{h}^{(M)}(\mathbf{p}_{j,t}^{(F)}) = \mathbf{p}_{j,t}^{(F)} \tag{27}$$

$$\mathbf{R}_{j,t}^{(M)} = \begin{bmatrix} \delta_{map}^2 & 0 \\ 0 & \delta_{map}^2 \end{bmatrix} \tag{28}$$

4.3. Optimized Variable Allocation and Data Association

The optimized variables are allocated to observations within the demanding space of perception. Unlike in a traditional multi-vehicle cooperative system, barring objects and features captured by vehicles, features that are not seen by any vehicle but are within the demanding space are also included in the optimized variables, and further optimized to yield the results of perception.

Observations that are associated are merged into existing variables and form constraints in the process of optimization. There are many methods that can be applied to our framework [44,45]. In this study, we assume that the vehicle's on-board sensors and HD maps can provide enough semantic clue to identify objects. The association algorithm itself is beyond the scope of this article.

4.4. Optimization Problem Solving

In this study, the nonlinear optimization problem in Equation (17) is solved via the Levenberg-Marquardt method [46]. We reorganize the residuals of time t into one vector, as expressed in Equation (29).

$$\mathbf{e} = \begin{pmatrix} \mathbf{e}_{ij}^{(S)} & \mathbf{e}_{kp}^{(S)} & \mathbf{e}_{m}^{(G)} & \mathbf{e}_{q}^{(M)} \end{pmatrix}^T \tag{29}$$

where $\mathbf{e}_{ij}^{(S)}$ is the residual of measurement from the on-board sensor of vehicle i to vehicle j. $\mathbf{e}_{kp}^{(S)}$ is the residual of measurement from the on-board sensor of vehicle k to feature or object p. $\mathbf{e}_{m}^{(G)}$ is the residual of GNSS/INS measurement of vehicle m, and $\mathbf{e}_{q}^{(M)}$ is the residual of measurement from HD map to feature q. The optimized variable at time t can then be rewritten as,

$$\mathbf{X} = \left(\mathbf{x}_i^{(V)}\ \mathbf{x}_j^{(V)}\ \mathbf{x}_k^{(V)}\ \mathbf{x}_p^{(\Xi)}\ \mathbf{x}_m^{(V)}\ \mathbf{x}_q^{(F)}\right), \tag{30}$$

where $\Xi \in \{F, O\}$. Let $\mathbf{R}$ be the overall covariance matrix such that

$$\mathbf{R} = diag\left(\mathbf{R}^{(S)}\ \mathbf{R}^{(S)}\ \mathbf{R}^{(G)}\ \mathbf{R}^{(M)}\right). \tag{31}$$

The cost function can be rewritten as,

$$\mathbf{f}(\mathbf{X}) = (\mathbf{R}^{-\frac{1}{2}}\mathbf{e})^T(\mathbf{R}^{-\frac{1}{2}}\mathbf{e}). \tag{32}$$

We can get the Jacobian matrix (see Equation (33)).

$$\mathbf{J}(\mathbf{X}) = \frac{\partial(\mathbf{R}^{-\frac{1}{2}}\mathbf{e})}{\partial\mathbf{X}} = \mathbf{R}^{-\frac{1}{2}} \begin{bmatrix} \frac{\partial \mathbf{e}_{ij}^{(S)}}{\partial \mathbf{x}_i^{(V)}} & \frac{\partial \mathbf{e}_{ij}^{(S)}}{\partial \mathbf{x}_j^{(V)}} & 0 & 0 & 0 & 0 \\ 0 & 0 & \frac{\partial \mathbf{e}_{kp}^{(S)}}{\partial \mathbf{x}_k^{(V)}} & \frac{\partial \mathbf{e}_{kp}^{(S)}}{\partial \mathbf{x}_p^{(\Xi)}} & 0 & 0 \\ 0 & 0 & 0 & 0 & \frac{\partial \mathbf{e}_{m}^{(G)}}{\partial \mathbf{x}_m^{(V)}} & 0 \\ 0 & 0 & 0 & 0 & 0 & \frac{\partial \mathbf{e}_{q}^{(M)}}{\partial \mathbf{x}_q^{(F)}} \end{bmatrix} \tag{33}$$

The initial values of the optimization iterations are given as follows. The vehicular position and attitude are calculated by the measurements of the GNSS/INS. The positions of features are determined by the map, and the initial positions of objects are calculated by converting the positions measured by the on-board sensors to the geodetic coordinate system according to the initial vehicle position and attitude. The cost function (17) can be minimized towards zero by iterations:

$$\mathbf{X}_{k+1} \leftarrow \mathbf{X}_k - \left(\mathbf{J}^T\mathbf{J} + \lambda diag\left(\mathbf{J}^T\mathbf{J}\right)\right)^{-1}\mathbf{J}^T\mathbf{f}\left(\mathbf{X}_{k+1}\right) \tag{34}$$

where λ is determined by the Levenberg-Marquardt method, and $\mathbf{J}$ and $\mathbf{f}$ are defined in Equations (32) and (33), respectively.

5. Theoretical Analysis on the Framework Performance

The proposed multiple-target positioning framework aims to solve a parameter estimation problem. Its performance can be evaluated either numerically or theoretically. In this section, the lower bounds on the estimation errors are determined from theoretical studies. As one of the most widely used lower bounds, the Cramér-Rao lower bound (CRLB) is chosen as the performance benchmark. The framework is performance-bound in terms of the minimum achievable variance provided by any unbiased estimators.

Assume that a deterministic signal $\boldsymbol{s}_t(\boldsymbol{\theta})$ with an unknown vector parameter $\boldsymbol{\theta}$ is observed in white Gaussian noise as Equation (35).

$$\mathbf{z}_t = \boldsymbol{h}_t(\boldsymbol{\theta}) + \mathbf{v}_t \tag{35}$$

where $\mathbf{v}_t \sim \mathcal{N}(\mathbf{0}, \mathbf{C}_t)$. We wish to estimate $\boldsymbol{\theta}$ from $\mathbf{z}$. The Fisher information matrix [47] is given by Equation (36).

$$[\mathbf{I}(\boldsymbol{\theta})]_{m,n} = \left[\frac{\partial \boldsymbol{h}_t(\boldsymbol{\theta})}{\partial \theta_m}\right]^T \mathbf{C}_t^{-1} \left[\frac{\partial \boldsymbol{h}_t(\boldsymbol{\theta})}{\partial \theta_n}\right] \tag{36}$$

Taking the inverse of $\mathbf{I}(\boldsymbol{\theta})$, the CRLB for the parameters is then obtained from its diagonal elements. The CRLB for θ_m is the (m, m) entry of $\mathbf{I}^{-1}(\boldsymbol{\theta})$.

For the proposed framework, the following measurements are considered.

$\mathbf{z}_{i,t}^{(V)}$ vehicle $i \in \mathcal{V}$, measured from GNSS/INS;

$\mathbf{z}_{il,t}^{(V2V)}$ measured from vehicle i to vehicle l, where $i \in \mathcal{V}$ and $l \in \mathcal{V}$;

$\mathbf{z}_{j,t}^{(F)}$ feature $j \in \mathcal{F}$, measured from the HD map;

$\mathbf{z}_{ij,t}^{(V2F)}$ vehicle $i \in \mathcal{V}$ to feature $j \in \mathcal{F}$, measured from the vehicle's on-board sensors; and

$\mathbf{z}_{ik,t}^{(V2O)}$ vehicle i to object $k \in \mathcal{O}$.

For convenience, all the measurements available are reformulated to the following compact form:

$$\mathbf{z}_t = \boldsymbol{h}_t(\boldsymbol{\theta}) + \mathbf{v}_t = \begin{bmatrix} z_{i,t}^G \\ z_{j,t}^M \\ z_{i,j,t}^\Xi \end{bmatrix} + \mathbf{v}_t \tag{37}$$

where $z_{i,t}^G$, $z_{j,t}^M$ and $z_{i,j,t}^\Xi$ are defined in Equations (6)–(8), respectively. The unknown parameters are obtained from Equation (38).

$$\boldsymbol{\theta}^T = \big[\underbrace{x_{i,t}^{(V)} \;\; y_{i,t}^{(V)} \;\; \theta_{i,t}}_{i\in\mathcal{V}} \;\; \underbrace{x_{j,t}^{(F)} \;\; y_{j,t}^{(F)}}_{j\in\mathcal{F}} \;\; \underbrace{x_{k,t}^{(O)} \;\; y_{k,t}^{(O)}}_{k\in\mathcal{O}}\big] \tag{38}$$

We observe that $\mathbf{z}_t$ is Gaussian distributed with mean $\boldsymbol{h}_t(\boldsymbol{\theta})$ and covariance matrix $\mathbf{C}_t$:

$$\mathbf{z}_t \sim \mathcal{N}\Big(\boldsymbol{h}_t(\boldsymbol{\theta}), \mathbf{C}_t\Big) \tag{39}$$

The CRLB for $\boldsymbol{\theta}$ is obtained by substituting $\boldsymbol{h}_t(\boldsymbol{\theta})$ and $\mathbf{C}_t$ into Equation (36).

6. Numerical Results

In this section, we discuss the simulation experiments conducted under typical vehicular network scenarios to verify the localization and perception capacity results of the proposed algorithm. We also demonstrate its environmental adaptability in subsequent discussions on factors that influence the final performance by considering different scene configurations.

As shown in Figure 4, we build an intersection with 2 two-way two-lane roads. This scenario consists of a busy urban area and a suburban area. The trajectories of all the vehicles and objects as well as the traffic scene configuration come from VISSIM, a behavior-based traffic flow simulator [48]. Each road is 330 m long. In the middle, until approximately 200 m from the intersection, we simulate a busy urban scenario with lamps, traffic lights, and traffic signs located randomly on the roadside. Pedestrians walking around the road and across the intersection are generated. Outside of and far from the intersection, nothing is placed on the roadside, which simulates the scenario of a suburban area. In the simulation, connected and disconnected vehicles start from one end of the road simultaneously, then travel straight or turn left or right at the intersection, and exit the scene almost simultaneously. Therefore, vehicles are in the urban area in the middle section of the simulation steps, and the starting and ending segments correspond to suburban scenes.

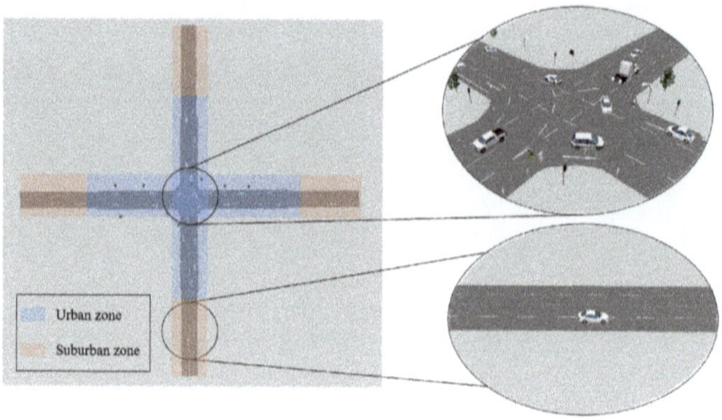

Figure 4. The considered intersection simulated in VISSIM.

6.1. Performance in a Typical Scenario

First, we validate the effectiveness of our algorithm in a fixed scenario and compare it with the method proposed by Gloria et al. [19], as well as the theoretical bound CRLB. We set the accuracy of each measurement to that of low-cost devices. The configuration for the scenario and each measurement are as follows.

- $N_v = 6$
- $N_f = 23$ (15 lamps, 4 traffic lights, and 4 traffic signs)
- $N_o = 18$ (10 pedestrians, and 8 disconnected vehicles)
- $\delta_{sensor} = 0.25$ m, and $\delta_{map} = 0.05$ m
- $\delta_{GNSS,l} = 2.5$ m, and $\delta_{GNSS,\theta} = 0.1$ rad
- $\theta_{FOV} = 70$ m, $R_s = 80$ m, and $\theta_b = 2°$
- $L_f = 100$ m, $L_r = 30$ m, and $W_d = 60$ m

We run the simulation 200 times, and noise is added to the measurements independently for each iteration. One localization result of the vehicles, objects, and features, and their true positions in the urban area is shown in Figure 5. As the 6 vehicles face similar scenes in every simulation step, we statistically analyze the positioning errors of all the vehicles. The root-mean-square error (RMSE) of self-positioning for all 6 vehicles at simulation time t is calculated using Equation (40).

$$\mathrm{RMSE}_t^{(V)} = \sqrt{\frac{1}{MN}\sum_{j=1}^{N}\sum_{i=1}^{M}\left\|\hat{\mathbf{p}}_{i,j,t}^{(V)} - \mathbf{p}_{i,j,t}^{(V)}\right\|_2^2} \tag{40}$$

where $\hat{\mathbf{p}}_{i,j,t}^{(V)}$ is the self-positioning result of vehicle i at the jth run at simulation step t, and $\mathbf{p}_{i,j,t}^{(V)}$ is its corresponding ground truth. The self-positioning mean-square error MSE bound is calculated using Equation (41)

$$\mathrm{CRLB}_t = \frac{1}{M}\sum_{i=1}^{M}\left(\mathrm{CRLB}(x_{i,t}^{(V)}) + \mathrm{CRLB}(y_{i,t}^{(V)})\right) \tag{41}$$

where $\mathrm{CRLB}(x_{i,t}^{(V)})$ and $\mathrm{CRLB}(y_{i,t}^{(V)})$ are the CRLBs of the x and y coordinates of vehicle i at simulation step t.

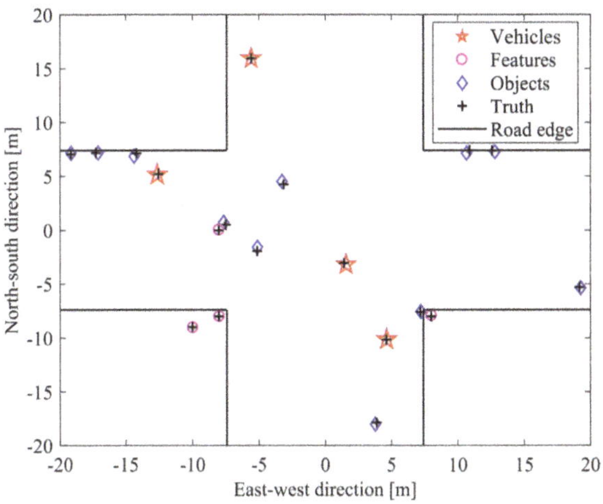

Figure 5. Multiple-target position performance in the urban area.

As shown in Figure 6, the proposed method is compared with that of Gloria et al. [19], as well as the RMSE bound $\sqrt{\mathrm{CRLB}_t}$ (see Equation (41)).

It is obvious that compared with the original GNSS measurements, we obtain a significantly improved positioning result by using the information from V2X and the HD map. In particular, in the urban area (simulation steps 21–55), our algorithm achieves high positioning accuracy (0.16 m), which is also lower than that of the method in [19] (1.79 m). Our positioning accuracy is close to the theoretical lower bound given by CRLB, which shows that we have effectively used all valuable information.

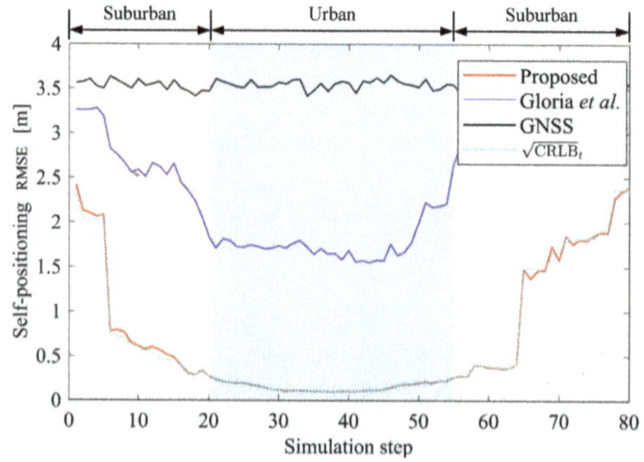

Figure 6. Self-localization error.

We also give the number of constraints at each simulation time in Figure 7. Overall, the greater the number of constraints available, the better our positioning results are. In fact, the study in [19] only uses GNSS and V-O constraints, while we have used additional constraints including V-F, V-V, and prior constraints of the HD map.

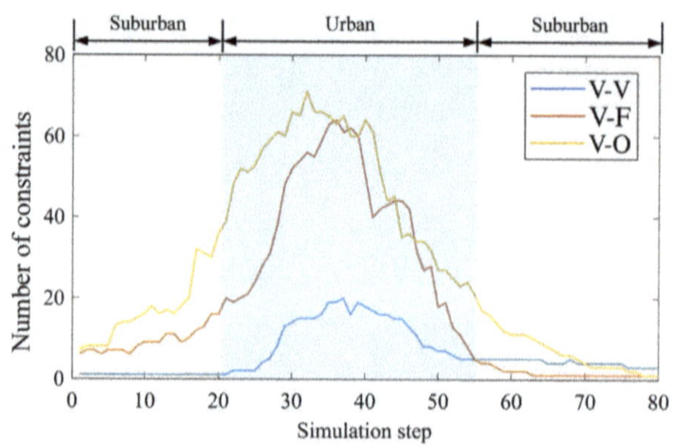

Figure 7. Amount of constraints used at every simulation step.

In the suburban area where the sensing ranges of different vehicles have little overlap, the method proposed in [19] acts ineffectively as few constraints are available. However, our positioning is still a significant improvement in these challenging areas. Environmental adaptability will be further discussed in the next section.

In terms of the positioning of other targets, we compared both the target location precision and sensing integrity. We transform the positioning results of these targets into the body coordinate system (i.e., the vehicle coordinate system shown in Figure 3 for analysis, as this analysis is consistent with the vehicle sensing system. The target positioning accuracy of a vehicle is evaluated by the RMSE of all targets within the demanding space (see Equation (42)).

$$\text{RMSE}_{i,t}^{(T)} = \sqrt{\frac{1}{ON}\sum_{k=1}^{O}\sum_{j=1}^{N}\left\|\hat{\mathbf{p}}_{k,j,t}^{(T)} - \mathbf{p}_{k,j,t}^{(T)}\right\|_2^2} \tag{42}$$

where $\hat{\mathbf{p}}_{k,j,t}^{(T)}$ is the localization result of the *k*th target in the vehicle coordinate system at the *j*th runtime, and $\mathbf{p}_{k,j,t}^{(T)}$ is its truth position. The RMSE of relative localization i.e., $\text{RMSE}_t^{(T)}$ is defined as the root mean square of the location of all vehicles. In fact, such a result is affected by both the absolute positioning and vehicular self-positioning, which makes our analysis more rigorous.The result is shown in Figure 8. In the urban area, the RMSE is 0.17 m; it is much smaller than that (0.24 m) obtained by Gloria's method as well as that (0.32 m) provided by the vehicles' on-board sensors. Higher perception accuracy is also achieved in suburban areas.

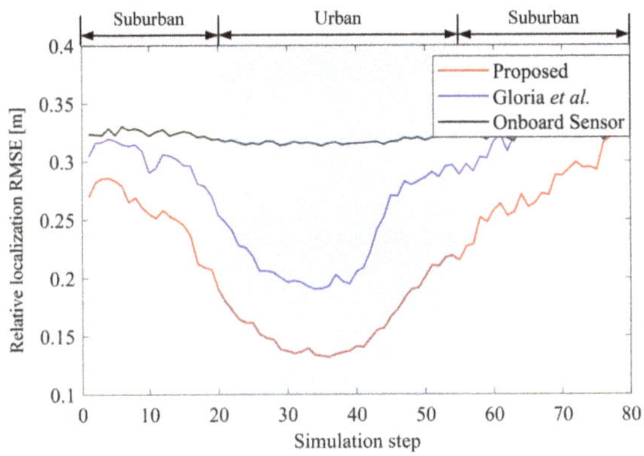

Figure 8. Relative positioning error of surrounding targets.

Figure 9 shows the improvement in sensing integrity. The blue line is the true value of the number of targets within the demanding space. Based on the raw data of the on-board sensors, only 41.81% of targets are captured in the urban area, owing to occlusion or limited sensing range, while the proposed method enables 90.42% of the targets to be captured. The improvement comes from the sharing of information between the connected vehicles, and the information provided by the real-time dynamic map.

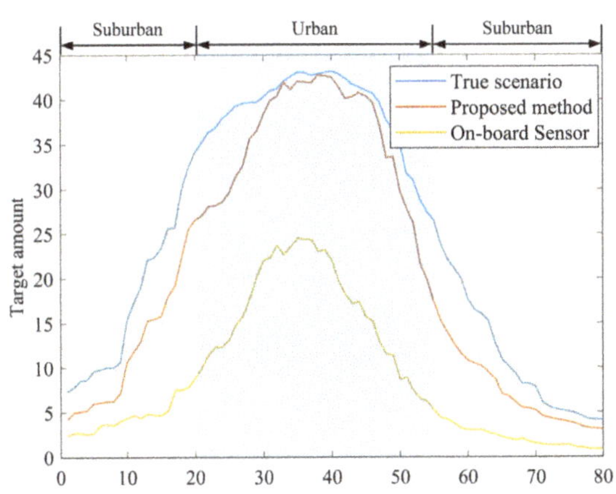

Figure 9. Perception integrity in terms of the number of targets captured.

In summary, our approach significantly improves multiple-target positioning in terms of accuracy and integrity over that achieved using the original measurements, and is also more effective than other methods.

6.2. *Adaptability to Different Scenarios*

In the following section, we analyze the impact of different elements on the results of multiple-target positioning to demonstrate the adaptability of our method to different environments. Simultaneously, we discuss the contributions of different constraint types to the results.

6.2.1. Number of Connected Vehicles

The vehicular positioning and relative localization of the targets in terms of the number of connected vehicles are shown in Figures 10 and 11, respectively. Except for the number of connected vehicles, the configurations of the scene are identical, with 6 features and 3 objects. The accuracy of sensing and GNSS/INS are the same as those in previous experiments.

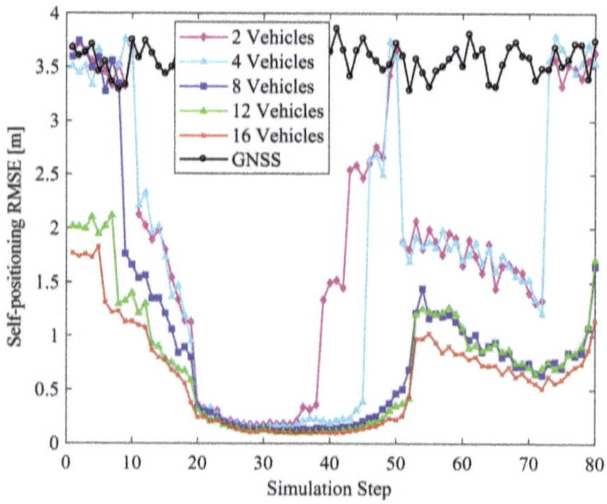

Figure 10. Effect of the number of connected vehicles on localization accuracy.

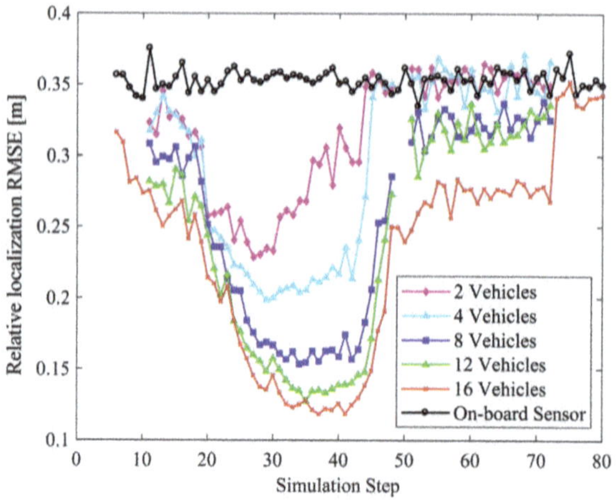

Figure 11. Effect of the number of connected vehicles on perception accuracy.

As can be seen from the figures, although there is a limited number of connected vehicles (only 2 vehicles), the positioning capacities of the vehicles and other targets are significantly improved in the urban areas. In general, the greater the number of connected vehicles, the smaller the corresponding positioning errors, as more V-V constraints are imposed between the vehicles. It is worth noting that there is a trend that the RMSEs of self-positioning increase as the simulation step increases from 70 to 80, as the vehicles are heading the end of the roads, where features and objects are becoming increasingly sparse. In theory, the lower bound of positioning error of 12 vehicles is lower than 8 vehicles. Due to the limited simulation times, the RMSE fluctuates near the theoretical bounds. Therefore, the RMSE of 12 vehicles seems close to those of 8 vehicles. However, based on the existing results, we cannot say that there is a trend that they will exceed those of 8 vehicles.

Another interesting observation is that for a given number of connected vehicles, the positioning errors in the suburban areas are larger than those in the urban areas, but it is worth noting that the decreasing trend of RMSE with increasing the number of connected vehicles is more significant. This is because, unlike the urban area with sufficient types of constraints, the constraints in suburbs are mainly of V-V. Therefore, we argue that the number of connected vehicles is important for improving location accuracy in the suburban area, which is consistent with the results shown in Figure 10.

As for the perception accuracy, in suburban area, with the increase of the number of vehicles, there are more constraints which benefits the positioning of connected vehicles and other objects. The underlying mechanism can be explained by Equation (36). As the constraints increase, the dimension of $h_t(\theta)$ increases, which leads to an increase in the value of the information matrix $\mathbf{I}(\theta)$. The CRLB for θ_m, which can be calculated with the diagonal element of $\mathbf{I}(\theta))^{-1}$ will decrease, which leads to the reduction of the absolute positioning error for each object. As the perception result is gained by projecting the absolute of other targets to the vehicle-body coordinate system based on the self-positioning, the reduction of the absolute positioning error finally improves the perception accuracy.

6.2.2. Number of Features

The effect of the number of features are shown in Figures 12 and 13. There are four connected vehicles running on the road with four objects and different numbers of features. It is obvious that increasing the number of features improves the accuracy of vehicle positioning and relative localization of the targets. Compared to raw measurements, even with a few features, the positioning and perception errors are reduced by exploiting the vehicle-to-target constraints and HD map information. It is noteworthy that at the intersection, the positioning accuracy is high when the number of features is 5, 10, or 30. However, if there is no feature i.e., no map information is used, the positioning error is obviously higher. This reflects the contribution of the HD map to vehicle positioning.

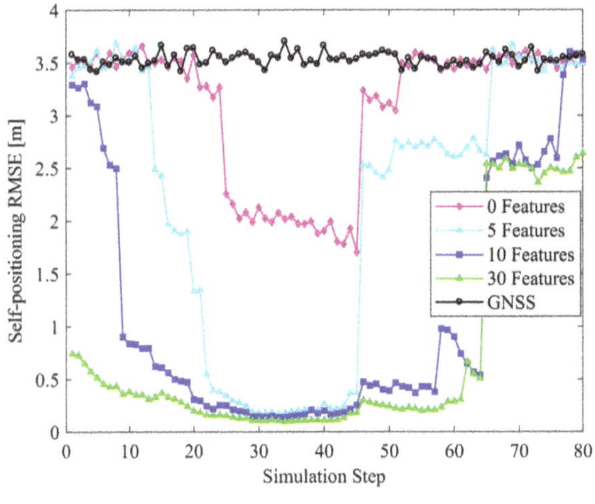

Figure 12. Effect of the number of features on localization accuracy.

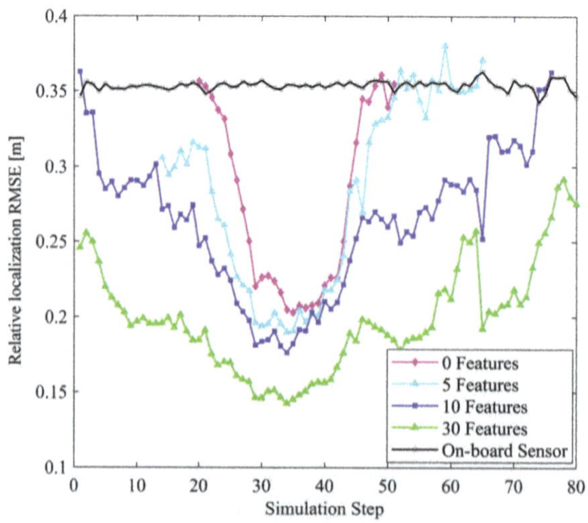

Figure 13. Effect of the number of features on perception accuracy.

6.2.3. Number of Objects

In comparing the results obtained when the number of objects varies, we set $N_v = 8$ and $N_f = 0$. The corresponding results are given in Figures 14 and 15. In the suburban zones, the increase in the number of objects improves the positioning. However, it is noteworthy that such an improvement in the intersection zone is not obvious. The reason is that in the former zone, there are very few V-V constraints, and V-O constraints play the main role in improving the results. Hence, adding objects can effectively improve the positioning. However, at the intersections, the V-V constraints formed by 8 vehicles are dominant and the results approach the theoretical bounds achievable. There is no significant improvement in accuracy with the addition of objects.

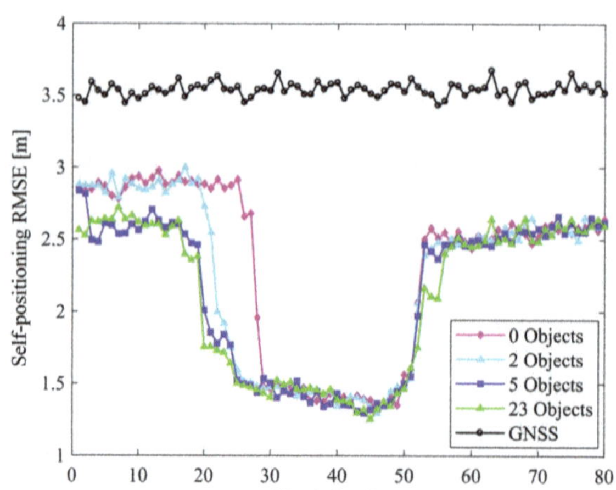

Figure 14. Effect of the number of objects on localization accuracy.

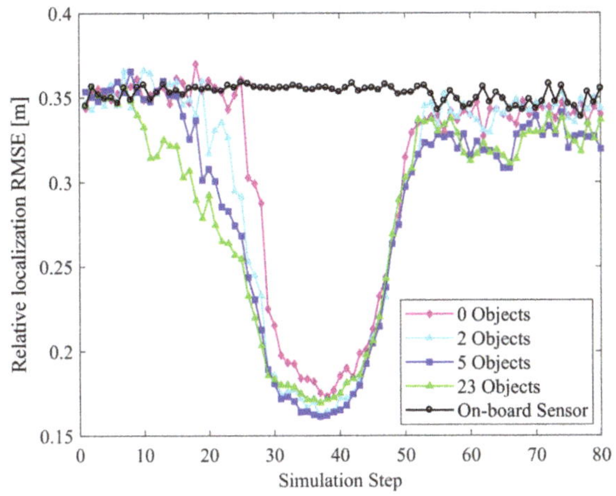

Figure 15. Effect of the number object on perception accuracy.

In the case of relative localization, increasing the number of objects can reduce the overall perception accuracy, instead. As the number of objects increases, the proportion of objects among all the perceived targets participating in the perceptual precision calculation increases. The objects are less constrained relative to other targets, and the overall clarity of perception declines. Considering this and the former discussion, we argue that V-O constraints are less effective than V-V and V-F constraints in improving multiple-target positioning accuracy.

7. Conclusions

This study focuses on the problem of multiple-target positioning for ICVs. We propose a unified theoretical framework for positioning both vehicles and other targets, wherein sensor data from V2X and HD map data are effectively fused with GNSS/INS and on-board sensors. By jointly exploiting the vehicle-to-target constraints and HD map information, the vehicular localization accuracy can be enhanced to meet the requirements of high-level automated driving by using low-cost GNSS/INS and on-board sensors in urban areas. Meanwhile, the confidence and integrity of the results of relative localization of targets are significantly improved, realizing sensing beyond line of sight and field of view, which can improve the transportation efficiency and safety. Furthermore, the proposed framework is applicable to more challenging scenarios entailing fewer connected vehicles and sparse features and objects. In future research, we plan to remove the limiting assumption of data association employed in this study by applying association methods in the process of optimization. We will also study the formulation of communication delay of V2X in the data fusion framework.

Author Contributions: Conceptualization, Z.X., D.Y. and F.W.; methodology, Z.X. and F.W.; software, Z.X.; validation, K.J. and F.W.; formal analysis, D.Y.; writing—original draft preparation, Z.X.; writing—review and editing, D.Y.; visualization, F.W.

Funding: This work was supported in part by the International Science and Technology Cooperation Program of China (2016YFE0102200), in part by the National Key Research and Development Program of China (2018YFB0105000), in part by the National Natural Science Foundation of China (61773234 and U1864203), in part by the Project of Tsinghua University and Toyota Joint Research Center for AI Technology of Automated Vehicle (TT2018-02), and in part by the software developed in the Beijing Municipal Science and Technology Program (D171100005117002 and Z181100005918001). This work was also supported by European Commission H2020 Marie Sklodowska-Curie project under the grant agreement No. 700044 and the State Key Laboratory of Automotive Safety and Energy under Project No. KF1804.

Conflicts of Interest: The authors declare no conflict of interest.

References

1. Wang, X.; Ning, Z.; Hu, X.; Ngai, E.C.; Wang, L.; Hu, B.; Kwok, R.Y.K. A city-wide real-time traffic management system: Enabling crowdsensing in social internet of vehicles. *IEEE Commun. Mag.* **2018**, *56*, 19–25. [CrossRef]
2. Usman, M.; Asghar, M.R.; Ansari, I.S.; Granelli, F.; Qaraqe, K.A. Technologies and solutions for location-based services in smart cities: Past, present, and future. *IEEE Access* **2018**, *6*, 22240–22248. [CrossRef]
3. Brummelen, J.V.; O'Brien, M.; Gruyer, D.; Najjaran, H. Autonomous vehicle perception: The technology of today and tomorrow. *Transp. Res. Part C Emerg. Technol.* **2018**, *89*, 384–406. [CrossRef]
4. Yang, D.; Jiang, K.; Zhao, D.; Yu, C.; Cao, Z.; Xie, S.; Xiao, Z.; Jiao, X.; Wang, S.; Zhang, K. Intelligent and connected vehicles: Current status and future perspectives. *Sci. China Technol. Sci.* **2018**, *61*, 1446–1471. [CrossRef]
5. Vivacqua, R.P.D.; Bertozzi, M.; Cerri, P.; Martins, F.N.; Vassallo, R.F. Self-localization based on visual lane marking maps: An accurate low-cost approach for autonomous driving. *IEEE Trans. Intell. Transp. Syst.* **2018**, *19*, 582–597. [CrossRef]
6. Kim, S.; Liu, W.; Ang, M.H.; Frazzoli, E.; Rus, D. The Impact of Cooperative Perception on Decision Making and Planning of Autonomous Vehicles. *IEEE Intell. Transp. Syst. Mag.* **2015**, *7*, 39–50. [CrossRef]
7. Bresson, G.; Alsayed, Z.; Yu, L.; Glaser, S. Simultaneous Localization and Mapping: A Survey of Current Trends in Autonomous Driving. *IEEE Trans. Intell. Veh.* **2017**, *2*, 194–220. [CrossRef]
8. Kuutti, S.; Fallah, S.; Katsaros, K.; Dianati, M.; Mccullough, F.; Mouzakitis, A. A Survey of the State-of-the-Art Localization Techniques and Their Potentials for Autonomous Vehicle Applications. *IEEE Internet Things J.* **2018**, *5*, 829–846. [CrossRef]
9. Liu, J.; Liu, J. Intelligent and connected vehicles: Current situation, future directions, and challenges. *IEEE Commun. Stand. Mag.* **2018**, *2*, 59–65. [CrossRef]
10. Skog, I.; Handel, P. In-car positioning and navigation technologies-A survey. *IEEE Trans. Intell. Transp. Syst.* **2009**, *10*, 4–21. [CrossRef]
11. Jackson, J.; Davis, B.; Gebre-Egziabher, D. A performance assessment of low-cost RTK GNSS receivers. In Proceedings of the IEEE/ION Position, Location and Navigation Symposium (PLANS), Monterey, CA, USA, 23–26 April 2018; pp. 642–649.
12. Ahmed, E.; Gharavi, H. Cooperative vehicular networking: A survey. *IEEE Trans. Intell. Transp. Syst.* **2018**, *19*, 996–1014. [CrossRef] [PubMed]
13. Nam, S.; Lee, D.; Lee, J.; Park, S. CNVPS: Cooperative Neighboring Vehicle Positioning System Based on Vehicle-to-Vehicle Communication. *IEEE Access* **2019**, *7*, 16847–16857. [CrossRef]
14. Jeong, H.Y.; Nguyen, H.H.; Bhawiyuga, A. Spatiotemporal Local-Remote Senor Fusion (ST-LRSF) for Cooperative Vehicle Positioning. *Sensors* **2018**, *18*, 1092. [CrossRef] [PubMed]
15. De, P.M.F. Survey on Ranging Sensors and Cooperative Techniques for Relative Positioning of Vehicles. *Sensors* **2017**, *17*, 271.
16. Severi, S.; Wymeersch, H.; Härri, J.; Ulmschneider, M.; Denis, B.; Bartels, M. Beyond GNSS: Highly accurate localization for cooperative-intelligent transport systems. In Proceedings of the IEEE Wireless Communications and Networking Conference, Barcelona, Spain, 15–18 April 2018; pp. 1–6.
17. Hobert, L.; Festag, A.; Llatser, I.; Altomare, L.; Visintainer, F.; Kovacs, A. Enhancements of V2X communication in support of cooperative autonomous driving. *IEEE Commun. Mag.* **2015**, *53*, 64–70. [CrossRef]
18. Shen, X.; Andersen, H.; Leong, W.K.; Kong, H.X.; Ang, M.H., Jr.; Rus, D. A General Framework for Multi-vehicle Cooperative Localization Using Pose Graph. *arXiv* **2017**, arXiv:1704.01252.
19. Soatti, G.; Nicoli, M.; Garcia, N.; Denis, B.; Raulefs, R.; Wymeersch, H. Implicit Cooperative Positioning in Vehicular Networks. *IEEE Trans. Intell. Transp. Syst.* **2018**, *19*, 3964–3980. [CrossRef]
20. Soatti, G.; Nicoli, M.; Garcia, N.; Denis, B.; Raulefs, R.; Wymeersch, H. Enhanced vehicle positioning in cooperative ITS by joint sensing of passive features. In Proceedings of the IEEE 20th International Conference on Intelligent Transportation Systems (ITSC), Yokohama, Japan, 16–19 October 2017; pp. 1–6.
21. Fascista, A.; Ciccarese, G.; Coluccia, A.; Ricci, G. Angle of arrival-based cooperative positioning for smart vehicles. *IEEE Trans. Intell. Transp. Syst.* **2018**, *19*, 2880–2892. [CrossRef]
22. Seif, H.G.; Hu, X. Autonomous Driving in the iCity—HD Maps as a Key Challenge of the Automotive Industry. *Engineering* **2016**, *2*, 159–162. [CrossRef]

23. Durrant-Whyte, H.; Bailey, T. Simultaneous localization and mapping: Part I. *IEEE Robot. Autom. Mag.* **2006**, *13*, 99–110. [CrossRef]
24. Javanmardi, E.; Javanmardi, M.; Gu, Y.; Kamijo, S. Factors to Evaluate Capability of Map for Vehicle Localization. *IEEE Access* **2018**, *6*, 49850–49867. [CrossRef]
25. Quack, T.M.; Reiter, M.; Abel, D. Digital map generation and localization for vehicles in urban intersections using LiDAR and GNSS data. *IFAC-PapersOnLine* **2017**, *50*, 251–257. [CrossRef]
26. Xiao, Z.; Jiang, K.; Xie, S.; Wen, T.; Yu, C.; Yang, D. Monocular Vehicle Self-localization method based on Compact Semantic Map. In Proceedings of the 21st International Conference on Intelligent Transportation Systems (ITSC), Maui, HI, USA, 4–7 November 2018; pp. 3083–3090.
27. Hsu, C.M.; Shiu, C.W. 3D LiDAR-Based Precision Vehicle Localization with Movable Region Constraints. *Sensors* **2019**, *19*, 942. [CrossRef]
28. Levinson, J.; Montemerlo, M.; Thrun, S. Map-based precision vehicle localization in urban environments. In Proceedings of the Robotics: Science and Systems III, Atlanta, GA, USA, 27–30 June 2007.
29. Wang, C.; Huang, H.; Ji, Y.; Wang, B.; Yang, M. Vehicle localization at an intersection using a traffic light map. *IEEE Trans. Intell. Transp. Syst.* **2019**, *20*, 1432–1441. [CrossRef]
30. Atia, M.M.; Hilal, A.R.; Stellings, C.; Hartwell, E.; Toonstra, J.; Miners, W.B.; Basir, O.A. A Low-Cost Lane-Determination System Using GNSS/IMU Fusion and HMM-Based Multistage Map Matching. *IEEE Trans. Intell. Transp. Syst.* **2017**, *18*, 3027–3037. [CrossRef]
31. Oguz-Ekim, P.; Ali, K.; Madadi, Z.; Quitin, F.; Tay, W.P. Proof of concept study using DSRC, IMU and map fusion for vehicle localization in GNSS-denied environments. In Proceedings of the IEEE 19th International Conference on Intelligent Transportation Systems (ITSC), Rio de Janeiro, Brazil, 1–4 November 2016; pp. 841–846.
32. Campbell, S.; O'Mahony, N.; Krpalcova, L.; Riordan, D.; Walsh, J.; Murphy, A.; Ryan, C. Sensor Technology in Autonomous Vehicles: A review. In Proceedings of the 29th Irish Signals and Systems Conference, Belfast, UK, 21–22 June 2018; pp. 1–4.
33. Kim, S.W.; Wei, L. Cooperative Autonomous Driving: A Mirror Neuron Inspired Intention Awareness and Cooperative Perception Approach. *IEEE Intell. Transp. Syst. Mag.* **2016**, *8*, 23–32. [CrossRef]
34. Kim, S.; Qin, B.; Chong, Z.J.; Shen, X.; Liu, W.; Ang, M.H.; Frazzoli, E.; Rus, D. Multivehicle Cooperative Driving Using Cooperative Perception: Design and Experimental Validation. *IEEE Trans. Intell. Transp. Syst.* **2015**, *16*, 663–680. [CrossRef]
35. Rauch, A.; Klanner, F.; Rasshofer, R.; Dietmayer, K. Car2X-based perception in a high-level fusion architecture for cooperative perception systems. In Proceedings of the IEEE Intelligent Vehicles Symposium, Alcala de Henares, Spain, 3–7 June 2012; pp. 270–275.
36. Xiao, Z.; Mo, Z.; Jiang, K.; Yang, D. Multimedia Fusion at Semantic Level in Vehicle Cooperactive Perception. In Proceedings of the IEEE International Conference on Multimedia Expo Workshops (ICMEW), San Diego, CA, USA, 23–27 July 2018; pp. 1–6.
37. Paden, B.; Čáp, M.; Yong, S.Z.; Yershov, D.; Frazzoli, E. A survey of motion planning and control techniques for self-driving urban vehicles. *IEEE Trans. Intell. Veh.* **2016**, *1*, 33–55. [CrossRef]
38. Yang, B.; Liang, M.; Urtasun, R. HDNET: Exploiting HD Maps for 3D Object Detection. In Proceedings of the 2nd Conference on Robot Learning, Zurich, Switzerland, 29–31 October 2018; Volume 87, pp. 146–155.
39. Kurdej, M.; Moras, J.; Cherfaoui, V.; Bonnifait, P. Map-Aided Evidential Grids for Driving Scene Understanding. *IEEE Intell. Transp. Syst. Mag.* **2015**, *7*, 30–41. [CrossRef]
40. Wang, S.; Fidler, S.; Urtasun, R. Holistic 3D scene understanding from a single geo-tagged image. In Proceedings of the IEEE Conference on Computer Vision and Pattern Recognition (CVPR), Boston, MA, USA, 7–12 June 2015; pp. 3964–3972.
41. Kschischang, F.R.; Frey, B.J.; Loeliger, H.A. Factor graphs and the sum-product algorithm. *IEEE Trans. Inf. Theory* **2001**, *47*, 498–519. [CrossRef]
42. He, Z.; Hu, Y.; Wu, J.; Wang, J.; Kang, W. A comprehensive method for multipath performance analysis of GNSS navigation signals. In Proceedings of the IEEE International Conference on Signal Processing, Xi'an, China, 14–16 September 2011.
43. Hamza, B.; Nebylov, A. Robust Nonlinear Filtering Applied to Integrated Navigation System INS/GNSS under Non Gaussian Measurement noise effect. *IFAC Proc. Vol.* **2012**, *45*, 202–207. [CrossRef]

44. Rauch, A.; Maier, S.; Klanner, F.; Dietmayer, K. Inter-vehicle object association for cooperative perception systems. In Proceedings of the 16th International IEEE Conference on Intelligent Transportation Systems, The Hague, The Netherlands, 6–9 October 2013; pp. 893–898.
45. Thomaidis, G.; Tsogas, M.; Lytrivis, P.; Karaseitanidis, G.; Amditis, A. Multiple hypothesis tracking for data association in vehicular networks. *Inf. Fusion* **2013**, *14*, 374–383. [CrossRef]
46. Marquardt, D. An Algorithm for Least-Squares Estimation of Nonlinear Parameters. *J. Soc. Ind. Appl. Math.* **1963**, *11*, 431–441. [CrossRef]
47. Kay, S.M. *Fundamentals of Statistical Signal Processing, Volume I: Estimation Theory*; Prentice Hall: Upper Saddle River, NJ, USA, 1993.
48. Lownes, N.E.; Machemehl, R. VISSIM: A multi-parameter sensitivity analysis. In Proceedings of the Simulation Conference, Monterey, CA, USA, 3–6 December 2006; pp. 1406–1413.

MDPI
St. Alban-Anlage 66
4052 Basel
Switzerland
Tel. +41 61 683 77 34
Fax +41 61 302 89 18
www.mdpi.com

Sensors Editorial Office
E-mail: sensors@mdpi.com
www.mdpi.com/journal/sensors

www.ingramcontent.com/pod-product-compliance
Lightning Source LLC
Chambersburg PA
CBHW041749220926
51847CB00014B/152
9783039283026